Ferrocenes

Edited by
A. Togni and T. Hayashi

Distribution:

VCH, P. O. Box 101161, D-69451 Weinheim, Federal Republic of Germany

Switzerland: VCH, P. O. Box, CH-4020 Basel, Switzerland

United Kingdom and Ireland: VCH, 8 Wellington Court, Cambridge CB1 1HZ, United Kingdom

USA and Canada: VCH, 220 East 23rd Street, New York, NY 10010–4606, USA

Japan: VCH, Eikow Building, 10-9 Hongo 1-chome, Bunkyo-ku, Tokyo 113, Japan

ISBN 3-527-29048-6

Ferrocenes

- Homogeneous Catalysis
- Organic Synthesis
- Materials Science

Edited by
Antonio Togni and Tamio Hayashi

Weinheim · New York
Basel · Cambridge · Tokyo

Professor Dr. Antonio Togni
Laboratorium für Anorganische Chemie
Eidgenössische Technische Hochschule
Universitätsstrasse 6
CH-8092 Zürich
Switzerland

Professor Dr. Tamio Hayashi
Department of Chemistry
Faculty of Science
Kyoto University
Sakyo, Kyoto 606-01

Published jointly by
VCH Verlagsgesellschaft, Weinheim (Federal Republic of Germany)
VCH Publishers, New York, NY (USA)

Editorial Directors: Dr. Thomas Mager, Dr. Ute Anton
Assistant Editor: Eva Schweikart
Production Manager: Claudia Grössl

Library of Congress Card No. applied for

A catalogue record for this book is available from the British Library.

Die Deutsche Bibliothek – CIP-Einheitsaufnahme
Ferrocenes : homogeneous catalysis, organic synthesis, materials science / ed. by Antonio Togni and Tamio Hayashi. – Weinheim ; New York ; Basel ; Cambridge ; Tokyo : VCH, 1995
ISBN 3-527-29048-6
NE: Togni, Antonio [Hrsg.]

Printed on acid-free and low-chlorine paper

Composition, Printing and Bookbinding: Druckhaus „Thomas Müntzer“ GmbH,
D-99947 Bad Langensalza
Printed in the Federal Republic of Germany

Preface

More than forty years after its serendipitous discovery in 1951 [1], ferrocene still enjoys a great deal of interest from scientists in many areas of research. Due to its high stability and the well-established methods for its incorporation into more complex structures, ferrocene has become a versatile building block for the synthesis of compounds with tailor-made properties.

The wealth of derivatives known is documented by an enormous number of publications. Despite progress in the use of electronic databases, systematic literature searches on ferrocenes still remain difficult. The most viable, useful, and comprehensive source of information about ferrocene is the ten volumes of Gmelin's *Handbook of Inorganic and Organometallic Chemistry* [2] covering the literature up to 1986. The annual reviews on ferrocene, appearing in the *Journal of Organometallic Chemistry* also provide fairly systematic and up-to-date accounts [3]. However, there is no single-volume source available delineating ferrocene chemistry and its applications.

This book is intended to provide an overview of the main areas of research where ferrocene plays a key role, because of both its chemical and physical properties. No attempt has been made to present a comprehensive work, but those areas of ferrocene research that have not been included (e.g., nonlinear-optical materials and flame retardants) either do not bear sufficient material to justify a review, or are no longer of general interest. The title of the book has been adapted from the title of a talk one of us (A. T.) gave in 1991 at an international conference and reflects the development of his own research interests. For this reason the arrangement of the chapters is somewhat arbitrary. However, the book is subdivided into three main sections: Homogeneous Catalysis, Organic Synthesis, and Materials Science, thus conveying its own logic. In order to provide, whenever possible, new perspectives in the different areas treated in the book, the majority of the authors have been recruited from the younger generation of internationally recognized authorities in their specific field. Chapter 1 (A. Hor and K. S. Gan) describes the coordination chemistry and catalytic applications of the unique ligand 1,1′-bis(diphenylphosphino)ferrocene (dppf) and serves as a background for the impressive series of successful application of chiral, enantiomerically pure ferrocenylphosphines in asymmetric catalysis, reported in Chapter 2 (T. Hayashi). In Chapter 3, Y. Butsugan and coworkers demonstrate that chiral ferrocene derivatives can not only be incorporated into phosphine ligands for late transition metals, but also into aminoalcohols, which, coordinated to zinc, can be used in catalytic reactions. More about optically active ferrocenes is to be found in Chapter 4, where

R. Herrmann and G. Wagner describe general methods for the preparation of such compounds and their application as templates in synthesis. After these first four chapters, which should be of great interest, in particular to the synthesis-oriented reader, the next three chapters mainly address coordination and electrochemical aspects. Thus, in Chapter 5, M. Herberhold reports on the synthesis and properties of ferrocenes containing heteroelements (mainly of group 16) directly attached to the metallocene core. Such compounds display interesting ligand properties. The incorporation of the redox-active ferrocene fragment into macrocyclic hosts is treated in Chapter 6 (C. D. Hall), whereas structural and electrochemical properties of ferrocene-containing ligands and their complexes are reviewed comprehensively by P. Zanello in Chapter 7. The last three chapters are more interdisciplinary in character, as they address the incorporation of ferrocenes in materials explicitly designed to possess specific physical properties. Thus, the combination of the well-known redox and structural properties of simple, for the most part highly symmetric, ferrocene derivatives allows the preparation of charge-transfer complexes having interesting magnetic behaviors (Chapter 8, A. Togni). A fairly new area of ferrocene research deals with liquid-crystalline materials, as presented in Chapter 9 by R. Deschenaux and J. W. Goodby. Last but not least, a not so novel, but still active area of research is the synthesis of polymers containing ferrocene units. This is reviewed by K. E. Gonsalves and X. Chen in Chapter 10.

We gratefully acknowledge the work done by all authors in presenting up-to-date and well-referenced contributions. Without their effort this volume would not have been possible. Furthermore, it was a pleasure to collaborate with the VCH "crew" in Weinheim. They not only did an excellent job in producing the book, but also helped us in a competent manner in all phases of its preparation. Finally, we are grateful to Dr. Peter Gölitz of *Angewandte Chemie* who originally encouraged the idea of making a book about ferrocene.

Zürich and Kyoto
October 1994

Antonio Togni
Tamio Hayashi

References

[1] (a) T. J. Kealey, P. L. Pauson, *Nature* **1951,** *168*, 1039–1040; (b) S. A. Miller, J. A. Tebboth, J. F. Tremaine, *J. Chem. Soc.* **1952**, 632–635.

[2] *Gmelin Handbook of Inorganic and Organometallic Chemistry, 8th Ed., Fe Organic Compounds*, Vols. A1 (1974), A2 (1977), A3 (1978), A4 (1980), A5 (1981), A6 (1977), A7 (1980), A8 (1985), A9 (1989), A10 (1991); Springer, Berlin.

[3] B. W. Rockett, G. Marr, *J. Organomet. Chem.* **1991**, *416*, 327–398; **1990**, *392*, 93–160; **1988**, *357*, 247–318; **1988**, *343*, 79–146; **1987**, *318*, 231–296; **1986**, *298*, 133–205; **1984**, *278*, 255–330; **1983**, *257*, 209–274; **1982**, *227*, 373–440; **1981**, *211*, 215–278; **1980**, *189*, 163–250; **1979**, *167*, 53–154; **1978**, *147*, 273–334; **1976**, *123*, 205–302; **1976**, *106*, 259–336; **1974**, *79*, 223–303.

Contents

Part 2 Organic Synthesis – Selected Aspects

4 Chiral Ferrocene Derivatives. An Introduction

G. Wagner and R. Herrmann

5 Ferrocene Compounds Containing Heteroelements

M. Herberhold

6 Macrocycles and Cryptands Containing the Ferrocene Unit

C. D. Hall

7 Electrochemical and X-ray Structural Aspects of Transition Metal Complexes Containing Redox-Active Ferrocene Ligands

P. Zanello

Part 3 Materials Science

8 Ferrocene-Containing Charge-Transfer Complexes. Conducting and Magnetic Materials

A. Togni

9 Ferrocene-Containing Thermotropic Liquid Crystals

R. Deschenaux and J. W. Goodby

List of Contributors

Shuki Araki
Department of Applied Chemistry
Nagoya Institute of Technology
Gokiso-cho, Showa-ku
Nagoya 466
Japan

Yasuo Butsugan
Department of Applied Chemistry
Nagoya Institute of Technology
Gokiso-cho, Showa-ku
Nagoya 466
Japan

Xiaohe Chen
Polymer Science Program
Institute of Materials Science, U-136
Department of Chemistry
University of Connecticut
Storrs, CT 06269
USA

Robert Deschenaux
Institut de Chimie
Université de Neuchâtel
Avenue de Bellevaux 51
2000 Neuchâtel
Switzerland

Kim-Suan Gan
Department of Chemistry
Faculty of Science
National University of Singapore
Kent Ridge
Singapore 0511

Kenneth E. Gonsalves
Polymer Science Program
Institute of Materials Science, U-136
Department of Chemistry
University of Connecticut
Storrs, CT 06269
USA

John W. Goodby
School of Chemistry
University of Hull
Hull, HU6 7RX
United Kingdom

C. Dennis Hall
Department of Chemistry
King's College
University of London, Strand
London WC2R 2LS
United Kingdom

Tamio Hayashi
Department of Chemistry
Faculty of Science
Kyoto University
Sakyo, Kyoto 606-01
Japan

Max Herberhold
Laboratorium für Anorganische Chemie
Universität Bayreuth
Universitätsstraße 30
D-95440 Bayreuth
Germany

Rudolf Herrmann
Organisch Chemisches Institut
Technische Universität München
Lichtenbergstraße 4
D-85747 Garching
Germany

T. S. Andy Hor
Department of Chemistry
Faculty of Science
National University of Singapore
Kent Ridge
Singapore 0511

Antonio Togni
Laboratorium für Anorganische Chemie
Eidgenössische Technische Hochschule
Universitätsstrasse 6
8092 Zürich
Switzerland

Gabriele Wagner
Organisch Chemisches Institut
Technische Universität München
Lichtenbergstraße 4
D-85747 Garching
Germany

Makoto Watanabe
Department of Applied Chemistry
Nagoya Institute of Technology
Gokiso-cho, Showa-ku
Nagoya 466
Japan

Piero Zanello
Dipartimento di Chimica
dell'Universita di Siena
Piano dei Mantellini, 44
53100 Siena
Italy

List of Abbreviations

A	acceptor
AIBN	azodi(isobutyronitrile)
Bct	benzene chromium tricarbonyl
BINAP	2,2′-bis(diphenylphosphino)-1,1′-binaphthyl, binaphthyl
BPPFA	(*R*)-*N*,*N*-dimethyl-1-[(*S*)-1′,2-bis(diphenylphosphino)-ferrocenyl]ethylamine
cis-dppet	*cis*-1,2-bis(diphenylphosphino)ethylene
chiraphos	2,3-bis(diphenylphosphino)butane
CIP	Cahn-Ingold-Prelog rules
cod	1,5-cyclooctadiene
COSY	correlated spectroscopy
Cp	η^5-cyclopentadienyl
Cp*	η^5-pentamethyl cyclopentadienyl
CT	charge transfer
ctb	capped trigonal biprism
CV	Cyclic voltammetry
D	donor
DAG	diacetone-2-keto-1-gluconic acid
dba	1,5-diphenylpenta-1,4-dien-3-one
DBU	1,8-diazabicyclo[5.4.0]undec-1-ene
DCNQI	*N*,*N*′-dicyanoquinonediimine
DDQ	2,3-dichloro-5,6-dicyanobenzoquinone
DDQH	2,3-dichloro-5,6-dicyanobenzohydroquinone
DFPE	(−)-(*R*,*S*)-1-[2-(diphenylhydroxymethyl)-ferrocenyl]-1-piperidino ethane
DIOP	*P*,*P*′-[2,2-dimethyl-1,3-dioxolane-4,5-diylbis(methylene)]-bis[diphenylphosphane]

DIPAMP	bis[(2-methoxyphenyl)phenylphosphino]ethane
DMAP	4-dimethylaminopyridine
DME	dimethoxyethane
DMF	*N*,*N*-dimethylformamide
dmit	1,3-dithiole-2-thione-4,5-dithiolate
DMSO	dimethylsulfoxide
dppb	1,4-bis(diphenylphosphino)butane
dppe	1,2-bis(diphenylphosphino)ethane
dppf	1,1′-bis(diphenylphosphino)ferrocene
dppm	bis(diphenylphosphino)methane
dppp	1,3-bis(diphenylphosphino)propane
DSC	differential scanning calorimetry
dtpe	1,2-bis(di-*p*-tolylphosphino)ethane
ESR	electron spin resonance
FAB MS	fast-atom bombardment mass spectrometry
fc	1,1′-ferrocenediyl
Fc	ferrocenyl
Fc^+	ferrocenium
FcCNP	1-[(dimethylamino)methyl]-2-(diphenylphosphino)ferrocene
GPC	gel permeation chromatography
$HB(pz)_3$	hydrido-tris(*N*-pyrazolyl)borate
$HB(pz^*)_3$	hydrido-tris(3,5-dimethyl-*N*-pyrazolyl)borate
HLADH	horse liver alcohol dehydrogenase
HOMO	highest occupied molecular orbital
I	isotropic liquid
L	ligand
MDI	4,4′-methylene-bis(phenylisocyanate)
Me_2DCNQI	2,5-dimethyl-*N*,*N*′-dicyanoquinonediimine
mim	*N*-methylimidazole
M_n	number average molecular weight
MWD	molecular weight distribution
N	nematic phase
nbd	norbornadiene
NLO	non-linear optical

NOE	nuclear Overhauser effect
oc	1,1′-osmocenediyl
odppf	1,1′-bis(oxodiphenylphosphoranyl)-ferrocene
PDCI	5,12-bis(cyanoimine)pentacene-7,14-dione
PPFA	(*R*)-*N*,*N*-dimethyl-1-[(*S*)-2-(diphenylphosphino)ferrocenyl]-ethylamine
PPG	poly(propylene glycol)
prophos	1,2-bis(diphenylphosphino)propane
PTCI	5,7,12,14-tetrakis(cyanoimine)pentacene
py	pyridine
QS	quadrupole splitting
rc	1,1′-ruthenocenediyl
Rc	η^1-ruthenocenyl
ROMP	ring opening metathesis polymerization
SCE	standard calomel electrode
SHG	second harmonic generator [efficiency] (sic)
SmA	smectic-A phase
SmC	smectic-C phase
SUMO	singly unoccupied molecular orbital
tbp	trigonal bipyramid
TCIDBT	5,7,12,14-tetrakis(cyanoimine)dibenzo[b,i]thianthrene
TCNE	tetracyanoethylene
TCNQ	tetracyano-*p*-quinodimethane
$TCNQF_4$	tetracyanoperfluoro-*p*-quinodimethane
T_g	glass transition temperature
T_m	melt transition temperature
TMEDA	*N*,*N*,*N*′,*N*′-tetramethylethylenediamine
TMTSF	tetramethyltetraselenofulvalene
triphos	1,1,1-tris(diphenylphosphino)methylethane
TTF	tetrathiafulvalene
UPS	ultraviolet photoelectron spectroscopy
XPS	X-ray photoelectron spectroscopy
9-BBN	borabicyclo[3.3.1]nonane

Part 1. Homogeneous Catalysis

1 1,1′-Bis(diphenylphosphino)ferrocene – Coordination Chemistry, Organic Syntheses, and Catalysis

Kim-Suan Gan and T. S. Andy Hor

1.1 Introduction

The use of ferrocenyl phosphines as ligands in coordination chemistry is well known [1 – 12]. The ability of these ligands to convey the ferrocenyl qualities to the resultant complexes without disturbing the inherent characteristics of the latter has widened the scope of metal complexes in the design of catalysts, drugs and materials. As diphosphine complexes metamorphose from their monophosphine analogues and begin to develop their individual traits [13 – 16], it is inevitable that ferrocenyl diphosphines must play a special role in the development of applied organometallic chemistry. Although the metalloligand 1,1′-bis(diphenylphosphino)ferrocene (dppf), which is probably the best developed ferrocenyl diphosphine, was first synthesized more than two decades ago, its chemical uniqueness and industrial importance were not fully appreciated until recently. In this chapter we attempt to trace its development as a coordination ligand and relate its chemistry to a variety of applications. The catalytic potential is emphasized in view of the ever growing influence of homogeneous catalysis in organic synthesis, manipulation of materials and production of fine chemicals.

Derivatives of dppf are beyond the scope of the present discussion and will be excluded. Chiral syntheses using dppf derivatives are covered in Chapter 2 and hence are also omitted. The coordination mode of dppf is usually specified in the complex formulas in the descriptions on coordination chemistry but ignored elsewhere for reasons of clarity. When reference is made to the C_5 ring of dppf, it is often, for the sake of simplicity, referred to as a cyclopentadienyl (Cp) ring, although, strictly speaking, it refers to a phosphinated cyclopentadienyl ring ($C_5H_4PPh_2$). Catalytic discussions are biased towards the complex $PdCl_2$(dppf), as it is with this catalyst that most coupling reactions are reported. Catalysts based on other phosphines are generally overlooked unless a comparison with dppf analogues merits their inclusion.

1.2 Preparation and Complexation

The synthesis of dppf was first recorded in 1965 by the lithiation of ferrocene with *n*-butyllithium, followed by condensation with chlorodiphenylphosphine [17]. A higher yield can be obtained in the presence of *N*, *N*, *N′*, *N′*-tetramethylethylene-

Table 1-1. List of ^{31}P NMR chemical and coordination shifts of some dppf complexes and their coordination geometries [dppf = $Fe(C_5H_4PPh_2)_2$] (blank entries denote unavailable data)

Complex	Coordination geometry	$\delta(^{31}P)$, ppm [a]	Δ ppm [b]	Ref.
$[Fe(C_5H_4PPh_2)_2]$ (dppf)	sandwich	−17.2(s)	–	[1]
		−16.8(s)	–	[2]
$Fe[C_5H_4P(O)Ph_2]_2$	sandwich	28.3(s)	45.5	[3]
Group 5				
$[NEt_4][V(CO)_4$(dppf-*P*, *P′*)]	octahedral	–	–	[4]
$[NEt_4][Ta(CO)_4$(dppf-*P*, *P′*)]	octahedral	–	–	[4]
Group 6				
$Cr(CO)_4$(dppf-*P*, *P′*)	octahedral	52.6(s)	69.8	[4, 5]
$Cr(CO)_5$(dppf-*P*)	octahedral	47.3(s); −17.2(s)	64.5	[5]
$[Cr(CO)_5]_2$(μ-dppf)	octahedral	47.5(s)	64.7	[5]
$Cr(CO)_5$(μ-dppf)$Mo(CO)_5$	octahedral	47.5(s,P_{Cr}); 28.5(s,P_{Mo})	64.7(P_{Cr}); 45.7 (P_{Mo})	[6]
$Cr(CO)_5$(μ-dppf)$W(CO)_5$	octahedral	47.5(s,P_{Cr}); 11.3(t,P_W), J(PW) 247 Hz	64.7(P_{Cr}); 28.5(P_W)	[6]
trans-$[Cr(CO)_4\{$(μ-dppf)$Cr(CO)_5\}_2]$	octahedral	64.8(s,$P_{Cr'}$); 46.8(s,$P_{Cr''}$) [c]	82.0($P_{Cr'}$); 64.0($P_{Cr''}$) [c]	[7]
trans-$[Cr(CO)_4\{$(μ-dppf)$Mo(CO)_5\}_2]$	octahedral	64.8(s,P_{Cr}); 27.9(s,P_{Mo})	82.0(P_{Cr}); 45.1(P_{Mo})	[7]
trans-$[Cr(CO)_4\{$(μ-dppf)$W(CO)_5\}_2]$	octahedral	64.8(s,P_{Cr}); 10.9(t,P_W), J(PW) 247 Hz	82.0(P_{Cr}); 28.1(P_W)	[7]
trans-$[Cr(CO)_4\{$(μ-dppf)$Fe(CO)_4\}_2]$	octahedral	67.0(s,P_{Fe}); 65.2(s,P_{Cr})	84.2(P_{Fe}); 82.4(P_{Cr})	[7]
trans-$[Cr(CO)_4\{$(μ-dppf)$Mn_2(CO)_9\}_2]$	octahedral	65.6(s,br,P_{Mn}); 64.6(s,P_{Cr})	82.8(P_{Mn}); 81.8(P_{Cr})	[7]
$Mo(CO)_4$(dppf-*P*, *P′*)	octahedral	33.9(s)	51.1	[4, 5, 8]
$Mo(CO)_5$(dppf-*P*)	octahedral	28.3(s); −17.10(s)	45.5	[5]
$[Mo(CO)_5]_2$(μ-dppf)	octahedral	28.5(s)	45.7	[5]
$Mo(CO)_5$(μ-dppf)$W(CO)_5$	octahedral	28.5(s,P_{Mo}); 11.3(t,P_W), J(PW) 247 Hz	45.7(P_{Mo}); 28.5(P_W)	[6]
cis-$[Mo(CO)_4\{$(μ-dppf)$Mo(CO)_5\}_2]$	octahedral	27.9(s,$P_{Mo''}$); 26.8(s,$P_{Mo''}$) [c]	45.1($P_{Mo''}$); 44.0($P_{Mo'}$) [c]	[7]
cis-$[Mo(CO)_4\{$(μ-dppf)$Cr(CO)_5\}_2]$	octahedral	46.9(s,P_{Cr}); 26.8(s,P_{Mo})	64.1(P_{Cr}); 44.0(P_{Mo})	[7]

cis-$[Mo(CO)_4\{(\mu\text{-dppf})Fe(CO)_4\}_2]$	octahedral	66.4(s,P_{Fe}); 26.7(s,P_{Mo})	83.6(P_{Fe}); 43.9(P_{Mo})	[7]
cis-$[Mo(CO)_4\{(\mu\text{-dppf})Mn_2(CO)_9\}_2]$	octahedral	65.6(s,P_{Mn}); 26.8(s,P_{Mo})	82.8(P_{Mn}); 44.0(P_{Mo})	[7]
$MoI_2(CO)_2$(dppf-*P*, *P′*)	octahedral	–	–	[9]
$MoI_2(CO)_3$(dppf-*P*, *P′*)	capped octahedral	–	–	[9]
cis-$[MoCl_2(CO)_2$(dppf-*P*, *P′*)(dppf-*P*)]	capped octahedral	–	–	[9]
$Mo_2I_4(CO)_6(\mu\text{-dppf})(XPh_3)_2$ (X = P, As, Sb)	capped octahedral	–	–	[10]
fac-$[Mo(N_2)$(dppf-*P*, *P′*)](triphos)[d]	octahedral	–19.7(d,2P),[e] $^2J(PP)$ 14 Hz	–2.5	[11]
fac-[Mo(dppf-P_a, P'_b)(dppf-P_c)(triphos)][d]	octahedral	37.0(dt,P_a); –20.9(s,P_d);[f] –21.9(d,P_b,P_c), $^2J(P_bP_c)$ 3.7 Hz	54.2(P_a); –3.7(P_d); –4.7(P_b,P_c)	[11]
$W(CO)_4$(dppf-*P*, *P′*)	octahedral	18.4(t), J(PW) 239 Hz	35.6	[4, 5]
$W(CO)_5$(dppf-*P*)	octahedral	11.3(t), J(PW) 249 Hz; –17.1(s)	28.5	[5]
$[W(CO)_5]_2(\mu\text{-dppf})$	octahedral	11.3(t), J(PW) 244 Hz	28.5	[5]
trans-$[W(CO)_4\{(\mu\text{-dppf})W(CO)_5\}_2]$	octahedral	16.2(t,$P_{W'}$), J(PW′) 283 Hz; 11.2(t,$P_{W''}$), J(PW″) 249 Hz[c]	33.4($P_{W'}$); 28.4($P_{W''}$)[c]	[7]
trans-$[W(CO)_4\{(\mu\text{-dppf})Cr(CO)_5\}_2]$	octahedral	15.6(t,P_W), J(PW) 282 Hz; 46.8(s,P_{Cr})	32.8(P_W); 64.0(P_{Cr})	[7]
cis-$[W(CO)_4\{(\mu\text{-dppf})Cr(CO)_5\}_2]$	octahedral	20.6(t,P_W), J(PW) 244 Hz; ≈47(s,br,P_{Cr})	37.8(P_W); ≈64.2(P_{Cr})	[7]
$[Et_4N][W_2(\mu\text{-H})(CO)_8(\mu\text{-dppf})]$	octahedral	10.1, J(PW) 219 Hz	27.3	[12]
$WI_2(CO)_3$(dppf-*P*, *P′*)	capped octahedral	–	–	[9]
$[W_2I_2(CO)_2(\mu\text{-dppf})$(dppm-*P*, *P′*)$_2(\eta^2\text{-MeC}_2\text{Me})_2][BF_4]_2$	octahedral	–	–	[13]
$W_2I_4(CO)_6(\mu\text{-dppf})(XPh_3)_2$ (X = P, As, Sb)	capped octahedral	–	–	[10]
Group 7				
$[Mn_2(CO)_9]_2(\mu\text{-dppf})$	octahedral	66.4(s,br)	83.6	[14]
$Mn_2(CO)_9(\mu\text{-dppf})Cr(CO)_5$	octahedral	66.2(s,br,P_{Mn}); 47.5(s,P_{Cr})	83.4(P_{Mn}); 64.7(P_{Cr})	[6]
$Mn_2(CO)_9(\mu\text{-dppf})Mo(CO)_5$	octahedral	66.2(s,br,P_{Mn}); 28.5(s,P_{Mo})	83.4(P_{Mn}); 45.7(P_{Mo})	[6]
$Mn_2(CO)_9(\mu\text{-dppf})W(CO)_5$	octahedral	66.2(s,br,P_{Mn}); 11.3(t,P_W), J(PW) 244 Hz	83.4(P_{Mn}); 28.5(P_W)	[6]

Table 1-1. (continued)

Complex	Coordination geometry	$\delta(^{31}P)$, ppm[a]	Δ ppm[b]	Ref.
$Mn(\eta^5\text{-MeCp})(CO)_2$(dppf-*P*)	3-legged piano-stool	84.7(s); −17.2(s)	101.9	[15]
$Mn(\eta^5\text{-MeCp})(CO)$(dppf-*P*, *P'*)	3-legged piano-stool	not observed due to exchange	–	[15–17]
cis-$[Mn_2Cl_2(CO)_8(\mu\text{-dppf})]$	octahedral	40.3(s)	57.5	[15, 18]
$[Re_2(CO)_9]_2(\mu\text{-dppf})$	octahedral	5.8(s)	23.0	[3]
$Re_2(CO)_9$(dppf-*P*)	octahedral	5.7(s,P_{Re}); −17.0(s,P_{free})	22.9(P_{Re})	[3]
$Re_2(CO)_9$(dppfO-*P*)[g]	octahedral	28.4(s,P=O); 5.7(s,P_{Re})	45.6(P=O); 22.9(P_{Re})	[3]
fac-$[ReBr(CO)_3$(dppf-*P*, *P'*)]	octahedral	0.27(s)	17.47	[3]
fac-$[ReCl(CO)_3$(dppf-*P*, *P'*)]	octahedral	2.8	20.0	[19]
cis-$[Re_2Cl_2(CO)_8(\mu\text{-dppf})]$	octahedral	−3.4(s)	13.8	[20]
cis-$[Re_2Br_2(CO)_8(\mu\text{-dppf})]$	octahedral	−7.4(s)	9.8	[20]
(*cis*-Br)$(OC)_4Re(\mu\text{-dppf})Re$(*trans*-Br)$(CO)_4$	octahedral	5.1(s,*trans*-P); −7.7(s,*cis*-P)	22.3(*trans*-P), 24.9(*cis*-P)	[20]
$Re_2(\mu\text{-OMe})_2(CO)_6(\mu\text{-dppf})$	octahedral	5.5(s)	22.5	[21]
$Re(MeCO_2\text{-}O)(CO)_3$(dppf-*P*, *P'*)	octahedral	7.9(s)	25.1	[21]
$[Re_2(CO)_9(Ph_2P(C_5H_4)Fe(C_5H_4)PMePh_2)]I$	octahedral	24.4(s,P_{Me}); 5.5(s,P_{Re})	41.6(P_{Me}); 22.7(P_{Re})	[3]
Group 8				
$Fe(CO)_3$(dppf-*P*, *P'*)	trigonal bipyramidal	62.0(s)	79.3	[22, 23]
$Fe(CO)_4$(dppf-*P*)	trigonal bipyramidal	66.8(s); −17.1(s)	84.0	[22, 23]
$[Fe(CO)_4]_2(\mu\text{-dppf})$	trigonal bipyramidal	67.0(s)	84.2	[22, 23]
$Fe_2(CO)_7$(dppf-*P*, *P'*)	octahedral	73.5(s); 64.6(s)	90.7; 81.8	[24]
$Fe(CO)_4(\mu\text{-dppf})Cr(CO)_5$	Fe: trigonal bipyramidal Cr: octahedral	67.0(s,P_{Fe}); 47.4(s,P_{Cr})	84.2(P_{Fe}); 64.6(P_{Cr})	[14]

$Fe(CO)_4(\mu\text{-dppf})Mo(CO)_5$	Fe: trigonal bipyramidal Mo: octahedral	$67.0(s,P_{Fe})$; $28.4(s,P_{Mo})$	$84.2(P_{Fe})$; $45.6(P_{Mo})$	[22]
$Fe(CO)_4(\mu\text{-dppf})W(CO)_5$	Fe: trigonal bipyramidal W: octahedral	$67.0(s,P_{Fe})$; $11.2(t,P_W)$, J(PW) 244 Hz	$84.2(P_{Fe})$; $28.4(P_W)$	[14]
$Fe(CO)_4(\mu\text{-dppf})Mn_2(CO)_9$	Fe: trigonal bipyramidal Mn: octahedral	$67.1(s,P_{Fe})$; $66.3(s,P_{Mn})$	$84.3(P_{Fe})$; $83.5(P_{Mn})$	[14]
$RuCl(\eta^5\text{-}C_5H_5)(\text{dppf-}P,P')$	3-legged piano-stool	–	–	[25]
$Ru(\eta^5\text{-}C_5H_5)H(\text{dppf-}P,P')$	3-legged piano-stool	–	–	[25]
$[Ru(\eta^5\text{-}C_5H_5)(\text{dppf-}P,P')(CH_3CN)]X$ $[X = PF_6, BF_4, BPh_4]$	3-legged piano-stool	–	–	[26]
$[Ru(\eta^5\text{-}C_5H_5)(CO)(\text{dppf-}P,P')]X$ $[X = BF_4, PF_6]$	3-legged piano-stool	–	–	[26]
$[Ru(C{=}CHR)(\eta^5\text{-}C_5H_5)(\text{dppf-}P,P')][PF_6]$ [R = Ph, tBu, Fc(ferrocenyl)]	3-legged piano-stool	–	–	[26]
$[Ru(C{\equiv}CR)(\eta^5\text{-}C_5H_5)(\text{dppf-}P,P')]$ $[R = Ph, tBu, C_6H_{13}, Fc, CO_2Me]$	3-legged piano-stool	–	–	[26]
$Ru\{C(OMe){=}CH(CO_2Me)\}(\eta^5\text{-}C_5H_5)(\text{dppf-}P,P')$	3-legged piano-stool	–	–	[26]
$Ru\{C(CO_2Me){=}CH(CO_2Me)\}(\eta^5\text{-}C_5H_5)(\text{dppf-}P,P')$	3-legged piano-stool	–	–	[25]
$Ru(\eta^3\text{-}C_3H_5)\{\eta^3\text{-}[OC(CF_3)]_2CH\}(\text{dppf-}P,P')$ [h]	pseudo-octahedral	58.5(d); 24.5(d), $^2J(PP)$ 34 Hz	75.7; 41.7	[27]
$[RuH(\text{dppf-}P,P')_2(\eta^2\text{-}H_2)][PF_6]$	distorted octahedral	38.2(br)	55.4	[28]
$Ru_2(\mu\text{-pz})_2(CO)_4(\text{dppf-}P)_2$ [i]	octahedral	–	–	[29]

Table 1-1. (continued)

Complex	Coordination geometry	$\delta(^{31}P)$, ppm[a]	Δ ppm[b]	Ref.
Group 9				
$CoCl_2$(dppf-*P*, *P'*)	tetrahedral	–	–	[2, 4]
[Co(acac)$_2$(dppf-*P*, *P'*)][BF_4]	octahedral	–	–	[30]
[Co(dtc)$_2$(dppf-*P*, *P'*)][BF_4][j]	octahedral	–	–	[30]
[Rh(dppf-*P*, *P'*)$_2$][BPh_4]	distorted square planar	22.2(d), J(RhP) 144 Hz	39.4	[31]
[Rh(dppf-*P*, *P'*)(nbd)][ClO_4]	square planar	14.8(d), J(RhP) 161 Hz	32.0	[8, 32, 33]
[Rh(dppf-*P*, *P'*)(cod)][ClO_4]	square planar	22.0(d), J(RhP) 149 Hz	39.2	[32, 34]
Rh{(η^6-C_6H_5)B(C_6H_5)$_3$}(dppf-*P*, *P'*)	2-legged piano-stool	43.0(d), J(RhP) 212 Hz	60.2	[32]
Rh_2(μ-S-*t*Bu)$_2$(CO)$_2$(μ-dppf)	square planar	27.0(d), J(Rh-P) 156 Hz	44.2	[34]
Rh_2H_2(CO)$_2$(dppf-*P*, *P'*)$_2$(μ-dppf)	trigonal bipyramidal	–	–	[35]
[Ir(dppf-*P*, *P'*)$_2$][BPh_4]	distorted square planar	7.7	24.9	[31]
Group 10				
NiX_2(dppf-*P*, *P'*) (X = Cl, Br)	tetrahedral	–	–	[2, 4, 8, 36]
cis-[$PdCl_2$(dppf-*P*, *P'*)]	square planar	34.0(s)	51.2	[2, 8, 37]
1-{(dppf-*P*, *P'*)Pd}B_3H_7	square planar	12.5(s)	29.7	[38]
[Pd(μ-Cl)(dppf-*P*, *P'*)]$_2$[BF_4]$_2$	square planar	46.5(s)	63.7	[39]
[Pd(μ-OH)(dppf-*P*, *P'*)]$_2$[BF_4]$_2$	square planar	38.7(s)	55.9	[39]
[Pd(dppf-*P*, *P'*)(PPh_3)][BF_4]$_2$	square planar	–	–	[40, 41]
trans-[$PdCl_2${(μ-dppf)Cr(CO)$_5$}$_2$]	Pd: square planar	47.5(s,P_{Cr}); 14.7(s,P_{Pd})	64.7(P_{Cr}); 31.9(P_{Pd})	[42]
trans-[$PdCl_2${(μ-dppf)Mo(CO)$_5$}$_2$]	Pd: square planar	28.5(s,P_{Mo}); 14.7(s,P_{Pd})	45.7(P_{Mo}); 31.9(P_{Pd})	[42]
trans-[$PdCl_2${(μ-dppf)W(CO)$_5$}$_2$]	Pd: square planar	14.7(s,P_{Pd}); 11.2(t,P_W), J(PW) 247 Hz	31.9(P_{Pd}); 28.4(P_W)	[42]
cis-[$PtCl_2$(dppf-*P*, *P'*)]	square planar	13.1, J(PtP) 3769 Hz	30.3	[1, 2, 43, 44]

trans-$[PtCl_2\{(\mu\text{-dppf})Cr(CO)_5\}_2]$	square planar	47.3(s,P_{Cr}); 10.0(t,P_{Pt}), J(PPt) 2630 Hz	64.5(P_{Cr}); 27.2(P_{Pt})	[42]
trans-$[PtCl_2\{(\mu\text{-dppf})Mo(CO)_5\}_2]$	square planar	28.4(s,P_{Mo}); 10.2(t,P_{Pt}), J(PPt) 2632 Hz	45.6(P_{Mo}); 27.4(P_{Pt})	[42]
trans-$[PtCl_2\{(\mu\text{-dppf})W(CO)_5\}_2]$	square planar	11.0(t,P_W), J(PW) 244 Hz; 10.1(t,P_{Pt}), J(PPt) 2632 Hz	28.2(P_W); 27.3(P_{Pt})	[42]
1-{(dppf-*P*, *P'*)Pt}B_3H_7	square planar	28.0, J(PtP) 2790 Hz	45.2	[45]
$[Pt_2H(\mu\text{-H})_2$(dppf-*P*, *P'*)$_2$]X [X = Cl, BF_4]	Pt(1): square planar Pt(2): trigonal bipyramidal	29.3, J(PtP) 3280 Hz	46.5	[1, 45]
[Pt(μ-Cl)(dppf-*P*, *P'*)]$_2$[BF_4]$_2$	square planar	18.3, J(PtP) 3987 Hz	35.5	[39, 46]
[Pt(μ-OH)(dppf-*P*, *P'*)]$_2$[BF_4]$_2$	square planar	6.5, J(PtP) 3857 Hz	23.7	[39]
[Pt(dppf-*P*, *P'*)(dmf)$_2$][BF_4]$_2$	square planar	7.0, J(PtP) 4143 Hz	24.2	[39]
[PtCl(dppf-*P*, *P'*)(dmso)][BF_4]	square planar	19.0(dt), J(PtP) 3913 Hz; 5.2(dt), J(PtP) 4086 Hz, ^{2}J(PP) 17 Hz	36.2; 22.4	[46]
[Pt{Ac_2(dT)}(dppf-*P*, *P'*)(dmso)][BF_4][k]	square planar	8.4(dt,P_A), J(PtP) 3452 Hz; 6.5(dt,P_B), J(PtP) 4394 Hz, ^{2}J(PP) 19 Hz[l]	25.6; 23.7	[46]
Pt(pz)$_2$(dppf-*P*, *P'*)[i]	square planar	7.2, J(PtP) 3198 Hz	24.4	[1]
Pt(3.5-Me_2pz)$_2$(dppf-*P*, *P'*)[m]	square planar	5.5, J(PtP) 3166 Hz	22.7	[1]
[Pt(dppf-*P*, *P'*)(3,5-Me_2pzH)$_2$][BF_4]$_2$[m]	square planar	−0.1, J(PtP) 3500 Hz	17.1	[1]
[Pt(dppf-*P*, *P'*)(H_2O)$_2$][BF_4]$_2$	square planar	18.4, J(PtP) 3973 Hz	35.6	[1]
[Pt(μ-OH)(dppf-*P*, *P'*)]$_2$[BF_4]$_2$	square planar	9.4, J(PtP) 3855 Hz	26.6	[1]
[Pt_2(μ-H)(μ-CO)(dppf-*P*, *P'*)$_2$][BF_4]	distorted square planar	25.5, J(PtP) 3923 Hz	42.7	[1]
Pt(S_2N_2)(dppf-*P*, *P'*)	square planar	20.9(*trans* to S), J(PtP) 2903 Hz; 9.4(*trans* to N), J(PtP) 3008 Hz, J(PP) 24 Hz[n]	38.1(*trans* to S); 26.6(*trans* to N)	[47]
Pt($CH_2SiMe_2CH{=}CH_2$)$_2$(dppf-*P*, *P'*)	square planar	20.0, J(PtP) 2051 Hz	37.2	[48]
cis-[Pt(1-MeTy(-H))(dppf-*P*, *P'*)(dmf)][BF_4][o]	square planar	8.0, J(PtP) 3375 Hz; 6.1, J(PtP) 4419 Hz	25.2; 23.3	[49]
cis-[Pt(1-MeTy(-H))(dppf-*P*, *P'*)(dmso)][BF_4][o]	square planar	8.3, J(PtP) 3427 Hz; 6.6, J(PtP) 4418 Hz	25.5	[49]

Table 1-1. (continued)

Complex	Coordination geometry	$\delta(^{31}P)$, ppm[a]	Δ ppm[b]	Ref.
cis-[Pt(1-MeTy(-H))(dppf-*P*, *P'*)(CH_3CN)][BF_4][o]	square planar	9.2, J(PtP) 4303 Hz; 4.9, J(PtP) 3165 Hz	26.4; 22.1	[49]
cis-[Pt(1-MeTy(-H))(dppf-*P*, *P'*)(1-MeCy)][BF_4][o,p]				
Isomer A	square planar	3.7, J(PtP) 3508 Hz; −0.9, J(PtP) 3718 Hz	20.9; 18.1	[49]
Isomer B	square planar	11.5, J(PtP) 3604 Hz; 4.7, J(PtP) 3340 Hz	28.7; 21.9	[49]
[Pt(1-MeCy(-H))(dppf-*P*, *P'*)][BF_4][p]	square planar	11.1, J(PtP) 3557 Hz; 7.6, J(PtP) 3545 Hz, J(PP) 25 Hz	28.3; 24.8	[49]
Pt(n-C_4H_9)$_2$(dppf-*P*, *P'*)	square planar	–	–	[44]
[$Pt_2Tl(\mu_3$-$S)_2$(dppf-*P*, *P'*)$_2$][PF_6]	Pt: square planar	18.0, J(PtP) 3160 Hz	35.2	[50]
Group 11				
CuI(dppf-*P*, *P'*)[q]	trigonal planar	–	–	[4]
[Cu_2(dppf-*P*, *P'*)$_2$(μ-dppf)]X_2 [X = ClO_4, BF_4]	trigonal planar	−7.8(br)[r]	9.4	[51, 52]
[Cu_2(dppf-*P*, *P'*)$_2$(μ-bpym)][BF_4]$_2$[s]	distorted tetrahedral	–	–	[53]
[Cu(μ-HCO_2)(dppf-*P*, *P'*)]$_2$	tetrahedral	−18.0	−0.8	[54]
[Ag(dppf-*P*, *P'*)$_2$][NO_3]	tetrahedral	3.5(d), J(AgP) 248 Hz	20.7	[55]
[Ag_2(dppf-*P*, *P'*)$_2$(μ-dppf)][PF_6]$_2$	trigonal planar	1.4(dt,2P), J(AgP) 328 Hz, J(PP) 21 Hz; 2.2(dd,4P), J(AgP) 328 Hz, J(PP) 23 Hz	18.6; 19.4	[55]
[Ag(NO_3)(μ-dppf)]$_2$	planar	5.8(s,br)	23.0	[56]
[Ag_2(μ-CH_3CO_2)(μ_3-CH_3CO_2)(μ-dppf)]$_2$	tetrahedral	5.0(s,br); 4.8(d,br), J(AgP) 630 Hz	22.2; 22.0	[56]
Ag_2(μ-$C_6H_5CO_2$)$_2$(μ-dppf)	trigonal planar	6.3(s,br); 6.2(d,br), J(AgP) 703 Hz	23.5; 23.4	[56]
Ag_2(HCO_2-*O*)$_2$(dppf-*P*, *P'*)$_2$(μ-dppf)	tetrahedral	0.3(br); −4.7(br)	17.5; 12.5	[56]
[Ag(μ-NCO)(μ-dppf)]$_2$	–	1.0(s,br)	18.2	[55]
[Ag(μ-Cl)(μ-dppf)]$_4$	–	−3.4(s,v.br)	13.8	[55]
[Ag(μ-SCN-*S*,*N*)(dppf-*P*, *P'*)]$_2$	tetrahedral	−0.9(s,v.br)	16.3	[55]

$[Ag(S_2CNEt_2)(\mu\text{-dppf})]_2$	tetrahedral	−6.7(s,br)	10.5	[55]
$Au_2Cl_2(\mu\text{-dppf})$	linear	28.9	46.1	[57, 58]
$Au_2(NO_3)_2(\mu\text{-dppf})$	linear	22.5	39.7	[59]
$Au_2(CN)_2(\mu\text{-dppf})$	linear	34.6(s)	51.8	[60]
$Au_2(C_6H_5CO_2\text{-}O)_2(\mu\text{-dppf})$	linear	22.8(s)	40.0	[59]
$Au_2Cl_2(\mu\text{-dppf})_2$	trigonal planar	30.1	47.3	[57]
$[AuCl(\mu\text{-dppf})]_n$	trigonal planar	29.2(s,br); 28.5(br)	46.4; 45.7	[59]
$[Au(\text{dppf-}P,P')(\text{dppf-}P)]Cl$	trigonal planar	≈26(br)[t]	≈43.2	[57]
$[Au_2(\text{dppf-}P,P')_2(\mu\text{-dppf})][NO_3]_2$	trigonal planar	38.8(d,4P), J(PP) 126 Hz; 38.3(t,2P), J(PP) 120 Hz	56.0; 55.5	[59]
$AuCl(\mu\text{-dppf})Cr(CO)_5$	Au: linear	47.5(P_{Cr}); 28.3(P_{Au})	64.7(P_{Cr}); 55.5(P_{Au})	[42]
$AuCl(\mu\text{-dppf})Mo(CO)_5$	Au: linear	28.5(P_{Mo}); 28.3(P_{Au})	45.7(P_{Mo}); 45.5(P_{Au})	[42]
$AuCl(\mu\text{-dppf})W(CO)_5$	Au: linear	28.3(P_{Au}); 11.3(P_W), J(PW) 244 Hz	45.5(P_{Au}); 28.5(P_W)	[42]
$[AuRe(CO)_5]_2(\mu\text{-dppf})$	Au: linear	53.6(s)	70.8	[60]
$[AuMn(CO)_5]_2(\mu\text{-dppf})$	Au: linear	42.1(s)	59.3	[60]
$AuCl(\mu\text{-dppf})Re_2(CO)_9$	Au: linear	28.3(s,P_{Au}); 5.8(s,P_{Re})	45.5(P_{Au}); 23.0(P_{Re})	[60]
Group 12				
$ZnCl_2(\text{dppf-}P,P')$	tetrahedral	−21.3	−4.1	[2]
$CdCl_2(\text{dppf-}P,P')$	tetrahedral	−8.7	8.5	[2]
$HgCl_2(\text{dppf-}P,P')$	tetrahedral	19.2, J(HgP) 4108 Hz	36.4	[2, 61]
$Hg_2X_4(\text{dppf-}P,P')$ [X = Cl, Br, I]	tetrahedral	–	–	[61]
$[Hg(\text{dppf-}P,P')_2]X_2$ [X = BF_4, PF_6]	tetrahedral	–	–	[61]
Group 14				
$Sn_2Cl_4(\text{dppf-}P,P')$	tetrahedral	–	–	[61]
Clusters				
$Fe_3(CO)_{10}(\mu\text{-dppf})$	triangular	39.0(s)	56.2	[24]
$Fe_3(\mu_3\text{-S})_2(CO)_7(\mu\text{-dppf})$	isosceles triangular	68.4(s)	85.6	[62]

Table 1-1. (continued)

Complex	Coordination geometry	$\delta(^{31}P)$, ppm [a]	Δ ppm [b]	Ref.
$Fe_3(\mu_3\text{-}S)_2(CO)_8(dppf\text{-}P)$	isosceles triangular	56.1(br); −18.0(s)	73.3	[62]
$[Fe_3(\mu_3\text{-}S)_2(CO)_8]_2(\mu\text{-}dppf)$	linked isosceles triangular	56.0(br)	73.2	[62]
$Ru_3(CO)_{10}(\mu\text{-}dppf)$	triangular	–	–	[63]
$Ru_3(CO)_8(\mu\text{-}dppf)_2$	triangular	–	–	[63]
$Ru_4(\mu\text{-}H)_4(CO)_{10}(\mu\text{-}dppf)$	tetrahedral	–	–	[63]
$Ru_3(\mu_3\text{-}S)_2(CO)_7(\mu\text{-}dppf)$	isosceles triangular	53.0(s)	70.3	[62]
$[Ru_3(\mu_3\text{-}S)_2(CO)_8]_2(\mu\text{-}dppf)$	linked isosceles triangular	50.1(s)	67.3	[62]
$Co_3(\mu_3\text{-}CMe)(CO)_7(\mu\text{-}dppf)$	triangular	25.4	42.6	[15, 18]
$Co_3(\mu_3\text{-}CPh)(CO)_7(\mu\text{-}dppf)$	triangular	23.9	41.1	[64]
$Fe_4(\mu\text{-}H)(CO)_{12}(\mu_6\text{-}B)Au_2(\mu\text{-}dppf)$	bridged butterfly	45.3	62.5	[65]
$Ru_4(\mu\text{-}H)(CO)_{12}(\mu_6\text{-}B)Au_2(\mu\text{-}dppf)$	bridged butterfly	42.8	60.0	[65]
$[Ru_4(\mu\text{-}H)(CO)_{12}(\mu\text{-}BH)Au]_2(\mu\text{-}dppf)$	twin bridged butterfly	48.0	65.2	[65]

a External standard 85% H_3PO_4.
b $\Delta = \delta$(complex) − δ(dppf); δ(dppf) −17.2 ppm (ref. 4).
c $M' = [M(CO)_4]$; $M'' = [M(CO)_5]$.
d triphos = $PhP(CH_2CH_2PPh_2)_2$.
e ^{31}P signals of dppf only.
f P_d refers to the pendant phosphorus of dppf.
g dppfO = $Ph_2P(C_5H_4)Fe\{(C_5H_4)P(O)Ph_2\}$.
h $\{[OC(CF_3)]_2CH\}$ (hexafluoropentanedionate).
i pz = pyrazolato.
j dtc = dimethyldithiocarbamato.
k $[Ac_2(dT)]$ = 3′,5′-diacetylthymidine.
l P_A *trans* to DMSO, P_B *trans* to $Ac_2(dT)$.
m 3,5-Me_2pzH = 3,5-dimethylparazole.
n Peak assignment deduced from ref. 66.
o 1-MeTy(-H) = 1-methylthyminato-N^3.
p 1-MeCy = 1-methylcytosine-N^3.
q Nuclearity in structure is unspecified in source of reference.
r Data obtained at 300 K; at 233 K, δ −6.0, −8.5; J(PP) 95 Hz.
s bypm = 2,2′-bypyrimidine.
t In the solid state ^{31}P spectrum, δ −14.9, ≈37 ppm.

References to Table 1-1

[1] A. L. Bandini, G. Banditelli, M. A. Cinellu, G. Sanna, G. Minghetti, F. Demartin, M. Manassero, *Inorg. Chem.* **1989**, *28*, 404–410.
[2] B. Corain, B. Longato, G. Favero, D. Ajò, G. Pilloni, U. Russo, F. R. Kreissl, *Inorg. Chim. Acta* **1989**, *157*, 259–266.
[3] T. S. A. Hor, H. S. O. Chan, K.-L. Tan, L.-T. Phang, Y. K. Yan, L.-K. Liu, Y.-S. Wen, *Polyhedron* **1991**, *10*, 2437–2450.
[4] A. W. Rudie, D. W. Lichtenberg, M. L. Katcher, A. Davison, *Inorg. Chem.* **1978**, *17*, 2859–2863.
[5] T. S. A. Hor, L.-T. Phang, *J. Organomet. Chem.* **1989**, *373*, 319–324.
[6] T. S. A. Hor, L.-T. Phang, *Polyhedron* **1990**, *9*, 2305–2308.
[7] L.-T. Phang, K.-S. Gan, H. K. Lee, T. S. A. Hor, *J. Chem. Soc., Dalton Trans.* **1993**, 2697–2702.
[8] I. R. Butler, W. R. Cullen, T.-J. Kim, S. J. Rettig, J. Trotter, *Organometallics* **1985**, *4*, 972–980.
[9] P. K. Baker, S. G. Fraser, P. Harding, *Inorg. Chim. Acta* **1986**, *116*, L5–L6.
[10] P. K. Baker, M. V. Kampen, D. A. Kendrick, *J. Organomet. Chem.* **1991**, *421*, 241–246.
[11] T. A. George, R. C. Tisdale, *J. Am. Chem. Soc.* **1985**, *107*, 5157–5159.
[12] J. T. Lin, Y. M. Shiao, *J. Organomet. Chem.* **1987**, *334*, C31–C34.
[13] P. K. Baker, K. R. Flower, *Polyhedron* **1990**, *9*, 2507–2510.
[14] T. S. A. Hor, L.-T. Phang, *J. Organomet. Chem.* **1990**, *390*, 345–350.
[15] S. Onaka, T. Moriya, S. Takagi, A. Mizuno, H. Furuta, *Bull. Chem. Soc. Jpn.* **1992**, *65*, 1415–1427.
[16] S. Onaka, H. Furuta, S. Takagi, *Angew. Chem. Int. Ed. Engl.* **1993**, *32*, 87–88.
[17] S. Onaka, *Bull. Chem. Soc. Jpn.* **1986**, *59*, 2359–2361.
[18] S. Onaka, A. Mizuno, S. Takagi, *Chem. Lett.* **1989**, 2037–2040.
[19] T. M. Miller, K. J. Ahmed, M. S. Wrighton, *Inorg. Chem.* **1989**, *28*, 2347–2355.
[20] Y. K. Yan, *M. Sc. Thesis*, National University of Singapore, **1992**.
[21] Y. K. Yan, H. S. O. Chan, T. S. A. Hor, K.-L. Tan, L.-K. Liu, Y.-S. Wen, *J. Chem. Soc. Dalton Trans.* **1992**, 423–426.
[22] T. S. A. Hor, L.-T. Phang, *J. Organomet. Chem.* **1990**, *381*, 121–125.
[23] T.-J. Kim, K.-H. Kwon, S.-C. Kwon, J.-O. Baeg, S.-C. Shim, D.-H. Lee, *J. Organomet. Chem.* **1990**, *389*, 205–217.
[24] T.-J. Kim, S.-C. Kwon, Y.-H. Kim, N. H. Heo, M. M. Teeter, A. Yamano, *J. Organomet. Chem.* **1991**, *426*, 71–86.
[25] M. I. Bruce, I. R. Butler, W. R. Cullen, G. A. Koutsantonis, M. R. Snow, E. R. T. Tiekink, *Aust. J. Chem.* **1988**, *41*, 963–969.
[26] M. Sato, M. Sekino, *J. Organomet. Chem.* **1993**, *444*, 185–190.
[27] N. W. Alcock, J. M. Brown, M. Rose, A. Wienand, *Tetrahedron: Asymmetry* **1991**, *2*, 47–50.
[28] M. Saburi, K. Aoyagi, T. Kodama, T. Takahashi, Y. Uchida, K. Kozawa, T. Uchida, *Chem. Lett.* **1990**, 1909–1912.
[29] F. Neumann, G. Süss-Fink, *J. Organomet. Chem.* **1989**, *367*, 175–185.
[30] M. Adachi, M. Kita, K. Kashiwabara, J. Fujita, N. Iitaka, S. Kurachi, S. Ohba, D.-M. Jin, *Bull. Chem. Soc. Jpn.* **1992**, *65*, 2037–2044.
[31] U. Casellato, B. Corain, R. Graziani, B. Longato, G. Pilloni, *Inorg. Chem.* **1990**, *29*, 1193–1198.
[32] B. Longato, G. Pilloni, R. Graziani, U. Casellato, *J. Organomet. Chem.* **1991**, *407*, 369–376.
[33] W. R. Cullen, T.-J. Kim, F. W. B. Einstein, T. Jones, *Organometallics* **1985**, *4*, 346–351.
[34] P. Kalck, C. Randrianalimanana, M. Ridmy, A. Thorez, H. T. Dieck, J. Ehlers, *New. J. Chem.* **1988**, *12*, 679–686.
[35] J. D. Unruh, W. J. Wells, III, DE 2617306 **1976** [*Chem. Abstr.* **1977**, *86*, 22369e].
[36] U. Casellato, D. Ajò, G. Valle, B. Corain, B. Longato, R. Graziani, *J. Crystallogr. Spectrosc. Res.* **1988**, *18*, 583–590.
[37] T. Hayashi, M. Konishi, Y. Kobori, M. Kumada, T. Higuchi, K. Hirotsu, *J. Am. Chem. Soc.* **1984**, *106*, 158–163.
[38] C. E. Housecroft, S. M. Owen, P. R. Raithby, B. A. M. Shaykh, *Organometallics* **1990**, *9*, 1617–1623.
[39] B. Longato, G. Pilloni, G. Valle, B. Corain, *Inorg. Chem.* **1988**, *27*, 956–958.

[40] H. Sato, H. Shigeta, M. Sekino, S. Akabori, *J. Organomet. Chem.* **1993**, *458*, 199–204.
[41] M. Sato, M. Sekino, S. Akabori, *J. Organomet. Chem.* **1988**, *344*, C31–C34.
[42] L.-T. Phang, S. C. F. Au-Yeung, T. S. A. Hor, S. B. Khoo, Z.-Y. Zhou, T. C. W. Mak, *J. Chem. Soc., Dalton Trans.* **1993**, 165–172.
[43] D. A. Clemente, G. Pilloni, B. Corain, B. Longato, M. Tiripicchio-Camellini, *Inorg. Chim. Acta* **1986**, *115*, L9–L11.
[44] G. M. Whitesides, J. F. Gaasch, E. R. Stedronsky, *J. Am. Chem. Soc.* **1972**, *94*, 5258–5270.
[45] B. S. Haggerty, C. E. Housecroft, A. L. Rheingold, B. A. M. Shaykh, *J. Chem. Soc., Dalton Trans.* **1991**, 2175–2184.
[46] B. Longato, G. Pilloni, G. M. Bonora, B. Corain, *J. Chem. Soc., Chem. Commun.* **1986**, 1478–1479.
[47] P. F. Kelly, A. M. Z. Slawin, D. J. Williams, J. D. Woollins, *Polyhedron* **1988**, *7*, 1925–1930.
[48] R. D. Kelly, G. B. Young, *Polyhedron* **1989**, *8*, 433–445.
[49] G. Bandoli, G. Trovó, A. Dolmella, B. Longato, *Inorg. Chem.* **1992**, *31*, 45–51.
[50] M. Zhou, Y. Xu, L.-L. Koh, K. F. Mok, P. H. Leung, T. S. A. Hor, unpublished results.
[51] U. Casellato, R. Graziani, G. Pilloni, *J. Crystallogr. Spectrosc. Res.*, **1993**, *23*, 571–575.
[52] G. Pilloni, R. Graziani, B. Longato, B. Corain, *Inorg. Chim. Acta* **1991**, *190*, 165–167.
[53] C. Vogler, W. Kaim, *J. Organomet. Chem.* **1990**, *398*, 293–298.
[54] S.-P. Neo, Z.-Y. Zhou, T. C. W. Mak, T. S. A. Hor, *J. Chem. Soc., Dalton Trans.* **1994**, in press.
[55] S.-P. Neo, T. S. A. Hor, Z.-Y. Zhou, T. C. W. Mak, *J. Organomet. Chem.* **1994**, *464*, 113–119.
[56] T. S. A. Hor, S. P. Neo, C. S. Tan, T. C. W. Mak, K. W. P. Leung, R.-J. Wang, *Inorg. Chem.* **1992**, *31*, 4510–4516.
[57] A. Houlton, R. M. G. Roberts, J. Silver, R. V. Parish, *J. Organomet. Chem.* **1991**, *418*, 269–275.
[58] D. T. Hill, G. R. Girard, F. L. McCabe, R. K. Johnson, P. D. Stupik, J. H. Zhang, W. M. Reiff, D. S. Eggleston, *Inorg. Chem.* **1989**, *28*, 3529–3533.
[59] L.-T. Phang, T. S. A. Hor, Z.-Y. Zhou, T. C. W. Mak, *J. Organomet. Chem.* **1994**, *469*, 253–261.
[60] P. M. N. Low, Y. K. Yan, H. S. O. Chan, T. S. A. Hor, *J. Organomet. Chem.* **1993**, *454*, 205–209.
[61] K. R. Mann, W. H. Morrison, Jr., D. N. Hendrickson, *Inorg. Chem.* **1974**, *13*, 1180–1185.
[62] Z.-G. Fang, Y.-S. Wen, R. K. L. Wong, S.-C. Ng, L.-K. Liu, T. S. A. Hor, *J. Cluster Sci.* **1994**, *5*, 327–340.
[63] S. T. Chacon, W. R. Cullen, M. I. Bruce, O. Shawkataly, F. W. B. Einstein, R. H. Jones, A. C. Willis, *Can. J. Chem.* **1990**, *68*, 2001–2010.
[64] W. H. Watson, A. Nagl, S. Hwang, M. G. Richmond, *J. Organomet. Chem.* **1993**, *445*, 163–170.
[65] S. M. Draper, C. E. Housecroft, A. L. Rheingold, *J. Organomet. Chem.* **1992**, *435*, 9–20.

diamine (TMEDA) [18, 19]. In spite of its early discovery, the development of dppf chemistry took off in a significant way only in the 1980s. Since then, a multitude of dppf complexes has been synthesized (Table 1-1) and the conformational flexibility and catalytic activity of dppf itself have been widely explored. Coordination complexes of dppf are generally prepared by the direct reaction of dppf with binary compounds [20, 21] or other primary forms of Lewis acids [22]. For carbonyl complexes, substitution reactions by photolysis, thermolysis and chemically-induced decarbonylation are usually employed.

The ligand dppf is capable of undergoing coordination to a variety of transition metals, for example the halo complexes of the late metals, carbonyl complexes of Group 6, 7 and 8 metals, and the Group 5 metalates, notably $[NEt_4][M(CO)_4$(dppf-

P, *P'*)] (M = V, Ta) [20]. Dppf usually functions as a phosphine donor, e.g., HgX_2(dppf-*P*, *P'*) (X = Cl, Br, I, SCN) [21], although occasionally direct Fe $\rightarrow$ M donation is observed, e.g., $Fe(Cp)_2 \cdot 7\,HgCl_2$ [23].

The steric bulk of the dppf ligand is best illustrated in the coordination geometry of NiX_2(dppf-*P*, *P'*) (X = Cl, Br, I). Due to the high steric requirements and bite angle of dppf, NiX_2(dppf-*P*, *P'*) prefers to adopt a high-spin tetrahedral geometry [20, 24] rather than the planar diamagnetic form [25]. In spite of this, seven-coordinated dppf complexes are known, such as [$WI_2(CO)_3$(dppf-*P*, *P'*)] [26], *cis*-[$MoCl_2(CO)_2$(dppf-*P*, *P'*)(dppf-*P*)] [26] and the bimetallic phosphine-bridged [$M_2I_4(CO)_6$(μ-dppf)L_2] (M = Mo, W; L = PPh_3, $AsPh_3$ or $SbPh_3$) [27]. At the other extreme, lower coordination geometries such as trigonal planar can be stabilized by dppf (e.g., [Au(dppf-*P*, *P'*)(dppf-*P*)]Cl [28]. This complex readily rearranges to give [Au(dppf-*P*, *P'*$)_2$]Cl in solution, presumably via a dibridged intermediate [Au(dppf-*P*, *P'*)(μ-dppf)$]_2{}^{2+}$. The monobridged relatives of the latter, [M_2(dppf-*P*, *P'*$)_2$(μ-dppf)$]^{2+}$ (M = Cu, **1a** [29, 30], Ag, **1b** [31], Au, **1c** [32] Fc = ferrocenyl), have recently been characterized structurally. The isolation of these complexes of varying geometries clearly demonstrates the flexibility of dppf as a ligand.

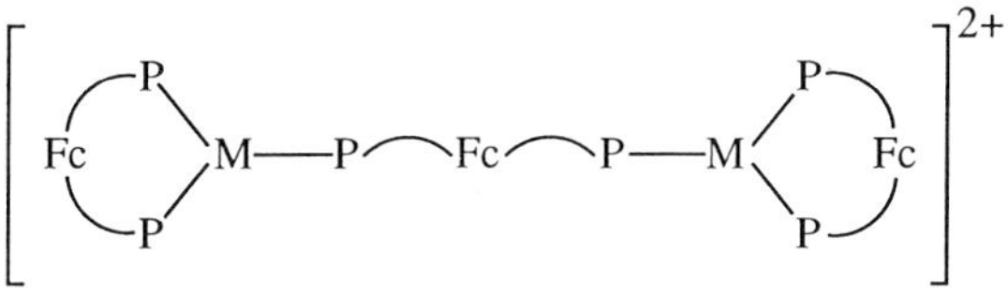

1a M = Cu, **1b** M = Ag, **1c** M = Au

1

Dppf is commonly seen in its chelating or bridging state. Of special interest is the quasi-closed bridging system of dppf complexes, $Re_2(\mu\text{-}OMe)_2(CO)_6$($\mu$-dppf) [33] and $Rh_2(\mu\text{-}S\text{-}t\text{Bu})_2(CO)_2$($\mu$-dppf), **2** [34], in which the sterically demanding dppf ligand coexists with two bridging ligands of much smaller size. Comparison of the

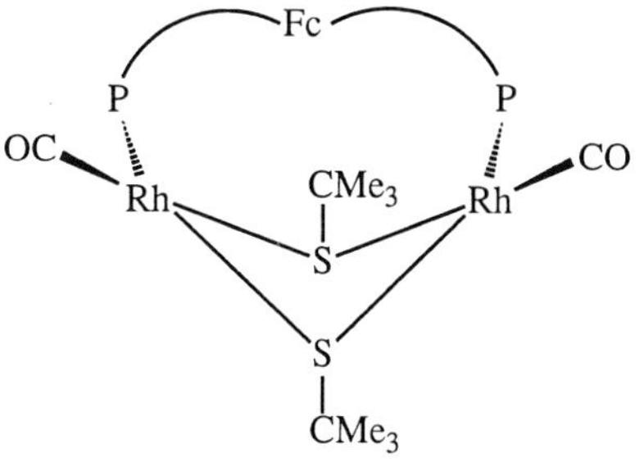

2

former with $Re_2(\mu\text{-OMe})_2(CO)_6(\mu\text{-dppm})$ [35] shows some resemblances despite the vast skeletal differences between dppf and dppm. Whether the "A-frame" chemistry of dppf can be developed remains an interesting prospect.

Bridging dppf is known to span a wide range of metal—metal distances. Strong M-M interactions are observed in some clusters supported by dppf bridges (see Sect. 1.6). The greatest M ... M separations are observed in singly-bridging complexes, e.g., $[M(CO)_5]_2(\mu\text{-dppf})$ (M = Cr, Mo) (>7 Å) [36]. These distances are significantly reduced when the metals are locked by two bridging links, e.g., $[Ag(NO_3)(\mu\text{-dppf})]_2$ (Ag ... Ag 3.936(2) Å) [8]. The ability of the dppf ligand to accommodate a wide range of M ... M distances distinguishes it from other common diphosphines.

Homoleptic complexes have emerged in recent years (e.g., $[M(dppf\text{-}P,P')_2]^+$ (M = Rh, Ir) and $[M_2(dppf\text{-}P,P')_2(\mu\text{-dppf})]^{2+}$ (M = Cu [29, 30], Ag [31], Au [32]) **1**. These expectedly crowded complexes often show the coexistence of two coordinating modes of dppf, notably the bridging and chelating modes, e.g., $Ag_2(HCO_2\text{-}O)_2(dppf\text{-}P,P')_2(\mu\text{-dppf})$ [8] and $Rh_2H_2(CO)_2(dppf\text{-}P,P')_2(\mu\text{-dppf})$ [37]. In other instances, e.g., *cis*-$[MoCl_2(CO)_2(dppf\text{-}P,P')(dppf\text{-}P)]$ [26] and *fac*-$[Mo(dppf\text{-}P,P')(dppf\text{-}P)(triphos)]$ (triphos = $PhP(CH_2CH_2PPh_2)_2$) [38], the presence of both chelating and monodentate modes is necessary to satisfy the electron count. Within the common structural framework of $[M_2(dppf)_3]^{2+}$ (M = Cu, Ag, Au), the dppm ligand opts for a triply-bridged species $[M(\mu\text{-dppm})_3M]^{n+}$ [39], whereas dppf prefers a bridge-chelate combination, $[(dppf\text{-}P,P')M(\mu\text{-dppf})M(dppf\text{-}P,P')]^{2+}$. These basic differences between the dinuclear chemistry of dppm and dppf have been discussed [32].

Dppf functions chiefly as a diphosphine ligand and usually no formal ferrocenyl-metal interaction occurs. Direct interaction between the ferrocenyl iron and the phosphine-bound metal center was first found in $[Pd(dppf\text{-}P,P')(PPh_3)][BF_4]_2$ **3** [40]. This complex is prepared from ligand addition to $[Pd(CH_3CN)_4][BF_4]_2$

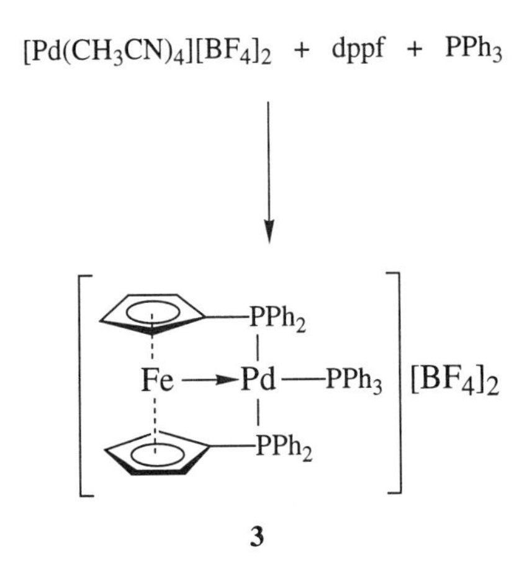

Fig. 1-1. Structure of $[Pd(dppf\text{-}P,P')(PPh_3)][BF_4]_2$ [40] showing an Fe → Pd donor/acceptor bond (dppf = $Fe(C_5H_4PPh_2)_2$).

(Fig. 1-1). The Fe → Pd dative bond is obviously promoted by the electron deficiency of the Pd center. This unusual structure was verified by its electronic spectrum, which is similar to that of $[Pd(PPh_3)\{(C_5H_4SR)_2Fe\}][BF_4]_2$ (R = Me, *i*Pr, *i*Bu, CH_2Ph, Ph) [40], and confirmed by X-ray crystallography [41]. Another rare case of Fe → M bonding is found in the cluster $Ru_3(\mu\text{-H})\{\mu_3\text{-PPh}_2(\eta^1,\eta^5\text{-C}_5H_3)Fe(\eta^5\text{-C}_5H_4PPh_2)\}(CO)_8$ **4**, which is obtained from the pyrolysis of $Ru_3(CO)_{10}(\mu\text{-dppf})$ under mild conditions [6].

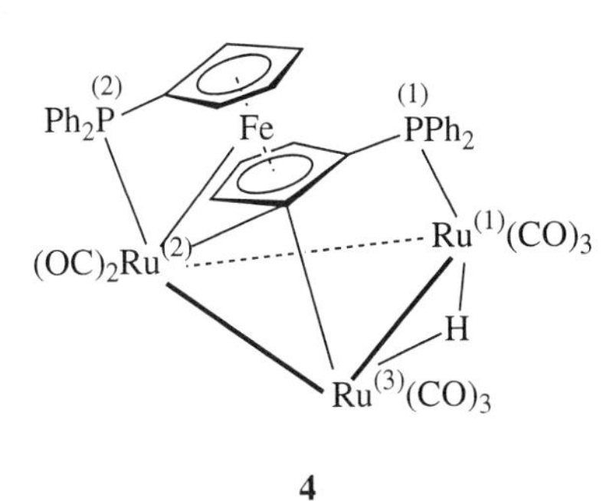

4

Bearing a ferrocenyl moiety, all dppf complexes are in principle redox active even though not many of them have been successfully demonstrated (see Chapter 7). Chemical oxidation of dppf or its complexes gives rise to two possibilities. In the presence of strong oxidizing agents, e.g. H_2O_2 or Br_2, complex degradation to free phosphine (di)oxide $Fe[C_5H_4P(O)Ph_2]_2$ [18, 42] is possible. Under more controlled conditions, a pendant phosphine site can be selectively oxidized whilst the coordinated site is unaffected. This case is illustrated in the H_2O_2 oxidation of $Re_2(CO)_9$(dppf-*P*) to $Re_2(CO)_9$(dppfO-*P*) [42]. The same complex can be conveniently prepared from $[Re_2(CO)_9]_2(\mu\text{-dppf})$ and Me_3NO. It is interesting that this oxidative bridge-scission via oxygen transfer from amine oxide to a stable bridging phosphine should proceed preferentially to a nucleophilic attack on a carbonyl carbon. Oxidation can also occur at the ferrocenyl iron center. With a one-electron oxidant ($NO^+BF_4^-$), *cis*-MCl_2(dppf-*P*, *P'*) (M = Pd, Pt) can be oxidized to green paramagnetic *cis*-$[MCl_2$(dppf-*P*, *P'*)$][BF_4]$ [43]. From ^{57}Fe Mössbauer spectroscopy, the $[dppf]^+$ ion is shown to behave like a typical ferrocenium species [28].

In recent years, dppf chemistry has witnessed enormous progress in its development from mononuclear complexes [43–45] to heteropolymetallic aggregates [7, 42, 45–47] and oligomers [48]. In line with the recent increase in interest in electroactive polymers, it seems natural that dppf polymers should play a key role in the contribution from organometallic chemistry. One recent example of this is the isolation of polymeric $[AuCl(\mu\text{-dppf})]_n$ **5** [32]. Further development in organometallic oligomers and polymers based on dppf chemistry is likely to find some useful applications in materials science. Another area of promise in dppf research is homogeneous catalysis. The variety of organic reactions catalyzed by dppf complexes is phenomenal. The future direction of coordination chemistry lies very

5

much in the interaction between [M(dppf)$_n$] and organic substrates. The relevance of such interactions to the design of catalysis and materials will be explored further in Section 1-5.

1.3 Structural Properties

1.3.1 Modes of Coordination

As expected for a diphosphine ligand, dppf can coordinate to a metal center in several ways (Fig. 1-2). To adjust its mode of coordination or to relieve the strain imposed by complex formation, the Cp rings of dppf can twist about the Cp(centroid)-Fe-Cp(centroid) axis (Fig. 1-3) and tilt towards or away from the Fe center (Fig. 1-4). The phosphorus atoms provide added flexibility by diverging from coplanarity with the Cp rings (Fig. 1-5). With this motive freedom, the bite angle and transannular distance between the two donor atoms are adaptable to the

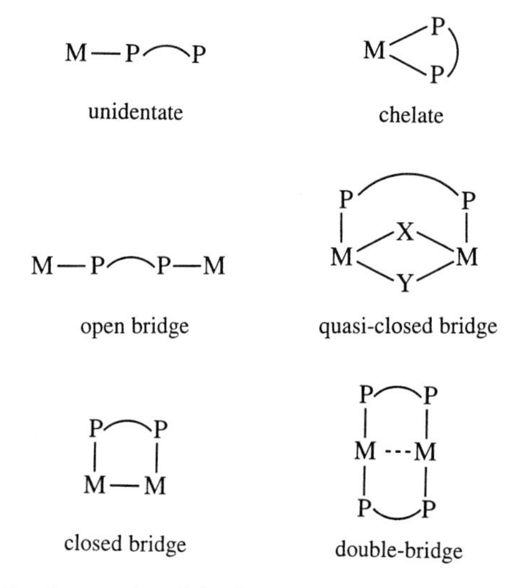

Fig. 1-2. Common coordination modes of dppf.

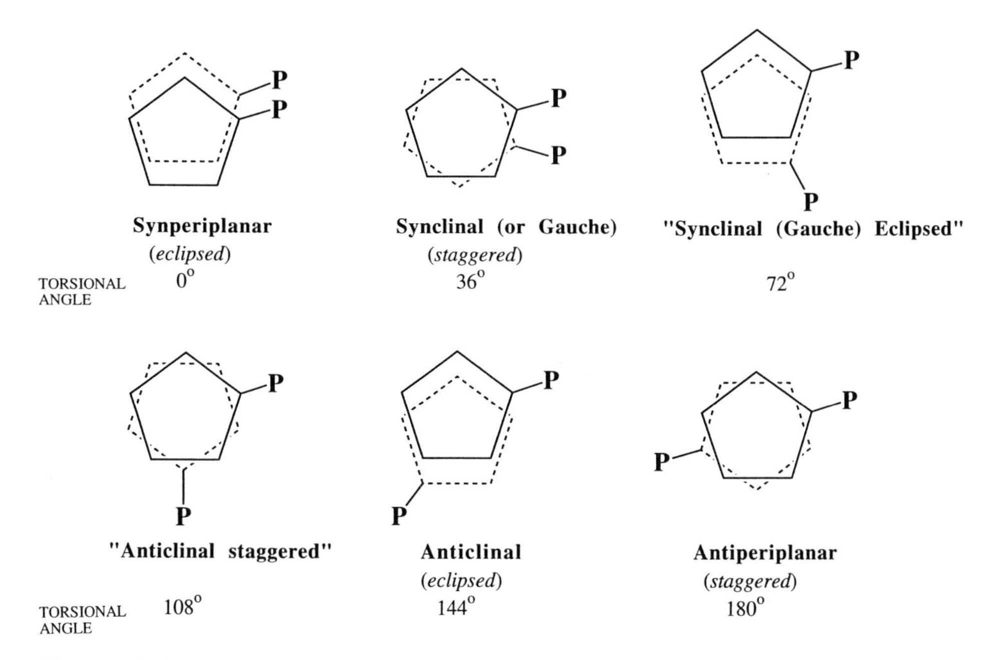

Fig. 1-3. Six ideal conformations of dppf arising from Cp ... Fe ... Cp torsional twist.

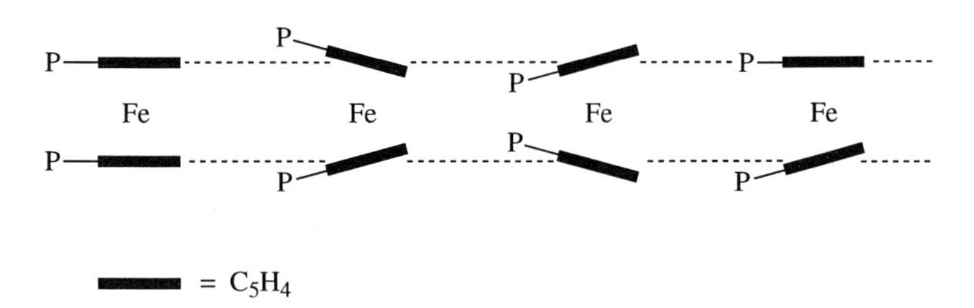

Fig. 1-4. Possible ring tilts of dppf.

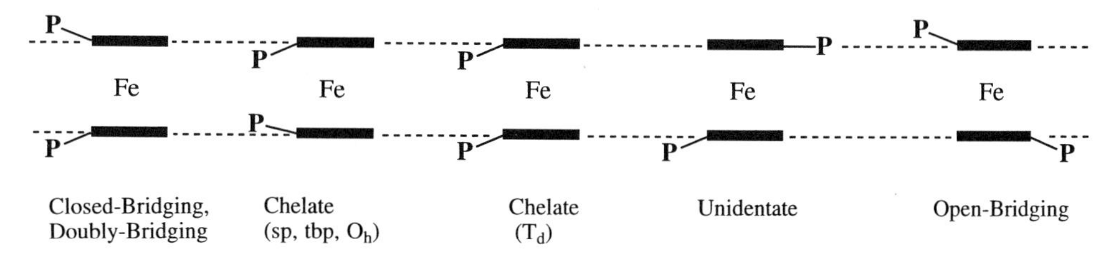

Fig. 1-5. A simplistic correlation of coordination modes of dppf with direction of displacement of phosphorus atom from coplanarity with cyclopentadienyl ring.

Table 1-2. Some structural parameters pertaining to the ferrocenyl skeleton of the crystallographically characterized complexes of dppf (blank entries denote unavailable data)

Complex	Coordination mode of dppf	Conformation	τ, °[a]
Sandwiched			
$Fe(C_5H_4PPh_2)_2$ *(dppf)*	–	antiperiplanar	180
$Fe\{C_5H_4P(S)Ph_2\}_2$	–	antiperiplanar	180
$[Fe\{C_5H_4P(O)Ph_2\}_2] \cdot 2H_2O$	–	antiperiplanar	180
Linear			
$Au_2Cl_2(\mu\text{-dppf}) \cdot CHCl_3$[f]	open bridge		
Molecule I		antiperiplanar	180
Molecule II		anticlinal (eclipsed)	150
Trigonal planar			
$[AuCl(\mu\text{-dppf})]_n \cdot nCH_2Cl_2$	open bridge	anticlinal (eclipsed)	153.1
$[AuCl(\mu\text{-dppf})]_n$[f]	open bridge		
Molecule I		antiperiplanar	≈180
Molecule II		anticlinal (eclipsed)	≈130
$[Cu_2(\text{dppf-}P,P')_2(\mu\text{-dppf})][ClO_4]_2$[f]	chelate	gauche (staggered)	
Molecule I			40.0(19)[h]
Molecule II			41.0(20)[h]
	open bridge	antiperiplanar	180
$[Ag_2(\text{dppf-}P,P')_2(\mu\text{-dppf})][PF_6]_2$	chelate	gauche (staggered)	44.1[i]
	open bridge	antiperiplanar	180[i]
$[Ag(NO_3)(\mu\text{-dppf})]_2$	doubly bridging	"gauche eclipsed" ≈ "anticlinal staggered"	89.6
$[Ag(NO_3)(\mu\text{-dppf})]_2 \cdot C_2H_4Cl_2$	doubly bridging	"gauche eclipsed" ≈ "anticlinal staggered"	90.7
$Ag_2(\mu\text{-}C_6H_5CO_2)_2(\mu\text{-dppf})$	quasi-closed bridge	anticlinal (staggered)[j]	–
$[Au_2(\text{dppf-}P,P')_2(\mu\text{-dppf})][NO_3]_2 \cdot 2\,H_2O$	chelate	gauche (staggered)	43.3[i]
	open bridge	antiperiplanar	180
Square planar			
cis-$[Rh(\text{dppf-}P,P)(\text{nbd})][ClO_4]$	chelate	gauche (staggered)[j]	–
$Rh_2(\mu\text{-S-}t\text{Bu})_2(CO)_2(\mu\text{-dppf})$	quasi-closed bridge	"gauche eclipsed"	72
$[Ir(\text{dppf-}P,P')_2][BPh_4]$	chelate	gauch (staggered)	34.8(12)[h]
		gauche (staggered)	25.1(12)[h]
cis-$[PdCl_2(\text{dppf-}P,P')] \cdot CH_2Cl_2$	chelate	gauche (staggered)	39.5
cis-$[PdCl_2(\text{dppf-}P,P')] \cdot CHCl_3$	chelate	gauche (staggered)	39.5[m]
cis-$[PdCl_2(\text{dppf-}P,P')] \cdot 0.25\text{thf}$	chelate	gauche (staggered)	
Molecule I			36.1

θ, °[b]	X_A-Fe-X_B, °	P-Fe-P, °	P ... P, Å	δ_P, Å[c]	δ_C, Å × 10^4[d]	Ref.
≈0	–	–	6.920[e]	–	–	[1]
0.04(19)	180	–	–	0.037(7)	22	[2]
0	180	–	–	−0.044	38	[2]
0[g]	180[g]	180	6.86(1)[g]	0.068[g]	0[g]	[3]
3.6	177	–	6.28(1)	0.029, 0.041	6, 5	
–	–	–	–	–	–	[4]
						[5]
–	–	–	–	–	–	
–	–	–	–	–	–	
0.8(8)[h]	178.4(10)[h]	66.9(1)[h]	3.762(12)[h]	0.092(18)[h], 0.013(18)	5[h], 4	[6]
0.7(9)[h]	178.2[h]	67.0(1)[h]	3.760(11)[h]	0.073(20)[h], 0.016(20)	−4[h], 0	
≈0	–	–	–	−0.115(22)[h], −0.090(20)	1[h], −3	
4.6[i]	175.5[i]	68.6	3.982	0.128[i], 0.033[i]	83[i]	[7]
0[i]	180[i]	–	–	0.026[i]	69[i]	
4.4	176.0	–	–	−0.138, −0.134	80, 57	[8]
1.5	177.9	–	–	−0.053, −0.105	44, 33	[9]
–	–	–	–	–	–	[8]
3.5[i]	177.2[i]	69.6[i]	3.895[i]	0.0024[i], 0.049	32[i], 88	[4]
–	–	–	–	–	–	
–	179.6	62.0[k]	3.479[k]	0.028, 0.107	–	[10]
–	–	–	5.7	–	–	[11]
9.4(5)[h]	176.1(7)[h]	61.1(1)[h]	3.468(8)[h]	−0.069(6)[h], −0.026(5)	−1[h], 0	[12]
4.9(4)[h]	179.9(7)[h]	59.3(1)[h]	3.410(6)[h]	0.125(4)[h], 0.030(5)	−6[h], −6	
6.2[l]	177.3	–	3.447(1)	0.062(1), 0.035(1)	–	[13]
–	–	–	3.487(2)	–	–	[14]
						[15]
4.8	178.5	–	3.469	0.032, 0.039	48, 22	

Table 1-2. (continued)

Complex	Coordination mode of dppf	Conformation	τ, °[a]
Molecule II			34.1
1-{dppf-*P*, *P'*)Pd}B_3H_7	chelate	gauche (staggered)	45
cis-[$PtCl_2$(dppf-*P*, *P'*)] · 0.5$(CH_3)_2CO$	chelate	gauche (staggered)	32.3(15)[h]
cis-[Pt(S_2N_2)(dppf-*P*, *P'*)]	chelate	gauche (staggered)	40
[Pt_2(μ-H)(μ-CO)(dppf-*P*, *P'*$)_2$][BF_4] · 0.5H_2O	chelate	gauche (staggered)	41.1[o]
		gauche (staggered)	42.6
[Pt(μ-OH)(dppf-*P*, *P'*)]$_2$[BF_4]$_2$	chelate	gauche (staggered)	36.0(15)[h]
[Pt_2Tl(μ_3-S)$_2$(dppf-*P*, *P'*)$_2$][PF_6]	chelate	gauche (staggered)	36.2
[Pt_2H(μ-H)$_2$(dppf-*P*, *P'*)$_2$]Cl	chelate	gauche (staggered)[i]	–
	chelate	gauche (staggered)[i]	–
cis-[Pt(1-MeTy(-H))(dppf-*P*, *P'*)(dmf)] [BF_4] · CH_2Cl_2[p]	chelate	gauche (staggered)[i]	–
cis-[Pt(1-MeTy(-H))(dppf-*P*, *P'*)(1-MeCy)] [BF_4][p,r]	chelate	gauche (staggered)[i]	–
trans-[$PtCl_2${(μ-dppf)W(CO)$_5$}$_2$]	open bridge	anticlinal (eclipsed)	153.9
Tetrahedral			
$NiCl_2$(dppf-*P*, *P'*)	chelate	synperiplanar	9[s]
$NiBr_2$(dppf-*P*, *P'*)	chelate	synperiplanar	6.5
[Ag_2(HCO_2-*O*)$_2$(dppf-*P*, *P'*)$_2$(μ-dppf)] · 2CH_2Cl_2	open bridge	antiperiplanar	180
	chelate	gauche (staggered)	56.5
[Cu(μ-HCO_2)(dppf-*P*, *P'*)]$_2$	chelate	gauche (staggered)[i]	–
[Cu(μ-I)(dppf-*P*, *P'*)]$_2$ · 2 CH_2Cl_2	chelate	gauche (staggered)	41.8
[Ag_2(μ-CH_3CO_2)(μ_3-CH_3CO_2)(μ-dppf)]$_2$	quasi-closed bridge	"gauche eclipsed"	88.8
Trigonal bipyramidal			
ax, *eq*-Fe(CO)$_3$(dppf-*P*, *P'*)	chelate	gauche (staggered)	28.5(3)
Fe(CO)$_4$(dppf-*P*)	unidentate	anticlinal (eclipsed)	132.8
[Fe(CO)$_4$]$_2$(μ-dppf)	open bridge	anticlinal (eclipsed) ~ antiperiplanar	162.8
Octahedral			
[Cr(CO)$_5$]$_2$(μ-dppf)	open bridge	antiperiplanar	171.6(3)
[Mo(CO)$_5$]$_2$(μ-dppf)	open bridge	antiperiplanar	173.2(3)
cis-[Mo(CO)$_4$(dppf-*P*, *P'*)] · C_6H_6	chelate	gauche (staggered)	41.9
Mo(CO)$_5$(dppf-*P*)	unidetante	anticlinal (eclipsed)	132.5
{[Mn_2(CO)$_9$]$_2$(μ-dppf)} · thf	open bridge	anticlinal (eclipsed)	131.4
cis-[Mn_2Cl_2(CO)$_8$(μ-dppf)]	open bridge	antiperiplanar	180
{[Re_2(CO)$_9$]$_2$(μ-dppf)} · solvate	open bridge	anticlinal (eclipsed)	131.3
[Re_2(CO)$_9$(dppfO-*P*)] · H_2O[f,x]	unidentate	anticlinal (eclipsed)	

θ, ° [b]	X_A-Fe-X_B, °	P-Fe-P, °	P ... P, Å	δ_P, Å [c]	δ_C, Å × 10^4 [d]	Ref.
4.4	178.2	–	–	0.017, 0.110	105, 12	
≈0	–	–	3.73	–	–	[16]
5.9(6)[h,l]	178.3(9)[h]	60.9(1)[h]	3.438(6)[h]	−0.010(4)[h]	58[h]	[17]
				−0.0004(40)	−26	
4.4	–	–	3.496[n]	–	–	[18]
2.1[o]	178.9[o]	65.5[o]	3.663[o]	0.028(2)[o], 0.044(2)	0[o], 0	[19]
4.4	178.3	65.1	3.658	0.062(2),	0,	
				−0.019(2)	0	
5.0(6)[h,l]	177.4(9)[h]	61.2(1)[h]	3.387(12)[h]	0.153(7)[h],	−12[h],	[20]
				−0.088(7)	28	
4.9	178.4	61.1	3.441	−0.0319,	84,	[21]
				0.0396	31	
0	–	64.4	3.620	–	–	[22]
5[l]	–	62.9	3.571	–	–	
6[l]	–	61.1[q]	3.426[q]	–	–	[23]
–	–	–	–	–	planar	[23]
5.1	175.7	–	–	–	planar	[24]
4.5[t]	–	–	3.668[u]	–	–	[1]
6.2[t]	173.1	–	3.573(9)	−0.157(6),	–	[13]
				0.087(6)		
0	180	–	–	−0.022	29	[8]
1.0	178.8	–	4.133	−0.053, 0.025	32, 52	
–	–	–	3.720	–	–	[9]
2.3	179.7	–	3.767	0.099, 0.049	242, 251	[9]
5.1	175.6	–	–	0.103, −0.024	35, 31	[8]
2.2	–	61.6[v]	3.439[v]	0.042(2), 0.034(2)	planar	[25]
2.64	–	–	–	–	planar	[26]
1.65	–	–	–	–	planar	[26]
1.3	–	174.7(1)	7.073(3)	−0.189(9)	≦50	[27]
1.3	–	174.8(1)	7.062(3)	−0.198(9)	≦50	[27]
2.2[l]	179.0	–	3.783(1)	0.017(1), 0.026(1)	–	[13]
2.3	177.4(2)	–	–	–	planar	[24]
–	–	140.3(1)	6.697(4)	−0.235(13)	–	[28]
0	–	–	–	−0.22(2),	4.4[w]	[29]
				−0.22(2)		
–	–	140.6(1)	6.689(4)	−0.22(3)	–	[28]
						[28]

Table 1-2. (continued)

Complex	Coordination mode of dppf	Conformation	τ, °[a]
Molecule I			127.0(12)
Molecule II			122.1(13)
fac-$(ReCl(CO)_3(dppf\text{-}P,P')]$	chelate	gauche (staggered)	14.8
$Re_2(\mu\text{-}OMe)_2(CO)_6(\mu\text{-}dppf)$	quasi-closed bridge	"gauche eclipsed"	85.9(5)
$[RuH(dppf\text{-}P,P')_2(\eta^2\text{-}H_2)][PF_6]$	chelate	"gauche eclipsed"[i]	–
$Ru(\eta^3\text{-}C_3H_5)(\eta^3\text{-}C_5HF_6O_2)(dppf\text{-}P,P')$[a']	chelate	gauche (staggered)[i]	–
2-legged piano-stool			
$Rh\{(\eta^6\text{-}C_6H_5)B(C_6H_5)_3\}(dppf\text{-}P,P')$	chelate	gauche (staggered)	32.2(8)[h]
3-legged piano-stool			
$Mn(\eta^5\text{-}MeCp)(CO)(dppf\text{-}P,P')\cdot CHCl_3$	chelate	synperiplanar	3.1(9)
$Mn(\eta^5\text{-}MeCp)(CO)_2(dppf\text{-}P)$	unidentate	anticlinal (eclipsed)	128(1)
$Ru(\eta^5\text{-}C_5H_5)H(dppf\text{-}P,P')$[f,g]	chelate	gauche (staggered)	
Molecule I			37[c']
Molecule II			29
Clusters			
$Fe_3(CO)_{10}(\mu\text{-}dppf)$	closed bridge	"gauched eclipsed"	61.6
$Fe_3(\mu_3\text{-}S)_2(CO)_7(\mu\text{-}dppf)$	quasi-closed bridge	"gauched eclipsed"	80.4(5)
$Ru_3(CO)_{10}(\mu\text{-}dppf)$	closed bridge	"gauche eclipsed"	≈72[d']
$Ru_4(\mu\text{-}H)(CO)_{12}(\mu_6\text{-}B)Au_2(\mu\text{-}dppf)$	closed bridge	"gauche ecliped"	80
$Co_3(\mu_3\text{-}CMe)(CO)_7(\mu\text{-}dppf)$	closed bridge	"gauche eclipsed"	69.6(6)
$Co_3(\mu_3\text{-}CPh)(CO)_7(\mu\text{-}dppf)$	closed bridge	"gauche eclipsed"[j]	–

a The torsion angle τ is defined as $C_A \ldots X_A \ldots X_B \ldots C_B$, where C_A is the carbon atom in Cp ring A that is bonded to a P atom (likewise for C_B) and X_A, X_B are the centroids of the two Cp rings.
b θ is the dihedral angle between the two Cp rings.
c δ_P is the deviation of the linked P atom from the same plane. A positive sign means that the P atom is on the same side of the Cp ring as the Fe atom.
d δ_C is the mean deviation of the C atoms in a Cp rings from their least-squares planes.
e Calculated based on centrosymmetry of the molecule (= 2 × Fe ... P).
f Two crystallographically independent molecules are present per unit cell.
g D. S. Eggleston, private communication.
h G. Pilloni and D. Clemente, private communication.
i T. S. A. Hor and T. C. W. Mak, unpublished results.
j Conformation deduced from X-ray diagram.
k Data calculated from [10].
l Attractive tilt between the two P atoms.
m Data obtained from [34].
n Data calculated from [18].
o F. Demartin, private communication.
p 1-MeTy(-H) = 1-methylthyminato-N^3.
q Data calculated from [23].
r 1-MeCy = 1-methylcytosine-N^3.

θ, °[b]	X_A-Fe-X_B, °	P-Fe-P, °	P ... P, Å	δ_P, Å[c]	δ_C, Å × 10^{4}[d]	Ref.
–	–	133.5(2)	6.369(7)	−0.22(3), 0.00	–	
–	–	131.7(2)	6.342(8)	−0.24(4), 0.00	–	
–	177.2	63.2[y]	3.644[y]	–	–	[30]
0.9(4)	–	–	–	0.016(16), 0.034(15)	–	[31]
–	–	–	3.850[z], 3.819	–	–	[32]
–	–	–	–	–	–	[33]
4.4(4)[h]	178.6(4)[h]	59.2(1)[h]	3.342(3)[h]	0.078(3)[h], 0.070(3)	−4[h], −5	[34]
4.3(6)	–	58.8[b′]	3.400[b′]	0.04(1), −0.04(1)	0[w], 6	[35, 36]
2.3(5)	–	–	–	−0.21(2), 0.00(2)	–	[35]
						[37]
4.9	177.4[c′]	60.6[c′]	3.43	0.064[c′], 0.091	88[c′], 83	
6.3	177.4	58.6	3.35	0.045, 0.018	73, 58	
3.07	–	–	–	–	–	[26]
3.2(4)	176.1(4)	–	–	−0.070(13), −0.054(14)	2, 0	[38]
5.4	174.6	93.1[c′]	5.16[c′]	−0.072[c′]	12[c′]	[39]
4.0	176.0(5)	–	–	–	–	[40]
2.3(3)	–	–	4.871(3)	−0.15(1), −0.16(1)	0.02, 0.01	[29, 35]
–	–	–	–	–	–	[41]

s Data obtained from [35].
t Repulsive tilt between the two P atoms.
u Data calculated from [1].
v Data calculated from [25].
w S. Onaka, private communication.
x dppfO = η^1-$Ph_2P(C_5H_4)Fe(C_5H_4)P(O)Ph_2$.
y Data calculated from [30].
z Data calculated from [32].
a′ $C_5HF_6O_2$ (hexafluoropentanedionate).
b′ Data calculated from [35, 36].
c′ M. I. Bruce and E. R. T. Tiekink, private communication.
d′ Data obtained from [40].

References to Table 1-2

[1] U. Casellato, D. Ajó, G. Valle, B. Corain, B. Longato, R. Graziani, *J. Crystallogr. Spectrosc. Res.* **1988**, *18*, 583–590.
[2] T. S. A. Hor, L.-K. Liu, T. C. W. Mak, unpublished results.
[3] D. T. Hill, G. R. Girard, F. L. McCabe, R. K. Johnson, P. D. Stupik, J. H. Zhang, W. M. Reiff, D. S. Eggleston, *Inorg. Chem.* **1989**, *28*, 3529–3533.

[4] L.-T. Phang, T. S. A. Hor, Z.-Y. Zhou, T. C. W. Mak, *J. Organomet. Chem.* **1994**, *469*, 253–261.
[5] A. Houlton, D. M. P. Mingos, D. M. Murphy, D. J. Williams, L.-T. Phang, T. S. A. Hor, *J. Chem. Soc., Dalton Trans.* **1993**, 3629.
[6] U. Casellato, R. Graziani, G. Pilloni, *J. Crystallogr. Spectrosc. Res.* **1993**, *23*, 571–575.
[7] S.-P. Neo, T. S. A. Hor, Z.-Y. Zhou, T. C. W. Mak, *J. Organomet. Chem.* **1994**, *464*, 113–119.
[8] T. S. A. Hor, S. P. Neo, C. S. Tan, T. C. W. Mak, K. W. P. Leung, R.-J. Wang, *Inorg. Chem.* **1992**, *31*, 4510–4516.
[9] S. P. Neo, T. S. A. Hor, T. C. W. Mak, unpublished results.
[10] W. R. Cullen, T.-J. Kim, F. W. B. Einstein, T. Jones, *Organometallics* **1985**, *4*, 346–351.
[11] P. Kalck, C. Randrianalimanana, M. Ridmy, A. Thorez, H. T. Dieck, J. Ehlers, *New. J. Chem.* **1988**, *12*, 679–686.
[12] U. Casellato, B. Corain, R. Graziani, B. Longato, G. Pilloni, *Inorg. Chem.* **1990**, *29*, 1193–1198.
[13] I. R. Butler, W. R. Cullen, T.-J. Kim, S. J. Rettig, J. Trotter, *Organometallics* **1985**, *4*, 972–980.
[14] T. Hayashi, M. Konishi, Y. Kobori, M. Kumada, T. Higuchi, K. Hirotsu, *J. Am. Chem. Soc.* **1984**, *106*, 158–163.
[15] L.-T. Phang, T. S. A. Hor, T. C. W. Mak, unpublished results.
[16] C. E. Housecroft, S. M. Owen, P. R. Raithby, B. A. M. Shaykh, *Organometallics* **1990**, *9*, 1617–1623.
[17] D. A. Clemente, G. Pilloni, B. Corain, B. Longato, M. Tiripicchio-Camellini, *Inorg. Chim. Acta* **1986**, *115*, L9–L11.
[18] P. F. Kelly, A. M. Z. Slawin, D. J. Williams, J. D. Woollins, *Polyhedron* **1988**, *7*, 1925–1930.
[19] A. L. Bandini, G. Banditelli, M. A. Cinellu, G. Sanna, G. Minghetti, F. Demartin, M. Manassero, *Inorg. Chem.* **1989**, *28*, 404–410.
[20] B. Longato, G. Pilloni, G. Valle, B. Corain, *Inorg. Chem.* **1988**, *27*, 956–958.
[21] M. Zhou, Y. Xu, L.-L. Koh, K. F. Mok, P.-H. Leung, T. S. A. Hor, unpublished results.
[22] B. S. Haggerty, C. E. Housecroft, A. L. Rheingold, B. A. M. Shaykh, *J. Chem. Soc., Dalton Trans.* **1991**, 2175–2184.
[23] G. Bandoli, G. Trovò, A. Dolmella, B. Longato, *Inorg. Chem.* **1992**, *31*, 45–51.
[24] L.-T. Phang, S. C. F. Au-Yeung, T. S. A. Hor, S. B. Khoo, Z.-Y. Zhou, T. C. W. Mak, *J. Chem. Soc., Dalton Trans.* **1993**, 165–172.
[25] T.-J. Kim, K.-H. Kwon, S.-C. Kwon, J.-O. Baeg, S.-C. Shim, D.-H. Lee, *J. Organomet. Chem.* **1990**, *389*, 205–217.
[26] T.-J. Kim, S.-C. Kwon, Y.-H. Kim, N. H. Heo, M. T. Teeter, A. Yamano, *J. Organomet. Chem.* **1991**, *426*, 71–86.
[27] T. S. A. Hor, L.-T. Phang, L.-K. Liu, Y.-S. Wen, *J. Organomet. Chem.* **1990**, *397*, 29–39.
[28] T. S. A. Hor, H. S. O. Chan, K.-L. Tan, L.-T. Phang, Y. K. Yan, L.-K. Liu, Y.-S. Wen, *Polyhedron* **1991**, *10*, 2437–2450.
[29] S. Onaka, A. Mizuno, S. Takagi, *Chem. Lett.* **1989**, 2037–2040.
[30] T. M. Miller, K. J. Ahmed, M. S. Wrighton, *Inorg. Chem.* **1989**, *28*, 2347–2355.
[31] Y. K. Yan, H. S. O. Chan, T. S. A. Hor, K.-L. Tan, L.-K. Liu, Y.-S. Wen, *J. Chem. Soc., Dalton Trans.* **1992**, 423–426.
[32] M. Saburi, K. Aoyagi, T. Kodama, T. Takahashi, Y. Uchida, K. Kozawa, T. Uchida, *Chem. Lett.* **1990**, 1909–1912.
[33] N. W. Alcock, J. M. Brown, M. Rose, A. Wienand, *Tetrahedron: Asymmetry* **1991**, *2*, 47–50.
[34] B. Longato, G. Pilloni, R. Graziani, U. Casellato, *J. Organomet. Chem.* **1991**, *407*, 369–376.
[35] S. Onaka, T. Moriya, S. Takagi, A. Mizuno, H. Furuta, *Bull. Chem. Soc. Jpn.* **1992**, *65*, 1415–1427.
[36] S. Onaka, *Bull. Chem. Soc. Jpn.* **1986**, *59*, 2359–2361.
[37] M. I. Bruce, I. R. Butler, W. R. Cullen, G. A. Koutsantonis, M. R. Snow, E. R. T. Tiekink, *Aust. J. Chem.* **1988**, *41*, 963–969.
[38] Z.-G. Fang, Y.-S. Wen, R. K. L. Wong, S.-C. Ng, L.-K. Liu, T. S. A. Hor, *J. Cluster Sci.* **1994**, *5*, 327–340.
[39] S. T. Chacon, W. R. Cullen, M. I. Bruce, O. Shawkataly, F. W. B. Einstein, R. H. Jones, A. C. Willis, *Can. J. Chem.* **1990**, *68*, 2001–2010.
[40] S. M. Draper, C. E. Housecroft, A. L. Rheingold, *J. Organomet. Chem.* **1992**, *435*, 9–20.
[41] W. H. Watson, A. Nagl, S. Hwang, M. G. Richmond, *J. Organomet. Chem.* **1993**, *445*, 163–170.

requirements of the metal. In the hope of relating various coordination modes of dppf to its structural conformations, many X-ray crystal structures of dppf complexes have been published. Significant variations in the torsional angle (τ), the dihedral angle (θ) between the two Cp rings and the deviation (δp) of the phosphorus atom from the Cp ring plane are eminent (Table 1-2). These structural parameters are a sensitive function of the metal geometry and its steric requirements.

In the free state, the dppf molecule is centrosymmetric with an inversion center at the iron atom [24]. This ideal conformation can be described as antiperiplanar when the Cp rings are parallel and staggered with a torsional angle of 180°. This *anti* arrangement of the PPh_2 moieties is understood based on its lowest conformational energy. Bond lengths and angles associated with the phosphorus atoms are reminiscent of those observed in free triphenylphosphine [49]. When dppf is oxidized to $[(C_5H_4)P(X)Ph_2]_2Fe$ (X = O, S), [50] no conformational change is observed.

In the open bridging mode, dppf spans two otherwise unbridged metal atoms. To minimize any steric hindrance, an ideal *anti* geometry for the bridge is expected. This conformation is frequently observed (e.g., $[Fe(CO)_4]_2(\mu$-dppf) [51], $[M(CO)_5]_2(\mu$-dppf) (M = Cr, Mo, W) [36, 44], *cis*-$[Mn_2Cl_2(CO)_8(\mu$-dppf)] [52, 53], $Au_2Cl_2(\mu$-dppf) [54], $Ag_2(HCO_2$-*O*$)_2$(dppf-*P*, *P'*$)_2(\mu$-dppf) [8], *trans*-$[PtCl_2\{(\mu$-dppf)$W(CO)_5\}_2]$ [47] and $[M_2$(dppf-*P*, *P'*$)_2(\mu$-dppf)$]^{2+}$ (M = Cu [29], Ag [31], Au [32]), **1**). Notable exceptions are $[M_2(CO)_9]_2(\mu$-dppf) (M = Mn (τ = 131.4°), Re (131.3°)) [42], both of which have a conformation closer to anticlinal (eclipsed). An interplay of lattice and intramolecular forces is probably responsible for this behavior. The former influence appears to gain significance in some "linear propagated" molecules [32]. One variation of an open bridge occurs when both phosphine sites are bonded to one metal while the Fe center interacts with the other (Fig. 1-6). Hitherto there has been no report of this bridging mode.

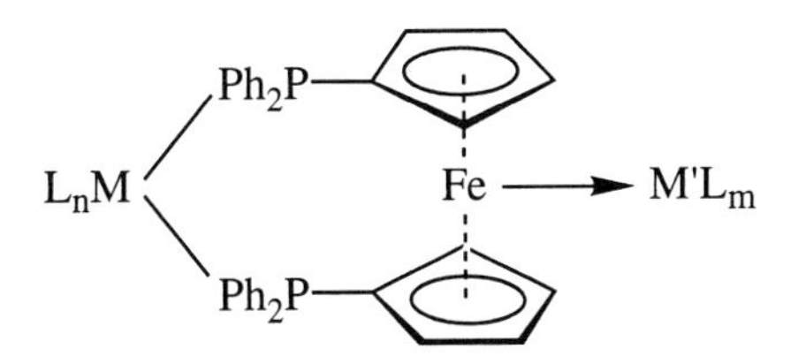

Fig. 1-6. A hitherto unknown bridging mode of dppf with an Fe → M bond.

When other ligands share the bridging responsibility with dppf, the complex may be defined as a quasi-closed system. The Cp rings in these cases commonly assume a fully or partially eclipsed conformation ("gauche eclipsed") with torsional angles of approximately 70–90° (e.g. $Rh_2(\mu$-S-*t*Bu$)_2(CO)_2(\mu$-dppf) **2** [34], $Re_2(\mu$-OMe$)_2$ · $(CO)_6(\mu$-dppf) [33] and $[Ag_2(\mu$-$CH_3CO_2)(\mu_3$-$CH_3CO_2)(\mu$-dppf$)]_2$ [8]). The ability of the sterically demanding dppf to exist with auxiliary bridging ligands of much smaller bite angle clearly illustrates its skeletal flexibility, which is largely attributed to the ease of ring twisting. In contrast to the open bridging systems, a relatively

Table 1-3. Metal-phosphorus bond lengths and chelate angles of metal chelates of dppf

Complex	Coordination geometry	Average M-P, Å	Bite angle P-M-P, °	Ref.
$[Cu_2(dppf\text{-}P,P')_2(\mu\text{-dppf})][ClO_4]_2$[a]	trigonal planar	2.279(6), 2.280(5)	111.3(2), 111.1(2)	[1]
$[Ag_2(dppf\text{-}P,P')_2(\mu\text{-dppf})][PF_6]_2$	trigonal planar	2.500(7)	105.6(2)	[2]
$[Au_2(dppf\text{-}P,P')_2(\mu\text{-dppf})][NO_3]_2 \cdot 2\,H_2O$	trigonal planar	2.389(3)	109.2(1)	[3]
cis-$[Rh(dppf\text{-}P,P')(nbd)][ClO_4]$	square planar	2.326(2)	96.8(1)	[4]
$[Ir(dppf\text{-}P,P')_2][BPh_4]$	distorted square planar	2.353(6), 2.340(6)	95.0(2), 93.6(2)	[5]
cis-$[PdCl_2(dppf\text{-}P,P')] \cdot CH_2Cl_2$	square planar	2.284(6)	98.0(1)	[6]
cis-$[PdCl_2(dppf\text{-}P,P')] \cdot CHCl_3$	square planar	2.292(1)	99.1(1)	[7]
cis-$[PdCl_2(dppf\text{-}P,P')] \cdot 0.25thf$[a]	square planar			
Molecule I		2.275(4)	99.4(1)	[8]
Molecule II		2.294(4)	97.8(1)	
1-$\{(dppf\text{-}P,P')Pd\}B_3H_7$	square planar	2.365(1)	104.2(1)	[9]
cis-$[PtCl_2(dppf\text{-}P,P')] \cdot 0.5(CH_3)_2CO$	square planar	2.256(4)	99.3(1)	[10]
cis-$[Pt(S_2N_2)(dppf\text{-}P,P')]$	square planar	2.289(3)	99.6(1)	[11]
$(Pt_2(\mu\text{-H})(\mu\text{-CO})(dppf\text{-}P,P')_2][BF_4] \cdot 0.5\,H_2O$	distorted square planar	2.329(2), 2.326(2)	103.7(1), 103,6(1)	[12]
$[Pt(\mu\text{-OH})(dppf\text{-}P,P')]_2[BF_4]_2$	square planar	2.242(4)	98.1(1)	[13]
$[Pt_2Tl(\mu_3\text{-S})_2(dppf\text{-}P,P')_2][PF_6]$	square planar	2.289(2)	97.5(1)	[14]
$[Pt_2H(\mu\text{-H})_2(dppf\text{-}P,P')_2]Cl$	Pt(1): square planar	2.291(6)	104.4(2)	[15]
	Pt(2): trigonal bipyramidal	2.283(6)	102.9(2)	
cis-$[Pt(1\text{-MeTy(-H)})(dppf\text{-}P,P')(dmf)][BF_4]$[b]	square planar	2.251(7)	99.1(3)	[16]
cis-$[Pt(1\text{-MeTy(-H)})(dppf\text{-}P,P')(1\text{-MeCy})][BF_4]$[b,c]	square planar	2.274(4)	99.8(2)	[16]
$[NiCl_2(dppf\text{-}P,P')]$	tetrahedral	2.312(2)	105.0(1)	[17]
$[NiBr_2(dppf\text{-}P,P')]$	tetrahedral	2.290(9)	102.5(2)	[6]
$[Cu(\mu\text{-HCO}_2)(dppf\text{-}P,P')]_2$	tetrahedral	2.260(2)	110.8(1)	[18]
$[Cu(\mu\text{-I})(dppf\text{-}P,P')]_2 \cdot 2\,CH_2Cl_2$	tetrahedral	2.283(5)	111.2(4)	[18]
$[Ag_2(HCO_2\text{-}O)_2(dppf\text{-}P,P')_2(\mu\text{-dppf})] \cdot 2\,CH_2Cl_2$	tetrahedral	2.534(5)	109.3(2)	[19]
ax,eq-$Fe(CO)_3(dppf\text{-}P,P')$	trigonal bipyramidal	2.250(3)	99.7/1)	[20]
cis-$[Mo(CO)_4(dppf\text{-}P,P')] \cdot C_6H_6$	octahedral	2.560(16)	95.3(1)	[6]
fac-$[ReCl(CO)_3(dppf\text{-}P,P')]$	octahedral	2.500(1)	93.6(1)	[21]
$[RuH(dppf\text{-}P,P')_2(\eta^2\text{-}H_2)][PF_6]$	distorted octahedral	2.410(7), 2.376(7)	106.0(2), 106.9(2)	[22]
$Ru(\eta^3\text{-}C_3H_5)(\eta^3\text{-}\{[OC(CF_3)]_2CH\})(dppf\text{-}P,P')$[d]	pseudooctahedral	—	98	[23]
$Rh\{(\eta^6\text{-}C_6H_5)B(C_6H_5)_3\}(dppf\text{-}P,P')$	2-legged piano-stool	2.248(3)	96.0(1)	[24]
$Mn(\eta^5\text{-MeCp})(CO)(dppf\text{-}P,P') \cdot CHCl_3$	3-legged piano-stool	2.216(3)	99.3(12)	[25]
$Ru(\eta^5\text{-}C_5H_5)H(dppf\text{-}P,P')$	3-legged piano-stool	2.255(4), 2.262(3)	99.1(1), 95.5(1)	[26]

a Two crystallographically independent molecules per unit cell.
b 1-MeTy(-H) = 1-methylthyminato-N^3.
c 1-MeCy = 1-methylcytosine-N^3.
d $\{[OC(CF_3)]_2CH\}$ (hexafluoropentanedionate).

References to Table 1-3

[1] U. Casellato, R. Graziani, G. Pilloni, *J. Crystallogr. Spectrosc. Res.* **1993**, *23*, 571–575.
[2] S.-P. Neo, T. S. A. Hor, Z.-Y. Zhou, T. C. W. Mak, *J. Organomet. Chem.* **1994**, *464*, 113–119.
[3] L.-T. Phang, T. S. A. Hor, Z.-Y. Zhou, T. C. W. Mak, *J. Organomet. Chem.* **1994**, *469*, 253–261.
[4] W. R. Cullen, T.-J. Kim, F. W. B. Einstein, T. Jones, *Organometallics* **1985**, *4*, 346–351.
[5] U. Casellato, B. Corain, R. Graziani, B. Longato, G. Pilloni, *Inorg. Chem.* **1990**, *29*, 1193–1198.
[6] I. R. Butler, W. R. Cullen, T.-J. Kim, S. J. Rettig, J. Trotter, *Organometallics* **1985**, *4*, 972–980.
[7] T. Hayashi, M. Konishi, Y. Kobori, M. Kumada, T. Higuchi, K. Hirotsu, *J. Am. Chem. Soc.* **1984**, *106*, 158–163.
[8] L.-T. Phang, T. S. A. Hor, T. C. W. Mak, unpublished results.
[9] C. E. Housecroft, S. M. Owen, P. R. Raithby, B. A. M. Shaykh, *Organometallics* **1990**, *9*, 1617–1623.
[10] D. A. Clemente, G. Pilloni, B. Corain, B. Longato, M. Tiripicchio-Camellini, *Inorg. Chim. Acta* **1986**, *115*, L9–L11.
[11] P. F. Kelly, A. M. Z. Slawin, D. J. Williams, J. D. Woollins, *Polyhedron* **1988**, *7*, 1925–1930.
[12] A. L. Bandini, G. Banditelli, M. A. Cinellu, G. Sanna, G. Minghetti, F. Demartin, M. Manassero, Inorg. Chem. **1989**, *28*, 404–410.
[13] B. Longato, G. Pilloni, G. Valle, B. Corain, *Inorg. Chem.* **1988**, *27*, 956–958.
[14] M. Zhou, Y. Xu, L.-L. Koh, K. F. Mok, P.-H. Leung, T. S. A. Hor, unpublished results.
[15] B. S. Haggerty, C. E. Housecroft, A. L. Rheingold, B. A. M. Shaykh, *J. Chem. Soc., Dalton Trans.* **1991**, 2175–2184.
[16] G. Bandoli, G. Trovò, A. Dolmella, B. Longato, *Inorg. Chem.* **1992**, *31*, 45–51.
[17] U. Casellato, D. Ajó, G. Valle, B. Corain, B. Longato, R. Graziani, *J. Crystallogr. Spectrosc. Res.* **1988**, *18*, 583–590.
[18] S.-P. Neo, Z.-Y. Zhou, T. C. W. Mak, T. S. A. Hor, *J. Chem. Soc., Dalton Trans.* **1994**, in press.
[19] T. S. A. Hor, S. P. Neo, C. S. Tan, T. C. W. Mak, K. W. P. Leung, R.-J. Wang, *Inorg. Chem.* **1992**, *31*, 4510–4516.
[20] T.-J. Kim, K.-H. Kwon, S.-C. Kwon, J.-O. Baeg, S.-C. Shim, D.-H. Lee, *J. Organomet. Chem.* **1990**, *389*, 205–217.
[21] T. M. Miller, K. J. Ahmed, M. S. Wrighton, *Inorg. Chem.* **1989**, *28*, 2347–2355.
[22] M. Saburi, K. Aoyagi, T. Kodama, T. Takahashi, Y. Uchida, K. Kozawa, T. Uchida, *Chem. Lett.* **1990**, 1909–1912.
[23] N. W. Alcock, J. M. Brown, M. Rose, A. Wienand, *Tetrahedron: Asymmetry* **1991**, *2*, 47–50.
[24] B. Longato, G. Pilloni, R. Graziani, U. Casellato, *J. Organomet. Chem.* **1991**, *407*, 369–376.
[25] S. Onaka, T. Moriya, S. Takagi, A. Mizuno, H. Furuta, *Bull. Chem. Soc. Jpn.* **1992**, *65*, 1415–1427.
[26] M. I. Bruce, I. R. Butler, W. R. Cullen, G. A. Koutsantonis, M. R. Snow, E. R. T. Tiekink, *Aust. J. Chem.* **1988**, *41*, 963–969.

---

small torsional twist ensures a shortened P ... P distance, which would be more compatible with the auxiliary ligands.

For the closed-bridging dppf complexes in which there is direct M-M bonding, the rings are "gauche eclipsed" with the C(Cp)-P vectors spanning an angle of about 70–80°. Among the few known examples, the tilt angle between the Cp rings is smallest for $Co_3(\mu_3\text{-CMe})(CO)_7(\mu\text{-dppf})$ **6** (2.3°) [52, 53] (Fig. 1-7). When compared with $Ru_4(\mu\text{-H})(CO)_{12}(\mu_6\text{-B})Au_2(\mu\text{-dppf})$ **7** (4.0°) [55] and $Ru_3(CO)_{10}(\mu\text{-dppf})$ (5.4°), [9, 55] there appears to be a trend in the M-M distances spanned by dppf: Co-Co (2.520(1) Å) < Au-Au (2.812(2) Å) < Ru-Ru (2.9284(5) Å). No conclusive statement, however, should be made until more data are available. The sign of the tilt

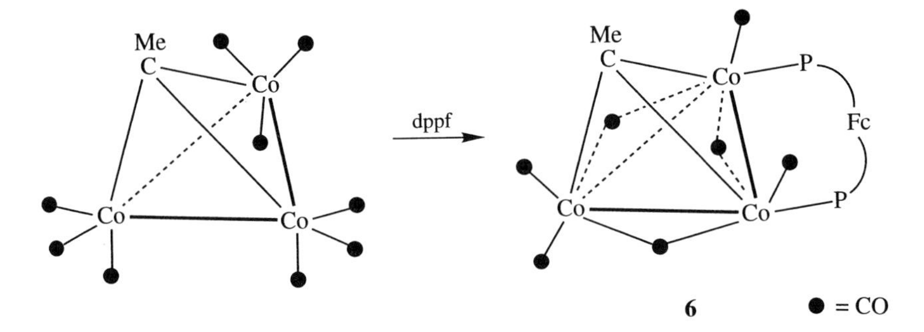

Fig. 1-7. Dppf substitution of a triangular cluster $Co_3(\mu_3\text{-}CMe)(CO)_9$.

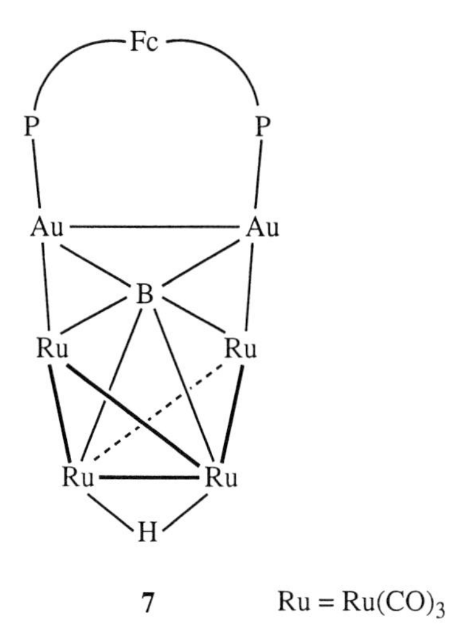

angle is often not clearly specified in the literature (positive for an attractive tilt between the phosphorus atoms).

Figure 1-8 summarizes the dependence of the torsional twist on the type of bridging. A large twist angle (180°) with an *anti*-configuration is only associated with an open-bridge, while a "gauche eclipsed" conformation (τ 72°) permits M-M bonding. Intermediate torsional twists could support weak M-M interactions. Twists significantly less than 72° have only been found in the chelating mode of dppf.

There are emerging examples in which dppf functions as a unidentate ligand. This coordination mode has been postulated as the key intermediate in some catalytic processes (see Sect. 1.5.2.1). Diphosphines acting as a unidentate ligand were once thought to be unstable based on entropy arguments. In this mode of coordination, the expected antiperiplanar orientation of the pendant phosphine is rarely observed. The approximate anticlinal (eclipsed) conformation of most known unidentate complexes (τ 122–123°) [42, 47, 51, 53] possibly reflects the higher influence of the

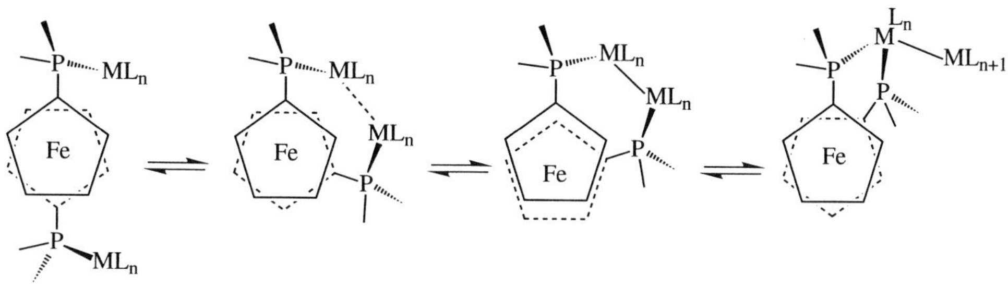

Fig. 1-8. Effects of torsional twist of dppf on metal-metal interaction.

lattice effect when the steric difference between the two phosphino moieties is quite high. Such lattice dependency of the conformational parameters (e.g. Cp twist angle and orientation of the phenyl rings) is best illustrated by Au_2Cl_2(μ-dppf), [54], [Cu_2(dppf-*P*, *P'*)$_2$(μ-dppf)][ClO_4]$_2$ **1a**, [29], $Re_2(CO)_9$(dppfO-*P*) [42] and Ru(η^5-Cp)H(dppf-*P*, *P'*), [56] in which two crystallographically independent molecules of (slightly) different conformations are located in the unit cell.

As most of the early complexes of dppf are chelates, the wealth of their X-ray crystal data affords a more thorough comparison (Tables 1-2 and 1-3). The Cp rings of these chelates generally adopt a gauche (staggered) conformation with a torsional angle of 25 – 45°. Among these are a few unusual cases with an eclipsed structure brought about by a very acute twist (e.g., Mn(η^5-MeCp)(CO)(dppf-*P*, *P'*) (3.1°) [53] and $NiBr_2$(dppf-*P*, *P'*) (6.5°) [22]). The resultant synperiplanar conformation confines the bite angles to a narrow range of ca. 98 – 102° and evidently restricts the coordination flexibility of the ligand. In contrast, a larger torsional twist of about 36° can subtend a wider bite range of ca. 95 – 110°. The largest torsional angle reported for dppf chelates is found in the crowded $Ag_2(HCO_2$-*O*)$_2$(dppf-*P*, *P'*)$_2$(μ-dppf) (τ 56.5°) [8], which displays a bite angle of 109.3°. In this complex, the Cp rings of the chelating ferrocenyl unit are virtually parallel but the phosphorus atoms are displaced slightly towards and away from the Fe center.

The skeletal flexibility of dppf allows a wide range of ligand bites (93 – 112°) and confers the ability to stabilize complexes of various geometries. These complexes encompass the linear dimeric species, Au_2Cl_2(μ-dppf) [54], chelates of trigonal planar ([M_2(dppf-*P*, *P'*)$_2$(μ-dppf)]$^{2+}$ (M = Cu [29], Ag [31], Au [32], **1**), square planar (e.g., $PdCl_2$(dppf-*P*, *P'*) [22, 57], Pt(S_2N_2)(dppf-*P*, *P'*) [58]), tetrahedral (e.g. NiX_2(dppf-*P*, *P'*) (X = Cl [24], Br [22]), $Ag_2(HCO_2$-*O*)$_2$(dppf-*P*, *P*)$_2$(μ-dppf) [8]), trigonal bipyramidal (e.g., $Fe(CO)_3$(dppf-*P*, *P'*) [59]) and octahedral geometry (e.g., $M(CO)_4$(dppf-*P*, *P'*) (M = Cr, Mo, W) [20, 22, 44], *fac*-[ReCl$(CO)_3$(dppf-*P*, *P'*)] [60]), and the 7-coordinate complex *cis*-[Mo$Cl_2(CO)_2$(dppf-*P*, *P'*)(dppf-*P*)] [26]. Many of these geometries are understandably non-ideal because of different ligand demands. Consequently, the chelate bites fluctuate over a range of nearly 20° (Table 1-3). The greatest angle deviation from the idealized 90° in the square planar geometry is observed in 1-{(dppf-*P*, *P'*)Pd}B_3H_7 **8** (104.2°) [61], [Pt_2(μ-H)(μ-CO)(dppf-*P*, *P'*)$_2$][BF_4] **9** (103.7°) [62] and [Pt_2H(μ-H)$_2$(dppf-*P*, *P'*)$_2$]Cl **10** (104.4°) [63]. This angle distortion provides a mechanism to relieve the geometrical constraints that would

8 **9**

have been imposed on the ferrocenyl backbone for a 90° bite. A plot of Pt-P bond length against bite angles (Fig. 1-9) using the limited available data indicates an interesting phenomenon – the chelation angle of Pt(II) planar complexes is confined to an exceptionally narrow range of 98 – 100°. Within this region, the bond generally weakens as the angle widens. Outside this domain, weak Pt-P bonds ($\approx$2.29 Å) are found. The reliability of this domain will be tested as more structural data are obtained. From the plot, the complex $[Pt_2Tl(\mu_3\text{-}S)_2(dppf\text{-}P,P')_2][PF_6]$ [64] appears to have an unusually weak Pt-P bond for an angle of 97.5°. The high *trans*-labilizing effect of the sulfido ligand is probably responsible. The smallest bite angle registered among the chelates is 93.6° in $ReCl(CO)_3(dppf\text{-}P,P')$ [60] and $[Ir(dppf\text{-}P,P')_2][BPh_4]$ [65]. In the latter, the ferrocenyl unit with a smaller chelate

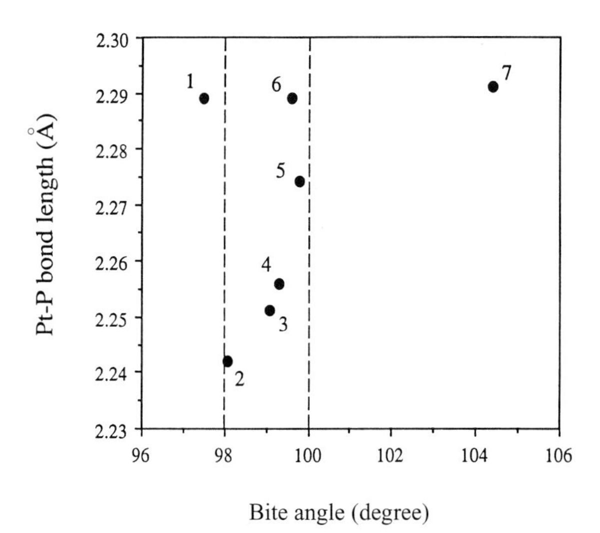

Fig. 1-9. A plot of Pt-P bond lengths against bite angles for the known structures of Pt chelates of dppf. [Data from [64] ($[Pt_2Tl(\mu_3\text{-}S)_2(dppf\text{-}P,P')_2][PF_6]$ **1**), [315] ($[Pt(\mu\text{-}OH)(dppf\text{-}P,P')]_2[BF_4]_2$ **2**), [331] (*cis*-$[Pt(1\text{-}MeTy(\text{-}H))(dppf\text{-}P,P')(dmf)][BF_4]$, **3**), [79] (*cis*-$[PtCl_2(dppf\text{-}P,P')] \cdot 0.5(CH_3)_2CO$, **4**), [331] (*cis*-$[Pt(1\text{-}MeTy(\text{-}H))(dppf\text{-}P,P')(1\text{-}MeCy)][BF_4]$, **5**), [58] ($Pt(S_2N_2)(dppf\text{-}P,P'$, **6**), [63] ($[Pt_2H(\mu\text{-}H)_2(dppf\text{-}P,P')_2]Cl$, **7**).].

angle (93.6°) is coordinated more strongly (Ir-P 2.340 Å) than its neighbor (95.0°, 2.353 Å).

All the trigonal planar chelates show a bite angle substantially lower than the ideal 120°. A bite of 111 – 112°, found in $[Cu_2(dppf\text{-}P,P')_2(\mu\text{-}dppf)][ClO_4]_2$ **1a** [29] and $[Cu(\mu\text{-}I)(dppf\text{-}P,P')]_2$, is hitherto the largest reported for dppf. The longer M-P link in chelating (2.279 Å) compared with open-bridging dppf (2.264 Å) in **1a** [29] is a general phenomenon when both modes are compared.

The structural data of tetrahedral chelates indicate minimal geometric or angle distortion and possibly reflect a "natural" chelate environment for the ligand. A perfect tetrahedral angle is in fact observed in $Ag_2(HCO_2\text{-}O)_2(dppf\text{-}P,P')(\mu\text{-}dppf)$ [8]. The few reported tetrahedral chelates do not deviate by more than 10° from the ideal bite of 109°. That a tetrahedral angle is most favored by dppf chelates is also exemplified in many trigonal planar complexes $[Au_2(dppf\text{-}P,P')_2(\mu\text{-}dppf)][NO_3]_2$ **1c** (109°) [32] and $[Cu_2(dppf\text{-}P,P')_2(\mu\text{-}dppf)][ClO_4]_2$ **1a** (111°) [29] as well as square planar complexes 1-{(dppf-P,P')Pd}B_3H_7 **8** (104°) [61] and $[Pt_2(\mu\text{-}H)(\mu\text{-}CO)(dppf\text{-}P,P')_2][BF_4]$ **9** (104°) [62].

A number of dppf complexes (e.g., *cis*-$PdCl_2$(dppf-P,P') [22, 57], $Au_2Cl_2(\mu\text{-}dppf)$ [54], $[Re_2(CO)_4]_2(\mu\text{-}dppf)$ [42] and $Mn(\eta^5\text{-}MeCp)(CO)(dppf\text{-}P,P')$ [53, 66]) tend to entrap solvent molecules in their crystal lattices. This clathrating property of dppf and its complexes provides an opportunity for the study of inclusion chemistry of organometallics.

1.3.2 Geometrical Distortions

The ability of dppf to rotate about the Cp-Fe-Cp axis affords a series of staggered or eclipsed conformations (Fig. 1-3) that are dependent, among other factors, on the ligand coordination mode. Such conformational variation is supplemented by other intrinsic geometrical distortions. These distortions are brought about by Cp ring tilting and displacement of phosphorus or Cp carbon atoms from the mean Cp ring plane. Other fluxional features include limited rotation of the C(Cp)-P bonds and skewing of the ligand with respect to the M ... M axis (Fig. 1-10). These atomic and molecular motions appear to be related to the coordination modes of the ligand and to the metal geometry.

In contrast to the unidentate and open-bridging dppf complexes, the chelates generally exhibit a greater extent of ring tilting but minimal divergence of the phosphorus atoms. Mean values of θ of known chelates: 4.1°; δ_P 0.05 Å, compared with values for unidentates of 2.4° and 0.11 Å and for open-bridged complexes of 1.4° and 0.13 Å (Table 1-2). Other restrained systems like the doubly, quasi-closed and closed bridging complexes also exhibit similar tilts to those of the chelates.

There is also evidence that Cp ring tilting is most marked among the square-planar chelates. This tilt could provide a facile mechanism to release the strain that is brought about by a small bite. A good example is illustrated in $[Ir(dppf\text{-}P,P')_2][BPh_4]$

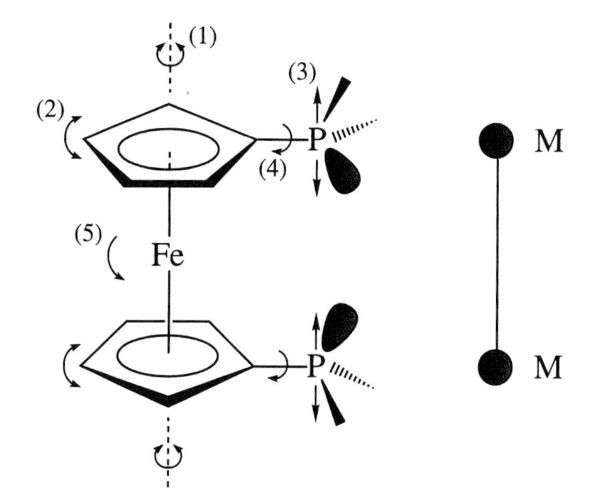

Fig. 1-10. Skeletal freedom of dppf in its bridging state: (1) Cp ... Fe ... Cp torsional twist; (2) Cp ring tilting; (3) displacement of phosphorus atom from coplanarity with Cp ring; (4) rotation of C(Cp)-P bond; (5) skewing of ferrocenyl backbone with respect to the M-M bonding axis.

[65], which shows the greatest distortion (θ 9.4°) at a very small bite of 95.0°. Conversely, near-parallel alignment of two Cp rings is only observed in chelates with larger bite angles (ca. 104°), e.g., 1-{(dppf-*P*, *P'*)Pd}B_3H_7, **8** [61] and $[Pt_2H(\mu\text{-}H)_2(\text{dppf-}P, P')_2]Cl$, **10** [63]. In a tetrahedral environment, the chelates usually display an approximate synperiplanar conformation. To accommodate an idealized bite of $\approx 109°$, which approaches the upper limit of the dppf bites, the two Cp rings are usually inclined at an angle of 4–6°. One notable exception is found in $[Ag_2(HCO_2\text{-}O)_2(\text{dppf-}P, P')_2(\mu\text{-dppf})]$ [8], in which a large torsional twist (56.5°) alleviates the ring tilt to a negligible 1°. Minimal ring tilt is observed in all known trigonal bipyramidal and octahedral complexes (θ 0–3°). It is not well understood why a 90° bite would induce less ring tilt in an octahedral environment compared with that of square planar.

10

In the open-bridging complexes, the more obvious distortion lies in the displacement of phosphorus donors from coplanarity with the Cp rings. To maintain an ideal centrosymmetry, the two C(Cp)-P vectors tend to diverge oppositely and equally away from the iron center. That this effect is coordination-induced is supported by a similar trend in the unidentate complexes in which the pendant phosphine is not displaced. Generally, the chelating phosphorus atoms are slightly displaced towards the ferrocenyl iron in the square planar, trigonal bipyramidal, and octahedral geometries. In the tetrahedral case, one of the phosphorus atoms is displaced slightly away from and the other towards the Fe center. Figure 1-5 summarizes this apparent relationship between the displacement of phosphorus and the modes of coordination. These simplistic generalizations are based on the limited available published data. A more reliable in detailed analysis would only be possible when more data are forthcoming.

1.4 Spectroscopic Characteristics

1.4.1 Techniques

As with other organophosphorus compounds, NMR spectroscopy represents the most useful diagnostic tool in the characterization of dppf complexes. From ^{31}P and ^{1}H NMR spectroscopic data, information can be obtained on the coordination site, the symmetry and the fluxional behavior of these complexes. Possible iron-metal interaction and ring conformational changes of the ferrocenyl moieties can also be studied using ^{57}Fe Mössbauer spectroscopy.

1.4.1.1 ^{31}P NMR Spectroscopy

On coordination to a metal, dppf almost invariably gives a downfield shift in the ^{31}P NMR spectrum. The coordination shift, Δ [δ (complex) – δ (free ligand)], however, depends on the metal, its coordination geometry and its auxiliary ligands (see Table 1-1).

In all the three common coordination modes, i.e., bridging, unidentate and chelating, the coordination shift of Group 6 dppf carbonyl complexes decreases in the expected order Cr > Mo > W [67]. Trends in the shift magnitudes with respect to both metals and ligands suggest that dppf behaves similarly to other common diphosphines (Fig. 1-11) [44, 68–70]. A plot of the chelates (Fig. 1-11c) [44, 68] reveals a clear decreasing trend in the order dppe > dppf > dppb > dppp > dppm. The pattern generally reflects the chelate ring size. The high chemical stability of the dppe chelates presumably relates to the exceptionally high shift. A similar trend is evident in the plot of the chelation shift [δ (chelate) – δ (unidentate)] (Fig. 1-12) [44, 68, 69]. This plot also reveals some anomalies, namely, (a) the 4-(dppm) and 6-(dppp) membered rings induce a shielding effect on the chelation shift, whereas dppf is associated with dppe and dppb in having the opposite effect; (b) the shifts of dppf chelates are significantly greater than those of dppm but most similar to those of dppb; this appears to correlate to the ring size;

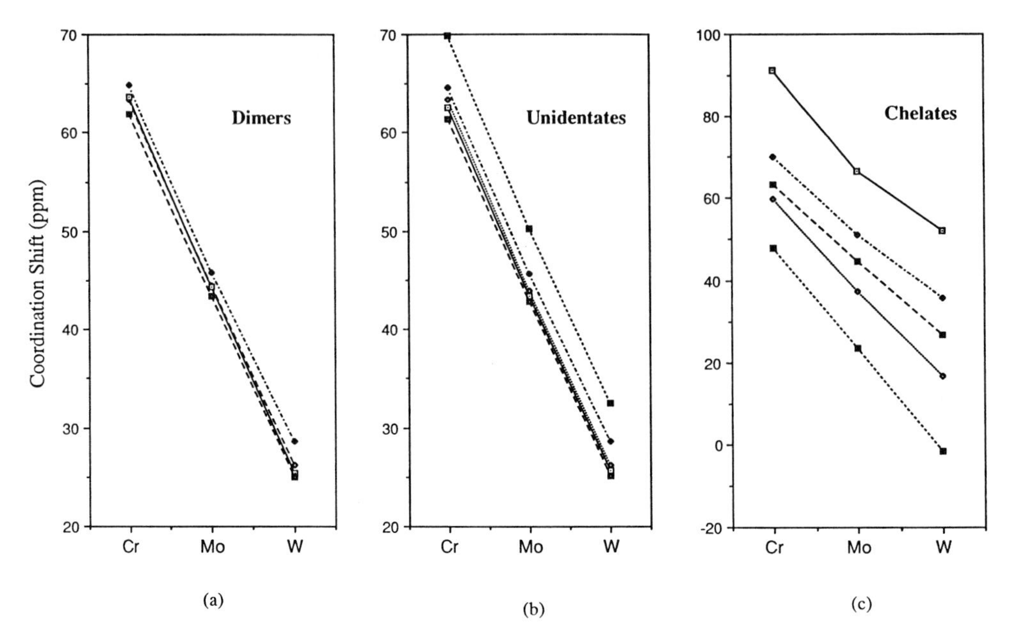

Fig. 1-11. ^{31}P NMR coordination shifts [δ(complex) − δ(free ligand)] of (a) $[M(CO)_5]_2(\mu$-P-P) (dimers); (b) $M(CO)_5(\eta^1$-P-P) (unidentates); (c) $M(CO)_4(\eta^2$-P-P) (chelates) [M = Cr, Mo, W; P-P = dppm (– – – – – –), dppe (———), dppp (· · · · · ·), dppb (— — —), dppf (—·—·—)]. [Data from [68] ($M(CO)_4(\eta^2$-P-P) (P-P = dppm, dppe, dppp, dppb)), [69] ($M(CO)_5(\eta^1$-P-P) (P-P = dppm, dppe, dppp, dppb)), [70] [$[M(CO)_5]_2(\mu$-P-P) (P-P = dppm, dppe, dppp, dppb)) and [44] ($M(CO)_4(\eta^2$-dppf), $M(CO)_5(\eta^1$-dppf), $[M(CO)_5]_2(\mu$-dppf))].

(c) only in dppf chelates can one find a highest shift in the heaviest congener, i.e., tungsten. Further research in this area will help to understand the implications of these trends.

Most dppf complexes known to date are those of Pt(II) and many of these are chelates. Their ^{31}P NMR resonances cover a wide spectrum, ranging from −0.9 ppm in *cis*-[Pt(1-MeTy(-H))(dppf-*P*, *P*')(1-MeCy)]$^+$ to 29.3 ppm in $[Pt_2H(\mu$-H$)_2$(dppf-*P*, *P*'$)_2$]Cl (Table 1-1). This clearly illustrates the sensitivity of ^{31}P chemical shift to the neighboring ligands and to the coordination geometry. In view of the importance of *cis*-$PdCl_2$(dppf-*P*, *P*') in catalytic reactions, it would be of interest to relate spectroscopic shifts to catalytic activity. The effect of the ligand on the latter is a subject of great interest. There are many examples of dppf complexes that show higher catalytic rates among their diphosphine analogues. A plot of coordination shift [δ (chelate) − δ (free ligand)] or "chelation shift" [δ (chelate)−δ(*cis*-MCl_2-$(PPh_3)_2$)] *vs.* the diphosphines (Pt is included for comparison) (Fig. 1-13) [43, 44, 71, 72] shows an overall trend of shifts increasing with chelate ring size. The only exception is MCl_2(dppe-*P*, *P*'), which shows remarkably high coordination and chelation shifts thus affirming the uncanny stability of a 5-membered ring. Similar to that observed in the Group 6 chelates, the close shift values of dppf and dppb complexes appear to be related to their similar ring sizes. A similar rela-

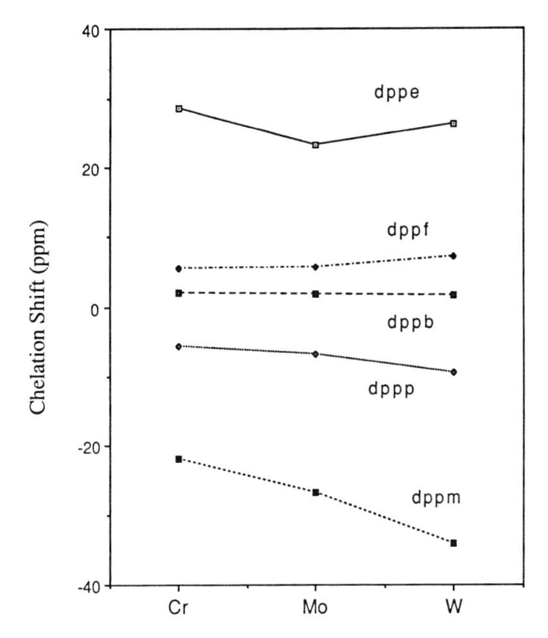

Fig. 1-12. ^{31}P NMR chelation shift [δ(chelate) − δ(unidentate)] of $M(CO)_4(\eta^2$-P-P) [M = Cr, Mo, W; P-P = dppm (– – – – – –), dppe (———), dppp (· · · · · ·), dppb (— — —), dppf (—·—·—)]. [Data from [68] ($M(CO)_4(\eta^2$-P-P) (P-P = dppm, dppe, dppp, dppb)), [69] ($M(CO)_5(\eta^1$-P-P) (P-P = dppm, dppe, dppp, dppb)) and [44] ($M(CO)_4(\eta^2$-dppf), $M(CO)_5(\eta^1$-dppf))].

tionship is also evident in the Group 6 $M(CO)_4(\eta^2$-P-P) complexes (Fig. 1-14) [44, 68]. A linear relationship is observed when the data from $M(CO)_4$(dppe-*P*, *P'*) are excluded.

1.4.1.2 ^1H NMR Spectroscopy

The ^1H NMR spectrum of the free dppf ligand displays two sets of multiplets at 3.99 and 4.26 ppm [54], attributed to α and β protons [with respect to $C(PPh_2)$] of the Cp rings. On complex formation, a downfield shift of these protons is generally observed. For complexes where assignment is given, H_α protons usually resonate at a higher field than those of H_β. Some notable exceptions include $WI_2(CO)_3$(dppf-*P*, *P'*) [26], $Pt(CH_2SiMe_2CH{=}CH_2)_2$(dppf-*P*, *P'*) **11** [73], $[Fe(CO)_4]_2(\mu$-dppf) [45], $Fe(CO)_4(\mu$-dppf)$M(CO)_5$ (M = Cr, Mo, W) [45, 46], $Fe(CO)_4(\mu$-dppf)$Mn_2(CO)_9$ [46], and $Re_2(\mu$-$OMe)_2(CO)_6(\mu$-dppf) [74]. Conformational analysis of $Re_2(\mu$-$OMe)_2(CO)_6(\mu$-dppf) using the molecular modelling software package CHEM-X [74] demonstrates that the unusually large downfield shift of the H_α protons arises from their proximity to the deshielding region of the phenyl rings.

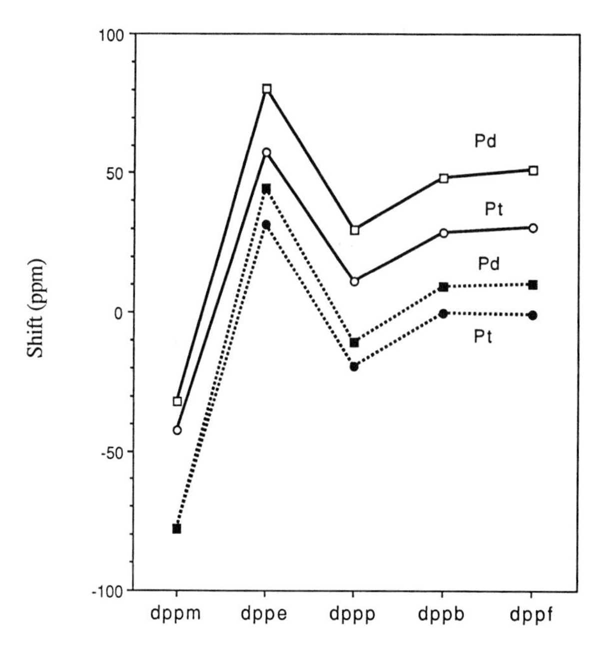

Fig. 1-13. ^{31}P NMR coordination shifts [δ(complex) – δ(free ligand)] (———) and "chelation shifts" [δ(chelate) – δ(*cis*-$MCl_2(PPh_3)_2$)] (------) of *cis*-$MCl_2(\eta^2$-P-P). [Data from [44] ($PtCl_2(PPh_3)_2$), [71] ($PdCl_2(PPh_3)_2$, $PdCl_2$(dppm-P,P')), [72] ($PtCl_2$(dppm-P,P'), $MCl_2(\eta^2$-P-P) (M = Pd, Pt; P-P = dppe, dppp, dppb)) and [43] (MCl_2(dppf-P,P') (M = Pd, Pt))].

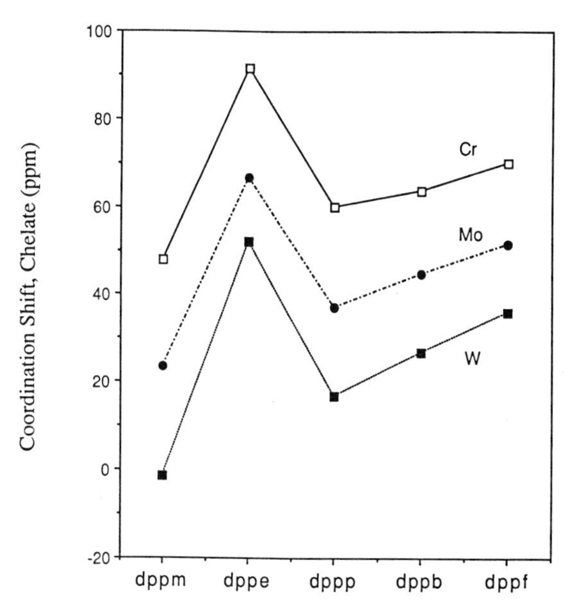

Fig. 1-14. ^{31}P NMR coordination shifts [δ(complex) – δ(free ligand)] of $M(CO)_4(\eta^2$-P-P) chelates (M = Cr, Mo, W; P-P = diphosphines) (data from [44, 68]).

1.4.1.3 Mössbauer Spectroscopy

The Mössbauer parameters of dppf complexes are collected in Table 1-4. Consistent with other ferrocenylphosphines [75], dppf shows a reduced quadrupole splitting compared with that of ferrocene (2.33 *vs.* 2.42 mm s^{-1}). On complexation, the isomer

Table 1-4. ^{57}Fe Mössbauer spectral data of some dppf complexes

Complex	δ, mm $s^{-1\,a}$	Δ, mm $s^{-1\,a}$	Ref.*
Sandwiched			
ferrocene	0.53	2.42	[1]
dppf	0.52	2.33	[1]
Linear			
$Au_2Cl_2(\mu$-dppf)	0.53	2.36	[2, 3]
Trigonal Planar			
$Au_2Cl_2(\mu\text{-dppf})_2$	0.52	2.31	[3]
[Au(dppf-*P*, *P'*)(dppf-*P*)]Cl	0.51	2.30	[3]
Tetrahedral			
$Mn(\eta^5\text{-MeCp})(CO)_2$(dppf-*P*)	0.51	2.29	[4]
$Mn(\eta^5$-MeCp)(CO)(dppf-*P*, *P'*)	0.50	2.28	[4]
$FeCl_2$(dppf-*P*, *P'*)	0.56	2.29	[5]
	0.70^b	2.79^b	
$CoCl_2$(dppf-*P*, *P'*)	0.56	2.35	[5]
$NiCl_2$(dppf-*P*, *P'*)	0.57	2.29	[5]
$NiBr_2$(dppf-*P*, *P'*)	0.55	2.31	[5]
NiI_2(dppf-*P*, *P'*)	0.58	2.33	[5]
$ZnCl_2$(dppf-*P*, *P'*)	0.55	2.35	[1]
$CdCl_2$(dppf-*P*, *P'*)	0.51	2.27	[1]
$HgCl_2$(dppf-*P*, *P'*)	0.56	2.36	[1]
$[Hg(\text{dppf-}P, P')_2][BF_4]$	0.44	2.26	[6]
Square Planar			
cis-$PdCl_2$(dppf-*P*, *P'*)	0.50	2.14	[5]
cis-$PtCl_2$(dppf-*P*, *P'*)	0.50	2.23^c	[1]
Octahedral			
$Cr(CO)_4$(dppf-*P*, *P'*)	0.52	2.23	[5]
$Mo(CO)_4$(dppf-*P*, *P'*)	0.52	2.27	[5]
$W(CO)_4$(dppf-*P*, *P'*)	0.54	2.23	[5]
cis-$[Mn_2Cl_2(CO)_8(\mu\text{-dppf})]$	0.52	2.35	[4]
Cluster			
$Co_3(\mu_3\text{-CMe})(CO)_7(\mu\text{-dppf})$	0.50	2.22	[4]

a Data measured at 77 K in [1], 300 K in [3], 50 K in [2] and 80 K in [4, 5].
b Referring to chloride-bound iron center. c Data measured at 293 K.
* References see page 40.

References to Table 1-4

[1] B. Corain, B. Longato, G. Favero, D. Ajò, G. Pilloni, U. Russo, F. R. Kreissl, *Inorg. Chim. Acta* **1989**, *157*, 259–266.
[2] D. T. Hill, G. R. Girard, F. L. McCabe, R. K. Johnson, P. D. Stupik, J. H. Zhang, W. M. Reiff, D. S. Eggleston, *Inorg. Chem.* **1989**, *28*, 3529–3533.
[3] A. Houlton, R. M. G. Roberts, J. Silver, R. V. Parish, *J. Organomet. Chem.* **1991**, *418*, 269–275.
[4] S. Onaka, T. Moriya, S. Takagi, A. Mizuno, H. Furuta, *Bull. Chem. Soc. Jpn.* **1992**, *65*, 1415–1427.
[5] A. Houlton, S. K. Ibrahim, J. R. Dilworth, J. Silver, *J. Chem. Soc., Dalton Trans.* **1990**, 2421–2424.
[6] K. R. Mann, W. H. Morrison, Jr., D. N. Hendrickson, *Inorg. Chem.* **1974**, *13*, 1180–1185.

---

shift (δ) and, to an even larger degree the quadrupole splitting (Δ), are metal-dependent. This behavior reflects the π-interaction of the phosphine groups without any significant perturbance of the σ-electrons [76]. These Mössbauer parameters are generally influenced by (a) direct iron-to-metal bonding or interaction; (b) ring substituent effects; (c) ring-tilt or variations in Fe-Cp distances; and (d) ring conformation [76]. The presence of iron-to-metal interaction in ferrocenyl complexes is usually diagnosed by large Δ values, >3.0 mm s^{-1} [77, 78]. The dppf complexes that have been studied by Mössbauer spectroscopy generally exhibit a narrow range of Δ (2.14–2.36 mm s^{-1}). This observation suggests no Fe → M bonding and hence agrees with the results of studies on solid state crystal structures [22, 53, 54, 79].

According to Silver et al. [76], the hyperfine interactions are influenced by the metal geometry. The latter imparts specific geometric demands on the ligand, which in turn affects the electronic environment of the ferrocenyl-iron atom. A plot of quadrupole splitting Δ *vs.* isomer shifts δ for dppf chelates gives a reasonably linear correlation with well-defined domains according to the type of metal coordination geometry (Fig. 1-15). The tetrahedral complexes show a tendency towards the largest values of δ and Δ, followed by the octahedral and square planar complexes. When more complexes [e.g., *cis*-$Mn_2Cl_2(CO)_8$(μ-dppf), Mn(η^5-MeCp)$(CO)_2$(dppf-*P*), and Mn(η^5-MeCp)(CO)(dppf-*P*, *P'*)] [53] are included in the analysis, they do not appear to fall into their respective geometric domains. These findings seem to indicate that other contributions such as the coordination mode of the ligand should also be taken into account.

Other spectroscopic techniques that have been used in the study of dppf complexes are XPS, UPS [43] and UV/VIS spectroscopy. XPS provides useful complementary information on the differentiation of pendant and coordinated phosphines [42, 80]. Caution needs to be exercised, however, as P(2p) core binding energy is not very sensitive towards changes in the ligand coordination mode or the oxidation state of the metal [42]. The electronic absorption spectra of dppf and MCl_2(dppf-*P*, *P'*) complexes (M = Co, Ni, Pd, Pt, Zn, Cd, Hg) exhibit an easily detected band at 405–465 nm, attributed to a transition in the ferrocenyl moiety [43]. The presence of any iron-to-metal bonding is usually indicated by an intense σ → σ* transition [21], as evident in the electronic spectrum of [Pd(dppf-*P*, *P'*)(PPh_3)][BF_4]$_2$, **3** [40].

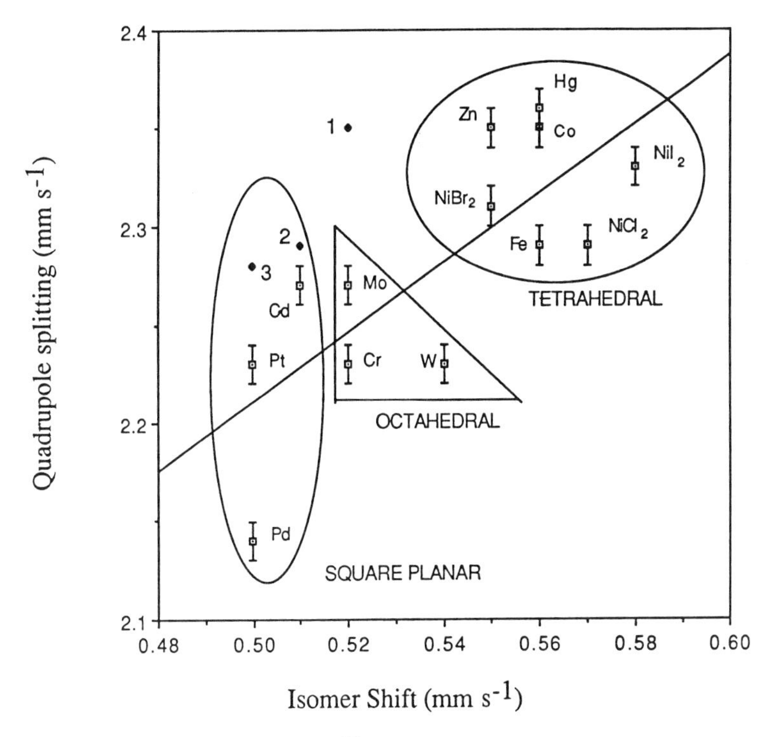

Fig. 1-15. Different geometrical domains of ^{57}Fe Mössbauer parameters (quadrupole splitting *vs.* isomer shift) in some dppf complexes (*cis*-$Mn_2Cl_2(CO)_8$(μ-dppf), **1**, Mn(η^5-MeCp)$(CO)_2$(dppf-*P*), **2**, Mn(η^5-MeCp)(CO)(dppf-*P*, *P'*), **3**, (adapted from [76], data from Table 1-4).

1.4.2 NMR Fluxionality

Being a flexible ligand that stabilizes complexes of varied metal geometry under different modes of coordination, dppf confers to many of its complexes some fluxionality in solution. Structural deformation relating to the Cp ring twist and tilt and the displacement of phosphorus from the ring planes provides an additional mechanism for stereo non-rigidity. Much of this fluxional behavior is found in the chelating and the bridging systems. The observed patterns may closely resemble those of ferrocenophanes or, in general, cyclophanes.

In the solid state, the Cp rings of 1-{(dppf-*P*, *P'*)Pd}B_3H_7 **8** are in a near-staggered conformation with each of the Cp protons in its own unique chemical environment [61]. Eight proton signals are thus expected in the Cp region of the ^{1}H NMR spectrum. This is, however, seldom observed and in this case only two resonances are visible at ambient temperature. On lowering the temperature to 203 K, four proton signals are resolved (Fig. 1-16). These spectral changes typify the general fluxional behavior of dppf complexes. The ABCD spin pattern at 203 K is consistent with a mutual twisting of the Cp rings, giving rise to an average eclipsed conformation

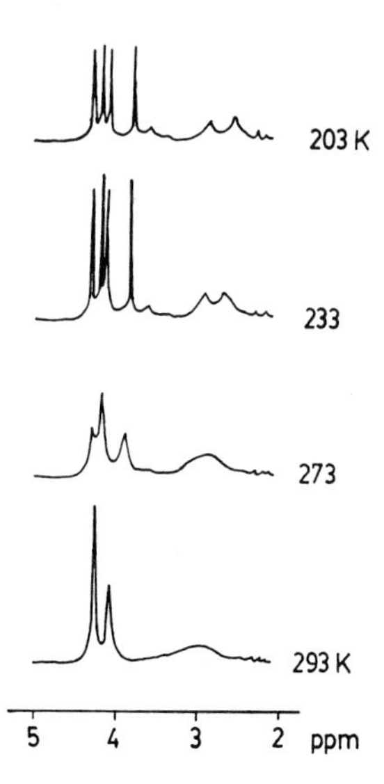

Fig. 1-16. Variable temperature ^{1}H NMR spectra of 1-{(dppf-*P*, *P'*)Pd}B_3H_7 (adapted from [61]).

in solution (Fig. 1-17a). If rapid inversion at each phosphorus atom is allowed, the pairs of Cp protons in the α and β positions will be time-averaged (Fig. 1-17b). This phenomenon mirrors that of other [3]-ferrocenophanes. Such a bridge-reversal process is indeed observed in **8** whereby the four Cp signals coalesce at 273 K and sharpen to give two peaks at 293 K (see Fig. 1-16). Further support of bridge-reversal is obtained from the resonances of the terminal borane protons. It is noteworthy that the two fluxional processes (i.e. Cp ring twisting and bridge-reversal) can be distinguished. Since inversion at the phosphorus atoms, but not ring twisting,

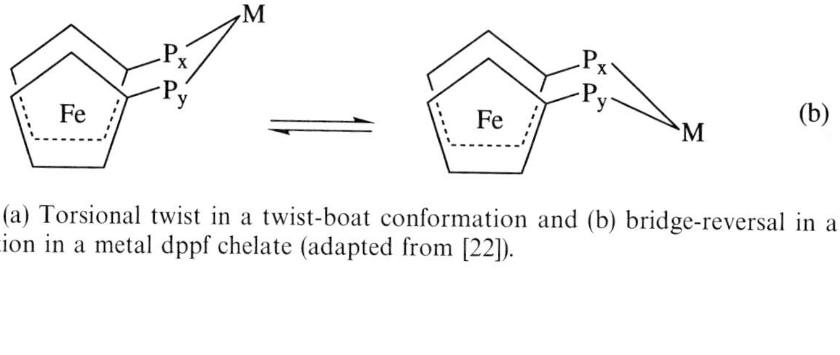

Fig. 1-17. (a) Torsional twist in a twist-boat conformation and (b) bridge-reversal in a half-chair conformation in a metal dppf chelate (adapted from [22]).

is frozen out at 253 K, it indicates that the energy barrier to the former is higher than that of the latter. This can be reasoned using purely steric grounds. In general, torsional twist of the Cp rings is facile and rarely static even at low temperature. Another typical example is $M(CO)_4$(dppf-*P*, *P'*)(M = Cr, Mo) [22]. Comparison of the fluxional properties of 1-{(dppf-*P*, *P'*)Pd}B_3H_7 **8** with those of *cis*-$PdCl_2$(dppf-*P*, *P'*) reveals that chloride replacement by the rigid borane fragment increases the barrier to bridge-reversal [61]. Similar fluxional processes are observed in the ^1H NMR spectra of Rh{(η^6-C_6H_5)B(C_6H_5)$_3$}(dppf-*P*, *P'*) **12** [81] and Pt($CH_2SiMe_2CH{=}CH_2$)$_2$(dppf-*P*, *P'*) **11** [73]. A single broad signal of **12** at 300 K is resolved into four sharp signals at 208 K. The H_α and H_β protons of **11** display slightly broadened signals at 298 K. On lowering the temperature, multiplets arising from the enantiomeric and magnetically nonequivalent Cp protons are obtained. Signal coalescence is also observed for the magnetically nonequivalent phenyl rings. It is, however, not always possible to freeze a bridge-reversal process. [$Pt_2H(\mu$-$H)_2$(dppf-*P*, *P'*)$_2$]Cl **10**, for example, exhibits two sharp Cp signals that persist from 293 K to 210 K [63]. Lack of any signal broadening indicates the very low energy barrier to inversion and that at least one of the ferrocenyl units has an average eclipsed conformation. Mechanisms and energy barriers to bridge-reversal process in [3]-ferrocenophanes have been discussed [82–84].

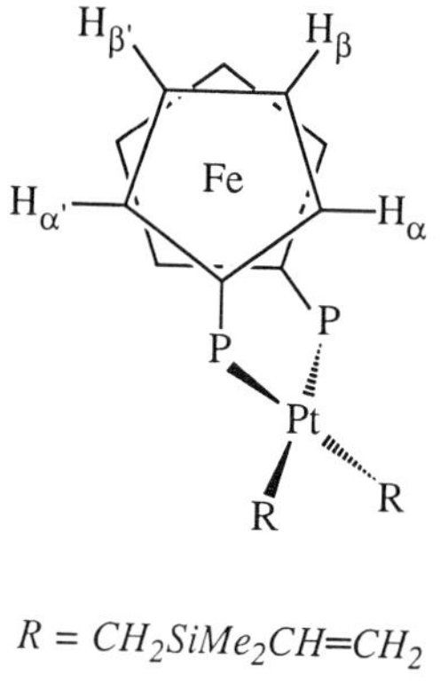

R = $CH_2SiMe_2CH{=}CH_2$

11 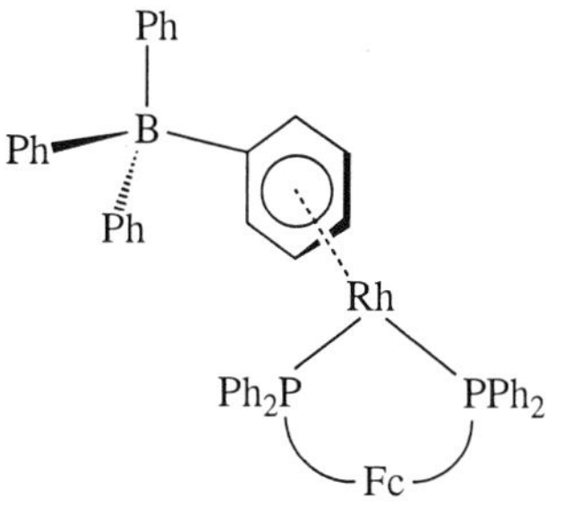 **12**

Dynamic exchange between chemically non-equivalent phosphorus donor sites is noted in the variable temperature $^{31}P\{^1H\}$ spectra of [$Pt_2(\mu$-H)(μ-CO)(dppf-*P*, *P'*)$_2$] · [BF_4], **9** [62]. At 308 K, a solitary signal is obtained at 25.5 ppm. The spectrum at 223 K, which consists of two signals at 31.0 and 16.7 ppm, agrees with the hetero-bridged solid state structure. In $Rh_2(\mu$-S-*t*Bu)$_2$(CO)$_2$(μ-dppf), **2** [34], the dppf ligand is not symmetrically disposed with respect to the Rh_2S_2 core. This results in two non-equivalent phosphorus sites in an enantiomer. Conformational interchange between the two enantiomers is facile at 312 K. The doublet arising from direct Rh-P coupling is neatly split into a doublet of doublets at 260 K (Fig. 1-18), a temperature at which the solid-state structure exists.

The temperature-dependent ^{31}P NMR spectrum of [Rh(dppf-*P*, *P'*)$_2$][BPh_4] reveals a different kind of solution fluxionality [65], which invokes changes in both

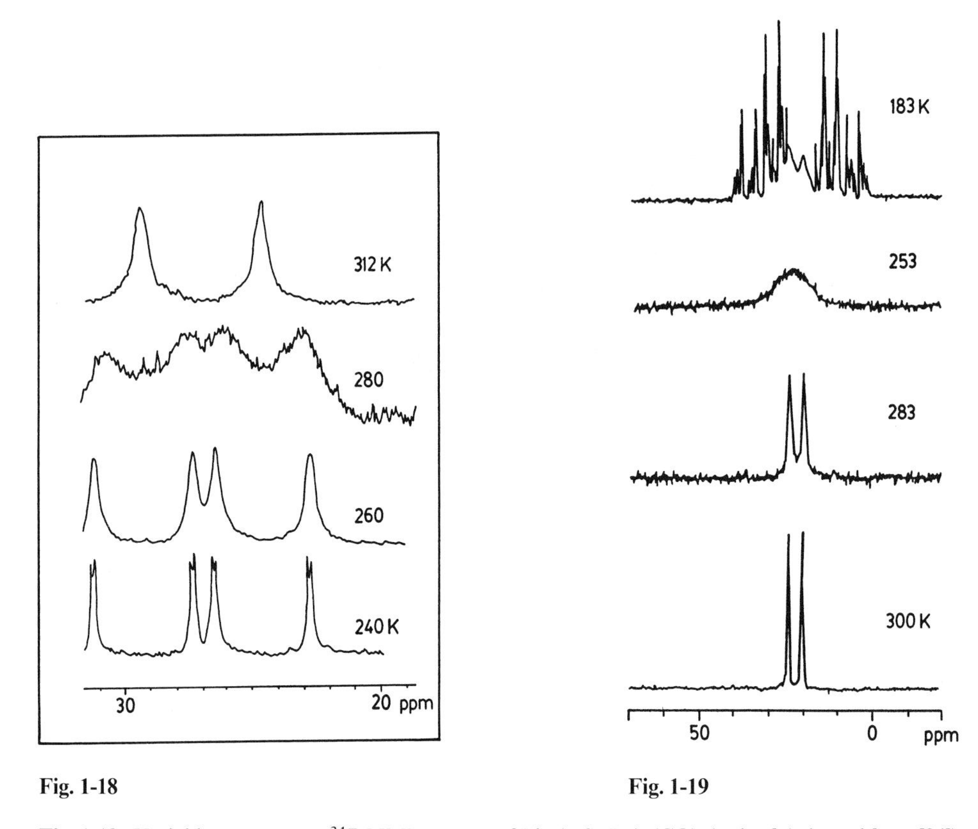

Fig. 1-18 **Fig. 1-19**

Fig. 1-18. Variable temperature ^{31}P NMR spectra of $Rh_2(\mu\text{-S-}t\text{Bu})_2(CO)_2(\mu\text{-dppf})$ (adapted from [34]).

Fig. 1-19. Variable temperature ^{31}P NMR spectra (36.23 MHz) of $[Rh(\text{dppf-}P,P')_2][BPh_4]$ (adapted from [65]).

the coordination geometry of the metal and the coordination mode of the dppf ligand. At room temperature the chemical equivalence of the phosphine ligands is characterized by a sharp doublet with Rh-P coupling. This doublet coalesces at 253 K and evolves to an AA'BB'X multiplet and a very broad doublet at 183 K on the NMR timescale (36.23 MHz, Fig. 1-19). This intermediate multiplet pattern is consistent with the existence of a dimeric species, $[Rh(\text{dppf-}P,P')(\mu\text{-dppf})]_2^{2+}$. The 162 MHz spectrum at 183 K gives a clean A_2M_2X multiplet (Fig. 1-20), which is explained by the emergence of a pentacoordinated solvated complex $[Rh(\text{dppf-}P,P')_2(\text{solv})]^+$. Quantitative conversion of the dimeric species to this complex is perceived (Scheme 1-1), followed by rapid interconversion between the phosphines at the equatorial and axial positions in the complex. This would explain the signal due to the indistinguishable phosphorus atoms observed in the room-temperature ^{31}P NMR spectrum. Such fluxional behavior has also been observed for $[Rh(\text{dppp-}P,P')_2]^+$ [85] and $[Ir(\text{dppf-}P,P')_2][BPh_4]$ [65].

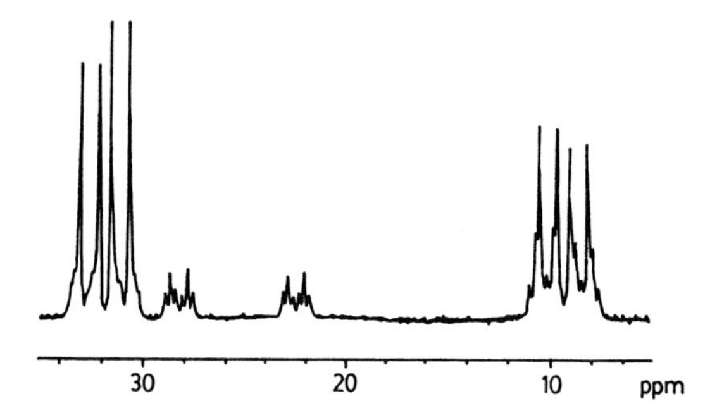

Fig. 1-20. 162 MHz ^{31}P NMR spectrum of $[Rh(dppf\text{-}P,P')_2][BPh_4]$ at 183 K showing an A_2M_2X spin pattern (adapted from [65]).

A similar dissociation process is displayed in another homoleptic complex $[Cu_2(dppf\text{-}P,P')_2(\mu\text{-dppf})][BF_4]_2$ **1a** [30]. A single broad resonance at 300 K is resolved into an AB_2 multiplet at 233 K. An exchange mechanism involving limited dissociation of the dimeric complex to two monomeric species via bridge splitting has been suggested. The monomers could then undergo facile mutual interconversion.

$$[Rh(dppf\text{-}P,P')(\mu\text{-dppf})]_2^{2+} + 2\ \text{solv} \Longleftrightarrow 2\ [Rh(dppf\text{-}P,P')_2(\text{solv})]^+$$

Scheme 1-1

A special fluxionality occurs in the dppf-bridged cluster complex $Ru_4(\mu\text{-H})(CO)_{12} \cdot (\mu_6\text{-B})Au_2(\mu\text{-dppf})$ **7**, in which a concerted mechanism composed of three mutually dependent processes is found: (a) the rocking motion of two Au atoms above the wing tips of the Ru_4B butterfly core; (b) inversion at the phosphorus atoms; and (c) Cp ring twisting [55]. These concomitant molecular motions lead to the interconversion of two enantiomers (Fig. 1-21), which provides a channel for the exchange of the Cp protons at the 2,5(α) and 3,4(β) positions. This result justifies the splitting of two Cp resonances at 298 K into four signals at 215 K.

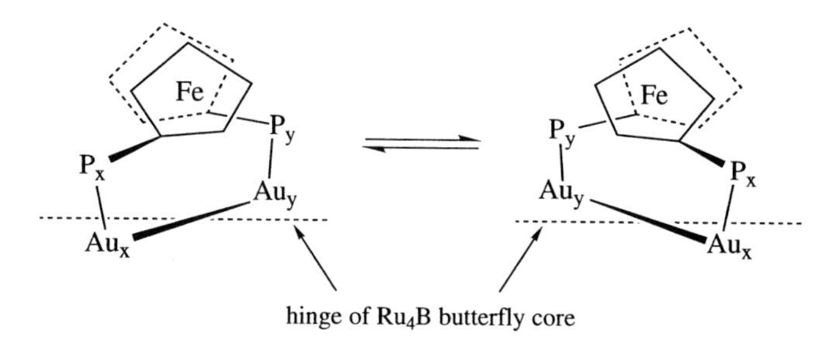

Fig. 1-21. Interconversion of two enantiomers in $Ru_4(\mu\text{-H})(CO)_{12}(\mu_6\text{-B})Au_2(\mu\text{-dppf})$ cluster complex [55].

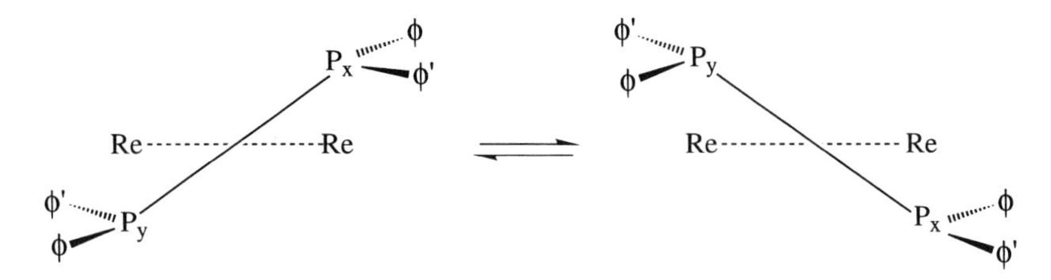

Fig. 1-22. Rocking motion of the dppf ligand with respect to the Re ... Re axis in $Re_2(\mu\text{-}OMe)_2(CO)_6(\mu\text{-dppf})$ (adapted from [74]).

A similar fluxional process is noted in $Re_2(\mu\text{-}OMe)_2(CO)_6(\mu\text{-dppf})$, in which dppf is in its quasi-closo bridging state [74]. The rocking motion of the diphosphine about the Re ... Re axis (Fig. 1-22) is a direct result of the opposing rotation of the

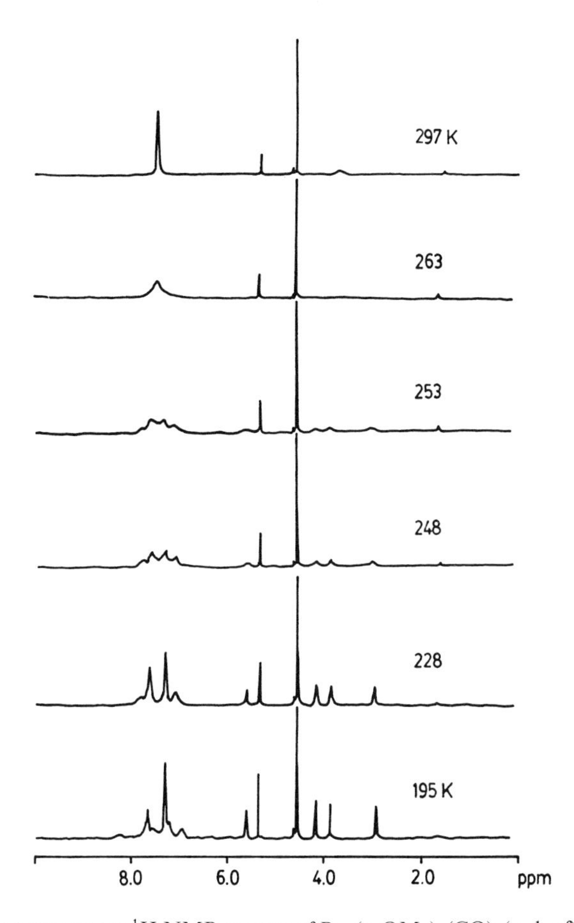

Fig. 1-23. Variable temperature ^1H NMR spectra of $Re_2(\mu\text{-}OMe)_2(CO)_6(\mu\text{-dppf})$ (adapted from [74]).

two Cp rings about the Cp-Fe-Cp axis. This motion causes chemical equivalence of both the phenyl protons and the proton pairs at the α and β positions (Fig. 1-23). The methoxy resonance remains unperturbed throughout the temperature range shown, which demonstrates the rigidity of the $[Re_2(\mu\text{-}OMe)_2]$ fragment even though it is being traversed by a fluxional dppf ligand.

1.5 Catalysis

1.5.1 Cross Coupling

1.5.1.1 Organic Electrophile and Organometallic Coupling

Work by Kumada et al. in the late 1970s on the use of $PdCl_2$(dppf) as a catalyst in the cross coupling of Grignard reagents and organic bromides [86] (Scheme 1-2) and allylic alcohols [87] (Scheme 1-3) created a new dimension in research into the homogeneous catalysis of coupling reactions. The superiority of $PdCl_2$(dppf) as a catalyst over other bidentate phosphine complexes or the classical $PdCl_2(PPh_3)_2$ is attributed to a delicate balance between steric and electronic influences of the phosphine moieties on the catalytic activity. This discovery had a significant bearing on the catalytic mechanism and also facilitated a series of elegant organic syntheses that had constantly troubled organic chemists but which could now be solved by this "magic" catalyst. Some recent representative developments in this ever-growing area of research are described below.

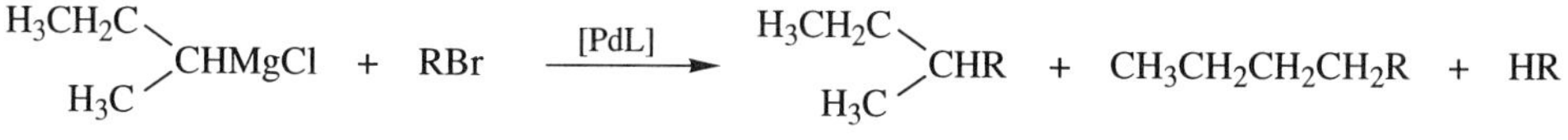

$$(H_3CH_2C)(H_3C)CHMgCl + RBr \xrightarrow{[PdL]} (H_3CH_2C)(H_3C)CHR + CH_3CH_2CH_2CH_2R + HR$$

R = Ph-, (E)-PhCH=CH-, H_2C=CMe-

Scheme 1-2

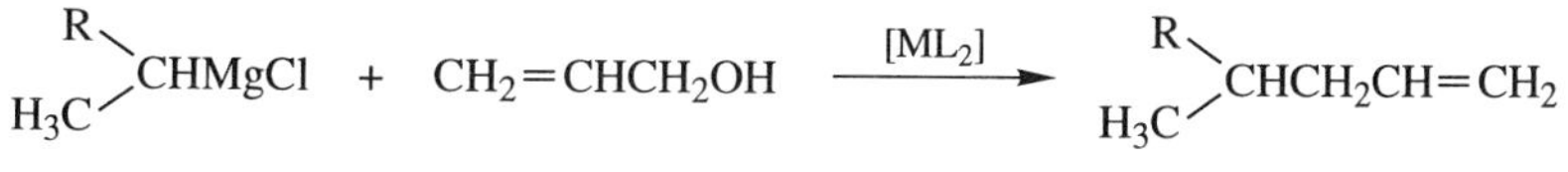

$$(R)(H_3C)CHMgCl + CH_2{=}CHCH_2OH \xrightarrow{[ML_2]} (R)(H_3C)CHCH_2CH{=}CH_2$$

R = n-C_6H_{13}, Ph

Scheme 1-3

Grignard and Organozinc Substrates

Mechanism

The wide-ranging applications of Grignard cross-coupling have generated a host of studies on its mechanistic details [88–90]. Although the precise mechanistic pathway is still uncertain, much synthetic and spectroscopic evidence points to an unsaturated Pd precursor that readily undergoes oxidative addition by an electrophile, followed by an alkyl transfer from a nucleophilic Grignard reagent, and subsequent reductive elimination to give the desired product (Fig. 1-24). According to this mechanism, the best catalyst obviously must be reduced and oxidized fairly easily, and the ligands must be of reasonable strength to support the stability of the catalyst and yet not be strong enough to over-saturate the active catalyst. An ideal catalyst should also suppress the unwanted β-elimination but favor rapid reductive elimination. Recent support for this mechanism comes from reports describing the coupling between *E*-2′-*p*-methoxystyryl halide **13** and *o*-methoxyphenyl Grignard (*o*-$MeOC_6H_4MgX$) to give 2,4′-dimethoxystilbene **14** catalyzed by [Pd(dppf)] (from *in situ* reduction of $PdCl_2$(dppf) by Li_2(cot) at −78 °C) [91–93]. There is spectroscopic evidence for the vinyl complex **15**, from the oxidative addition step, and the alkene complex **16**, from the reductive coupling, even though the arylation product **17** is too unstable to be detected. This observation thus provides concrete support for sequential addition of the electrophile and nucleophile to a *cis*-[Pd(dppf)] moiety in the catalytic cycle. No phosphine dissociation is invoked, nor is oxidative addition to give a Pd(IV) complex necessary. Similar arylation using Pt(dppf)(C_2H_4) as a catalyst allows direct observation of all the key intermediates [94]. Replacement of dppf by DIOP [95, 96] does not diminish the catalytic efficiency.

Ni(II) and Pd(II)/Pd(0) phosphine complexes were developed and modified as a result of the above considerations. The designed syntheses of dppf in the early

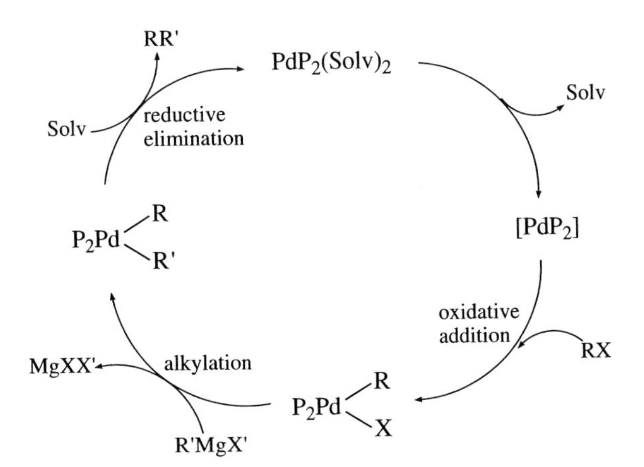

Fig. 1-24. Pd-catalyzed cross-coupling of alkyl halide and Grignard reagent. (Solv = coordinating solvent; P = monophosphine, PR''_3) (adapted from [264]).

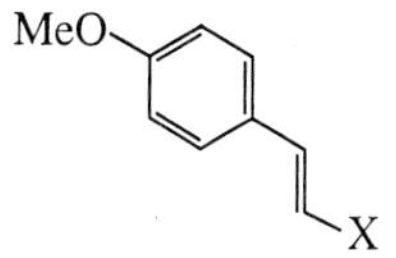

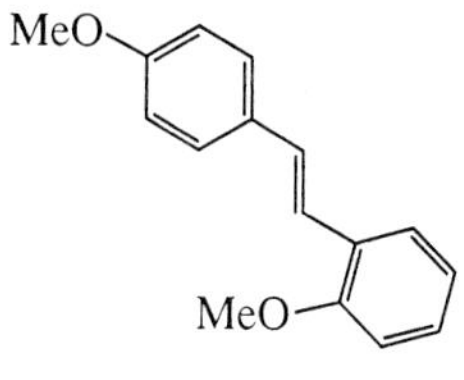

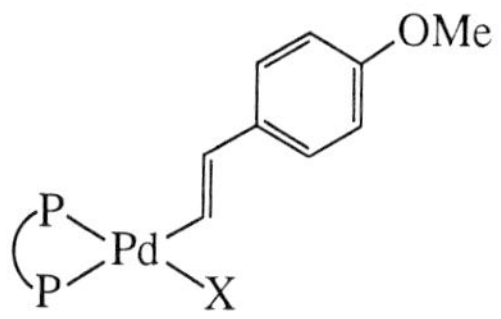

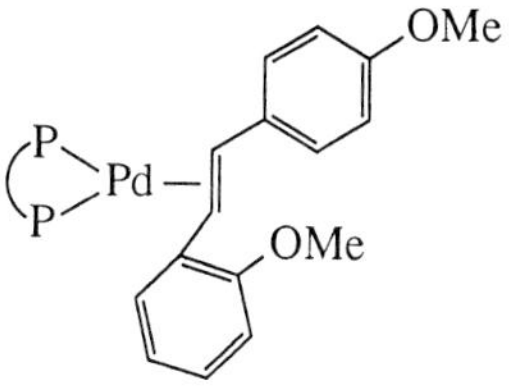

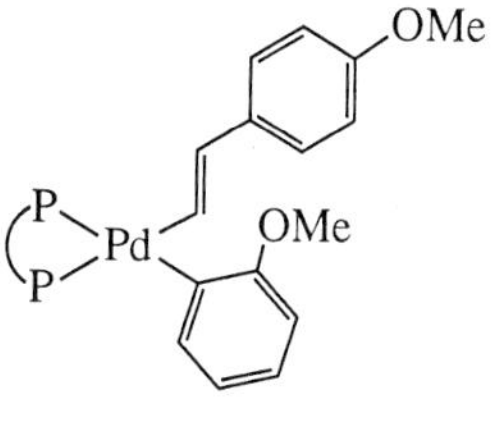

1970s and $PdCl_2$(dppf) in 1979 marked a significant breakthrough in this area of research. The role of this ferrocene-based bidentate ligand in the catalytic cycle is the subject of vigorous speculation in the literature. The use of mono-versus diphosphines in the catalyst system is also a subject of great interest. It remains unclear whether a diphosphine would significantly alter the mechanistic pathways indicated in Fig. 1-24, which are based on two monophosphine ligands. What seems certain is that a diphosphine such as the dppf ligand improves the catalytic activity and selectivity in many circumstances. This dppf ligand was first employed by Hayashi and Higuchi et al. [57] in the coupling reactions between primary and secondary Grignard or alkylzinc substrates with organic halides. The value of this work lies in the high efficiency of the catalyst, which suppresses the undesirable β-elimination products [97–99]. Its superiority over other traditional PPh_3 and diphosphine catalysts of Ni and Pd is summarized (Table 1-5) in the coupling of *n*- and *sec*-BuMgCl with RBr. The advantages of $PdCl_2$(dppf) over the other catalysts are ascribed to the large bite angle (99.1°) and small Cl-Pd-Cl angle (87.8°), which facilitate the reductive elimination step. A chelating phosphine is clearly superior to a monophosphine in this respect as it favors the *cis* disposition of the coupling fragments. The *trans* product is also more susceptible to β-hydride elimination [100, 101]. The longer Pd-P bonds (2.292 Å) compared with those of the other diphosphine complexes also suggest a more facile phosphine dissociation [100, 102–104], which has been proposed as a prerequisite in the elimination process. Another advantage offered by dppf over other common diphosphines is its flexible bite size and angle, which could stabilize the 14-electron species [Pd(P-P)] formed on elimination of the coupled product.

Table 1-5. Reaction conditions and catalyst effects on Grignard (*n*- and *sec*-BuMgCl) coupling with alkyl and aryl bromides (data from [57])

	Catalyst	RBr	Reaction conditions[a]		Product distribution	
			T, °C	time, h	*sec*BuR	*n*BuR
(a) *sec*-BuMgCl	$PdCl_2(dppf)$	PhBr	r.t.	1	95	0
	$Pd(PPh_3)_4$		r.t.	24	4	6
	$PdCl_2(PPh_3)_2$		r.t.	24	5	6
	$PdCl_2(dppe)$		r.t.	48	0	0
	$PdCl_2(dppe)$		reflux	8	4	1
	$PdCl_2(dppp)$		r.t.	24	43	19
	$PdCl_2(dppb)$		r.t.	8	51	25
	$NiCl_2(PPh_3)_2$		r.t.	23	3	5
	$NiCl_2(dppp)$		r.t.	23	29	3
	$PdCl_2(dppf)$	Ph-C=C-Br	0	2	97	0
	$PdCl_2(dppf)$		r.t.	20	93	0
	$Pd(PPh_3)_4$		0	3	33	36
	$PdCl_2(dppe)$		0	5	3	3
	$PdCl_2(dppp)$		0	5.5	76	5
	$PdCl_2(dppb)$		0	3.5	53	25
	$NiCl_2(PPh_3)_2$		r.t.	24	56	3
	$NiCl_2(dppp)$		0	5	88	1
	$PdCl_2(dppf)$	MeO-C6H4-Br	r.t.	19	75	1
	$PdCl_2(PPh_3)_2$		r.t.	18	3	2
	$PdCl_2(dppf)$	2-Me-C6H4-Br	r.t.	19	58	0
	$PdCl_2(PPh_3)_2$		r.t.	20	0	2
	$PdCl_2(dppf)$	$CH_2{=}CMeBr$	0	8	80	0
(b) *n*-BuMgCl	$PdCl_2(dppf)$	PhBr	r.t.	24		92
	$PdCl_2(dppe)$		reflux	7		3
	$PdCl_2(dppp)$		reflux	8		12
	$PdCl_2(dppb)$		r.t.	24		54
	$NiCl_2(PPh_3)_2$		r.t.	24		14
	$NiCl_2(dppp)$		r.t.	24		42
	$PdCl_2(dppf)$	Ph-C=C-Br	r.t.	20		90

a r.t. = room temperature.

Reactions

A recent effect of the use of dppf in a Pd catalyst is seen in an efficient and selective mono-alkylation and -arylation of dichlorobenzenes (Scheme 1-4) [105, 106]. An elevated temperature is used to overcome the sluggish oxidative addition known for aryl halides. Among the diphosphine ligands in the catalyst $PdCl_2$(P-P) (P-P = diphosphine), dppf is again found to be the most efficient. The activity of $PdCl_2(PPh_3)_2$ is poor. Introduction of free dppf to the catalytic mixture exerts two opposing effects, as it seems to suppress both the oxidative addition of the aryl halide and the β-elimination.

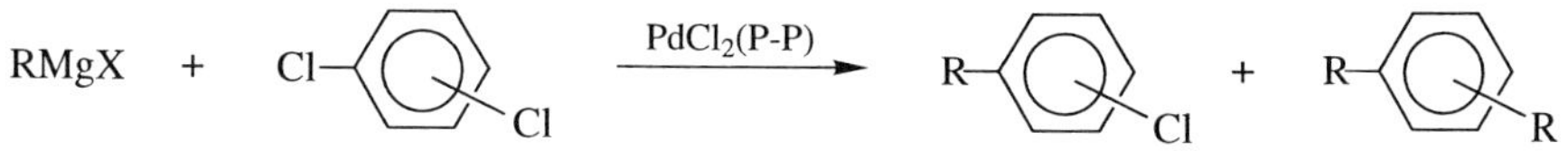

RMgX + Cl–C_6H_4–Cl $\xrightarrow{PdCl_2(P\text{-}P)}$ R–C_6H_4–Cl + R–C_6H_4–R

Scheme 1-4

This Grignard coupling can be extended to the allylic systems [107, 108]. Cross coupling of allylic ethers such as (*E*)-1-triethylsiloxy-2-butene **18** or 3-triethylsiloxy-1-butene **19** with aryl Grignard reagents gives regioselective products **20** or **21** (Scheme 1-5). Interestingly, the use of $PdCl_2$(dppf) as a catalyst favors C-C

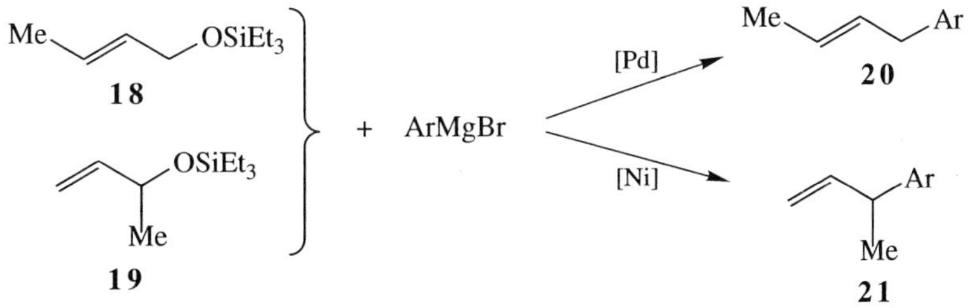

Scheme 1-5

bond formation at the less substituted position (**20**), whereas the Ni analogue gives the more substituted product **21**. The reason for the opposite regioselectivity is unclear but it is possible that a steric effect is operational, which exemplifies the differences between tetrahedral Ni(II) and planar Pd(II) in the intermediates. Among the Ni diphosphine catalysts $NiCl_2$(dppf) also exhibits the highest regioselectivity. The high activity and regioselectivity of the dppf catalysts is a feature in this type of aryl-allyl synthesis (Table 1-6). Arylation of *cis*-**22** and *trans*-5-methyl-2-cyclohexenyl silyl ethers with both Pd and Ni catalysts proceeds with inversion of configuration. A π-allyl M(II) complex **23** is believed to be a key intermediate. The mechanism follows the general pathway described in Fig. 1-24 except that the active catalytic species **24** is cationic. Nucleophilic attack by the Grignard reagent gives **25** which reverts back to **24**. There are occasions when the use of a Grignard reagent would yield the homo- rather than hetero-coupling product. The *N*-substituted isocyanide dichloride **26** reacts catalytically with tributyl(trimethylsilylethynyl)tin to give *N*-substituted dialkynylimines **27**, which are readily hydrolyzed to dialkynylketones **28**. The use of a Grignard reagent instead of the organostannyl substrate would give high yields of dialkynes by oxidative homo-coupling of the Grignard [109]. The homo-coupling is catalyzed by $PdCl_2$(dppf) and liberates isocyanide as the byproduct. Synthesis of alcohols can also be achieved by coupling a vinyl bromide with a silyl-substituted Grignard reagent **29** [110]. Oxidation of the coupling product **30** gives the corresponding alcohol **31** directly. Grignard coupling can also be applied to heterocyclics, as illustrated in the bromobithiophene synthesis at 0 °C [111] and arylation at the C-2 position of 2-(methylthio)-

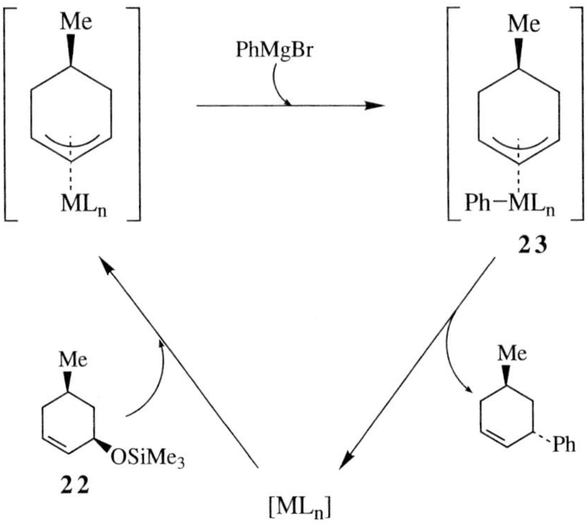

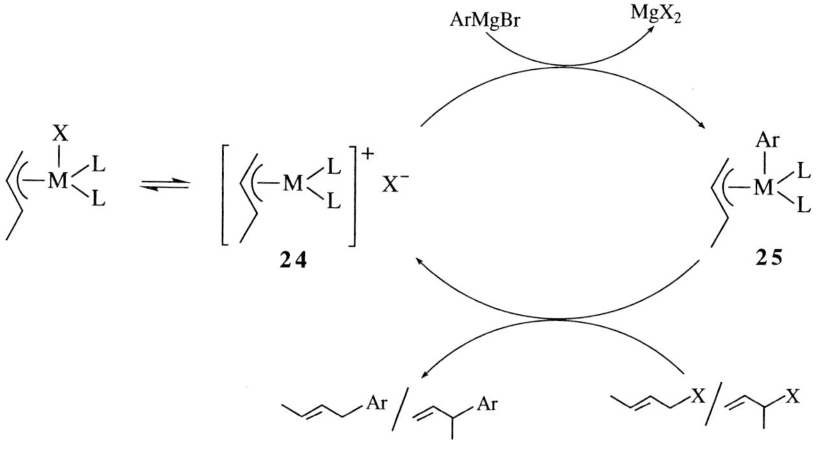

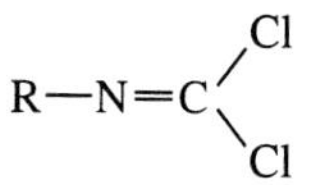

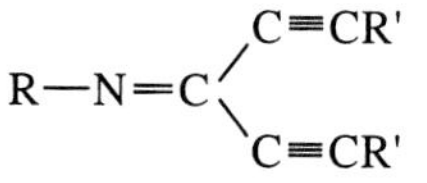

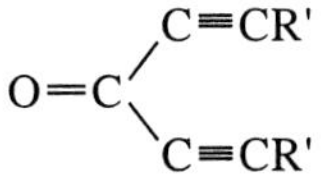

Table 1-6. Catalyst effect on the Grignard (PhMgBr) coupling with allyl ether (data from [107])

ArMgBr	Allyl ether	Catalyst	Reaction time, h	Total yield, %	Product ratio (*E*/*Z*) **20**:**21**
PhMgBr	Me~~OSiEt$_3$ **18**	$NiCl_2$(dppf)	4	100	12(11/1)/88
		$NiCl_2$(dppp)	40	44	59(59/0)/41
		$NiCl_2(PPh_3)_2$	4	100	67(66/1)/33
		$PdCl_2$(dppf)	4	100	96(92/4)/4
		$PdCl_2$(dppe)	40	3	–
		$PdCl_2$(dppp)	40	68	93(80/13)/7
		$Pd(PPh_3)_4$	20	52	90(85/5)/10
PhMgBr	CH$_2$=CH–CH(Me)OSiEt$_3$ **19**	$NiCl_2$(dppf)	4	91	19(10/9)/81
		$NiCl_2$(dppp)	40	81	58(53/5)/42
		$NiCl_2(PPh_3)_2$	4	93	55(39/16)/45
		$PdCl_2$(dppf)	4	83	91(75/16)/9
		$PdCl_2$(dppp)	40	60	88(80/8)/12

SiMe(OEt)$_2$ R MgX **29** SiMe(OEt)$_2$ R Ph **30** OH R Ph **31**

CH$_3$ H$_3$C N SCH$_3$ O **32** + H$_3$C–C$_6$H$_4$–MgBr $\xrightarrow{MCl_2P_2}$ CH$_3$ H$_3$C N O –C$_6$H$_4$–CH$_3$ **33**

Scheme 1-6

4,4-dimethyl-2-oxazoline **32** to give **33** (Scheme 1-6) [112]. The catalytic merit of $PdCl_2$(dppf) is clearly seen in the latter reaction when activities of common catalysts are compared (Table 1-7). This synthetic strategy of removing a thiolato substituent from a 2-oxazoline ring has found some use in the preparation of aromatic acids, aldehydes, esters and ketones. Another means of facilitating substitution of a heterocyclic ring is exemplified in the coupling between a 1-methyl-2-pyrrolyl Grignard reagent or a Zn substrate with aryl iodide at room temperature [113]. The activity of $PdCl_2$(P-P) (P-P = dppb, dppf) as catalyst is higher than that of $NiCl_2$(dppp). A detailed account on catalytic couplings in heterocycles has recently been published by Kalinin [114].

The use of organozinc substrates in coupling reactions is often supported by $PdCl_2$(dppf). On the other hand, the use of the PPh_3 derivative does not always give the desired products [115]. This is best illustrated in the coupling between 1,2-dihaloalkenes and 1-alkynylzinc reagents (Scheme 1-7), in which $Pd(PPh_3)_4$

Table 1-7. Catalyst effect on the Grignard (*p*-tolylmagnesium bromide) coupling with 2-(methylthio)-4,4-dimethyl-2-oxazoline (data from [112])

Catalyst	*T*, °C	Product distribution, %	
		33	**32**
$NiCl_2(PPh_3)_2$	35	2	25
$NiCl_2$(dppe)	35	64	10
$NiCl_2$(dppe)	20	69	17
$NiCl_2$(dppp)	35	21	49
$NiCl_2$(dppf)	35	21	59
$PdCl_2(PPh_3)_2$	35	24	26
$PdCl_2$(dppe)	35	9	74
$PdCl_2$(dppb)	35	79	–
$PdCl_2$(dppb)	20	87	–
$PdCl_2$(dppf)	35	61	12
$PdCl_2$(dppf)	20	99	–

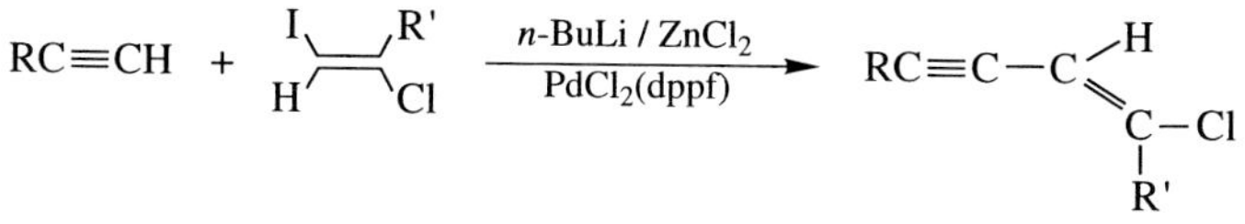

RC≡CH + I(H)C=C(R')Cl —[*n*-BuLi / $ZnCl_2$; $PdCl_2$(dppf)]→ RC≡C–CH=C(Cl)R'

R = n-C_4H_9, n-C_5H_{11}, Me_3Si, Ph
R' = H, n-C_5H_{11}

Scheme 1-7

performs poorly but $PdCl_2$(dppf) is by far the most effective. Again, the presence of the dppf ligand in the metal core is thought to promote the reductive elimination step by a steric effect. Another fine example can be drawn from a comparative study among different catalysts for coupling between vinyl- or aryl-triflates with the Reformatsky reagent $tC_4H_9CO_2CH(CH_3)ZnBr$ [116]. The desired coupling product is obtained only when [Pd(dppf)] (obtained by reducing $PdCl_2$(dppf) with $i$$Bu_2AlH$) is used as the catalyst and 2-naphthyl triflate the substrate (Scheme 1-8). Other catalysts tend to give reduction products arising from β-elimination or coupling products with the reductant. These observations support the common theory that phosphine dissociation promotes β-elimination, and hence formation of the reduced byproducts, and that the dppf ligand facilitates reductive elimination to give the desired coupling products.

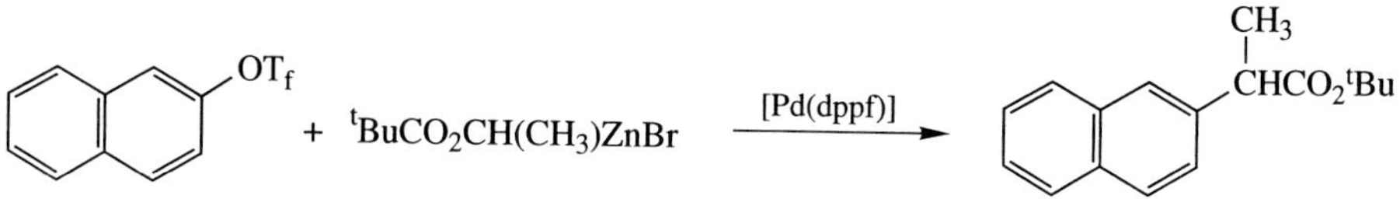

2-naphthyl-OTf + $^tBuCO_2CH(CH_3)ZnBr$ —[Pd(dppf)]→ 2-naphthyl-$CH(CH_3)CO_2{}^tBu$

Scheme 1-8

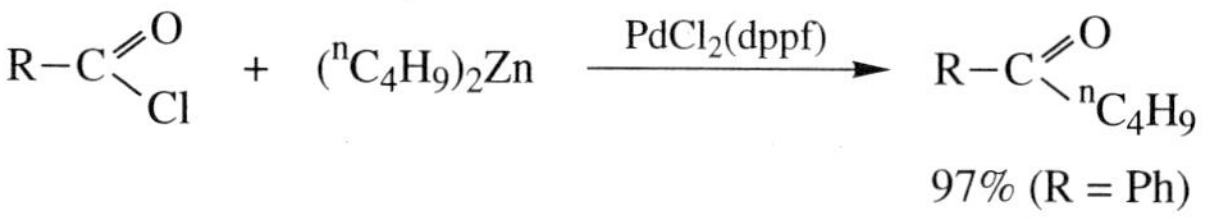

$$R{-}C(=O)Cl + ({}^{n}C_4H_9)_2Zn \xrightarrow{PdCl_2(dppf)} R{-}C(=O){}^{n}C_4H_9$$

97% (R = Ph)

Scheme 1-9

Dialkylzinc is another common substrate in Pd-catalyzed couplings. An efficient ketone synthesis uses acid chlorides as starting material (Scheme 1-9). The use of $PdCl_2$(dppf) offers high yields and is best in restricting the formation of benzaldehyde as a β-elimination byproduct [117]. *N*-Aryl aromatic imine **34** can be obtained from aromatic iodides and aryl isocyanide **35** [118]. The proposed intermediate, 1-[(arylimino)alkyl]zinc **36**, which arises from a 1,1-insertion of isocyanide into the Zn-Et bond, is an organometallic mimic of an acyl anion. Other applications of organozinc substrates are found in the $PdCl_2$(dppf)-catalyzed vinyl-alkyl coupling of the terpenic chains (Scheme 1-10). This process facilitates total synthesis of a physiologically active constituent of *Angelas* sponges, (±)-Ageline A [119], and the cross-coupling between bromoacetophenones or -benzaldehydes with RZnX (Scheme 1-11) [120].

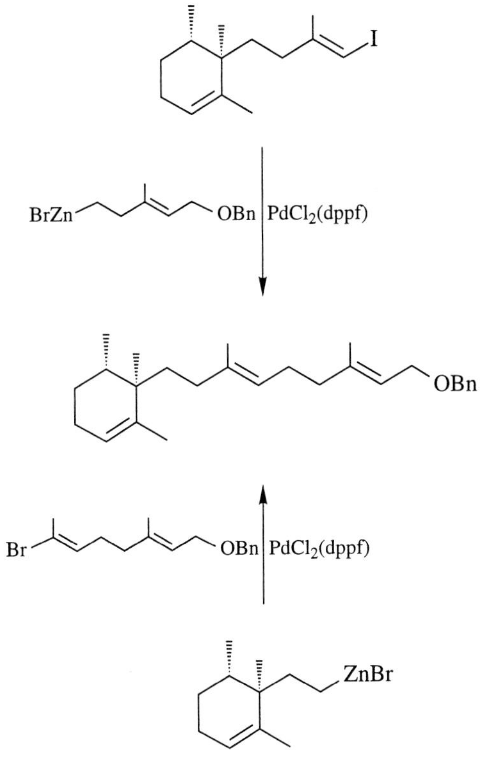

I

BrZn OBn $PdCl_2$(dppf)

OBn

Br OBn $PdCl_2$(dppf)

ZnBr

Scheme 1-10

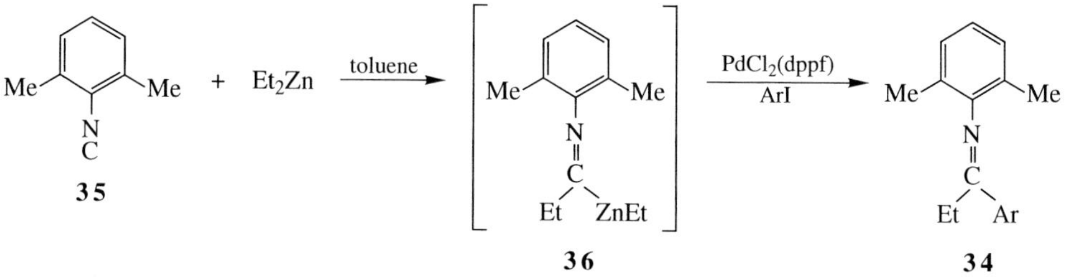

Me Me + Et_2Zn toluene → [Me Me N=C Et ZnEt] → $PdCl_2(dppf)$ ArI → Me Me N=C Et Ar

35 36 34

Ar = p-MeO-C_6H_4, α-naphthyl

$$R'C_6H_4Br + RZnX \xrightarrow{PdCl_2(dppf)} R'C_6H_4R + R'C_6H_4\text{-}C_6H_4R'$$

R' = p-MeCO, m-MeCO, p-HCO;
R = Bu, Ph; X = Cl or Br

Scheme 1-11

Organoboranes

Catalytic coupling reactions of alkyl- [121, 122] and 1-alkenyl- [123 – 125] boronates usually proceed with a base. Cross-coupling between R_3B or B-alkyl-9-BBN (BBN = borabicyclo[3.3.1]nonane) with aryl or 1-alkenyl halides, for example,

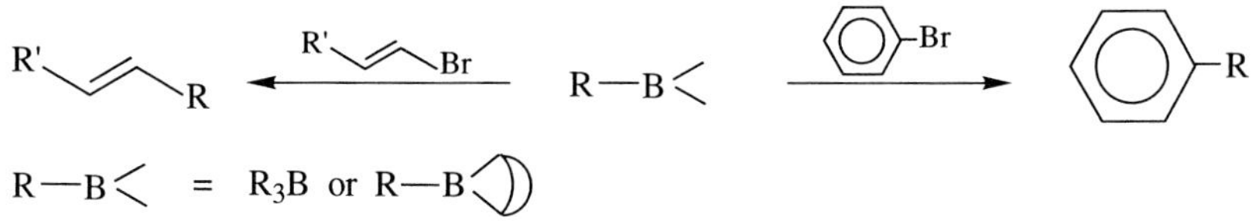

R' R ← R' Br — R—B< — PhBr → Ph-R

R—B< = R_3B or R—B(9-BBN)

Scheme 1-12

$(CH_2)_nCH=CH_2$ Br → HB(9-BBN) → $(CH_2)_{n+2}B$ Br → $PdCl_2(dppf)$/ NaOH → $(CH_2)_{n+2}$

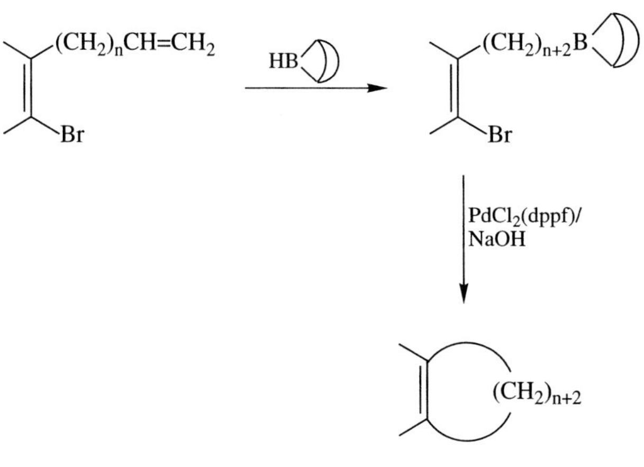

Scheme 1-13

$$\mathrm{\underset{R'}{\overset{R}{>}}C{=}CH_2 \;+\; HB\text{(9-BBN)} \longrightarrow \underset{R'}{\overset{R}{>}}CHCH_2B\text{(9-BBN)} \xrightarrow[\text{PdCl}_2\text{(dppf)/base}]{\text{XCH=CHR''/}} \underset{R'}{\overset{R}{>}}CHCH_2CH{=}CHR''}$$

Scheme 1-14

proceeds smoothly with $PdCl_2$(dppf) but only in the presence of a strong base (Scheme 1-12) [126]. Coupling with haloalkenes can occur either intra- (Scheme 1-13) or intermolecularly (Scheme 1-14). Using this method, stereodefined exocyclic alkenes can be synthesized conveniently [127]. Similar intramolecular coupling using ω-alkenyl triflates gives indan [128]. The use of catalytic $PdCl_2$(dppf), which functions together with a base, ensures that only fairly mild conditions are necessary. Triflates are also used in this cross-coupling (Table 1-8) in which geometrically pure alkylated arenes and alkenes are obtained. An interesting combination of haloarenes and triflates in two consecutive couplings is demonstrated in the synthesis of unsymmetrically disubstituted naphthalene from 1-iodo-2-naphthol (Scheme 1-15). The same catalytic mixture has found a recent application in the condensation of

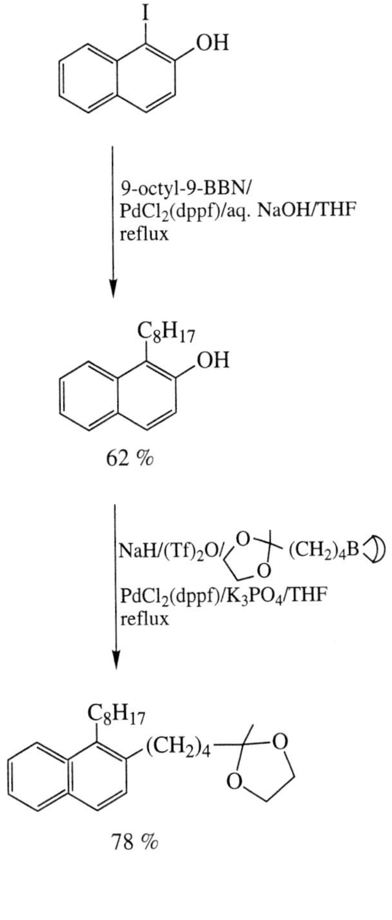

1-iodo-2-naphthol (I, OH)

↓ 9-octyl-9-BBN/ $PdCl_2$(dppf)/aq. NaOH/THF reflux

1-octyl-2-naphthol (C_8H_{17}, OH)

62 %

↓ NaH/$(Tf)_2O$/ 2-methyl-1,3-dioxolan-2-yl-$(CH_2)_4B$(9-BBN); $PdCl_2$(dppf)/K_3PO_4/THF reflux

1-C_8H_{17}-2-[$(CH_2)_4$-(2-methyl-1,3-dioxolan-2-yl)]naphthalene

Scheme 1-15 78 %

Table 1-8. Cross-coupling between triflates and 9-alkyl-9-BBN derivatives (from alkenes and 9-BBN) (BBN = borabicyclo[3.3.1]nonane) (data from [128])

Triflate	Alkene	Reaction time, h A[a]	Reaction time, h B[a]	Product	GC yield, % A	GC yield, % B
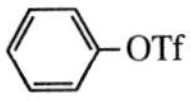 PhOTf	$CH_2{=}CH(CH_2)_8CO_2Me$	5	5	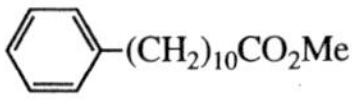 $Ph(CH_2)_{10}CO_2Me$	87	99
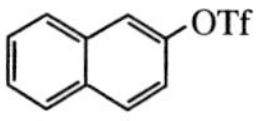 2-naphthyl-OTf	1-octene	5	5	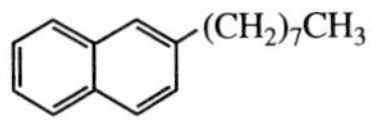 2-naphthyl-$(CH_2)_7CH_3$	82	97
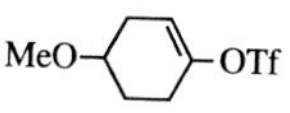 $MeO-C_6H_4-OTf$	$CH_2{=}CHCH_2COPh$	5	5	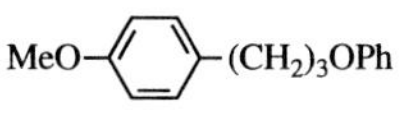 $MeO-C_6H_4-(CH_2)_3OPh$	92	76
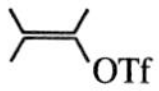 $Me_2C{=}C(Me)OTf$	1-octene		5	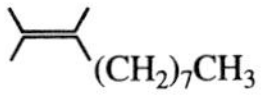 $Me_2C{=}C(Me)(CH_2)_7CH_3$		91
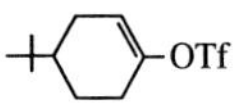 4-tBu-cyclohexenyl-OTf	$CH_2{=}CH(CH_2)_8CO_2Me$	5	5	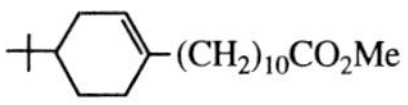 4-tBu-cyclohexenyl-$(CH_2)_{10}CO_2Me$	89	99
 camphor-derived enol triflate	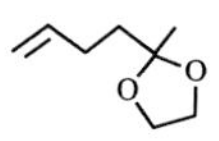 alkenyl dioxolane	18	18	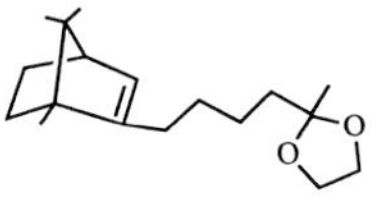 coupled dioxolane product	57	74

a Procedure A: $Pd(PPh_3)_4$ in dioxane at 85 °C.
Procedure B: $PdCl_2(dppf)$ in refluxing THF.

2,5-dibromoaniline with B-*n*-dodecyl-9-BBN leading ultimately to a diimide monomer. The latter undergoes condensation with an aryl diboronic acid to give a soluble rigid-rod polyimide [129]. Boronic esters also fail to react smoothly under similar conditions unless $PdCl_2(dppf)/Tl_2CO_3$ is used as the catalytic mixture [130]. Numerous boronyl ketones and arylboronic esters can be used in these couplings under mild conditions (Table 1-9). The stereoselectivity promoted by the dppf ligand is best reflected in the synthesis of conjugated 2,4-alkadienoates [131]. The use of $Pd(OAc)_2(dppf)$ as a catalyst precursor is witnessed in the synthesis of a variety of pharmaceutically important pyrazines [132]. A typical example is represented by the coupling between halo pyrazines (e.g., methyl 5-bromonicotinate **37**) with areneboronic acids (e.g., 3-pyridineboronic acid **38**) to give substituted bipyridines **39** [133].

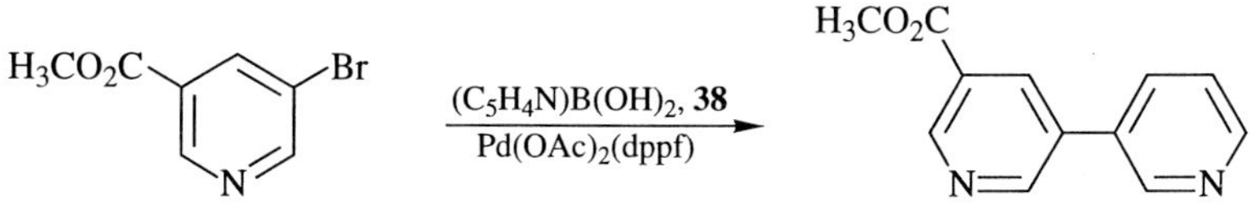

H_3CO_2C, Br, N — **37** —[$(C_5H_4N)B(OH)_2$, **38** / $Pd(OAc)_2(dppf)$]→ H_3CO_2C, N, N — **39**

Table 1-9. Cross-coupling reaction of boronic esters and halides catalyzed by $PdCl_2(dppf)$ in THF at 50 °C for 16 h in the presence of a base (Tl_2CO_3) (data from [130])

Boronic ester	Aryl halide	Product	Yield, %
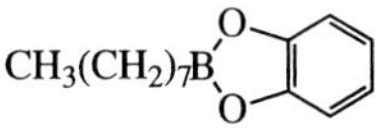 $CH_3(CH_2)_7B$ (catechol boronate)	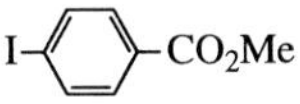 I–C_6H_4–CO_2Me	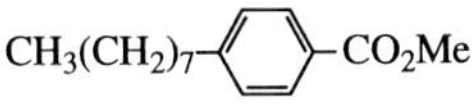 $CH_3(CH_2)_7$–C_6H_4–CO_2Me	88
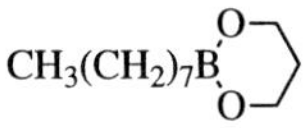 $CH_3(CH_2)_7B$ (1,3,2-dioxaborinane)	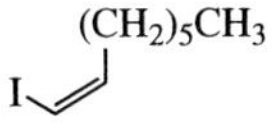 I–CH=CH–$(CH_2)_5CH_3$	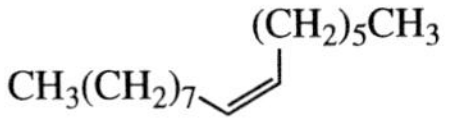 $CH_3(CH_2)_7$–CH=CH–$(CH_2)_5CH_3$	15
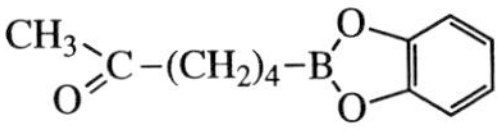 $CH_3C(=O)$–$(CH_2)_4$–B (catechol boronate)	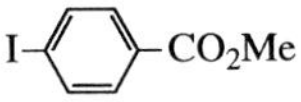 I–C_6H_4–CO_2Me	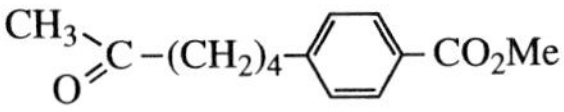 $CH_3C(=O)$–$(CH_2)_4$–C_6H_4–CO_2Me	62
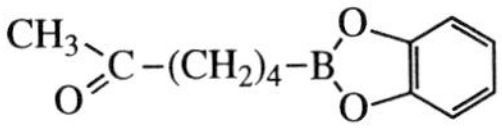 $CH_3C(=O)$–$(CH_2)_4$–B (catechol boronate)	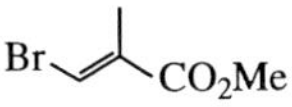 Br, CO_2Me	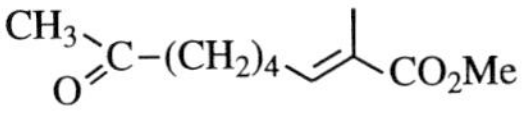 $CH_3C(=O)$–$(CH_2)_4$, CO_2Me	68
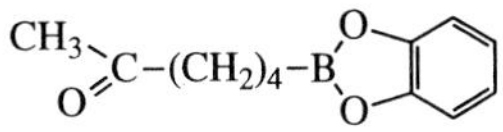 $CH_3C(=O)$–$(CH_2)_4$–B (catechol boronate)	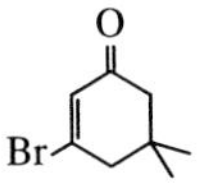 O, Br	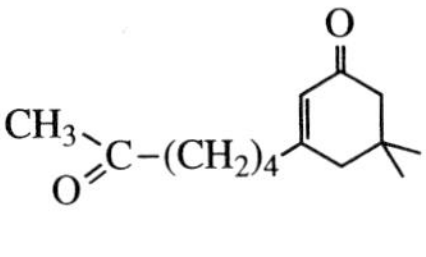 $CH_3C(=O)$–$(CH_2)_4$, O	66
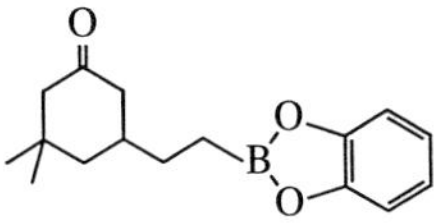 O, B (catechol boronate)	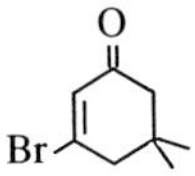 O, Br	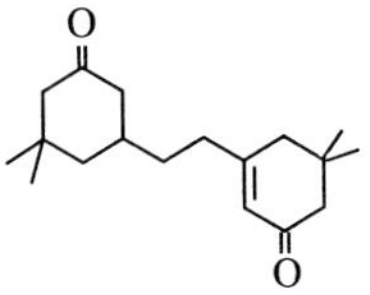 O, O	63

Organosilicon and Stannanes

Mechanism

Since Stille reported Pd-catalyzed coupling reactions between unsaturated halides and sulfonates with organostannanes (Scheme 1-16) [134], there have been many mechanistic studies on these and related reactions [135–137]. A commonly accepted mechanistic pathway resembles that of Grignard cross-coupling, which requires an

$$RX + R'SnR''_3 \xrightarrow{Pd(0)} RR' + XSnR''_3$$

R, R' = aryl, vinyl, silyl;
X = Br, I, OTf

Scheme 1-16

oxidative addition of the organic halide to an unsaturated Pd(o) catalyst, followed by alkylation (sometimes called transmetalation) from the organometallic substrate and reductive elimination to give the desired coupling product. Although the oxidative step is slow for allylic acetates and some alkyl halides in Grignard coupling [138], it has been shown recently in some typical Stille reactions that alkylation is the rate determining step. The slow step requires ligand (usually phosphine) dissociation and an incoming nucleophile to give an olefinic intermediate **40** [139].

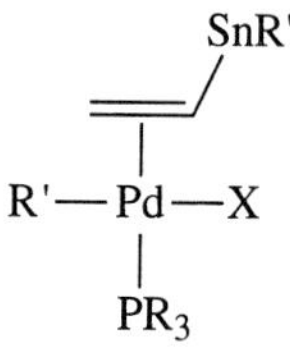

SnR''_3

R'—Pd—X

PR_3

40

Reactions

The use of organotin substrates in organic syntheses is well-known [140, 141]. Metal-catalyzed ketone syntheses in the presence of organometallics, for example, have met with varying degree of success [142–144], but one of the best routes appears to be the use of organotin compounds catalyzed by Pd complexes [145]. Organostannanes are widely used in Stille coupling reactions [146, 147]. The use of 1,4-naphthoquinones as electrophiles in cross-coupling with Bu_4Sn (Scheme 1-17) [148] has found application in the synthesis of antibiotics WS 5995 A and C [149–151]. Another recent example employs internal N → Sn coordination of 1-aza-5-stannabicyclo[3.3.3]undecane **41** in a selective alkylation of aryl bromides catalyzed by $PdCl_2$(dppf) (Table 1-10) [152]. This coordination accelerates the usually sluggish alkyl transfer in the Stille reaction [153–156] and restricts the transfer of undesirable organic moieties from the Sn substrate. The long exocyclic Sn-C bond is possibly responsible for the enhanced reactivity through a four-centered Sn ... C ... Pd ... Y (Y = halide) bond breaking bond making mechanism or electron transfer from Sn to Pd. A more traditional coupling is found in the inter-

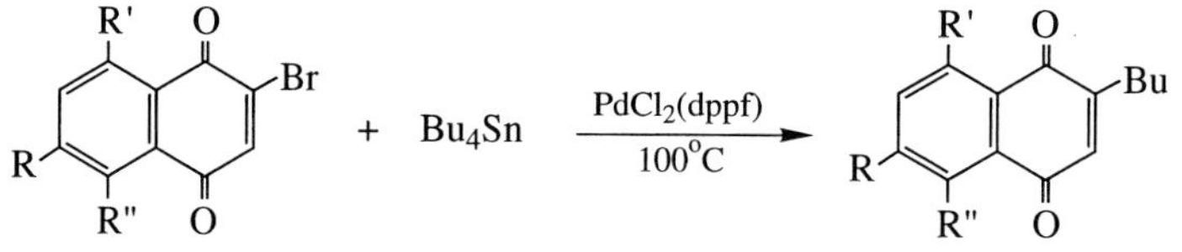

R' O Br R R'' O + Bu_4Sn $\xrightarrow[100^{\circ}C]{PdCl_2(dppf)}$ R' O Bu R R'' O

R	R'	R''	Yield (%)
H	H	OH	82
OMe	OH	H	74

Scheme 1-17

Table 1-10. Alkylation of aryl bromides by 1-aza-5-stannabicyclo[3.3.3]undecane catalyzed by $PdCl_2(dppf)$ (data from [152])

Reactant, **41**	X (XC_6H_4Br)	Conditions, °C, h	Yield, %
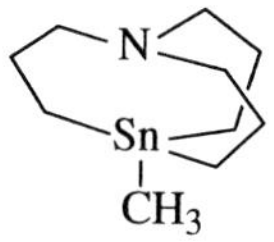 1-aza-5-stannabicyclo[3.3.3]undecane, Sn–CH_3	*p*-CH_3O	75, 2	94
	p-$(CH_3)_2N$	105, 30	56
	m-NO_2	105, 2	93
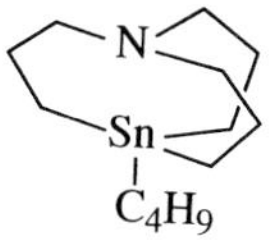 1-aza-5-stannabicyclo[3.3.3]undecane, Sn–C_4H_9	*p*-CH_3O	105, 48	64
	m-NO_2	105, 12	86

action between phenyl(trimethylstannyl)acetylene and aryl halides catalyzed by $PdCl_2(PPh_3)_2$ or $PdCl_2(dppf)$ (Scheme 1-18) [157]. An electron withdrawing moiety on the aryl group enhances the coupling rate, presumably due to acceleration of the oxidative addition step [158]. This reaction is one of the many Stille-type processes that are better catalyzed by monophosphine complexes. A similar observation is made in the ketimine synthesis from imidoyl chlorides and an organostannane (Scheme 1-19) [159]. A notable exception is found in the reaction between hydroxy-anthraquinone triflates and stannanes. The difficulties experienced with $Pd(PPh_3)_4$ or $PdCl_2(PPh_3)_2$ can be overcome with $PdCl_2(dppf)$, which readily gives the anthraquinone derivatives **42** in good yield [160]. This coupling is effective even for aryl substrates with a free *o*-hydroxyl group. There are occasions when the use of $PdCl_2(dppf)$ as a catalyst is satisfactory but the regioselectivity is not better than that of the PPh_3 analogue. An example of this is found in the coupling between acetylenic aryl triflates **43** and vinyltributyltin or 1-(trimethylsilyl)-2-(tributylstannyl)acetylene $Me_3SiCCSnBu_3$, which gives (*Z*)-indanylidene derivatives **44** by cyclization [161].

Interesting coupling reactions can be carried out with alkenyl- and alkynylstannanes. Reductive coupling of acid chlorides with (*E*)-1,2-bis(tri-*n*-butylstannyl)-

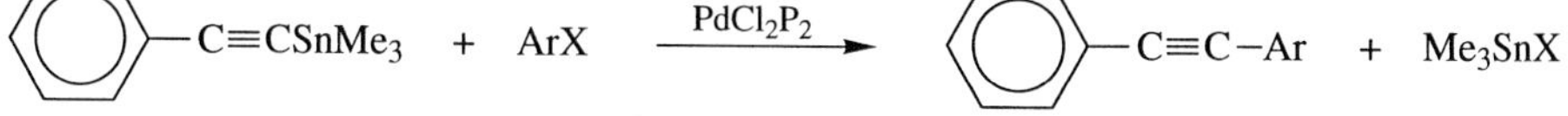

$$C_6H_5-C{\equiv}CSnMe_3 + ArX \xrightarrow{PdCl_2P_2} C_6H_5-C{\equiv}C-Ar + Me_3SnX$$

Scheme 1-18

$$R'\text{-}N{=}C(Cl)R + R''SnR_3 \xrightarrow{\text{Pd cat.}} R'\text{-}N{=}C(R'')R + SnClR_3$$

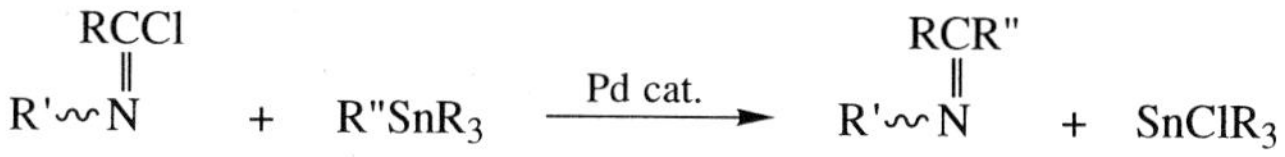

Scheme 1-19

$RSnR'_3$, $PdCl_2(dppf)$ → **42**

43 → Pd cat., $RSnBu_3$ → **44**

Bu_3Sn–CH=CH–$SnBu_3$ → RCOCl, [Pd] → **45a**

RCO–CH=CH–$SnBu_3$ → R'COCl, [Pd] → **45b**

ethene or β-stannyl enones give 1,4-diketones **45(a/b)** [162]. The catalyst of choice is $Pd(PPh_3)_4$ but other Pd complexes such as $PdCl_2$(dppf) are also active. The stannylalkynyl coupling to organic halides has been elaborated elegantly by Ito et al. in the synthesis of unsymmetrical dialkynyl imines by a stepwise coupling mechanism of *N*-phenyl phenylthioimidoyl chloride **46** with alkynyltin substrates [163]. Interestingly, while $PdCl_2$(dppf) functions efficiently in the attachment of the first alkynyl group, introduction of the second alkynyl substituent is best carried out with $Pd(PPh_3)_4$. The steric effect inherent in the dppf group may be responsible for the sterically demanding disubstitution. Selective formation of *N,S*-diphenyl alkynecarbothioimidates **47** as the intermediate allows a general synthesis of *N*-phenyl alkynyl imines **48** by Grignard coupling. Coupling of **46** with stannyl diacetylenes **49** provides a route to polyacetylenic imines **50**. Both reactions are again best catalyzed by $PdCl_2$(dppf). These coupling reactions can be extended to the synthesis of functionalized organotin derivatives. In a rare use of two different organometallic substrates, hydroboration of ω-stannyl-1-alkenes **51** with 9-BBN gives an interesting heterometal substrate that couples chemoselectively at the B-C site with aryl or 1-alkenyl halides to give the desired tin derivatives (Scheme 1-20) [164]. Both Pd(0) ([$Pd(PPh_3)_4$]) and Pd(II) ([$PdCl_2$(dppf)]) can be used as the coupling catalyst.

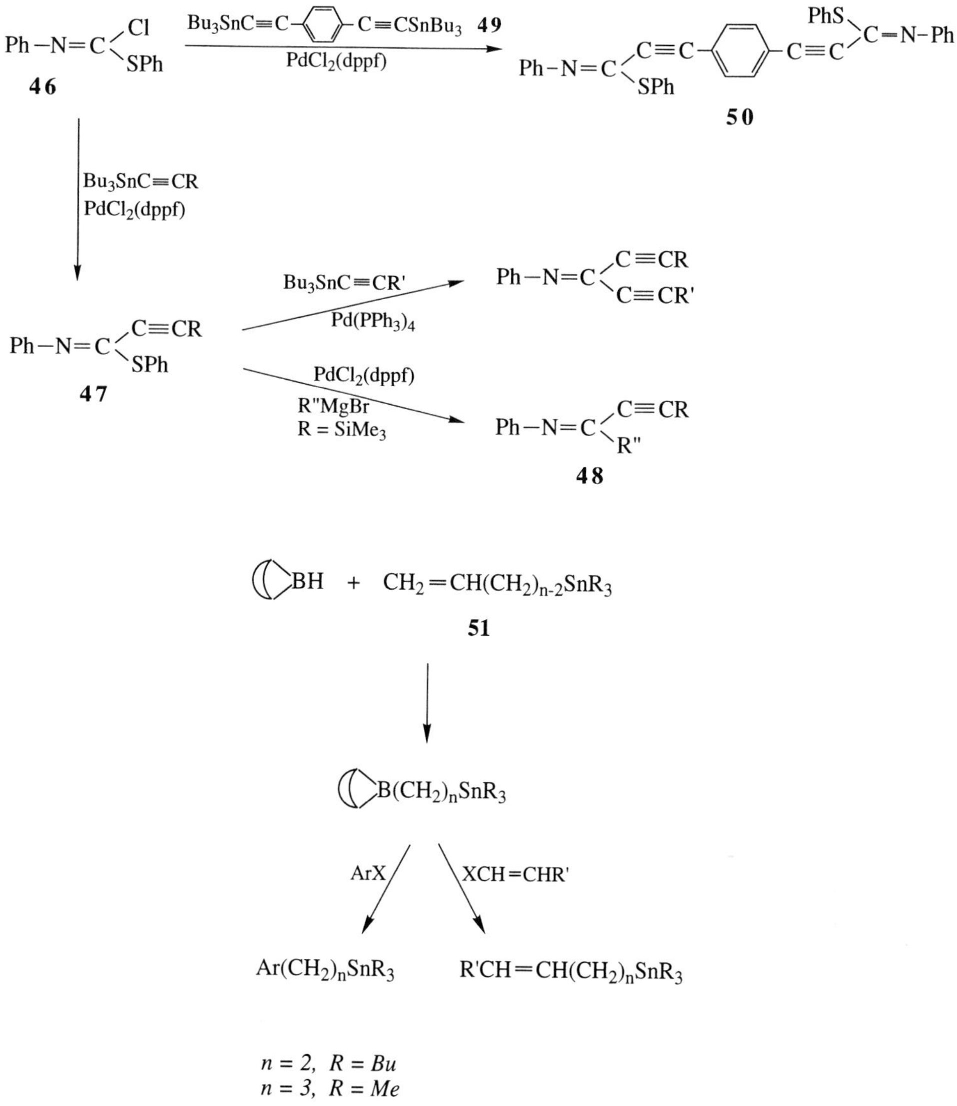

$$Ph{-}N{=}C(Cl)(SPh) \ (\mathbf{46}) \xrightarrow[PdCl_2(dppf)]{Bu_3SnC{\equiv}C{-}C_6H_4{-}C{\equiv}CSnBu_3\ \mathbf{49}} Ph{-}N{=}C(SPh){-}C{\equiv}C{-}C_6H_4{-}C{\equiv}C{-}C(SPh){=}N{-}Ph \ (\mathbf{50})$$

$$\mathbf{46} \xrightarrow[PdCl_2(dppf)]{Bu_3SnC{\equiv}CR} Ph{-}N{=}C(C{\equiv}CR)(SPh) \ (\mathbf{47})$$

$$\mathbf{47} \xrightarrow[Pd(PPh_3)_4]{Bu_3SnC{\equiv}CR'} Ph{-}N{=}C(C{\equiv}CR)(C{\equiv}CR')$$

$$\mathbf{47} \xrightarrow[R''MgBr,\ R = SiMe_3]{PdCl_2(dppf)} Ph{-}N{=}C(C{\equiv}CR)(R'') \ (\mathbf{48})$$

$$(9\text{-BBN})BH + CH_2{=}CH(CH_2)_{n-2}SnR_3 \ (\mathbf{51}) \longrightarrow (9\text{-BBN})B(CH_2)_nSnR_3$$

$$(9\text{-BBN})B(CH_2)_nSnR_3 \xrightarrow{ArX} Ar(CH_2)_nSnR_3$$

$$(9\text{-BBN})B(CH_2)_nSnR_3 \xrightarrow{XCH{=}CHR'} R'CH{=}CH(CH_2)_nSnR_3$$

n = 2, R = Bu
n = 3, R = Me

Scheme 1-20

Others

Other miscellaneous syntheses that have benefited from the introduction of dppf to the catalytic mixture include the coupling reactions of aryl triflates or halides with ketene trimethylsilyl acetals [165]. The catalyst mixture is $[Pd(OAc)(\eta^3\text{-}C_4H_7)]_2$/LiOAc/phosphine. In this case, nucleophilic activation of the Si-O bond of the acetal enables an Si-Pd “transmetalation” to give a σ-complex **52** without necessitating a phosphine

P P Pd Ar C R' R'' C O OMe

52

dissociation step. Hydrolysis of the alkyl 2-arylalkanoates yields alkyl 2-arylalkanoic acids, which are anti-inflammatory pharmaceutics [166]. Using the coupling reaction between aryl triflate and (*E*)-1-methoxy-1-(trimethylsiloxy)propene as a model (Scheme 1-21 and Table 1-11), one can conclude from the yield of methyl 2-phenylpropanoate that dppf is the ligand of choice among the common diphosphines and PPh_3.

$$C_6H_5OSO_2CF_3 + (E)\text{-}CH_3CH{=}C(OCH_3)[OSi(CH_3)_3] \xrightarrow{[Pd(OAc)(\eta^3\text{-}C_4H_7)]_2} C_6H_5(CH_3)CH\text{-}C(=O)OCH_3$$

Scheme 1-21

Cyanation of aryl halides to form nitriles is generally robust unless catalyzed by Co, Ni or Pd complexes [167–171]. A useful catalytic mixture comprises $NiBr_2$/Zn/phosphine/KCN. The use of PPh_3 is sufficient to achieve good conversion of chlorobenzene to benzonitrile. But for the more resistant *o*-dichlorobenzene and *o*-chlorobenzonitrile, dppf is reported to show better conversion and selectivity [172]. The choice of phosphine is obviously crucial as $P(o\text{-}Tol)_3$ and dppe are distinctly ineffective. The superiority of dppf over many other phosphines is remarkably evident in the cyanation of aryl triflates and halides when catalyzed by phosphine-doped $Pd_2(dba)_3$ [173].

Table 1-11. Phosphine ligand effect in the regioselectivity of $Pd(OAc)_2$-catalyzed coupling between butyl vinyl ether and 1-naphthyl triflate (data from [181])

L	Pd/L	Triflate/Pd	Yield, %
dppf	1:2	50	73
dppf	1:2	100	72
dppf	1:2	200	70
dppf	1:1	50	13
dppb	1:2	50	24
dppp	1:2	50	10
dppe	1:2	50	trace
PPh_3	1:3	50	15

1.5.1.2 Arylation and Vinylation of Alkenes

Pioneering work by Heck [174, 175] in the arylation of olefins demonstrated the importance of Pd catalysts in the control of stereo- and regioselectivity. The most commonly used catalyst mixture contains $Pd(OAc)_2$, PPh_3 and NEt_3. The reducing property of the amine enables generation of $[Pd(0)(PPh_3)_2]$, the active catalyst, to take place, oxidative addition of which occurs to give $Pd(II)(Ar)X(PPh_3)_2$, which triggers off a fruitful catalytic cycle. A crucial step in this cycle requires olefin coordination, which would promote an intramolecular insertion of the olefin into the M-Ar bond. Owing to the reluctance of Pd(II) to give 18-electron 5-coordinate species, this olefin coordination is generally believed to occur via (phosphine) dissociative attack. The ready lability of PPh_3 is hence long thought to be an ideal candidate for Heck-type catalysts, and diphosphines with good chelating ability are a poorer choice. Recent results, especially from the laboratory of Cabri, have challenged this conventional view. In using the triflate **53** as an electrophile and methyl acrylate **54** as a weakly-donating olefin, two alternative mechanisms were

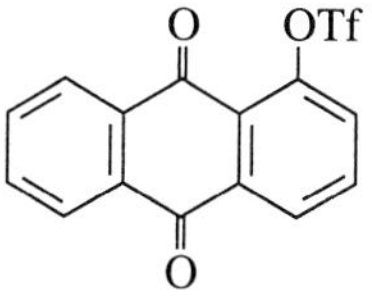

O OTf

O

O O

53 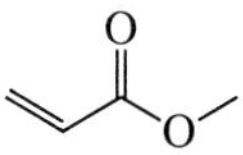 **54**

proposed: the olefin coordination can be initiated either by ionization (**A**) or phosphine dissociation (**B**) from a chelating to a monodentate state (Fig. 1-25) [176]. Triflate, being a weakly coordinating ligand, and dppf, a bulky but flexible difunctional phosphine, are ideal for this purpose. Similar proposals had also been made by Ozawa and Hayashi in an asymmetric arylation of 2,3-dihydrofuran with aryl triflates [177]. This form of phosphine dissociation prior to olefin attack has precedence in the literature [178, 179]. A recent result in a direct vinylation of aryl chlorides illustrates the importance of the chelate effect in catalysis and supports the idea of chelate opening as a key catalytic step [180]. In suggesting the regioselectivity control by the phosphine and anion dissociation, Cabri et al. also argued that the cationic intermediate favors α-substitution and the neutral intermediate β-substitution. This argument is based on the higher polarization of the olefinic moiety, which leads to a lower charge density of the α-carbon in the cationic complex [181]. This accounts for the regioselective α-arylation of butyl vinyl ether (Scheme 1-22 and Table 1-12) and unsymmetrical olefins [182] when bidentate phosphines such as dppf are used. The authors also traced the competing reduction pathway to an amine coordination. This competes with the olefin especially when the Pd(II) center is electron-deficient and, through a β-elimination mechanism, gives a reducing hydrido complex. Such β-elimination is undesirable in most coupling reactions. One notable exception is the glycal (1,1-anhydro sugar) synthesis from a carbohydrate (Scheme 1-23) [183]. Mesylation of the latter gives an α-alkoxy electrophile with an anomeric center,

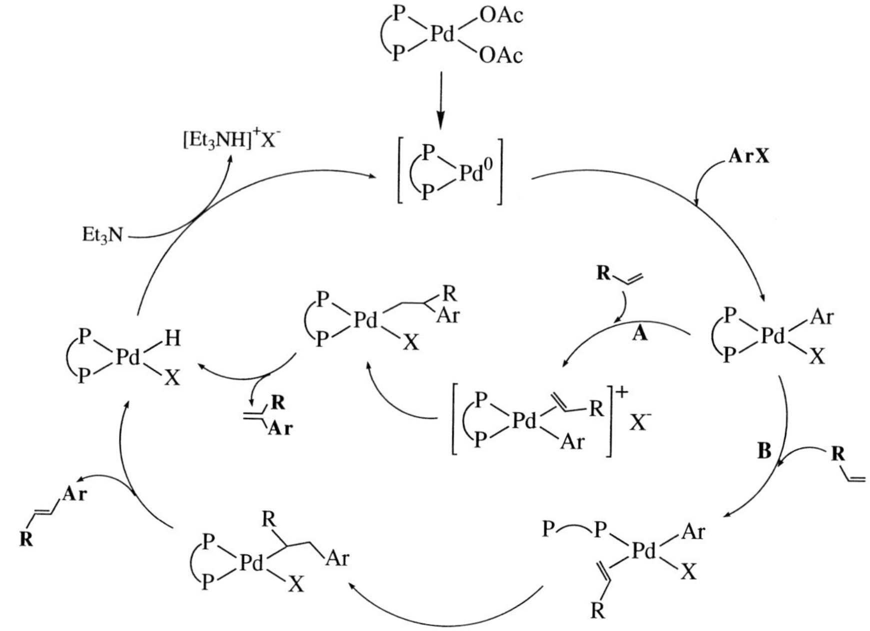

Fig. 1-25. Two mechanistic pathways for olefinic arylation mediated by [Pd^0(P-P)]. (P-P = diphosphine) (adapted from [182]).

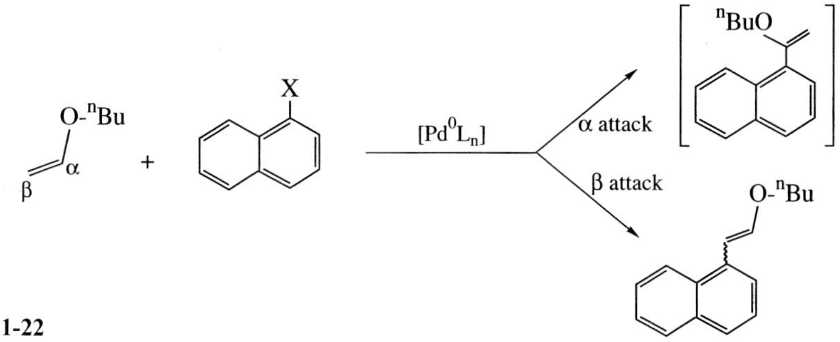

Scheme 1-22

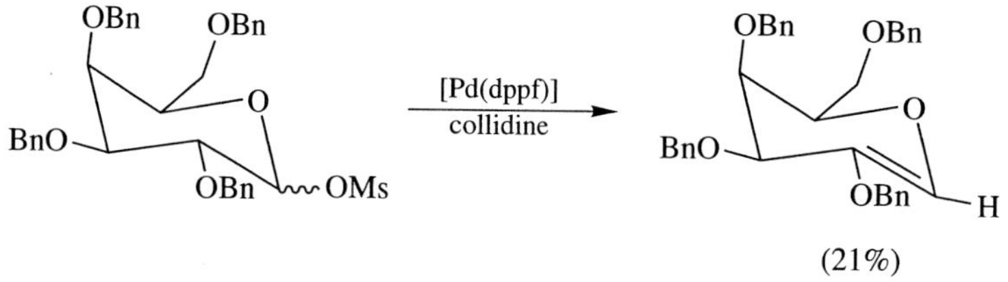

Scheme 1-23

Table 1-12. Effects of ligand (L) and substrate ratios on the coupling between phenyl triflate and (*E*)-1-methoxy-1-(trimethylsiloxy)propene (L = phosphine ligand) (data from [165])

Ligand (L)	*T*, °C	Time, h	Conversion, %	α/β
none	100	24	8	55/45
PPh_3	100	1.5	100	63/37
P(*p*-tolyl)$_3$	100	5	100	61/39
P(*o*-tolyl)$_3$	100	5	100	63/37
$P(CH_3)Ph_2$	80	5	100	>99/1
$P(CH_3)_2Ph$	80	6	100	>99/1
dppm	100	24	60	80/20
dppe	80	24	95	>99/1
dtpe[a]	80	24	93	>99/1
dppp	60	0.5	100	>99/1
dppb	60	1.5	100	>99/1
dppf	60	2	100	>99/1
c-dppet[a]	80	12	100	>99/1

a dtpe = 1,2-bis(di-*p*-tolylphosphino)ethane; *c*-dppet = *c*-1,2-bis(diphenylphosphino)ethylene.

which can oxidatively add to [Pd(dppf)] (from reduction of $PdCl_2$(dppf) with *n*-BuLi). Subsequent β-hydride elimination would then provide a route to oxyglycals.

Most of the Pd-catalyzed syntheses are carried out in the presence of phosphine although there are exceptions when this is not necessary. One such example is the coupling of aryl triflate with methyl α-acetamidoacrylate catalyzed by a mixture containing $Pd(OAc)_2$/dppf/$n$$Bu_3N$/LiF (Scheme 1-24) [184]. A mixture containing $Pd(OAc)_2/NaHCO_3/nBu_4NCl$ is also catalytically active in most cases.

ArOTf + $CH_2{=}C(CO_2Me)(NHCOMe)$ —Pd(0)→ $ArCH{=}C(CO_2Me)(NHCOMe)$

Scheme 1-24

1.5.1.3 Carbonylation and Carbonylative Coupling

Palladium complexes provide a catalytic means for introducing a carbonyl functionality to many organic substrates [185]. The unique value of $PdCl_2$(dppf) as a carbonylation catalyst was realized over a decade ago in the synthesis of acetylenic ketones from terminal acetylenes and organic halides (Scheme 1-25) [186]. Catalytic carbonylation of arylacetylenes in the presence of HI in methanol gives methyl

$$RX + CO + HC{\equiv}CR' \xrightarrow{NR''_3 \;\to\; [NR''_3H]^+X^-} R(C{=}O){-}C{\equiv}CR'$$

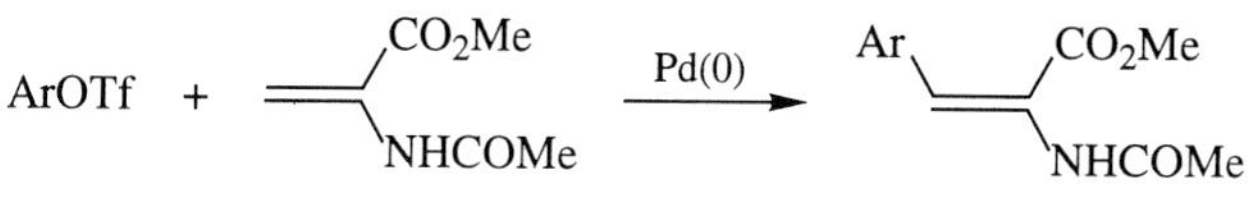

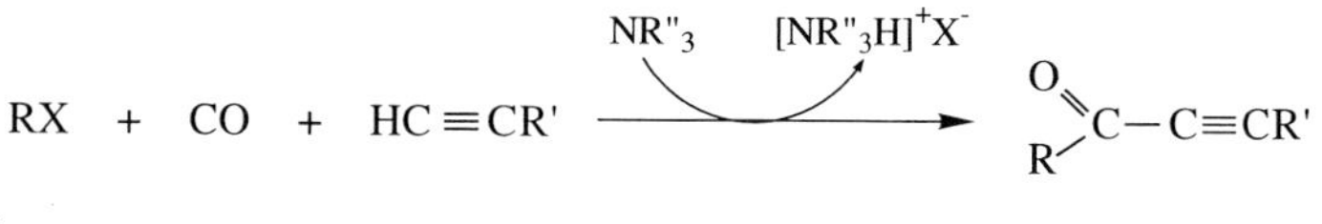

Scheme 1-25

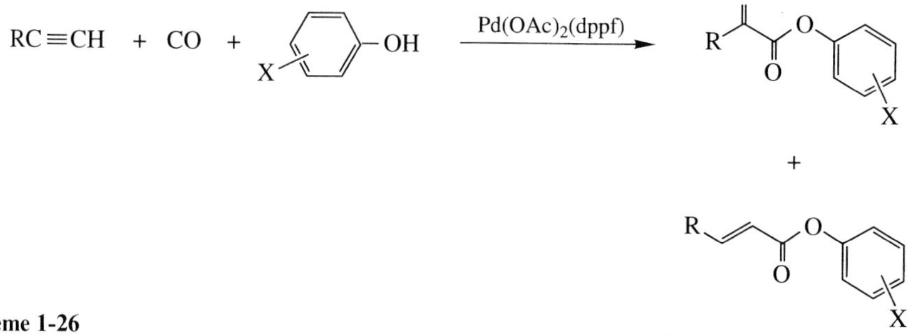

RC≡CH + CO + X-C6H4-OH —Pd(OAc)2(dppf)→ products

Scheme 1-26

2-arylpropenoates, which are useful precursors for the synthesis of optically active anti-inflammatory 2-arylpropanoic acids [187–189]. Terminal alkynes undergo carbonylative coupling with phenols to give 2-substituted-2-propenoic acid aryl esters when catalyzed by equimolar quantities of $Pd(OAc)_2$ and dppf (Scheme 1-26) [190]. It is noteworthy that this catalysis can be carried out at atmospheric pressure whereas the analogous PPh_3 system requires CO gas at high pressure. Another example of the effective use of $PdCl_2$(dppf) as a catalyst in alkynyl coupling is the synthesis of the antibacterial 6-substituted methyl 4-pyridone-3-carboxylates **55** from 2-bromo-3-aminoacrylates **56** [191]. Formation of the pyridone derivatives is accomplished by cyclization. Although the choice of the Pd catalyst is not crucial, the use of $PdCl_2$(dppf) generally gives the best results. This coupling followed by a cyclization is also used in the synthesis of flavones and chromones and forms the basis of the synthesis of many biological and medicinal products [192, 193] from *o*-iodophenols and terminal acetylenes (Scheme 1-27) [194]. Another effective catalyst contains $Pd(OAc)_2$(dppf) supported by 1,8-diazabicyclo[5.4.0]undec-7-ene (DBU) and a base [195]. Carbonylative coupling is generally believed to occur via oxidative

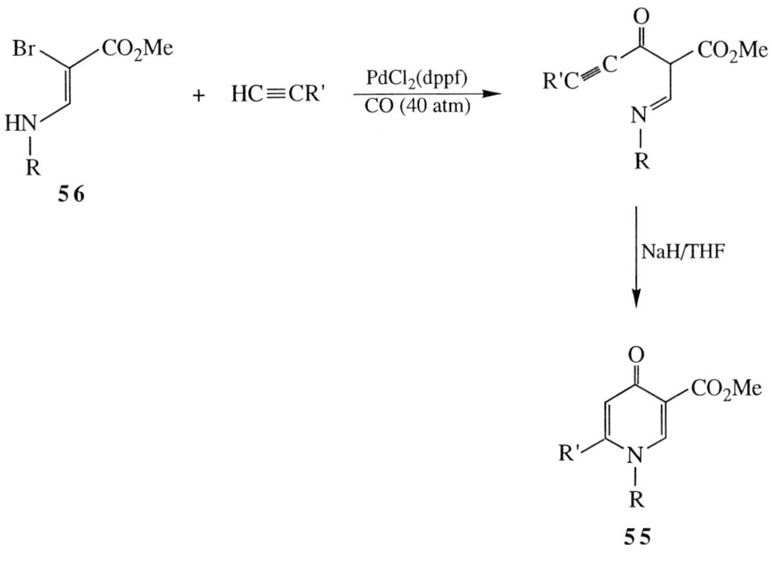

56 + HC≡CR' —$PdCl_2$(dppf), CO (40 atm)→ intermediate —NaH/THF→ **55**

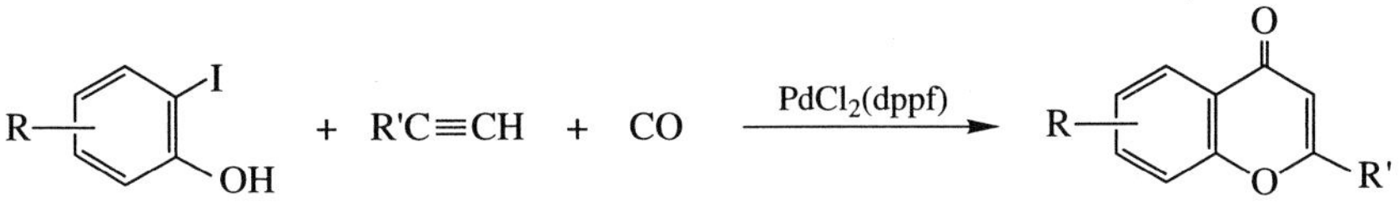

Scheme 1-27

addition of the aryl halide, CO insertion to give an aroyl functionality, nucleophilic attack from the acetylide, and reductive elimination to give the products (Fig. 1-26) [196, 197]. The carbonyl insertion step could involve one or more of the following: opening up of the chelate ring; carbonyl complex formation; and ionic dissociation [198]. Introduction of the alkynyl group may also proceed via insertion for example into Pd-H bond. There is evidence that the cyclization step described above is also Pd-catalyzed [199]. The use of $Pd(OAc)_2(dppf)$ again increases the rate and yield of the cyclized product compared to $Pd(PPh_3)_4$. When *o*-iodoanilines are used as the substrates, 2-aryl-4-quinolones **57** are formed in 62–84% yield [200]. Substituted 4-quinolones are known to be pharmacologically active [201, 202]. The catalytic activity of $PdCl_2(dppf)$ is also the highest among the Pd complexes tested.

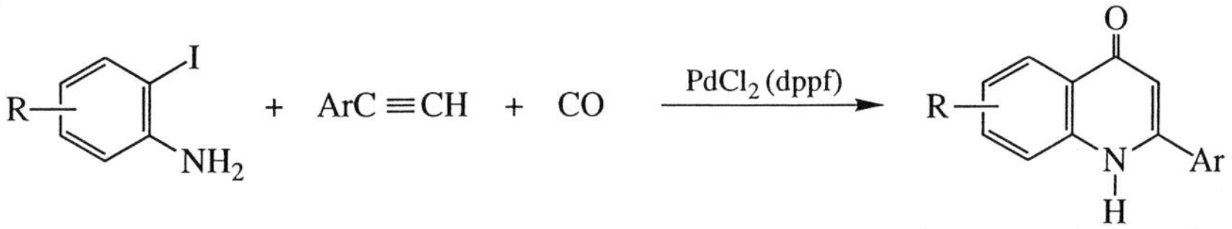

57

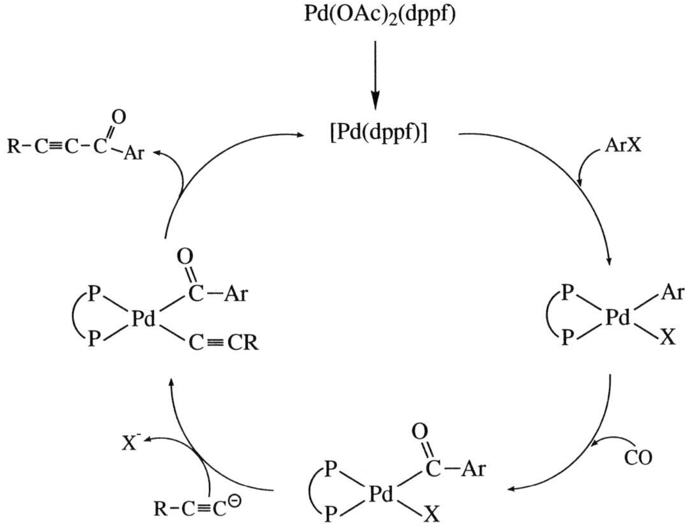

Fig. 1-26. Carbonylative aryl-acetylide coupling catalyzed by [Pd(dppf)].

Numerous aryl bromides, iodides [203], borates [204] and triflates [205, 206] have been successfully carbonylated. Triflates could serve as a route for the synthesis of arenecarboxylic acid derivatives from phenols. This carbonylation using dppf in a catalytic mixture generally shows higher efficiency than PPh_3 or $P(o\text{-Tol})_3$ [207]. Poor performance is also noted for PPh_3 in a Pd-catalyzed vinyl substitution of aryl bromides [208]. Side-reactions involving the formation of $[PPh_3Ar]Br$ and ArH are responsible. A system which is catalyzed effectively by $PdCl_2$(dppf) under 10 atm CO is the desulfonylation of 1-naphthalenesulfonyl chloride **58** in the presence of $Ti(OiPr)_4$. Formation of isopropyl 1-naphthoate **59** can be explained in a sequence of oxidative addition, SO_2 extrusion, carbonylation and reductive elimination (Fig. 1-27) [209]. A notable side-product is di-1-naphthyl disulfide.

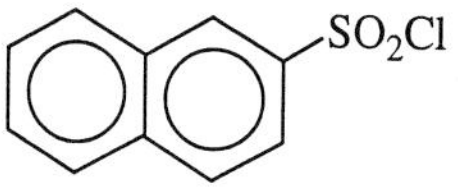

SO_2Cl

58

A useful synthesis of (*E*)-*β*-ethoxycarbonylvinylsilanes by palladium-catalyzed regio- and stereospecific hydroesterification (EtOH + CO) (or carboethoxylation) of trimethylsilylacetylenes has been reported recently [210]. Alkoxycarbonyl or carbonyl functionalization of vinylsilanes are useful synthetic intermediates [211, 212]. The use of $PdCl_2$(dppf) as a catalyst (with $SnCl_2 \cdot 2\,H_2O$ as cocatalyst) is found to be superior and gives excellent yields. A key step in the reaction is thought to involve hydropalladation to give **60** or **61**. The preference for **60** to **61** is understood

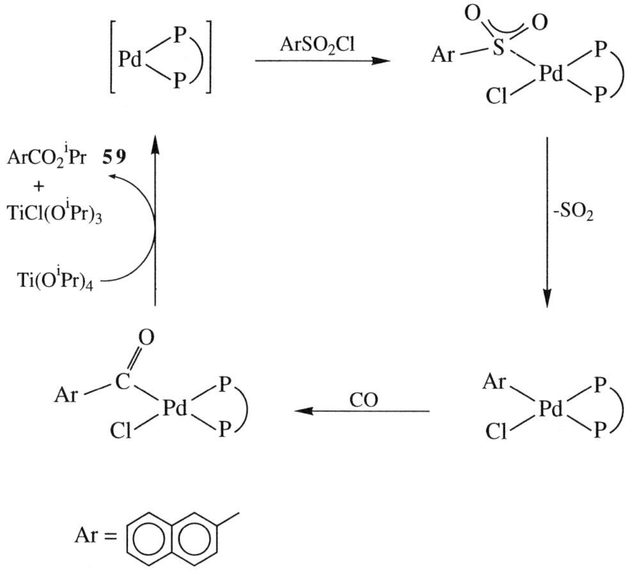

$ArSO_2Cl$

$-SO_2$

CO

$ArCO_2{}^iPr$ **59** + $TiCl(O^iPr)_3$

$Ti(O^iPr)_4$

Ar =

Fig. 1-27. Formation of isopropyl 1-naphthoate in a carbonylative coupling (adapted from [209]).

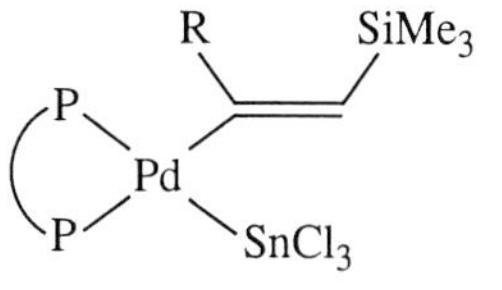

60 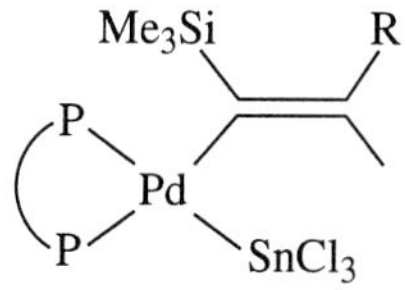 **61**

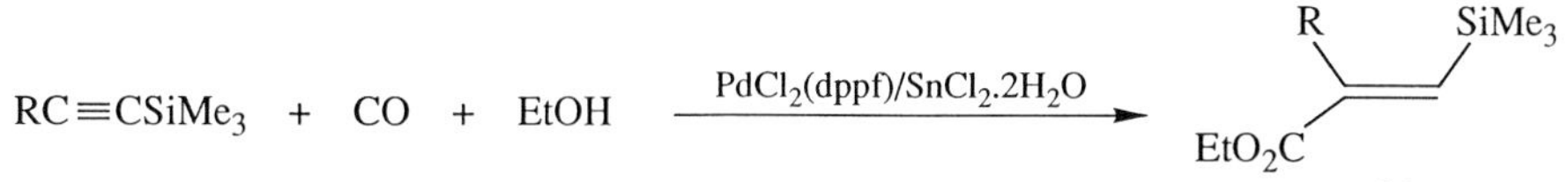

$RC{\equiv}CSiMe_3 + CO + EtOH \xrightarrow{PdCl_2(dppf)/SnCl_2.2H_2O}$ **62**

Me_3Si–CH=CH$_2$ + CO + EtOH ⟶ **63** + **64**

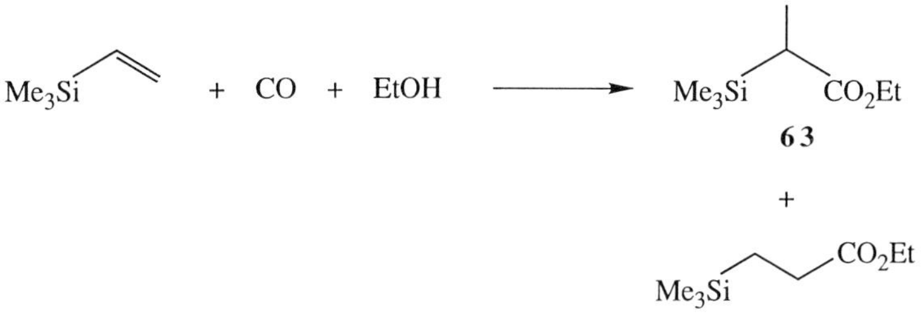

in terms of the repulsive effects between the dppf ligand and the silyl group. This geometrical preference dictates the stereospecificity thus giving the *syn* isomer of **62**. Formation of **62** from **60** is explained by CO insertion into the Pd-vinyl link to give an acyl complex followed by nucleophilic attack from EtOH. Similar hydroesterification of trimethylvinylsilane gives the α-**63** or β-**64** silyl esters [213]. The regioselectivity depends on the catalysts used, with the Pd-catalysts again showing high β-selectivity. Interestingly, the catalyst with monodentate phosphine, $PdCl_2(PPh_3)_2$, is more effective than its dppf analogue. Other bidentate complexes such as dppe or dppb are inactive. This contrast is attributed to the varying extent of phosphine dissociation to give the active substrate. The dppe or dppb chelates may be too stable compared with the dppf ring. The larger bite angle of dppf [57] appears to promote the catalytic effect of the corresponding complex. The presence of phosphine in the catalyst is important, as shown in the lack of activity of $PdCl_2(PhCN)_2$. A similar disparity in the catalytic effect between the monodentate and bidentate phosphine is also observed in a similar hydrocarboxylation (H_2O + CO). A general mechanism is shown in Fig. 1-28. The nature of the phosphine catalyst can also influence the isomeric distribution in the products. This is best illustrated in the hydroesterification and hydrocarboxylation of 3,3,3-trifluoropropene and pentafluorostyrene, which give the branched and straight-chain acids (Scheme 1-28) [214]. The hydrocarboxylation mechanism is illustrated in Fig. 1-29, which involves a hydroxycarbonyl complex [M-COOH] **65** and a β-(hydroxycarbonyl)alkyl complex [M-C-C-COOH] **66** as intermediates. Since the insertion of an olefin into the former to give the latter is regiodetermining, the structure of the

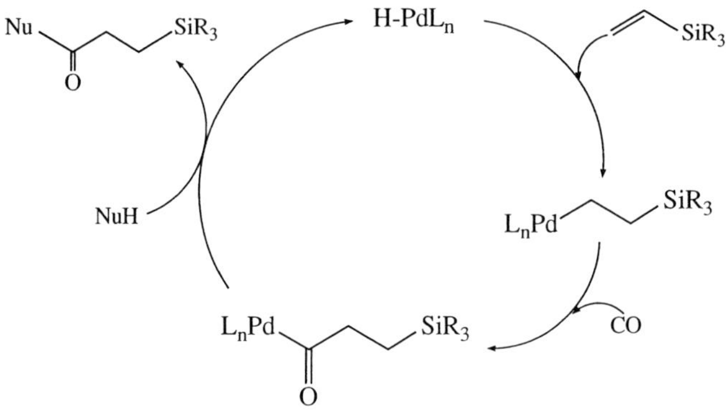

Nu = OEt
OH

Fig. 1-28. Hydrocarboxylation and hydroesterification of trialkylvinylsilane (adapted from [213]).

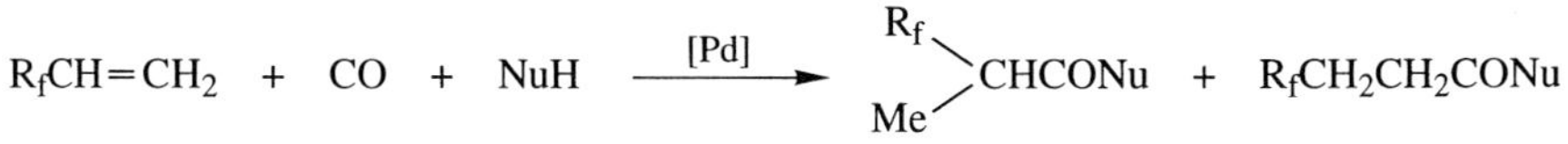

$$R_fCH{=}CH_2 + CO + NuH \xrightarrow{[Pd]} (R_f)(Me)CHCONu + R_fCH_2CH_2CONu$$

Nu = RO
HO

Scheme 1-28

transition state that leads to **66** would influence the branch/chain product ratio. A diphosphine such as dppf or dppb could remain chelating throughout the catalytic cycle but a monophosphine such as PPh_3 can dissociate easily. The high *trans* effect of the diphosphine tends to stabilize the intermediate with the electron-withdrawing group at the α carbon **67** and leads to the straight chain acids. On the other hand, displacement of PPh_3 would exert a less distinctive *trans* effect and explains the minor quantities of branched isomers formed when $PdCl_2(PPh_3)_2$ is used as catalyst. Similar behavior is observed in the hydroesterification, although the mechanism is modified to incorporate hydrido and acyl complexes as intermediates. Another route for carboxylic acids is found in the hydroxycarbonylation of triflates in the presence of KOAc (Scheme 1-29) [215]. Vinyl triflates are conveniently catalyzed at room temperature by $Pd(OAc)_2(PPh_3)_2$ whereas aryl triflates are best promoted by $Pd(OAc)_2$/dppf.

Carbocarbonylation [216] of alkyl halide with a perfluoro substituent leads directly to the analogous carboxylic acids and esters, provided the β-hydride elimination is slower than CO insertion. The use of $PdCl_2$(dppf) in the esterification of **68** gives

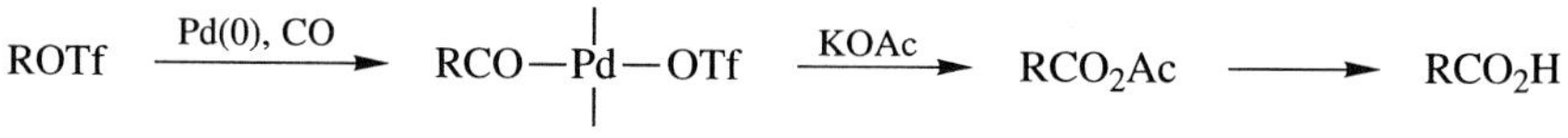

$$ROTf \xrightarrow{Pd(0),\ CO} RCO{-}Pd{-}OTf \xrightarrow{KOAc} RCO_2Ac \longrightarrow RCO_2H$$

Scheme 1-29

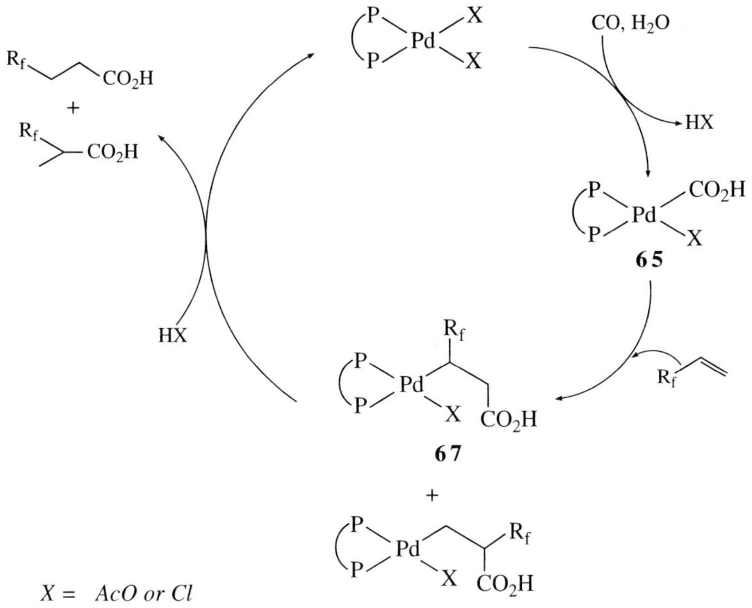

Fig. 1-29. A mechanism of hydrocarboxylation with hydroxycarbonyl and β-(hydroxycarbonyl)-alkyl complexes as intermediates (R_f = fluoroalkyl) (adapted from [214]).

an 83% yield of **69** at 80 °C under CO at 30 atm [217]. It is clearly evident that alcohol as solvent in carbonylative coupling of aryl halide must be used with caution.

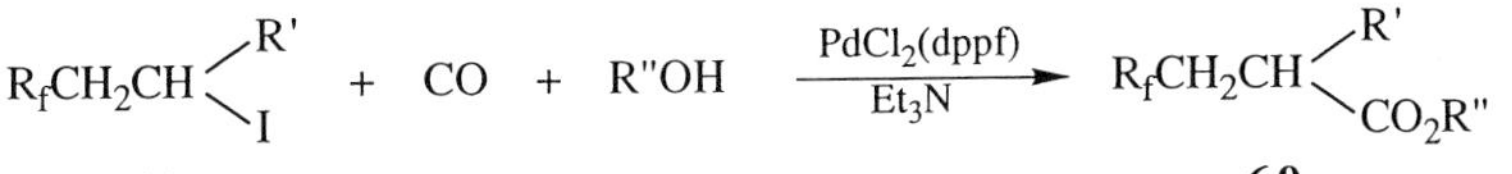

68 **69**

α-Keto esters are known to be produced in a double carbonylation catalyzed by Pd complexes (Scheme 1-30) [218].

$$2ArI + 3CO + 2ROH + 2Et_3N \xrightarrow{[Pd]} ArCOCO_2R + ArCO_2R + 2[Et_3NH]^+I^-$$

Scheme 1-30

On a similar note, the use of a secondary amine as a reductant in the catalytic mixture could also yield competitive products such as α-keto amides (Scheme 1-31) [219–221]. Although these carbonylations are chiefly catalyzed by PPh_3 complexes

$$2ArX + 3CO + 4HNR_2 \xrightarrow{[Pd]} ArCOCONR_2 \; (+ \; ArCONR_2) + 2[R_2NH_2]^+X^-$$

Scheme 1-31

of Pd, there is a clear possibility that dppf could perform similarly. Other forms of halides can also be carbocarboxylated readily. Bis(chloromethyl)arenes **70** can be quantitatively converted to 1,4-bis(alkoxycarbonylmethyl)benzene **71**, although the use of dppf and other bidentate chelates is found to be less desirable than the monophosphine analogues [222]. The inability of diphosphines to vacate a coordination site is held responsible. In these carbonylations, the nucleophiles need not be ROH or H_2O. For example, $PdCl_2$(dppf) has also found its use in catalyzing the coupling of PhCH=CHBr with Et_3N to give the amide PhCH=$CHCONEt_2$ [223] and of carboxylic acids **72** with aryl halides [224]. Another interesting example is the coupling of an organoborane (9-alkyl-9-BBN) with vinyl halides, with the help of K_3PO_4 as a cocatalyst (Scheme 1-32).

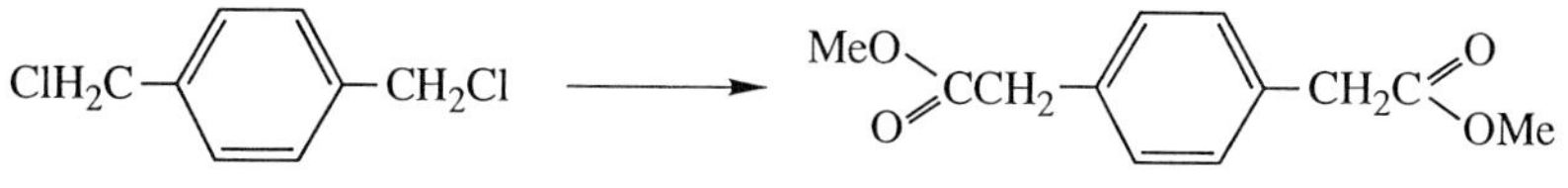

ClH_2C–C_6H_4–CH_2Cl → $MeO(O=)CCH_2$–C_6H_4–$CH_2C(=O)OMe$

70 **71**

RX + CO + $R'HC(Z)CO_2R''$ —(Pd cat.; NEt_3 → $[Et_3NH]^+X^-$)→ $R-C(=O)-C(R')(Z)CO_2R''$

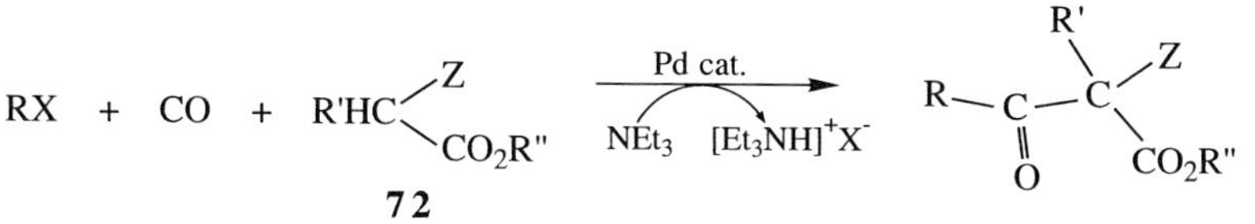

72

Z = CO_2R'', CN

This catalytic carbonylation can be carried out at room temperature and atmospheric pressure, but the yield in the reaction using $PdCl_2$(dppf) as catalyst is somewhat lower than that obtained using $Pd(PPh_3)_4$ or $PdCl_2(PPh_3)_2$ [225]. In a rare use of CO_2 in coupling reactions, $Ni(dppf)_2$ (from $Ni(cod)_2$ and dppf) promotes cycloaddition of diynes with CO_2 to give bicyclic α-pyrones **73** [226]. A reverse process is the decarbonylation of *N*-protected aspartic anhydride to give α- and β-alanine when catalyzed by NiL_2(cod) complexes, such as Ni(dppf)(cod) [227].

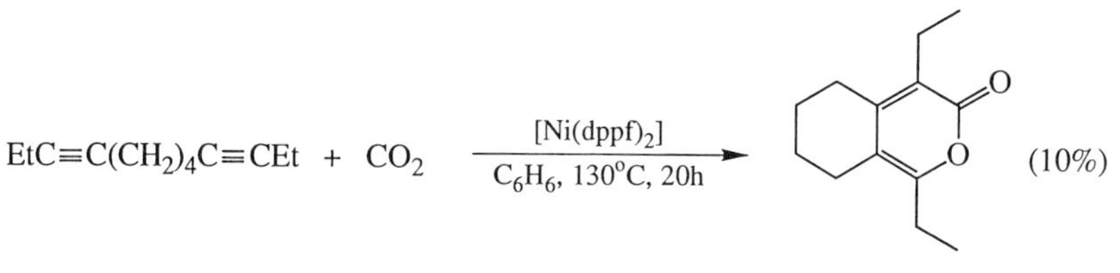

$EtC{\equiv}C(CH_2)_4C{\equiv}CEt + CO_2$ —($[Ni(dppf)_2]$; C_6H_6, 130°C, 20h)→ **73** (10%)

9-R-9-BBN + CO + X–C(R')=CH–R'' —($PdCl_2$(dppf), K_3PO_4)→ R–C(=O)–C(R')=CH–R''

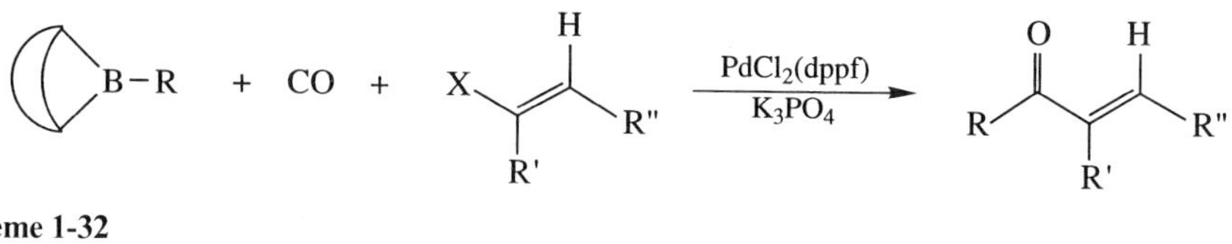

Scheme 1-32

1.5.1.4 Nucleophilic Substitution

Many nucleophilic substitutions in organic syntheses are catalyzed by Pd. For example, azidation of (*E*)-2-hexen-1-yl derivatives does not take place with NaN_3 unless catalyzed by $Pd(PPh_3)_4$ [228]. In many cases, free diphosphines are added to improve the stereoselectivity. Azidation of allyl esters is a good example. Mechanistically, since both the first step, i.e. oxidative addition of allyl esters, and the second step, i.e., direct nucleophilic attack on the allyl ligand, occur with inversion of configuration, overall configuration retention is achieved (Scheme 1-33). Monophosphine allyl complexes, however, tend to lose this stereospecificity because of faster isomerization of the allyl intermediate. In the azidation of (*Z*)-5-(methoxycarbonyl)-2-cyclohexen-1-yl esters with $Pd_2(dba)_3$ as catalyst (Scheme 1-34), the conversion rate increases with increasing chain length of the hydrocarbon-based diphosphine. The reactivity of dppf is higher than that of any of the alkyl diphosphines tested, although its stereoselectivity is lower than that of dppb. Stereochemical retention is also observed in nucleophilic substitution of (*S*)-(*E*)-allyl acetate with sodium acetylacetonate (Scheme 1-35) [229]. The use of dppf in place of dppe in this Pd-catalyzed substitution improves the yield of [1-((*E*)-styryl)ethyl]-acetylacetone **74**. A similar advantage is also achieved in the allylic alkylation of α-isocyanoacetate in which higher retention is achieved when $Pd(PPh_3)_4$ is replaced with $[Pd(\mu\text{-}Cl)(\eta^3\text{-}C_3H_5)]_2$/dppf as a catalyst (Scheme 1-36) [230].

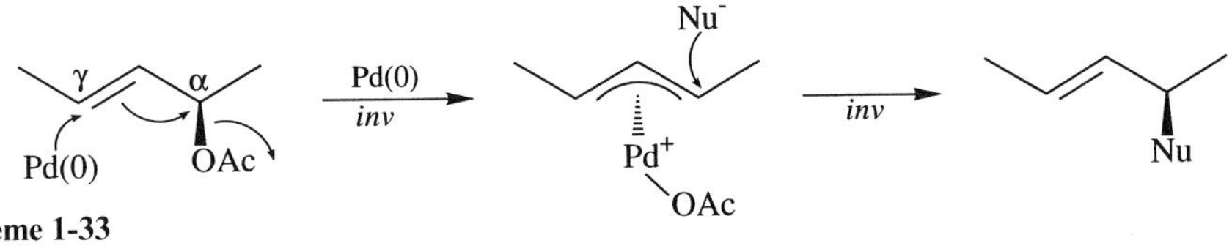

Scheme 1-33

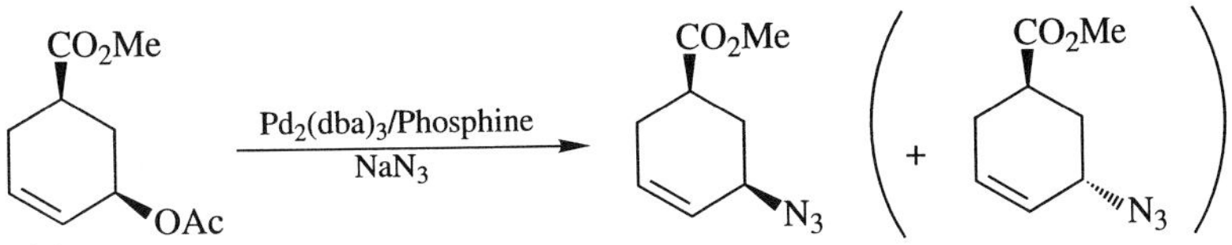

Scheme 1-34

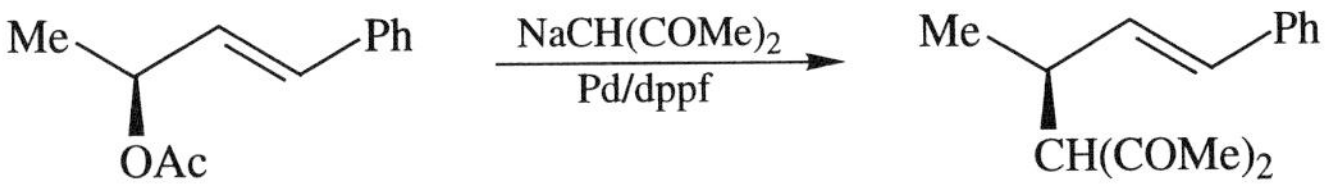

Scheme 1-35

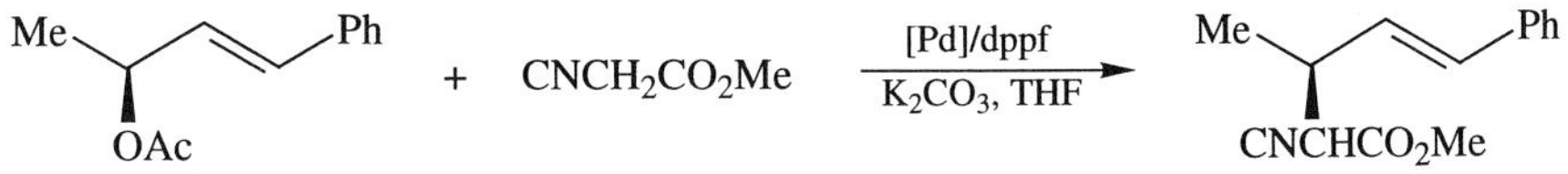

Scheme 1-36

Trost and Scanlan reported a Pd-catalyzed condensation of a vinyl epoxide **75** and an allyl sulfone **76** in the presence of dppf under neutral conditions [231]. This alkylation allows a room temperature entry to a basic indolizidine ring system as a step towards the synthesis of (+)-*ayllo*-Pumiliotoxin 339B [232]. The modification of allylic alkylations by condensation of a diene **77** with a pronucleophile **78** also leads to C-C bond formation at the allylic position in both 1:1 (**79** and **80**) and 2:1 (**81** and **82**) products [233]. Reactions between ketene silyl acetals **83** with allyl

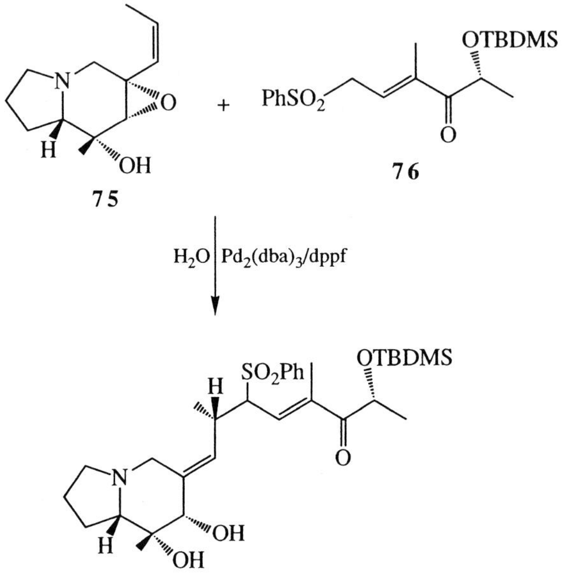

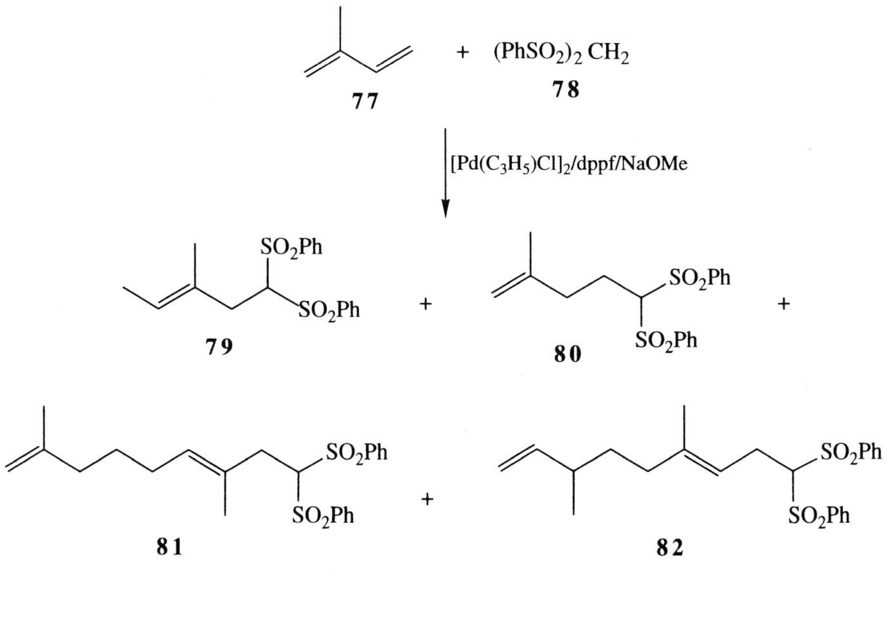

N O OH H **75** + $PhSO_2$ OTBDMS O **76**

H_2O | $Pd_2(dba)_3$/dppf

H SO_2Ph OTBDMS O N H OH OH

77 + $(PhSO_2)_2CH_2$ **78**

$[Pd(C_3H_5)Cl]_2$/dppf/NaOMe

SO_2Ph SO_2Ph **79** + SO_2Ph SO_2Ph **80** +

SO_2Ph SO_2Ph **81** + SO_2Ph SO_2Ph **82**

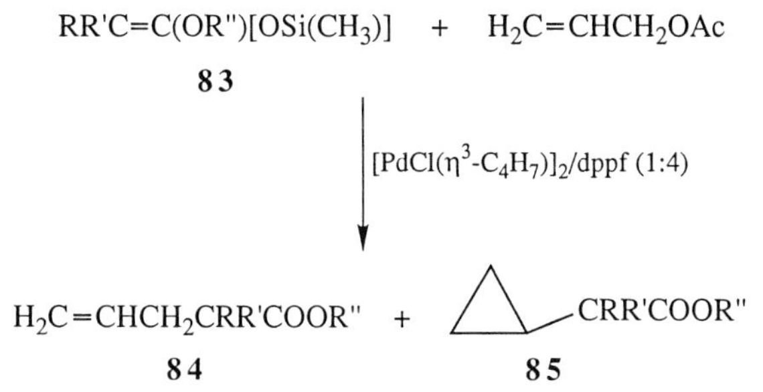

$$RR'C{=}C(OR'')[OSi(CH_3)] + H_2C{=}CHCH_2OAc$$
83

$[PdCl(\eta^3\text{-}C_4H_7)]_2$/dppf (1:4)

$$H_2C{=}CHCH_2CRR'COOR'' + \triangle\text{-}CRR'COOR''$$
84 **85**

acetates give, besides the usual alkylation products **84**, cyclopropane derivatives **85** by nucleophilic attack of the silyl enolate on the central carbon of the allyl group [234, 235]. Dppf, among many other phosphines, supports these coupling reactions

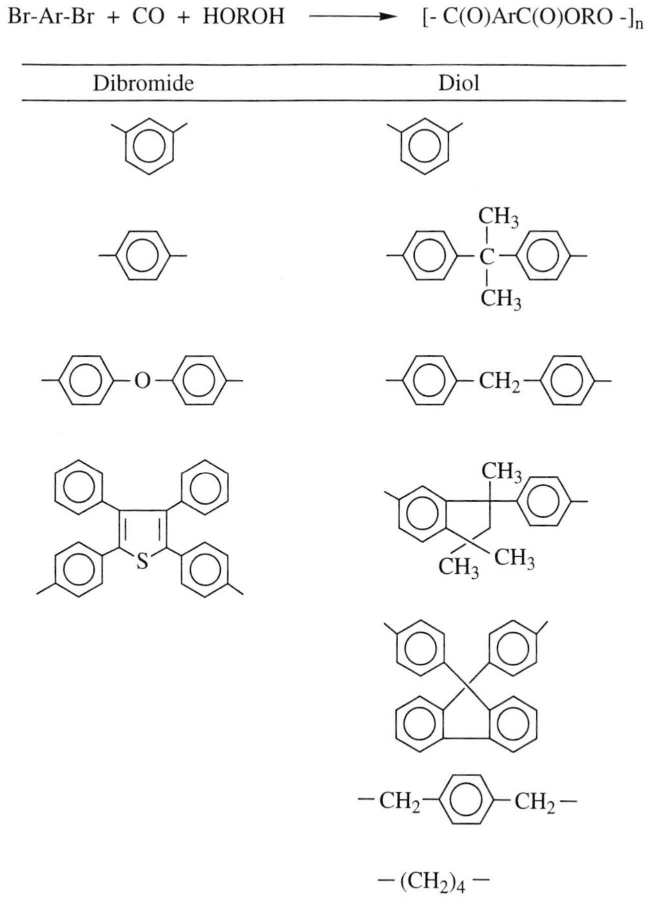

Br-Ar-Br + CO + HOROH ⟶ [- C(O)ArC(O)ORO -]$_n$

Dibromide	Diol
1,3-phenylene	1,3-phenylene
1,4-phenylene	$-C_6H_4-C(CH_3)_2-C_6H_4-$
$-C_6H_4-O-C_6H_4-$	$-C_6H_4-CH_2-C_6H_4-$
tetraphenylthiophene	indane (CH_3, CH_3, CH_3)
	fluorene-bis(phenyl)
	$-CH_2-C_6H_4-CH_2-$
	$-(CH_2)_4-$

Fig. 1-30. Polyester synthesis from the carbonylative polycondensation of aromatic dibromides and diols (adapted from [237]).

best. Catalysts based on other metals have also been used successfully in this coupling. Allylation of enolates by allylic ethers can be initiated by $NiCl_2$(dppf) in the presence of PhMgBr [236].

1.5.1.5 Polycondensation and Polymerization

Some industrially important polymeric materials can be prepared using the basic strategies discussed earlier. A representative example can be found in the synthesis of polyesters using the carbonylative polycondensation of aromatic dibromides and diols (Fig. 1-30) [237]. The underlying principle is no different from the fundamentals of carbonylative coupling presented earlier in Section 1.5.1.3. Replacement of the diols with hydrazides **86** similarly yields poly(acylhydrazide)s **87** [238]. The catalytic

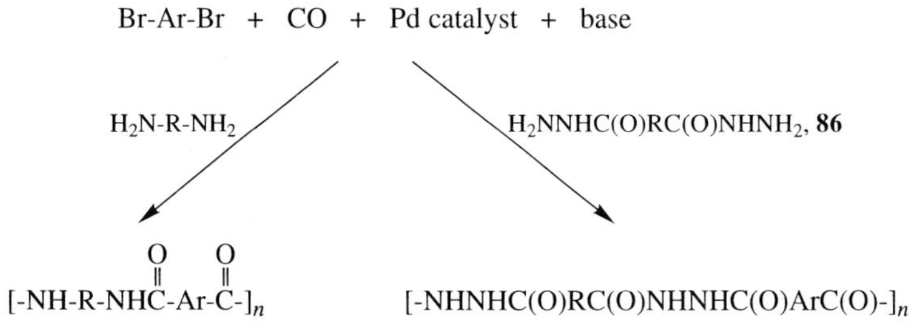

Br-Ar-Br + CO + Pd catalyst + base

H_2N-R-NH_2 → [-NH-R-NHC(=O)-Ar-C(=O)-]$_n$ **88**

$H_2NNHC(O)RC(O)NHNH_2$, **86** → [-NHNHC(O)RC(O)NHNHC(O)ArC(O)-]$_n$ **87**

activity of $PdCl_2$(dppf) is significantly higher than that of the PPh_3 analogue although $Pd(PPh_3)_4$ is the catalyst of choice because of the higher viscosity of the polymers generated. Similar polycondensation using diamines to give polyamides **88** is also catalyzed best by $PdCl_2$(dppf) in terms of reaction duration and yield although the quality of the polymer is lower [239]. These couplings using aryl dihalides can be extended to the synthesis of conducting polyheteroarenediylviny-

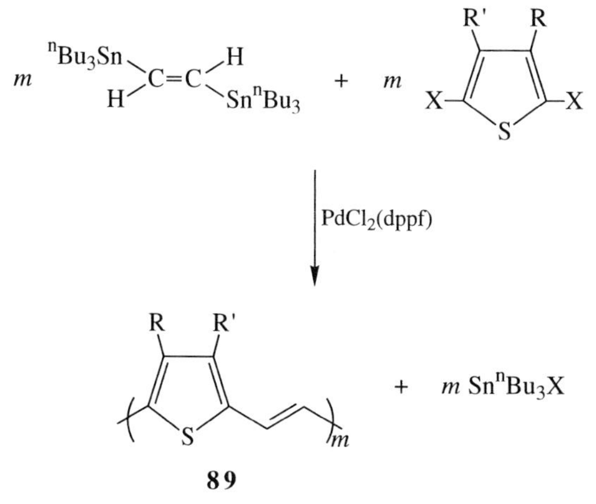

m $^nBu_3$Sn(H)C=C(H)Sn$^nBu_3$ + *m* 2,5-X-3-R-4-R'-thiophene

$PdCl_2$(dppf)

→ poly(3-R-4-R'-thiophene-2,5-diyl-vinylene)$_m$ **89** + *m* Sn$^nBu_3$X

X = Br: R = R' = H
X = I: R = n-decyl, R' = H;
R = R' = ^{n}Bu

lenes **89** when 1,2-bis(tributylstannyl)ethylene is used [240]. This polymer synthesis employs the Stille-type coupling between arylhalides and organostannanes [146]. Organoboranes are also useful substrates. A fine example is found in the hydroboration of a diolefin with 9-BBN and subsequent intermolecular coupling of the α,ω-bis(B-alkanediyl-9-borabicyclo)[3.3.1]nonanes with dihaloarenes (Fig. 1-31) [241]. This is reminiscent of the work of Suzuki et al. in the cross-coupling reactions of B-alkyl-9-borabicyclo[3.3.1]nonanes with monohaloarenes (inset in Fig. 1-31). Similar hydroboration can be carried out on a difunctional monomer such as *p*-bromostyrene (Fig. 1-32). Oxidation of the boron derivatives at the chain ends can yield functional polymers such as **90** and **91**. The catalyst used in these phase-transfer base-catalyzed couplings, $PdCl_2$(dppf), gives satisfactory yields.

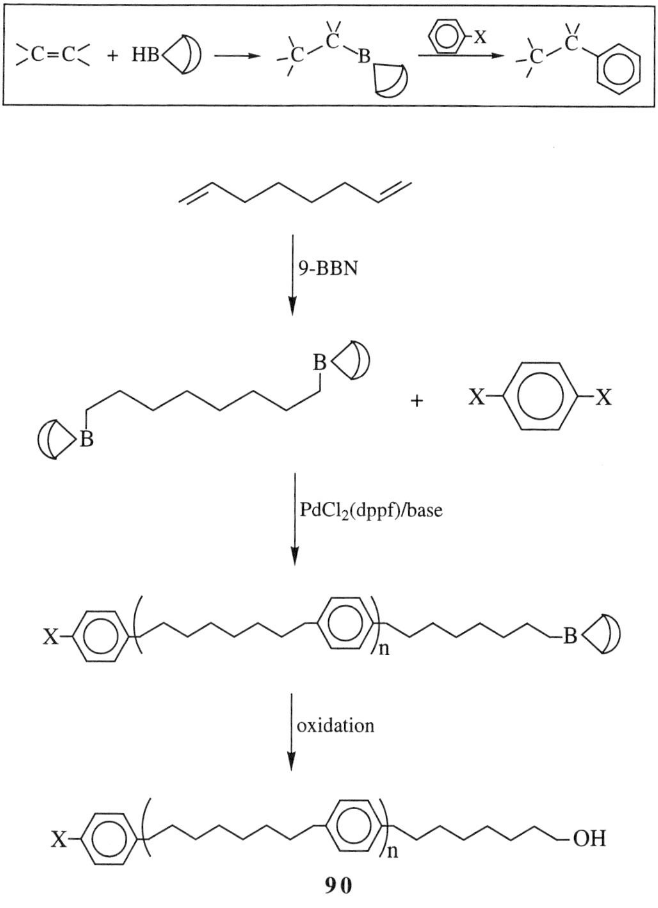

9-BBN

PdCl$_2$(dppf)/base

oxidation

90

Fig. 1-31. Double hydroboration of a diolefin with 9-BBN followed by coupling with aryl halide. (Inset showing a simple hydroboration and arylation) (9-BBN = 9-borabicyclo[3.3.1]nonane) (adapted from [241]).

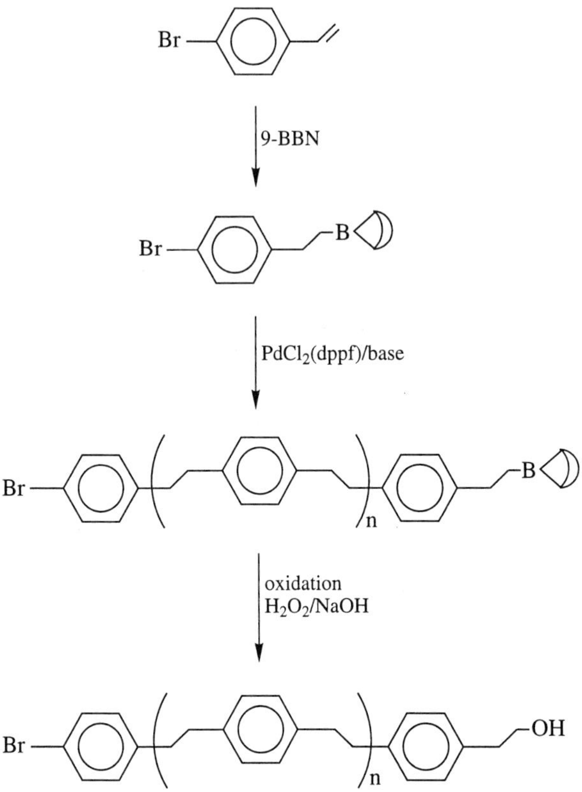

Br — C₆H₄ — CH=CH₂

9-BBN

$PdCl_2(dppf)$/base

oxidation
H_2O_2/NaOH

91

Fig. 1-32. Synthesis of functional polymers from hydroboration of *p*-bromostyrene and coupling reaction catalyzed by $PdCl_2(dppf)$ (adapted from [241]).

1.5.2 Olefin Functionalization

1.5.2.1 Hydroformylation

One major advantage offered by the dppf ligand in Rh-catalyzed olefin hydroformylation is exemplified in its higher linear aldehyde selectivity when present in a dppf : Rh ratio of 1.5 or higher [37, 242]. This result leads to the proposed key intermediate of a Rh dimer with both chelating and bridging phosphine in the catalytic cycle. It also confirms the significance of the tris(phosphine) moieties at the point when the aldehyde selectivity is determined, i.e., the step in which the hydride is inserted into the M-olefin bond. This involvement of a dinuclear or tris(phosphine) intermediate appears to differ from the intermediate $RhH(CO)(PR'_3)_2$(olefin) (which is converted into the square planar $Rh(R)(CO)(PR'_3)_2$ by hydride insertion) commonly accepted for hydroformylation catalyzed by monophosphine complexes. ^{31}P NMR studies also established the existence of the equilibrium in which the disphosphine can be

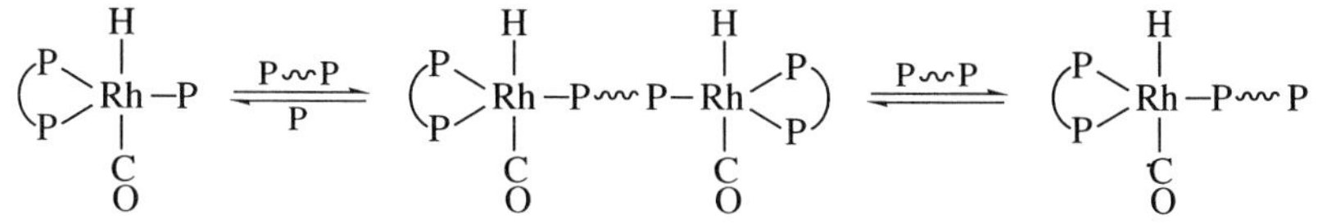

Scheme 1-37

in a chelating, unidentate, or bridging state (Scheme 1-37) and that the active species under hydroformylative conditions may contain a combination of these ligand modes within the same complex [243]. Crystal structures of the tbp mononuclear complex with both chelating and monodentate phosphines have been reported [244]. In this context, the isolation and complete characterization of many di- and polynuclear dppf complexes that show the coexistence of various coordination modes are encouraging (see Sect. 1.3.1) [8, 32, 245]. In order to account for the enhanced selectivity at a higher dppf:Rh ratio, a mechanism that encompasses chelating phosphine and dinuclear complexes has been proposed [246]. An alternative proposal that does not invoke dimer formation has been proposed by Casey and Whiteker [247]. This suggestion conjures a bite-dependence site selectivity for a diphosphine ligand in a 5-coordinate complex as the key issue. The authors suggested a natural bite angle of about 120° to be the most favorable for high regioselectivity for *n*-aldehyde formation. That dppf shows a higher selectivity towards unbranched aldehyde compared with PPh_3 is also seen in the phosphine-doped $RhH(CO)(PPh_3)_3$-catalyzed hydroformylation of methyl methacrylate (Scheme 1-38) [248], and 1-octene [249]. Activities with dienes, e.g., 1,3-butadiene, have also been documented.

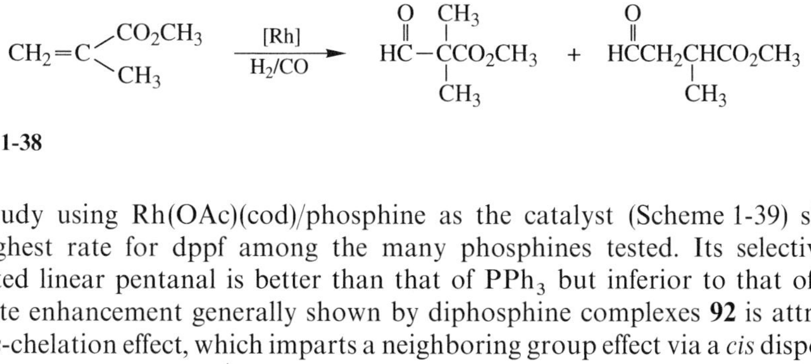

$$CH_2{=}C(CO_2CH_3)(CH_3) \xrightarrow[H_2/CO]{[Rh]} HC(O){-}C(CH_3)_2CO_2CH_3 + HC(O)CH_2CH(CH_3)CO_2CH_3$$

Scheme 1-38

A study using Rh(OAc)(cod)/phosphine as the catalyst (Scheme 1-39) showed the highest rate for dppf among the many phosphines tested. Its selectivity to saturated linear pentanal is better than that of PPh_3 but inferior to that of dppe. The rate enhancement generally shown by diphosphine complexes **92** is attributed to it *cis*-chelation effect, which imparts a neighboring group effect via a *cis* disposition of the carbonyl to the η^1-allyl. This proximity effect facilitates subsequent acyl formation whereas the *trans* configuration **93** would be unfavorable [250]. Recently, the idea of diphosphine-enrichment has been extended to asymmetric hydroformyla-

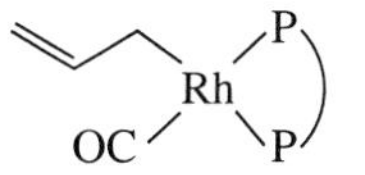

92 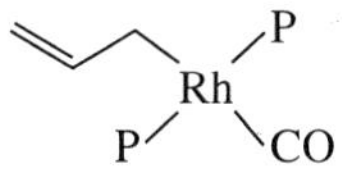 **93**

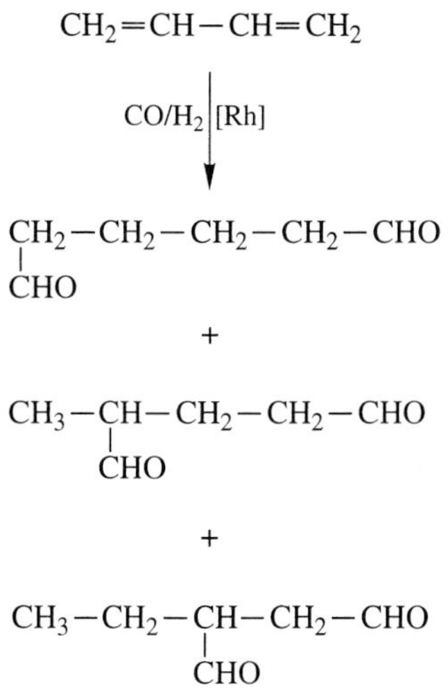

$CH_2{=}CH{-}CH{=}CH_2$

CO/H_2 [Rh]

$CH_2{-}CH_2{-}CH_2{-}CH_2{-}CHO$ (with CHO on the first carbon)

+

$CH_3{-}CH{-}CH_2{-}CH_2{-}CHO$ (with CHO on the second carbon)

+

$CH_3{-}CH_2{-}CH{-}CH_2{-}CHO$ (with CHO on the third carbon)

Scheme 1-39

$CH_2{=}C(CO_2CH_3)(NHCOCH_3)$ $\xrightarrow[H_2/CO]{RhH(CO)(PPh_3)_3/Diphosphine}$ $H{-}C(CO_2CH_3)(NHCOCH_3)(CH_3)$ + $H{-}C(CO_2CH_3)(CH_2CHO)(NHCOCH_3)$ + $OHC{-}C(CO_2CH_3)(NHCOCH_3)(CH_3)$

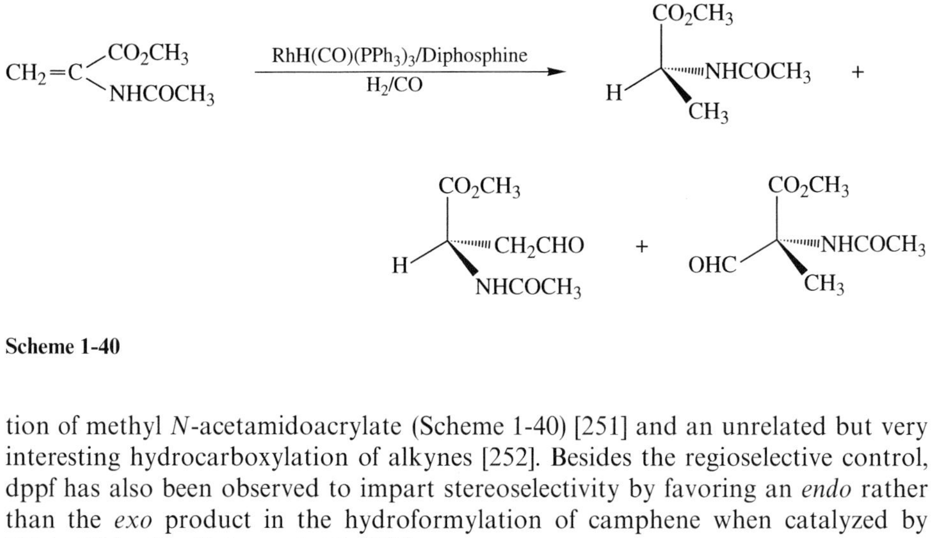

Scheme 1-40

tion of methyl *N*-acetamidoacrylate (Scheme 1-40) [251] and an unrelated but very interesting hydrocarboxylation of alkynes [252]. Besides the regioselective control, dppf has also been observed to impart stereoselectivity by favoring an *endo* rather than the *exo* product in the hydroformylation of camphene when catalyzed by $[Rh(\mu\text{-}Cl)(cod)]_2$ (Scheme 1-41) [253].

These olefin hydroformylations are not confined to Rh catalysts. Higher linear selectivity has also been reported in a Ru system when the catalyst, $Ru(CO)_3(PPh_3)_2$ is doped with a ten-fold excess of dppf; the rate, however, is lower than that of the unassisted catalyst [254]. In another study of 1-octene using a "melt" catalyst containing RuO_2 and $Ru_3(CO)_{12}$, both nonanal and nonanol can be formed depending on the conditions. Dppf gives a higher formation rate for the aldehyde and linear alcohol selectivity than PPh_3 but does not appear to offer any advantages over other diphosphines [255].

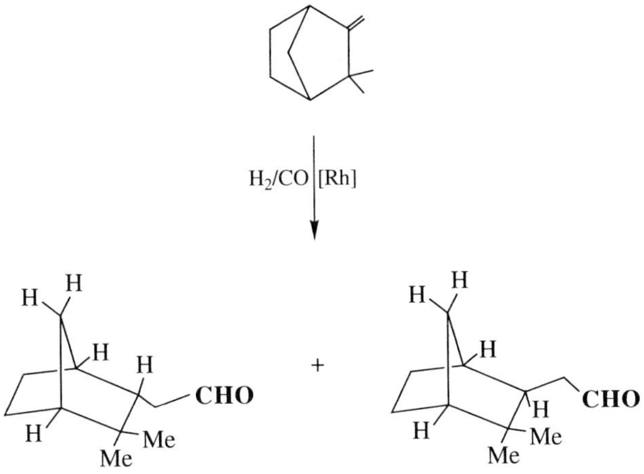

H_2/CO [Rh]

endo- exo-

Scheme 1-41

1.5.2.2 Hydrogenation and Reduction

Olefin hydrogenation catalyzed by $[Pd(dppf)(solv)_2][ClO_4]_2$ (solv = DMF or py) [256], $Ru(C_3H_5)\{[OC(CF_3)]_2CH\}(dppf)$ [257] or the Pd(II) derivatives of dppf supported on polystyrene [258] has been reported. A catalyst precursor $[Rh(P\text{-}P)(nbd)]^+$ can be hydrogenated under very mild conditions to give a trihydride-bridged Rh(III) complex $[Rh_2H_2(\mu\text{-}H)_3(FcPP)_2]^+$ (FcPP = ferrocenylphosphine) [259]. Using a series of ferrocene-based diphosphine complexes $[Rh(P\text{-}P)(nbd)][ClO_4]$ as model catalyst, Cullen et al. demonstrated that the hydrogenation proceeds faster when there is less steric hindrance on the ligand and that the *t*-butyl group is preferred to the phenyl substituent on the phosphine [260]. Polynuclear heteroaromatic compounds can also be hydrogenated regioselectively when catalyzed by $[Rh(dppf)(nbd)][ClO_4]$ or $[Rh_2(\mu\text{-}Cl)_2(dppf)_2]$ [261]. The observed turnover rates are significantly higher than those of $RhCl(PPh_3)_3$ [262] or $RuClH(PPh_3)_3$ [263].

Reduction of primary halides with Grignard reagents is significantly accelerated when catalyzed by $PdCl_2(dppf)$ or [Pd(dppf)] (generated by reducing $PdCl_2(dppf)$ with DIBAL [264, 265]. The reduction products may originate simply from proton (or hydride) transfer or cross-coupling, as discussed earlier in Section 1.5.1.1. In an elegant study using deuterium labeling, Yuan and Scott suggested two reduction mechanisms with or without direct participation of a Pd intermediate. The former pathway is based on the concept of equilibrating Grignard reagents and two competing processes, hydride transfer and transmetalation. The latter relies on oxidative addition of the electrophile, *β*-hydride transfer from the Grignard reagent and reductive elimination. Non-organometallics can also be used as reductants. Reduction of aryl triflates by $[Et_3NH]^+[HCO_2]^-$ (from $Et_3N + HCO_2H$) occurs readily when catalyzed by $Pd(OAc)_2$ in the presence of phosphines (Scheme 1-42) [266]. A variety of mono- and bidentate phosphines can be used. Generally, dppf

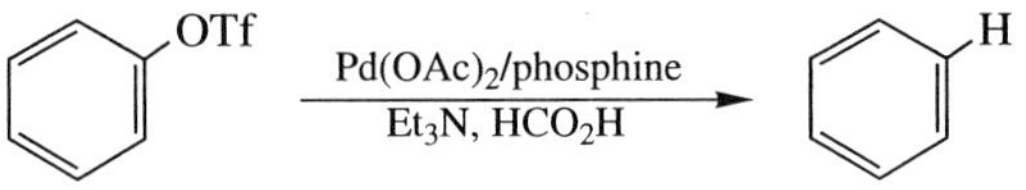

OTf $\xrightarrow[Et_3N,\ HCO_2H]{Pd(OAc)_2/phosphine}$ H

Scheme 1-42

functions better in reactions that are poorly promoted by PPh_3. This catalytically reducing mixture containing dppf is responsible for the synthesis of (+)-4-demethoxydaunomycinone **94**, which is pharmacologically important [267]. A postulated mechanism requires initial reduction of Pd(II) to Pd(0), followed by oxidative addition from the triflate, formato substitution, CO_2 extrusion and, finally,

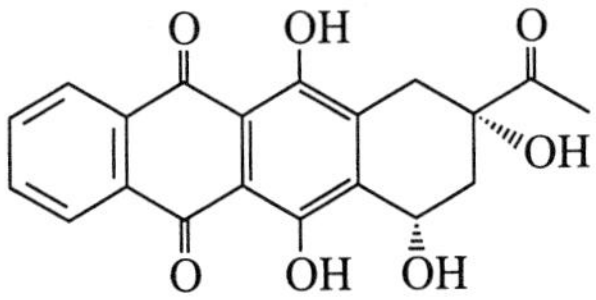

94

reductive elimination to the desired arene (Fig. 1-33). A variation of this reduction employs $Pd(PPh_3)_4$ as the catalytic precursor and $NaBH_4$ as the hydrogen source [156]. $PdCl_2$(dppf) has also been found to be effective in reducing a tricyclic triflate (Scheme 1-43) when the other catalytic substrates gave disappointing results. This

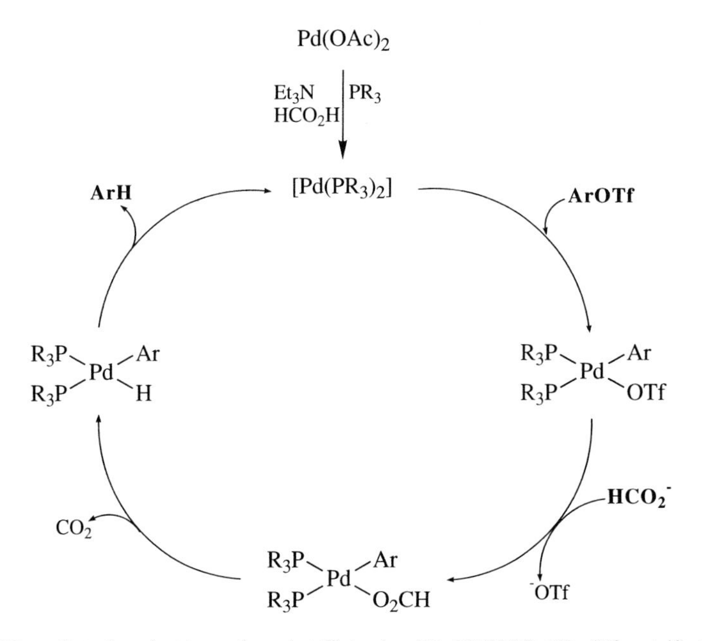

Fig. 1-33. Pd-mediated reduction of aryl triflate by $[Et_3NH][CO_2H]$. (Tf = triflate, $CF_3SO_3^-$) (adapted from [176]).

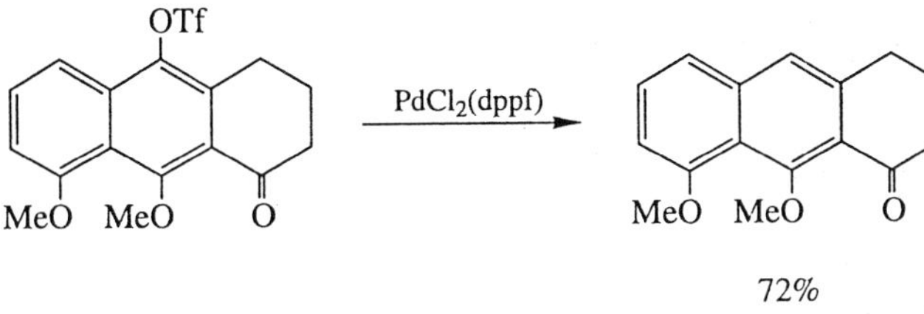

OTf

MeO MeO O

$PdCl_2(dppf)$

MeO MeO O

72%

Scheme 1-43

synthesis of a tricyclic ketone from an aryl carbene **95** via a phenol **96** is relevant to the synthesis of olivin **97** and chromomycinone **98**, which are relatives of some antitumor antibiotics. This reduction also applies to the sulfonates [268]. Reduction of 4-fluorobenzenesulfate **99** is favored over hydrolysis. The reduction is promoted by monophosphines of higher basicity, which lends some support to the oxidative addition step, and by diphosphines of larger chelate angles such as dppp, dppb and dppf. Remarkably, dppm chemoselectively yields the hydrolytic product; the mechanistic implication would be of some significance.

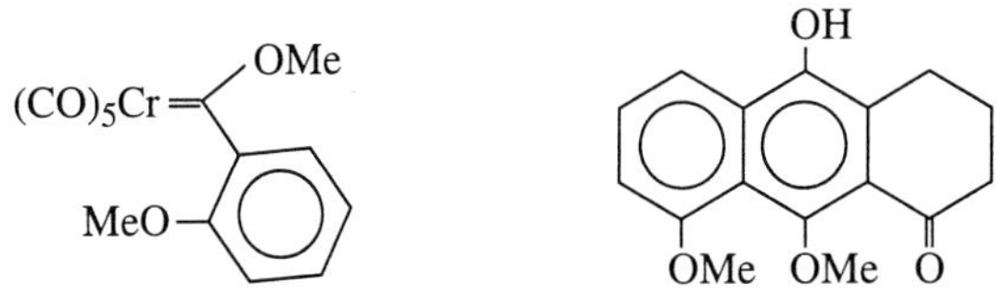

$(CO)_5Cr$ OMe MeO

95

OH OMe OMe O

96

O $O\text{-}SO_2\text{-}(4\text{-}C_6H_4F)$ MeO O

99

90°C $Pd(OAc)_2/dppf/HCO_2H/Et_3N$

HO R OH OH O H OMe OH Me O OH OH

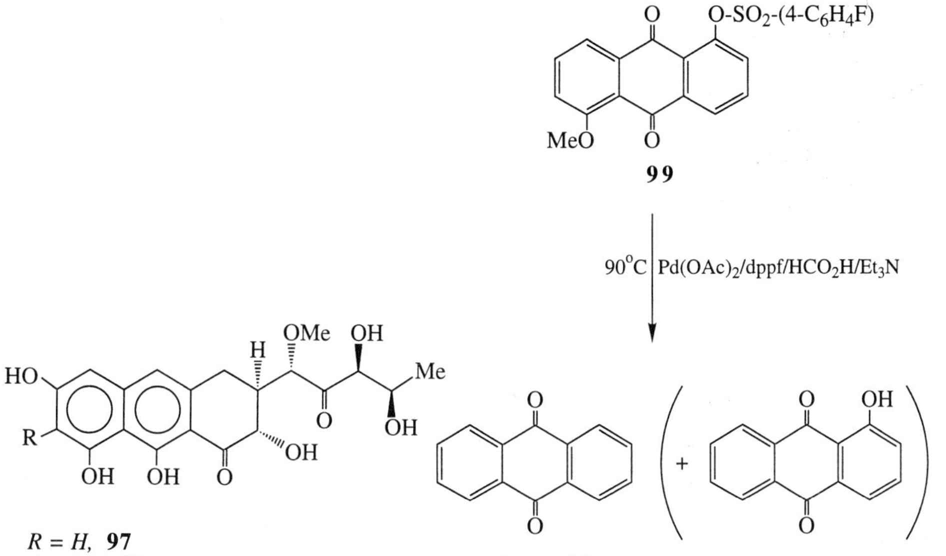

R = H, **97**
R = Me, **98**

O O + O OH O

95 - 100%

1.5.2.3 Hydroboration and Hydrosilylation

Cross-coupling between 1-alkenyl- and arylboron compounds with organic electrophiles have found wide application (see Sect. 1.5.1.1) in organic synthesis [125, 132, 269–271]. The value of this methodology is further realized with the use of alkylboranes, such as *B*-alkyl-9-borabicyclo[3.3.1]nonanes (*B*-R-9-BBN), which can be conveniently prepared by hydroboration of olefins [272]. The hydroboration proceeds with high stereoselectivity and chemoselectivity. The choice of phosphines in a catalytic system sometimes affects the chemo- and regioselectivity of the hydroboration. Hydroboration of 1-(ethylthio)-1-propyne with catecholborane can be satisfactorily carried out with $PdCl_2$(dppf) but the regioselectivity is best for the dppe and dppp complexes of Ni [273]. Notably, PPh_3 complexes perform poorly.

Me HB ½ $[Pd_2(dba)_3]$/L Me H Me B–O O 100 + Me B–O O 101 Me O B–O 102

A similar example is seen in the $[Pd_2(dba)_3]$-catalyzed hydroboration of 2-methyl-1-buten-3-ynes [274]. While PPh_3 and $PPh_2(C_6F_5)$ favor the 1,4-addition product allenylborane **100** all diphosphines yield the 1,2-addition product (*E*)-dienylborane **102** exclusively (Table 1-13). This remarkable difference in selectivity is explained based on an 1,3-enyne monophosphine complex **103** and an alkynyl diphosphine complex **104** as intermediates. Dppf exhibits the best product yield among the phosphines tested. Similar observation was noted in the asymmetric hydroboration (Scheme 1-44) [275]. The action of catecholborane on 1-phenyl-1,3-butadiene also proceeds regioselectively to give, after oxidation, *anti*-1-phenyl-1,3-butanediol

Table 1-13. Ligand (L) effect on the Pd-catalyzed hydroboration of 1-buten-3-ynes with catecholborane (data from [274])

L	Time, h	Yield, %	**100/101/102**
none	5	0	–
2 PPh_3	0.5	70	62/38/0
2 $PPh_2(C_6F_5)$	0.5	73	83/17/0
dppe	2	39	0/0/100
dppb	2	61	0/0/100
dppf	0.5	89	0/0/100

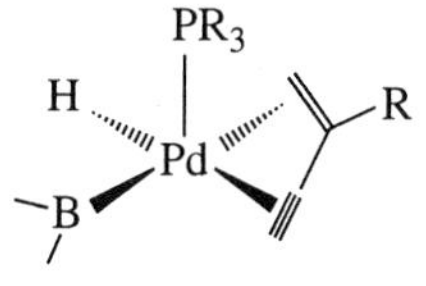

103 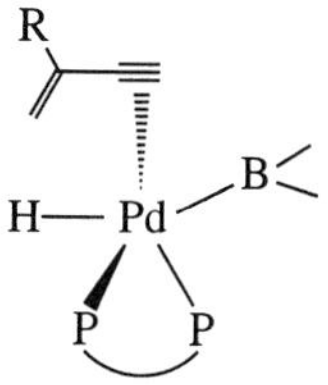 104

(Scheme 1-45) [276]. This double-hydroboration is Rh-catalyzed in the presence of PPh_3 or dppf. Again, the use of other phosphines would diminish the chemo- or regioselectivity or both.

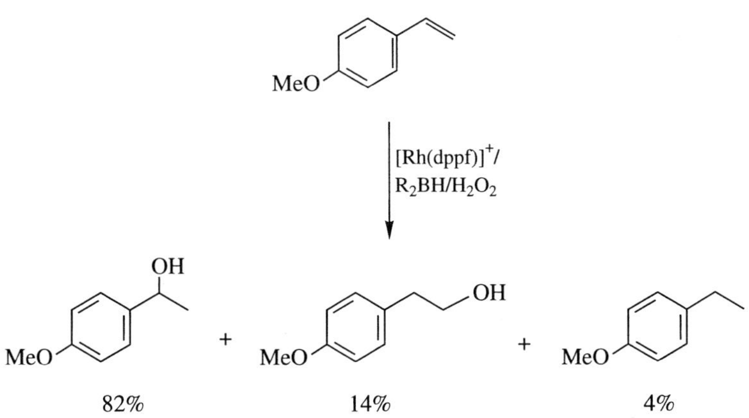

$[Rh(dppf)]^+$/ R_2BH/H_2O_2

82% 14% 4%

Scheme 1-44

$[Rh]/PPh_3$ or dppf

H_2O_2 NaOH

anti

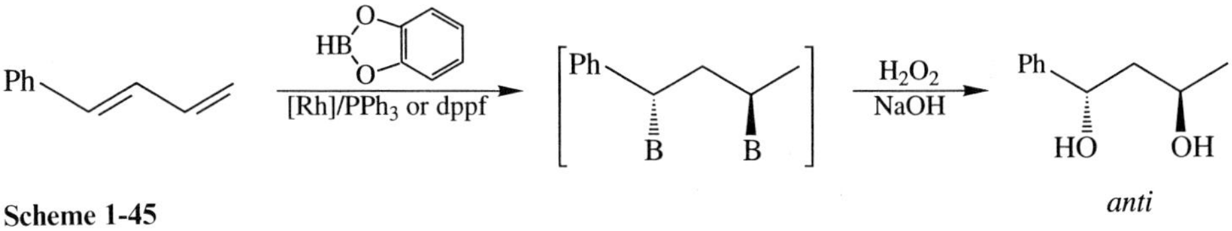

Scheme 1-45

Olefin hydrosilylation is classically carried out using Speier's catalyst (H_2PtCl_6-$6\,H_2O$) [277] in a heterogeneous-like mechanism [278–280]. Homogeneous catalysis by metal complexes [281–283], for example ferrocene derivatives [1, 284–286], is gaining importance especially in asymmetric induction (described by Hayashi, Chapter 2). The use of supported catalysts has also been reported. The $[PdCl_2]$ complex of dppf-functionalized polystyrene conveniently catalyzes the hydrosilylation of styrene by $HSiCl_3$ (Scheme 1-46) [287]. The regioselectivity is increased to 100% in favor of the α-adduct during the recycling runs.

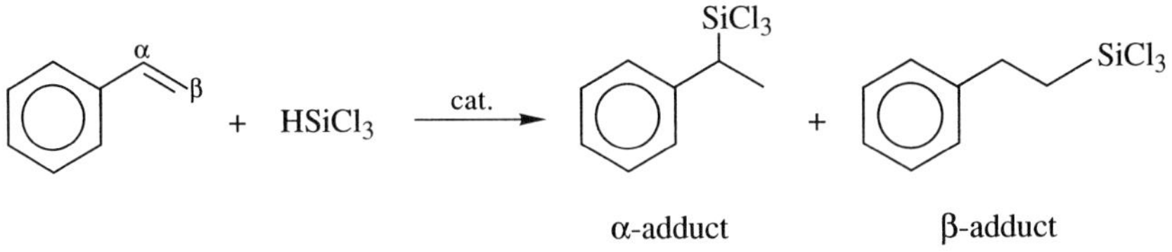

α β + $HSiCl_3$ cat. → α-adduct + β-adduct ($SiCl_3$)

Scheme 1-46

1.5.2.4 Isomerization

Olefin isomerization can be promoted by a variety of metal complexes [288–294]. One of the best known examples is [Rh(P-P)][ClO_4] (produced in situ by hydrogenation of [Rh(P-P)(cod)][ClO_4]), which converts an (*Z*)-allylamine **105** to a racemic (*E*)-enamine **106** with high chemical selectivity and conversion rate; the best phosphine ligand for this purpose is BINAP, followed by dppe and dppf [295, 296].

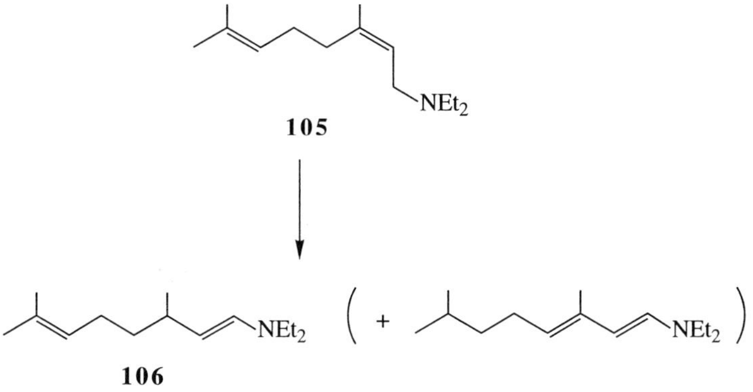

NEt_2

105

NEt_2 (+ NEt_2)

106

This isomerization is enantioselective when optically active BINAP is used and provides practical access to optically active aldehydes and alcohols such as L-menthol, which is a key fragrance chemical [297–299]. The proposed mechanism involves amine, iminium, and enamine as complex intermediates [300]. Extension of this olefin isomerization is realized in the isomerization of an alkyne to a conjugated diene (Scheme 1-47) [301]. High chemoselectivity is achieved when $Pd(OAc)_2$ or [$Pd_2(dba)_3$]/HOAc, in the presence of phosphine, is used as catalyst (Table 1-14). The phosphine of choice is dppb although dppf could give a similar yield.

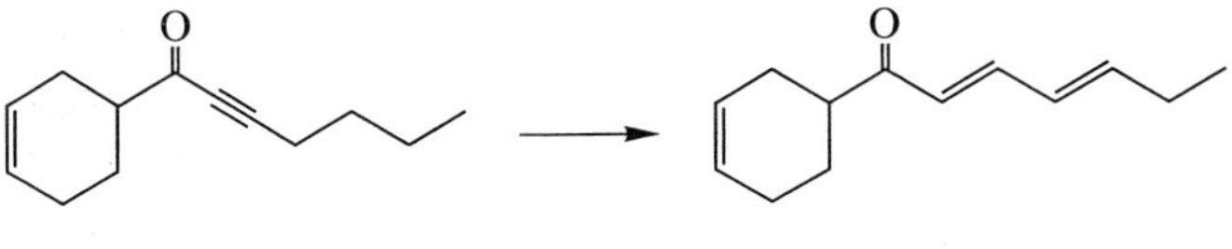

O → O

Scheme 1-47

Table 1-14. Effect of catalyst on the isomerization of an alkynone to dienone in toluene at 100 °C (data from [301])

Catalyst	Time, h	Isolated yield, %
$Pd(OAc)_2/PPh_3$	4.5	74
$Pd(OAc)_2$/dppb	5	81
$Pd(OAc)_2$/dppf	20	77
$Pd_2(dba)_3/PPh_3$/HOAc	1	53
$Pd_2(dba)_3$/dppb/HOAc	2.5	82

1.6 Cluster Complexes

The use of dppf as a flexible and versatile ligand with catalytic potential is too important to be left out of a study of metal cluster chemistry. Its variable coordination modes in adapting to metals of different steric and electronic demands and its ability to span a wide range of M-M distances make it a very attractive stabilizing ligand for metal clusters.

Almost all known dppf clusters are prepared by carbonyl substitution of the parent clusters. One of the earlier examples of dppf clusters, $Co_3(\mu_3\text{-CMe})(CO)_7(\mu\text{-dppf})$, **6** [52, 53], is obtained from the thermal reaction of $Co_3(\mu_3\text{-CMe})(CO)_9$ with dppf (Fig. 1-7). The steric bulk of dppf forces a terminal-to-bridging carbonyl rearrangement, with dppf itself edge-bridging the metal triangle. In this closed-bridging system, although the Co-Co bond spanned by dppf (2.520(1) Å) is longer than the other two Co-Co bonds (2.479(1), 2.513(1) Å) [53], it is the shortest ever reported for a dppf-bridged M-M bond. It also provides the most compelling evidence that dppf-bridged cluster chemistry can be developed. In the dppm analogue of the Co_3 cluster [302], the M-M bond elongation is smaller and the seven carbonyls remain at their terminal positions. A related cluster, $Co_3(\mu_3\text{-CPh})(CO)_7(\mu\text{-dppf})$ [303], can be similarly prepared. It exhibits an identical dppf-bridged Co-Co bond (2.519(2) Å) as that of the CMe analogue.

Other instances of dppf-supported M-M bonds in triangular clusters are $M_3(CO)_{10}(\mu\text{-dppf})$ (M = Fe [51], Ru [9]), which is best prepared from PPN^+-catalyzed carbonyl replacement [9]. The structures of $Fe_3(CO)_{10}(\mu\text{-dppf})$ **107** and $Ru_3(CO)_{10}(\mu\text{-dppf})$ **108** resemble those of the parent carbonyls $M_3(CO)_{12}$ and are thus different. In the former, there are eight terminal and two bridging carbonyls. The Fe-Fe bond supported by the two bridging carbonyls and a bridging dppf is significantly shorter (2.553(2) Å) compared with the other two unbridged bonds (2.728(2) and 2.730(2) Å). In contrast, the Ru cluster has no bridging carbonyls. The bridged Ru-Ru distance is 2.9284(5) Å while the unsupported bond length is 2.8600(4) Å. The elongation of the bridged bond in the Ru cluster is not entirely due to the repulsive *cis*-effect but the steric demands and the bite of the dppf ligand. Purely on electronic reasoning, long Ru-Ru bonds should be associated with short Ru-P bonds. This is indeed observed in $Ru_3(CO)_{10}(\mu\text{-dppf})$ with the

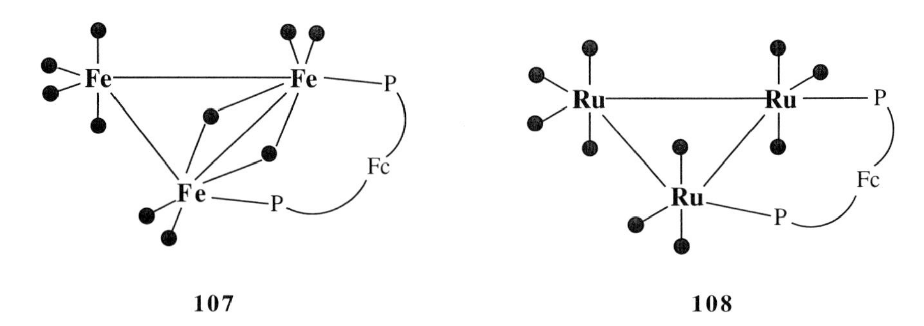

107 **108**

Ru-P distance (average bond length 2.3503(8) Å) being marginally shorter than that in $Ru_3(CO)_{11}(PFcPh_2)$ (2.369(1) Å), which has a shorter Ru-Ru bond (average bond length 2.88 Å) [9].

Higher substitution of $Ru_3(CO)_{12}$ with a two-fold excess of dppf affords $Ru_3(CO)_8(\mu\text{-dppf})_2$ [9]. Hydrogenation of $Ru_3(CO)_{10}(\mu\text{-dppf})$ gives a major product $Ru_4(\mu\text{-H})_4(CO)_{10}(\mu\text{-dppf})$ [9], which can also be prepared from the ketyl-catalyzed reaction of $Ru_4(\mu\text{-H})_4(CO)_{12}$ with dppf. Cullen et al. have investigated extensively the pyrolytic products of ferrocenyl clusters [3, 6, 304–308]. Pyrolysis of $Ru_3(CO)_{10}(\mu\text{-dppf})$ under mild conditions affords five characterizable products among which three show skeletal disintegration of the dppf ligand by C-H and P-C cleavage, $Ru_3\{\mu\text{-P}(C_6H_4)(\eta^5\text{-}C_5H_4)Fe(\eta^5\text{-}C_5H_4PPh_2)\}(\mu\text{-CO})(CO)_8$ **109**,

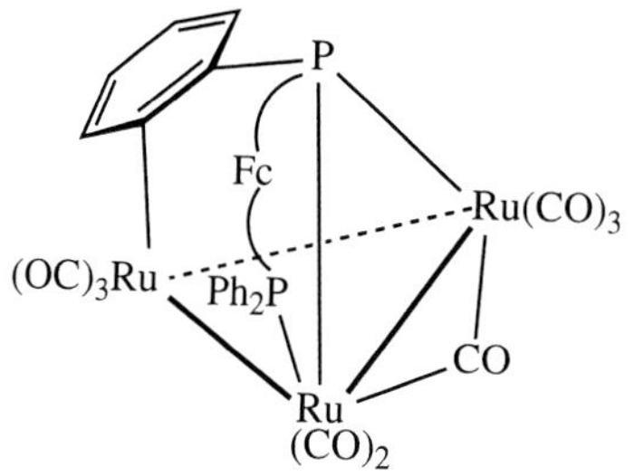

109 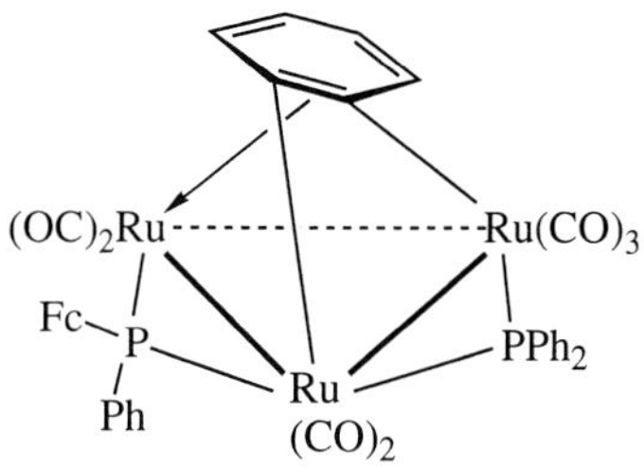 **110**

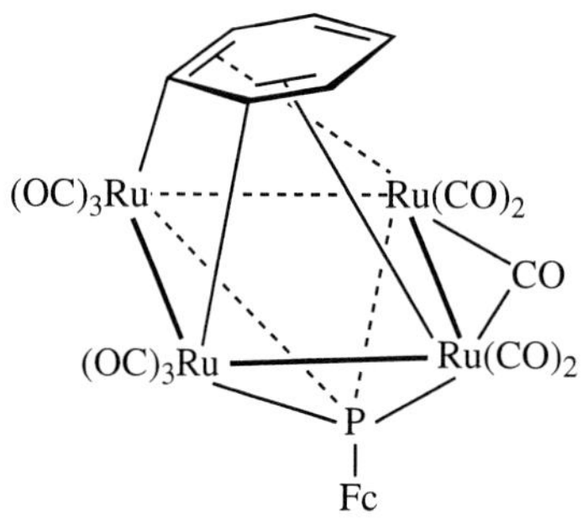

111

$Ru_3(\mu_3\text{-}\eta^1,\eta^1,\eta^2\text{-}C_6H_4)(\mu\text{-}PPh_2)(\mu\text{-}PPhFc)(CO)_7$ **110** and $Ru_4(\mu_4\text{-}PFc)(\mu_4\text{-}C_6H_4)\text{-}(\mu\text{-}CO)(CO)_{10}$ **111** [6]. This collapse of the ligand, with is doubtlessly initiated by the decarbonylation, provides a channel for cluster stabilization in the absence of added ligand. Ortho-metalation of the Cp or phenyl rings is an obvious means of such stabilization [1, 2]. This phenomenon is observed in $Ru_3(\mu\text{-}H)\{\mu_3\text{-}PPh_2(\eta^1,\eta^5\text{-}C_5H_3)Fe(\eta^5\text{-}C_5H_4PPh_2)\}(CO)_8$ **4**, in which ortho-metalation of a Cp ring with Ru(2) and Ru(3) brings the iron atom to within bonding distance of Ru(2) (3.098(3) Å). As a result, this weak Fe → Ru(2) donation strengthens the Ru(2)-P(2) bond (2.306(5) Å) compared with Ru(1)-P(1) (2.366(4) Å). Other instances of ferrocenyl iron-metal interactions in clusters include $Os_3(\mu\text{-}H)_2(CO)_8PAr(C_6H_5)(\eta^5\text{-}C_5H_3)\text{-}Fe(\eta^5\text{-}C_5H_4)$ (Ar = C_6H_5, $Fe(\eta^5\text{-}C_5H_5)(\eta^5\text{-}C_5H_4)$) and $Os_3(\mu\text{-}H)(CO)_7(\mu_3\text{-}P\textit{i}Pr)\text{-}\{Fe(\eta^5\text{-}C_5H_4P\textit{i}Pr_2)(\eta^5\text{-}C_5H_4)\}$ [3]. Double-metalation of a phenyl ring is observed in $Ru_3(\mu\text{-}H)\{\mu_3\text{-}PPh(\eta^1,\eta^2\text{-}C_6H_4)(\eta^5\text{-}C_5H_4)Fe(\eta^5\text{-}C_5H_4PPh_2)\}(CO)_8$ **112** [6].

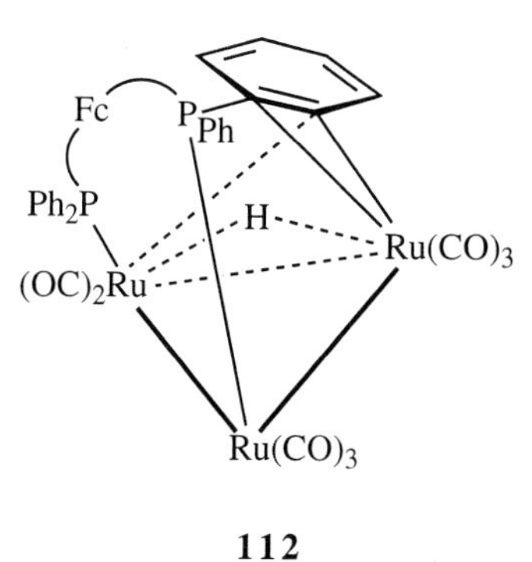

112

This ring thereby constitutes an unusual $\mu\text{-}\eta^1,\eta^2$-vinyl group whereby the phosphino carbon is also metalated. Intramolecular oxidative addition of the *ortho* C-H bond of the phenyl substituents is a common feature in crowded iridium phosphine complexes [65, 309, 310].

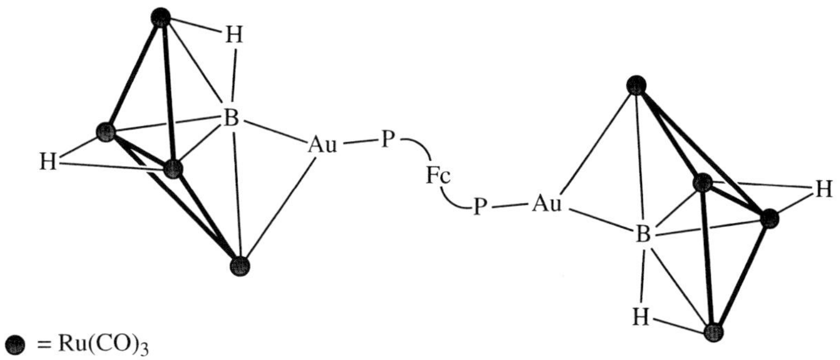

● = Ru(CO)3

113

Mixed-metal clusters of dppf have also been reported. One special example is $M_4(\mu\text{-H})(CO)_{12}(\mu_6\text{-B})Au_2(\mu\text{-dppf})$ (M = Fe, Ru, 7) [55], whereby a butterfly cluster is capped with a heterometal moiety $\{Au_2(dppf)\}$ with an Au-Au length of 2.818(2) Å. This Au-Au interaction is in contrast to the large Au-Au separation in $\{Ru_4(\mu\text{-H})(CO)_{12}(\mu\text{-BH})Au\}_2(\mu\text{-dppf})$ **113** [55], which can be considered as two clusters openly bridged by a dppf ligand. This form of bicluster linked by dppf is also observed in $[Fe_3(\mu_3\text{-S})_2(CO)_8]_2(\mu\text{-dppf})$ **114**, which is produced together with $Fe_3(\mu_3\text{-S})_2(CO)_8(\text{dppf-}P)$ **115** and $Fe_3(\mu_3\text{-S})_2(CO)_7(\mu\text{-dppf})$ **116** from $Fe_3(\mu_3\text{-S})_2(CO)_9$ [311]. This reaction illustrates again the alternative coordination modes of dppf. In these complexes, the Fe triangle is capped by two 6-electron sulfide bridges. The absence of direct bonding between the two Fe core atoms spanned by dppf in **116** is in accordance with the 18-electron rule. The analogous reaction of the congeneric $Ru_3(\mu_3\text{-S})_2(CO)_9$ with dppf gives the quasi-closed bridge and open bridge clusters as the major products [311].

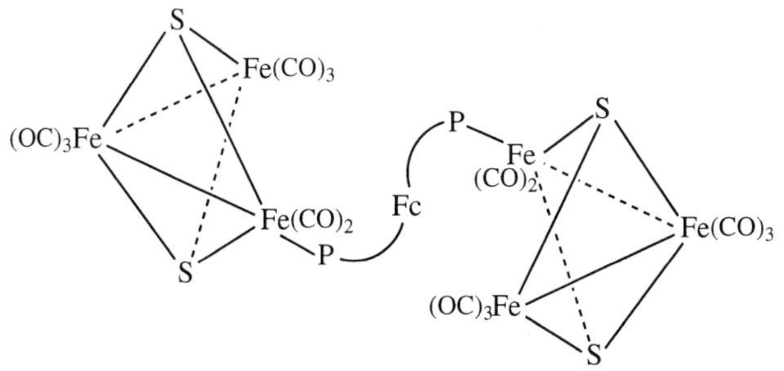

114

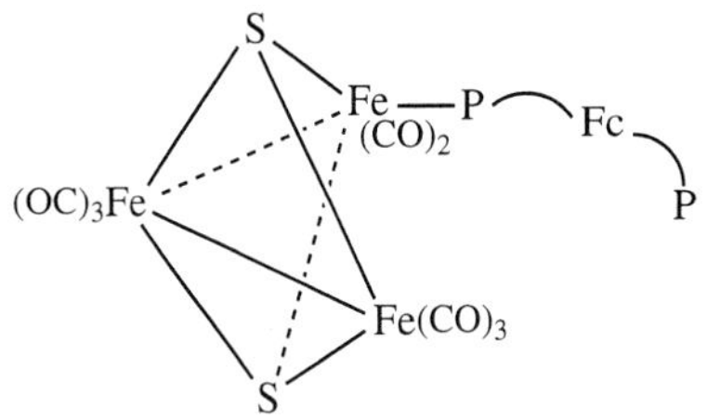

115

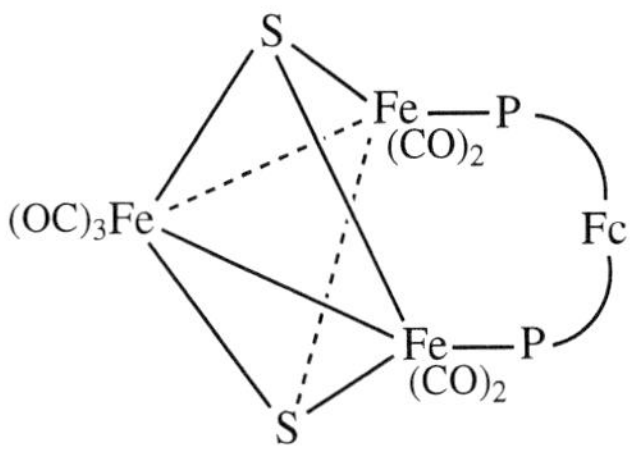

116

The bridging form of dppf is the predominant mode of coordination observed among its cluster complexes. The metal proximity effect largely explains this preference. Similar to other bridging or capping ligands [312], dppf can contribute to the stabilization of the cluster core. The ability of dppf to hold the metals together at a distance comfortable for the metals is an advantage. It is possible but not common for dppf to span a direct metal-metal bond. An open bridge or quasi-closed bridge on the other hand could permit a greater degree of freedom and hence less conformational strain on the dppf skeleton.

1.7 Electrochemistry

The electrochemistry of ferrocene-type ligands and their complexes is reviewed in detail by Zanello in Chapter 7. Hence the present description discusses only briefly some unique features of dppf complexes. These complexes are generally expected to exhibit a ferrocene-centered oxidation process. The general interest lies in the modification of the redox potential of the ferrocene/ferrocenium couple on phosphination of the Cp rings, complexation of the resultant dppf ligand, and variations among the various known coordination modes of the ligand.

A marked positive shift (0.27 V) [60] is observed on phosphination of the Cp rings of ferrocene. The redox behavior of dppf has been studied by several groups. Although Wrighton et al. [60] reported the oxidation of dppf in CH_2Cl_2 to be irreversible, DuBois et al. [313] observed quasi-reversibility (+0.51 V *vs.* SCE) in CH_3CN followed by an irreversible chemical reaction. Support of the latter came from Corain et al. [43] who found it to be essentially reversible in 1,2-$C_2H_4Cl_2$, and Housecroft et al. [61] who concurred with a single reversible oxidation in CH_3CN. Pilloni et al. [314] subsequently confirmed the reversibility of the oxidation in 1,2-$C_2H_4Cl_2$ and proposed an intramolecular electron transfer between the ferrocene and phosphine moieties to give a phosphine radical that rapidly dimerizes and then hydrolyzes to give the oxidized (dppfO and $dppfO_2$) and protonated ($[dppfH]^+$ and $[dppfH_2]^{2+}$) products. The overall reactions can thus be represented as follows:

$$3\,\text{dppf} + H_2O \rightarrow \text{dppfO} + 2\,[\text{dppfH}]^+ + 2\,e^-,$$

$$3\,\text{dppf} + 2\,H_2O \rightarrow \text{dppfO}_2 + 2\,[\text{dppfH}_2]^{2+} + 4\,e^-$$

Scheme 1-48

These decomposition species are sometimes liberated from the oxidized precursor. Irreversible ferrocene-based oxidation has been observed in e.g., $Mo(CO)_4$(dppf-*P*, *P*′) [313], 1-{Pd(dppf-*P*, *P*′)}(B_3H_7) **8** [61] and $Co_3(\mu\text{-CPh})(CO)_7(\mu\text{-dppf})$ [303].

Electrochemical study of a series of MCl_2(dppf-*P*, *P*′) complexes (M = Co, Ni, Zn, Cd, Hg) suggests high lability of the oxidized complex [43]. Oxidation of the ligand in MCl_2(dppf-*P*, *P*′) (M = Zn, Cd, Hg) significantly weakens the P → M bond (which is purely σ in character), destabilizes the cationic $[MCl_2(\text{dppf-}P,P')]^+$, and results in the absence of a cathodic peak. This facile decomposition of the oxidized species appears to be fairly common among dppf coordination complexes.

$$MCl_2(\text{dppf}) - e^- \rightarrow [MCl_2(\text{dppf})]^+ \rightarrow MCl_2 + [\text{dppf}]^+$$
$$\rightarrow \text{rapid decomposition } (M = \text{Zn, Cd, Hg})$$

Scheme 1-49

In the case where the weakening of the P-M bonds is compensated by M → P backbonding [e.g., in MCl_2(dppf-*P*, *P*′) (M = Pd, Pt)], stable oxidized species {e.g.,

[MCl_2(dppf-*P*, *P'*)]$^+$} can result. A reversible one-electron oxidation is commonly observed in many dppf complexes, e.g., MCl_2(dppf-*P*, *P'*) [61, 79], [M(μ-X)(dppf-*P*, *P'*)]$_2$[BF_4]$_2$ (M = Pd, Pt; X = Cl, OH) [315], [Pt(dppf-*P*, *P'*)(dmf)$_2$][BF_4]$_2$ [315], ReCl(CO)$_3$(dppf-*P*, *P'*) [60], Ru_2Cl_4(η^6-*p*-cym)$_2$(μ-dppf) (*p*-cym = *p*-MeC_6H_4-$CHMe_2$) [316], [Cu(dppf-*P*, *P'*)(CH_3CN)$_2$]$^+$ [317]. Complexation inevitably leads to anodic (positive) shift of the redox potential (e.g. ≈0.15 V for ReCl(CO)$_3$(dppf-*P*, *P'*), ≈0.2 V for Mo(CO)$_4$(dppf-*P*, *P'*) [313], ≈0.3 V for [Cu_2(dppf-*P*, *P'*)$_2$(μ-dppf)]$^{2+}$, ≈0.4 V for MCl_2(dppf-*P*, *P'*) (M = Pd [61], Pt [79]) and ≈0.5 V in 1-{Pd(dppf-*P*, *P'*)}(B_3H_7) (irreversible) [61]. The presence of non-communicating ferrocenyl groups within a complex usually results in either a multi-step one-electron (e.g., [M(μ-X)(dppf-*P*, *P'*)]$_2$[BF_4]$_2$, (M = Pd, Pt; X = Cl, OH)) or single-step multi-electron (e.g., [Cu_2(dppf-*P*, *P'*)$_2$(μ-dppf)][BF_4]$_2$ [317]) charge transfer. The latter oxidation provides a means for the electrochemical synthesis of [Cu(dppf)(dppfO)]$^+$. A one-electron transfer is occasionally observed, (e.g., M'Cl_2{(μ-dppf)M(CO)$_5$}$_2$ (M = Cr, Mo, W; M' = Pd or Pt)), although there are two dppf bridges in the complex [47]. Some dppf complexes of Au(*I*) are even electrochemically inactive [28]. Structurally related monomers and dimers such as in MCl_2(dppf-*P*, *P'*) and [M(μ-Cl)(dppf-*P*, *P'*)]$_2$[BF_4]$_2$ (M = Pd, Pt) [315] may also show similar cyclic voltammetric responses (in terms of its shape and current).

The high resistance of dppf complexes towards reduction compared with other alkyl-chained diphosphine analogues is best illustrated in the electrochemical reduction of Ru_2Cl_4(η^6-*p*-cym)$_2$(μ-P-P). The registered reduction potential increases in the order dppe < dppm ≪ dppf, which is consistent with the stronger donating ability of dppf [316]. A similar effect can be obtained by replacing the phenyl group with an electron-donating substituent on the phosphine. The opposite trend is observed for [Co(dtc)$_2$(P-P)][BF_4], which follows the order of dppf ≈ dppb < dppp < dppe [318]. This result may be explained by steric effects in which a larger chelate ring imparts a destabilizing effect on the complex. The interplay of these electronic and steric effects of diphosphines on complex reduction is significant but presently not well understood.

1.8 Biomedical Applications

The interest in diphosphine ligands and diphosphine complexes as antitumor drugs or other therapeutic agents [319–325] has recently been extended to the complexes of ferrocenyl phosphines such as dppf. This extension is inevitable as many cyclopentadienyl complexes display antitumor activity and are cytotoxic [326–329]. Variations of the diphosphine backbone and the spatial separation between the two phosphine moieties are important considerations in their pharmacological and therapeutic value. The high affinity to pyrimidyl nucleobases shown by [PtCl(dppf)(solv)]$^+$ and [Pt(dppf)(solv)$_2$]$^{2+}$ [330, 331] prompted some in vitro tests to be conducted on the cytostatic ability of [Pt(μ-Cl)(dppf)]$_2$$^{2+}$, which is active in inhibiting the growth of tumor cells [315]. In general, the complexes [M(dppf)L_2]$^{2+}$

(M = Pd, Pt) with labile ligands L and free dppf are active antiproliferating agents [332]. The antitumor activity of $[Cu_2(dppf)_2(\mu\text{-dppf})][BF_4]_2$ is also reported to be comparable to that of cisplatin [30]. By using mice-bearing ip P388 leukemia as model, both dppf and $Au_2Cl_2(\mu\text{-dppf})$ were shown to exhibit antitumor effects [54]. In general, however, the activity of free dppf and its complexes is lower than that of their dppe analogues. The activity of free dppe towards Eagle's KB cell line is comparable to that of cisplatin. The exact reason for this lower activity of dppf compared with dppe complexes and the cytostatic mechanism are unknown. A wider separation between the phosphine moieties in the former may be responsible. There is also evidence that suggests that the free ligand is the active substrate. The interplay of ligand lability, ionic dissociation, solvolytic effects, lipophilicity and transportation characters is a subject for future work.

1.9 Summary

The results described in this article form the tip of the iceberg of what could be done with the complex 1,1′-bis(diphenylphosphino)ferrocene. It is not an ordinary diphosphine ligand, as exemplified by many of the descriptions given. The recent evolution of dppf chemistry underscores the significance of other less developed ferrocenyl diphosphines. The full potential of these diphosphines will only be realized as more of their complexes are synthesized and their applications examined. The current flux of works on dppf should be extended to its close relatives, such as 1,1′-bis(dialkylarsino)ferrocenes and 1,1′-bis(dialkylphosphino)ruthenocenes. The latter in particular should have some special characteristics in view of the wider Cp ... Cp separation and different chelation bite. An examination of the recent literature on dppf chemistry springs a few surprises: (a) an Fe → M coordination bond is possible; (b) dppf can bridge some short metal-metal distances; (c) dppf displays a variety of coordination modes under very similar circumstances; and, (d) many complexes of dppf are significantly more active catalytically than other diphosphine analogues. We can expect more surprises to surface as other metallocenyl diphosphines are developed. In the catalytic reactions, the mechanisms proposed usually do not take into consideration the special traits of dppf as compared with other diphosphines. This flaw should be rectified as the coordination chemistry of dppf matures further. As one cannot realistically expect dppf to adopt a *trans* orientation relative to a metal center, the assumption of similar behavior between $[M(PPh_3)_2]$ and [M(dppf)] seems questionable. Similarly, one needs to re-examine the analogy between [M(dppf)] and [M(dppe)] or [M(dppp)] as the chelating ability of the latter two can be significantly higher. Most reported mechanisms also ignore the possibilities for the involvement of polynuclear dppf intermediates. Further studies of dppf aggregates and clusters may shed more light on this matter. Many Pd-catalyzed C-C bond formations in organic syntheses currently involve $PdCl_2(dppf)$ and $Pd(OAc)_2(dppf)$ and the assumption is that the active species generated by reducing these species with either NEt_3 or RMgX are identical. These assumptions are risky as so little is known about the structural properties of the dppf complexes of low valent palladium. The coordination and structural chemistry of $Pd(OAc)_2(PPh_3)_2$ has been surprisingly little developed. It is therefore equally speculative to draw an analogy between this complex and its dppf counterpart. The known lability of monophosphine ligands, metal association of carboxylates, and the variable coordination modes of carboxylates only make this scenario more complicated. The immediate future of catalyst design in many organic syntheses lies in a better understanding of their mechanistic pathways. This understanding will not be possible unless more fundamental coordination chemistry of dppf complexes is forthcoming.

Acknowledgements

The preparation of this manuscript was funded by the Lee Foundation and the Shaw Foundation of Singapore. The authors' research work was generously supported by the National University of Singapore (RP 850030). Contributions from our collaborators and the many able students in our laboratory, whose names appear in the cited references, have been invaluable. Our grateful thanks to the many experts in various disciplines on their work. The materials (reprints, personal communications, etc.) provided by them have been most useful; we are also grateful for their encouragement. Extra crystallographic data from Professors M. I. Bruce, D. Clemente, F. Demartin, D. S. Eggleston, S. Onaka, G. Pilloni and E. R. T. Tiekink were particularly helpful. We thank Prof. G. Pilloni for a preprint of his paper. We are honored by the invitation from the editors to make this contribution. H. K. Lee helped us to remove many chemical and linguistic ambiquities. Prof. T. Hayashi provided some invaluable comments on our draft. S. P. Neo and M. Zhou assisted in the calculations of some structural data. Efforts from B. H. Aw, H. K. Lee and P. M. N. Low in the proof-reading are gratefully acknowledged. This manuscript could not have been completed without the selfless and tireless effort of Y. P. Leong on the figures, data compilation and stenographic work.

References

[1] W. R. Cullen, J. D. Woollins, *Coord. Chem. Rev.* **1981**, *39*, 1–30.
[2] T. Hayashi, M. Kumada, *Acc. Chem. Res.* **1982**, *15*, 395–401.
[3] W. R. Cullen, S. J. Rettig, T.-C. Zheng, *Organometallics* **1992**, *11*, 277–283.
[4] A. Togni, G. Rihs, R. E. Blumer, *Organometallics* **1992**, *11*, 613–621.
[5] C. R. S. M. Hampton, I. R. Butler, W. R. Cullen, B. R. James, J. P. Charland, J. Simpson, *Inorg. Chem.* **1992**, *31*, 5509–5520.
[6] M. I. Bruce, P. A. Humphrey, O. Shawkataly, M. R. Snow, E. R. T. Tiekink, W. R. Cullen, *Organometallics* **1990**, *9*, 2910–2919.
[7] T. S. A. Hor, L.-T. Phang, *Polyhedron* **1990**, *9*, 2305–2308.
[8] T. S. A. Hor, S. P. Neo, C. S. Tan, T. C. W. Mak, K W. P. Leung, R.-J. Wang, *Inorg. Chem.* **1992**, *31*, 4510–4516.

[9] S. T. Chacon, W. R. Cullen, M. I. Bruce, O. Shawkataly, F. W. B. Einstein, R. H. Jones, A. C. Willis, *Can. J. Chem.* **1990**, *68*, 2001–2010.
[10] M. Sato, M. Sekino, *J. Organomet. Chem.* **1993**, 444, 185–190.
[11] J. A. Adeleke, L. K. Liu, *J. Chinese Chem. Soc.* **1992**, *39*, 61–65.
[12] T. Hayashi, A. Yamamoto, Y. Ito, E. Nishioka, H. Miura, K. Yanagi, *J. Am. Chem. Soc.* **1989**, *111*, 6301–6311.
[13] A. N. Hughes in *Phosphorus Chemistry*, (Eds.: E. N. Walsh, E. J. Griffith, R. W. Parry, L. D. Quin), ACS Symposium Series no. 486, ACS, Washington, DC **1992**, Chap. 14, p. 173–185.
[14] A. L. Balch in *Catalytic Aspects of Metal Phosphine Complexes* (Eds.: E. C. Alyea, D. W. Meek), Advances in Chemistry Series, ACS, Washington, DC **1982**, no. 196, Chap. 14, p. 243–256.
[15] R. J. Puddephatt, *Chem. Soc. Rev.* **1983**, *12*, 99–127.
[16] C. A. McAuliffe, *Phosphine, Arsine and Stibene Complexes of Transition Elements*, Elsevier, Amsterdam, **1979**.
[17] G. P. Sollot, J. L. Snead, S. Portnoy, W. R. Peterson, Jr., H. E. Mertwoy, *Chem. Abstr.* **1965**, *63*, 18147b.
[18] J. J. Bishop, A. Davison, M. L. Katcher, D. W. Lichtenberg, R. E. Merrill, J. C. Smart, *J. Organomet. Chem.* **1971**, *27*, 241–249.
[19] G. Marr, T. Hunt, *J. Chem. Soc. (C)* **1969**, 1070–1072.
[20] A. W. Rudie, D. W. Lichtenberg, M. L. Katcher, A. Davison, *Inorg. Chem.* **1978**, *17*, 2859–2863.
[21] K. R. Mann, W. H. Morrison, Jr., D. N. Hendrickson, *Inorg. Chem.* **1974**, *13*, 1180–1185.
[22] I. R. Butler, W. R. Cullen, T.-J. Kim, S. J. Rettig, J. Trotter, *Organometallics* **1985**, *4*, 972–980.
[23] W. H. Morrison, Jr., D. N. Hendrickson, *Inorg. Chem.* **1972**, *11*, 2912–2917.
[24] U. Casellato, D. Ajò, G. Valle, B. Corain, B. Longato, R. Graziani, *J. Crystallogr. Spectrosc. Res.* **1988**, *18*, 583–590.
[25] G. R. van Hecke, W. D. Horrocks, Jr., *Inorg. Chem.* **1966**, *5*, 1968–1974.
[26] P. K. Baker, S. G. Fraser, P. Harding, *Inorg. Chim. Acta* **1986**, *116*, L5–L6.
[27] P. K. Baker, M. V. Kampen, D. A. Kendrick, *J. Organomet. Chem.* **1991**, *421*, 241–246.
[28] A. Houlton, R. M. G. Roberts, J. Silver, R. V. Parish, *J. Organomet. Chem.* **1991**, *418*, 269–275.
[29] U. Casellato, R. Graziani, G. Pilloni, *J. Crystallogr. Spectrosc. Res.* **1993**, *23*, 571–575.
[30] G. Pilloni, R. Graziani, B. Longato, B. Corain, *Inorg. Chim. Acta* **1991**, *190*, 165–167.
[31] S.-P. Neo, T. S. A. Hor, Z.-Y. Zhou, T. C. W. Mak, *J. Organomet. Chem.* **1994**, *464*, 113–119.
[32] L.-T. Phang, T. S. A. Hor, Z.-Y. Zhou, T. C. W. Mak, *J. Organomet. Chem.* **1994**, *469*, 253–261.
[33] Y. K. Yan, H. S. O. Chan, T. S. A. Hor, K.-L. Tan, L.-K. Liu, Y.-S. Wen, *J. Chem. Soc., Dalton Trans.* **1992**, 423–426.
[34] P. Kalck, C. Randrianalimanana, M. Ridmy, A. Thorez, H. T. Dieck, J. Ehlers, *New J. Chem.* **1988**, *12*, 679–686.
[35] K.-W. Lee, W. T. Pennington, A. W. Cordes, T. L. Brown, *Organometallics* **1984**, *3*, 404–413.
[36] T. S. A. Hor, L.-T. Phang, L.-K. Liu, Y.-S. Wen, *J. Organomet. Chem.* **1990**, *397*, 29–39.
[37] J. D. Unruh, W. J. Wells, III, DE 2617306, **1976** [*Chem. Abstr.* **1977**, *86*, 22369e].
[38] T. A. George, R. C. Tisdale, *J. Am. Chem. Soc.* **1985**, *107*, 5157–5159.
[39] P. A. W. Dean, J. J. Vittal, R. S. Srivastava, *Can. J. Chem.* **1987**, *65*, 2628–2633.
[40] M. Sato, M. Sekino, S. Akabori, *J. Organomet. Chem.* **1988**, *344*, C31–C34.
[41] M. Sato, H. Shigeta, M. Sekino, S. Akabori, *J. Organomet. Chem.* **1993**, *458*, 199–204.
[42] T. S. A. Hor, H. S. O. Chan, K.-L. Tan, L.-T. Phang, Y. K. Yan, L.-K. Liu, Y.-S. Wen, *Polyhedron* **1991**, *10*, 2437–2450.
[43] B. Corain, B. Longato, G. Favero, D. Ajò, G. Pilloni, U. Russo, F. R. Kreissl, *Inorg. Chim. Acta* **1989**, *157*, 259–266.
[44] T. S. A. Hor, L.-T. Phang, *J. Organomet. Chem.* **1989**, *373*, 319–324.
[45] T. S. A. Hor, L.-T. Phang, *J. Organomet. Chem.* **1990**, *381*, 121–125.
[46] T. S. A. Hor, L.-T. Phang, *J. Organomet. Chem.* **1990**, *390*, 345–350.

[47] L.-T. Phang, S. C. F. Au-Yeung, T. S. A. Hor, S. B. Khoo, Z.-Y. Zhou, T. C. W. Mak, *J. Chem. Soc., Dalton Trans.* **1993**, 165–172.
[48] L.-T. Phang, K.-S. Gan, H. K. Lee, T. S. A. Hor, *J. Chem. Soc., Dalton Trans.* **1993**, 2697–2702.
[49] J. J. Daly, *J. Chem. Soc.* **1964**, 3799–3810.
[50] T. S. A. Hor, L.-K. Liu, T. C. W. Mak, unpublished results.
[51] T.-J. Kim, S.-C. Kwon, Y.-H. Kim, N. H. Heo, M. T. Teeter, A. Yamano, *J. Organomet. Chem.* **1991**, *426*, 71–86.
[52] S. Onaka, A. Mizuno, S. Takagi, *Chem. Lett.* **1989**, 2037–2040.
[53] S. Onaka, T. Moriya, S. Takagi, A. Mizuno, H. Furuta, *Bull. Chem. Soc. Jpn.* **1992**, *65*, 1415–1427.
[54] D. T. Hill, G. R. Girard, F. L. McCabe, R. K. Johnson, P. D. Stupik, J. H. Zhang, W. M. Reiff, D. S. Eggleston, *Inorg. Chem.* **1989**, *28*, 3529–3533.
[55] S. M. Draper, C. E. Housecroft, A. L. Rheingold, *J. Organomet. Chem.* **1992,** *435*, 9–20.
[56] M. I. Bruce, I. R. Butler, W. R. Cullen, G. A. Koutsantonis, M. R. Snow, E. R. T. Tiekink, *Aust. J. Chem.* **1988**, *41*, 963–969.
[57] T. Hayashi, M. Konishi, Y. Kobori, M. Kumada, T. Higuchi, K. Hirotsu, *J. Am. Chem. Soc.* **1984**, *106*, 158–163.
[58] P. F. Kelly, A. M. Z. Slawin, D. J. Williams, J. D. Woollins, *Polyhedron* **1988**, *7*, 1925–1930.
[59] T.-J. Kim, K.-H. Kwon, S.-C. Kwon, J.-O. Baeg, S.-C. Shim, D.-H. Lee, *J. Organomet. Chem.* **1990**, *389*, 205–217.
[60] T. M. Miller, K. J. Ahmed, M. S. Wrighton, *Inorg. Chem.* **1989**, *28*, 2347–2355.
[61] C. E. Housecroft, S. M. Owen, P. R. Raithby, B. A. M. Shaykh, *Organometallics* **1990**, *9*, 1617–1623.
[62] A. L. Bandini, G. Banditelli, M. A. Cinellu, G. Sanna, G. Minghetti, F. Demartin, M. Manassero, *Inorg. Chem.* **1989**, *28*, 404–410.
[63] B. S. Haggerty, C. E. Housecroft, A. L. Rheingold, B. A. M. Shaykh, *J. Chem. Soc., Dalton Trans.* **1991**, 2175–2184.
[64] M. Zhou, Y. Xu, L.-L. Koh, K. F. Mok, P.-H. Leung, T. S. A. Hor, unpublished results.
[65] U. Casellato, B. Corain, R. Graziani, B. Longato, G. Pilloni, *Inorg. Chem.* **1990**, *29*, 1193–1198.
[66] S. Onaka, *Bull. Chem. Soc. Jpn.* **1986**, *59*, 2359–2361.
[67] J. A. Connor, J. P. Day, E. M. Jones, G. K. McEwen, *J. Chem. Soc., Dalton Trans.* **1973**, 347–354.
[68] G. T. Andrews, I. J. Colquhoun, W. McFarlane, *Polyhedron* **1983**, *2*, 783–790.
[69] T. S. A. Hor, K.-S. Gan, Y. L. Yong, unpublished results.
[70] K. C. Tan, B. Sc. Honours Thesis **1991**, National University of Singapore.
[71] T. S. A. Hor, M. O. Wong, C. H. Chin, unpublished results.
[72] A. R. Sanger, *J. Chem. Soc., Dalton Trans.* **1977**, 1971–1976.
[73] R. D. Kelly, G. B. Young, *Polyhedron* **1989**, *8*, 433–445.
[74] S.-L. Lam, Y.-X. Cui, S. C. F. Au-Yeung, Y.-K. Yan, T. S. A. Hor, *Inorg. Chem.* **1994**, *33*, 2407–2412.
[75] A. Houlton, P. T. Bishop, R. M. G. Roberts, J. Silver, M. Herberhold, *J. Organomet. Chem.* **1989**, *364*, 381–389.
[76] A. Houlton, S. K. Ibrahim, J. R. Dilworth, J. Silver, *J. Chem. Soc., Dalton Trans.* **1990**, 2421–2424.
[77] R. M. G. Roberts, J. Silver, I. E. G. Morrison, *J. Organomet. Chem.* **1981**, *209*, 385–391.
[78] M. Watanabe, H. Ichikawa, I. Motoyama, H. Sano, *Bull. Chem. Soc. Jpn.* **1983**, *56*, 3291–3293.
[79] D. A. Clemente, G. Pilloni, B. Corain, B. Longato, M. Tiripicchio-Camellini, *Inorg. Chim. Acta* **1986**, *115*, L9–L11.
[80] H. S. O. Chan, T. S. A. Hor, L.-T. Phang, K.-L. Tan, *J. Organomet. Chem.* **1991**, *407*, 353–357.
[81] B. Longato, G. Pilloni, R. Graziani, U. Casellato, *J. Organomet. Chem.* **1991**, *407*, 369–376.
[82] E. W. Abel, M. Booth, K. G. Orrell, *J. Organomet. Chem.* **1981**, *208*, 213–224.
[83] E. W. Abel, M. Booth, C. A. Brown, K. G. Orrell, R. L. Woodford, *J. Organomet. Chem.* **1981**, *214*, 93–105.

[84] A. Davison, J. C. Smart, *J. Organomet. Chem.* **1979**, *174*, 321–334.
[85] M. P. Anderson, L. H. Pignolet, *Inorg. Chem.* **1981**, *20*, 4101–4107.
[86] T. Hayashi, M. Konishi, M. Kumada, *Tetrahedron Lett.* **1979**, 1871–1874.
[87] T. Hayashi, M. Konishi, M. Kumada, *J. Organomet. Chem.* **1980**, *186*, C1–C4.
[88] A. Yamamoto, *Organotransition Metal Chemistry*, John Wiley, New York, **1986**, p. 374–390.
[89] J. K. Kochi, *Organometallic Mechanism and Catalysis*, Academic Press, New York, **1987**.
[90] G. W. Parshall, S. D. Ittel, *Homogeneous Catalysis*, 2nd ed., John Wiley, New York, **1992**, Chap. 7, p. 177–180.
[91] J. M. Brown, N. A. Cooley, *J. Chem. Soc., Chem. Commun.* **1988**, 1345–1347.
[92] J. M. Brown, N. A. Cooley, *Organometallics* **1990**, *9*, 353–359.
[93] J. M. Brown, N. A. Cooley, *Phil. Trans. R. Soc. Lond.* **1988**, *A 326*, 587–594.
[94] J. M. Brown, N. A. Cooley, D. W. Price, *J. Chem. Soc., Chem. Commun.* **1989**, 458–460.
[95] J. M. Brown, S. J. Cook, S. J. Kimber, *J. Organomet. Chem.* **1984**, *269*, C58–C60.
[96] D. Parker, R. J. Taylor, *J. Chem. Soc., Chem. Commun.* **1987**, 1781–1783.
[97] D. E. Bergstrom, M. K. Ogawa, *J. Am. Chem. Soc.* **1978**, *100*, 8106–8112.
[98] R. H. Grubbs, A. Miyashita, M. Liu, P. Burk, *J. Am. Chem. Soc.* **1978**, *100*, 2418–2425.
[99] D. L. Reger, E. C. Culbertson, *J. Am. Chem. Soc.* **1976**, *98*, 2789–2794.
[100] F. Ozawa, T. Ito, Y. Nakamura, A. Yamamoto, *Bull. Chem. Soc. Jpn.* **1981**, *54*, 1868–1880.
[101] F. Ozawa, T. Ito, A. Yamamoto, *J. Am. Chem. Soc.* **1980**, *102*, 6457–6463.
[102] M. K. Loar, J. K. Stille, *J. Am. Chem. Soc.* **1981**, *103*, 4174–4181.
[103] K. Tatsumi, R. Hoffmann, A. Yamamoto, J. K. Stille, *Bull. Chem. Soc. Jpn.* **1981**, *54*, 1857–1867.
[104] A. Gillie, J. K. Stille, *J. Am. Chem. Soc.* **1980**, *102*, 4933–4941.
[105] T. Katayama, M. Umeno, *Chem. Lett.* **1991**, 2073–2076.
[106] T. Katayama, Y. Uchibori, M. Umeno, *Jpn. Kokai Tokkyo Koho JP 04356431*, **1992 Heisei** 6 pp.
[107] T. Hayashi, M. Konishi, K.-I. Yokota, M. Kumada, *J. Organomet. Chem.* **1985**, *285*, 359–373.
[108] T. Hayashi, M. Konishi, K.-I. Yokota, M. Kumada, *J. Chem. Soc., Chem. Commun.* **1981**, 313–314.
[109] Y. Ito, M. Inouye, M. Murakami, *Tetrahedron Lett.* **1988**, *29*, 5379–5382.
[110] K. Tamao, T. Iwahara, R. Kanatani, M. Kumada, *Tetrahedron Lett.* **1984**, *25*, 1909–1912.
[111] N. Jayasuriya, J. Kagan, *Heterocycles* **1986**, *24*, 2901–2904.
[112] L. N. Pridgen, L. B. Killmer, *J. Org. Chem.* **1981**, *46*, 5402–5404.
[113] A. Minato, K. Tamao, T. Hayashi, K. Suzuki, M. Kumada, *Tetrahedron Lett.* **1981**, *22*, 5319–5322.
[114] V. N. Kalinin, *Synthesis (J. Synth. Org. Chem.)* **1992**, *5*, 413–432.
[115] F.-T. Luo, J.-H. Lai, J.-C. Shaeh, *Bull. Inst. Chem., Academia Sinica* **1987**, 17–22.
[116] F. Orsini, F. Pelizzoni, L. M. Vallarino, *J. Organomet. Chem.* **1989**, *367*, 375–382.
[117] R. A. Grey, *J. Org. Chem.* **1984**, *49*, 2288–2289.
[118] M. Murakami, H. Ito, W. A. W. A. Bakar, A. B. Baba, Y. Ito, *Chem. Lett.* **1989**, 1603–1606.
[119] K. Asao, H. Iio, T. Tokoroyama, *Tetrahedron Lett.* **1989**, *30*, 6401–6404.
[120] Y. Okamoto, K. Yoshioka, T. Yamana, H. Mori, *J. Organomet. Chem.* **1989**, *369*, 285–290.
[121] N. Miyaura, T. Ishiyama, H. Sasaki, M. Ishikawa, M. Satoh, A. Suzuki, *J. Am. Chem. Soc.* **1989**, *111*, 314–321.
[122] Y. Hoshino, T. Ishiyama, N. Miyaura, A. Suzuki, *Tetrahedron Lett.* **1988**, *29*, 3983–3986.
[123] M. Satoh, N. Miyaura, A. Suzuki, *Chem. Lett.* **1986**, 1329–1332.
[124] N. Miyaura, M. Satoh, A. Suzuki, *Tetrahedron Lett.* **1986**, *27*, 3745–3748.
[125] N. Miyaura, K. Yamada, H. Suginome, A. Suzuki, *J. Am. Chem. Soc.* **1985**, *107*, 972–980.
[126] N. Miyaura, T. Ishiyama, M. Ishikawa, A. Suzuki, *Tetrahedron Lett.* **1986**, *27*, 6369–6372.
[127] N. Miyaura, M. Ishikawa, A. Suzuki, *Tetrahedron Lett.* **1992**, *33*, 2571–2574.
[128] T. Oh-e, N. Miyaura, A. Suzuki, *J. Org. Chem.* **1993**, *58*, 2201–2208.
[129] L. Schmitz, M. Rehahn, M. Ballauff, *Polymer* **1993**, *34*, 646–649.
[130] M. Sato, N. Miyaura, A. Suzuki, *Chem. Lett.* **1989**, 1405–1408.
[131] T. Yanagi, T. Oh-e, N. Miyaura, A. Suzuki, *Bull. Chem. Soc. Jpn.* **1989**, *62*, 3892–3895.
[132] W. J. Thompson, J. Gaudino, *J. Org. Chem.* **1984**, *49*, 5237–5243.
[133] W. J. Thompson, J. H. Jones, P. A. Lyle, J. E. Thies, *J. Org. Chem.* **1988**, *53*, 2052–2055.

[134] J. Godschalx, J. K. Stille, *Tetrahedron Lett.* **1980**, *21*, 2599–2602; D. Milstein, J. K. Stille, *J. Am. Chem. Soc.* **1978**, *100*, 3636–3638; W. J. Scott, G. T. Crisp, J. K. Stille, *J. Am. Chem. Soc.* **1984**, *106*, 4630–4632.
[135] R. F. Heck, *Palladium Reagents in Organic Syntheses*, Academic Press, New York, **1985**, Chap. 6, p. 293.
[136] B. M. Trost, T. R. Verhoeven in *Comprehensive Organometallic Chemistry* (Eds.: G. Wilkinson, F. G. A. Stone, E. W. Abel), Pergamon Press, Oxford, **1982**, Vol. 8, Chap. 57, p. 799–938.
[137] A. Sekiya, N. Ishikawa, *J. Organomet. Chem.* **1976**, *118*, 349–354.
[138] P. Fitton, M. P. Johnson, J. E. McKeon, *J. Chem. Soc., Chem. Commun.* **1968**, 6–7.
[139] V. Farina, B. Krishnan, *J. Am. Chem. Soc.* **1991**, *113*, 9585–9595.
[140] A. W. Parkins, R. C. Poller, *An introduction to Organometallic Chemistry*, MacMillan, London, **1986**, Chap. 8, p. 182–185.
[141] I. P. Beletskaya, N. A. Bumagin in *Fundamental Research in Homogeneous Catalysis: Proceedings of the 4th International Symposium on Homogeneous Catalysis, Leningrad, U.S.S.R., Sept. 24–28, 1984* (Ed.: A. E. Shilov), Gordon and Breach, New York, **1986**, p. 281–296.
[142] J. R. Stille, R. H. Grubbs, *J. Am. Chem. Soc.* **1983**, *105*, 1664–1665
[143] S. Inaba, R. D. Rieke, *Tetrahedron Lett.* **1983**, *24*, 2451–2452.
[144] T. Sato, T. Itoh, T. Fujisawa, *Chem. Lett.* **1982**, 1559–1560.
[145] D. Milstein, J. K. Stille, *J. Org. Chem.* **1979**, *44*, 1613–1618.
[146] J. K. Stille, *Angew. Chem. Int. Ed. Engl.* **1986**, *25*, 508–524.
[147] J. K. Stille, *Pure Appl. Chem.* **1985**, *57*, 1771–1780.
[148] N. Tamayo, A. M. Echavarren, M. C. Paredes, *J. Org. Chem.* **1991**, *56*, 6488–6491.
[149] M. Watanabe, M. Date, S. Furukawa, *Chem. Pharm. Bull.* **1989**, *37*, 292–297.
[150] H. Ikushima, S. Takase, Y. Kawai, Y. Itoh, M. Okamoto, H. Tanaka, H. Imanaka, *Agri. Biol. Chem.* **1983**, *47*, 2231–2235.
[151] H. Ikushima, M. Okamoto, H. Tanaka, O. Ohe, M. Kohsaka, H. Aoki, H. Imanaka, *J. Antibiot.* **1980**, *33*, 1107–1113.
[152] E. Vedejs, A. R. Haight, W. O. Moss, *J. Am. Chem. Soc.* **1992**, *114*, 6556–6558.
[153] A. M. Echavarren and J. K. Stille, *J. Am. Chem. Soc.* **1987**, *109*, 5478–5486.
[154] D. R. McKean, G. Parrinello, A. F. Renaldo, J. K. Stille, *J. Org. Chem.* **1987**, *52*, 422–424.
[155] E. Nakamura, S. Aoki, K. Sekiya, H. Oshino, I. Kuwajima, *J. Am. Chem. Soc.* **1987**, *109*, 8056–8066.
[156] G. A. Peterson, F.-A. Kunng, J. S. McCallum, W. D. Wulff, *Tetrahedron Lett.* **1987**, *28*, 1381–1384.
[157] M. E. Wright, *Organometallics* **1989**, *8*, 407–411.
[158] D. Milstein, J. K. Stille, *J. Am. Chem. Soc.* **1979**, *101*, 4992–4998.
[159] T. Kobayashi, T. Sakakura, M. Tanaka, *Tetrahedron Lett.* **1985**, *26*, 3463–3466.
[160] N. Tamayo, A. M. Echavarren, M. C. Paredes, F. Fariña, P. Noheda, *Tetrahedron Lett.* **1990**, *31*, 5189–5192.
[161] F.-T. Luo, R.-T. Wang, *Tetrahedron Lett.* **1991**, *32*, 7703–7706.
[162] M. Pérez, A. M. Castaño, A. M. Echavarren, *J. Org. Chem.* **1992**, *57*, 5047–5049.
[163] Y. Ito, M. Inouye, M. Murakami, *Chem. Lett.* **1989**, 1261–1264.
[164] T. Ishiyama, N. Miyaura, A. Suzuki, *Synlett* **1991**, 687–688.
[165] C. Carfagna, A. Musco, G. Sallese, R. Santi, T. Fiorani, *J. Org. Chem.* **1991**, *56*, 261–263.
[166] T. Y. Shen, *Angew. Chem. Int. Ed. Engl.* **1972**, *11*, 460–472.
[167] L. Cassar, M. Foá, F. Montanari, G. P. Marinelli, *J. Organomet. Chem.* **1979**, *173*, 335–339.
[168] A. Sekiya, N. Ishikawa, *Chem. Lett.* **1975**, 277–278.
[169] K. Takagi, T. Okamoto, Y. Sakakibara, S. Oka, *Chem. Lett.* **1973**, 471–474.
[170] K. Takagi, T. Okamoto, Y. Sakakibara, A. Ohno, S. Oka, N. Hayama, *Bull. Chem. Soc. Jpn.* **1976**, *49*, 3177–3180.
[171] G. P. Ellis, T. M. Romney-Alexander, *Chem. Rev.* **1987**, *87*, 779–794.
[172] Y. Sakakibara, F. Okuda, A. Shimobayashi, K. Kirino, M. Sakai, N. Uchino, K. Takagi, *Bull. Chem. Soc. Jpn.* **1988**, *61*, 1985–1990.
[173] K. Takagi, K. Sasaki, Y. Sakakibara, *Bull. Chem. Soc. Jpn.* **1991**, *64*, 1118–1121.

[174] R. F. Heck, *Pure Appl. Chem.* **1981**, *53*, 2323–2332.
[175] R. F. Heck, *Acc. Chem. Res.* **1979**, *12*, 146–151.
[176] W. Cabri, I. Candiani, S. DeBernardinis, F. Francalanci, P. Sergio, *J. Org. Chem.* **1991**, *56*, 5796–5800.
[177] F. Ozawa, A. Kubo, T. Hayashi, *J. Am. Chem. Soc.* **1991**, *113*, 1417–1419.
[178] J. S. Brumbaugh, R. R. Whittle, M. Parvez, A. Sen, *Organometallics* **1990**, *9*, 1735–1747.
[179] E. G. Samsel, J. R. Norton, *J. Am. Chem. Soc.* **1984**, *106*, 5505–5512.
[180] Y. Ben-David, M. Portnoy, M. Gozin, D. Milstein, *Organometallics* **1992**, *11*, 1995–1996.
[181] W. Cabri, I. Candiani, A. Bedeschi, S. Penco, R. Santi, *J. Org. Chem.* **1992**, *57*, 1481–1486.
[182] W. Cabri, I. Candiani, A. Bedeschi, R. Santi, *J. Org. Chem.* **1992**, *57*, 3558–3563.
[183] G. S. Jones, W. J. Scott, *J. Am. Chem. Soc.* **1992**, *114*, 1491–1492.
[184] A. Arcadi, S. Cacchi, F. Marinelli, E. Morera, G. Ortar, *Tetrahedron* **1990**, *46*, 7151–7164.
[185] H. M. Colquhoun, J. Holton, D. J. Thompson, M. V. Twigg, *New Pathways for Organic Synthesis*, Plenum, New York, **1984**, p. 195.
[186] T. Kobayashi, M. Tanaka, *J. Chem. Soc., Chem. Commun.* **1981**, 333–334.
[187] T. Hiyama, N. Wakasa, T. Ueda, T. Kusumoto, *Bull. Chem. Soc. Jpn.* **1990**, *63*, 640–642.
[188] T. Ohta, H. Takaya, M. Kitamura, K. Nagai, R. Noyori, *J. Org. Chem.* **1987**, *52*, 3174–3176.
[189] K. Mori, T. Mizoroki, A. Ozaki, *Chem. Lett.* **1975**, 39–42.
[190] K. Itoh, M. Miura, M. Nomura, *Tetrahedron Lett.* **1992**, *33*, 5369–5372
[191] S. Torii, L. H. Xu, H. Okumoto, *Synlett* **1991**, 695–696.
[192] E. Wollenweber, M. Jay in *The Flavonoids. Advances in Research Since 1980* (Ed.: J. B. Harborne), Chapman & Hall, London, **1988**, Chap. 7, p. 233–302.
[193] J. D. Hepworth in *Comprehensive Heterocyclic Chemistry* (Eds.: A. R. Katritzky, C. W. Rees, A. J. Boulton, A. McKillop), Pergamon Press, Oxford, **1984**, Vol. 3, part 2B, Chap. 2.24, p. 874.
[194] V. N. Kalinin, M. V. Shostakovsky, A. B. Ponomaryov, *Tetrahedron Lett.* **1990**, *31*, 4073–4076.
[195] P. G. Ciattini, E. Morera, G. Ortar, S. S. Rossi, *Tetrahedron* **1991**, *47*, 6449–6456.
[196] R. F. Heck, *Palladium Reagents in Organic Synthesis*, Academic Press, New York, **1985**, Chap. 8, p. 359.
[197] L. S. Hegedus in *The Chemistry of the Metal-Carbon Bond* (Ed.: F. R. Hartley and S. Patai), John Wiley & Sons, New York, **1985**, p. 431.
[198] G. P. C. M. Dekker, A. Buijs, C. J. Elsevier, K. Vrieze, P. W. N. M. van Leeuwen, W. J. J. Smeets, A. L. Spek, Y. F. Wang, C. H. Stam, *Organometallics* **1992**, *11*, 1937–1948.
[199] Z.-W. An, M. Catellani, G. P. Chiusoli, *J. Organomet. Chem.* **1990**, *397*, 371–373.
[200] V. N. Kalinin, M. V. Shostakovsky, A. B. Ponomaryov, *Tetrahedron Lett.* **1992**, *33*, 373–376.
[201] T. Osawa, H. Ohta, K. Akimoto, K. Harada, H. Soga, Y. Jinno, EP 343574, **1989** [*Chem. Abstr.* **1990**, *112*, 235197q].
[202] H. Ueda, H. Miyamoto, H. Yamashita, H. Tone, EP 287951, **1988** [*Chem. Abstr.* **1989**, *110*, 173109k].
[203] R. F. Heck, *Palladium Reagents in Organic Syntheses*, Academic Press, New York, **1985**, Chap. 8, p. 348.
[204] O. Ohe, K. Ohe, S. Uemura, N. Sugita, *J. Organomet. Chem.* **1988**, *344*, C5–C7.
[205] A. M. Echavarren, J. K. Stille, *J. Am. Chem. Soc.* **1988**, *110*, 1557–1565.
[206] R. E. Dolle, S. J. Schmidt, L. I. Kruse, *J. Chem. Soc., Chem. Commun.* **1987**, 904–905.
[207] S. Cacchi, P. G. Ciattini, E. Morera, G. Ortar, *Tetrahedron Lett.* **1986**, *27*, 3931–3934.
[208] C. B. Ziegler, Jr., R. F. Heck, *J. Org. Chem.* **1978**, *43*, 2941–2946.
[209] K. Itoh, H. Hashimoto, M. Miura, M. Nomura, *J. Mol. Cat.* **1990**, *59*, 325–332.
[210] R. Takeuchi, M. Sugiura, N. Ishii, N. Sato, *J. Chem. Soc., Chem. Commun.* **1992**, 1358–1359.
[211] M. R. Najafi, M.-L. Wang, G. Zweifel, *J. Org. Chem.* **1991**, *56*, 2468–2476.
[212] M. P. Cooke, Jr., *J. Org. Chem.* **1987**, *52*, 5729–5733.
[213] R. Takeuchi, N. Ishii, M. Sugiura, N. Sato, *J. Org. Chem.* **1992**, *57*, 4189–4194.
[214] T. Fuchikami, K. Ohishi, I. Ojima, *J. Org. Chem.* **1983**, *48*, 3803–3807.
[215] S. Cacchi, A. Lupi, *Tetrahedron Lett.* **1992**, *33*, 3939–3942.
[216] H. Urata, H. Yugari, T. Fuchikami, *Chem. Lett.* **1987**, 833–836.

[217] H. Urata, O. Kosukegawa, Y. Ishii, H. Yugari, T. Fuchikami, *Tetrahedron Lett.* **1989**, *30*, 4403–4406.
[218] F. Ozawa, N. Kawasaki, H. Okamoto, T. Yamamoto, A. Yamamoto, *Organometallics* **1987**, *6*, 1640–1651.
[219] L. Huang, F. Ozawa, A. Yamamoto, *Organometallics* **1990**, *9*, 2603–2611.
[220] L. Huang, F. Ozawa, A. Yamamoto, *Organometallics* **1990**, *9*, 2612–2620
[221] F. Ozawa, H. Soyama, H. Yanagihara, I. Aoyama, H. Takino, K. Izawa, T. Yamamoto, A. Yamamoto, *J. Am. Chem. Soc.* **1985**, *107*, 3235–3245.
[222] T. Kobayashi, F. Abe, M. Tanaka, *J. Mol. Cat.* **1988**, *45*, 91–109.
[223] T. Kobayashi, M. Tanaka, *J. Organomet. Chem.* **1982**, *231*, C12–C14.
[224] T. Kobayashi, M. Tanaka, *Tetrahedron Lett.* **1986**, *27*, 4745–4748.
[225] T. Ishiyama, N. Miyaura, A. Suzuki, *Bull. Chem. Soc. Jpn.* **1991**, *64*, 1999–2001.
[226] T. Tsuda, R. Sumiya, T. Saegusa, *Synth. Commun.* **1987**, *17*, 147–154.
[227] A. M. Castaño, A. M. Echavarren, *Tetrahedron Lett.* **1990**, *31*, 4783–4786.
[228] S.-I. Murahashi, Y. Taniguchi, Y. Imada, Y. Tanigawa, *J. Org. Chem.* **1989**, *54*, 3292–3303.
[229] T. Hayashi, A. Yamamoto, T. Hagihara, *J. Org. Chem.* **1986**, *51*, 723–727.
[230] Y. Ito, M. Sawamura, M. Matsuoka, Y. Matsumoto, T. Hayashi, *Tetrahedron Lett.* **1987**, *28*, 4849–4852.
[231] B. M. Trost, T. S. Scanlan, *J. Am. Chem. Soc.* **1989**, *111*, 4988–4990.
[232] T. Tokuyama, J. W. Daly, R. J. Highet, *Tetrahedron* **1984**, *40*, 1183–1190.
[233] B. M. Trost, L. Zhi, *Tetrahedron Lett.* **1992**, *33*, 1831–1834.
[234] C. Carfagna, R. Galarini, A. Musco, R. Santi, *J. Mol. Cat.* **1992**, *72*, 19–27.
[235] C. Carfagna, L. Mariani, A. Musco, G. Sallese, R. Santi, *J. Org. Chem.* **1991**, *56*, 3924–3927.
[236] E. Alvarez, T. Cuvigny, M. Julia, *J. Organomet. Chem.* **1988**, *339*, 199–212.
[237] M. Yoneyama, M. Kakimoto, Y. Imai, *Macromolecules* **1989**, *22*, 2593–2596.
[238] M. Yoneyama, M. Kakimoto, Y. Imai, *Macromolecules* **1989**, *22*, 4152–4155.
[239] M. Yoneyama, M. Kakimoto, Y. Imai, *J. Polymer Sci.: Part A: Polymer Chem.* **1989**, *27*, 1985–1992.
[240] R. Galarini, A. Musco, R. Pontellini, A. Bolognesi, S. Destri, M. Catellani, M. Mascherpa, G. Zhuo, *J. Chem. Soc., Chem. Commun.* **1991**, 364–365.
[241] E. Cramer, V. Percec, *J. Polymer Sci.: Part A: Polymer Chem.* **1990**, *28*, 3029–3046.
[242] O. R. Hughes, J. D. Unruh, *J. Mol. Catal.* **1981**, *12*, 71–83.
[243] O. R. Hughes, D. A. Young, *J. Am. Chem. Soc.* **1981**, *103*, 6636–6642.
[244] C. P. Casey, G. T. Whiteker, M. G. Melville, L. M. Petrovich, J. A. Gavney, Jr., D. R. Powell, *J. Am. Chem. Soc.* **1992**, *114*, 5535–5543.
[245] T. S. A. Hor, L.-T. Phang, *Bull. Sing. N. I. Chem.* **1990**, *18*, 29–46, and references therein.
[246] J. D. Unruh, J. R. Christenson, *J. Mol. Cat.* **1982**, *14*, 19–34.
[247] C. P. Casey, G. T. Whiteker, *Isr. J. Chem.* **1992**, *114*, 5535–5543.
[248] C. U. Pittman, Jr., W. D. Honnick, J. J. Yang, *J. Org. Chem.* **1980**, *45*, 684–689.
[249] C. U. Pittman, Jr., W. D. Honnick, *J. Org. Chem.* **1980**, *45*, 2132–2139.
[250] P. W. N. M. van Leeuwen, C. F. Roobeek, *J. Mol. Cat.* **1985**, *31*, 345–353.
[251] S. Gladiali, L. Pinna, *Tetrahedron: Asymmetry* **1990**, *1*, 693–696.
[252] D. Zargarian, H. Alper, *Organometallics* **1993**, *12*, 712–724.
[253] Y. Kou, Y. Yin, *J. Mol. Cat. (China)* **1989**, *8*, 262–267.
[254] C. U. Pittman, Jr., G. M. Wilemon, *J. Org. Chem.* **1981**, *46*, 1901–1905.
[255] J. F. Knifton, *J. Mol. Cat.* **1988**, *47*, 99–116.
[256] W. R. Cullen, N. F. Han, *Appl. Organomet. Chem.* **1987**, *1*, 1–6.
[257] N. W. Alcock, J. M. Brown, M. Rose, A. Wienand, *Tetrahedron: Asymmetry* **1991**, *2*, 47–50.
[258] I. R. Butler, W. R. Cullen, N. F. Han, F. G. Herring, N. R. Jagannathan, J. Ni, *Appl. Organomet. Chem.* **1988**, *2*, 263–275.
[259] I. R. Butler, W. R. Cullen, T.-J. Kim, F. W. B. Einstein, T. Jones, *J. Chem. Soc., Chem. Commun.* **1984**, 719–721.
[260] W. R. Cullen, T.-J. Kim, F. W. B. Einstein, T. Jones, *Organometallics* **1985**, *4*, 346–351.
[261] T.-J. Kim, K.-C. Lee, *Bull. Korean Chem. Soc.* **1989**, *10*, 279–282.
[262] R. H. Fish, J. L. Tan, A. D. Thormodsen, *J. Org. Chem.* **1984**, *49*, 4500–4505.
[263] R. H. Fish, J. L. Tan, A. D. Thormodsen, *Organometallics* **1985**, *4*, 1743–1747.

[264] P. L. Castle, D. A. Widdowson, *Tetrahedron Lett.* **1986**, *27*, 6013–6016.
[265] K. Yuan, W. J. Scott, *J. Org. Chem.* **1990**, *55*, 6188–6194.
[266] S. Cacchi, P. G. Ciattini, E. Morera, G. Ortar, *Tetrahedron Lett.* **1986**, *27*, 5541–5544.
[267] W. Cabri, S. De Bernardinis, F. Francalanci, S. Penco, *J. Chem. Soc., Perkin Trans.* **1990**, 428–429.
[268] W. Cabri, S. De Bernardinis, F. Francalanci, S. Penco, R. Santi, *J. Org. Chem.* **1990**, *55*, 350–353.
[269] A. Suzuki, *Pure Appl. Chem.* **1985**, *57*, 1749–1758.
[270] A. Suzuki, *Acc. Chem. Res.* **1982**, *15*, 178–184.
[271] R. B. Miller, S. Dugar, *Organometallics* **1984**, *3*, 1261–1263.
[272] N. Miyaura, T. Ishiyama, H. Sasaki, M. Ishikawa, M. Satoh, A. Suzuki, *J. Am. Chem. Soc.* **1989**, *111*, 314–321.
[273] I. D. Gridnev, N. Miyaura, A. Suzuki, *Organometallics* **1993**, *12*, 589–592.
[274] Y. Matsumoto, M. Naito, T. Hayashi, *Organometallics* **1992**, *11*, 2732–2734.
[275] J. M. Brown, G. C. Lloyd-Jones, *Tetrahedron: Asymmetry* **1990**, *1*, 869–872.
[276] Y. Matsumoto, T. Hayashi, *Tetrahedron Lett.* **1991**, *32*, 3387–3390.
[277] J. C. Saam, J. L. Speier, *J. Am. Chem. Soc.* **1958**, *80*, 4104–4106.
[278] L. N. Lewis, *J. Am. Chem. Soc.* **1990**, *112*, 5998–6004.
[279] L. N. Lewis, N. Lewis, *Chem. Mater.* **1989**, *1*, 106–114.
[280] L. N. Lewis, N. Lewis, *J. Am. Chem. Soc.* **1986**, *108*, 7228–7231.
[281] I. Ojima in *The Chemistry of Organic Silicon Compounds*, Part 2 (Eds.: S. Patai, Z. Rappoport), John Wiley, New York, **1989**, Chap. 25, p. 1479–1526.
[282] I. Ojima, *Pure and Appl. Chem.* **1984**, *56*, 99–110.
[283] J. L. Speier, *Adv. Organomet. Chem.* **1979**, *17*, 407–447.
[284] W. R. Cullen, S. V. Evans, N. F. Han, J. Trotter, *Inorg. Chem.* **1987**, *26*, 514–519.
[285] T. Hayashi, K. Tamao, Y. Katsuro, I. Nakae, M. Kumada, *Tetrahedron Lett.* **1980**, 1871–1874.
[286] Y. Kiso, M. Kumada, K. Tamao, M. Umeno, *J. Organomet. Chem.* **1973**, *50*, 297–310.
[287] W. R. Cullen, N. F. Han, *J. Organomet. Chem.* **1987**, *333*, 269–280.
[288] G. W. Parshall, S. D. Ittel, *Homogeneous Catalysis*, 2nd ed., John Wiley, New York, **1992**, Chap. 2, p. 9–24.
[289] P. N. Rylander, *Organic Syntheses with Noble Metal Catalysts*, Academic Press, New York, **1973**, Chap. 5, p. 145–147.
[290] J. K. Stille, Y. Becker, *J. Org. Chem.* **1980**, *45*, 2139–2145.
[291] H. Alper, K. Hachem, *J. Org. Chem.* **1980**, *45*, 2269–2271.
[292] D. Bingham, B. Hudson, D. Webster, P. B. Wells, *J. Chem. Soc., Dalton Trans.* **1974**, 1521–1524.
[293] F. G. Cowherd, J. L. von Rosenberg, *J. Am. Chem. Soc.* **1969**, *91*, 2157–2158.
[294] A. J. Birch, G. S. R. Subba Rao, *Tetrahedron Lett.* **1968**, 3797–3798.
[295] K. Tani, T. Yamagata, S. Akutagawa, H. Kumobayashi, T. Taketomi, H. Takaya, A. Miyashita, R. Noyori, S. Otsuka, *J. Am. Chem. Soc.* **1984**, *106*, 5208–5217.
[296] S. Otsuka, K. Tani, T. Yamagata, S. Akutagawa, H. Kumobayashi, M. Yagi, EP 68506, **1983**.
[297] R. Noyori, H. Takaya, *Acc. Chem. Res.* **1990**, *23*, 345–350.
[298] K. Tani, T. Yamagata, S. Otsuka, H. Kumobayashi, S. Akutagawa, *Org. Synth.* **1989**, *67*, 33–43.
[299] H. Kumobayashi, S. Akutagawa, S. Otsuka, *J. Am. Chem. Soc.* **1978**, *100*, 3949–3950.
[300] S. Inoue, H. Takaya, K. Tani, S. Otsuka, T. Sato, R. Noyori, *J. Am. Chem. Soc.* **1990**, *112*, 4897–4905.
[301] B. M. Trost, T. Schmidt, *J. Am. Chem. Soc.* **1988**, *110*, 2301–2303.
[302] G. Balavoin, J. Collin, J. J. Bonnet, G. Lavigne, *J. Organomet. Chem.* **1985**, *280*, 429–439.
[303] W. H. Watson, A. Nagl, S. Hwang, M. G. Richmond, *J. Organomet. Chem.* **1993**, *445*, 163–170.
[304] W. R. Cullen, S. J. Rettig, T. C. Zheng, *Organometallics* **1993**, *12*, 688–696.
[305] W. R. Cullen, S. T. Chacon, M. I. Bruce, F. W. B. Einstein, R. H. Jones, *Organometallics* **1988**, *7*, 2273–2278.
[306] W. R. Cullen, S. J. Rettig, T.-C. Zheng, *Organometallics* **1992**, *11*, 924–935.

[307] W. R. Cullen, S. J. Rettig, T.-C. Zheng, *Organometallics* **1992**, *11*, 853–858.
[308] W. R. Cullen, S. J. Rettig, T.-C. Zheng, *Can. J. Chem.* **1993**, *71*, 399–409.
[309] M. I. Bruce, *Angew. Chem. Int. Ed. Engl.* **1977**, *16*, 73–86.
[310] F. Morandini, B. Longato, S. Bresadola, *J. Organomet. Chem.* **1977**, *132*, 291–299.
[311] Z.-G. Fang, Y.-S. Wen, R. K. L. Wong, S.-C. Ng, L.-K. Liu, T. S. A. Hor, *J. Cluster Sci.* **1994**, *5*, 327–340.
[312] W. L. Gladfelter, K. J. Roesselet in *The Chemistry of Metal Cluster Complexes* (Eds. D. F. Shriver, H. D. Kaesz, R. D. Adams), VCH, New York, **1990**, Chap. 7, p. 329–365.
[313] D. L. DuBois, C. W. Eigenbrot, Jr., A. Miedaner, J. C. Smart, R. C. Haltiwanger, *Organometallics* **1986**, *5*, 1405–1411.
[314] G. Pilloni, B. Longato, B. Corain, *J. Organomet. Chem.* **1991**, *420*, 57–65.
[315] B. Longato, G. Pilloni, G. Valle, B. Corain, *Inorg. Chem.* **1988**, *27*, 956–958.
[316] F. Estevan, P. Lahuerta, J. Latorre, A. Sanchez, C. Sieiro, *Polyhedron* **1987**, *6*, 473–478.
[317] G. Pilloni, B. Longato, *Inorg. Chim. Acta* **1993**, *208*, 17–21.
[318] M. Adachi, M. Kita, K. Kashiwabara, J. Fujita, N. Iitaka, S. Kurachi, S. Ohba, D.-M. Jin, *Bull. Chem. Soc. Jpn.* **1992**, *65*, 2037–2044.
[319] S. J. Berners-Price, P. J. Sadler, *Struct. Bonding (Berlin)* **1988**, *70*, 27–102, and references therein.
[320] S. J. Berners-Price, P. J. Sadler in *Platinum and Other Metal Coordination Compounds in Cancer Chemotherapy* (Ed.: M. Nicolini), Martinus Nijhoff, Boston, **1988**, p. 527, and references therein.
[321] C. K. Mirabelli, D. T. Hill, L. F. Faucette, F. L. McCabe, G. R. Girard, D. B. Bryan, B. M. Sutton, J. O. Bartus, S. T. Crooke, R. K. Johnson, *J. Med. Chem.* **1987**, *30*, 2181–2190.
[322] C. K. Mirabelli, B. D. Jensen, M. R. Mattern, C.-M. Sung, S.-M. Mong, D. T. Hill, S. W. Dean, P. S. Schein, R. K. Johnson, S. T. Crooke, *Anti-Cancer Drug Des.* **1986**, *1*, 223–234.
[323] P. Köpf-Maier in *Platinum and Other Metal Coordination Compounds in Cancer Chemotherapy* (Ed.: M. Nicolini), Martinus Nijhoff, Boston, **1988**, p. 601, and references therein.
[324] S. J. Berners-Price, R. K. Johnson, C. K. Mirabelli, L. F. Faucette, F. L. McCabe, P. J. Sadler, *Inorg. Chem.* **1987**, *26*, 3383–3387.
[325] J. E. Schurig, H. A. Meinema, K. Timmer, B. H. Long, A. M. Casazza, *Prog. Clin. Biochem. Med.* **1989**, *10*, 205–216.
[326] P. Köpf-Maier, H. Köpf, E. W. Neuse, *J. Cancer Res. Clin. Oncol.* **1984**, *108*, 336.
[327] P. Köpf-Maier, H. Köpf, *Chem. Rev.* **1987**, *87*, 1137–1152.
[328] P. Köpf-Maier, H. Köpf, E. W. Neuse, *Angew. Chem. Int. Ed. Engl.* **1984**, *23*, 456–457.
[329] P. Köpf-Maier, H. Köpf, *Struct. Bonding (Berlin)* **1988**, *70*, 103–194.
[330] B. Longato, G. Pilloni, G. M. Bonora, B. Corain, *J. Chem. Soc., Chem. Commun.* **1986**, 1478–1479.
[331] G. Bandoli, G. Trovó, A. Dolmella, B. Longato, *Inorg. Chem.* **1992**, *31*, 45–51.
[332] V. Scarcia, A. Furlani, B. Longato, B. Corain, G. Pilloni, *Inorg. Chim. Acta* **1988**, *153*, 67–70.

Received: December 21, 1993

2 Asymmetric Catalysis with Chiral Ferrocenylphosphine Ligands

Tamio Hayashi

2.1 Introduction

Among various types of asymmetric reactions, reaction with a chiral catalyst is obviously the best choice, provided that the catalytic asymmetric reactions proceed with high stereoselectivity producing the desired enantiomeric isomer in high yield [1]. Catalytic reactions by homogeneous transition metal complexes have been rapidly developed recently and now a wide variety of reactions can be effected by transition-metal catalysts [2]. Many of the transition-metal complexes used for catalytic reactions contain tertiary phosphines as ligands, so that it is convenient to use optically active phosphine ligands to make the metal complexes function as chiral catalysts. Thus, the most significant point for obtaining high efficiency in the transition metal-catalyzed asymmetric reactions is the design and preparation of a chiral phosphine ligand that will bring about high enantioselectivity as well as high catalytic activity in a given reaction.

A number of chiral phosphine ligands have been developed in order to achieve high enantioselectivity in catalytic asymmetric reactions and some of these ligands have been found to be effective for the rhodium-catalyzed asymmetric hydrogenation of α-(acylamino)acrylic acids or their analogs giving the hydrogenation products of over 90% *ee* [3]. Nevertheless, few of them can exhibit this high enantioselectivity in other types of catalytic asymmetric reactions. The exceptions are exemplified by 2,2′-bis(diphenylphosphino)-1,1′-binaphthyl (BINAP) [4] and a series of ferrocenylphosphines. The former can generate chiral surroundings in close proximity to a catalyst that is efficient enough to control the enantioface of reacting substrates by steric repulsion, giving high values of enantiomeric excess in various types of catalytic asymmetric reactions including rhodium-catalyzed hydroboration, palladium-catalyzed Heck-type reaction, rhodium-catalyzed isomerization, and the rhodium- and ruthenium-catalyzed hydrogenation of olefins and ketones [1]. Typical chiral ferrocenylphosphines [5] (Fig. 2-1) have the following unique features that are hardly found in other chiral phosphine ligands: (a) they possess functional groups (X) at the ferrocenylmethyl position on the side chain; (b) they contain the ferrocene planar chirality that never undergoes racemization; (c) both monophosphine and bisphosphines can be prepared from one chiral source, namely, *N*,*N*-dimethyl-1-ferrocenylethylamine; (d) they posses the orange color characteristic of ferrocene and are readily located during the column chromatographic work-up. Of these features, the most significant is the presence of functional groups on the side chain.

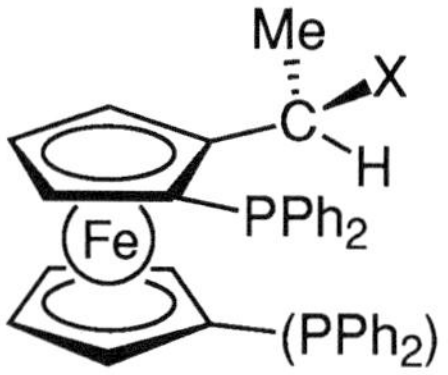

Fig. 2-1. Chiral ferrocenylphosphines.

These groups are controlled by the ferrocenyl and methyl groups on the chiral carbon center to face towards the reaction site on the catalyst coordinated with phosphorus atoms on the ferrocenylphosphine ligand and are expected to interact with a functional group on a substrate in a catalytic asymmetric reaction. By the secondary interaction between functional groups on the phosphine ligand and the reacting substrate [6], the ferrocenylphosphines bring about high enantioselectivity in a variety of asymmetric catalytic reactions. According to the demand of the substrate and reaction type, desired functional groups can be readily introduced on the side chain. The high efficiency of the ferrocenylphosphine ligands is highlighted by nickel- or palladium-catalyzed asymmetric cross-coupling, palladium-catalyzed allylic substitution reactions, and gold-catalyzed asymmetric aldol type reaction of isocyanocarboxylates, in which high selectivity is achieved only with the well-designed ferrocenylphosphine ligand.

The main purpose of this article is to survey the asymmetric reactions catalyzed by chiral ferrocenylphosphine-metal complexes, focusing attention on the design of the chiral ligand.

2.2 Preparation of Chiral Ferrocenylphosphines

Chiral ferrocenylphosphines were first prepared by Hayashi and Kumada in 1974 [7]. The asymmetric ortho-lithiation of optically resolved *N*,*N*-dimethyl-1-ferrocenylethylamine **1** with butyllithium reported by Ugi and coworkers [8] (see Chapter 4) was conveniently used for their preparation. The addition of diphenylchlorophosphine to the ortho-lithiated ferrocene **2** generated from (*R*)-**1** gave (*R*)-*N*,*N*-dimethyl-1-[(*S*)-2-(diphenylphosphino)ferrocenyl]ethylamine ((*R*)-(*S*)-PPFA **3a**) in 60–70% yield (Scheme 2-1) [9]. The first (*R*) designates the carbon central chirality

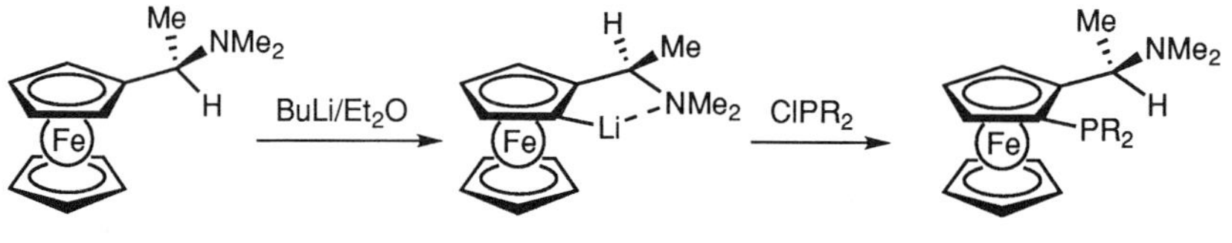

(*R*)-**1** → BuLi/Et_2O → **2** → $ClPR_2$ →

(*R*)-(*S*)-**3a**: R = Ph (PPFA)
(*R*)-(*S*)-**3b**: R = Me

Scheme 2-1

originated from **1** and the second (*S*) designates the ferrocene planar chirality generated at the stereoselective lithiation. The enantiomeric ferrocenylphosphine (*S*)-(*R*)-PPFA can also be obtained starting with (*S*)-**1**. Several bis(substituted phenyl)phosphino groups and dimethylphosphino group were introduced into (*R*)-**1** in a similar manner using the corresponding chlorophosphines [9].

The stepwise lithiation of (*R*)-**1** with butyllithium in ether and then with butyllithium and *N*,*N*,*N'*,*N'*-tetramethylethylenediamine (TMEDA) in the same solvent followed by diphenylphosphination with diphenylchlorophosphine led to the introduction of two diphenylphosphino groups, one onto each of the cyclopentadienyl rings, to give (*R*)-*N*,*N*-dimethyl-1-[(*S*)-1',2-bis(diphenylphosphino)ferrocenyl]-ethylamine ((*R*)-(*S*)-BPPFA **4a**) in 60–70% yield (Scheme 2-2) [9]. The analogous ferrocenylbisphosphines **4** containing bis(substituted phenyl)phosphino groups or a dialkylphosphino group were prepared in a similar manner [10–12].

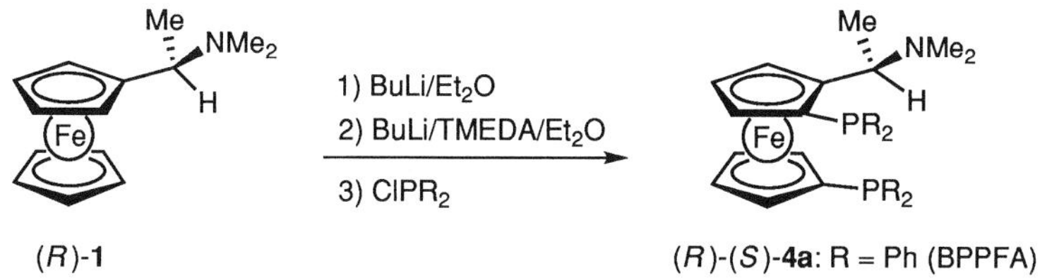
Me
NMe2
H
Fe
1) BuLi/Et2O
2) BuLi/TMEDA/Et2O
3) ClPR2
Me
NMe2
H
PR2
Fe
PR2
(R)-1
(R)-(S)-4a: R = Ph (BPPFA)

Scheme 2-2

The ferrocenylphosphines with (*R*)-(*R*) or (*S*)-(*S*) configuration, which are diastereomeric to (*R*)-(*S*) or (*S*)-(*R*) isomers and are readily obtained as shown in Schemes 2-1 and 2-2, are prepared by protection of the ring hydrogen liable to lithiation with a trimethylsilyl group followed by lithiation of the other less reactive hydrogen (Scheme 2-3) [9, 13, 14]. After introduction of one or two diphenylphosphino groups the silyl group is removed with potassium *t*-butoxide.

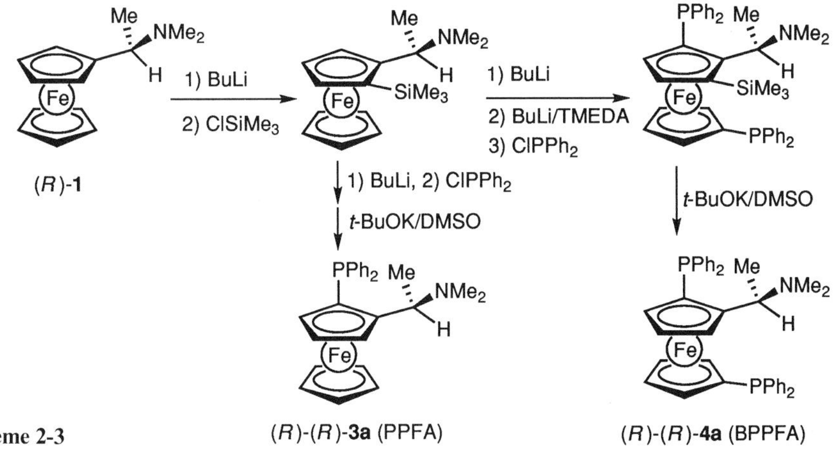
Me
NMe2
H
Fe
1) BuLi
2) ClSiMe3
(R)-1
Me
NMe2
H
SiMe3
Fe
1) BuLi
2) BuLi/TMEDA
3) ClPPh2
PPh2 Me
NMe2
H
SiMe3
Fe
PPh2
1) BuLi, 2) ClPPh2
t-BuOK/DMSO
t-BuOK/DMSO
PPh2 Me
NMe2
H
Fe
PPh2 Me
NMe2
H
Fe
PPh2
(R)-(R)-3a (PPFA)
(R)-(R)-4a (BPPFA)

Scheme 2-3

The dimethylamino group on the ferrocenylmethyl position of (*R*)-(*S*)-**3** and -**4** can be replaced by other functional groups by way of acetates (*R*)-(*S*)-**5** and -**6**, respectively, which are obtained in epimerically pure form by treatment of the dimethylamines **3** and **4** with acetic anhydride (Scheme 2-4) [9]. The substitution of acetate with amines takes place with retention of configuration [15] in refluxing methanol or ethanol to give the corresponding new ferrocenylphosphines **7** or **8** in high yield. Modification of the side chain of the ferrocenylphosphines by this method is very important for the design of the chiral ligand for some catalytic asymmetric reactions *(see below)*. Ferrocenylphosphines containing other functional groups such as methoxy **9** and **10**, hydroxyl **11**, or hydrothio **12** are also prepared from acetates (*R*)-(*S*)-**5** and -**6** (Schemes 2-5 and 2-6) [9, 16]. Introduction of a diphenyl- or dicyclohexylphosphino group at the ferrocenylmethyl position is also effected by the substitution reaction of (*R*)-(*S*)-**5** with the corresponding dialkylphosphine, or more conveniently, by reaction of the dimethylamine (*R*)-(*S*)-**3** with dialkylphosphine in acetic acid [9, 17] (Scheme 2-7), which produces a new type of chiral ferrocenylbisphosphine **13**.

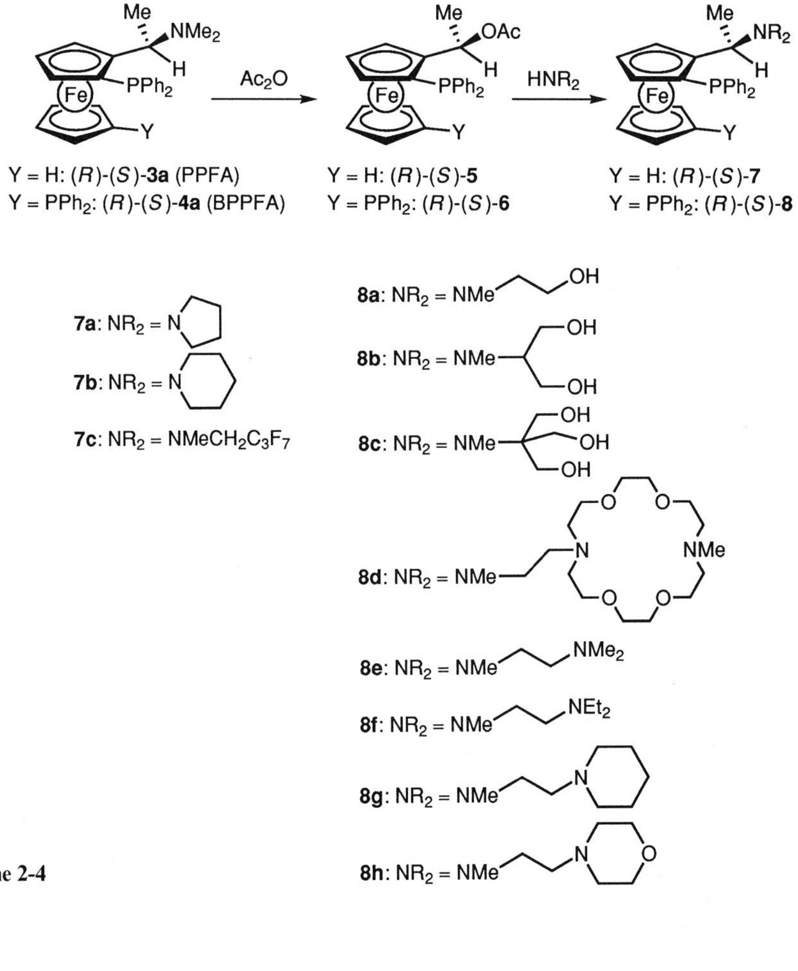

Me, NMe_2, H, PPh_2, Fe, Y — Ac_2O → Me, OAc, H, PPh_2, Fe, Y — HNR_2 → Me, NR_2, H, PPh_2, Fe, Y

Y = H: (*R*)-(*S*)-**3a** (PPFA)
Y = PPh_2: (*R*)-(*S*)-**4a** (BPPFA)

Y = H: (*R*)-(*S*)-**5**
Y = PPh_2: (*R*)-(*S*)-**6**

Y = H: (*R*)-(*S*)-**7**
Y = PPh_2: (*R*)-(*S*)-**8**

7a: NR_2 = N(pyrrolidinyl)
7b: NR_2 = N(piperidinyl)
7c: NR_2 = $NMeCH_2C_3F_7$

8a: NR_2 = NMe–CH₂CH₂–OH
8b: NR_2 = NMe–CH(CH₂OH)₂
8c: NR_2 = NMe–C(CH₂OH)₃
8d: NR_2 = NMe–CH₂CH₂–N(aza-crown, O, O, NMe, O, O)
8e: NR_2 = NMe–CH₂CH₂–NMe_2
8f: NR_2 = NMe–CH₂CH₂–NEt_2
8g: NR_2 = NMe–CH₂CH₂–N(piperidinyl)
8h: NR_2 = NMe–CH₂CH₂–N(morpholinyl, O)

Scheme 2-4

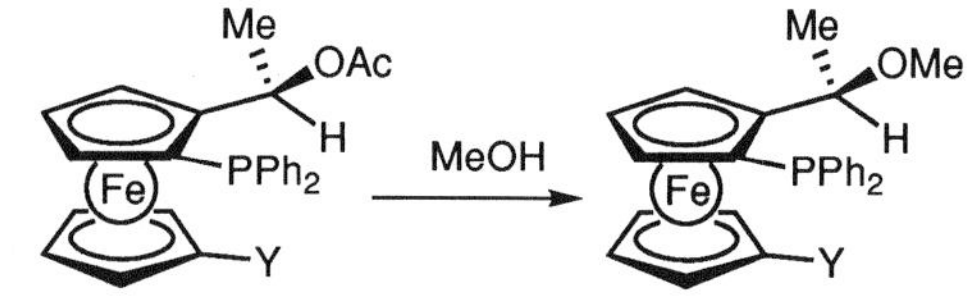

MeOH

Y = H: (R)-(S)-**5**
Y = PPh_2: (R)-(S)-**6**

Y = H: (R)-(S)-**9**
Y = PPh_2: (R)-(S)-**10**

Scheme 2-5

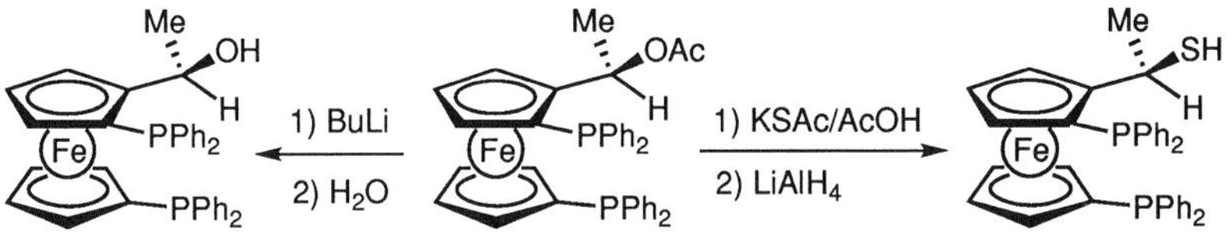

1) BuLi
2) H_2O

1) KSAc/AcOH
2) $LiAlH_4$

(R)-(S)-**11** (R)-(S)-**6** (R)-(S)-**12**

Scheme 2-6

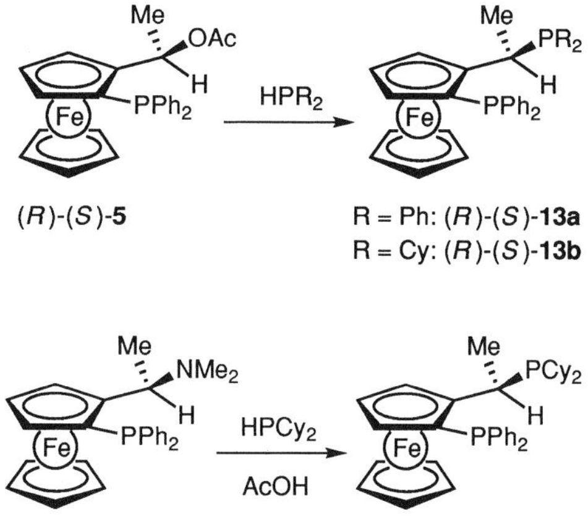

HPR_2

(R)-(S)-**5**

R = Ph: (R)-(S)-**13a**
R = Cy: (R)-(S)-**13b**

$HPCy_2$
AcOH

(R)-(S)-**3a** (R)-(S)-**13b**

Scheme 2-7

An optically active ferrocenylbisphosphine **14**, which is C_2 symmetric and has two 1-(dimethylamino)ethyl side chains, one on each of the two cyclopentadienyl rings, has been obtained by the optical resolution of its racemate (Scheme 2-8) [18, 19].

A chiral bisphosphine containing two ferrocene molecules has been prepared using Ugi's lithiation as a key step (Scheme 2-9) [20]. Thus, the diphenylphosphinyl group is introduced at the ferrocenylmethyl position of the iodide obtained from lithiated ferrocene **2**. Oxidative coupling of the iodoferrocene followed by reduction of the phosphine oxide gives the bisferrocene-bisphosphine **15**. The bisphosphine is unique in that a *trans*-chelate is formed on its coordination to a metal.

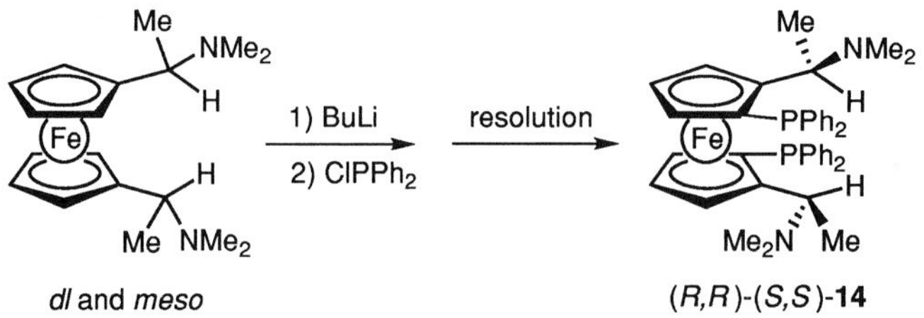

Scheme 2-8

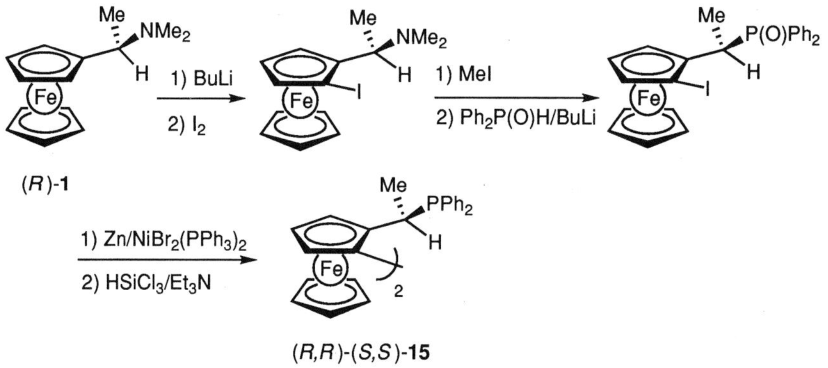

Scheme 2-9

Ferrocenylmonophosphines **16**, which are analogous to **3** but possess an isopropyl or phenyl group in place of the methyl group at the ferrocenylmethyl position, are prepared starting with the corresponding optically resolved *N*,*N*-dimethyl-1-ferrocenylalkylamines in a similar manner to that used for the preparation of **3** (Scheme 2-10) [21].

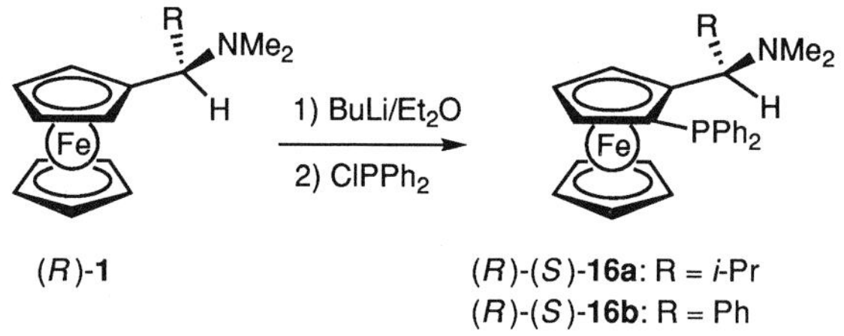

Scheme 2-10

The optically active ferrocenylmonophosphine and bisphosphine that have dimethylamino functionality but do not possess the chiral carbon center on the side chain are obtained by optical resolution of the racemic ferrocenylphosphine sulfide **17** (Scheme 2-11) [9]. The ferrocenylphosphines **18** and **19** are important for studies on the role of the central chirality on asymmetric induction in some catalytic asymmetric reactions.

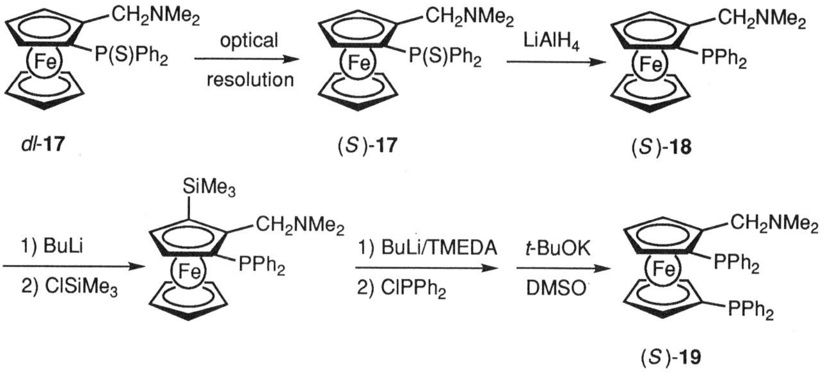

CH_2NMe_2 $P(S)Ph_2$ optical resolution CH_2NMe_2 $P(S)Ph_2$ $LiAlH_4$ CH_2NMe_2 PPh_2

dl-**17** (*S*)-**17** (*S*)-**18**

1) BuLi 2) $ClSiMe_3$ $SiMe_3$ CH_2NMe_2 PPh_2 1) BuLi/TMEDA 2) $ClPPh_2$ *t*-BuOK DMSO CH_2NMe_2 PPh_2 PPh_2

(*S*)-**19**

Scheme 2-11

A chiral bisphosphine that is analogous to BINAP but that contains a biferrocenyl backbone has been obtained by optical resolution of the corresponding phosphine oxide [22].

2.3 Structure of Chiral Ferrocenylphosphines and their Transition-Metal Complexes

Several reports have appeared on X-ray crystal structure analysis of chiral ferrocenylphosphine-transition metal complexes. The ferrocenylmonophosphine PPFA coordinates to palladium to form a stable complex $PdCl_2$(PPFA) [23], which is an effective catalyst for asymmetric cross-coupling and asymmetric hydrosilylation of olefins *(see below)*. The palladium complex adopts a square planar structure in which both phosphorus and nitrogen atoms in PPFA coordinate to palladium forming a six-membered ring chelate. The P-Pd-N angle is 96.1°. A similar square planar structure has been reported for a rhodium-PPFA complex, [Rh(PPFA)(NBD)]PF_6, in which PPFA also forms a chelate with phosphorus and nitrogen atoms [24].

The palladium complex, $PdCl_2$(BPPFA), where BPPFA is a ferrocenylbisphosphine ligand containing dimethylamino group on the side chain, adopts a square planar geometry with two *cis* chlorine and two phosphorus atoms, and the nitrogen atom that is not bound to palladium (Figure 2-2) [25]. The conformation of the four phenyl rings around the phosphorus atoms is nearly the same as that in an achiral analog, $PdCl_2$(dppf) (dppf = 1,1′-bis(diphenylphosphino)ferrocene) [26]. The phenyl rings are oriented in the so-called face and edge manner. The bite angle (P-Pd-P) of the bisphosphine ligand in this bisphosphine complex is 100.9°, being very large compared with that of other bisphosphine-palladium complexes. The large bite angle is proposed to make the phenyl rings come close to the reaction site of the central metal during the catalytic reactions taking place, thereby increasing enantioselectivity.

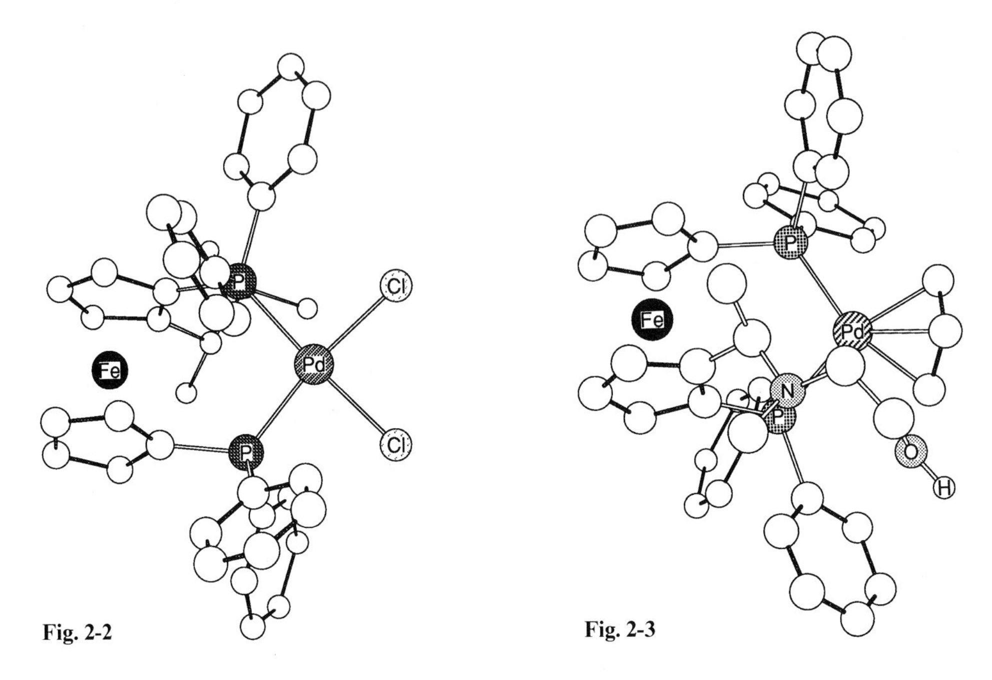

Fig. 2-2. X-Ray crystal structure of $PdCl_2$[(*R*)-(*S*)-**4a** (BPPFA)].

Fig. 2-3. X-Ray crystal structure of $Pd(\pi\text{-}C_3H_5)$[(*R*)-(*S*)-**8a**]$^+$.

Another important point of the chiral ferrocenylbisphosphine ligand is the location of the side chain attached to the ferrocene moiety. A functional group on the terminal of the pendant side chain is directed towards the reaction site of the catalyst (Fig. 2-3), which has been demonstrated by the X-ray structure analysis of a π-allylpalladium complex that is complexed with a ferrocenylbisphosphine bearing a 2-hydroxyethylamino group on the ferrocenylmethyl position [27]. The functional groups on the pendant side chain, which are readily introduced by nucleophilic substitution, are expected to bring about high enantioselectivity by attractive interactions with a functional group on the reacting substrate. The ferrocenylphosphines can be designed and modified by the introduction of a suitable functional group according to the demand of the reaction type and the substrate.

2.4 Catalytic Asymmetric Reactions with Chiral Ferrocenylphosphine Ligands

2.4.1 Cross-Coupling of Organometallics with Halides

Nickel and palladium complexes catalyze the reaction of organometallic reagents (R-m) with alkenyl or aryl halides and related compounds (R′-X) to give cross-coupling products (R-R′), which provides one of the most useful synthetic methods for making a carbon-carbon bond (Scheme 2-12) [2]. The catalytic cycle of the reaction is generally accepted to involve an unsymmetrical diorganometal complex LnM(II)(R)R′ as a key intermediate. From this intermediate the product R-R′ is released by reductive elimination to leave an LnM(0) species that undergoes oxidative addition to R′-X generating an intermediate LnM(II)(X)R′. Transfer of an alkyl group from R-m to this intermediate reproduces the diorganometal complex.

$$\text{R-m} + \text{R'-X} \xrightarrow{[\text{M}]} \text{R-R'} + \text{mX}$$

M = Ni, Pd
m = Mg, Zn, Al, Zr, Sn, B, etc.
R' = aryl, alkenyl
X = Cl, Br, I, OSO_2CF_3, $OPO(OR)_2$, etc.

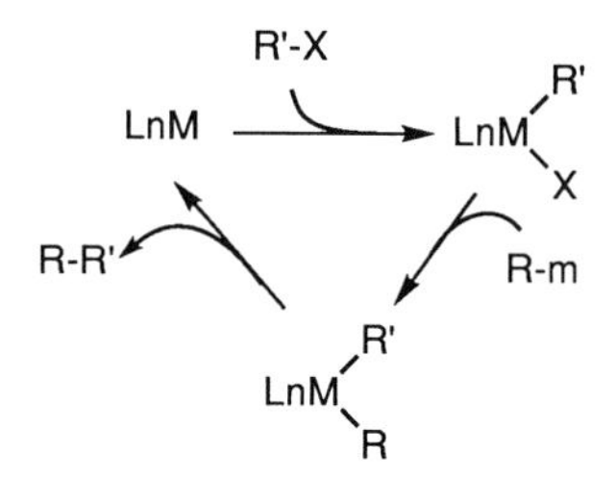

Scheme 2-12

Asymmetric synthesis by the catalytic cross-coupling reaction has been most extensively studied with secondary alkyl Grignard reagents. The asymmetric cross-coupling with chiral catalysts allows transformation of a racemic mixture of the secondary alkyl Grignard reagent into an optically active product by a kinetic resolution of the Grignard reagent. Since the secondary alkyl Grignard reagents usually undergo racemization at a rate comparable to that of the cross-coupling, optically active coupling product is formed even at 100% conversion of the Grignard reagent (Scheme 2-13).

The asymmetric cross-coupling of 1-phenylethylmagnesium chloride **20a** with vinyl bromide **21a** has been reported by Hayashi and Kumada to proceed in the presence of 0.5 or 1 mol% of nickel or palladium catalyst coordinated with a chiral ferrocenylphosphine ligand to give optically active 3-phenyl-1-butene **22** [28].

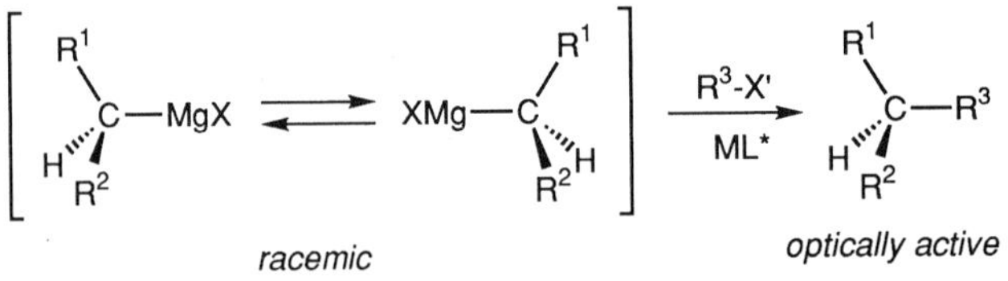

R^1 C—MgX H R^2 ⇌ XMg—C R^1 H R^2 — R^3-X', ML* → R^1 C—R^3 H R^2

racemic — optically active

Scheme 2-13

Scheme 2-14, in which some of the representative results are summarized, shows that the ferrocenylphosphines containing (dialkylamino)alkyl group on the side chain are enantioselective ligands for the cross-coupling. (*S*)-(*R*)-PPFA **3a** and (*S*)-(*R*)-BPPFA **4a** give the coupling product **22** with 68% *ee* (*R*) and 65% *ee* (*R*), respectively. Comparable enantioselectivity to (*S*)-(*R*)-PPFA is observed with (*R*)-(*R*)-PPFA, which is the diastereoisomer with opposite configuration at the chiral carbon center and with (*S*)-**18**, which lacks the central carbon chirality. The results demonstrate that ferrocene planar chirality, rather than the carbon central chirality, plays an important role in the asymmetric cross-coupling. On the other hand, a dramatic decrease in the enantioselectivity is observed with (*R*)-**23**, which does not contain a dimethylamino group. Thus, the presence of the (dialkylamino)alkyl side

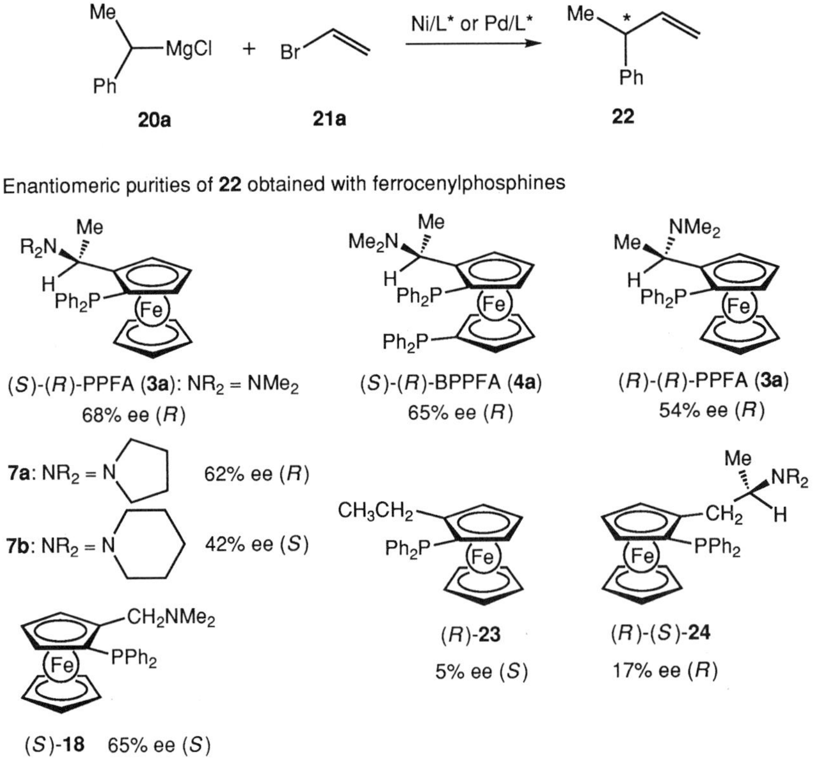

20a + **21a** —Ni/L* or Pd/L*→ **22**

Enantiomeric purities of **22** obtained with ferrocenylphosphines

(*S*)-(*R*)-PPFA (**3a**): $NR_2 = NMe_2$ 68% ee (*R*)

7a: NR_2 = N(pyrrolidine) 62% ee (*R*)

7b: NR_2 = N(piperidine) 42% ee (*S*)

(*S*)-(*R*)-BPPFA (**4a**) 65% ee (*R*)

(*R*)-(*R*)-PPFA (**3a**) 54% ee (*R*)

(*S*)-**18** 65% ee (*S*)

(*R*)-**23** 5% ee (*S*)

(*R*)-(*S*)-**24** 17% ee (*R*)

Scheme 2-14

chain is of primary importance for the high selectivity and the enantioselectivity is strongly affected by the structure of the dialkylamino group (**7a** and **7b**). A ferrocenylphosphine **24** that contains a dimethylamino group at a position one carbon removed from the ferrocene moiety gives **22** of lower enantiomeric excess (17% *ee*) [29]. The high enantioselectivity is not obtained with other chiral phosphine ligands such as DIOP and chiraphos. The amino group is proposed to coordinate with the magnesium atom in the Grignard reagent at the transmetallation step in the catalytic cycle, where the coordination occurs selectively with one of the enantiomers of the racemic Grignard reagent to bring about high selectivity [28], although the coordination has not been supported by NMR studies of palladium complex [30].

The asymmetric cross-coupling of the Grignard reagent catalyzed by PPFA-nickel has also been examined in the combination of several other secondary alkyl Grignard reagents with alkenyl bromides. For example, the reaction of 1-phenylethylmagnesium chloride with (*E*)-1-phenyl-2-bromoethene [28b] and that of 2-butylmagnesium chloride with (*E*)-1-phenylthio-2-bromoethene [31] gives the corresponding coupling products of around 50% *ee*. Optically active α-curcumene has been prepared by use of the asymmetric cross-coupling of 1-(*p*-tolyl)-1-ethyl magnesium chloride with vinyl bromide producing 3-(*p*-tolyl)-1-butene (66% *ee*) as a key step [32]. The reaction of 4-bromo-1-butene with phenylmagnesium bromide in the presence of PPFA-nickel catalyst proceeds through alkyl group isomerization from primary to secondary to produce a moderate yield of 3-phenyl-1-butene **22** with 38% *ee* [33].

The use of 1-arylethylzinc reagents in place of the corresponding Grignard reagents sometimes increases the stereoselectivity. The reaction of zinc reagents **20b** prepared from the Grignard reagent **20a** with zinc halides in THF in the presence of a palladium catalyst coordinated with a chiral ferrocenylphosphine [(*R*)-(*S*)-PPFA **3a**] proceeded with 85–86% enantioselectivity (Scheme 2-15) [34]. A higher selectivity (93% *ee*) was obtained with a C_2-symmetric ferrocenylphosphine ligand **14** that has two phosphorus atoms and two aminoalkyl side chains on the ferrocene skeleton [18].

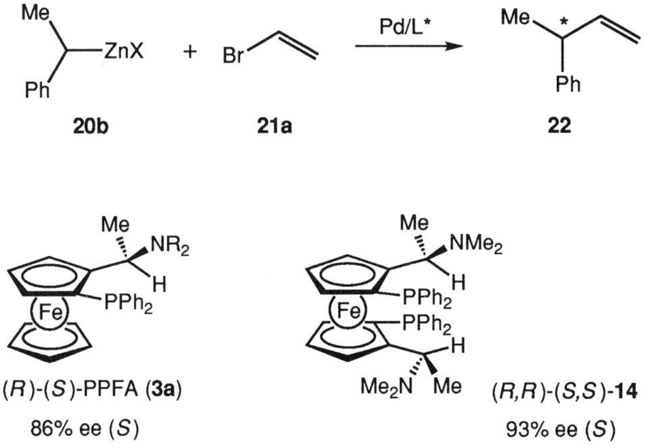

Scheme 2-15

The palladium-catalyzed coupling of silyl ketene acetals with aryl bromides in the presence of TlOAc has been applied to the asymmetric synthesis of 2-arylpropionic acids (40 – 50% *ee*) by the use of BPPFA and several other chiral phosphine ligands (Scheme 2-16) [35].

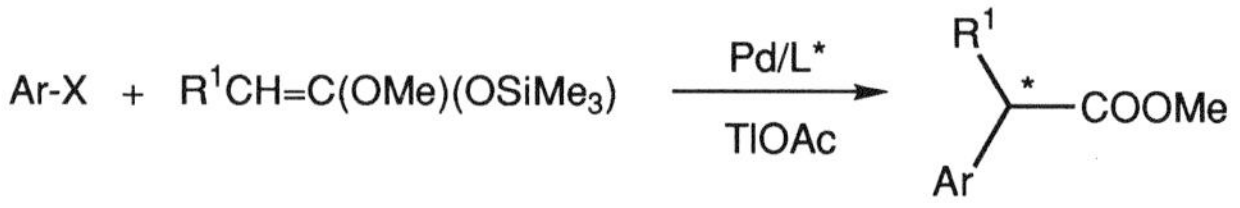

$$\text{Ar-X} + \text{R}^1\text{CH=C(OMe)(OSiMe}_3) \xrightarrow[\text{TlOAc}]{\text{Pd/L}^*} \text{Ar}\overset{*}{\text{C}}\text{H(R}^1)\text{COOMe}$$

Scheme 2-16

The asymmetric cross-coupling with the chiral ferrocenylphosphine ligand has been successfully applied by Hayashi to the synthesis of optically active allylsilanes [36]. The reaction of α-(trimethylsilyl)benzylmagnesium bromide **25** with vinyl bromide **21a**, (*E*)-bromostyrene **21b**, and (*E*)-bromopropene **21c** in the presence of 0.5 mol% of palladium complex coordinated with chiral ferrocenylmonophosphine, (*R*)-(*S*)-PPFA (**3a**), produces high yields of the corresponding (*R*)-allylsilanes **26a** – **c** with 95, 85, and 95% *ee*, respectively, which were substituted with phenyl group at the chiral carbon center bonded to the silicon atom (Scheme 2-17). With other chiral phosphine ligands such as DIOP or prophos, the optical purity of allylsilanes is not higher than 5% *ee*. Interestingly, the nickel/PPFA catalyst, which has been used successfully for the reaction of 1-arylethyl Grignard reagents, is much less catalytically active for this particular reaction. The optically active allylsilanes obtained here can be used for the $S_{E'}$-reactions forming optically active homoallyl alcohols

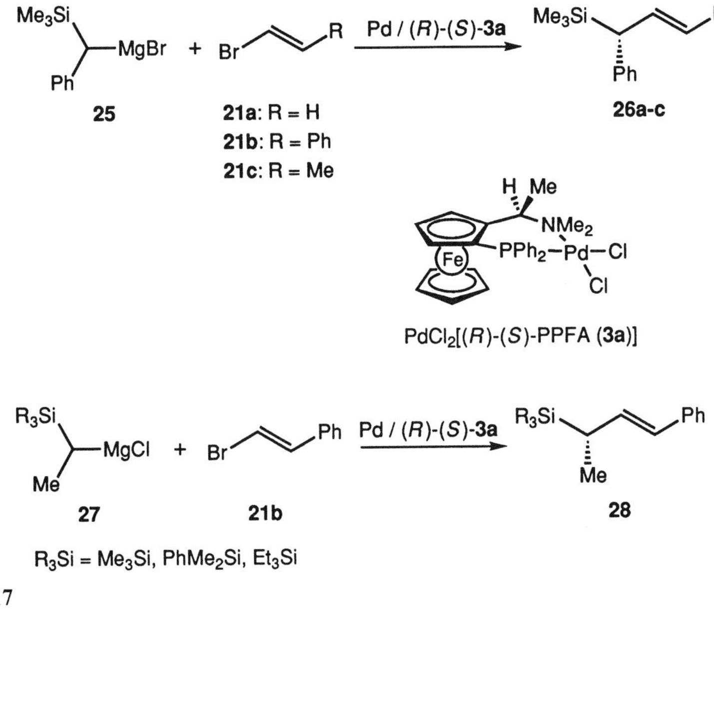

Me_3Si(Ph)CH–MgBr **25** + Br–CH=CH–R (**21a**: R = H; **21b**: R = Ph; **21c**: R = Me) → Pd / (*R*)-(*S*)-**3a** → Me_3Si–CH(Ph)–CH=CH–R **26a-c**

PdCl$_2$[(*R*)-(*S*)-PPFA (**3a**)]

R_3Si(Me)CH–MgCl **27** + Br–CH=CH–Ph **21b** → Pd / (*R*)-(*S*)-**3a** → R_3Si–CH(Me)–CH=CH–Ph **28**

R_3Si = Me_3Si, $PhMe_2Si$, Et_3Si

Scheme 2-17

and π-allylpalladium complexes [36c]. Lower enantioselectivity is observed with (*Z*)-alkenyl bromides. The palladium/PPFA catalyst is also effective for the reaction of 1-(trialkylsilyl)ethylmagnesium chlorides **27** with (*E*)-bromostyrene **21b** giving allylsilanes bearing a methyl group at the chiral carbon center. The enantioselectivity was dependent on the trialkylsilyl group, triethylsilyl being best to produce 1-phenyl-3-silyl-1-butene **28** with 93% *ee*.

The palladium/PPFA-catalyzed asymmetric cross-coupling of α-(trimethylsilyl)-benzylmagnesium bromide **25** has also been applied to the synthesis of optically active propargylsilane **29** (18% *ee*) by the use of 1-bromo-2-phenylacetylene as a coupling partner (Scheme 2-18) [37].

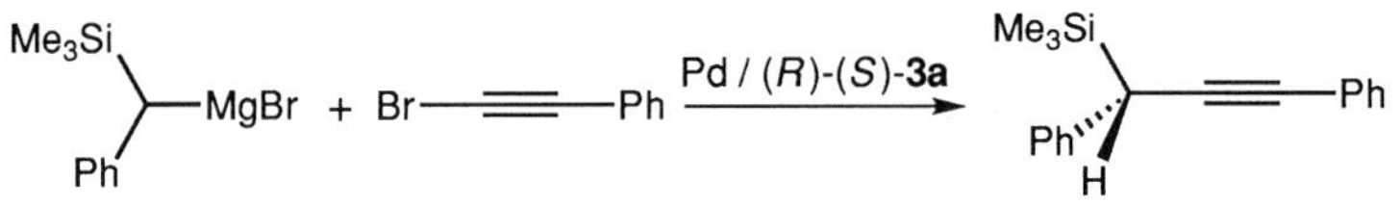

Me_3Si / Ph —MgBr **25** + Br—≡—Ph —Pd / (*R*)-(*S*)-**3a**→ Me_3Si / Ph / H —≡—Ph **29**

Scheme 2-18

Catalytic asymmetric induction of planar chirality in an (arene)chromium complex has been reported in the cross-coupling of tricarbonyl(*o*-dichlorobenzene)chromium **30** with vinylic metals, where one of the *meso* chlorine atoms undergoes the coupling to give the monovinylation product **31** with up to 44% *ee* (Scheme 2-19) [38].

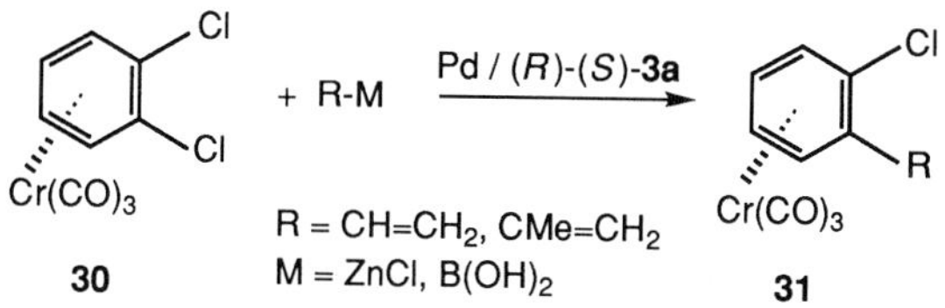

Cl, Cl, $Cr(CO)_3$ **30** + R-M —Pd / (*R*)-(*S*)-**3a**→ Cl, R, $Cr(CO)_3$ **31**

R = $CH{=}CH_2$, $CMe{=}CH_2$
M = ZnCl, $B(OH)_2$

Scheme 2-19

Preparation of axially chiral binaphthyls is another example of the exciting application of the catalytic asymmetric cross-coupling to organic synthesis. The cross-coupling between sterically hindered substrates, for example, 2-methyl-1-naphthylmagnesium bromide **32a** and 1-bromo-2-methylnaphthalene **33a** forming 2,2′-dimethyl-1,1′-binaphthyl **34a** (Scheme 2-20), is efficiently catalyzed only by nickel complexes coordinated with a monodentate phosphine ligand, chelating phosphine-nickel complexes or any palladium complexes being much less catalytically active. Initial attempts with (−)-DIOP and (*S*)-(*R*)-BPPFA **4a** have suffered from very poor enantioselectivity (<5%) as well as very low chemical yield [39]. The use of ferrocenylmonophosphine ligand (*S*)-(*R*)-**9**, which is a chiral monophosphine ligand containing methoxy group on the side chain, dramatically increases the selectivity to produce a high yield of (*R*)-**34a** with 95% *ee* [40]. High enantioselectivity is also attained in the reaction of **32a** with 1-bromonaphthalene **33b** catalyzed by the nickel/(*S*)-(*R*)-**9**, which gave (*R*)-2-methyl-1,1′-binaphthyl **34b** with 83% *ee*, but much lower selectivity (16% *ee*) is observed in the reaction of the other combination, that is, coupling of 1-naphthylmagnesium bromide **32b** with **33a**.

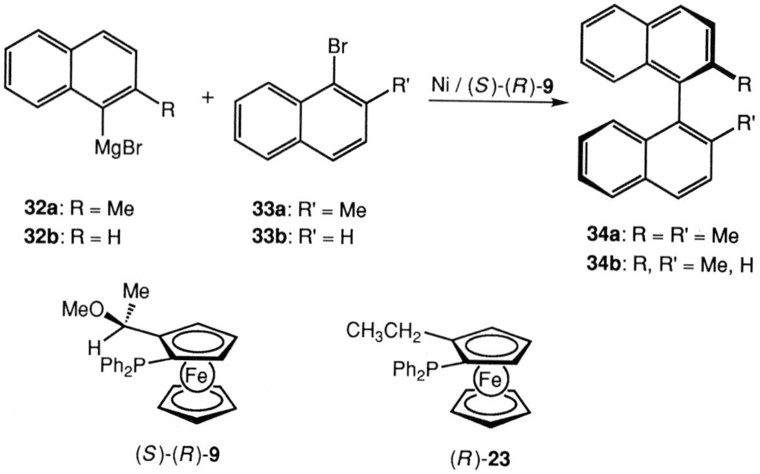

32a: R = Me
32b: R = H

33a: R' = Me
33b: R' = H

Ni / (S)-(R)-9

34a: R = R' = Me
34b: R, R' = Me, H

(S)-(R)-9

(R)-23

Scheme 2-20

The large difference suggests that the stereochemistry of binaphthyl is determined kinetically at the formation of diastereomeric diorganonickel(II) species NiL(Ar)Ar′, which has a chiral propeller structure, and hardly undergoes epimerization due to steric hindrance preventing rotation about the nickel-carbon bonds. Here again the chiral ferrocenylphosphine (*R*)-**23**, which does not contain the alkoxy functionality on the side chain, gives a racemic coupling product, indicating that the presence of the alkoxy group on the ferrocenylphosphine ligand is essential for high enantioselectivity.

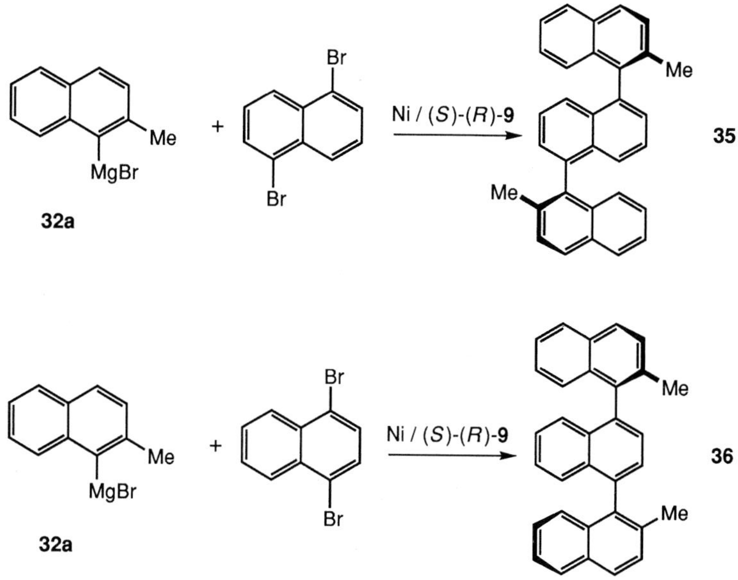

32a + dibromonaphthalene → Ni / (S)-(R)-9 → 35

32a + 1,4-dibromonaphthalene → Ni / (S)-(R)-9 → 36

Scheme 2-21

The nickel-catalyzed cross-coupling of 2-methyl-1-naphthylmagnesium bromide **32a** has been extended to the asymmetric synthesis of ternaphthalenes (Scheme 2-21) [41]. The reaction with 1,5- and 1,4-dibromonaphthalenes in the presence of the same catalyst, nickel/(*S*)-(*R*)-**9**, produces the corresponding optically active ternaphthalenes with 99% *ee* (**35**) and 95% *ee* (**36**), respectively, together with a small amount of *meso* compounds.

2.4.2 Allylic Substitution Reactions via π-Allyl Complexes

Substitution reactions of allylic substrates with nucleophiles have been shown to be catalyzed by certain palladium complexes [2, 42]. The catalytic cycle of the reactions involves π-allylpalladium as a key intermediate (Scheme 2-22). Oxidative addition of the allylic substrate to a palladium(0) species forms a π-allylpalladium(II) complex, which undergoes attack of a nucleophile on the π-allyl moiety to give an allylic substitution product. The substitution reactions proceed in an S_N or $S_{N'}$ manner depending on catalysts, nucleophiles, and substituents on the substrates. Studies on the stereochemistry of the allylic substitution have revealed that soft carbon nucleophiles represented by sodium dimethyl malonate attack the π-allyl carbon directly from the side opposite to the palladium (Scheme 2-23).

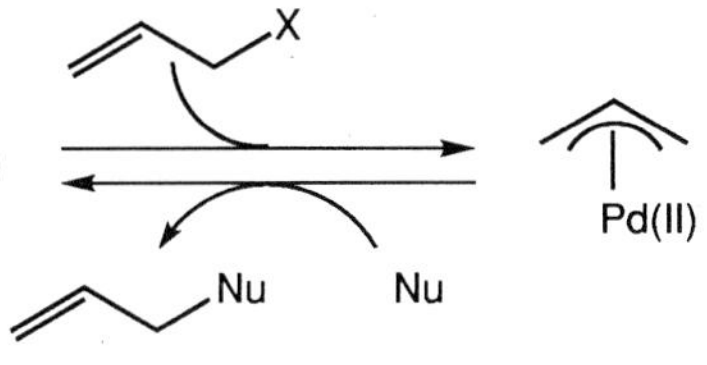

X
Pd(0)
Pd(II)
Nu
Nu

Scheme 2-22

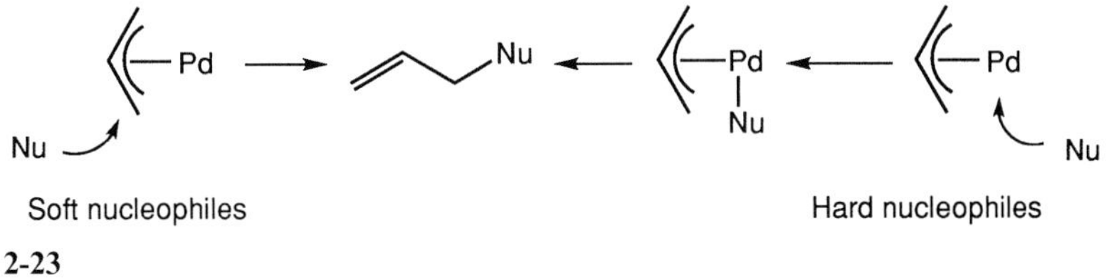

Pd
Nu
Nu
Pd
Nu
Pd
Nu

Soft nucleophiles

Hard nucleophiles

Scheme 2-23

Two types of palladium-catalyzed asymmetric reaction have been reported. One is the allylation of nucleophiles in which a new chiral carbon center is created in the nucleophile and the other is the allylic substitution reaction in which it is created in the allylic substrate (Scheme 2-24). Chiral ferrocenylbisphosphines designed and modified on the side chain have been successfully used for both of the two types of asymmetric reaction [5c, d].

Since the soft carbon nucleophiles attack the π-allyl group from the side opposite to palladium, the asymmetric induction at the carbon nucleophiles is difficult with

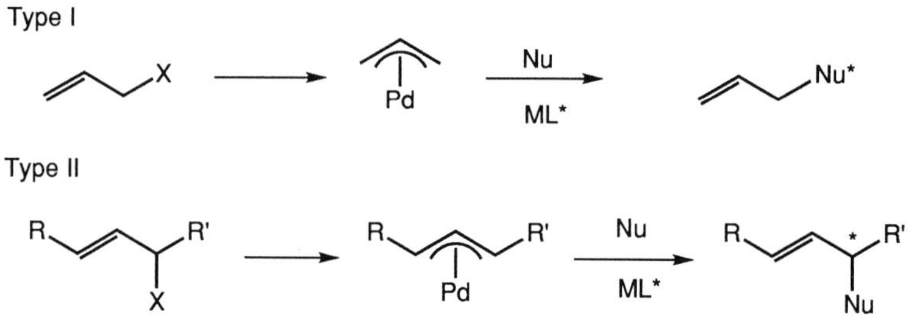

Type I

Type II

Scheme 2-24

conventional chiral phosphine ligands, whose chiral surroundings are due only to the orientation of the phenyl rings on the phosphorus atoms. The attacking prochiral nucleophiles are too remote from the chiral surroundings. Chiral ferrocenylbisphosphines **8a**–**c** bearing hydroxyl group(s) at the terminal of the pendant side chain are much more efficient than DIOP or prophos for the allylation of active methine compounds giving optically active ketones with a chiral quaternary carbon center (Scheme 2-25) [43]. The most effective ligand is (*R*)-(*S*)-**8a**, which contains a 2-hydroxyethylamino group on the side chain. Thus, the allylation of 2-acetylcyclohexanone **37a** with allyl acetate at −60 °C in the presence of a palladium catalyst generated from (*R*)-(*S*)-**8a** and $[PdCl_2(\pi\text{-}C_3H_5)]_2$ gives a high yield of (*S*)-2-allyl-acetylcyclohexanone **38a** with 81% *ee*. The replacement of the terminal hydroxyl group in (*R*)-(*S*)-**8a** by an amino or a methoxy group or removal of the whole pendant side chain results in the formation of **38a** with opposite configuration (*R*) in less than 22% *ee*. The distance between the terminal hydroxyl group and ferrocene moiety is also important, the 3-hydroxypropylamino side chain being less enantioselective (46% *ee*).

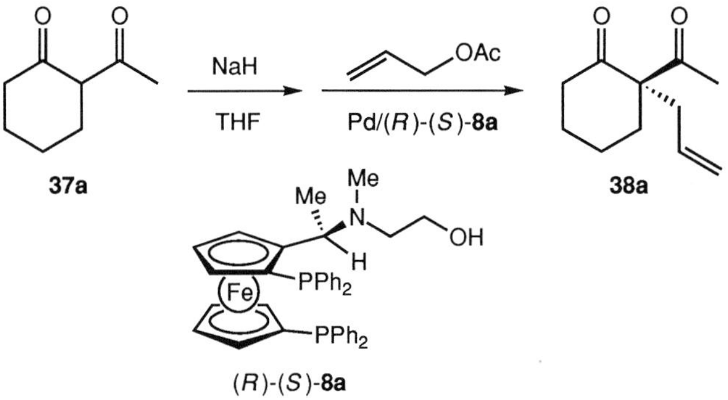

Scheme 2-25

The high enantioselectivity of the ligand **8a** is ascribed to the stereocontrol effected by attractive interactions probably due to hydrogen bonding between the terminal hydroxyl group on the ligand and the prochiral enolate of a *β*-diketone that is to attack the π-allyl carbon on the π-allylpalladium intermediate from the side opposite to palladium. The X-ray crystal structure of $(Pd(\pi\text{-}C_3H_5)(\mathbf{8a})]ClO_4$ [27] (see Fig. 2-3 and Scheme 2-31), in which the terminal hydroxyl group is located in close proximity to the π-allyl group, supports the interaction of the hydroxyl group with the prochiral nucleophile.

The ferrocenylphosphine (*R*)-(*S*)-**8a** is also effective for the asymmetric allylation of several other active methine compounds, including 2-acetyl-1-tetralone, 6,6-dimethyl-2-acetylcyclohexanone, 2-acetylcyclooctanone, 1-phenyl-2-methylbutane-1,3-dione, 2-phenylpropanal, and methyl α-isocyano(phenyl)acetate [43, 44]. The allylation products and the values of enantiomeric excess are shown in Scheme 2-26.

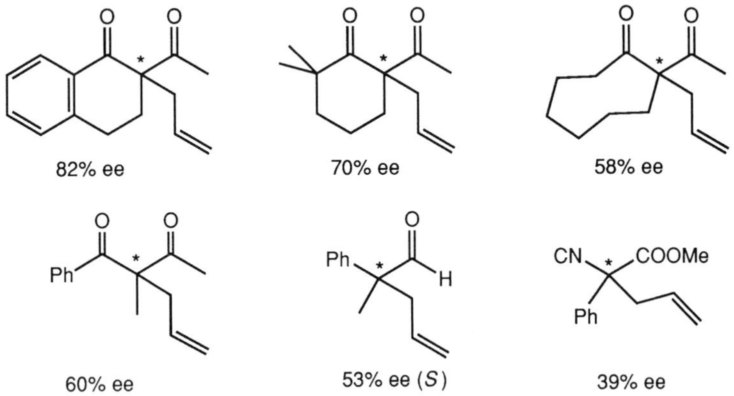

82% ee 70% ee 58% ee

60% ee 53% ee (*S*) 39% ee

Scheme 2-26

The chiral ferrocenylbisphosphine ligands that contain monoaza or diaza crown ethers of varying ring sizes have been prepared and used for the allylation of active methine compounds [45]. The most enantioselective is the ligand (*R*)-(*S*)-**8d**, bearing the 1,10-diaza-18-crown-6 moiety, which gives (*R*)-**38a** of 75% *ee* in the allylation of 2-acetylcyclohexanone at −40 °C. Interestingly, the absolute configu-

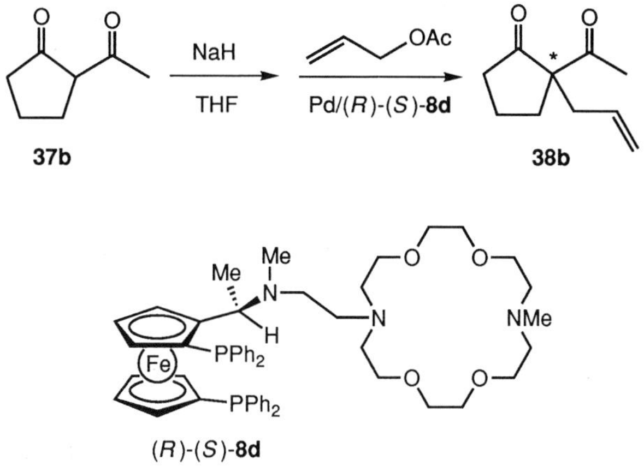

NaH, THF; OAc, Pd/(*R*)-(*S*)-**8d**

37b **38b**

(*R*)-(*S*)-**8d**

Scheme 2-27

ration of **38a** is opposite to that obtained with (*R*)-(*S*)-**8a**. The crown-modified ligand (*R*)-(*S*)-**8d** has a higher enantioselectivity than (*R*)-(*S*)-**8a** in the allylation of 2-acetylcyclopentanone **37b** to give the corresponding allylation product **38b** of 65% *ee* (Scheme 2-27). The aza crown ether moiety is proposed to form an inclusion complex with the nucleophiles. NOE studies on the π-allylpalladium complex coordinated with one of the crown-modified ligands suggest that the aza crown ether is in the proper orientation to interact with the incoming nucleophile.

In the allylic substitution of racemic 2-propenyl acetates or related substrates with the same substituents at 1 and 3 positions, the π-allylpalladium intermediate containing a *meso* type π-allyl group is formed from both enantiomers of the allylic substrate. Two π-allyl carbons at the 1- and 3-positions are diastereotopic on coordination of a chiral phosphine ligand to palladium. The asymmetric induction arises from preferential attack by the nucleophile on either of the two diastereotopic π-allyl carbon atoms (Scheme 2-28).

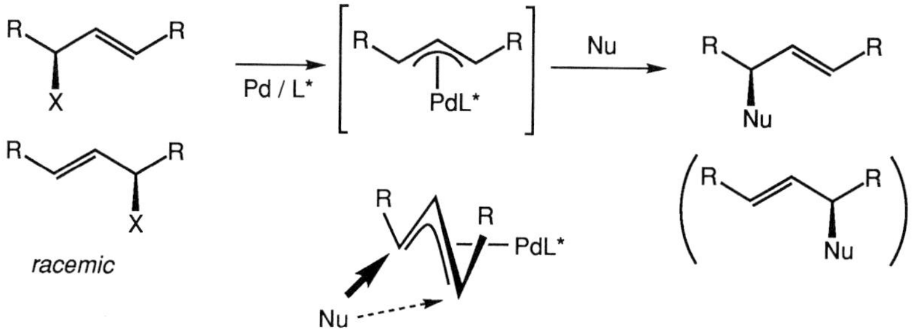

Scheme 2-28

For this type of allylic alkylation, 1,3-diphenyl-2-propenyl acetate **39a** have been often used as allylic substrates. Although an enantioselectivity of ca. 90% *ee* has been reported in the reaction of **39a** with acetamidomalonate with usual chiral phosphines such as BINAP or chiraphos [46], the ferrocenylphosphines containing a hydroxyl functionality on the pendant side chain are more effective ligands for the reaction with sodium acetylacetate (Scheme 2-29) [5d, 47]. The enantioselectivity increases as the number of hydroxyl groups on the side chain increases. Thus, the ferrocenylphosphines **8a**, **8b**, and **8c** give the alkylation product with 71, 90, and 96% *ee*, respectively. High enantioselectivity is also observed in the reaction of 1,3-diaryl-2-propenyl acetates (Ar = 1-Np **39b**: 92% *ee*, 3-$MeOC_6H_4$ **39c**: 86% *ee*) and *cis*-3-acetoxy-5-carbomethoxycyclohexene **40** (72% *ee*) in the presence of the Pd/(*R*)-(*S*)-**8b** catalyst.

The allylic amination with benzylamine or related amines such as (3,4-dimethoxyphenyl)methylamine catalyzed by palladium coordinated with the dihydroxylated ferrocenylphosphine **8b** takes place with high enantioselectivity (Scheme 2-30) [27]. Thus, reaction of 1,3-diphenyl-2-propenyl ethyl carbonate **41** with benzylamine gives a quantitative yield of allylic amination product **42** with >97% *ee*. High enantioselectivity is also obtained in the allylic amination of 2-propenyl esters **43** substituted

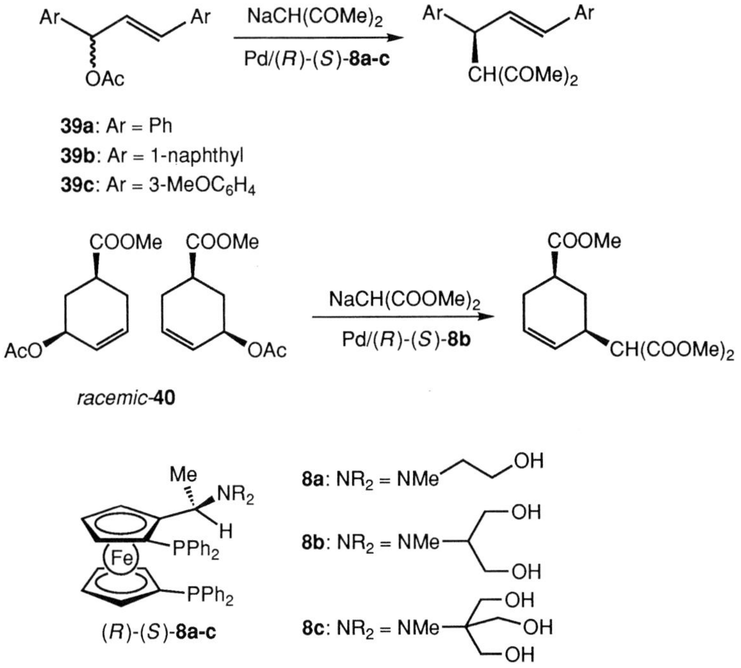

Ar, Ar, OAc — $NaCH(COMe)_2$ / Pd/(*R*)-(*S*)-**8a-c** → Ar, Ar, $CH(COMe)_2$

39a: Ar = Ph
39b: Ar = 1-naphthyl
39c: Ar = 3-$MeOC_6H_4$

COOMe, AcO, COOMe, OAc — $NaCH(COOMe)_2$ / Pd/(*R*)-(*S*)-**8b** → COOMe, $CH(COOMe)_2$

racemic-**40**

Me, NR_2, H, Fe, PPh_2, PPh_2

(*R*)-(*S*)-**8a-c**

8a: NR_2 = NMe–CH_2CH_2–OH
8b: NR_2 = NMe–CH(CH_2OH)$_2$
8c: NR_2 = NMe–C(CH_2OH)$_3$

Scheme 2-29

Ph, Ph, OCOOEt — $ArCH_2NH_2$ / Pd/(*R*)-(*S*)-**8b** → Ph, Ph, $NHCH_2Ar$

41 **42**

Ar = Ph, 3,4-dimethoxyphenyl (OMe, OMe)

Me, Me, N, OH, OH, H, Fe, PPh_2, PPh_2

(*R*)-(*S*)-**8b**

R, R, X — $PhCH_2NH_2$ / Pd/(*R*)-(*S*)-**8b** → R, R, $NHCH_2Ph$ → R, COOR, NHR

43a: R = *n*-Pr, X = OP(O)Ph_2
43b: R = *i*-Pr, X = OCOOR

82% ee (R = n-Pr)
97% ee (R = i-Pr)

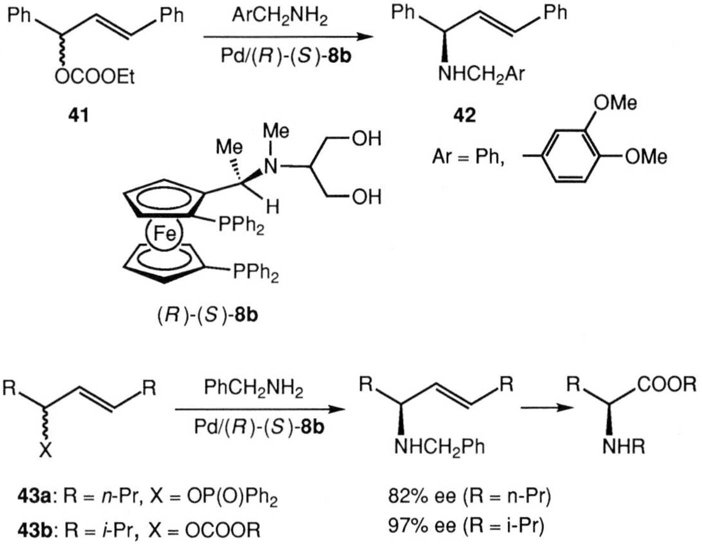

Scheme 2-30

with *n*Pr or *i*Pr groups at the 1- and 3-positions. The allylic amines produced can be converted into optically active α-amino acids. The high enantioselectivity is not observed with ferrocenylphosphines lacking the hydroxyl functionality. It is proposed that the hydroxyl group(s) on the ligand, which are located outside the π-allyl group of the π-allylpalladium intermediate, direct the incoming nucleophile preferentially to one of the two diastereotopic carbon atoms. ^{31}P NMR studies on the π-allylpalladium complex bearing the 1,3-diphenyl π-allyl and the phosphine ligand **8b** reveals that the palladium complex adopts one of the two possible conformational isomers, "W" shape π-allyl and "M" shape π-allyl, with high selectivity (20/1) in an equilibrium (Scheme 2-31).

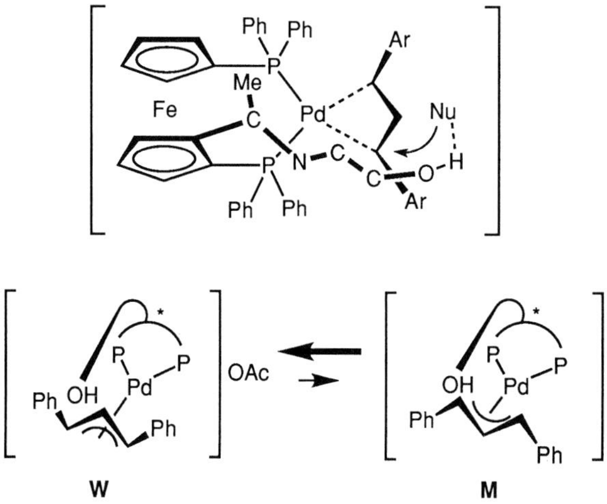

Scheme 2-31

Asymmetric allylic substitution reactions of the racemic substrates bearing two different substituents at 1- and 3-positions involve the chiral π-allylpalladium intermediates, which cannot undergo epimerization under the usual reaction conditions. Although the substitution reactions proceed with net retention of configuration, optically active products can be formed from racemic substrates by a regioselective nucleophilic attack on the π-allyl. Use of a dihydroxylated ferrocenylphosphine ligand **8b** for the alkylation of racemic allyl acetates **44** (which have slightly different aryl groups at the 1- and 3-positions) gives rise to high enantioselectivity in the allylic alkylation products **45** and **46** (Scheme 2-32) [48]. The high enantiomeric purity (up to 95% *ee*) of the products results from the regiocontrol by chiral ligand **8b**, which directs the nucleophilic attack onto one of the π-allyl carbons. A kinetic resolution of a racemic allyl acetate has been reported in the reaction of 3-acetoxy-4-methyl-1-phenylpentene **47** with sodium acetylacetate catalyzed by Pd/**8b** (Scheme 2-33) [49]. The relative rate ratio for the enantiomers is $k_{fast}/k_{slow} = 14$, practically enantiomerically pure (>99% *ee*) (*R*)-**47** being recovered at 80% conversion. Allylic alkylation product (*S*)-**48** of high purity (>98% *ee*) is also obtained at 40% conversion.

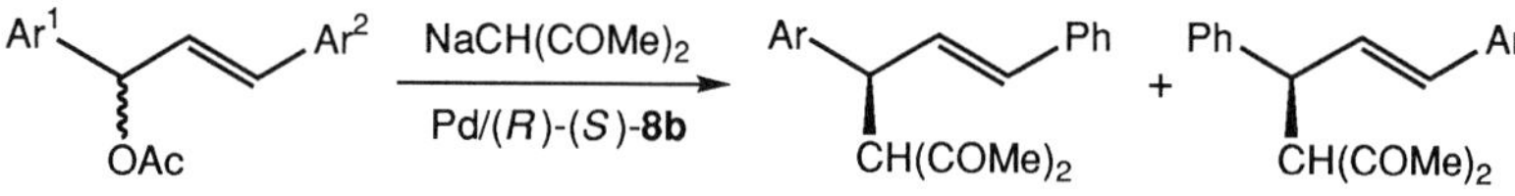

44a: Ar^1, Ar^2 = Ph, 3-$MeOC_6H_4$
44b: Ar^1, Ar^2 = Ph, 1-naphthyl

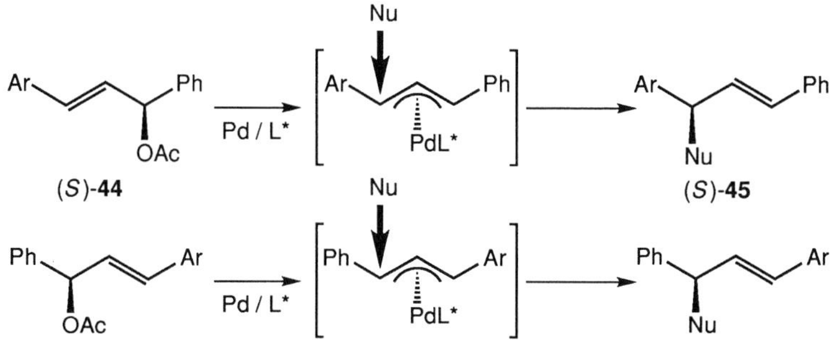

Scheme 2-32

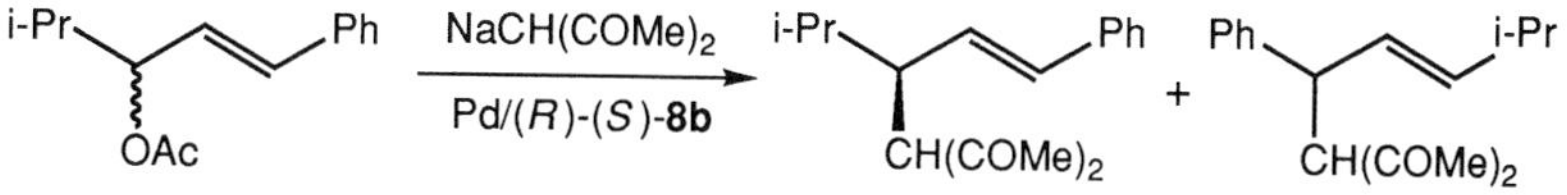

Scheme 2-33

For the asymmetric allylic substitution reactions that proceed via π-allylpalladium intermediates with monosubstitution at the 1-position of the π-allyl group, the dihydroxylated ferrocenylphosphine ligand **8b** also exhibits high enantioselectivity. Allylic amination of (*E*)-2-butenyl acetate **49** with benzylamine in the presence of 3 mol% of palladium catalyst coordinated with **8b** gives the amination product (*S*)-3-benzylamino-1-butene **50** with 84% *ee*, together with a small amount of achiral regioisomer (*E*)-**51**, the ratio being 97:3 (Scheme 2-34) [50]. The same amination starting with (*Z*)-**49** resulted in the formation of (*S*)-**50** with lower enantiomeric purity (53% *ee*) and achiral **51** with (*Z*) geometry, although the regioselectivity is still high (95:5). Here again ferrocenylphosphine BPPFA **4a**, which does not have the hydroxyl group, is less enantioselective (49% *ee* for (*E*)-**49**). In this allylic amination, π-allylpalladium intermediate undergoes the epimerization of monosubstituted π-allyl moiety which is fast compared with the nulceophilic attack of benzylamine, but the *syn-anti* isomerization of the π-allyl is slow.

Intramolecular asymmetric allylic substitution reactions have been applied to the synthesis of optically active cyclic compounds. The high efficiency of the dihydroxylated ferrocenylphosphine **8b** has been shown in the cyclization of 2-butenylene dicarbamates **52** to form optically active 4-vinyl-2-oxazolidones **53**, which are

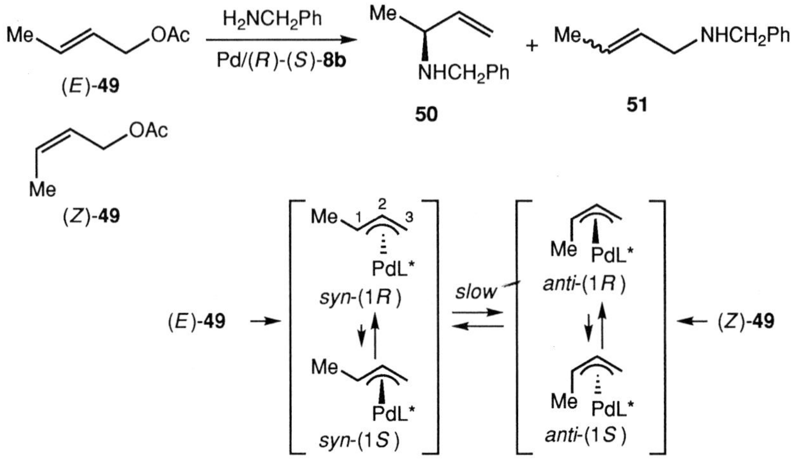

Scheme 2-34

hydrolyzed to 2-amino-3-butenols (Scheme 2-35) [51]. The enantiomeric purity of **53a** is 77% *ee* from (*Z*)-butenylene *N,N*-diphenylcarbamate **52a**. Higher stereoselectivity has been observed for the cyclization of 1,1-dimethyl-2-butenylene dicarbamate **54**, which gave oxazolidone **55** of 90% *ee*, although the regioselectivity is low (30%).

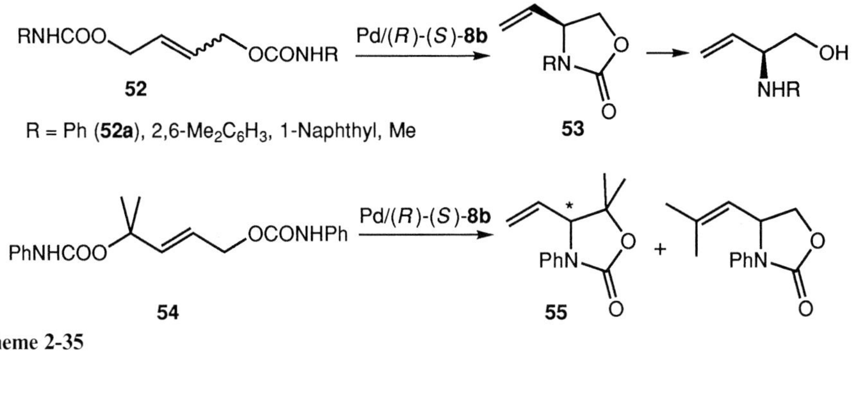

Scheme 2-35

Asymmetric [3 + 2] cycloaddition of 2-(sulfonylmethyl)-2-propenyl carbonate **56** with methyl acrylate or methyl vinyl ketone is also catalyzed by the palladium-ferrocenylphosphine **8b** complex to produce methylenecyclopentane derivatives with up to 78% *ee* (Scheme 2-36) [52].

The chiral ferrocenylbisphosphine BPPFA **4a** has been used for several types of asymmetric allylic substitution reactions, although the enantioselectivity is not always high enough. Five examples in which BPPFA gives higher enantiomeric excess than other chiral phosphine ligands are shown in the following Schemes. (a) The cyclization of an allylic carbonate with palladium/BPPFA **4a** catalyst gives an 83% yield of a cyclohexanone derivative with 48% *ee* (Scheme 2-37) [53]. (b) Reaction of

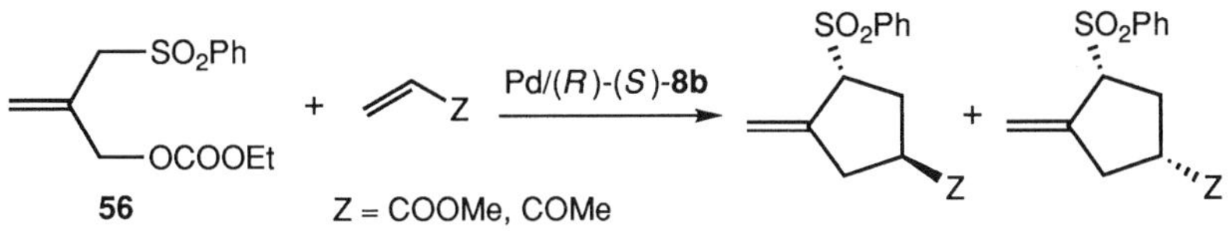

Scheme 2-36

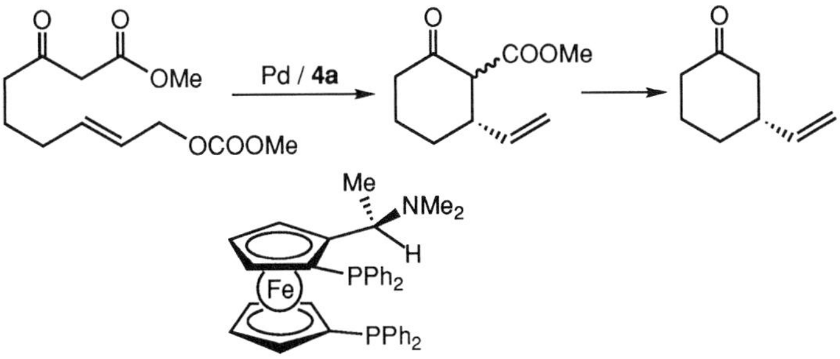

Scheme 2-37

(*Z*)-2-butenylene dicarbonate with dimethyl malonate gives a low yield (20–40%) of 2-vinylcyclopropane-1,1-dicarboxylate with up to 70% *ee* (Scheme 2-38) [54]. Reaction with methyl acetoacetate or acetylacetone takes place in a different manner to give a dihydrofuran derivative (59% *ee*), which results from nucleophilic attack of enolate oxygen at the cyclization step. (c) Asymmetric elimination of an acetylacetate ester gives (*R*)-4-*tert*-butyl-1-vinylcyclohexene of up to 44% *ee* (Scheme 2-39) [55]. (d) Palladium-catalyzed allylic silylation is also applied to asymmetric synthesis

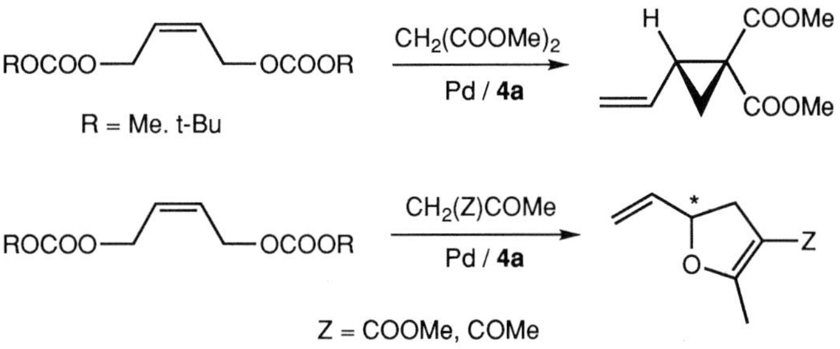

Scheme 2-38

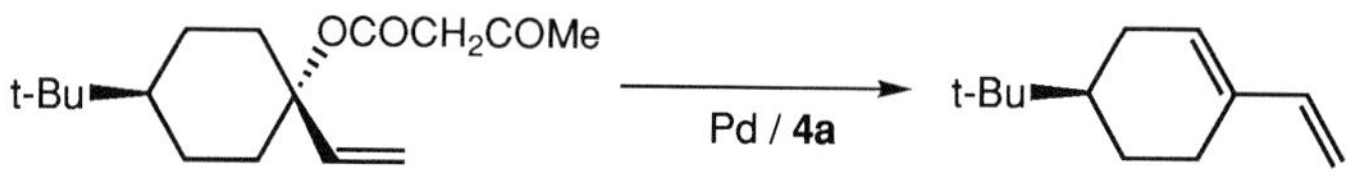

Scheme 2-39

of an optically active allylsilane (61% *ee*) (Scheme 2-40) [56]. (e) A *cis*-decalin derivative has been synthesized with up to 83% *ee* by the enantioposition-selective allylic alkylation of a prochiral allylic acetate (Scheme 2-41) [57].

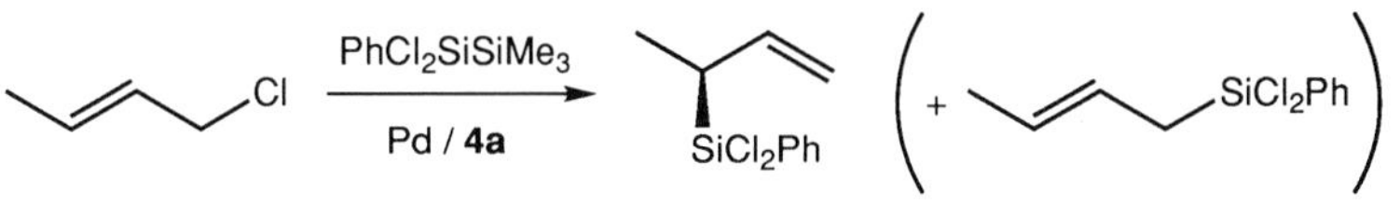

Scheme 2-40

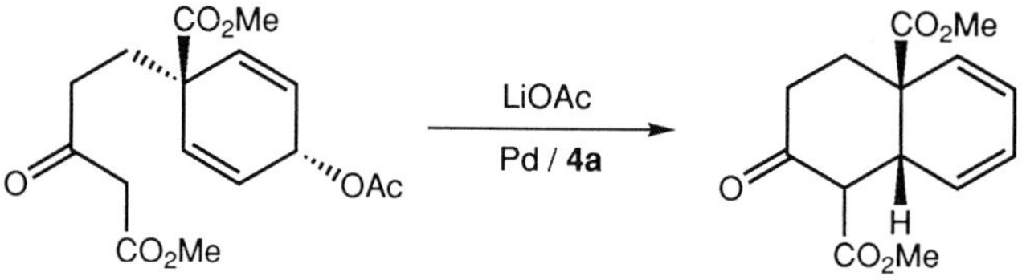

Scheme 2-41

2.4.3 Hydrogenation of Olefins and Ketones

Asymmetric hydrogenation of α-amidoacrylic acids and related olefins has been reported to proceed with high enantioselectivity in the presence of chiral ferrocenylphosphine-rhodium catalysts. Both ferrocenylmonophosphines [21, 24, 58] and ferrocenylbisphosphines [17, 59, 60] can produce *N*-acetylphenylalanine with >80% *ee* in the rhodium-catalyzed hydrogenation of α-acetoaminocinnamic acid or its esters (Scheme 2-42). The high selectivity is ascribed mainly to a characteristic structure of the olefinic substrate that has an amido group at the proper position [3], and the use of the ferrocenylphosphine ligands is not necessary for this high selectivity. Many other chiral phosphine ligands are known to be more effective than the ferrocenylphosphines for the hydrogenation of this particular substrate.

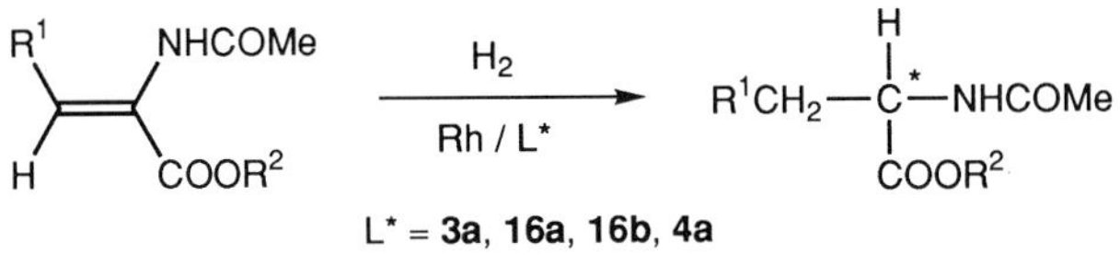

L* = **3a**, **16a**, **16b**, **4a**

Scheme 2-42

The ferrocenylbisphosphines **8f**–**h** bearing the amino pendant side chain are unique ligands that effect the rhodium-catalyzed asymmetric hydrogenation of tetrasubstituted olefins **57** (Scheme 2-43) [61]. Thus, the hydrogenation of α-arylacrylic acid **57a** in the presence of a cationic rhodium catalyst coordinated with **8g** gives a quantitative yield of carboxylic acid **58a** with 98.4% *ee*. Other tetra-

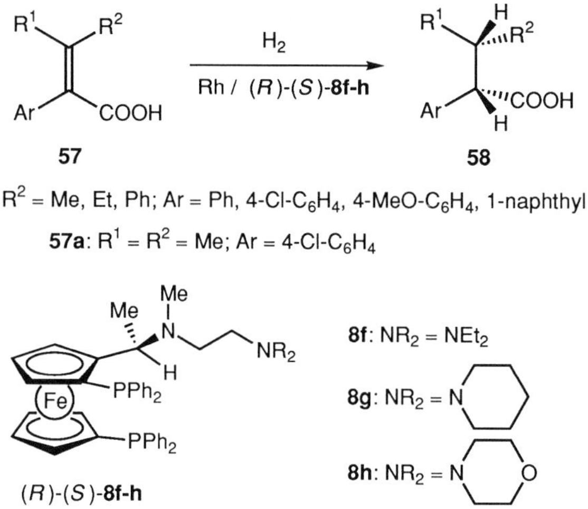

H_2

Rh / (*R*)-(*S*)-**8f-h**

57 **58**

R^1, R^2 = Me, Et, Ph; Ar = Ph, 4-Cl-C_6H_4, 4-MeO-C_6H_4, 1-naphthyl

57a: R^1 = R^2 = Me; Ar = 4-Cl-C_6H_4

(*R*)-(*S*)-**8f-h**

8f: NR_2 = NEt_2

8g: NR_2 = N

8h: NR_2 = N O

Scheme 2-43

substituted olefins **57** can also be converted into the corresponding hydrogenation products with over 95% enantioselectivity. The hydrogenation of the highly substituted olefins is very slow with other phosphine-rhodium catalysts. It is likely that the terminal amino group on the ferrocenylphosphine ligand forms an ammonium carboxylate with the olefinic substrate and consequently attracts the substrate to the coordination sphere of the catalyst to promote the hydrogenation, and the attractive interaction effects the selective enantioface differentiation of the olefin to bring about high enantioselectivity.

For the asymmetric hydrogenation of prochiral carbonyl compounds, a rhodium complex coordinated with ferrocenylbisphosphine **11**, which contains a hydroxyl group at the ferrocenylmethyl position, is one of the most active and enantioselective catalysts (Scheme 2-44) [62]. The rhodium complex catalyzes the hydrogenation of acetophenone and pinacolone, both of which are simple ketones lacking other functional groups, to give a quantitative yield of the corresponding secondary alcohols with 43% *ee*. The hydrogenation with BPPFA **4a** ligand is much slower and the selectivity is lower (15% *ee*). The high efficiency of ligand **11** can probably be ascribed to hydrogen bonding between the carbonyl group on the ketone and the hydroxyl group on the ligand, which will activate the carbonyl group towards hydrogenation and increase conformational rigidity in the diastereomeric transition states. The hydrogenation of functionalized carbonyl compounds such as pyruvic acid or aminomethyl aryl ketone hydrochlorides proceeds with higher enantioselectivity. Medicinally useful 2-amino-1-arylethanols are obtained in up to 95% *ee*. A mechanism for the asymmetric hydrogenation of aminoalkyl ketones has been proposed [63].

The rhodium/**11** complex is also an effective catalyst for the hydrogenation of enol phosphinates giving optically active secondary alkyl alcohols with up to 78% *ee* (Scheme 2-45) [64], although the role of the hydroxyl group on the ligand remains to be clarified.

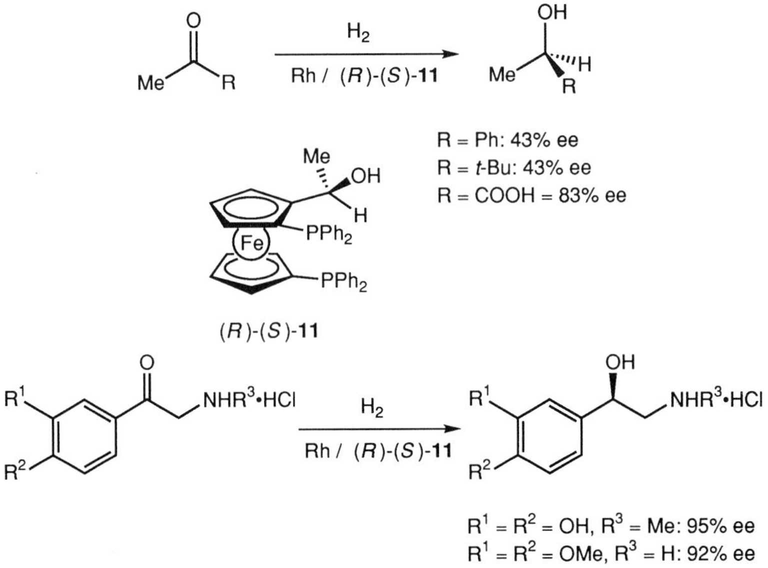

O
Me R
H_2
Rh / (R)-(S)-11
OH
Me H
R

R = Ph: 43% ee
R = t-Bu: 43% ee
R = COOH = 83% ee

Me
OH
H
Fe
PPh_2
PPh_2

(R)-(S)-11

O
R^1
NHR^3•HCl
R^2
H_2
Rh / (R)-(S)-11
OH
R^1
NHR^3•HCl
R^2

$R^1 = R^2 = OH$, $R^3 = Me$: 95% ee
$R^1 = R^2 = OMe$, $R^3 = H$: 92% ee

Scheme 2-44

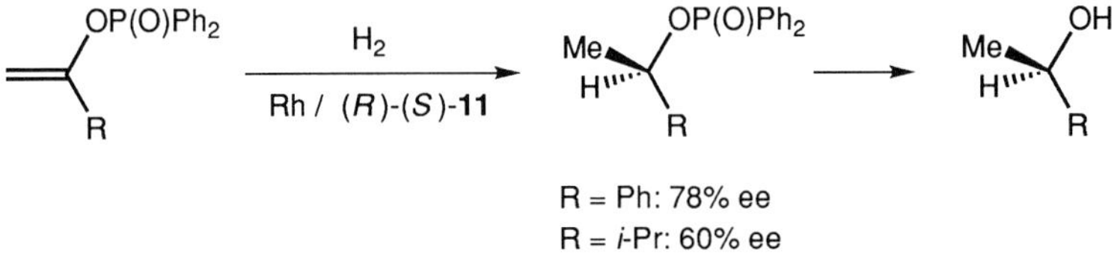

OP(O)Ph_2
R
H_2
Rh / (R)-(S)-11
Me
H
OP(O)Ph_2
R
Me
H
OH
R

R = Ph: 78% ee
R = i-Pr: 60% ee

Scheme 2-45

2.4.4 Hydrosilylation of Olefins and Ketones

For the hydrosilylation of olefins, palladium complexes coordinated with monodentate phosphine ligands are catalytically much more active than those coordinated with chelating bisphosphine ligands. A palladium-ferrocenylmonophosphine complex, $PdCl_2$[(*R*)-(*S*)-PPFA (**3a**)], catalyzes the hydrosilylation of prochiral olefins with trichlorosilane (Scheme 2-46) [65]. The optically active alkyltrichlorosilanes obtained from norbornene and styrene are converted into optically active alcohols and bromides (52–53% *ee*) via pentafluorosilicates. In this hydrosilylation, the dimethylamino group on the PPFA **3a**, which forms ammonium chloride with a trace amount of hydrogen chloride present in the trichlorosilane, is released from palladium and the PPFA coordinates to palladium as a monodentate phosphine

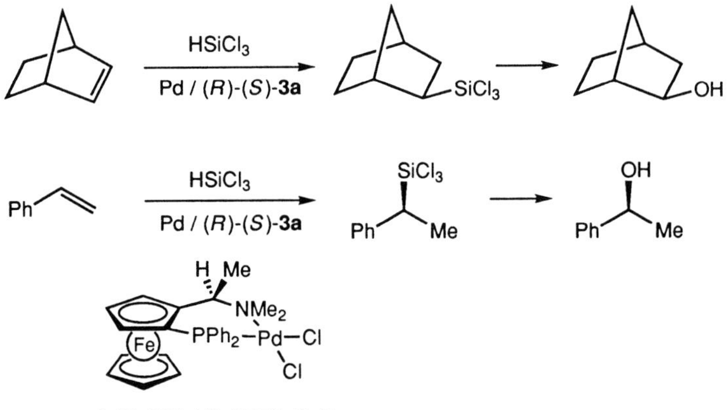

Scheme 2-46

ligand. The PPFA-palladium catalyst is also useful for the asymmetric hydrosilylation of 1,3-dienes, which provides an efficient route to optically active allylsilanes containing functional groups on the silicon atom (Scheme 2-47) [66]. Reaction of 1-phenyl-1,3-butadiene **59a** with trichlorosilane proceeds regioselectivity in a 1,4-fashion to give (*Z*)-1-phenyl-1-silyl-2-butene **60** of 64% *ee*. The hydrosilylation is proposed to proceed via the π-allylpalladium intermediate **61**, which is formed by migration of hydride in the hydrido(silyl)palladium species to the terminal carbon of the diene in a cisoid conformation. The asymmetric hydrosilylation of cyclopentadiene with the PPFA-palladium catalyst gives cyclic allylsilane in around 25% *ee*.

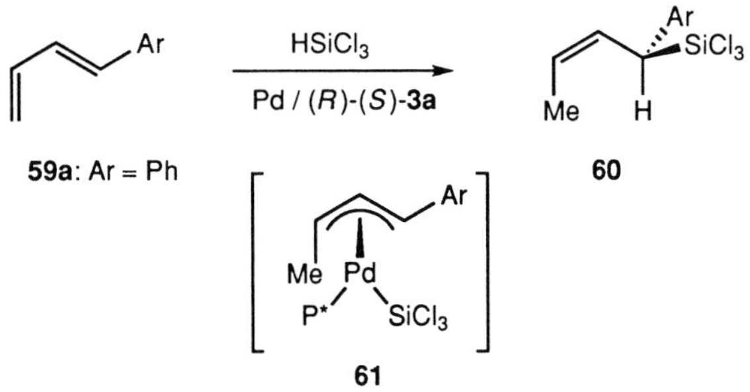

Scheme 2-47

The chiral ferrocenylmonophosphine ligand **7c**, which is analogous to PPFA **3a** but contains a perfluoroalkyl group on the aminoethyl side chain, is more enantioselective than PPFA (Scheme 2-48) [67]. The enantioselectivity in the palladium-catalyzed hydrosilylation of cyclopentadiene and 1-vinylcyclohexene is increased to 60% *ee* and 43% *ee*, respectively. The polymer-supported PPFA **62** has been used for the hydrosilylation of styrene (15% *ee*) [68].

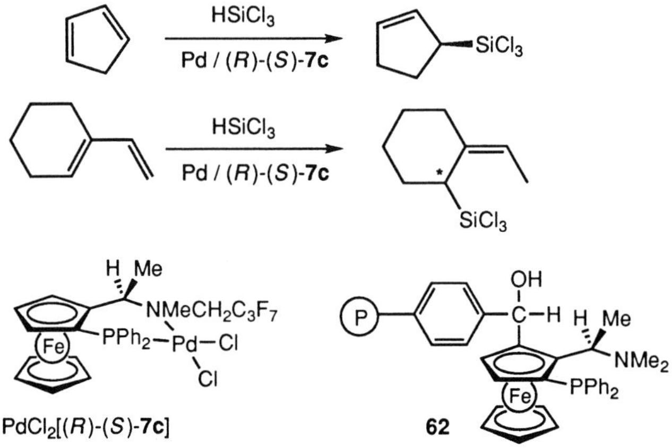

Scheme 2-48

For asymmetric hydrosilylation of ketones a rhodium complex coordinated with the ferrocenyl(dimethyl)phosphine **3b** has been reported to be more effective than other ferrocenylphosphines to give optically active alcohols (up to 49% *ee*) after hydrolysis (Scheme 2-49) [7].

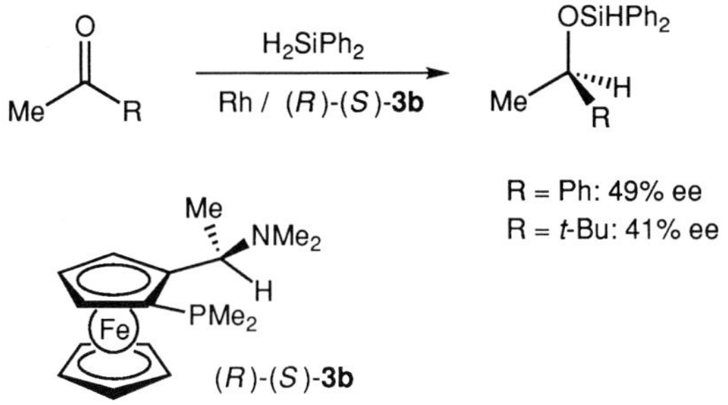

Scheme 2-49

2.4.5 Aldol Reaction of α-Isocyanocarboxylates

One of the most interesting and exciting uses of chiral ferrocenylphosphine ligands is the asymmetric aldol-type reaction of α-isocyanocarboxylates, which provides an efficient route to optically active β-hydroxy-α-amino acids and their derivatives. The first report by Ito and Hayashi appeared in 1986 [69] in which 1 mol% of gold(I) complexes generated from $[Au(c\text{-HexNC})_2]BF_4$ and ferrocenylbisphosphines containing 2-(dialkylamino)ethylamino group at the ferrocenylmethyl position catalyze the reaction of several kinds of aldehydes with methyl isocyanoacetate to give 5-alkyl-2-oxazoline-4-carboxylates **63** with high enantio- and *trans*-selectivity (Scheme 2-50). The *trans*-**63** can be readily hydrolyzed to the corresponding

R	trans	(trans/cis)
R = Ph:	93% ee	(89/11)
R = *i*-Pr:	90% ee	(98/2)
R = *t*-Bu:	97% ee	(100/0)

Scheme 2-50

threo-β-hydroxy-α-amino acids **64**. The *trans*-selectivity and the enantioselectivity for the *trans* isomer are both >90% for most aldehydes, except for small alkyl aldehydes with the ferrocenylphosphine ligand **8f** containing the diethylamino group at the terminal of the side chain. The aminoethyl side chain is essential for the high selectivity. Racemic products are formed with chiral phosphine ligands lacking the side chain and ferrocenylphosphines bearing a 3-aminopropyl side chain are much less enantioselective than those bearing the 2-aminoethyl side chain. It has been proposed that the ferrocenylbisphosphine coordinates to gold(I) with two phosphorus atoms and the dialkylamino group at the terminal of the pendant participates in the formation of enolate of isocyanoacetate coordinated with gold. The intramolecular formation of ammonium enolate may permit a favorable arrangement of the enolate and aldehyde on the gold at the diastereomeric transition state to bring about higher stereoselectivity. The crystal structure of an isolated gold(I)-**8e** complex has been reported to adopt a trimeric structure in which the nitrogen atoms on the ferrocenylbisphosphine **8e** do not coordinate to gold [70].

Both *trans*-selectivity and enantioselectivity depend on the structure of the terminal amino group, six-membered ring amines represented by piperidino **8g** and morpholino **8h** generally being most selective [71]. Substituted aromatic aldehydes, α,β-unsaturated aldehydes, and secondary and tertiary alkyl aldehydes can be converted into the corresponding *trans*-oxazolines with high enantioselectivity. Enantiomeric purities and *trans*/*cis* ratios obtained for the aldol reaction of several aldehydes in the presence of Au/(*R*)-(*S*)-**8h** are shown in Scheme 2-51. The gold-catalyzed aldol reaction of isocyanoacetate has been applied to the synthesis

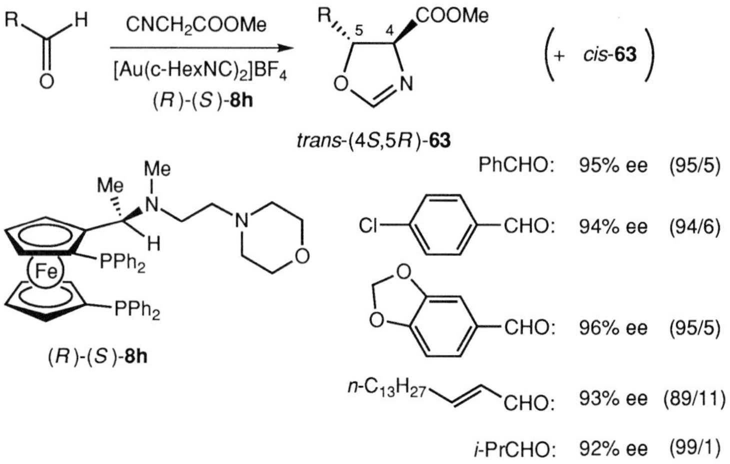

CNCH$_2$COOMe, [Au(c-HexNC)$_2$]BF$_4$, (*R*)-(*S*)-**8h** → *trans*-(4*S*,5*R*)-**63** (+ *cis*-**63**)

(*R*)-(*S*)-**8h**

PhCHO:	95% ee	(95/5)
Cl–C$_6$H$_4$–CHO:	94% ee	(94/6)
3,4-(methylenedioxy)C$_6$H$_3$–CHO:	96% ee	(95/5)
n-C$_{13}$H$_{27}$CH=CH–CHO:	93% ee	(89/11)
i-PrCHO:	92% ee	(99/1)

Scheme 2-51

of optically active sphingosines [72] and MeBmt, which is Cyclosporin's unusual amino acid [73].

The importance of the carbon central chirality at the ferrocenylmethyl position has been reported by Togni and Pastor (Scheme 2-52) [13, 74]. The ferrocenylphosphine ligand (*R*)-(*S*)-**8e**, where *R* and *S* refer to the carbon central chirality and ferrocene planar chirality, respectively, shows high enantioselectivity for the aldol reaction of benzaldehyde to give *trans*-oxazoline (*4S*, *5R*)-**63a** of high % *ee* (91% *ee*, *trans*/*cis* = 90/10), while the diastereomeric ligand (*S*)-(*S*)-**8e**, in which the carbon central chirality is changed to *S*, gives enantiomeric *trans*-oxazoline (*4R*, *5S*)-**63a** of much lower % *ee* (41% *ee*). NMR studies of (*R*)-(*S*)-**8e** and (*S*)-(*S*)-**8e** suggest that the direction of the aminoalkyl pendant side chain is not the same in both compound.

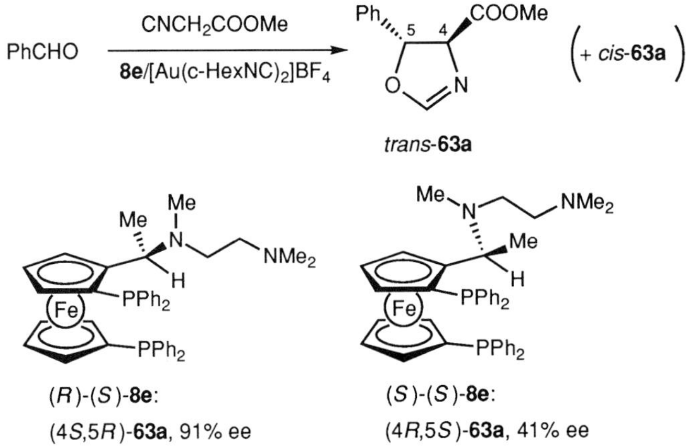

PhCHO, CNCH$_2$COOMe, **8e**/[Au(c-HexNC)$_2$]BF$_4$ → *trans*-**63a** (+ *cis*-**63a**)

(*R*)-(*S*)-**8e**: (4*S*,5*R*)-**63a**, 91% ee

(*S*)-(*S*)-**8e**: (4*R*,5*S*)-**63a**, 41% ee

Scheme 2-52

A number of ferrocenylphosphines containing chiral carbon center(s) on the side chain have been also prepared and examined for the gold-catalyzed aldol reaction [16, 75]. The enantioselectivity is strongly dependent on the configurations of the chiral carbons. A few of the good results obtained for the aldol reaction of benzaldehyde are shown in Scheme 2-53.

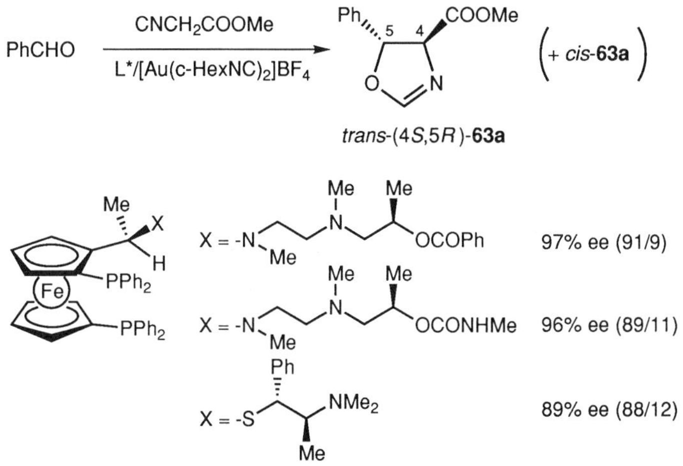

PhCHO → ($CNCH_2COOMe$, $L^*/[Au(c\text{-}HexNC)_2]BF_4$) *trans*-(4*S*,5*R*)-**63a** (+ *cis*-**63a**)

X = $-N(Me)CH_2CH_2N(Me)CH_2CH(Me)OCOPh$ — 97% ee (91/9)

X = $-N(Me)CH_2CH_2N(Me)CH_2CH(Me)OCONHMe$ — 96% ee (89/11)

X = $-SCH(Ph)CH(Me)NMe_2$ — 89% ee (88/12)

Scheme 2-53

For the aldol reaction of small alkyl aldehydes such as acetaldehyde, the enantioselectivity is improved by the use of *N*,*N*-dialkyl-α-isocyanoacetamides instead of isocyanoacetate esters (Scheme 2-54) [76]. For example, the reaction of acetaldehyde with *N*,*N*-dialkyl-α-isocyanoacetamides **64** in the presence of (*R*)-(*S*)-**8g**/gold catalyst gives the corresponding *trans*-oxazoline **65** of 99% *ee*, which is much higher than the enantioselectivity (85% *ee*) observed in the reaction with methyl isocyanoacetate under the same reaction conditions.

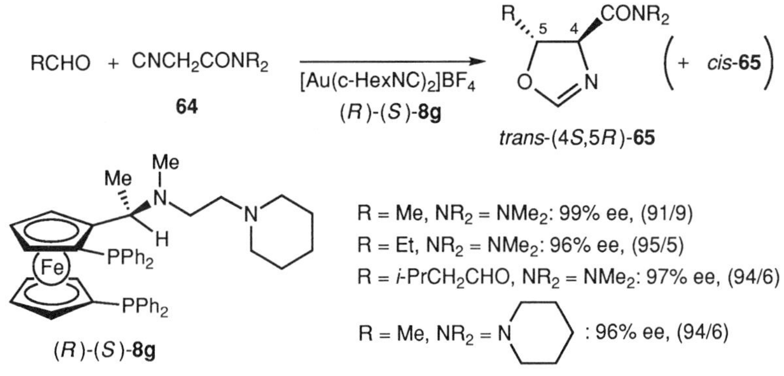

RCHO + $CNCH_2CONR_2$ (**64**) → ($[Au(c\text{-}HexNC)_2]BF_4$, (*R*)-(*S*)-**8g**) *trans*-(4*S*,5*R*)-**65** (+ *cis*-**65**)

(*R*)-(*S*)-**8g**

R = Me, $NR_2 = NMe_2$: 99% ee, (91/9)
R = Et, $NR_2 = NMe_2$: 96% ee, (95/5)
R = *i*-$PrCH_2CHO$, $NR_2 = NMe_2$: 97% ee, (94/6)
R = Me, NR_2 = N(piperidino) : 96% ee, (94/6)

Scheme 2-54

The gold-catalyzed aldol reaction of α-keto esters gives the corresponding oxazolines with up to 90% *ee*, which are converted into optically actice β-alkyl-β-hydroxyaspartic acid derivatives (Scheme 2-55) [77].

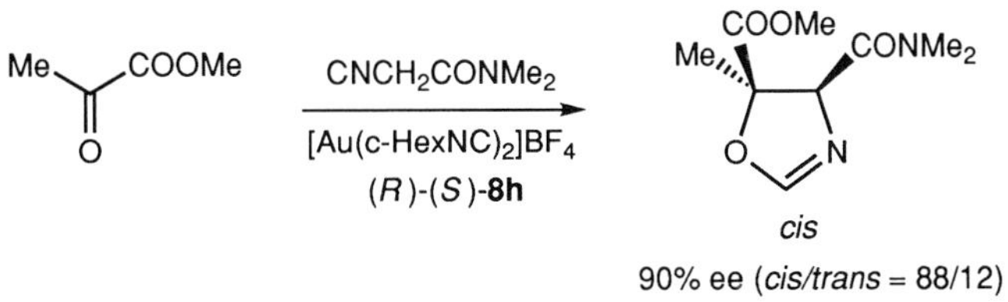

Me–C(=O)–COOMe + $CNCH_2CONMe_2$ → ([Au(c-HexNC)$_2$]BF_4, (*R*)-(*S*)-**8h**) → oxazoline (Me, COOMe, $CONMe_2$)

cis

90% ee (*cis/trans* = 88/12)

Scheme 2-55

α-Alkyl-substituted β-hydroxy-α-amino acids can be prepared by the asymmetric aldol reaction of α-alkyl-substituted isocyanocarboxylates, which are readily available from proteinogenic α-amino acids (Scheme 2-56) [78, 79]. Hydroxymethylation with paraformaldehyde, where the stereoselectivity is concerned only with the enantioface differentiation of the enolate of α-isocyanocarboxylates, proceeds with moderate enantioselectivity to give high yields of 4-alkyl-2-oxazoline-4-carboxylates, which are readily hydrolyzed to optically active α-alkylserines (52 – 81% *ee*). The combination of α-alkyl-substituted isocyanocarboxylates with prochiral aldehydes such as acetaldehyde or benzaldehyde gives the corresponding oxazolines with high enantiomerily purity, although the diastereoselectivity is not always high. The oxazolinecarboxylates obtained are converted by hydrolysis into α,β-dialkylserines.

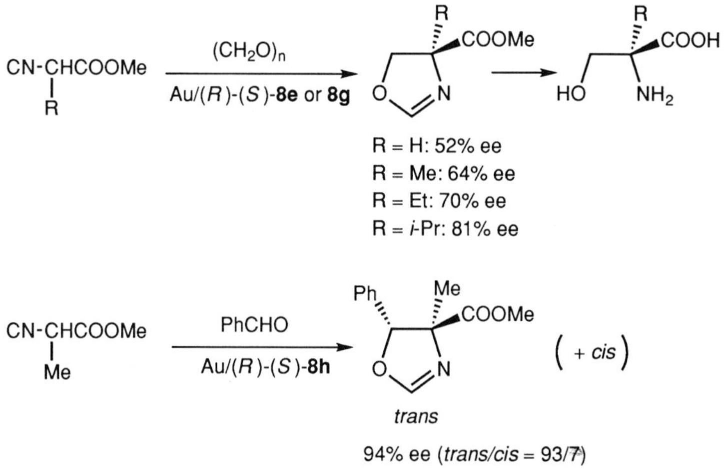

CN-CH(R)COOMe → ($(CH_2O)_n$, Au/(*R*)-(*S*)-**8e** or **8g**) → oxazoline (R, COOMe) → HO–CH$_2$–C(R)(NH$_2$)COOH

R = H: 52% ee
R = Me: 64% ee
R = Et: 70% ee
R = *i*-Pr: 81% ee

CN-CH(Me)COOMe → (PhCHO, Au/(*R*)-(*S*)-**8h**) → oxazoline (Ph, Me, COOMe) (+ *cis*)

trans

94% ee (*trans/cis* = 93/7)

Scheme 2-56

The gold-catalyzed aldol reaction has been successfully applied to the asymmetric synthesis of (1-aminoalkyl)phosphonic acids, phosphonic analogs of α-amino acids (Scheme 2-57) [80, 81]. The reaction of (isocyanomethyl)phosphonates **65** with aldehydes proceeds with high *trans*-selectivity (>98%) to give *trans*-5-alkyl-2-oxazoline-4-phosphonates **66** in high yield. The enantioselectivity observed with phenyl ester **66a** is over 95% *ee*.

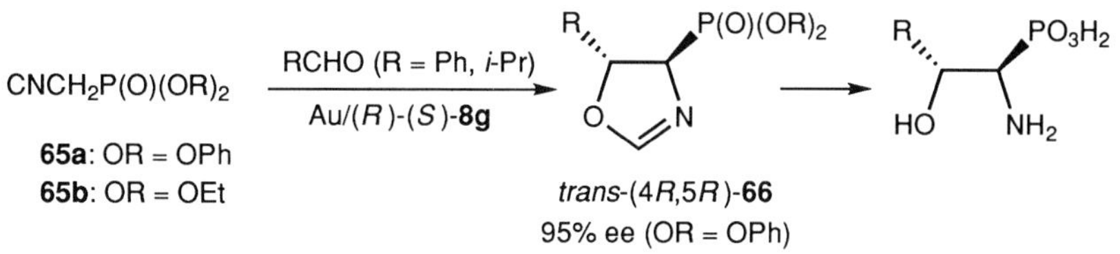

$CNCH_2P(O)(OR)_2$ → [RCHO (R = Ph, *i*-Pr); Au/(*R*)-(*S*)-**8g**] → *trans*-(4*R*,5*R*)-**66** (95% ee (OR = OPh)) → amino hydroxy phosphonic acid (PO_3H_2, HO, NH_2)

65a: OR = OPh
65b: OR = OEt

Scheme 2-57

Silver(I) complex coordinated with the ferrocenylbisphosphine ligand **8g** is also effective as a catalyst for the asymmetric aldol reaction of isocyanoacetate when the isocyanoacetate is kept in low concentration in the reaction system (Scheme 2-58) [82]. Thus, by the slow addition of isocyanoacetate over a period of 1 h to a solution of aldehyde and the silver catalyst, *trans*-oxazolines are formed in 80–90% *ee*, the enantioselectivity being only a little lower than that observed in the gold(I)-catalyzed

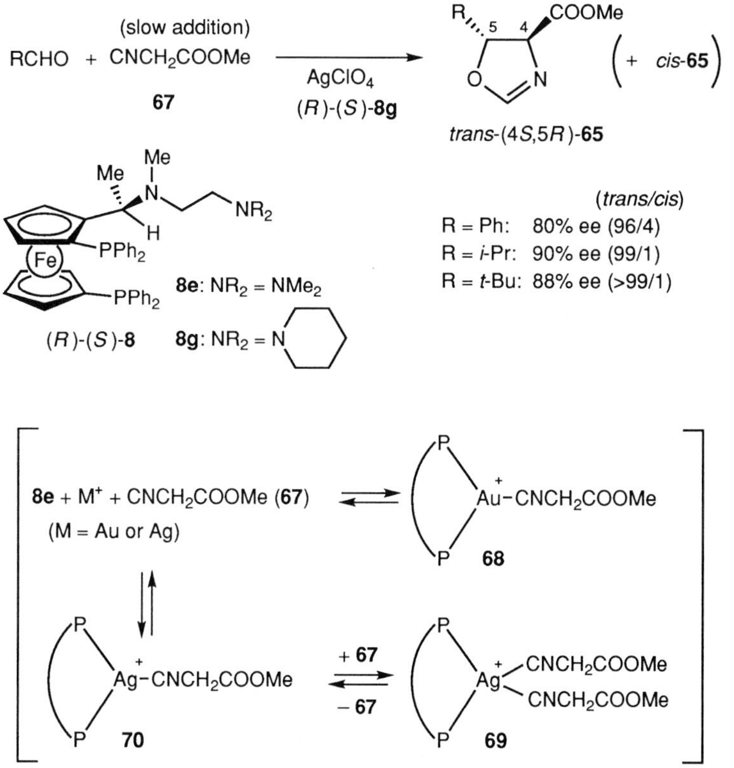

RCHO + $CNCH_2COOMe$ (**67**) (slow addition) → [$AgClO_4$, (*R*)-(*S*)-**8g**] → *trans*-(4*S*,5*R*)-**65** (+ *cis*-**65**)

R	ee	(*trans/cis*)
R = Ph:	80% ee	(96/4)
R = *i*-Pr:	90% ee	(99/1)
R = *t*-Bu:	88% ee	(>99/1)

(*R*)-(*S*)-**8**
8e: $NR_2 = NMe_2$
8g: NR_2 = N(piperidino)

8e + M^+ + $CNCH_2COOMe$ (**67**) (M = Au or Ag) ⇌ P–Au^+–$CNCH_2COOMe$ (**68**)

⇅ P–Ag^+–$CNCH_2COOMe$ (**70**) ⇌ (+ **67** / − **67**) P–Ag^+($CNCH_2COOMe$)$_2$ (**69**)

Scheme 2-58

reaction. Under the standard conditions where isocyanoacetate is present in a high concentration, silver is much less stereoselective than gold. On the basis of IR studies on the structure of gold(I) and silver(I) complexes coordinated with ferrocenylphosphine **8e** in the presence of methyl isocyanoacetate **67**, it has been proposed that the most significant difference between those metals is in the coordination number of the isocyanide to metal. The gold(I) adopts tri-coordinated structure **68** with two phosphorus atoms in **8e** and one molecule of **67**, even in the presence of a large excess of isocyanide. On the other hand, two molecules of isocyanide coordinate to silver(I) to form a tetra-coordinated complex **69**, which is in equilibrium with tri-coordinated species **70**. In the presence of 20 equiv. of **67**, only **69** is observed. The equilibrium constant for the formation of tetra-coordinated silver species **69** is larger at lower temperature. The high enantioselectivity observed in the slow addition method and studies on the temperature effects on the stereoselectivity suggest that the tri-coordinated species is the key intermediate for the high stereoselectivity. A conformation of tetra-coordinated silver complex in solution has been proposed by NOE studies where the terminal amino group of pendant side chain is located close to methylene hydrogens of isocyanoacetate coordinated to silver [83]. A transition-state model for the gold-catalyzed reaction has been proposed by Togni [13b].

The ferrocenylphosphine-silver complex catalyzes the aldol-type reaction of tosylmethyl isocyanide **71** with aldehydes with higher stereoselectivity than the gold complex (Scheme 2-59) [84]. The reaction with several aldehydes produces *trans*-4-tosyloxazolines **72** in up to 86% *ee*, which can be converted into optically active 1-alkyl-2-aminoethanols by reduction with $LiAlH_4$.

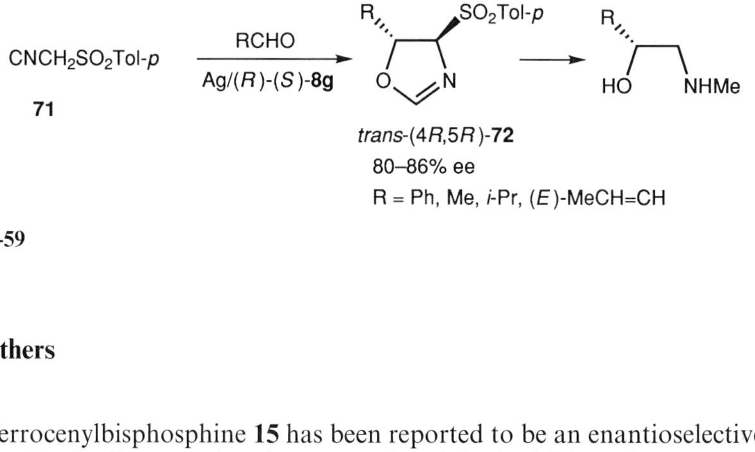

$CNCH_2SO_2Tol$-*p* **71**

RCHO, Ag/(*R*)-(*S*)-**8g**

R, SO_2Tol-*p*, O, N

trans-(4*R*,5*R*)-**72**
80–86% ee
R = Ph, Me, *i*-Pr, (*E*)-MeCH=CH

R, HO, NHMe

Scheme 2-59

2.4.6 Others

The bisferrocenylbisphosphine **15** has been reported to be an enantioselective chiral ligand for the rhodium-catalyzed asymmetric Michael addition of α-cyano carboxylates [85] to vinyl ketones (Scheme 2-60). The *trans*-chelation of **15** generates efficient chiral surroundings around the rhodium atom to produce a quaternary chiral carbon center with up to 89% *ee*.

For an intramolecular asymmetric Heck-type reaction of alkenyl iodide **73** to form an optically active indolizidine derivative, a palladium catalyst coordinated

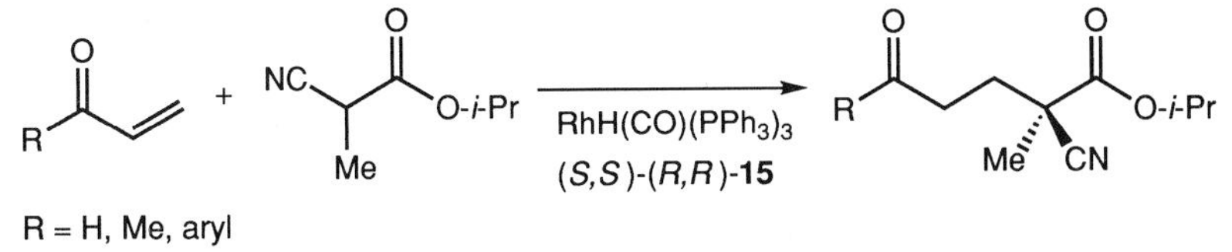

R = H, Me, aryl

Scheme 2-60

with hydroxyferrocenylbisphosphine **11** is more stereoselective than those with other chiral bisphosphine ligands giving the cyclization products as a mixture of **74** and **75** of up to 86% *ee* (Scheme 2-61) [86]. An interaction between the hydroxyl group on **11** and the carbonyl group on **73** has been proposed for the high enantioselectivity of **11.**

A new type of ferrocenylbisphosphine (*R*)-(*S*)-**13b** has been used for the rhodium-catalyzed hydroboration of styrene, which gives, after oxidation, (*R*)-1-phenylethanol of 92% *e* [17].

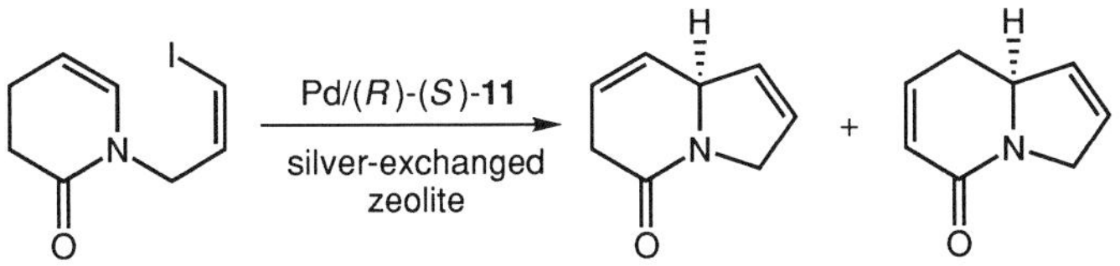

Scheme 2-61

References

[1] For recent reviews: a) H. Brunner, *Synthesis* **1988**, 645–654; b) H. Brunner, *Top. Stereochem.* **1988**, *18*, 129–247; c) G. Consiglio, R. M. Waymouth, *Chem. Rev.* **1989**, *89*, 257–276, d) R. Noyori, M. Kitamura, in *Modern Synthetic Methods* (Ed.: R. Scheffold), Springer-Verlag, New York**1989**, Vol. 5, p. 115; e) I. Ojima, N. Clos, C. Bastos, *Tetrahedron* **1989**, *45*, 6901–6939; f) I. Ojima, *Catalytic Asymmetric Synthesis*, VCH Publishers New York, **1993**.

[2] For reviews: a) B. M. Trost, T. R. Verhoeven, in *Comprehensive Organometallic Chemistry* (Eds.: G. Wilkinson, F. G. A. Stone, E. W. Abel) Pergamon, Oxford, **1982**, Vol. 8, p. 799; b) R. F. Heck, *Palladium Reagents in Organic Synthesis*, Academic Press, New York, **1985**; c) P. W. Jolly, in *Comprehensive Organometallic Chemistry* (Eds.: G. Wilkinson, F. G. A. Stone, E. W. Abel), Pergamon, Oxford, **1982**, Vol. 8, p. 713.

[3] Review: J. D. Morrison, *Asymmetric Synthesis*, Academic Press, London, **1985**, Vol. 5.

[4] R. Noyori, H. Takaya, *Acc. Chem. Res.* **1990,** *23*, 345–350, and references cited therein.

[5] For earlier reviews on chiral ferrocenylphosphines: a) T. Hayashi, M. Kumada, in *Fundamental Research in Homegeneous Catalysis* (Eds.: Y. Ishii, M. Tsutsui), Plenum, New York, **1978**, Vol. 2, p. 159; b) T. Hayashi, M. Kumada, *Acc. Chem. Res.* **1982,** *15*, 395–401; c) T. Hayashi, in *Organic Synthesis: An Interdisciplinary Challenge* (Eds.: J. Streith, H. Prinzbach, G. Schill), Blackwell Scientific Publishers, Boston, **1985**, p. 35; d) T. Hayashi, *Pure Appl. Chem.* **1988,** *60*, 7–12.

[6] For a pertinent review on the secondary interaction: M. Sawamura, Y. Ito, *Chem. Rev.* **1992**, *92*, 857–871.
[7] T. Hayashi, K. Yamamoto, M. Kumada, *Tetrahedron Lett.* **1974**, 4405–4408.
[8] D. Marquarding, H. Klusacek, G. Gokel, P. Hoffmann, I. Ugi, *J. Am. Chem. Soc.* **1970**, *92*, 5389–5393.
[9] T. Hayashi, T. Mise, M. Fukushima, M. Kagotani, N. Nagashima, Y. Hamada, A. Matsumoto, S. Kawakami, M. Konishi, K. Yamamoto, M. Kumada, *Bull. Chem. Soc. Jpn.* **1980**, *53*, 1138–1151.
[10] T. D. Appleton, W. R. Cullen, S. V. Evans, T.-J. Kim, J. Trotter, *J. Organomet. Chem.* **1985**, *279*, 5–21.
[11] R. Sihler, U. Werz, H.-A. Brune, *J. Organomet. Chem.* **1989**, *368*, 213–221.
[12] T. Hayashi, A. Yamazaki, *J. Organomet. Chem.* **1991**, *413*, 295–302.
[13] a) S. D. Pastor, A. Togni, *J. Am. Chem. Soc.* **1989**, *111*, 2333–2334; b) A. Togni, S. D. Pastor, *J. Org. Chem.* **1990**, *55*, 1649–1664.
[14] N. Deus, G. Hübener, R. Herrmann, *J. Organomet. Chem.* **1990**, *384*, 155–163.
[15] G. Gokel, D. Marquarding, I. K. Ugi, *J. Org. Chem.* **1972**, *37*, 3052–3058.
[16] a) A. Togni, R. Häusel, *Synlett.* **1990**, 633–635; b) A. Togni, G. Rihs, R. E. Blumer, *Organometallics* **1992**, *11*, 613–621.
[17] A. Togni, C. Breutel, A. Schnyder, F. Spindler, H. Landert, A. Tijani, *J. Am. Chem. Soc.* **1994**, *116*, 4062–4066.
[18] T. Hayashi, A. Yamamoto, M. Hojo, Y. Ito, *J. Chem. Soc., Chem. Commun.* **1989**, 495–496.
[19] T. Hayashi, A. Yamamoto, M. Hojo, K. Kishi, Y. Ito, E. Nishioka, H. Miura, K. Yanagi, *J. Organomet. Chem.* **1989**, *370*, 129–139.
[20] M. Sawamura, H. Hamashima, Y. Ito, *Tetrahedron Asymmetry* **1991**, *2*, 593–596.
[21] K. Yamamoto, J. Wakatsuki, R. Sugimoto, *Bull. Chem. Soc. Jpn.* **1980**, *53*, 1132–1137.
[22] M. Sawamura, A. Yamauchi, T. Takegawa, Y. Ito, *J. Chem. Soc., Chem. Commun.* **1991**, 874–875.
[23] F. H. van der Steen, J. A. Kanters, *Acta Cryst.* **1986**, *C42*, 547–550.
[24] W. R. Cullen, F. W. B. Einstein, C.-H. Huang, A. C. Willis, E.-S. Yeh, *J. Am. Chem. Soc.* **1980**, *102*, 988–993.
[25] T. Hayashi, M. Kumada, T. Higuchi, K. Hirotsu, *J. Organomet. Chem.* **1987**, *334*, 195–203.
[26] T. Hayashi, M. Konishi, Y. Kobori, M. Kumada, T. Higuchi, K. Hirotsu, *J. Am. Chem. Soc.* **1984**, *106*, 158–163.
[27] T. Hayashi, A. Yamamoto, Y. Ito, E. Nishioka, H. Miura, K. Yanagi, *J. Am. Chem. Soc.* **1989**, *111*, 6301–6311.
[28] a) T. Hayashi, M. Tajika, K. Tamao, M. Kumada, *J. Am. Chem. Soc.* **1976**, *98*, 3718–3719; b) T. Hayashi, M. Konishi, M. Fukushima, T. Mise, M. Kagotani, M. Tajika, M. Kumada, *J. Am. Chem. Soc.* **1982**, *104*, 180–186.
[29] T. Hayashi, M. Konishi, T. Hioki, M. Kumada, A. Ratajczak, H. Niedbala, *Bull. Chem. Soc. Jpn.* **1981**, *54*, 3615–3616.
[30] K. V. Baker, J. M. Brown, N. A. Cooley, G. D. Hughes, R. J. Taylor, *J. Organomet. Chem.* **1989**, *370*, 397–406.
[31] C. Cardellicchio, V. Fiandanese, F. Naso, *Gazz. Chim. Ital.* **1991**, *121*, 11–16.
[32] K. Tamao, T. Hayashi, H. Matsumoto, H. Yamamoto, M. Kumada, *Tetrahedron Lett.* **1979**, 2155–2156.
[33] M. Zembayashi, K. Tamao, T. Hayashi, T. Mise, M. Kumada, *Tetrahedron Lett.* **1977**, 1799–1802.
[34] T. Hayashi, T. Hagihara, Y. Katsuro, M. Kumada, *Bull. Chem. Soc. Jpn.* **1983**, *56*, 363–364.
[35] R. Galarini, A. Musco, R. Pontellini, R. Santi, *J. Mol. Cat.* **1992**, *72*, L11–L13.
[36] a) T. Hayashi, M. Konishi, H. Ito, M. Kumada, *J. Am. Chem. Soc.* **1982**, *104*, 4962–4963; b) T. Hayashi, M. Konishi, Y. Okamoto, K. Kabeta, M. Kumada, *J. Org. Chem.* **1986**, *51*, 3772–3781; c) T. Hayashi, *Chemica Scripta* **1985**, *25*, 61–70.
[37] T. Hayashi, Y. Okamoto, M. Kumada, *Tetrahedron Lett.* **1983**, *24*, 807–808.
[38] M. Uemura, H. Nishimura, T. Hayashi, *Tetrahedron Lett.* **1993**, *34*, 107–110.
[39] a) K. Tamao, H. Yamamoto, H. Matsumoto, N. Miyake, T. Hayashi, M. Kumada, *Tetrahedron Lett.* **1977**, 1389–1392; b) K. Tamao, A. Minato, N. Miyake, T. Matsuda, Y. Kiso, M. Kumada, *Chem. Lett.* **1975**, 133–136.

[40] T. Hayashi, K. Hayashizaki, T. Kiyoi, Y. Ito, *J. Am. Chem. Soc.* **1988**, *110*, 8153–8156.
[41] T. Hayashi, K. Hayashizaki, Y. Ito, *Tetrahedron Lett.* **1989**, *30*, 215–218.
[42] Review: C. G. Frost, J. Howarth, J. M. J. Williams, *Tetrahedron Asymmetry* **1992**, *3*, 1089–1122.
[43] T. Hayashi, K. Kanehira, T. Hagihara, M. Kumada, *J. Org. Chem.* **1988**, *53*, 113–120.
[44] Y. Ito, M. Sawamura, M. Matsuoka, M. Matsumoto, T. Hayashi, *Tetrahedron Lett.* **1987**, *28*, 4849–4852.
[45] M. Sawamura, H. Nagata, H. Sakamoto, Y. Ito, *J. Am. Chem. Soc.* **1992**, *114*, 2586–2592.
[46] a) M. Yamaguchi, T. Shima, T. Yamagishi, M. Hida, *Tetrahedron Lett.* **1990**, *31*, 5049–5052; b) M. Yamaguchi, T. Shima, T. Yamagishi, M. Hida, *Tetrahedron Asymmetry* **1991**, *2*, 663–666.
[47] T. Hayashi, A. Yamamoto, T. Hagihara, Y. Ito, *Tetrahedron Lett.* **1986**, *27*, 191–194.
[48] T. Hayashi, A. Yamamoto, Y. Ito, *Chem. Lett.* **1987**, 177–180.
[49] T. Hayashi, A. Yamamoto, Y. Ito, *J. Chem. Soc., Chem. Commun.* **1986**, 1090–1092.
[50] T. Hayashi, K. Kishi, A. Yamamoto, Y. Ito, *Tetrahedron Lett.* **1990**, *31*, 1743–1746.
[51] T. Hayashi, A. Yamamoto, Y. Ito, *Tetrahedron Lett.* **1988**, *29*, 99–102.
[52] A. Yamamoto, Y. Ito, T. Hayashi, *Tetrahedron Lett.* **1989**, *30*, 375–378.
[53] K. Yamamoto, J. Tsuji, *Tetrahedron Lett.* **1982**, *23*, 3089–3092.
[54] T. Hayashi, A. Yamamoto, Y. Ito, *Tetrahedron Lett.* **1988**, *29*, 669–672.
[55] T. Hayashi, K. Kishi, Y. Uozumi, *Tetrahedron Asymmetry* **1991**, 2, 195–198.
[56] Y. Matsumoto, A. Ohno, T. Hayashi, *Organomettalics* **1993**, *12*, 4051–4055.
[57] T. Takemoto, Y. Nishikimi, M. Sodeoka, M. Shibasaki, *Tetrahedron Lett.* **1992**, *33*, 3531–3532.
[58] W. R. Cullen, J. D. Woollins, *Can. J. Chem.* **1982**, *60*, 1793–1799.
[59] T. Hayashi, T. Mise, S. Mitachi, K. Yamamoto, M. Kumada, *Tetrahedron Lett.* **1976**, 1133–1134.
[60] H. Brunner, B. Schönhammer, B. Schönhammer, C. Steinberger, *Chem. Ber.* **1983**, *116*, 3529–3538.
[61] a) T. Hayashi, N. Kawamura, Y. Ito, *J. Am. Chem. Soc.* **1987**, *109*, 7876–7878; b) T. Hayashi, N. Kawamura, T. Ito, *Tetrahedron Lett.* **1988**, *29*, 5969–5972.
[62] a) T. Hayashi, T. Mise, M. Kumada, *Tetrahedron Lett.* **1976**, 4351–4354; b) T. Hayashi, A. Katsumura, M. Konishi, M. Kumada, *Tetrahedron Lett.* **1979**, 425–428.
[63] K. Inoguchi, S. Sakuraba, K. Achiwa, *Synlett* **1992**, 169–178.
[64] T. Hayashi, K. Kanehira, M. Kumada, *Tetrahedron Lett.* **1981**, *22*, 4417–4420.
[65] T. Hayashi, K. Tamao, Y. Katsuro, I. Nakae, M. Kumada, *Tetrahedron Lett.* **1980**, *21*, 1871–1874.
[66] a) T. Hayashi, K. Kabeta, T. Yamamoto, K. Tamao, M. Kumada, *Tetrahedron Lett.* **1983**, *24*, 5661–5664; b) T. Hayashi, K. Kabeta, *Tetrahedron Lett.* **1985**, *26*, 3023–3026.
[67] a) T. Hayashi, Y. Matsumoto, I. Morikawa, Y. Ito, *Tetrahedron Asymmetry* **1990**, *1*, 151–154; b) T. Hayashi, S. Hengrasmee, Y. Matsumoto, *Chem. Lett.* **1990**, 1377–1380.
[68] W. R. Cullen, N. F. Han, *J. Organomet. Chem.* **1987**, *333*, 269–280.
[69] Y. Ito, M. Sawamura, T. Hayashi, *J. Am. Chem. Soc.* **1986**, 108, 6405–6406.
[70] A. Togni, S. D. Pastor, G. Rihs, *J. Organomet. Chem.* **1990**, *381*, C21–C25.
[71] a) Y. Ito, M. Sawamura, T. Hayashi, *Tetrahedron Lett.* **1987**, *28*, 6215–6218; b) T. Hayashi, M. Sawamura, Y. Ito, *Tetrahedron* **1992**, *48*, 1999–2012.
[72] Y. Ito, M. Sawamura, T. Hayashi, *Tetrahedron Lett.* **1988**, *29*, 239–240.
[73] A. Togni, S. D. Pastor, G. Rihs, *Helv. Chim. Acta* **1989**, *72*, 1471–1478.
[74] A. Togni, R. E. Blumer, P. S. Pregosin, *Helv. Chim. Acta* **1991**, *74*, 1533–1543.
[75] a) S. D. Pastor, A. Togni, *Tetrahedron Lett.* **1990,** *31*, 839–840; b) S. D. Pastor, A. Togni, *Helv. Chim. Acta* **1991**, *74*, 905–933; c) S. D. Pastor, R. Kesselring, A. Togni, *J. Organomet. Chem.* **1992**, *429*, 415–420.
[76] Y. Ito, M. Sawamura, M. Kobayashi, T. Hayashi, *Tetrahedron Lett.* **1988**, *29*, 6321–6324.
[77] Y. Ito, M. Sawamura, H. Hamashima, T. Emura, T. Hayashi, *Tetrahedron Lett.* **1989**, *30*, 4681–4684.
[78] Y. Ito, M. Sawamura, E. Shirakawa, K. Hayashizaki, T. Hayashi, *Tetrahedron Lett.* **1988**, *29*, 235–238.
[79] Y. Ito, M. Sawamura, E. Shirakawa, K. Hayashizaki, T. Hayashi, *Tetrahedron* **1988**, *44*, 5253–5262.

[80] A. Togni, S. D. Pastor, *Tetrahedron Lett.* **1989**, *30*, 1071–1072.
[81] M. Sawamura, Y. Ito, T. Hayashi, *Tetrahedron Lett.* **1989**, *30*, 2247–2250.
[82] T. Hayashi, Y. Uozumi, A. Yamazaki, M. Sawamura, H. Hamashima, Y. Ito, *Tetrahedron Lett.* **1991**, *32*, 2799–2802.
[83] M. Sawamura, Y. Ito, T. Hayashi, *Tetrahedron Lett.* **1990**, *31*, 2723–2726.
[84] M. Sawamura, H. Hamashima, Y. Ito, *J. Org. Chem.* **1990**, *55*, 5935–5936.
[85] M. Sawamura, H. Hamashima, Y. Ito, *J. Am. Chem. Soc.* **1992**, *114*, 8295–8296.
[86] S. Nukui, M. Sodeoka, M. Shibasaki, *Tetrahedron Lett.* **1993**, *34*, 4965–4968.

Received: October 18, 1993

3 Enantioselective Addition of Dialkylzinc to Aldehydes Catalyzed by Chiral Ferrocenyl Aminoalcohols

Yasuo Butsugan, Shuki Araki and Makoto Watanabe

3.1 Introduction

Some kinds of optically active ferrocenyl compounds have been known as useful chiral auxiliaries in asymmetrically induced synthesis. Chiral α-ferrocenylalkylamines have been prepared [1–5] and used as chiral auxiliaries for steroselective peptide synthesis by four-component condensation [6], asymmetric transamination [7], and asymmetric condensation [8] (see Chap. 4). Chiral ferrocenylphosphines have been used as chiral ligands of transition metal complexes that catalyse asymmetric reactions, e.g. hydrogenation, Grignard cross-coupling, allylation, aldol reaction, and hydrosilylation (see Chap. 2). Thus, chiral ferrocenyl derivatives have been recognized as useful chiral ligands in many asymmetric reactions, but to the best of our knowledge no chiral ferrocenyl compounds have been used for the asymmetric alkylation of carbonyl compounds. Enantioselective addition of organometallic reagents to aldehydes is one of the most important and fundamental asymmetric reactions that afford optically active secondary alcohols. Optically active secondary alcohols are components of many naturally occurring compounds, various biologically active compounds, and industrial materials such as liquid crystals. Furthermore, they are also important as synthetic intermediates.

In 1984 Oguni and Omi reported that certain β-amino alcohols, e.g., (*S*)-leucinol, catalyse the addition of diethylzinc to benzaldehyde (Scheme 3-1) [9]. Although the observed *ee* value was not satisfactory and the reaction was slow, it was the first successful catalytic asymmetric alkylation. Since the report was disclosed, the catalytic asymmmetric addition of dialkyzincs to various aldehydes has been investigated intensively by a number of research groups [10]. From these investigations it was found that a bulky substituent near the C-O bond and a tertiary amine moiety in the amino alcohol catalyst are essential for attaining high enantioselectivity and for acceleration of the alkylation. In the meantime, the reaction mechanism involving asymmetric amplification has also been elucidated [11]. Typical highly enantioselective catalysts are shown in Fig. 3-1.

$$\mathrm{PhCHO} + \mathrm{Et_2Zn} \xrightarrow{\beta\text{-amino alcohol (2 mol \%)}} \mathrm{Ph{-}\overset{OH}{\underset{*}{C}H}{-}Et}$$

Scheme 3-1

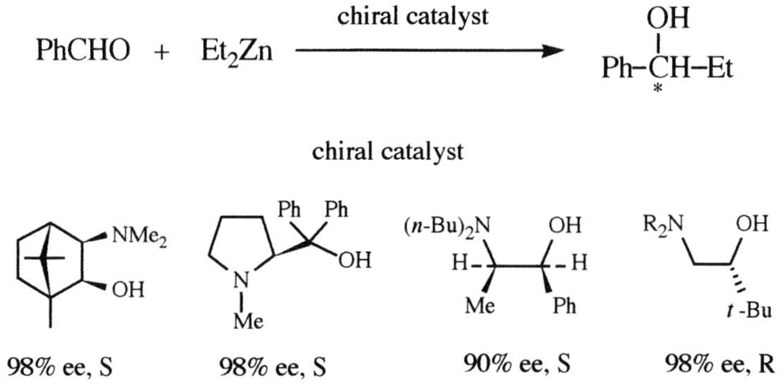

PhCHO + Et_2Zn —chiral catalyst→ Ph-CH(OH)-Et (*)

chiral catalyst

98% ee, S   98% ee, S   90% ee, S   98% ee, R

Fig. 3-1. Highly enantioselective catalysts used in diethylzinc addition.

We have tried here to incorporate the ferrocene moiety into amino alcohol catalysts and have synthesized the following four kinds of compounds; (a) chiral ferrocenyl zincs bearing an aminoethanol auxiliary [12]; (b) *N*-(1-ferrocenylalkyl)-*N*-alkylnorephedrines [13]; (c) chiral polymers bearing *N*-ferrocenylmethylephedrine [14]; and (d) chiral 1,2-disubstituted ferrocenyl amino alcohols (Fig. 3-2) [15–17].

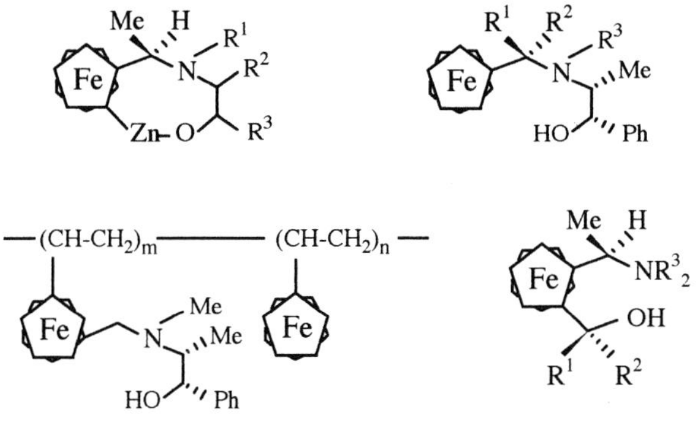

Fig. 3-2. Chiral ferrocenyl amino alcohol catalysts employed in alkylation of aldehydes with dialkylzincs.

3.2 Chiral Ferrocenylzincs Bearing an Aminoethanol Auxiliary [12]

3.2.1 Synthesis of the Catalysts

Ugi and coworkers investigated the reactivity of the lithiation of *N,N*-dimethyl-1-ferrocenylethylamine **1** at the ferrocene ring and found that the stereochemistry of the α-carbon is retained during the reaction [1]. Thus, the lithiation of (*R*)-**1** with *n*-butyllithium proceeded at room temperature to afford preferentially (*R*)-*N*,*N*-dimethyl-1-[(*R*)-2-lithioferrocenyl]ethylamine **2a** with a diastereomer ratio of 96:4 (Scheme 3-2).

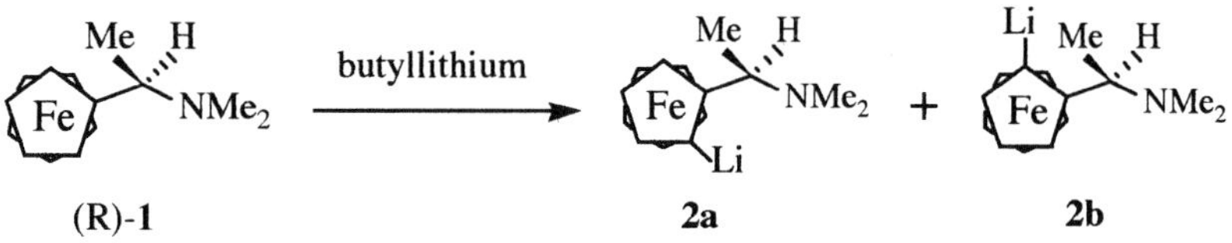

Me H, Fe, NMe_2 (R)-**1** —butyllithium→ Me H, Fe, NMe_2, Li **2a** + Li Me H, Fe, NMe_2 **2b**

Scheme 3-2

We investigated this lithiation reaction with *sec*- and *tert*-butyllithum and higher regioselectivity was attained with *sec*-butyllithium (**2a**:**2b** = 98.5:1.5) (with *tert*-butyllithium **2a**:**2b** = 96:4). The diastereoselectivity was measured by ^1H NMR analysis after treatment of the lithioferrocene **2** with trimethylchlorosilane [1]. Lithiation of (*R*)-**1** with *sec*-butyllithium in ether and subsequent iodination with iodine in THF gave (*R*)-*N*,*N*-dimethyl-1-[(*S*)-2-iodoferrocenyl]ethylamine (**3**) (82% yield, 97% *de*) (Scheme 3-3). Optically pure **3** (the precursor of **2a**) could be obtained by recrystallization from acetonitrile (*m.p.* 79 °C, $[\alpha]_D^{22}$ = 9.32 (*c* = 1.01, EtOH).

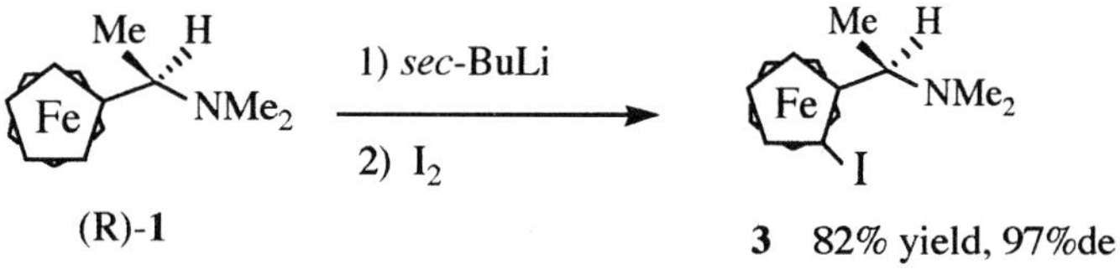

Me H, Fe, NMe_2 (R)-**1** —1) *sec*-BuLi, 2) I_2→ Me H, Fe, NMe_2, I **3** 82% yield, 97%de

Scheme 3-3

Chiral ferrocenyl derivatives with a trimethylammonium group in the α-position are known to undergo nucleophilic substitution with complete retention of configuration [2]. Quaternization of **3** with methyl iodide followed by reaction with *N*-methylethanolamine, (*S*)-prolinol, or (1*S*,2*R*)-ephedrine gave the corresponding (*R*,*S*)-2-iodoferrocenes **4a**, **4b**, and **4c** in 80–90% yield with complete retention of configuration (Scheme 3-4). Diastereomers **4d** and **4e** were similarly prepared starting with (*S*,*R*)-**3**.

3.2.2 Addition of Diethylzinc to Benzaldehyde

The ferrocenylzinc complexes **5a**–**e**, derived from **4a**–**e** by the treatment with two equivalents of *n*-butyllithium followed by zinc chloride in ether (Scheme 3-5), proved to be effective catalysts for asymmetric alkylation of benzaldehyde.

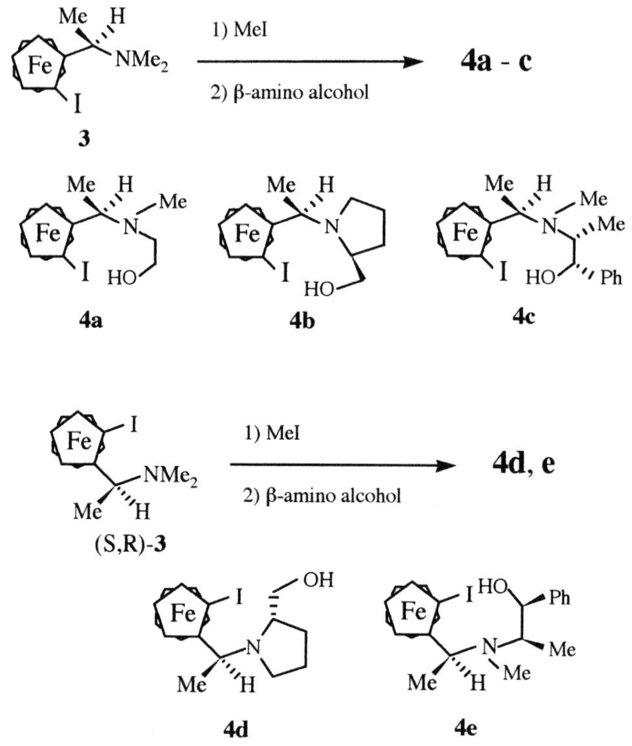

1) MeI
2) β-amino alcohol

3 → **4a - c**

4a **4b** **4c**

1) MeI
2) β-amino alcohol

(S,R)-**3** → **4d, e**

4d **4e**

Scheme 3-4

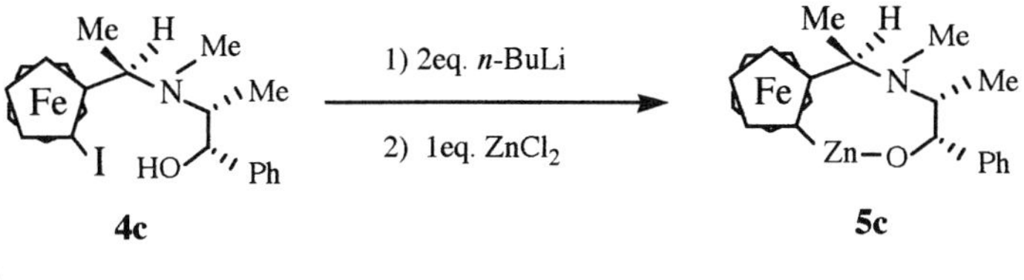

1) 2eq. *n*-BuLi
2) 1eq. $ZnCl_2$

4c → **5c**

Scheme 3-5

Asymmetric ethylation of benzaldehyde was conducted in the presence of 5 mol% of the ferrocenylzinc complex **5** in toluene or hexane (Scheme 3-6). Results are listed in Table 3-1. The iodoferrocene **4a** itself bearing a simple *N*-methylaminoalcohol side chain showed poor catalytic effect (entry 1); however, by chelation with zinc metal enantioselectivity was doubled (entry 2). With (*S*)-prolinol derivatives **4b** and **4d**, 1-phenylpropanol with *R* configuration was formed (entries

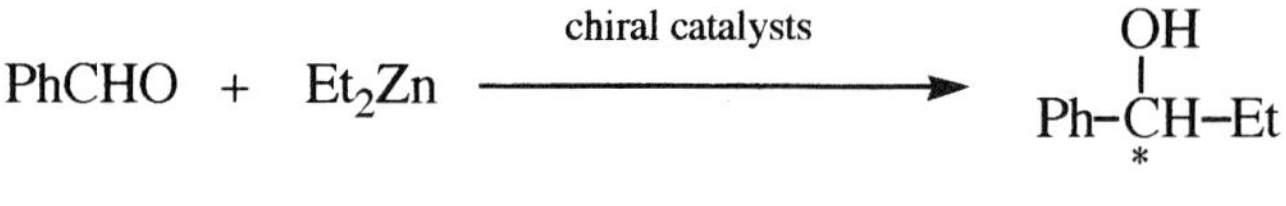

PhCHO + Et_2Zn —chiral catalysts→ Ph-C*H(OH)-Et

Scheme 3-6

Table 3-1. Asymmetric addition of diethylzinc to benzaldehyde[a]

Entry	Iodo ferrocene	Solvent	T, °C	Yield, %[b]	% *ee*[c]	Configuration of product
1[d]	**4a**	toluene	20	69	18	*S*
2	**4a**	toluene	20	66	37	*S*
3	**4a**	toluene-ether	20	60	21	*S*
4	**4b**	toluene	20	65	56	*R*
5	**4b**	hexane	20	77	59	*R*
6	**4d**	hexane	20	50	68	*R*
7	**4c**	hexane	20	74	84	*S*
8	**4c**	toluene	20	65	87	*S*
9	**4c**	toluene	0	44	86	*S*
10	**4c**	toluene	40	81	88	*S*
11	**4c**	toluene	60	81	88	*S*
12	**4e**	hexane	20	47	69	*S*

a The reaction was carried out with 5 mol% of catalyst and 1.6 equiv of diethylzinc, to benzaldehyde for 7 h.
b Isolated yield.
c Based on the reported value of $[\alpha]_D = 45.45$ ($c = 5.15$, $CHCl_3$) for (*S*)-1-phenylpropanol [19].
d *n*-Butyllithium and zinc chloride were not added.

4–6). Between these two compounds, (*S*)-ferrocenylethyl diastereomer **4d** gave a higher enantiomeric purity than (*R*)-ferrocenylethyl ligand **4b**. On the contrary, the catalysts **4c** and **4e** possessing a (1*S*, 2*R*)-ephedrine auxiliary afforded (*S*)-1-phenylpropanol (entries 7–12), and **4e** with (*R*)-ferrocenylethyl structure gave higher optical yields (entries 7 and 12). As (*S*)-prolinol and (1*R*,2*S*)-*N*-methylephedrine are known to catalyze the asymmetric ethylation of benzaldehyde to give (*R*)-1-phenylpropanol [18], the observed direction of enantioselectivity in our ferrocene-catalyzed reaction is considered to depend predominantly on the structure of the amino alcohol auxiliary, the configuration of the ferrocenylethyl moiety playing a minor role. Although direct evidence for the structure of the active catalytic species in our reaction is not available at present, chelated ferrocenylzinc structure **5** could be postulated. High enantioselectivity, even at elevated reaction temperature (60 °C) (entry 11), may be ascribed to the rigid nature of the catalyst **5**. In the ethylation of benzaldehyde with **5c**, only hydrolyzed ferrocene **6c** was recovered quantitatively after aqueous workup (Scheme 3-7).

Me H Me N Me Fe Zn–O Ph **5c** 1) PhCHO, Et_2Zn 2) H_3O^+ Me H Me N Me Fe HO Ph **6c** quant.

Scheme 3-7

3.3 *N*-(1-Ferrocenylalkyl)-*N*-alkylnorephedrines [13]

3.3.1 Synthesis of the Catalysts

Quaternization of readily accessible (*R*)- and (*S*)-*N*,*N*-dimethyl-1-ferrocenylethylamine ((*R*)-**1** and (*S*)-**1**) with methyl iodide followed by reaction with (1*S*,2*R*)-norephedrine in acetonitrile at room temperature gave the corresponding chiral ferrocenes **8a** and **8b**. A similar reaction of the methiodide derived from *N*,*N*-dimethylferrocenylmethylamine **7** with (1*S*,2*R*)-norephedrine under reflux in acetonitrile gave **8c** (Scheme 3-8).

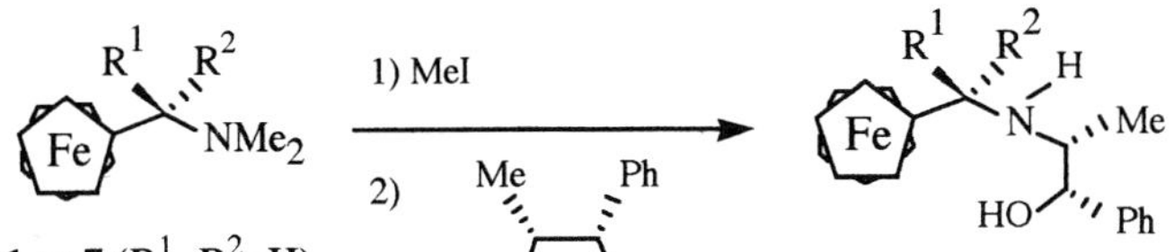

1 or **7** (R^1=R^2=H) → 1) MeI, 2) (1*S*,2*R*)-norephedrine → **8a-c**

	R^1	R^2
8a	Me	H
8b	H	Me
8c	H	H

Scheme 3-8

N-Methyl derivatives **9a**, **9b** and **9c** were obtained from **8a**, **8b** and **8c**, respectively, by treatment with formaldehyde and $NaBH_4$ in methanol (Scheme 3-9). The *N*-(*n*-butyl) derivative **9d** was synthesized from **8c** by treatment with *n*-butyl iodide and sodium carbonate under reflux in ethanol (Scheme 3-10).

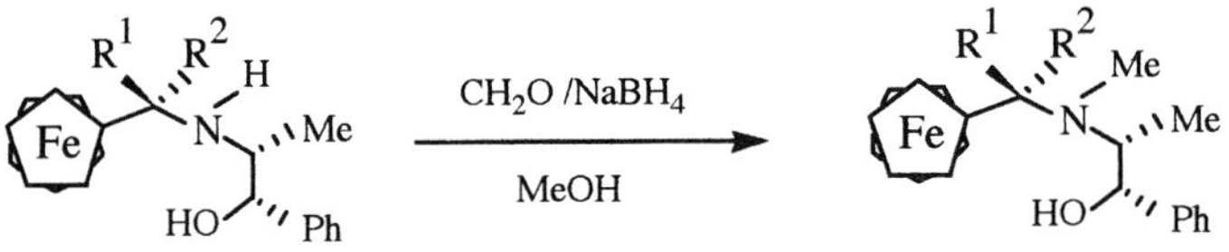

8a-c → CH_2O /$NaBH_4$, MeOH → **9a-c**

	R^1	R^2
9a	Me	H
9b	H	Me
9c	H	H

Scheme 3-9

Scheme 3-10

3.3.2 Addition of Diethylzinc to Aldehydes

Reactions of diethylzinc with benzaldehyde and heptanal were carried out in the presence of a catalytic amount (5 mol%) of **8** or **9**, in order to examine the effect of the substitution pattern (R^1, R^2) of catalysts **8** or **9** on the enantioselectivity. Results are summarized in Table 3-2. The secondary amine **8a** was catalytically less active and less enantioselective than tertiary amine catalysts (entry 1). Tertiary amines **9a**, **9b** and **9c** afforded (*S*)-1-phenylpropanol by reaction with benzaldehyde

Table 3-2. Enantioselective addition of diethylzinc to aldehydes in the presence of **8** or **9**[a]

Entry	Aldehyde	Catalyst	Solvent	*T*, °C	*t*, h	Yield, %[b]	% *ee*[c]
1	PhCHO	**8a**	hexane	rt[d]	8	58	47
2	PhCHO	**9a**	hexane	0	10	61	84
3	PhCHO	**9a**	hexane	rt	7	79	93
4	PhCHO	**9a**	hexane	40	3	92	95
5	PhCHO	**9a**	toluene	40	2	92	95
6	PhCHO	**9b**	hexane	rt	4	94	94
7	PhCHO	**9b**	hexane	40	2	92	94
8	PhCHO	**9c**	hexane	0	8	68	89
9	PhCHO	**9c**	hexane	40	2	91	94
10	PhCHO	**9c**	hexane	rt	4	93	94
11[e]	PhCHO	**9c**	hexane	rt	9	58	64
12[f]	PhCHO	**9c**	hexane	rt	2	94	93
13	PhCHO	**9c**	H-T[g]	40	3	78	94
14	PhCHO	**9c**	H-T[g]	60	2	88	91
15	PhCHO	**9d**	hexane	rt	3	85	99
16	n-$C_6H_{13}CHO$	**9a**	hexane	40	2	81	71
17	n-$C_6H_{13}CHO$	**9c**	hexane	rt	3	81	80
18	n $C_6H_{13}CHO$	**9d**	hexane	rt	3	75	78

a The reaction was carried out with 5 mol% of catalyst and 1.6 equiv. of diethylzinc to 1 equiv. aldehyde to afford (*S*)-alcohol in all entries.
b Isolated yield.
c Based on the reported values of $[\alpha]_D = 45.45$ ($c = 5.15$, $CHCl_3$) for (*S*)-1-phenylpropanol [19] and $[\alpha]_D^{24} = +9.6$ ($c = 8.3$, $CHCl_3$) for (*S*)-3-nonanol [20].
d Room temperature.
e 2 mol% of catalyst was used, based on benzaldehyde.
f 10 mol% of catalyst was used, based on benzaldehyde.
g Hexane-toluene.

in high optical yields, regardless of the configuration of the 1-ferrocenylethylamine moiety (**9a** *vs.* **9b**). Moreover, absence of the methyl group ($R^1 = R^2 = H$) did not cause any decrease in either the chemical or optical yields. Catalysts **9a**–**c** exhibited the highest enantioselectivity and the highest catalytic activity at 40 °C. A limitation of the turnover was noted in the presence of 2 mol% of catalyst (entry 11). Considering the alkyl group on the nitrogen atom, a butyl group **9d** (entry 15) was superior to a methyl group, at least in the reaction with benzaldehyde. The ethylation of heptanal, as a representative of aliphatic aldehydes, also proceeded to give (*S*)-3-nonanol in high enantioselectivity by using **9a**, **9c**, and **9d** (entries 16–18).

In summary, the effect of the *N*-alkyl and *N*-ferrocenyl alkyl substituents of **8** and **9** on enantioselectivity was studied and it was found that the calalyst **9d**, in which the nitrogen atom of norephedrine is substituted by a ferrocenylmethyl and an *n*-butyl group, afforded the highest enantioselectivity even at high reaction temperature. Since both enantiomers of norephedrine are readily availabe and an *N*-ferrocenylalkyl group is also easily introduced, the present method allows practical synthesis of both enantiomers of secondary alcohols with high optical purity more conveniently than previously reported catalytic systems.

Other *N*,*N*-disubstituted norephedrine catalysts such as *N*-isopropylephedrine **10** [21], *N*-(β-aminoethyl)ephedrine **11** [22], the *N*-hindered phenol derivative **12** [23], *N*,*N*-dibutylnorephedrine **13** [24], and arene chromium tricarbonyl complex **14** [25] have been reported (Scheme 3-11).

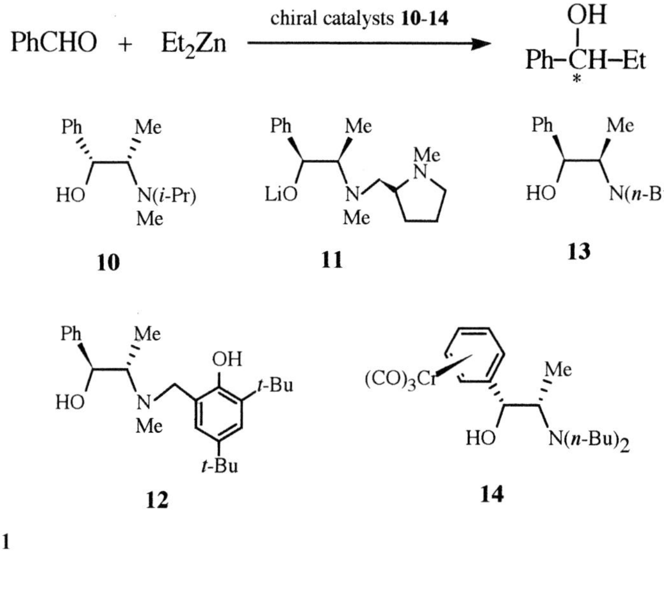

PhCHO + Et_2Zn —(chiral catalysts **10-14**)→ Ph–CH(OH)*–Et

10: Ph, Me, HO, N(*i*-Pr)Me

11: Ph, Me, LiO, N(Me), Me–N (pyrrolidine)

13: Ph, Me, HO, $N(n\text{-Bu})_2$

12: Ph, Me, HO, N(Me), OH, *t*-Bu, *t*-Bu

14: $(CO)_3Cr$, Me, HO, $N(n\text{-Bu})_2$

Scheme 3-11

3.4 Chiral Polymers Bearing *N*-Ferrocenylmethylephedrine [14]

3.4.1 Synthesis of the Catalysts

Chiral polymer-supported catalysts have been utilized in asymmetric addition of dialkylzinc to aldehydes because of the easy product isolation and workup [26]. In the previous section, a highly enantioselective addition of diethylzinc to aldehydes was described using *N*-(1-ferrocenylalkyl)-*N*-alkylnorephedrines as effective catalysts. We then examined incorporation of the catalyst into polymeric systems.

The chiral vinylferrocene monomer **18** bearing an ephedrine residue as a chiral auxiliary was prepared from *N*,*N*-dimethylferrocenylmethylamine (**7**) (Scheme 3-12).

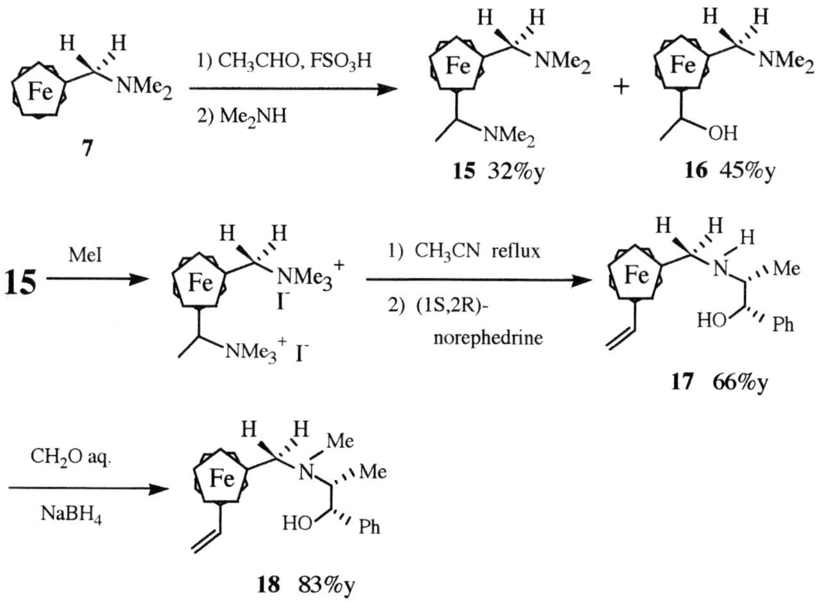

Scheme 3-12

The substituted diamine **15** was prepared directly from **7** in 32% yield according to the published procedure [27]. The diamine **15** was readily separated by column chromatography on alumina from the amino alcohol **16** byproduct (45% yield). The amino alcohol **16** could be converted into **15** by repeating the same procedure [27]. Compound **15** exhibited no absorption in the IR spectrum near 9 or 10 μm, indicating that **15** does not contain an unsubstituted cyclopentadienyl ring [28]. Diamine **15** was quarternized with methyl iodide to give the corresponding dimethiodide, which was refluxed in acetonitrile for 1 h and then treated with excess (1*S*,2*R*)-norephedrine in acetonitrile under reflux for 2 h. Thus, **15** was converted to the vinylferrocene derivative **17** bearing an *N*-substituted norephedrine in 66% overall yield in a one-pot process; **17**: *m.p.* 57–59 °C, $[\alpha]_D^{22} = \pm 20.9$ ($c = 1.12$,

EtOH). The vinylferrocene derivative **18** bearing an *N*-substituted ephedrine was synthesized from **17** by treatment with formaldehyde and $NaBH_4$ in methanol in 83% yield; oil, $[\alpha]_D^{22} = -10.8$ ($c = 0.508$, EtOH).

It has been reported that vinylferrocene is polymerized by using a radical, cationic, anionic, or Ziegler system initiator [29–34]. In particular, higher molecular weight products can be obtained using radical-initiated bulk polymerization [31]. Indeed, both bulk copolymerization and solution copolymerization (in benzene) of **18** with vinylferrocene by using a radical initiator (AIBN) afforded the chiral polymers **19a**–**e** (Scheme 3-13), which were purified by reprecipitation of the benzene solution with methanol. The ratios of the two comonomers were varied in copolymerization. The composition data of the copolymers obtained revealed nearly the same reactivity between **18** and vinylferrocene, which suggests that **19a**–**e** are random copolymers.

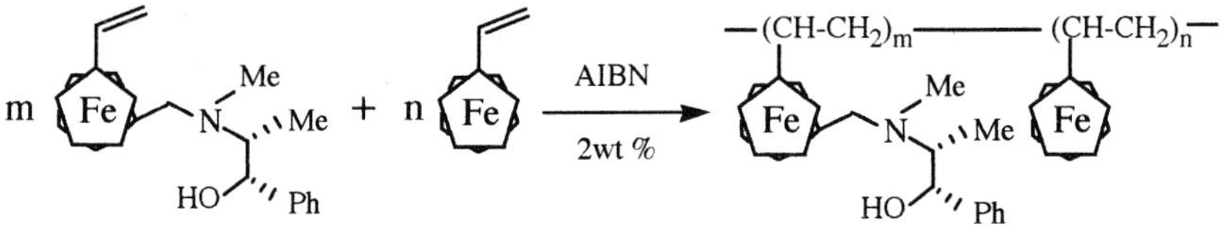

Scheme 3-13

3.4.2 Addition of Diethylzinc to Benzaldehyde

The polyvinylferrocene-supported ephedrines **19a**–**e** were applied as chiral catalysts to the asymmetric ethylation of benzaldehyde in a heterogeneous system (hexane, room temperature). A quantity of 5 mol% of total ephedrine unit per benzaldehyde was used and the effect of the content of ephedrine unit in the polymers on enantioselectivity was investigated (Table 3-3). All chiral polymers **19a**–**e** afforded (*S*)-1-phenylpropanol. Chiral polymer **19c**, containing 32.8 mol% ephedrine units,

Table 3-3. Asymmetric addition of diethylzinc to benzaldehyde[a]

Entry	Chiral polymer[b] m/n[c]	*t*, h	Yield %[d]	% *ee*[e]
1	**19a** 0.103(9.34)	9	70	55
2	**19b** 0.276(21.6)	8	71	51
3	**19c** 0.488(32.8)	8	85	72
4	**19d** 0.988(49.7)	9	75	63
5	**19e** 0.466(31.8)	15	67	58

a The reaction was carried out in hexane at room temperature with 5 mol% of total ephedrine units of **19** relative to benzaldehyde to afford (*S*)-1-phenylpropanol in all entries.
b Polymers **19a**–**d** were synthesized by bulk copolymerization, **19e** by solution copolymerization in benzene.
c Based on nitrogen analysis. The value in parentheses means mol% of ephedrine unit in the polymer.
d Isolated yield.
e Based on the reported value of $[\alpha]_D = 45.45$ ($c = 5.15$, $CHCl_3$) for (*S*)-1-phenylpropanol [19].

gave the best results (entry 3). When chiral catalysts **19a**, **19b**, and **19d**, were used, the enantioselectivities decreased (entries 1, 2, and 4). The difference in the copolymerization method of the chiral polymers apparently influenced the enantioselectivity (entry 5). These polymers, being insoluble in hexane, can be easily separated from the reaction mixture by filtration.

The chiral vinylferrocene monomer **18** and the chiral copolymers **19a**–**e** have been prepared for the first time. Insolubilization of the chiral ferrocene catalyst lowered the enantioselectivity to some extent compared with that of the soluble monomeric catalysts. Nevertheless, polymers **19a**–**e** are usefull for the enantioselective synthesis of secondary alcohols.

3.5 Chiral 1,2-Disubstituted Ferrocenyl Aminoalcohols [15]

3.5.1 Synthesis of the Catalysts

Chiral 1,2-disubstituted ferrocenyl amino alcohols **20** were synthesized as shown in Scheme 3-14. Optically pure **3** was readily obtained from (*R*)-*N*,*N*-dimethyl-1-ferrocenylethylamine ((*R*)-**1**), as described in Section 3.2.1. Quaternization of **3** with

Table 3-4. Preparation of (*R*,*S*)-1,2-disubstituted ferrocenyl amino alcohol **20**[a,b]

Entry	Starting material	Yield, %		*m.p.*, °C	$[\alpha]_D^{22}$, (*c*, EtOH)	
1	**3**, benzaldehyde	**20a**	37	oil	−120.6	(1.04)
		20b	24	oil	− 92.0	(0.760)
2	**3**, pivalaldehyde	**20c**	64	91–92	+130.2	(0.765)
		20d	9	121–124	− 44.2	(0.606)
3	**3**, diisopropyl ketone	**20e**	27	60–61	+ 50.8	(0.634)
4	**3**, benzophenone	**20f**	72	44–54	−182.8	(0.488)
5	**3**, cyclohexanone	**20g**	43	108–109	− 17.3	(0.550)
6[c]	**3**, 2,6-dimethylbenzaldehyde	**20h**	48	45–49	− 51.7	(0.704)
7	**3**, mesitaldehyde	**20i**	28	53–58	−135.2	(0.398)
		20j	39	136–139	+109.6	(0.356)
8	**3**, ferrocenecarboxaldehyde	**20k**	32	65–68	+ 45.2	(0.354)
9	**3**, 9-anthradehyde	**201**	27	97–106	+ 64.9	(0.276)
		20m	28	57–65	− 51.3	(0.372)
10	**21**, benzophenone	**20o**	83	65–69	−209.9	(0.486)
11	**21**, anthrone	**20p**	28	93–98	−190.6	(0.770)
12	**21**, pivalaldehyde	**20q**	48	oil	+131.3	(0.128)
		20r	6	oil	− 30.8	(0.120)
13	**22**, benzophenone	**20s**	52	50–55	−192.0	(0.762)
14	**23**, benzophenone	**20t**	56	46–54	−229.5	(0.346)
15	**25**, benzaldehyde	**20u**	53	oil	−125.5	(0.494)

a Except where otherwise noted, the reaction was carried out in ether at 0 °C – room temperature and quenched with aqueous H_3PO_4.

b The deiodinated ferrocenes were recovered for all entries.

c The reaction mixture was treated with aqueous H_3PO_4 for 5 h at room temperature.

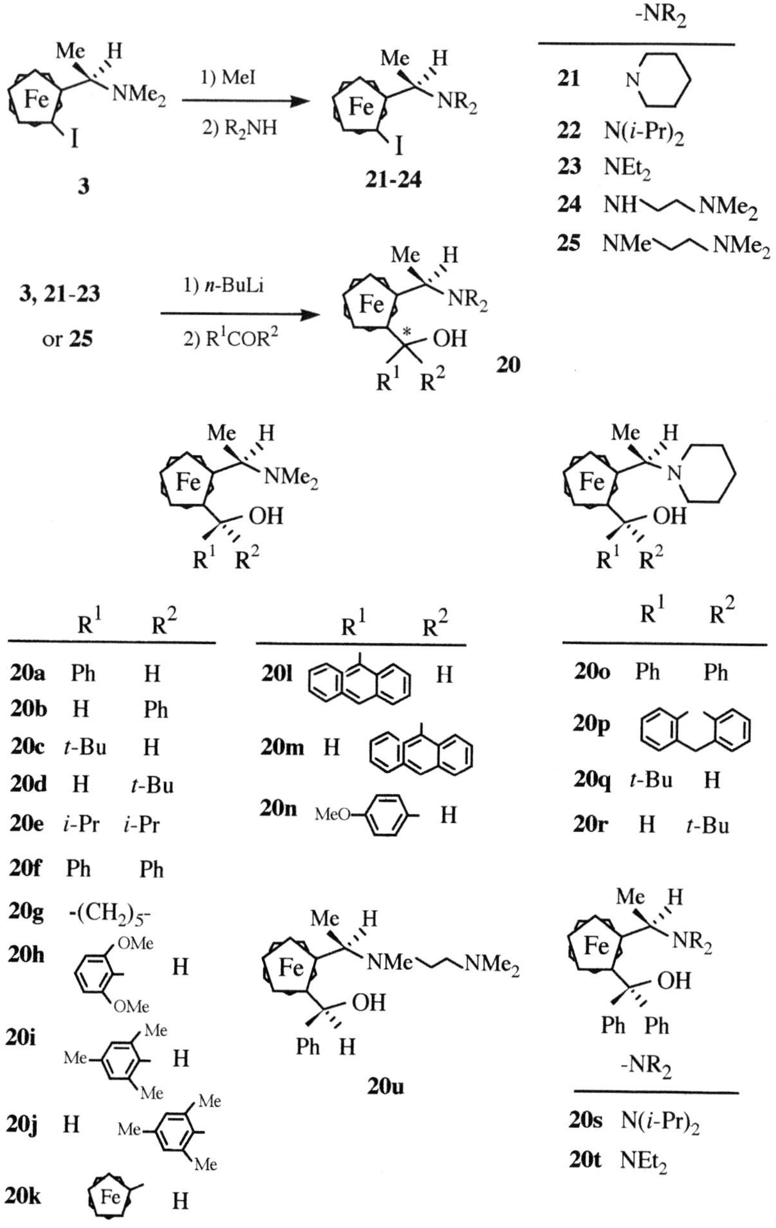

	$-NR_2$
21	N-piperidinyl
22	$N(i\text{-}Pr)_2$
23	NEt_2
24	$NH(CH_2)_2NMe_2$
25	$NMe(CH_2)_2NMe_2$

	R^1	R^2
20a	Ph	H
20b	H	Ph
20c	t-Bu	H
20d	H	t-Bu
20e	i-Pr	i-Pr
20f	Ph	Ph
20g	$-(CH_2)_5-$	
20h	2,6-(MeO)$_2C_6H_3$	H
20i	2,4,6-$Me_3C_6H_2$	H
20j	H	2,4,6-$Me_3C_6H_2$
20k	ferrocenyl	H

	R^1	R^2
20l	9-anthryl	H
20m	H	9-anthryl
20n	$MeO-C_6H_4$	H

	R^1	R^2
20o	Ph	Ph
20p	fluorenylidene	
20q	t-Bu	H
20r	H	t-Bu

	$-NR_2$
20s	$N(i\text{-}Pr)_2$
20t	NEt_2

20u

Scheme 3-14

methyl iodide followed by reaction with piperidine, diisopropylamine, diethylamine, or *N*,*N*-dimethylethylenediamine gave the respective (*R*,*S*)-2-iodoferrocenes **21**–**24** with complete retention of configuration. Reaction of **24** with formaldehyde and $NaBH_4$ in methanol gave **25**.

(*R*,*S*)-1,2-Disubstituted ferrocenyl amino alcohols **20a – u** were derived from **3**, **21 – 23**, and **25** by treatment with *n*-butyllithium followed by reaction with a carbonyl compound. When aldehydes were used as carbonyl component, two chromatographically separable diastereomers (**20a – b**, **20c – d**, **20i – j**, **20l – m**, **20q – r**) were obtained. In the cases of **20h**, **20k**, and **20u**, the diastereomeric mixtures were isomerized to single diastereomers without separation by treatment with aqueous phosphoric acid. The yields and the properties of **20** are summarized in Table 3-4. The absolute configuration of the two diastereomers was tentatively assigned on the basis of their ^{1}H NMR spectra, their stability to aqueous phosphoric acid, and spectral comparison with **20n**, absolute configuration of which was confirmed by single-crystal X-ray analysis [35].

The enantiomeric (*S*,*R*)-1,2-disubstituted ferrocenyl amino alcohols (*S*,*R*)-**20o** and **27 – 29** were similarly prepared from (*S*,*R*)-**21** and **26**, respectively (Scheme 3-15). 1-Ferrocenylisobutylamine derivative **27** was synthesized by lithiation of **26** with *n*-butyllithium and subsequent treatment with benzaldehyde. The diastereoselectivity of the lithiation was reported to be 97 : 3 [3].

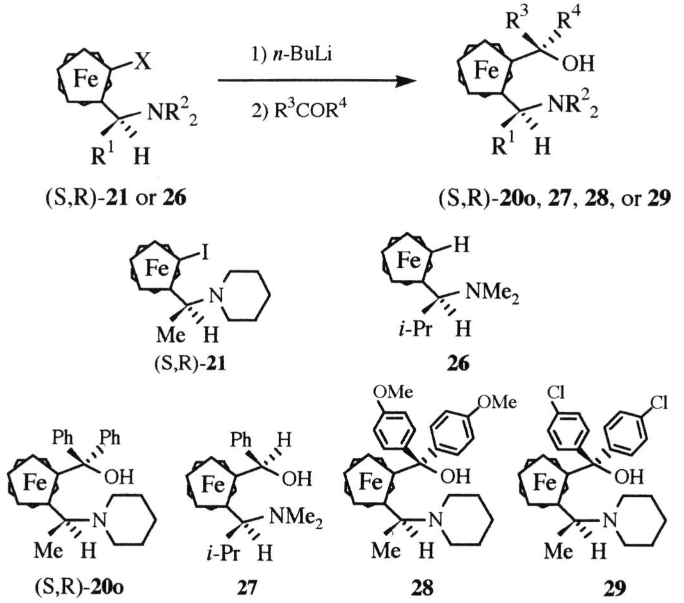

1) *n*-BuLi
2) R^3COR^4

X, NR^2_2, R^1 H; R^3 R^4 OH

(S,R)-**21** or **26** → (S,R)-**20o**, **27**, **28**, or **29**

(S,R)-**21** **26** (S,R)-**20o** **27** **28** **29**

Scheme 3-15

3.5.2 Addition of Dialkylzinc to Aldehydes

The reaction of diethylzinc with benzaldehyde was carried out in the presence of several (*R*,*S*)-1,2-disubstituted ferrocenyl amino alcohols **20** in order to examine the effect of the structure of **20** on the enantioselectivity. The reaction conditions and results are summarized in Table 3-5.

Table 3-5. Asymmetric addition of diethylzinc to benzaldehyde in the presence of (*R*,*S*)-catalyst **20**[a]

Entry	Catalyst	T, °C	t, h	Yield, %[b]	$[\alpha]_D^{22}$, (c, $CHCl_3$)[c]		% *ee*[d]
1	**20a**	rt[e]	3	95	−44.4	(5.28)	91
2	**20a**	0	6	80	−43.7	(4.19)	90
3	**20b**	rt	2	96	−45.3	(4.34)	93
4	**20c**	rt	2	95	−45.6	(4.09)	94
5	**20d**	rt	1	95	−45.7	(3.90)	94
6	**20e**	rt	2	92	−46.5	(4.08)	94
7	**20f**	rt	2	99	−47.0	(4.15)	97
8	**20o**	rt	1	99	−48.5	(3.94)	99
9	**20p**	rt	2	92	−47.5	(3.21)	98
10	**20s**	rt	22	51	−18.4	(3.97)	36
11[f]	**20o**	rt	3	95	−48.3	(3.28)	98
12[g]	**20o**	rt	22	72	−41.8	(2.90)	86
13[g]	**20o**	40	8	88	−43.1	(3.49)	89
14[g]	**20o**	60	5	78	−39.9	(3.09)	82

a Except where otherwise noted, the reaction was carried out in hexane using 5 mol% of catalyst and 1.6 equiv. of diethylzinc per equiv. benzaldehyde.
b Isolated yield.
c Reported value for (*S*)-1-phenylpropanol in 98% *ee* is $[\alpha]_D^{22} = 47.6$ ($c = 6.11$, $CHCl_3$) [36].
d Determined by HPLC using Daicel Chiralcel OB.
e Room temperature.
f 2 mol% of catalyst was used based on benzaldehyde.
g 0.5 mol% of catalyst relative to benzaldehyde was used.

At room temperature, all catalysts **20** (5 mol%) bearing a dimethylamino or piperidinyl group afforded (*S*)-1-phenylpropanol in high yield with >91% *ee*, regardless of the stereochemistry of the asymmetric carbon bearing the hydroxyl group. Among catalysts **20a**−**f** bearing a dimethylamino group, the benzhydrol derivative **20f** gave the highest *ee* (entry 7, 97% *ee*). The piperidinyl group (**20o**, (−)-(*R*,*S*)-1-[2-(diphenylhydroxymethyl)ferrocenyl]-1-piperidinoethane, (−)-DFPE) was superior to the dimethylamino group (entry 8, 99% *ee*). The 9-hydroxyanthracene derivative **20p**, a rigid analogue of **20o**, was almost as effective (entry 9) as **20o**. On the other hand, catalyst **20s**, bearing a diisopropylamino group, had both low catalytic activity and stereoselectivity (entry 10), probably owing to the bulky substituents that interfere with coordination of the nitrogen to the zinc alkoxide.

Reducing the catalyst: benzaldehyde ratio from 5 mol% to 2 mol% had no effect (entry 11), but reducing the ratio to 0.5 mol% reduced both yield and % *ee* (entry 12). The influence of reaction temperature on enantioselectivity was investigated with 0.5 mol% of **20o** (entries 12−14). Both chemical and optical yields reached maxima at 40 °C, although the reaction ran well at 60 °C.

The asymmetric addition of diethylzinc to heptanal, a straight-chain aliphatic aldehyde, was catalyzed by compound **20** (Table 3-6). All *R*,*S* catalysts **20**, except **20s** and **20u**, afforded (*S*)-3-nonanol in moderate enantiomeric excess, regardless of the stereochemistry of the asymmetric carbon bearing the hydroxyl group. Cyclohexanol derivative **20g**, which is less bulky around the carbinol center than catalysts

Table 3-6. Asymmetric addition of diethylzinc to heptanal in the presence of chiral 1,2-disubstituted ferrocenylamino alcohols[a]

Entry	Catalyst	T, °C	t, h	Yield, %[b]	$[\alpha]_D^{22}$, (c, $CHCL_3$)		% ee[c]	Config.
1	**20a**	rt[d]	3	86	+5.39	(7.84)	56	*S*
2[e]	**20a**	rt	28	69	+5.09	(3.86)	53	*S*
3	**20c**	rt	2	87	+5.32	(8.15)	55	*S*
4[f]	**20c**	rt	2	84	+5.28	(6.36)	55	*S*
5	**20c**	0	6	80	+5.41	(7.44)	56	*S*
6	**20f**	rt	2	87	+4.80	(7.08)	50	*S*
7	**20g**	rt	2	76	+3.26	(6.44)	34	*S*
8	**20h**	rt	2	86	+5.79	(7.21)	60	*S*
9	**20i**	rt	5	77	+4.67	(6.11)	49	*S*
10	**20j**	rt	7	50	+5.04	(3.83)	53	*S*
11	**20k**	rt	6	60	+3.72	(3.77)	39	*S*
12	**20m**	rt	3	70	+5.42	(5.87)	56	*S*
13	**20o**	rt	3	93	+5.63	(7.64)	59	*S*
14	**20q**	rt	2	72	+5.55	(5.60)	58	*S*
15	**20s**	rt	20	55	0	(2.54)	0	
16	**20t**	rt	3	62	+4.63	(4.08)	48	*S*
17	**20u**	rt	26	8[g]				
18	**27**	rt	2	94	−5.46	(7.98)	57	*R*

a Except where otherwise noted, the reaction was carried out in hexane using 5 mol% of catalyst and 1.6 equiv. of diethylzinc per equiv. heptanal.
b Isolated yield.
c Based on the reported value of $[\alpha]_D^{24} = +9.6$ ($c = 8.3$, $CHCl_3$), for (S)-3-nonanol [20].
d Room temperature.
e Toluene was used as solvent.
f 7 mol% of catalyst based on heptanal was used.
g Determined by GC analysis of the reaction mixture.

with an aryl or *tert*-butyl group, gave low selectivity (entry 7). Diferrocenylcarbinol **20k** also gave poor selectivity (entry 11), and diamine **20u** which has a second amino group on the side chain, gave a complex mixture (entry 17). On the other hand, **27**, which has an isopropyl group on the asymmetric carbon bearing the amino group, and **20o** were the most active catalyst and afforded good stereoselectivity (entries 18 and 13). 2,6-Dimethoxybenzenemethanol **20h** and *tert*-butyl derivative **20q** gave comparable selectivities (entries 8 and 14). We conclude that the enantioselective ethylation of heptanal with **20** as catalyst is unsatisfactory.

The effective catalyst **20o** ((−)-DFPE) was examined for the ethylation of other aldehydes (Table 3-7). *Para*-Substituted benzaldehydes, (E)-cinnamaldehyde, 2-furaldehyde, and 2-naphthaldehyde, which possess π electrons adjacent to the carbonyl group, afforded the corresponding secondary alcohols in high enantiomeric purity (entries 1−5). In addition, cyclohexanecarboxaldehyde, 2-ethylbutyraldehyde, and pivalaldehyde, which are branched α to the carbonyl group, were ethylated in >98% *ee* (entries 6−11). On the other hand, ethylation of isovaleraldehyde, 3-phenylpropionaldehyde, and n-butyraldehyde, which lack a substituent α to the carbonyl group, proceeded with low selectivity (entries 12−14).

Table 3-7. Asymmetic addition of diethylzinc to aldehydes in the presence of the catalyst **20o**[a]

Entry	Aldehyde	*T*, °C	*t*, h	Yield, %[b]	$[\alpha]_D^{22}$, (*c*, solvent)	% *ee*	Configuration of product
1	*p*-ClC_6H_4CHO	rt[c]	1	100	−28.2 (5.01, C_6H_6)[d]	100[e]	*S*
2	*p*-$MeOC_6H_4CHO$	rt	4	97	−32.8 (4.14, C_6H_6)[d]	90[e]	*S*
3	(*E*)-$C_6H_5CH = CHCHO$	rt	3	90	−6.30 (2.70, $CHCl_3$)[f]	100[g] (72)[h]	*S*
4	2-furaldehyde	rt	1	87	−16.2 (1.06, $CHCl_3$)[i]	87[j]	*S*
5	2-naphtaldehyde	rt	3	91	−26.6 (3.35, C_6H_6)[k]	97[e]	*S*
6	*c*-$C_6H_{11}CHO$	0	3	92	−8.02 (6.82, Et_2O)[l]	>98[m]	*S*[n]
7	*c*-$C_6H_{11}CHO$	rt	2	92	−7.96 (6.64, Et_2O)[l]	>98[m]	*S*[n]
8[o]	*c*-$C_6H_{11}CHO$	rt	4	92	−7.76 (6.64, Et_2O)[l]	97[p]	*S*[n]
9[q,r]	*c*-$C_6H_{11}CHO$	rt	3	91	+7.95 (7.11, Et_2O)[l]	>98[m]	*R*[n]
10	$(CH_3CH_2)_2CHCHO$	rt	3	83	−0.45 (4.00, $CHCl_3$)	>98[m]	*S*[s]
11	$(CH_3)_3CCHO$	rt	3	93[t]	−32.6 (2.38, $CHCl_3$)[u]	98[m]	*S*
12	$(CH_3)_2CHCH_2CHO$	rt	2	70	+12.6 (3.70, EtOH)[v]	62[g]	*S*
13[q]	$PhCH_2CH_2CHO$	rt	4	88	−16.8 (4.96, EtOH)[w]	63[g]	*R*
14	$CH_3(CH_2)_2CHO$	rt	4	88	−4.92 (7.62, EtOH)[x]	58[y]	*S*

a Unless otherwise noted, the reaction was carried out in hexane with 5 mol% of catalyst **20o** and 1.6 equiv. of diethylzinc per equiv. of aldehyde.
b Isolated yield.
c Room temperature.
d Reported values for (*S*)-1-(*p*-chlorophenyl)propanol in 43% *ee* and (*S*)-1-(*p*-methoxyphenyl)propanol in 51% *ee* are $[\alpha]_D^{22} = -10.4$ ($c = 5$, C_6H_6) and $[\alpha]_D^{22} = -17.2$ ($c = 5$, C_6H_6), respectively [37].
e Determined by HPLC analysis using Daicel Chiralcel OB.
f $[\alpha]_D^{22} = -5.7$ ($CHCl_3$) for (*S*)-1-phenylpent-1-en-3-ol in 96% *ee* determind by HPLC using a chiral column [36].
g Based on the reported values.
h The % *ee* in parenthesis is based on $[\alpha]_D^{22} = -6.6$ ($c = 3.2$, $CHCl_3$) in 75% *ee* [38].
i $[\alpha[_{578}^{22} = -17.9$ ($c = 1.75$, $CHCl_3$) for (*S*)-1-(2-furyl)propanol in 91% *ee* [39].
j Determined by HPLC analysis using Daicel Chiralcel OB of the corresponding (*S*)-MTPA ester.
k $[\alpha]_D^{20} = -18.81$ (C_6H_6) for (*S*)-1-(2-naphtyl)propanol in 44.7% *ee* [40].
l $[\alpha]_D^{20} = -9.60$ ($c = 12.3$, Et_2O) for (−)-1-cyclohexylpropanol [41].
m Determined by GC analysis of the corresponding (*S*)-MTPA ester.
n Assigned by the (−)-1-cyclohexylpropanol obtained by reduction of (−)-(*S*)-1-phenylpropanol [42].
o Catalyst **20a** was used.

p Based on the rotation value obtained in entry 6.
q Catalyst (*S,R*)-**20o**, representing the enantiomer of **20o**, was used.
r Aldehyde was added to a mixture of (*S,R*)-**20o**, Et_2Zn, and hexane.
s Tentatively assigned by analogy to the other compounds.
t Determined by GC analysis of the reaction mixture.
u $[\alpha]_D^{23} = +27.4$ (neat) for (*R*)-2,2-dimethyl-3-pentanol [43].
v $[\alpha]_D^{21} = -20.3$ ($c = 5.25$, EtOH) for (*R*)-5-methyl-3-hexanol [44].
w $[\alpha]_D = +26.8$ ($c = 5.0$, EtOH) for (*S*)-1-phenyl-3-pentanol [38].
x $[\alpha]_D^{20} = -8.21$ ($c = 11.5$, EtOH) for (*R*)-3-hexanol [44].
y Determind by GC analysis of the corresponding menthyl carbonate with (−)-MCF.

Table 3-8. Asymmetric addition of di-*n*-butylzinc to aldehydes in the presence of catalyst **20o**[a]

Entry	Aldehyde	*T*, °C	*t*, h	Yield, %[b]	$[\alpha]_D^{22}$ (*c*, solvent)	% *ee*	Configuration of product
1	PhCHO	rt[c]	5	92	−39.3 (3.08, C_6H_6)[d]	99[e]	*S*
2	$(CH_3)_2CHCHO$	rt	5	66	−27.1 (2.95, EtOH)[f]	>98[g]	*S*

a The reaction was carried out in hexane with 5 mol% of catalyst **20o** and 1.6 equiv. of di-*n*-butylzinc per equiv. aldehyde.
b Isolated yield.
c Room temperature.
d $[\alpha]_D^{25} = +35.7$ ($c = 3.00$, C_6H_6) for (*R*)-1-phenylpentanol [45].
e Determined by HPLC analysis using Daical Chiralcel OB.
f $[\alpha]_D = +27.67$ ($c = 10$, EtOH) for (+)-2-methyl-3-heptanol [46].
g Determined by GC analysis of the corresponding (*S*)-MTPA ester.

We examined the reaction of di-*n*-butylzinc with aldehydes in the presence (5 mol%) in hexane at room temperature (Scheme 3-16, Table 3-8). Benzaldehyde afforded (*S*)-1-phenylpentanol in 92% yield and 99% *ee*, and isobutyraldehyde gave (*S*)-2-methyl-3-heptanol in 66% yield and >98% *ee*.

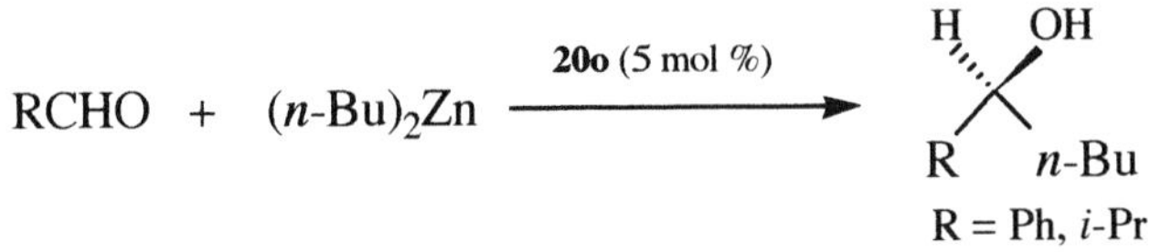

$$\text{RCHO} + (n\text{-Bu})_2\text{Zn} \xrightarrow{\textbf{20o}\ (5\ \text{mol}\ \%)} \text{R-CH(OH)-}n\text{-Bu} \quad \text{R = Ph, } i\text{-Pr}$$

Scheme 3-16

All chiral catalysts except **20b** and **20 m** were recovered unchanged in >90% yield after the reaction. Catalysts **20b** and **20m**, with an (*R*) carbinol configuration, were partially isomerized (<20%) to **20a** and **20l**, respectively, with total recovery >90%. In contrast, catalysts **20d** and **20j**, also with the (*R*) carbinol configuration, were recovered without isomerization.

The catalytic reaction with **20** is completely different from the system with ferrocenylzinc complex **5** in Section 3.2. In the ethylation of aldehyde with **5c**, only hydrolyzed ferrocene **6c** was recovered after aqueous workup (Scheme 3-7) in Section 3.2), which rules out the possbility that the intermediates in the reaction are compounds akin to **20**.

Other investigations on *β*-aminoalcohols as catalysts have shown that the degree of enantioselectivity depends on the bulk of the substituents on the hydroxy-bearing carbon and that the configuration of the alcohol moiety of the catalyst determines the configuration of the predominant enantiomer of the product. On the other hand, our work with catalysts **20** shows that the bulk around the alcohol moiety also contributes to asymmetric induction, but the sense of induction is independent of the configuration of the alcohol moiety.

A proposed mechanism for the enantioselective addition of diethylzinc to an aldehyde on the zinc alkoxides of **20a** and **20b** is shown in Fig. 3-3. Molecular models suggest that the zinc alkoxide forms a seven-membered ring with a chair conformation, regardless of the configuration of the alcohol moiety in the catalyst

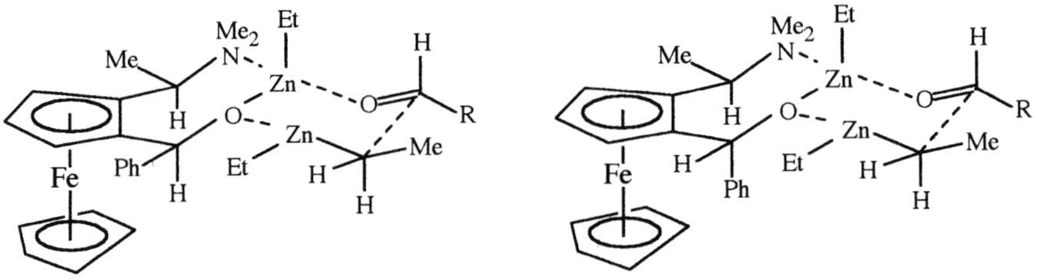

30 **31**

Fig. 3-3. Proposed mechanism for addition of chiral-ligand complexed diethylzinc to aldehydes.

(**30** and **31**). The alkylation reaction can be interpreted in terms of a six-membered cyclic transition state. Nucleophilic attack of the ethyl group from the *Si* face of the aldehyde leads to the (*S*) isomer. The high enantioselectivity in the ethylation of α-branched aldehydes results from repulsion between the α substituent of the aldehyde and the ethyl group on the bridging zinc alkoxide.

The ethylation of benzaldehyde with 50% *ee* of (*S*,*R*)-**20o** at 0 °C gave no asymmetric amplifiction. It thus appears that a single monomeric zinc alkoxide species is involved in the catalysis with **20**.

After we reported this enantioselective dialkylzinc addition using new chiral 1,2-disubstituted ferrocenyl amino alcohols, a similar reaction system was reported using chiral (η^6-arene)chromium complexes as a δ-amino alcohol analogue (Scheme 3-17) [47] and interesting results were obtained (Table 3-9). (*R*,*S*)-Chromium complexes **32** and **33** afforded (*S*)-alcohol, regardless of the stereochemistry of the carbinol moiety (entry 1 and 2). Chromium free δ-amino alcohol **34** gave (*S*)-alcohol in low selectivity (entry 3). On the other hand, **35**, which is the (*R*,*R*)-diastereomer of **32** with opposite planar chirality, gave (*R*)-alcohol in low *ee* (entry 4). These results mean that the degree and the sense of asymmetric induction

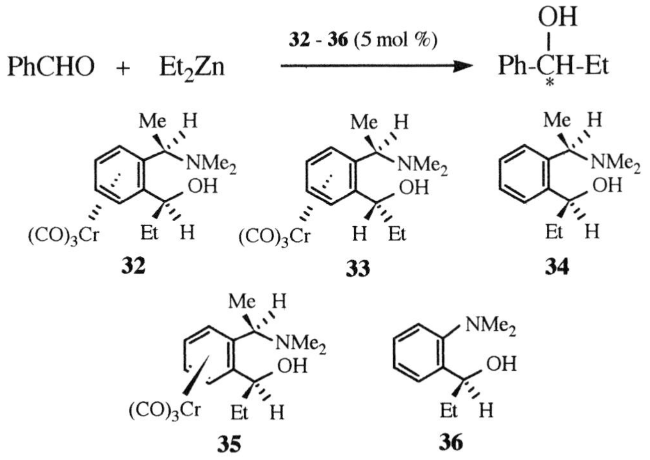

$$\text{PhCHO} + \text{Et}_2\text{Zn} \xrightarrow{\textbf{32}-\textbf{36}\ (5\ \text{mol}\ \%)} \text{Ph-}\overset{*}{\text{C}}\text{H(OH)-Et}$$

32 **33** **34** **35** **36**

Scheme 3-17

Table 3-9. Asymmetric addition of diethylzinc to benzaldehyde by chiral arenechromium complexes

Entry	Catalyst	Yield, %	% *ee*	Configuration of product
1	**32**	87	93	*S*
2	**33**	87	50	*S*
3	**34**	71	24	*S*
4	**35**	79	30	*R*
5	**36**	56	12	*S*

of the alcohol produced is strongly related to the planar chirality of the δ-amino alcohol catalyst. On the other hand, γ-amino alcohol **36** gave only poor stereoselectivity (entry 5).

3.5.3 Addition of Dialkylzincs to *o*-Phthalaldehyde: A Facile Synthesis of Optically Active 3-Alkylphthalides [16]

In the previous subsection we described the high enantioselectivity of chiral 1,2-disubstituted ferrocenyl amino alcohols as a chiral catalyst of asymmetric addition of dialkylzinc to aldehydes. In particular, (−)- and (+)-DFPE (**20o** and (*S*,*R*)-**20o**) gave the highest enantioselectivity and catalytic activity. Here a facile preparation of optically active 3-ethyl- and 3-*n*-butylphthalides **39** is described. Optically active phthalides are naturally occurring substances, many of which possess biological activity [48]. The approaches to the asymmetric synthesis of the phthalides can be classified into the following three procedures; (a) the addition of chiral *o*-substituted aryllithium reagents to carbonyl compounds [48c, 49]; (b) the addition of organometallics or metal hydrides to chiral (*o*-acylaryl)oxazolines [49b] and chiral (*o*-formylaryl)oxazolidines [50]; (c) the stoichiometric or catalytic asymmetric reduction of prochiral *o*-acylbenzoic esters [48c, 51]; and (d) the enantioselective addition of dialkylzincs to 2-bromobenzaldehyde using chiral β-amino alcohol [52]. The present procedure is a new one based on the highly enantioselective addition of dialkylzinc reagents to *o*-phthalaldehyde **37**, catalyzed by chiral 1,2-disubstituted ferrocenyl amino alcohols **20o**, (*S*,*R*)-**20o**, and **28**−**29**, followed by oxidation of the resulting lactols **38** (Scheme 3-18).

The reaction of diethyl- and di-*n*-butylzinc with *o*-phthalaldehyde **37** was carried out in the presence of **20o**, (*S*,*R*)-**20o**, or **28**−**29** (5−10 mol%) in hexane at room temperature to afford monoadducts **38** with high optical purity. Powdered *o*-phthalaldehyde is only slightly soluble in hexane, but gradually went into solution during the reaction, and an almost clear solution was obtained at the end point. The diastereomer ratio of lactols **38** at the 1-position was determined to be 1:1 by ^{1}H NMR spectroscopy. The reaction conditions and results are summarized in Table 3-10.

In the presence of 5 mol% of **20 o** ((−)-DFPE) the ethylation was sluggish at 0 °C (entry 1), but the reaction was smoothly complete in an hour at room temperature (entry 2). Comparable selectivity was obtained in toluene, in which **37** is easily soluble (entry 3). With the enantiomer of **20o**, (*S*,*R*)-**20o** ((+)-DFPE), the enantiomeric (*R*) lactol **38a** was obtained (entry 4). Of the two *para*-substituted analogues **28** and **29**, the *p*-chloro derivative **29** gave a slightly higher enantioselectivity (entry 6, 90% *ee*). As the reaction was found to proceed even without the catalyst in hexane at room temperature yielding the racemic lactol **38a** in a low yield (entry 7), we tried to reduce the uncatalyzed reaction by changing the physical form of **37** and the amount of catalyst. By using of a few pieces of solid mass (not powder) of **37**, the optical yield of **38a** was improved to 92% *ee* (entry 8). It seems

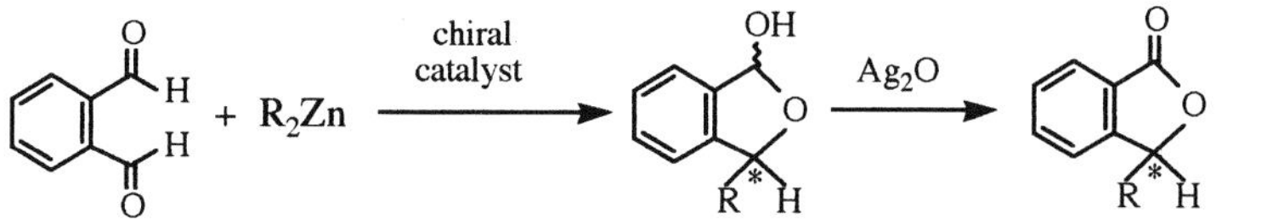

chiral catalyst

37 + R_2Zn → (chiral catalyst) → **38a** R = Et, **38b** R = *n*-Bu → (Ag_2O) → **39a** R = *n*-Bu, **39b** R = *n*-Bu

chiral catalyst

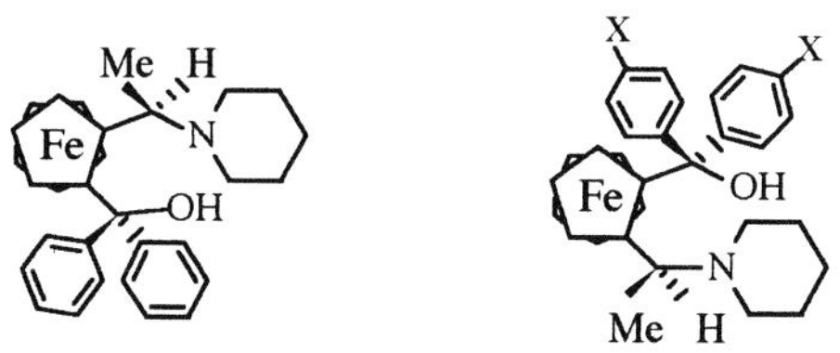

20o

(S,R)-**20o** X = H
28 X = OMe
29 X = Cl

Scheme 3-18

Table 3-10. Asymmetric addition of dialkylzincs to *o*-phthalaldehyde in the presence of chiral catalysts[a]

Entry	Catalyst	R_2Zn	*T*, °C	*t*, h	Yield, %[b]	% *ee*	Configuration of product[c]
1	**20o**	Et_2Zn	0	4	54	86	*S*
2	**20o**	Et_2Zn	rt	1	95	88	*S*
3[d]	**20o**	Et_2Zn	rt	1	92	87	*S*
4	(*S*,*R*)-**20o**	Et_2Zn	rt	1	92	88	*R*
5	**28**	Et_2Zn	rt	1	86	86	*R*
6	**29**	Et_2Zn	rt	0.5	85	90	*R*
7[e]	none	Et_2Zn	rt	1	30		
8[f]	(*S*,*R*)-**20o**	Et_2Zn	rt	3	85	92	*R*
9[f,g]	(*S*,*R*)-**20o**	Et_2Zn	rt	3	87	95	*R*
10[f,g]	**29**	Et_2Zn	rt	3	88	98	*R*
11	**20o**	$(n\text{-Bu})_2Zn$	rt	1	50	89	*S*[h]
12[f,g]	**29**	$(n\text{-Bu})_2Zn$	rt	3	57	94	*R*[h]

a Unless otherwise noted, the reaction was carried out in hexane with 5 mol% of catalyst and 1.2 equiv. of dialkylzinc was added to a suspension of powdered *o*-phthalaldehyde.
b Isolated yield.
c The configuration at the 3-position of lactol **38**. Unless otherwise noted, tentatively assigned in that (*R*,*S*)-catalyst afforded (*S*)-alcohols in the alkylation of simple aldehydes [15].
d Toluene as solvent.
e The reaction was conducted without catalyst.
f A solid mass of *o*-phthalaldehyde was used.
g 10 mol% of catalyst to aldehyde was used.
h Assigned in that (*S*)-(−)-3-*n*-butylphthalide **39b** was obtained by oxidation of the (−)-lactol **38b** [53].

that the small surface area of solid mass of **37** lowered the concentration of **37** in hexane and hence increased the relative catalyst concentration to **37**. Further, the use of 10 mol% of (*S*,*R*)-**20o** afforded a better result (entry 9, 95% *ee*). The bis(*p*-chlorophenyl) derivative **29** was superior to (*S*,*R*)-**20o** (entry 10, 98% *ee*). The electron-withdrawing *p*-chloro groups of **29** are considered to lower the electron density on the zinc metal in the complex formed from **29** and dialkylzinc. This increases the Lewis acidity and the reactivity of the complex, and consequently reduces the uncatalyzed reaction. The reaction of **37** with di-*n*-butylzinc in the presence of **29** (10 mol%) in hexane at room temperature afforded lactol **38b** in 94% *ee* (entry 12). The enantioselectivities of **38a** and **38b** were estimated by HPLC analysis of the corresponding diols **40a**, **b**, obtained by sodium borohydride reduction of **38a**, **b** (Scheme 3-19).

The lactols **38** obtained were oxidized with silver oxide to optically active 3-alkylphthalides **39** in 80–81% yield without racemization. The absolute configuration of 3-*n*-butylphthalide **39b** was determined by comparison with literature data [53].

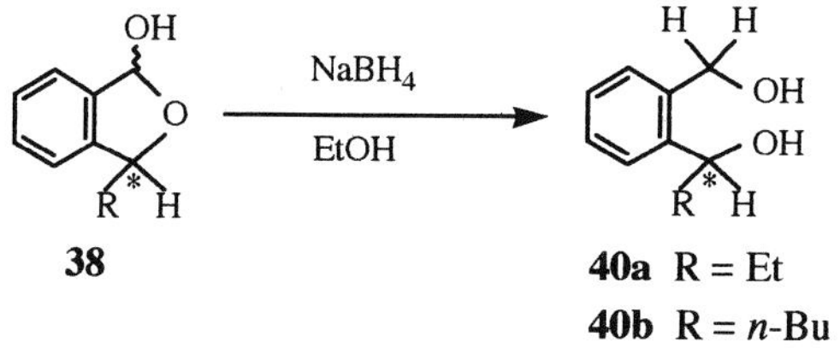

OH O R * H **38** —$NaBH_4$ / EtOH→ H H OH OH R * H

40a R = Et
40b R = *n*-Bu

Scheme 3-19

3.5.4 Enantio- and Diastereoselective Addition of Diethylzinc to Racemic α-Thio- and α-Selenoaldehydes [17]

The importance of optically active vicinal thio- and selenoalcohols has been recognized as potential intermediates for enantiomerically pure epoxides, which have been used in the synthesis of more complex enantiomerically enriched compounds [54]. The synthesis of racemic vicinal thio- and selenoalcohols via stereoselective reduction of α-thio- [55] and α-selenoketones [56] has been reported, whereas very few asymmetric syntheses of highly optically active vicinal thioalcohols are known [57].

In Section 3.5.2 we described the highly enantioselective addition of dialkylzinc to achiral aldehydes catalyzed by chiral 1,2-disubstituted ferrocenyl amino alcohols, which shows high stereoselectivity even in the alkylation of α-branched aliphatic aldehydes. In order to further develop the characteristics of our catalysts, the ethylation of aldehydes substituted with an α-thio- and seleno group was investigated.

In this section, the asymmetric synthesis of the vicinal thio- and selenoalcohols **42**–**45** is described based on the highly enantio- and diastereoselective addition of diethylzinc reagent to racemic α-thio- and selenoaldehydes **41**, catalyzed by **20o** ((−)-DFPE) and (*S*,*R*)-**20o** ((+)-DFPE) (Scheme 3-20). Although the enantioselective addition of dialkylzinc reagents to achiral aldehydes using chiral catalysts has been well investigated [10], there are no known catalytic enantio- and diastereoselective dialkylzinc additions to aldehydes with chiral centers, except for the alkylation of α-methyl- [58, 59], α-chloro- [59], and β-alkoxyaldehydes [60]. The reaction of diethylzinc with racemic α-thio- and selenoaldehydes **41** was carried out in the presence of **20o** or (*S*,*R*)-**20o** (5–50 mol%) in hexane at room temperature for 12–16 h. The results are summarized in Table 3-11.

We first examined the ethylation of α-phenylthiobutyraldehyde in the presence of various catalytic amounts of **20o** or (*S*,*R*)-**20o** (entries 1–7). The yield and the

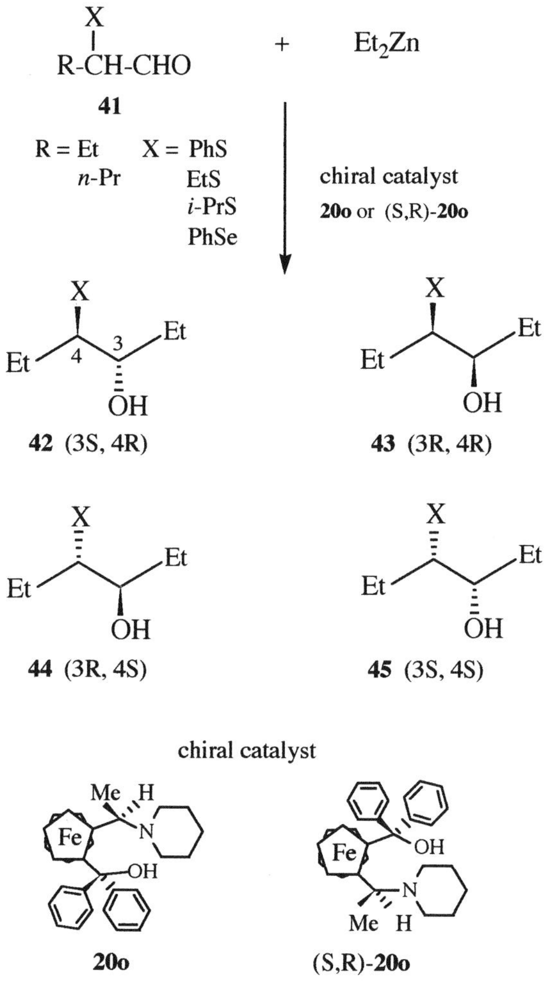

X
R-CH-CHO + Et_2Zn
41
R = Et, *n*-Pr X = PhS, EtS, *i*-PrS, PhSe
chiral catalyst
20o or (S,R)-**20o**

42 (3S, 4R) **43** (3R, 4R)

44 (3R, 4S) **45** (3S, 4S)

chiral catalyst

20o (S,R)-**20o**

Scheme 3-20

Table 3-11. Enantio- and diastereoselective addition of diethylzinc to racemic α-thio- and α-seleno-aldehydes **41** using (−)-DFPE **20o** or (+)-DFPE (*S*,*R*)-**20o**[a]

Entry	Aldehyde **41**		Catalyst (mol%)	Product			
	R	X		Yield, %[b]	**42/43**[c]	**44/45**[c]	(**42** + **43**)/(**44** + **45**)[c]
1	Et	PhS	**20o** (5)	23	96:4	61:39	82:18
2	Et	PhS	**20o** (10)	34	99:1	58:42	87:13
3	Et	PhS	**20o** (25)	49	99:1	46:54	87:13
4[d]	Et	PhS	**20o** (5)	30	96:4	55:45	80:20
5	Et	PhS	(*S*,*R*)-**20o** (5)	18	59:41	94:6	17:83
6	Et	PhS	(*S*,*R*)-**20o** (25)	53	35:65	98:2	18:82
7	Et	PhS	(*S*,*R*)-**20o** (50)	46	27:73	99:1	22:78
8[e]	Et	PhS	–	88	82:18	82:18	50:50
9	Et	EtS	(*S*,*R*)-**20o** (5)	39	63:37	96:4	19:81
10	Et	EtS	(*S*,*R*)-**20o** (10)	19	57:43	94:6	21:79
11	Et	EtS	(*S*,*R*)-**20o** (25)	24	57:43	96:4	14:86
12[e]	Et	EtS	–	90	80:20	80:20	50:50
13	Et	*i*-PrS	(*S*,*R*)-**20o** (10)	20	48:52	88:12	25:75
14	Et	*i*-PrS	(*S*,*R*)-**20o** (25)	14	41:59	93:7	17:83
15[e]	Et	*i*-PrS	–	85	80:20	80:20	50:50
16	Pr	PhS	**20o** (5)	15	98:2	56:44	84:16
17	Pr	PhS	**20o** (25)	38	100:0	45:55	90:10
18[e]	Pr	PhS	–	92	86:14	86:14	50:50
19	Et	PhSe	**20o** (5)	10	98:2	64:36	89:11
20	Et	PhSe	**20o** (25)	68	98:2	42:58	67:33
21[e]	Et	PhSe	–	93	92:8	92:8	50:50

a Unless otherwise noted, the reaction was carried out in hexane at room temperature with 1.6 equiv. of diethylzinc to aldehyde for 12–16 h.
b Isolated yield.
c Determined by ^{19}F NMR spectrum analysis of the corresponding (*S*)-MTPA ester.
d 2.5 equiv. of diethylzinc used per equiv. of aldehyde.
e The reaction was conducted with EtMgBr.

diastereoselectivity increased with increasing amount of the catalyst up to 25 mol% (entries 3 and 6), but the use of 50 mol% of the catalyst showed no enhanced activity (entry 7). When **20o** was used, the diastereomer ratio (**42/43**) obtained for ethylation of the (*R*)-enantiomer of **41** was much superior to that (**44/45**) for the (*S*)-enantiomer of **41**, as well as to the ratio of non-catalyzed ethyl Grignard reaction (entry 8). The (*R*)-enantiomer was consumed faster than the (*S*)-one (**42** + **43/44** + **45**), but the recovered aldehyde **41** was not (*S*) enriched but completely racemized owing to the presence of the highly acidic α-hydrogen. Similarly, the enantio- and diastereoselective addition of diethylzinc to racemic (α-ethylthio)- and (α-isopropylthio)butyraldehydes was examined (entries 9–14). The diastereoselectivity was somewhat lower than the reaction with the aldehyde with a phenylthio substitutent. The ethylation of (α-phenylthio)valeraldehyde in the presence of 25 mol% of **20o** afforded 100% *de* for the (*R*)-enantiomer and the largest reactivity ratio between the (*R*)- and (*S*)-enaniomers was attained (entry 17). The ethylation of (α-phenyl-seleno)butyraldehyde with **20o** showed similar high diastereoselectivity for the (*R*)-enantiomer and much lower selectivity for the (*S*)-enantiomer. The racemic aldehyde **41** was recovered through all reactions after workup. Although EtMgBr reaction afforded good yields of the thio- and selenoalcohols (entries 8, 12, 15, 18, 21), the addition of EtMgBr after diethylzinc reaction did not increase the yield of thioalcohol. Therefore, unchanged aldehyde **41** is considered to be converted to the enolate form during the diethylzinc reaction. In addition to a marked tendency of the substrate aldehydes to undergo enolization, the low product yield in the reactions under low catalyst concentration may be ascribed to the strong coordinating properties of the products as bidentate alcohols, thereby deactivating the diethyl-zinc-DFPE complex.

A plausible intermediate for the enantio- and diastereoselective addition of diethylzinc to α-thioaldehyde catalyzed by **20o** is shown in Fig. 3-4. When **20o** ((−)-DFPE) is used in the ethylation of the racemic aldehyde **41**, the (*R*)-enantiomer reacts faster than the (*S*)-enantiomer and the newly produced stereogenic center from (*R*)-**41** has the *S*-configuration. This stereochemical property can be reasonably explained by considering the 7/6-fused bicyclic intermediate depicted. Compound (*R*)-**41** has lower steric hindrance than (*S*)-**41** in the reaction complex and ethylzinc preferentially attacks from the *Si* face of (*R*)-**41** to afford the *S*-configuration. Therefore, (*R*)-**41** reacts faster than (*S*)-**41** to afford (3*S*,4*R*)-**42**.

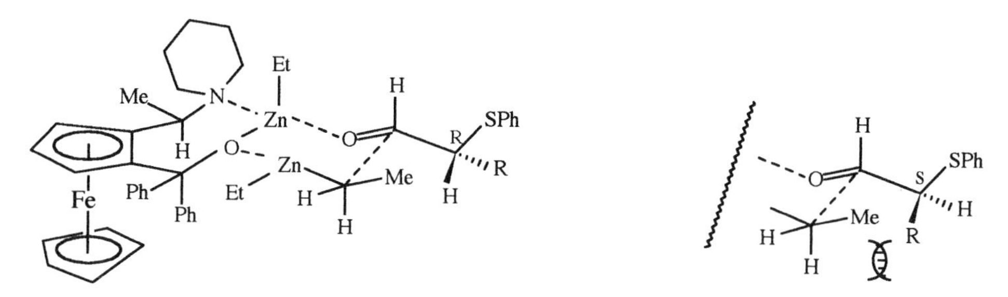

Fig. 3-4. Proposed mechanism for addition of chiral ligand complexed diethylzinc to (*R*)- and (*S*)-α-thioaldehydes.

3.6 Summary

Optically active ferrocene derivatives, particularly ferrocenyl phosphines, have hitherto been utilized as chiral ligands for a wide range of asymmetric synthesis. We have now revealed that the ferrocene moiety can easily be incorporated in amino alcohol ligands instead of phosphinic ligands. The preparative methods for several types of ferrocenylamino alcohols were developed and they were successfully used to catalyze enantioselective addition of dialkylzinc to aldehydes with high enantio-selectivity. In particular, 1,2-disubstituted ferrocenyl amino alcohols with planar chirality were found to possess high enantioselectivity and high catalytic efficiency, and were applied for the synthesis of optically active 3-alkylphthalides and vicinal thio- and selenoalcohols.

References

[1] D. Marquarding, H. Klusacek, G. Gokel, P. Hoffmann, I. Ugi, *J. Am. Chem. Soc.* **1970**, *92*, 5389–5393.
[2] G. Gokel, D. Marquarding, I. Ugi, *J. Org. Chem.* **1972**, *37*, 3052–3058.
[3] D. Marquarding, H. Burghard, I. Ugi, R. Urban, H. Klusacek, *J. Chem. Res., Synop.* **1977**, 82–83; *J. Chem. Res., Miniprint* **1977**, 915–958.
[4] G. Eberle, I. Lagerlund, I. Ugi, R. Urban, *Tetrahedron* **1978**, *34, 977–980.*
[5] R. Herrmann, G. Hubener, F. Siglmuller, I. Ugi, *Liebigs Ann. Chem.* **1986**, 251–268.
[6] R. Urban, *Tetrahedron* **1979**, *35*, 1841–1843; G. Eberle, I. Ugi, *Angew. Chem. Int. Ed. Engl.* **1976**, *15*, 492–493.
[7] A. Ratajczak, A. Czech, *Bull. Acad. Pol. Sci. Ser. Sci. Chim.* **1979**, *27*, 661–664.
[8] M. Cushman, J. Chen, *J. Org. Chem.* **1987**, *52*, 1517–1521.
[9] N. Oguni, T. Omi, *Tetrahedron Lett.* **1984**, *25*, 2823–2824.
[10] Reviews: R. Noyori, M. Kitamura, *Angew. Chem. Int. Ed. Engl.* **1991**, *30*, 49–69; K. Soai, S. Niwa, *Chem. Rev.* **1992**, *92*, 833–856.
[11] M. Kitamura, S. Okada, S. Suga, R. Noyori, *J. Am. Chem. Soc.* **1989**, *111*, 4028–4036.
[12] M. Watanabe, S. Araki, Y. Butsugan, M. Uemura, *Chem. Express* **1989**, *4*, 825–828.
[13] M. Watanabe, S. Araki, Y. Butsugan, M. Uemura, *Chem. Express* **1990**, *5*, 661–664.
[14] M. Watanabe, S. Araki, Y. Butsugan, M. Uemura, *Chem. Express* **1990**, *5*, 761–764.
[15] M. Watanabe, S. Araki, Y. Butsugan, M. Uemura, *J. Org. Chem.* **1991**, *56*, 2218–2224.
[16] M. Watanabe, N. Hashimoto, S. Araki, Y. Butsugan, *J. Org. Chem.* **1992**, *57*, 742–744.
[17] M. Watanabe, M. Komota, M. Nishimura, S. Araki, Y. Butsugan, *J. Chem. Soc. Perkin Trans. 1* **1993**, 2193–2196.
[18] G. Muchow, Y. Vannoorenberghe, G. Buono, *Tetrahedron Lett.* **1987**, *28*, 6163–6166.
[19] R. H. Pickard, J. Kenyon, *J. Chem. Soc.* **1914**, 1115–1131.
[20] T. Mukaiyama, K. Hojo, *Chem. Lett.* **1976**, 893–896.
[21] P. A. Chaloner, S. A. R. Perera, *Tetrahedron Lett.* **1987**, *28*, 3013–3014; P. A. Chaloner, E. Langadianou, S. A. R. Perera, *J. Chem. Soc. Perkin Trans. 1* **1991**, 2731–2735.
[22] E. J. Corey, F. J. Hannon, *Tetrahedron Lett.* **1987**, *28*, 5233–5236.
[23] E. J. Corey, F. J. Hannon, *Tetrahedron Lett.* **1987**, *28*, 5237–5240.
[24] K. Soai, S. Yokoyama, K. Ebihara, T. Hayasaka, *J. Chem. Soc. Chem. Commun.* **1987**, 1690–1691; K. Soai, S. Yokoyama, T. Hayasaka, *J. Org. Chem.* **1991**, *56*, 4246–4268.
[25] S. B. Heaton, G. B. Jones, *Tetrahedron Lett.* **1992**, *33*, 1693–1696.
[26] S. Itsuno, J. M. J. Frechet, *J. Org. Chem.* **1987**, *52*, 4140–4142; S. Itsuno, Y. Sakurai, K. Ito, T. Maruyama, S. Nakahara, J. M. J. Frechet, *J. Org. Chem.* **1990**, *55*, 304–310; K. Soai, S. Niwa, M. Watanabe, *J. Org. Chem.* **1988**, *53*, 927–928; *J. Chem. Soc. Perkins Trans. 1* **1989**, 109–113.
[27] R. Herrmann, I. Ugi, *Tetrahedron* **1981**, *37*, 1001–1009.
[28] M. Rosenblum, R. B. Woodward, *J. Am. Chem. Soc.* **1958**, *80*, 5443–5449.
[29] F. S. Arimoto, A. C. Heaven, Jr., *J. Am. Chem. Soc.* **1955**, *77*, 6295–6297.
[30] Y. H. Chen, M. F. Refojo, H. G. Cassidy, *J. Polym. Sci.* **1959**, *40*, 433–441.
[31] Y. Sasaki, L. L. Walker, E. L. Hurst, C. V. Pittman, Jr., *J. Polym. Sci. Polym. Chem. Ed.* **1973**, *11*, 1213–1224.
[32] C. Aso, T. Kunitake, T. Nakashima, *Makromol. Chem.* **1969**, *124*, 232–240.
[33] G. F. Hayes, R. N. Young, *Polymer* **1977**, *18*, 1286–1287.
[34] Cr. Simionescu, T. Lixandru, I. Negulescu, I. Mazilu, L. Tataru, *Makromol. Chem.* **1973**, *16*, 59–74.

[35] L. F. Battele, R. Bau, G. Gokel, R. T. Oyakawa, I. Ugi, *J. Am. Chem. Soc.* **1973**, *95*, 482–486.
[36] M. Kitamura, S. Suga, K. Kawai, R. Noyori, *J. Am. Chem. Soc.* **1986**, *108*, 6071–6072.
[37] J. Capillon, J. Guette, *Tetrahedron* **1979**, *35*, 1817–1820.
[38] T. Sato, Y. Gotoh, Y. Wakabayashi, T. Fujisawa, *Tetrahedron Lett.* **1983**, *24*, 4123–4126.
[39] A. V. Oeveren, W. Menge, B. L. Feringa, *Tetrahedron Lett.* **1989**, *30*, 6427–6430.
[40] N. Oguni, T. Omi, Y. Yamamoto, A. Nakamura, *Chem. Lett.* **1983**, 841–842.
[41] P. A. Levene, L. A. Mikeska, *J. Biol. Chem.* **1927**, *75*, 587–605.
[42] P. A. Levene, P. G. Stevevs, *J. Biol. Chem.* **1930**, *87*, 375–391.
[43] W. M. Foley, F. J. Welch, E. M. La Combe, H. S. Mosher, *J. Am. Chem. Soc.* **1959**, *81*, 2779–2784.
[44] J. L. Coke, R. S. Shue, *J. Org. Chem.* **1973**, *38*, 2210–2211.
[45] J. P. Mazaleyrat, D. J. Cram, *J. Am. Chem. Soc.* **1981**, *103*, 4585–4586.
[46] R. H. Pickard, J. Kenyon, *J. Chem. Soc.* **1913**, 1923–1959.
[47] M. Uemura, R. Miyake, M. Shiro, Y. Hayashi, *Tetrahedron Lett.* **1991**, *32*, 4569–4572; M. Uemura, R. Miyake, Y. Hayashi, *J. Chem. Soc. Chem. Commun.* **1991**, 1696–1697.
[48] a) D. H. R. Barton, J. X. de Vries, *J. Chem. Soc.* **1963**, 1916–1919. b) M. Elander, K. Leander, B. Luning, *Acta Chem. Scand.* **1969**, *23*, 2177–2178. c) Y. Ogawa, K. Hosaka, M. Chin, H. Mitsuhashi, *Heterocycles* **1989**, *29*, 865–872.
[49] a) M. Asami, T. Mukaiyama, *Chem. Lett.* **1980**, 17–20. b) A. I. Meyers, M. A. Hanagan, L. M. Trefonas, R. J. Baker, *Tetrahedron* **1983**, *39*, 1991–1999. c) S. Matsui, A. Uejima, Y. Suzuki, K. Tanaka, *J. Chem. Soc. Perkin Trans. 1* **1993**, 701–704.
[50] H. Takahashi, T. Tsubuki, K. Higashiyama, *Chem. Pharm. Bull.* **1991**, *39*, 3136–3139.
[51] T. Ohkuma, M. Kitamura, R. Noyori, *Tetrahedron Lett.* **1990**, *31*, 5509–5512.
[52] K. Soai, H. Hori, M. Kawahara, *Tetrahedron: Asymmetry* **1991**, *2*, 253–254.
[53] U. Nagai, T. Shishido, R. Chiba, H. Mitsuhashi, *Tetrahedron* **1965**, *21*, 1701–1709.
[54] W. H. Pirkle, P. L. Rinaldi, *J. Org. Chem.* **1978**, *43*, 3803–3807.
[55] M. Shimagaki, T. Maeda, Y. Matsuzaki, I. Hori, T. Nakata, T. Oishi, *Tetrahedron Lett.* **1984**, *25*, 4775–4778; M. Shimagaki, Y. Matsuzaki, I. Hori, T. Nakata, T. Oishi, *Tetrahedron Lett.* **1984**, *25*, 4779–4782.
[56] S. Uemura, K. Ohe, T. Yamauchi, S. Mizutaki, K. Tamaki, *J. Chem. Soc. Perkin Trans. 1* **1990**, 907–910.
[57] Asymmetric addition of phenylthiomethyllithium to aldehydes was reported in moderate % *ee*: K. Soai, T. Mukaiyama, *Bull. Chem. Soc. Jpn.* **1979**, *52*, 3371–3376.
[58] M. Hayashi, H. Miwata, N. Oguni, *Chem. Lett.* **1989**, 1969–1970; S. Niwa, T. Hatanaka, K. Soai, *J. Chem. Soc. Perkin Trans. 1* **1991**, 2025–2027.
[59] M. Hayashi, H. Miwata, N. Oguni, *J. Chem. Soc. Perkin Trans. 1* **1991**, 1167–1171.
[60] K. Soai, T. Hatanaka, T. Yamashita, *J. Chem. Soc. Chem. Commun.* **1992**, 927–929.

Received: November 2, 1993

Part 2. Organic Synthesis – Selected Aspects

4 Chiral Ferrocene Derivatives. An Introduction

Gabriele Wagner and Rudolf Herrmann

4.1 Central and Planar Chirality in Metallocenes

Soon after its discovery, it was found, that ferrocene behaves in many respects like an aromatic electron-rich organic compound, which is activated towards electrophilic reactions almost like phenol. As a consequence, a rich organic chemistry of ferrocene was developed, in which the organometallic moiety is treated like a simple phenyl group. Thus, Friedel – Crafts acylation leads to ferrocenyl ketones, which can be reduced to secondary ferrocenyl alcohols containing an asymmetric carbon atom. Compounds of this type can be resolved into the enantiomers by standard resolution techniques [1] and do not provide any problems with respect to chemistry and nomenclature. However, ferrocene is different from phenyl with respect to stereochemistry: a derivative with at least two different substituents in the same ring cannot be superimposed with its mirror image, i.e., it is chiral. The first compound of this type to be prepared was 3,1′-dimethyl-ferrocenecarboxylic acid, in which both enantiomers were obtained by resolution of the racemate with cinchonidine and quinidine (Fig. 4-1) [2]. Crystals of one enantiomer suitable for X-ray structure analysis could be grown; this result confirmed unequivocally the correlation with the true configuration of the molecule [3].

Schlögl has investigated the problem systematically and suggested this isomerism to be an example of *planar* chirality [1, 4]. A simple rule was developed to assign chirality descriptors to homoannular di- or polysubstituted ferrocenes: The observer regards the molecule from the side of the ring to be assigned (called the "upper" ring). The substituents are then analyzed for priority according to the Cahn – Ingold – Prelog (CIP) rules. If the shortest path from the substituent with highest priority to that following in hierarchy is clockwise, the chirality descriptor is (*R*), otherwise it is (*S*) [5]. If both rings bear at least two different substituents, the process is started with the ring containing the substituent of highest priority, the molecule is then turned upside down and the second ring subsequently assigned. In order to show clearly that the chirality descriptor belongs to a *planar* element of chirality, it is often written as ($_pR$) or ($_pS$), respectively. Another convention exists, which states that in ferrocene derivatives with both central and planar chirality, the descriptor of the central chirality is written first. Thus, (*R*, *S*) should mean that the compound contains an element of central chirality with (*R*) configuration and an element of planar chirality with (*S*) configuration. This order is used in this book. However, the opposite order of the descriptors was suggested in a recent monograph

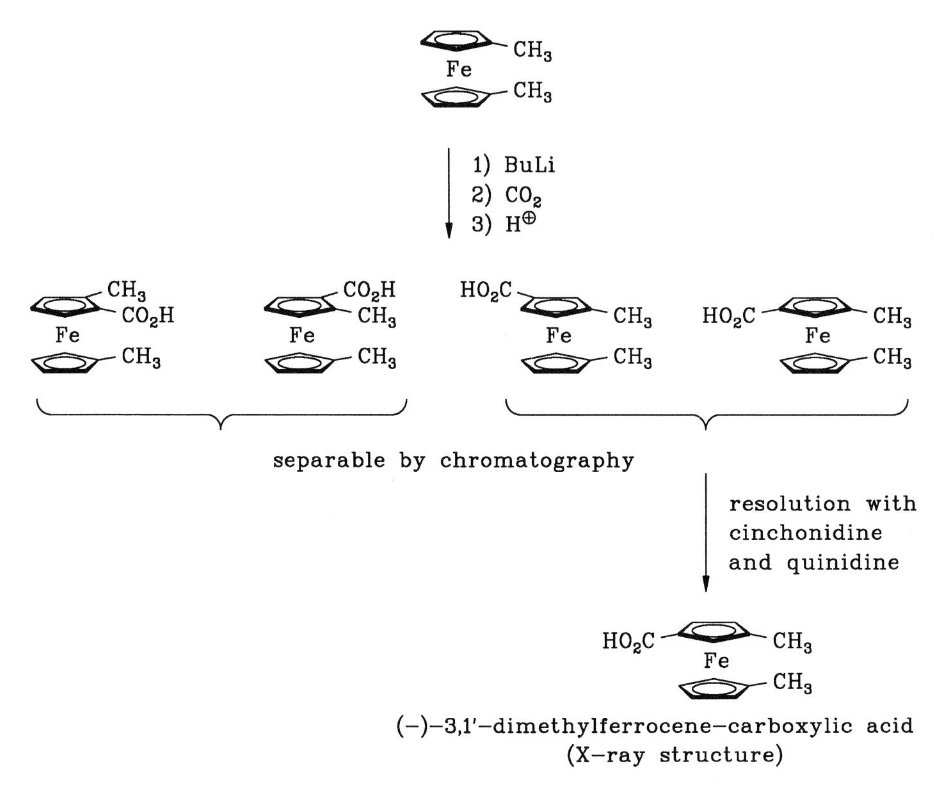

Fig. 4-1. Synthesis and resolution of 3,1′-dimethyl-ferrocenecarboxylic acid.

on chirality in organometallic compounds [6]. When writing an article, it is therefore desirable to specify clearly the descriptor for the planar chirality by using the subscript *p*, or at least the procedure as to how the descriptors were derived and arranged (if this is not done, severe problems may arise if the structures in the figures do not correspond to the text, as in a recent publication on catalytic applications of ferrocenes with both central and planar elements of chirality [7]; it is impossible for the reader to determine which compounds have actually been used).

These rules for determining chirality descriptors were so obvious a way to assign chirality descriptors to ferrocene derivatives that they were soon adopted by almost all ferrocene chemists. However, in their basic paper on stereochemical nomenclature, Cahn, Ingold, and Prelog stated that for metallocenes, the planar chirality can be "reduced" to central chirality, by considering single bonds between ring carbon atoms and iron [8]. This means that every ring carbon is now a (distorted) tetrahedron to which the rules for central chirality apply. The descriptor of the carbon atom bearing the substituent with highest priority then becomes the chirality descriptor of the whole molecule (Fig. 4-2).

The basic problem of this suggestion was that the chirality descriptors obtained by the "central" procedure are the opposite of those derived from the "planar" approach. Schlögl consequently changed his stereochemical nomenclature [5, 9] and,

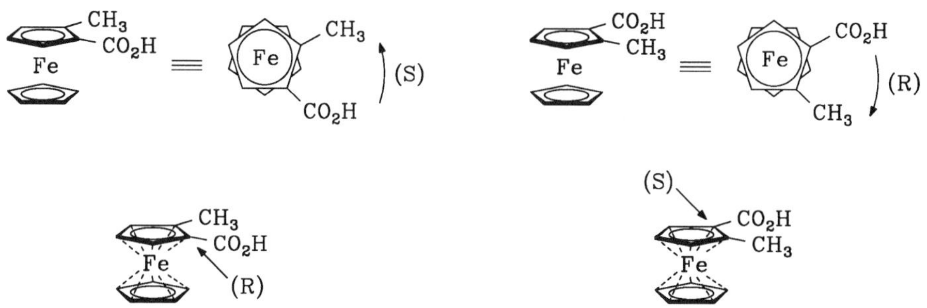

Fig. 4-2. Planar-chiral and central-chiral stereochemical descriptors for 2-methyl-ferrocenecarboxylic acid.

from 1965 to today, he and his Austrian school of ferrocene chemists use descriptors as suggested by CIP, in combination with the term "metallocene chirality". Only the French chemists in general followed Schlögl here and use, for example, (R_p) for the chirality descriptor in the "central" meaning of the CIP rules [10, 11]. According to the best knowledge of the authors, there exist only seven further publications in which this "central" nomenclature occurs [12–18]; even in a French–Russian "joint venture", the *planar* approach is used [19]. As in the vast majority of the publications, the assignment of chirality descriptors according to the *planar* nomenclature will be applied consequently in this book.

A word of caution should be added with respect to Chemical Abstracts, as far as chirality assignments of homoannular substituted ferrocene derivatives are concerned. Until the 8th collective index, only (+) and (−) are found as chirality indicators. For quite a long time, no descriptors were given at all, only the remark "stereoisomer", followed by the registry number, which does not allow identification of a compound easily. This fact is in sharp contrast to the claims of Chemical Abstracts Service authors that they would consequently use Schlögl's central descriptors [20, 21]. Since volume 114, the (R^*, S^*) nomenclature for ferrocene derivatives begins to appear, but its application is not very consequent, at least at the time where the book was written, and it is advisable to examine the orginal article rather than trust Chemical Abstract's descriptors.

4.2 α-Ferrocenylalkyl Carbocations

4.2.1 Structure and Stability

One fascinating aspect of ferrocene chemistry is its extraordinary ability to stabilize carbocations that formally should have their positive charge in a position adjacent to the cyclopentadienyl ring (α-ferrocenylalkyl carbocations). Such cations are so stable that they form quantitatively from appropriate precursors (e.g., alcohols) on treatment with acid and many of them remain unchanged in solution for days and

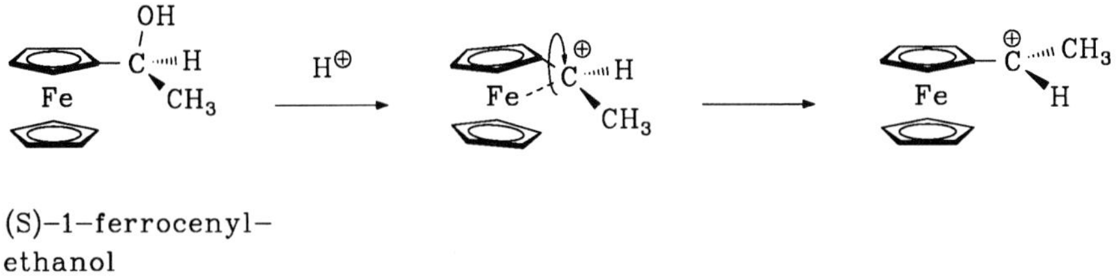

Fig. 4-3. Formation, structure, and racemization of the α-ferrocenylethyl carbocations.

even weeks (for a review on their chemistry, see [22]). It is now known that this stability is due to direct participation of the iron atom in charge delocalization, which causes the two cyclopentadienyl rings to deviate from being parallel (tilt angle 4 – 5°) and leads to a bending of the substituent (Fig. 4-3) bearing the cationic center out of the plane of the ring, in the direction of the iron atom (deviation 10 – 20°). This deviation was predicted by theoretical considerations [23] and confirmed by X-ray structures of crystalline salts of such cations [24 – 26]. These distortions seem to be a general feature of α-ferrocenylalkyl carbocations. For convenience, however, they will generally be drawn in a more conventional way, i.e., as if they were planar at the cationic center (see the right-hand part of Fig. 4-3).

As a consequence of the special structure of such cations, the rotation around the bond between the ring carbon and the cationic centre is hindered, due to its partial double bond character. Thus, the cations exist as distinct enantiomers with only a low tendency towards racemization; their optical rotations remain constant over long periods and the barrier for hindered rotation of the 1-ferrocenylethylium cation in trifluoroacetic acid was determined to have $\Delta H^{\ddagger} = 81.5$ kJ/mol and $T\Delta S^{\ddagger} = -22.2$ kJ/mol [27]; other cations have similar properties [28].

The stabilizing effect of ferrocene on positive charges in its neighbourhood is not limited to charges on carbon; it has also been observed for phosphenium ions [29].

4.2.2 Stereochemistry

The formation of a chiral carbocation by treatment of a chiral α-ferrocenylalkyl alcohol is stereoselective, without loss of chirality information. The same is true for the reaction of the cation with nucleophiles. An overall *retention* of the configuration for this S_N1 process is observed. This was demonstrated experimentally by the construction of synthetic cycles in which the product is identical with the starting material and shows the same enantiomeric excess (*ee*, Fig. 4-4) [30 – 34].

A plausible explanation for these observations is that the leaving group of any α-ferrocenylalkyl compound (e.g., water in the case of a protonated α-ferrocenylalkyl alcohol) will always depart in the direction away from the iron atom (*exo*), as there

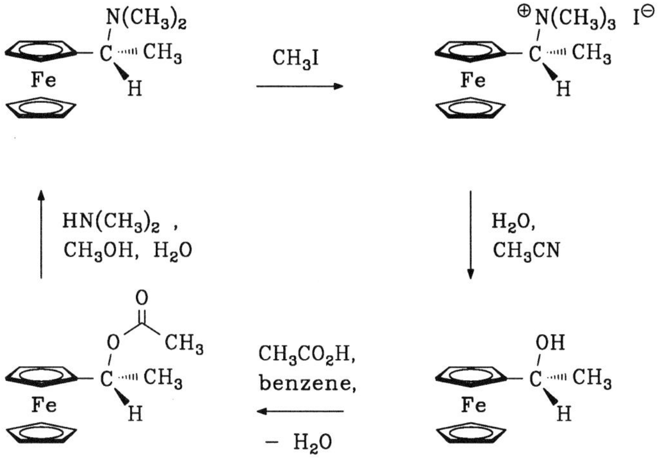

Fig. 4-4. Synthetic cycle showing retentive S_N1 reaction of α-ferrocenylalkyl compounds.

is not much space between the two rings and strong steric repulsion could be expected if the reaction were to occur at this side. As a result, any entering nucleophile will also attack from the *exo*-side as well, which leads to net retention of configuration provided that no rotation around the bond between the ring carbon and the formally cationic center occurs (Fig. 4-5) [35]. In addition, the same *exo* logic also applies to protonation reactions of α-ferrocenyl alkenes in which cations are formed, i.e., the steric influence of the ferrocene moiety reaches even longer distances [36, 37].

There is some evidence that the influence of the ferrocene extends even to cations in the β-position, so that retention during nucleophilic substitution is also found here [38, 39]. This chemistry has not been developed further, presumably because β-functionalized ferrocenes are less easily accessible.

As the enantiomeric α-ferrocenylalkyl carbocations are stable species, a stereochemical nomenclature is desirable. We have made a suggestion that extends the "planar" approach for 1,2-disubstituted ferrocenes to the adjacent cationic center.

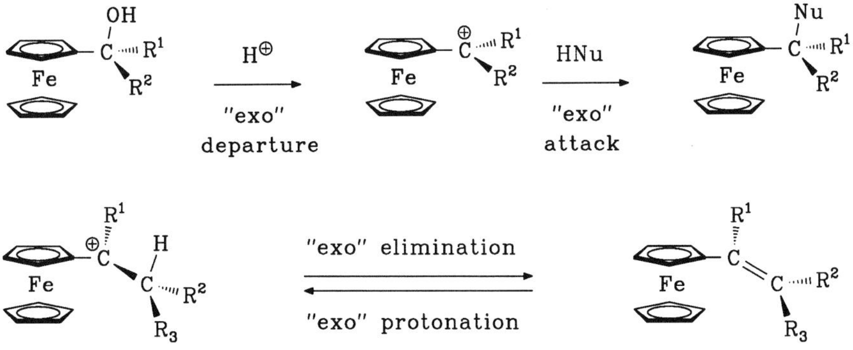

Fig. 4-5. *Exo*-type stereochemistry in the formation and reaction of α-ferrocenylalkyl carbocations.

Fc > R^1 > R^2 : (R)

Fig. 4-6. Stereochemical nomenclature of α-ferrocenylalkyl carbocations.

Thus, the cyclopentadienyl ring, the cationic centre and its substituents are considered to lie in the same plane. The observer then regards the molecule from the "upper" side, i.e., that which contains the ring with the cationic substituent (Fig. 4-6). The priority of the three substituents of the cationic center is then determined according to the CIP sequence rules, and the descriptor (*R*) corresponds to a clockwise arrangement of the substituents in descending priority, while (*S*) is used if the arrangement is counterclockwise [40]. This convention is not only simple to apply, but has the additional advantage that the chirality descriptors are not changed on going from an alcohol to the cation and during its reaction with most nucleophiles (which normally become the substituents with the highest priority according to the CIP rules), i.e., the *configurational* retention is reflected in the retention of the chirality descriptors. According to Sokolov [6], the situation can be seen as planar chiral at ferrocene where two nonidentical substituents are not directly attached to the cyclopentadienyl ligand, but are one atom distant. This obviously leads to the same chirality descriptor, provided that ferrocene (which is omitted from priority assignments in Sokolov's approach) is the substituent of highest priority at the cationic centre, which is normally true.

If other elements of chirality are present in the molecule, the descriptors should be arranged in the order central > carbocationic > axial > planar > torsional (note that Sokolov [6] uses a different order for the descriptors). The effects caused by such other types of chirality are discussed elsewhere in some detail [41], e.g., *S-cis*/*S-trans*-isomerism in ferrocenyl ketones and alkenes, as well as the chemistry of biferrocenyl (containing a direct bond between two ferrocenes).

4.3 Central Chiral Ferrocene Derivatives

4.3.1 Syntheses

4.3.1.1 By Resolution

Ferrocene behaves like an aromatic compound activated for electrophilic substitution reactions. Thus, only minor modifications of experimental procedures developed for aromatics are necessary to obtain ferrocene derivatives (a useful review on general methods is given by Schlögl and Falk [42]). For central chiral ferrocenes, resolution of the racemate is a frequently applied technique. Traditionally, resolutions are best achieved by salt formation between a chiral acid or base and the

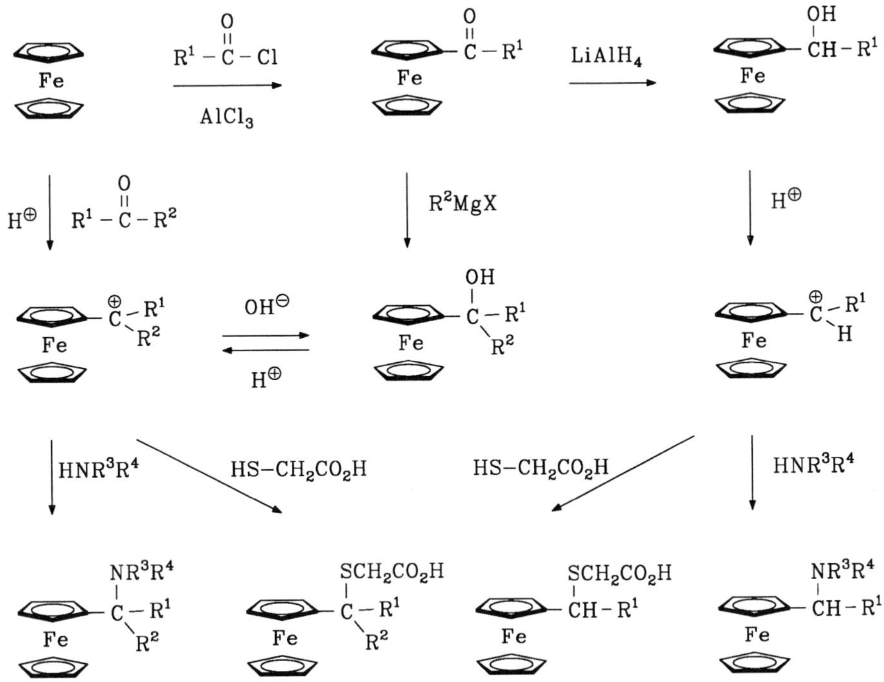

Fig. 4-7. Synthesis of racemic central chiral ferrocene derivatives.

racemic base or acid to be resolved, followed by separation of the diastereoisomeric salts by crystallization, provided that their solubilities are comparatively different. This approach is also valid for ferrocene derivatives. For the synthesis of the racemic compounds, Friedel–Crafts acylation of ferrocene gives α-ferrocenyl ketones, which are reduced to the corresponding secondary alcohols by complex hydrides [43, 44]. These alcohols are the key intermediates for functional group transformations (Fig. 4-7), as they readily form carbocations on treatment with acid, which may then react with nucleophiles to give acidic or basic compounds suitable for resolutions [31, 45–49]. It is also possible to form the carbocations directly from ferrocene and carbonyl compounds in acidic medium (also a Friedel–Crafts type reaction), which allows convenient one-pot syntheses of a variety of ferrocene derivatives [50–53].

As shown in Fig. 4-8, many amines can be resolved with the help of cheap acids such as (*R*,*R*)-tartaric acid [54] or (−)-diacetone-2-keto-l-gulonic acid [(−)-DAG; 2,3-4,6-di-*O*-(1-methylethylidene)-α-*L*-xylo-2-hexulofuranosonic acid] [40]. The resolution of the acidic compounds, e.g., the derivatives of thioglycolic acid (2-mercaptoacetic acid), is possible with chiral bases such as ephedrine [45] or phenylethylamine [55].

If the chiral auxiliary used for the formation of diastereoisomers is covalently attached to the ferrocene derivative, other techniques such as chromatography may be applied for diastereoisomer separation. 1-Phenylethylamine is a comparatively cheap reagent for this purpose. It reacts readily with racemic α-ferrocenylalkyl carbocations to give secondary amines separable by chromatography on silica gel [40, 56]. The same amines were also obtained by an asymmetric synthesis in 90%

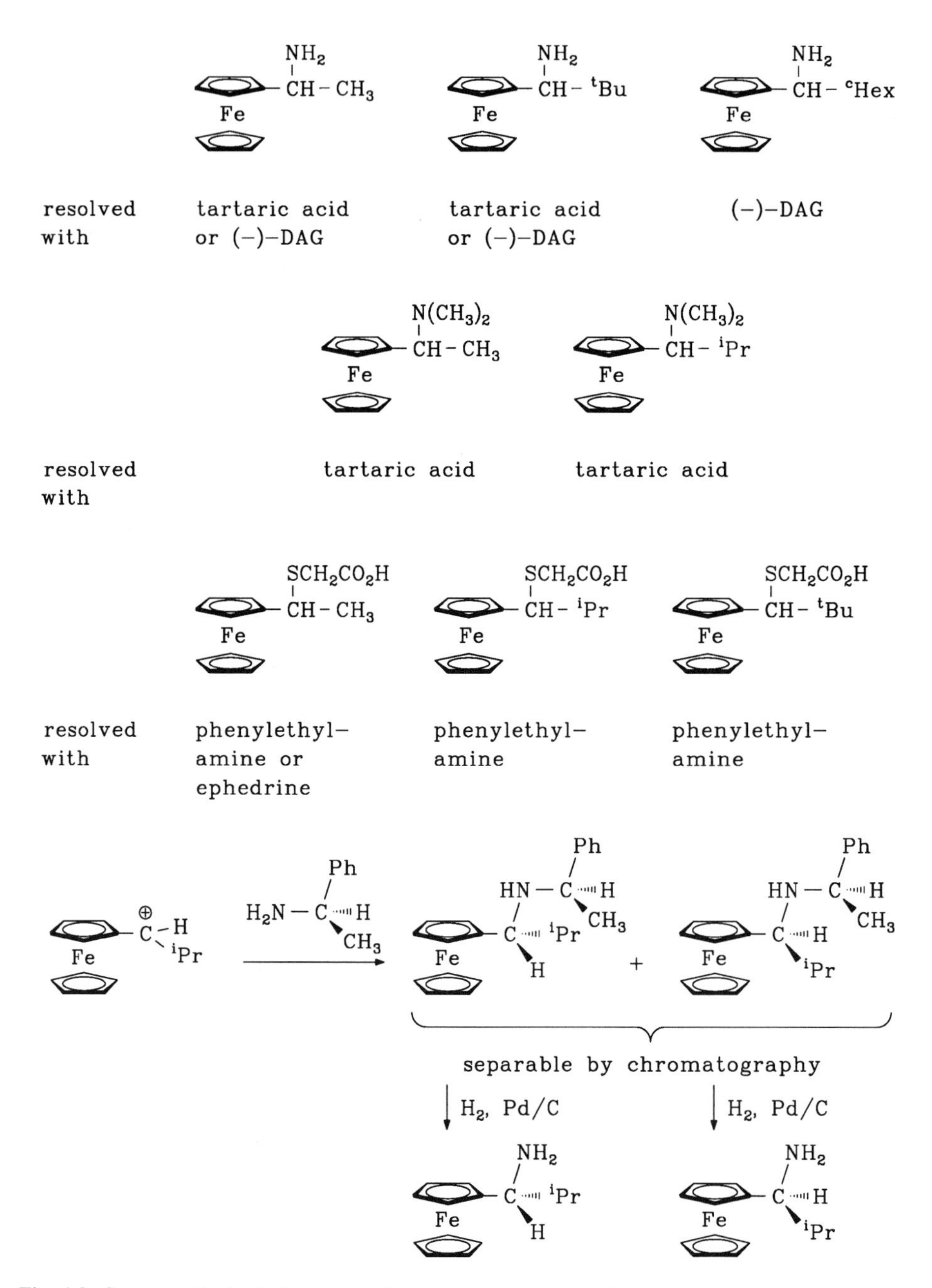

Fig. 4-8. Some synthetically important ferrocene derivatives with central chirality.

ee, by the reduction of the imine from phenylethylamine and acetylferrocene, using $NaBH_4$ in methanol [57]. The chiral auxiliary can then be cleaved by catalytic hydrogenation to the primary amine [40], but this is accompanied by loss of the chiral auxiliary. Chiral thiols, namely *N*-acetyl-*L*-cysteine [58] and dimethyl (*R*)-2-mercaptosuccinate [59], form separable diastereoisomeric sulfides with α-ferrocenylalkyl carbocations. The sulfides can be converted to chiral α-ferrocenylalkanols by the $HgCl_2$ method [45, 59].

With chiral stationary phases, chromatographic separation of enantiomeric ferrocene derivatives is possible. An apparatus for the resolution of ferrocenyl alcohols and other compounds on triacetylcellulose has been described [60]. Analytical enantiomer separation of ferrocenyl alcohols, ethers, sulfides, and amines for the determination of enantiomeric excesses is best achieved on cyclodextrin bonded phases [61].

Kinetic resolution of ferrocene derivatives, mainly alcohols, had an important place during the early stage of stereochemical investigations of ferrocene derivatives. The reaction of (partially) resolved ferrocenylalkyl alcohols and amines with racemic 2-phenylbutyric acid anhydride (Horeau's method) was the basis for the configurational assignment before the establishment of structures by X-ray crystallography [41]. There has been some debate on the reliability of the method [62, 63], and additional chirality information seems necessary for certainty. Recently, the kinetic resolution of 1-ferrocenylethanol by transesterification with vinyl acetate, catalyzed by a lipase from *Pseudomonas fluorescens*, led to an enantiomeric excess of 90 – 96% of both enantiomers [64], opening new preparative aspects.

The compounds obtained by the resolution procedures can often directly be used as chiral auxiliaries, but sometimes it is necessary to convert them to other products, e.g., planar chiral compounds. Frequent functional group transformations are summarized in Sections 4.3.3 and 4.3.3.2. The link to planar chiral compounds is described in Sections 4.3.3.3 and 4.4.1.

4.3.1.2 By Asymmetric Synthesis

If an achiral ferrocene derivative is converted to a chiral one by chiral reagents or catalysts, this may be called an asymmetric synthesis. All asymmetric syntheses of ferrocene derivatives known so far are reductions of ferrocenyl ketones or aldehydes to chiral secondary alcohols. Early attempts to reduce benzoylferrocene by the Clemmensen procedure in (*S*)-1-methoxy-2-methylbutane as chiral solvent led to complex mixtures of products with low enantiomeric excess [65]. With (2*S*,3*R*)-4-dimethylamino-1,2-diphenyl-3-methyl-2-butanol as chiral modifier for the $LiAlH_4$ reducing agent, the desired alcohol was formed with 53% *ee* (Fig. 4-9a) [66]. An even better chiral ligand for $LiAlH_4$ is natural quinine, which allows enantioselective reduction of several ferrocenyl ketones with up to 80% *ee* [67]. Inclusion complexes of ferrocenyl ketones with cyclodextrins can be reduced by $NaBH_4$ with up to 84% enantioselectivity (Fig. 4-9b) [68 – 70].

Microorganisms like baker's yeast are able to reduce some ferrocenyl ketones enantioselectively (for a review on microorganisms and enzymes for the reduction

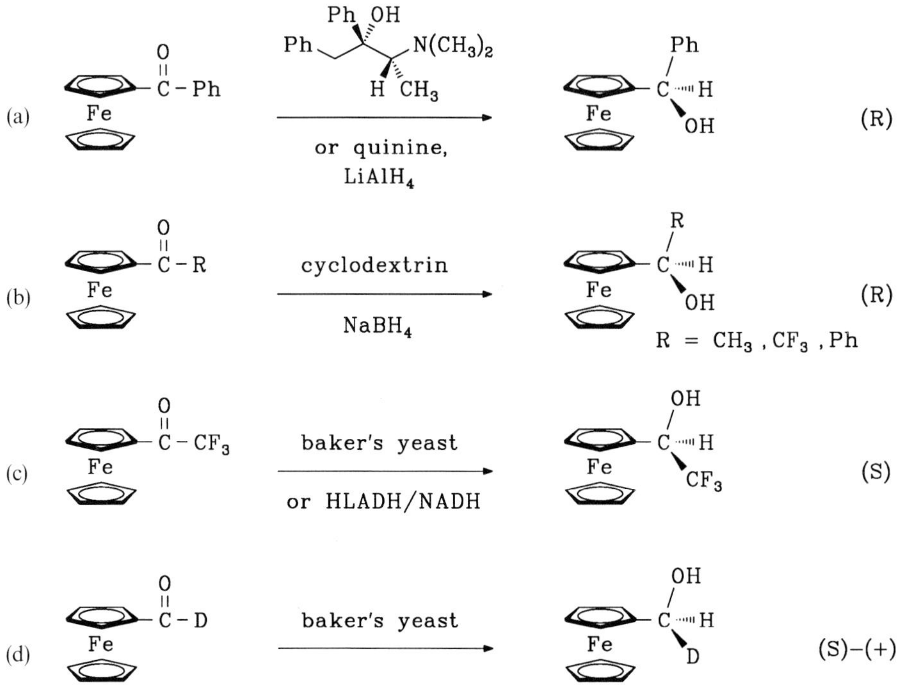

Fig. 4-9. Chiral α-ferrocenylalkanols by stereoselective reduction of ketones.

of organometallic compounds, see Ryabov [71]), in particular ketones with electron-withdrawing substituents [72], such as trifluoroacetylferrocene (Fig. 4-9c). An interesting application was the reduction of deutero-ferrocenealdehyde to optically active α-deutero-ferrocenemethanol (Fig. 4-9d) [73]. Isolated enzymes, such as horse liver alcohol dehydrogenase (HLADH), may also catalyze enantioselective reductions [74].

4.3.1.3 From the Chiral Pool

As resolution procedures are often tedious, and asymmetric synthesis provides chiral products with only limited enantiomeric excess, it seems an obvious strategy to use an enantiomerically pure material from the chiral pool to construct chiral ferrocenes by incorporating these compounds in the final product. As such chiral materials, cheap terpenes (menthone, α- and β-pinene, and camphor) were chosen. The reaction of ferrocene with carbonyl compounds under acidic conditions is a very convenient way to obtain directly α-ferrocenylalkyl carbocations. The starting materials were therefore converted to aldehydes or their enol ethers (menthone and camphor are too sterically hindered and do not react with ferrocene). Joint dissolution of the aldehydes and ferrocene in trifluoroacetic acid or in the trichloroacetic acid/fluorosulfonic acid system gives α-ferrocenylalkyl carbocations, which can either rearrange or be trapped by nucleophiles (typically azide or dimethylamine). Only the compounds obtained from (−)-menthone and some of those from (+)-α-pinene will be discussed here; the other systems are described in detail elsewhere [75].

When the enol ether derived from (−)-menthone and ferrocene are allowed to react in acidic medium at room temperature, a single cation with the ferrocenylalkyl substituent in the equatorial position at the cyclohexane ring is formed (configuration at the cyclohexane: (*R*)) (Fig. 4-10). By trapping the cation with a nucleophile (azide or dimethylamine), a stereochemically uniform product (with configuration at the center adjacent to the ferrocene (*R*)) is formed. If the cation is left to stand for a few hours at room temperature, it rearranges by rotation around the ferrocene–cation bond to give the thermodynamically more stable (*S*)-cation quantitatively, which on trapping with nucleophiles leads to another set of isomeric amines [76]. Deprotonation of the cations by base occurs in the *exo*-mode (see Sect. 4.2.2); the (*R*)-cation gives pure *trans*-alkene, and the more stable (*S*)-cation a mixture of *cis*- and *trans*-alkene (at low temperature, more *cis*-alkene is obtained). Interestingly, the rotation around the ferrocene–alkene bond is so hindered in the *cis*-alkene that a kind of atropisomerism with respect to the orientation of the iron and the unsubstituted cyclopentadienyl ring relative to the cyclohexane is observed by NMR spectroscopy ("up" and "down" isomers). On deprotonation of the corresponding cations, the "down" isomer of both *cis*- and *trans*-alkene is originally formed, but the "down" isomer of the *cis*-alkene is so sterically hindered that its characterization is difficult (already in the original mixture from the low temperature deprotonation, 20% of the "up" isomer is present). Rapid isomerization occurs first to the "up" isomer, which in turn gives the *trans*-alkene [77], and it can therefore be concluded that the *trans*-alkene is not a direct product of the deprotonation of the (*S*)-cation, but comes from a secondary thermal rearrangement of the *cis*-alkene. The less sterically hindered *trans*-alkene is a uniform isomer according to NMR. On reprotonation, it forms a single cation having the ferrocenylmethyl substituent in an *axial* position at the cyclohexane ring (configuration (*S*)); this means that protonation of the "up" isomer occurs exclusively. Subsequent trapping with azide and reduction gives an amine whose X-ray structure proved the structural assignments made above (configuration (*S*) at the carbon adjacent to the ferrocene) [78].

Thus, as shown in Fig. 4-10, three isomeric enantiomerically pure amines can be prepared from the same starting material by the appropriate choice of the reaction conditions [76].

An enantiomerically pure aldehyde, (1*R*,2*R*,3*R*)-2,7,7-trimethylbicyclo[3.1.1]heptane-2-aldehyde, is produced from α-pinene by rhodium-catalyzed hydroformylation [79, 80]. Initially, reaction with ferrocene under acidic conditions leads to a 1 : 1 mixture of diastereoisomeric cations, but on standing for a few hours at room temperature, isomerization by rotation around the ferrocene – cationic carbon bond to the thermodynamically more stable cation (with configuration (*R*) at the cationic center) occurs (Fig. 4-11). An enantiomerically pure amine is available by trapping of this cation by azide and reduction [75]. Analogously, the isomeric aldehyde with the bicyclo [2.2.1] heptane structure is formed by hydroformylation of α-pinene with cobalt catalysts [79, 80] and was used as the starting material for an isomeric series of chiral amines [75].

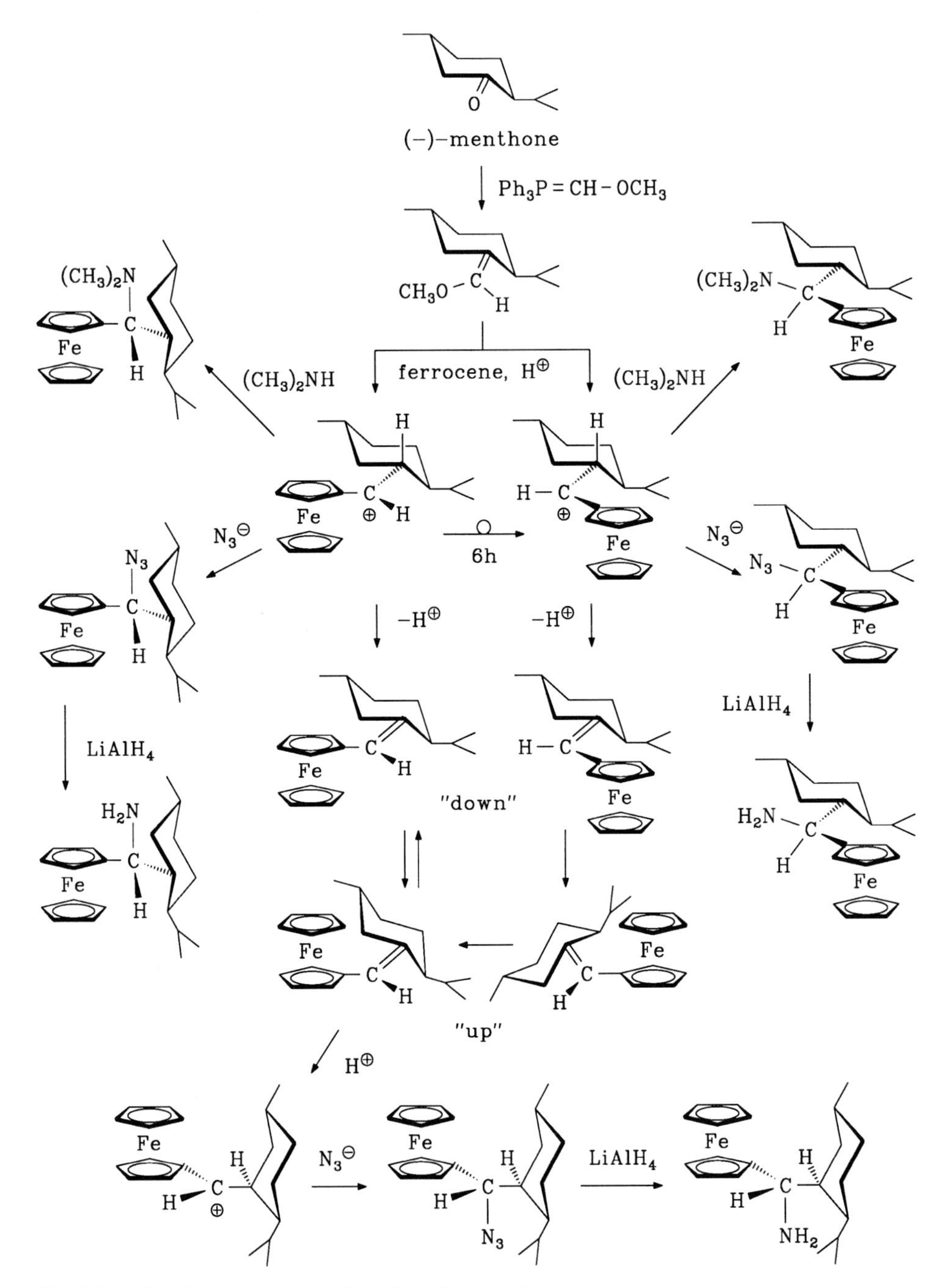

Fig. 4-10. Chiral ferrocene derivatives from (–)-menthone.

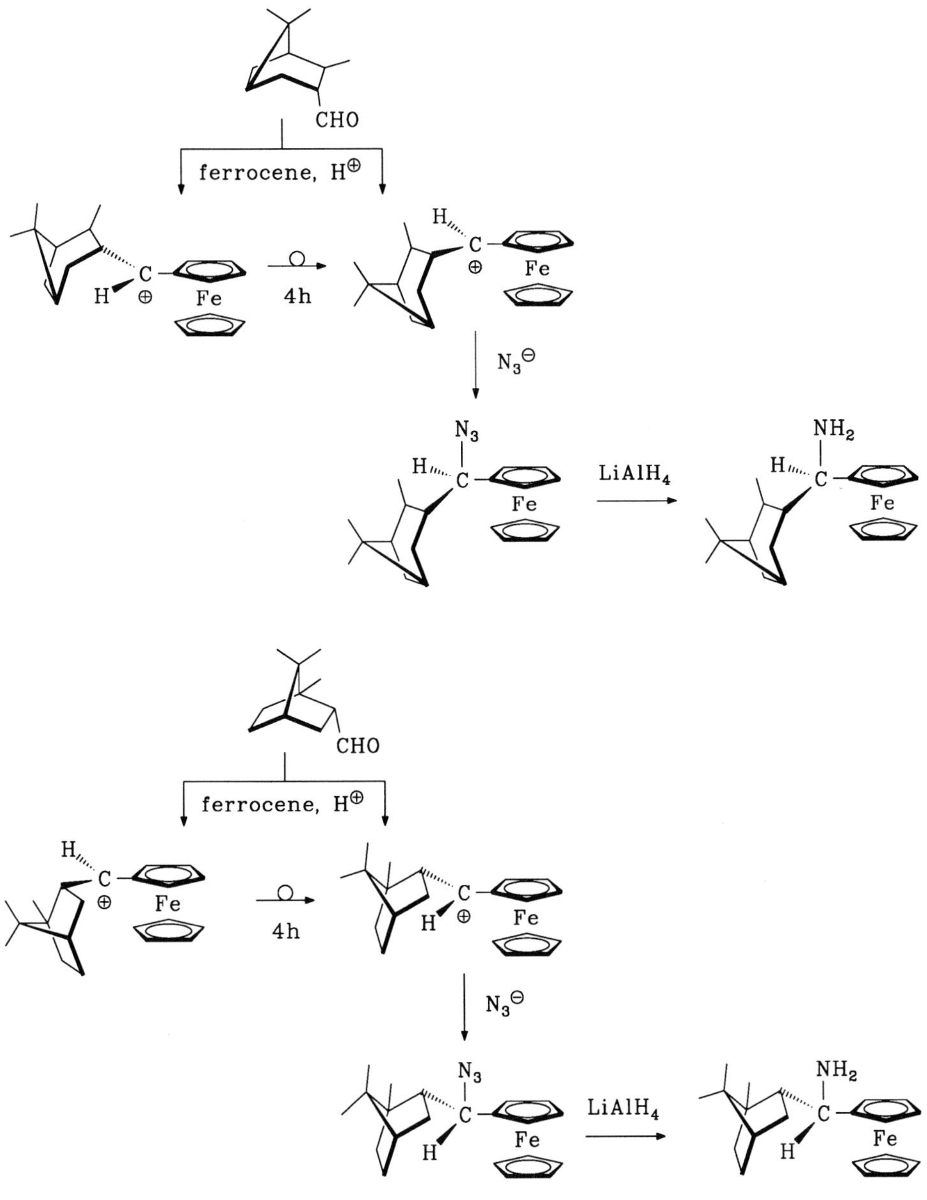

Fig. 4-11. Chiral ferrocene derivatives from α-pinene.

So far, all approaches to chiral ferrocenes have started with unsubstituted ferrocene and the chirality introduced later. If a chiral derivative of cyclopentadiene could be prepared, it would be possible to obtain a chiral ferrocene derivative by simple reaction with $FeCl_2$ and base. This concept has been realized by several groups.

β-Pinene was the starting material for the preparation of a chiral fulvene. Ligand exchange with [(cyclopentadienyl)Fe(*p*-xylidene)]$^+$ gave a mixture of diastereoisomeric chiral ferrocene carbaldehydes, due to incomplete stereocontrol in the formation of the planar chirality during the ligand exchange reaction (Fig. 4-12a). The isomers are separable by chromatography [81]. A similar ligand system from camphor was only applied for the synthesis of titanocene derivatives, not ferrocenes [82].

From tartaric acid, a chiral cyclopentadiene was derived that gave a ferrocene derivative having a phosphine-substituted dioxolane in each ring (Fig. 4-12b) [83].

Chiral starting materials are not only accessible from natural products, but also by asymmetric synthesis. Thus, as shown in Fig. 4-12c, a prochiral fulvene was

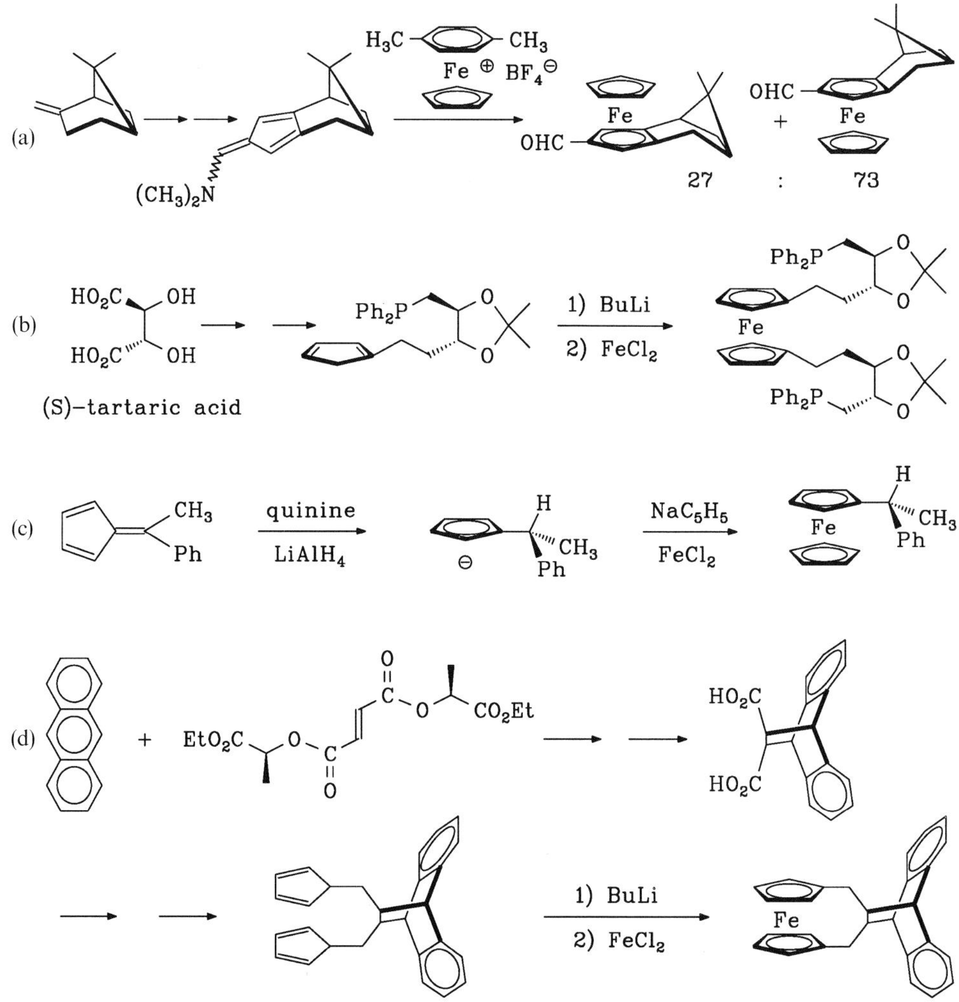

Fig. 4-12. Chiral ferrocene derivatives from chiral cyclopentadienes.

reduced with 17% *ee* by the $LiAlH_4$/quinine complex and the chiral cyclopentadiene anion obtained was treated with cyclopentadienyl sodium, followed by small portions of $FeCl_2$ in THF [84]. The configuration of the chiral 1-ferrocenyl-1-phenylethane could be determined by chemical correlation with chiral 1-ferrocenylethanol.

A chiral C_2-symmetric bridged ferrocene was constructed from a synthetic ligand containing two cyclopentadiene units (Fig. 4-12d). The key step in the synthesis of the ligand is a diastereoselective Diels–Alder reaction of anthracene with bis[(*S*)-1-ethoxycarbonylethyl]fumarate. When oxidized to the ferrocinium salt, the ferrocene derivative has Lewis acid properties and catalyzes Diels–Alder reactions with some enantioselectivity [85].

Another type of C_2-symmetric compound contains two bridged ferrocenes and has only elements of planar chirality. Although not strictly from the chiral pool, similarity with the compounds discussed here justifies inclusion in this section. The biferrocenyl shown in Fig. 4-13, top, was prepared from a bisindenyl ligand by reaction with butyl lithium and $FeCl_2$, and resolved by first introducing an aldehyde group and separating the diastereoisomeric acetals formed with chiral pentane-2,4-diol, and then cleaving the adehyde function with a Rh(I) catalyst [86].

Finally, some rather exotic ferrocenes must be mentioned that are formed from chiral helicenes having five-membered terminal rings by deprotonation and reaction with $FeCl_2$-THF complex (Fig. 4-13, bottom) [87, 88]. Such materials have unusual optical properties, e.g., extremely high optical rotations.

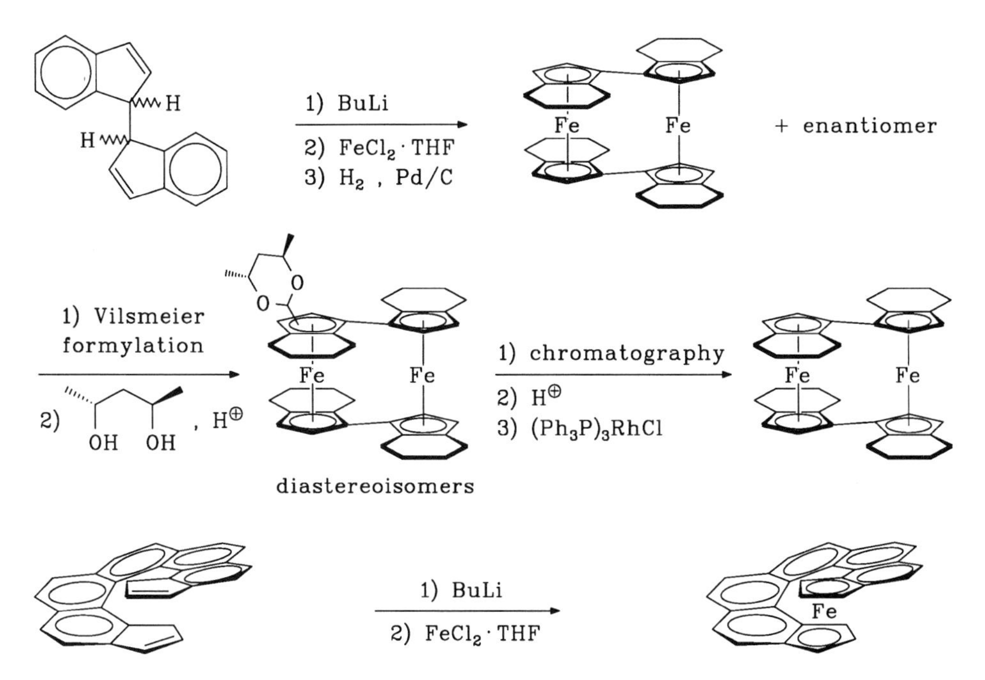

Fig. 4-13. Ferrocene derivatives by asymmetric synthesis or from the chiral pool.

4.3.2 Alkenes

Generally, all ferrocenylcarbocations containing a β-hydrogen atom can eliminate this as a proton by treatment with base to form alkenes, which are therefore interesting potential intermediates in further derivatization. Only the protonation of ferrocenylalkenes has earlier been studied in detail; it always occurs in an *exo*-mode [22, 37, 51, 75, 76]. As a first step for a more general investigation of their reactivity, we reasoned that, in analogy to protonation, alkylation of the ferrocenylalkenes by electrophilic carbon should also be feasible. This is indeed true for alkoxymethyl carbocations as alkylating agents. We have studied this reaction with many ferrocenylalkenes [89], including the menthone-derived chiral *trans*-alkene [76, 78] described above. The alkylation is stereoselective and occurs in the same *exo* manner as the protonation; for the sterically hindered menthone-derived *trans*-alkene, the "up" isomer is alkylated and leads to the cation with the ferrocenylmethyl substituent in the axial position. The cation was trapped with nucleophiles, e.g., azide and dimethylamine (Fig. 4-14), and the products were transformed to chiral amino alcohols, which may be useful as chelating ligands for catalytic applications [89].

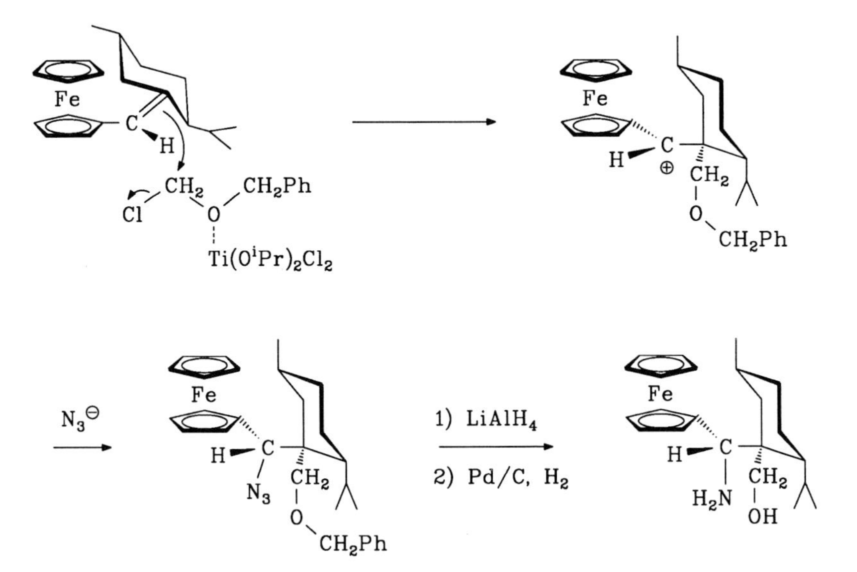

Fig. 4-14. Alkoxymethylation of chiral ferrocenylalkenes.

4.3.3 Amines

α-Ferrocenylalkylamines are by far the most important class of ferrocene derivatives. It is thus not surprising that there are many methods for their preparation. The direct resolution of the racemic amine (see Sect. 4.3.1.1) is not always feasible, so that functional group transformations were developed that rely on the retentive nucleophilic substitution described in Section 4.2.2. A summary of important techniques is shown in Fig. 4-15.

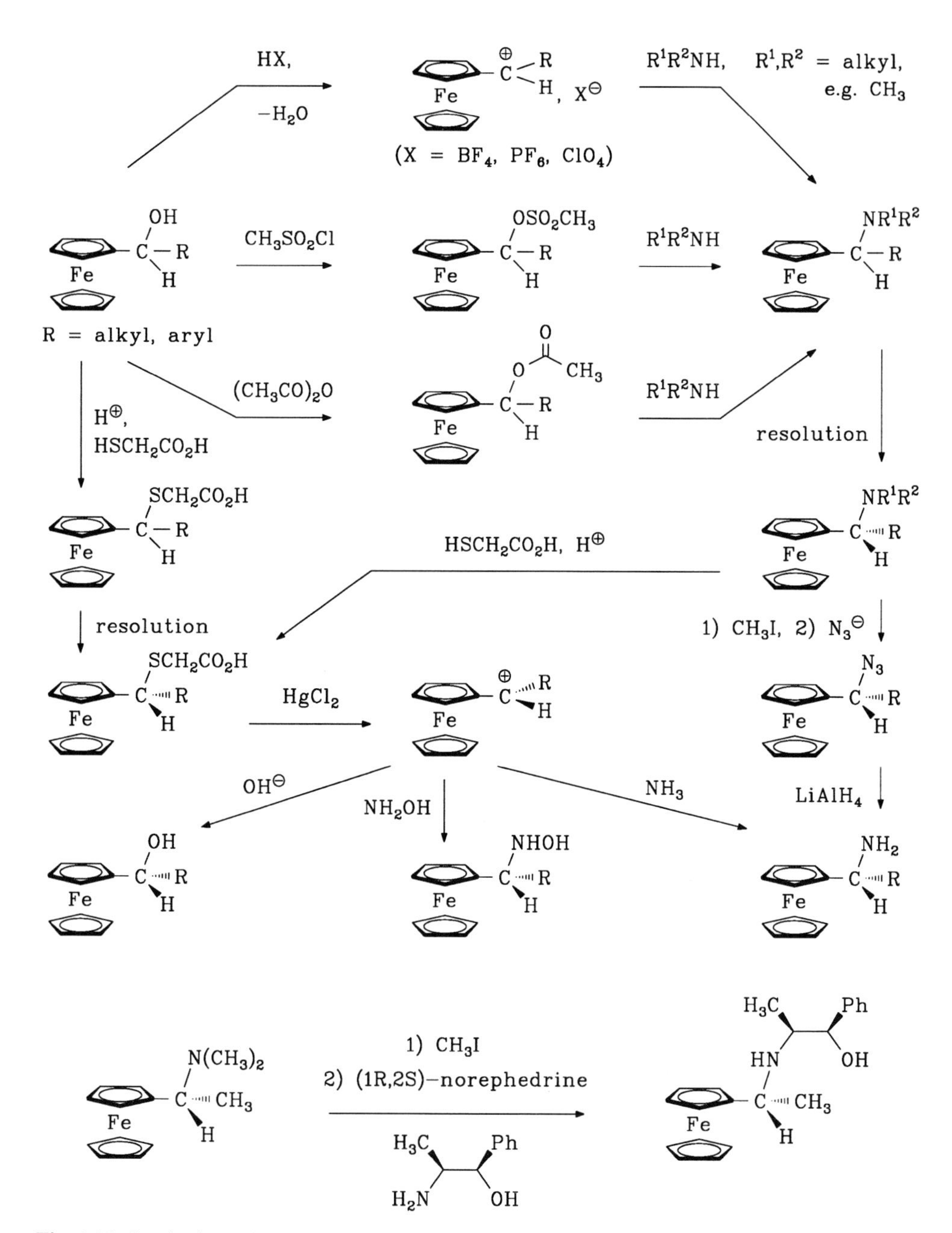

Fig. 4-15. Synthetic methods for the preparation of chiral α-ferrocenylalkyl amines.

Primary ferrocenylalkylamines are often used as chiral auxiliaries and are therefore the main goal of synthesis. As ammonia itself is not a good nucleophile, yields in the direct synthesis from the carbocations are not high. Secondary amines (if not too sterically hindered) [50], mercaptans [51, 52], and azide [30, 32, 40] have a much higher nucleophilicity, and yields are good to excellent. Generally, tertiary amines are easier to resolve than primary amines and during the early stage of the development of asymmetric synthesis by ferrocene derivatives, the preparation and resolution of compounds such as 1-ferrocenyl-*N*,*N*-2-trimethylpropylamine was the key step in the preparation of chiral auxiliaries [90]. The racemates were generally prepared from the corresponding alcohols via stable covalent intermediates like the acetate [31, 32] or the methanesulfonate [47], or salts like the tetrafluoroborate, the hexafluorophosphate, or the perchlorate [48, 49, 53, 91]. The resolved tertiary amines were converted to the primary amines either by quaternization, followed by substitution with azide and reduction [30, 32] or via the derivative of 2-mercaptoacetic acid formed on treatment with acid, which was then cleaved to the carbocation with $HgCl_2$. Trapping of the intermediate cation with hydroxide ions led to chiral alcohols [45], and with ammonia to the primary amine [46, 90]. Chiral *N*-(α-ferrocenylalkyl)hydroxylamines are the products when hydroxylamine is used as nucleophile [55]. Substitution reactions of the enantiomers of *N*,*N*-dimethyl-1-ferrocenylethylamine with norephedrine via the quaternary ammonium salt, leading to ferrocene-substituted amino alcohols with three centers of chirality, are discussed in the chapter on enantioselective addition of dialkyl zinc to carbonyl compounds (Chapter 3).

4.3.3.1 Reactions at Nitrogen

The nitrogen atom in α-ferrocenylalkylamines generally shows the same reaction pattern as that in other amines; alkylation and acylation do not provide synthetic problems. Due to the high stability of the α-ferrocenylalkyl carbocations, ammonium salts readily lose amine and are, therefore, important synthetic intermediates. Acylation of primary amines with esters of formic acid gives the formamides, which can be dehydrated to isocyanides by the standard $POCl_3$/diisopropylamine technique (Fig. 4-16) [92]. Chiral isocyanides are obtained from chiral amines without any racemization during the reaction sequence. The isocyanides undergo normal α-addition at the isocyanide carbon, but could not be deprotonated at the α-carbon by even strong bases. This deviation from the normal reactivity of isocyanides prompted us to study the electrochemistry of these compounds, but no abnormal redox behaviour, compared with that of other ferrocene derivatives, was detected [93]. The isocyanides form chromium pentacarbonyl complexes on treatment with $Cr(CO)_5(THF)$ (Fig. 4-16) and electrochemistry demonstrated that there is no electronic interaction between the two metal centres.

4.3.3.2 Reaction at Chiral Carbon

As described above, nucleophilic substitution at a chiral α-carbon atom normally occurs with retention of the configuration. Exceptions are possible only when the

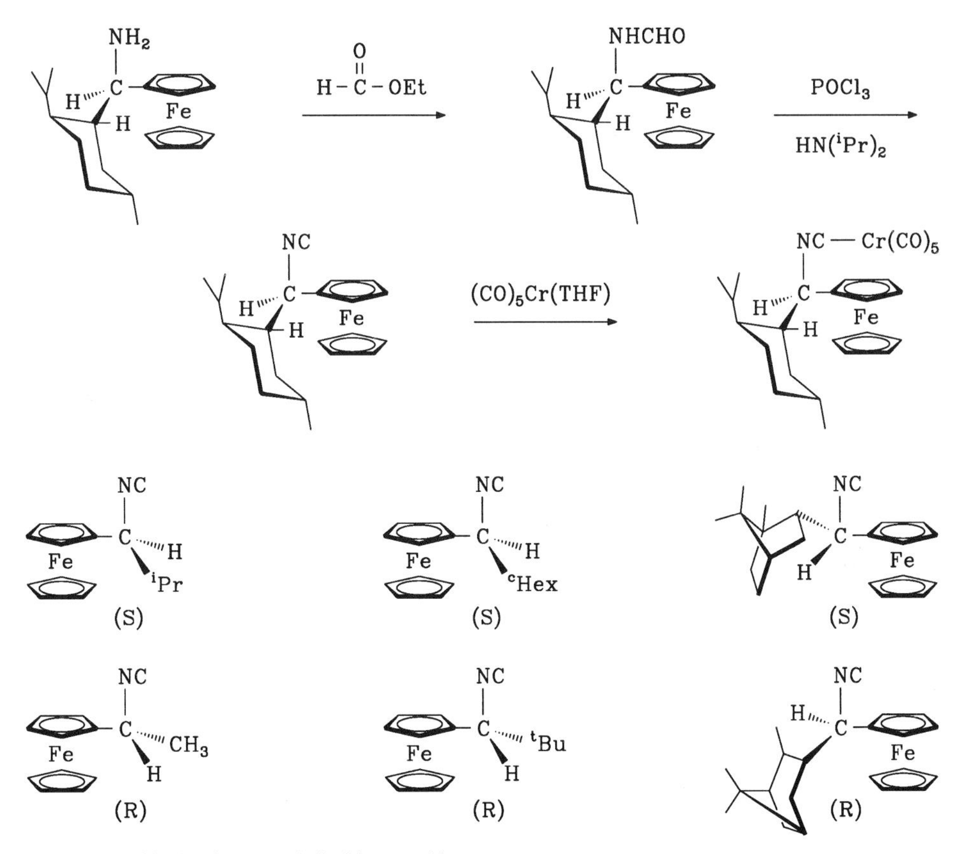

Fig. 4-16. Chiral α-ferrocenylalkyl isocyanides.

intermediate carbocation is stored for prolonged periods to allow rotation around the ferrocene–cation bond; most activation enthalpies for this process lie in the range 40–120 kJ/mol in acidic medium [22]. For nonpolar solvents, the barrier towards rotation may be lower and partial racemization may occur, although it is not observed if trapping with nucleophiles is rapid enough. Using functional group transformations at chiral carbon, together with intramolecular substitution reactions at the cyclopentadienyl rings, Schlögl and coworkers have correlated chemically several hundreds of ferrocene derivatives with compounds whose structure was known by X-ray crystallography, and have thus founded the basis for our precise knowledge on configurations and enantiomeric purity of both central chiral and planar chiral ferrocenes [5, 41, 42, 94–96].

As to the nucleophiles that can be applied in the nucleophilic reactions, ammonia and amines, water and alcohols, and mercaptans have been mentioned already. Sulfide and thiourea have been used only with the achiral ferrocenylmethylium ion [97]. To obtain chiral derivatives of 1-ferrocenyl-ethylmercaptan, substitution of the acetate with potassium thioacetate in acetic acid, followed by reduction with $LiAlH_4$, was found appropriate [98, 99]. Reaction of (*R*)-1-ferrocenylethanol with NaH and

CS_2, with subsequent alkylation with methyl iodide, led to (*R*)-1-ferrocenylethyl-*S*-methyltrithiocarbonate, which involves a rearrangement with retention (Fig. 4-17, top) [100]. Sulfides [101], tertiary phosphines [101], and tertiary amines [102] are also reasonably good nucleophiles and form reactive ionic products, e.g., pyridinium salts [103], but this has only been verified for achiral or racemic substrates. Pyridinium salts may be considered as a storage form of α-ferrocenylalkyl carbocations, and show almost the same behaviour towards nucleophiles [103, 104]. Primary and

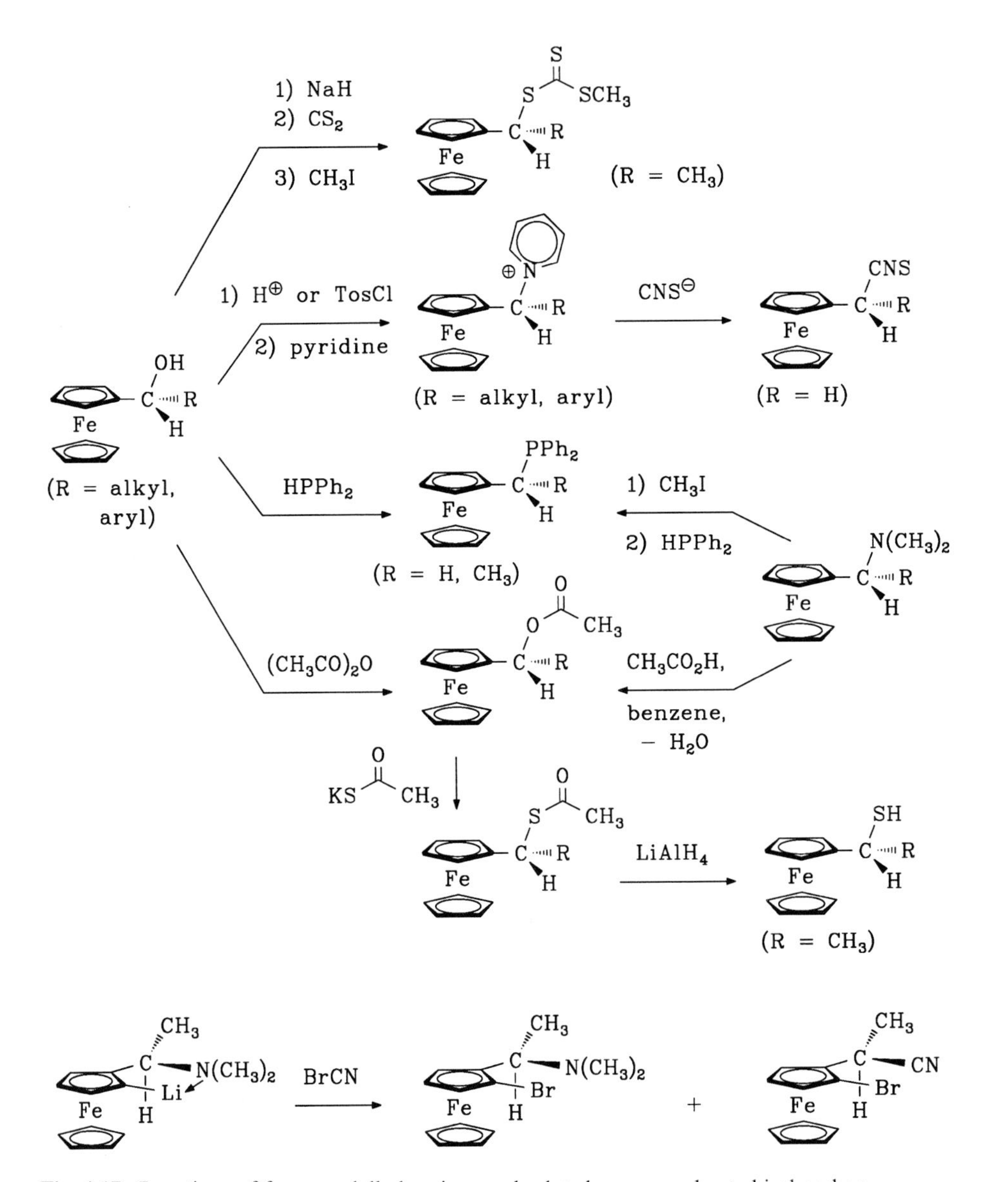

Fig. 4-17. Reactions of ferrocenylalkyl amines and related compounds at chiral carbon.

secondary phosphines are alkylated smoothly by ferrocenylalkyl carbocations, as demonstrated earlier with achiral or racemic compounds [105], and we have recently obtained optically active ferrocenylalkyl phosphines by this technique [106]. Trialkyl phosphites react with the ferrocenylmethylium ion under Arbusov rearrangement to give dialkyl ferrocenylmethylphosphonates [107]. In contrast to azide, other pseudohalogens (e.g., cyanide) are less nucleophilic and give lower yields [35]; from pyridinium salts, nitriles [103] and thiocyanates [104] could be prepared by the reaction with cyanide or thiocyanate ions, but only in racemic form. With compounds containing special structural features, chiral nitriles may be obtained in good yields, such as in the reaction of lithiated *N*,*N*-dimethyl-1-ferrocenylethylamine with BrCN shown on the bottom of Fig. 4-17 (see also Sects. 4.4.1 and 4.4.5) [106]. As α-ferrocenylalkyl halides are only of limited stability [108], the use of halide ions as nucleophiles is not a good synthetic method.

4.3.3.3 Stereoselective Metalation

Lithiation

The hydrogen atoms at the cyclopentadienyl rings of ferrocene exhibit some acidity and are therefore amenable towards metalation reactions [109, 110]. Commercial butyl lithium is a convenient reagent for this purpose. It was discovered soon that selective 1,1′-dilithiation of ferrocene can be achieved by the addition of *N*,*N*,*N*′,*N*′-tetramethylethane-1,2-diamine (TMEDA), and that a substituent at ferrocene derivative having a suitably located donor atom such as nitrogen or oxygen selectively directs lithiation in the adjacent position. The lithium derivatives of ferrocene react with many electrophiles in the same manner as other organometallic compounds to give stable 1,1′- or 1,2-disubstituted products. The state of the art shortly before the beginning of the "stereoselective era" in the lithiation field is covered by an excellent review [111].

A first attempt towards an asymmetric synthesis by the lithiation reaction used (−)-spartein as a chiral additive [12, 13], but the enantioselectivity was disappointing. With the chiral (*S*)-*N*-ferrocenylmethyl-2-methylpiperidine, a high diastereoselectivity was observed [12, 13, 112] (note that the polemic about the structure of the product is due to different stereochemical nomenclature, i.e., "central" *vs.* "planar", as discussed in Sect. 4.1). A breakthrough was achieved by the readily resolvable [54] enantiomerically pure *N*,*N*-dimethyl-1-ferrocenylethylamine as starting material for the lithiation reaction. The diastereoselectivity is very high (92% diastereomeric excess), and enantiomerically pure 1,2-disubstituted products are obtained in high yield by reaction of the lithiated intermediate with electrophiles [62]. The stereochemistry of the reaction sequence could be demonstrated unequivocally by an X-ray structure analysis of the product of the lithiation of (*R*)-*N*,*N*-dimethyl-1-ferrocenylethylamine, followed by reaction with 4-methoxybenzaldehyde [113, 114]. Both the configuration of the starting amine and the stereochemical course of the lithiation reaction where established by this method. Ethers can be used instead of amines for this technique, but with less stereocontrol [115].

The rationale for the high diastereoselectivity of the lithiation reaction is that the lone pair of the heteroatom (here nitrogen) is in a suitable position to form a donating bond to the lithium atom in position 2, stabilizing the 2-lithio-isomer by the chelating effect with respect to the 3-lithio-compound or the l′-isomer (having the lithium in the other cyclopentadienyl ring). If one now considers that there is not much space at the *endo*-side of the ferrocene moiety (i.e., between the two rings, Fig. 4-18) it seems convincing that the hydrogen atom and not the methyl group (which are the remaining substituents of the chiral carbon atom) will preferably occupy this space; the methyl group would have strong repulsing interactions with the hydrogen atoms of the second ring. Thus, (*R*,*R*)-1-(1-dimethylaminoethyl)-2-lithioferrocene is the more stable diastereoisomer and is obtained in a 96:4 ratio with respect to its (*R*,*S*)-isomer [62].

This explanation for the enantioselectivity has recently been questioned on the basis of an X-ray structure of a product formed from a lithiation reaction after standing for 4 weeks at 20 °C, after having filtered off the bulk of lithiated *N*,*N*-dimethyl-1-ferrocenylethylamine (presumably the racemate) as its TMEDA adduct [116]. In this product, lithium is coordinated in several different modes to the ferrocene ring, some of them stabilized by TMEDA molecules and by ethoxide ions formed by cleavage of the solvent diethyl ether. It seems, however, unlikely that conclusions can be drawn from the structure of a by-product with respect to the main reaction, more so when the prolonged storage of the reaction mixture is considered. In addition to this study, NOE measurements on 1-(1-dimethylamino-ethyl)-1′-diphenylphosphinoferrocene suggested that the methyl group at the chiral carbon should at least be partially in the space between the rings [117]. Again, it

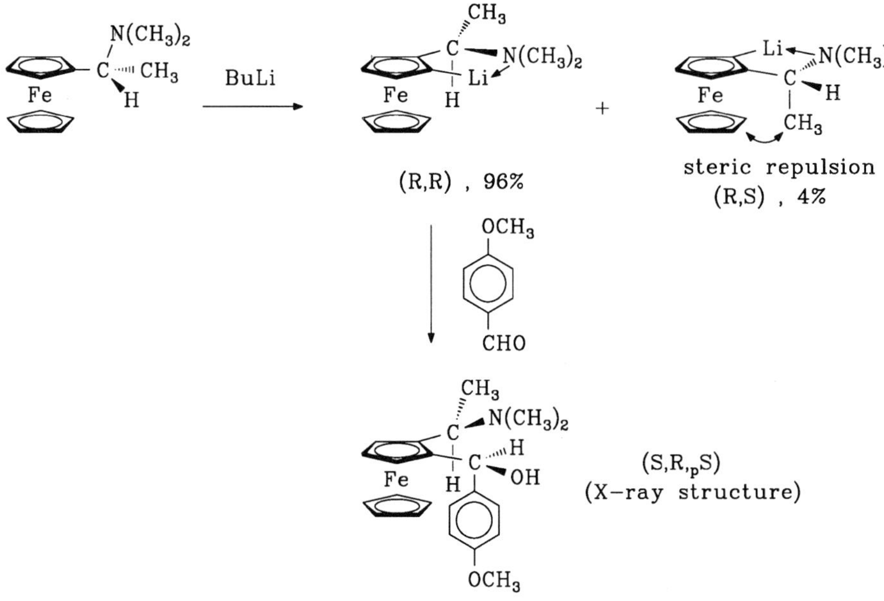

Fig. 4-18. Stereoselective lithiation of α-ferrocenylalkyl amines.

seems unlikely that the steric effect of a diphenylphosphino group at the other ring towards the dimethylaminoethyl substituent should be the same as that of a lithium atom in the same ring. Thus, the rationale for the diastereoselectivity of the lithiation reaction continues to be accepted by most chemists.

Ugi has coined the term "stereorelating synthesis" for the sequence lithiation/reaction with electrophiles [62, 118], and used this technique as a method for the chemical correlation of the structure and for the determination of the enantiomeric purity of many 1,2-disubstituted ferrocene derivatives obtained either by resolution or by asymmetric synthesis (for a compilation, see [118]). It is important to note that all stereochemical features discussed above for central chiral compounds, such as retentive nucleophilic substitution, remain valid when more substituents are present at the ferrocene ring and the conversion of functional groups in planar chiral ferrocenes can be achieved by the same methods as described.

Palladation

The metalation of ferrocene and its derivatives is not limited to lithium, but is also possible with transition metals, in particular mercury and palladium. As in the case of lithiation, the use of central chiral ferrocenes, such as *N*,*N*-dimethyl-1-ferrocenylethylamine, leads to diastereoselective metalation and forms stable chloro-bridged dimeric products. The nitrogen lone pair has the same chelating effect on the transition metal as it has on lithium, as shown by X-ray structures of palladated products [119], but the observed diastereoselectivity is lower (a 70:30 ratio was observed, compared with 96:4 for the lithiation) [120]. Palladium, platinum, and nickel derivatives of high optical purity are accessible from the lithiated compound by exchange of the metal with mercury, which is in turn replaced by the desired transition metal [119, 121, 122]. A higher selectivity of the palladation reaction was found for *N*,*N*-dimethyl-1-ferrocenyl-2,2,2-trifluoroethylamine (97:3 diastereomeric ratio) [123], and the asymmetric induction is much higher when conformationally rigid amines such as 1-(dimethylamino)-[3]-ferrocenophanes are palladated [19].

In contrast to lithiation, attempts to use the achiral *N*,*N*-dimethyl-ferrocenylmethylamine (Fig. 4-19, R = H) as starting material and introduce chirality by chiral additives have been successful in the case of the palladation reaction. Carboxylates are effective additives, and with the sodium salts of enantiomerically pure *N*-acetylvaline or *N*-acetyl-leucine about 90% *ee* could be obtained [124, 125]. The rationale for the asymmetric induction by the chiral additive is that palladium first coordinates to the nitrogen of the ferrocenylamine and to the carboxylate group of the amino acid derivative. The carbonyl oxygen of the carboxylate should then interact with the hydrogen in position 2 of the ferrocene ring, making it more acidic and thus prone to attack by palladium in the transition state. The preferred diastereoisomer is that in which the larger substituent of the chiral amino acid has less interaction with the ferrocene system, i.e., where it is directed to the *exo*-side of the ferrocene ring [124]. As the palladated products have higher stability than the lithiated compounds, crystallization is an attractive possibility for access to enantiomerically

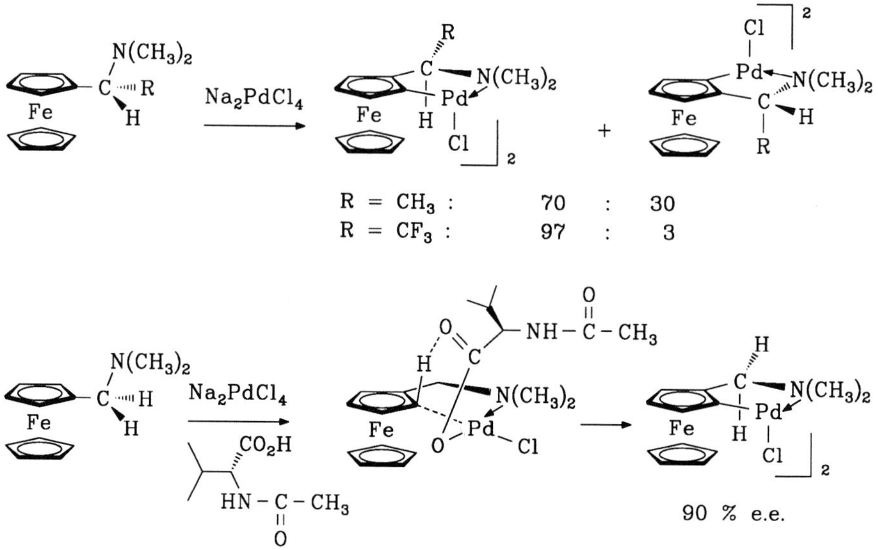

$R = CH_3$: 70 : 30
$R = CF_3$: 97 : 3

90 % e.e.

Fig. 4-19. Stereoselective palladation of ferrocenylalkyl amines.

pure products. This was successful in the case of proline as additive [126]. Analogous to the lithium compounds, the palladated ferrocenes react with electrophiles to form a variety of planar chiral products [124, 127].

4.4 Planar Chiral Ferrocene Derivatives

4.4.1 General Synthetic Methods

The best entrance to planar chiral ferrocene derivatives is obviously the stereoselective metalation discussed in the preceding section. Some additional features of these reactions shall be adressed which show the high degree of flexibility of the method. It was mentioned already that the addition of TMEDA to the lithiation of ferrocene leads to the formation of 1,1′-dilithioferrocene. If TMEDA and a second equivalent of butyl lithium are added to lithiated *N*,*N*-dimethyl-1-ferrocenylethylamine, a selective lithiation occurs in the second ring (Fig. 4-20, top). Subsequent reaction with an electrophile gives a pure diastereoisomer of the 2,1′-disubstituted amine in a convenient one-pot procedure [128, 129]. If the 2-position is blocked by a substituent (Fig. 4-20, middle) such as in (*R*,*S*)-2-(*N*,*N*-dimethyl-1-ethyl)-1-(4-methylphenylthio)ferrocene, the position 5 of the ferrocene ring is lithiated selectively (in the absence of TMEDA), leading to 1,2,3-trisubstituted ferrocenes [129–131]. Although the methyl group at the chiral carbon atom must point

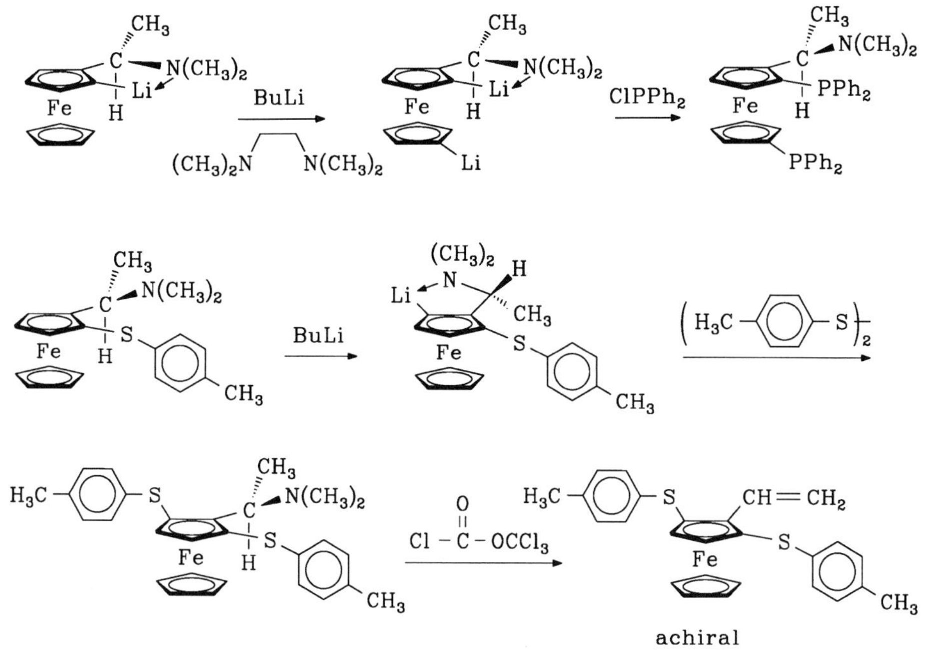

Fig. 4-20. Synthesis of higher substituted planar chiral ferrocenes.

downwards *(endo)* in order to allow for stabilization of the lithium in position 5 (see Sect. 4.3.3.3 for the less stable lithiation product of the simple amine), the chelating effect is still sufficiently strong to prevent lithiation in positions 3 or 1′. This effect was demonstrated by the introduction of two equal *p*-tolylthio substituents in the lithiated amine, and subsequent elimination of dimethylamine, which led to a substituted vinylferrocene devoid of any sign of optical activity (Fig. 4-20, bottom) [132]. Trisubstituted ferrocenes are also accessible via the palladated ferrocenylamines [6]. Individual compounds of these types will be described in the following sections.

In addition to stereoselective metalation, other methods have been applied for the synthesis of enantiomerically pure planar chiral compounds. Many racemic planar chiral amines and acids can be resolved by both classical and chromatographic techniques (see Sect. 4.3.1.1 for references on resolution procedures). Some enzymes have the remarkable ability to differentiate planar chiral compounds. For example, horse liver alcohol dehydrogenase (HLADH) catalyzes the oxidation of achiral ferrocene-1,2-dimethanol by NAD^+ to (*S*)-2-hydroxymethyl-ferrocenealdehyde with 86% *ee* (Fig. 4-21a) and the reduction of ferrocene-1,2-dialdehyde by NADH to (*R*)-2-hydroxymethyl-ferrocenealdehyde with 94% *ee* (Fig. 4-21b) [14]. Fermenting baker's yeast also reduces ferrocene-1,2-dialdehyde to (*R*)-2-hydroxymethyl-ferrocenealdehyde [17]. HLADH has been used for a kinetic resolution of 2-methyl-ferrocenemethanol, giving 64% *ee* in the product, (*S*)-2-methyl-ferrocenealdehyde

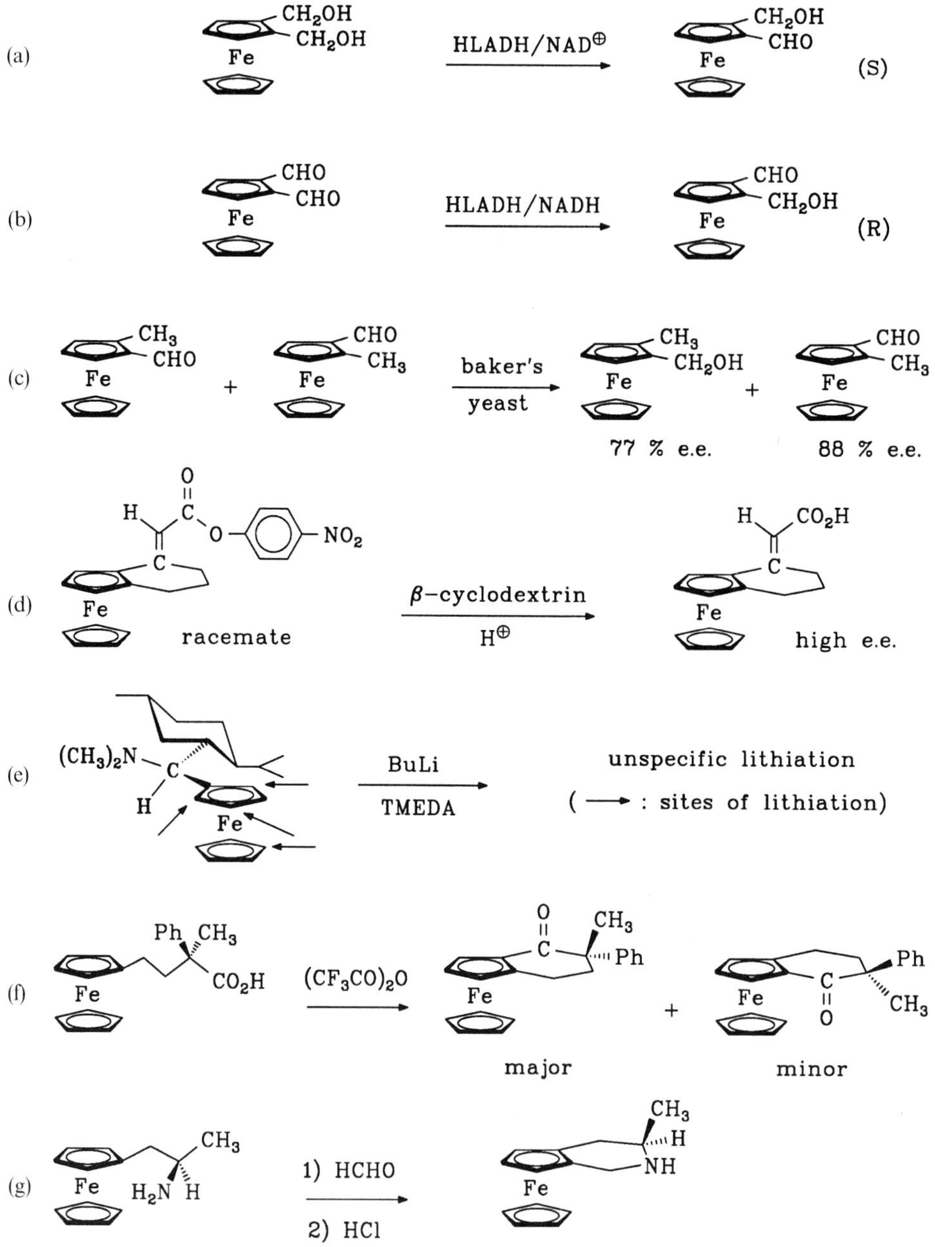

(a) CH_2OH, CH_2OH, Fe — HLADH/NAD$^{\oplus}$ → CH_2OH, CHO, Fe (S)

(b) CHO, CHO, Fe — HLADH/NADH → CHO, CH_2OH, Fe (R)

(c) CH_3, CHO, Fe + CHO, CH_3, Fe — baker's yeast → CH_3, CH_2OH, Fe + CHO, CH_3, Fe

77 % e.e. 88 % e.e.

(d) O, H, C, C, O, NO_2, Fe racemate — β-cyclodextrin, H$^{\oplus}$ → H, CO_2H, C, Fe high e.e.

(e) $(CH_3)_2N$, C, H, Fe — BuLi, TMEDA → unspecific lithiation

(→ : sites of lithiation)

(f) Ph, CH_3, CO_2H, Fe — $(CF_3CO)_2O$ → O, C, CH_3, Ph, Fe + Ph, C, O, CH_3, Fe

major minor

(g) CH_3, H_2N, H, Fe — 1) HCHO, 2) HCl → CH_3, H, NH, Fe

Fig. 4-21. Alternative approaches to planar chiral ferrocenes.

[15]. Intact baker's yeast resolves 2-methyl-ferrocenealdehyde kinetically, leading to 88% *ee* in the remaining (*R*)-aldehyde and 77% *ee* in (*S*)-2-methyl-ferrocenemethanol (Fig. 4-21c) [16]. A comparatively high enantiomeric excess is observed quite often in baker's yeast reductions of suitably constructed planar chiral ferrocene derivatives [18].

Considering the high efficiency of enzymes, model compounds have been developed that mimic their functions (Fig. 4-21). Among them, β-cyclodextrin occupies a prominent place, as it forms inclusion complexes with suitably constructed ferrocene derivatives that then may undergo kinetic resolution (Fig. 4-21d). An ester of an unsaturated acid derived from a 1,2-ferrocenophane is the best substrate for this technique; the rate of hydrolysis of one enantiomer is enhanced by 3.2×10^6 [133].

Planar chiral compounds should also be accessible from the chiral pool. An example (with limited stereoselectivity) of such an approach is the formation of a ferrocene derivative from a β-pinene-derived cyclopentadiene (see Sect. 4.3.1.3 [81]). A C_2-symmetric binuclear compound (although not strictly from the chiral pool, but obtained by resolution) has also been mentioned [86]. Another possibility should be to use the central chiral tertiary amines derived from menthone or pinene (see Sect. 4.3.1.3 [75, 76]) as starting materials for the lithiation reaction. In these compounds, the methyl group at the chiral carbon of *N*,*N*-dimethyl-1-ferrocenylethylamine is replaced by bulky terpene moieties, e.g., the menthane system (Fig. 4-21e). It was expected that the increase in steric bulk would also increase the enantioselectivity over the 96:4 ratio, as indicated by the results with the isopropyl substituent [118]. However, the opposite was observed: almost all selectivity was lost, and lithiation also occurred in the position 3 and in the other ring [134]. Obviously, there exists a limit in bulkiness, where blocking of the 2-position prevents the chelate stabilization of the lithium by the lone pair of the nitrogen.

Asymmetric induction in ring closure reactions of central chiral ferrocene derivatives has been reported. Moderate diastereoselectivity was found in the ring closure of the enantiomeric 4-ferrocenyl-2-methyl-2-phenyl-butanoic acids by treatment with trifluoroacetic anhydride (Fig. 4-21f) [10]. The diastereoisomeric ketones could be separated by chromatography. A higher induction was observed in an asymmetric Pictet–Spengler type cyclization of a reactive imine formed from enantiomerically pure 2-ferrocenyl-2-propylamine and formaldehyde, as only one isomer of the product was detected (Fig. 4-21g) [135, 136].

As many planar chiral ferrocenes are applied as chiral auxiliaries and catalysts, the question arises whether there is any possibility for racemization. It seems that under normal reaction conditions for chemical transformations this does not occur. However, several planar chiral compounds racemize quite rapidly when treated with strong acids, like $HClO_4$ or $AlCl_3$, particularly in polar solvents. Racemization is fastest in nitroalkanes as solvent [137, 138]. According to a thorough study of the reaction mechanism by deuteration experiments, the racemization starts with a decomplexation of the substituted cyclopentadienyl ring from the iron, leaving the ligand in the solvent cage, where it turns upside down (Fig. 4-22). When it now coordinates again to the iron, the product is the enantiomer of the initial ferrocene and when the mixture is left under these conditions for some time, complete

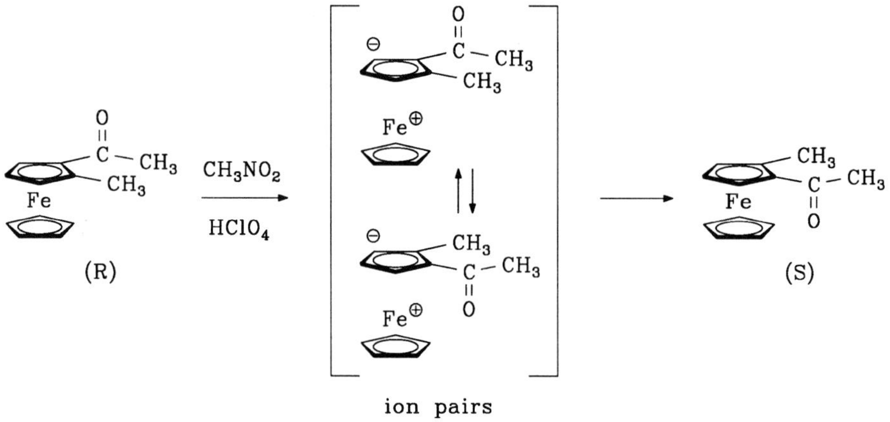

Fig. 4-22. Racemization of planar chiral ferrocenes.

racemization is observed [138]. Nitroalkanes should therefore be avoided as solvents and care should be taken when planar chiral ferrocenes are treated with strong acids.

4.4.2 Phosphines

Chiral phosphines are readily available from lithiated or palladated tertiary ferrocenylalkyl amines by reaction with chlorodiarylphosphine or chlorodialkylphosphine. As the chemistry of such phosphines is the topic of a separate chapter of this book (see Chapter 2), we will mention just a few aspects here. As with the phosphines, the corresponding derivatives of arsenic are obtained from chlorodiarylarsines [139].

All stereoisomers of planar chiral ferrocenylphosphines are accessible by stereoselective lithiation of the enantiomeric *N*,*N*-dimethyl-1-ferrocenyl-ethylamines, followed by reaction with either chlorodiphenylphosphine to give the (*R*,*S*)- or (*S*,*R*)-compounds or with chlorotrimethylsilane to the corresponding trimethylsilyl derivatives, which are blocked in the position 2. A second lithiation introduces the substituent therefore in position 5, and trapping with chlorodiphenylphosphine and subsequent cleavage of the trimethylsilyl substituent from the trisubstituted product with potassium *tert*-butoxide in DMSO leads to the (*R*,*R*)- or (*S*,*S*)-isomers, respectively [129, 130]. Stereoisomeric compounds with a phosphine and a sulfide as substituents in the positions 2 and 5 are conveniently obtained by simple reversal of the order of addition of the electrophiles to the lithiated intermediates, as shown in Fig. 4-23; the products are interesting ligands for transition metals, as they possess three different donor atoms with coordination ability. Rhodium forms a chelate with phosphorus and nitrogen, which has some catalytic activity in asymmetric hydrogenation of dehydroamino acids, while nickel prefers the sulfur rather than the phosphorus for coordination [131].

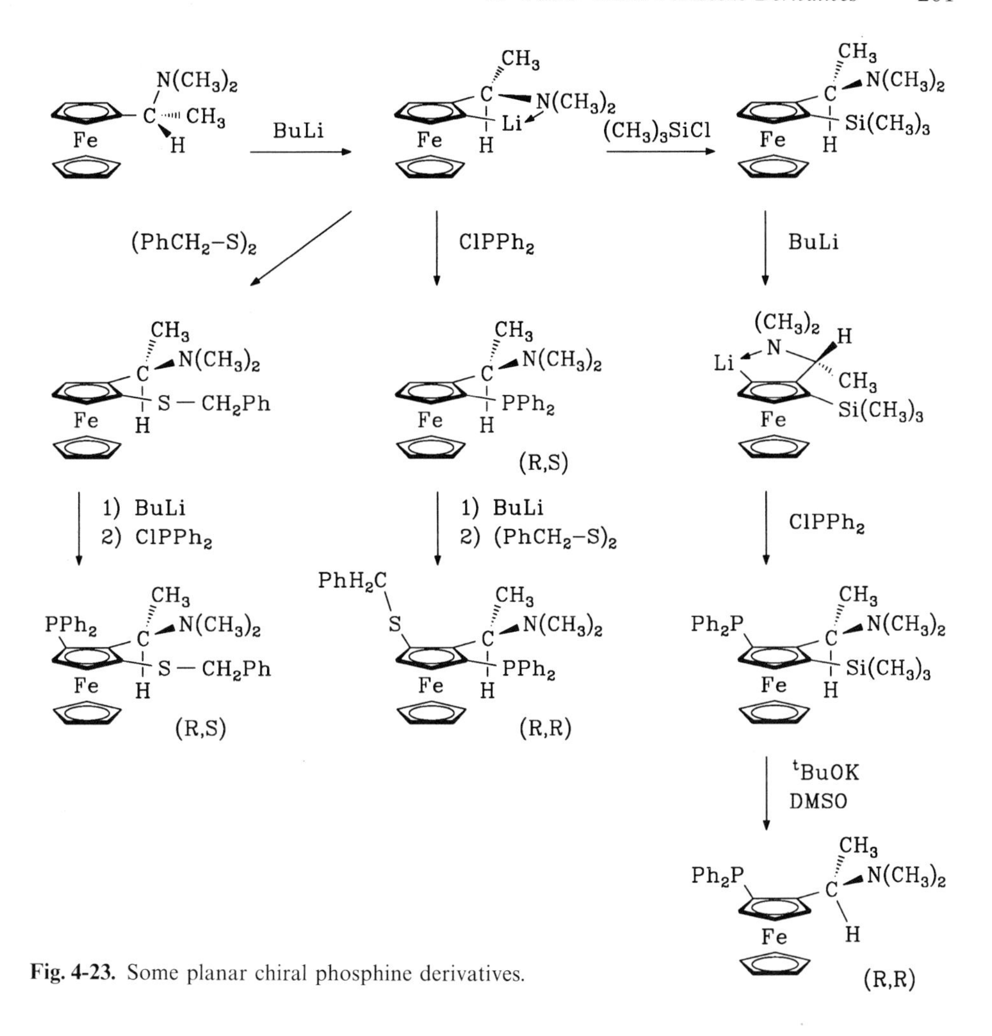

Fig. 4-23. Some planar chiral phosphine derivatives.

4.4.3 Sulfur and Selenium Compounds

The rapid development of chiral phosphine derivatives of ferrocene was undoubtedly due to their application in catalysis. In contrast, chiral sulfur compounds from lithiated *N,N*-dimethyl-1-ferrocenylethylamine were only prepared about 15 years later [140]. For the synthesis of such derivatives, the lithiated amine is treated with disulfides as shown in Fig. 4-24, top (and analogously, diselenides [141]). The sulfides obtained are easily oxidized by peracids or $NaIO_4$ to the corresponding sulfoxides. As sulfur becomes a new center of chirality by the oxidation, diastereoisomeric sulfoxides are formed in ratios depending on the oxidant [140]. If chiral oxaziridines [106, 142] are used as oxidizing agents, the diastereoisomeric ratio is appreciably

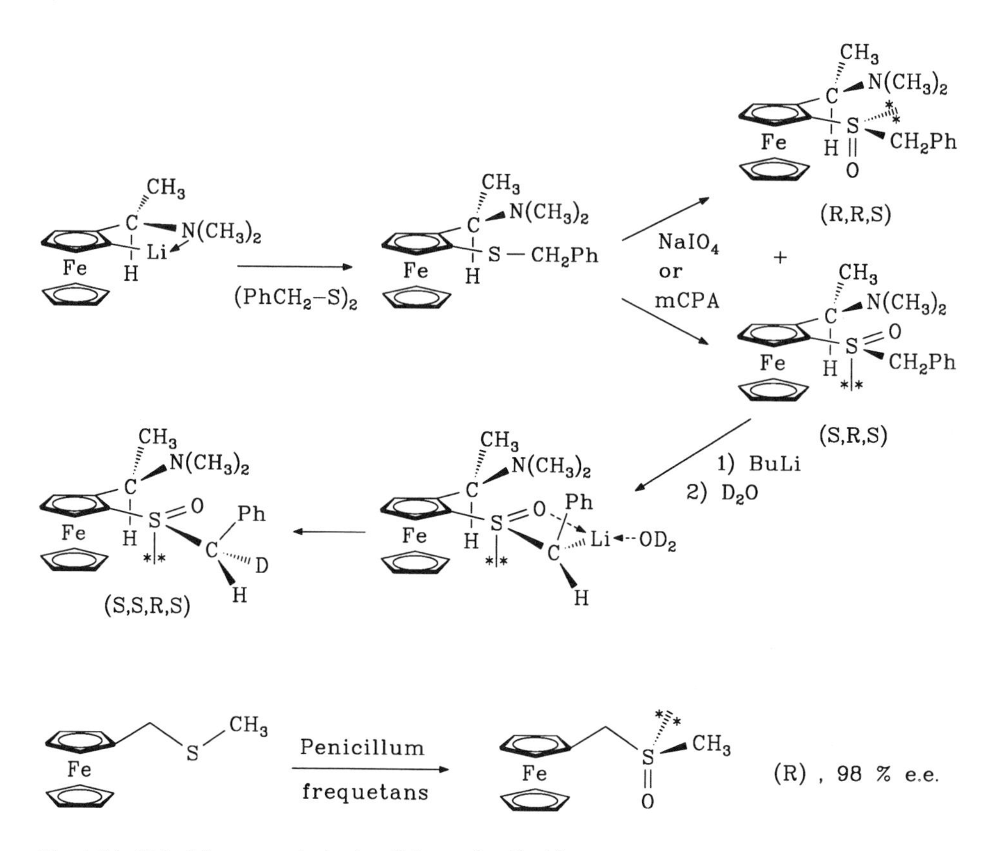

Fig. 4-24. Chiral ferrocene-derived sulfides and sulfoxides.

improved. The diastereoisomers can be separated by chromatography. The benzyl sulfoxides (from the oxidation of (*R*,*S*)-1-benzylthio-2-(1-dimethylaminoethyl)ferrocene) are deprotonated by butyl lithium. Addition of D_2O leads to 100% diastereoselective introduction of deuterium in the benzylic CH_2-group (configuration (*S*,*S*,*R*,*S*) means (*S*) at sulfur, (*S*) at the carbon atom bearing deuterium, (*R*) at the carbon atom of the dimethylaminoethyl group, and (*S*) for planar chirality). The anions do not, however, react with acylating agents [132]. A simpler ferrocene-derived chiral sulfoxide (shown in Fig. 4-24, bottom) was obtained by asymmetric oxidation of ferrocenylmethylmethylsulfide with *Penicillum frequetans* in 98% *ee* [143].

The reaction of chiral lithiated ferrocenylamines with dialkylthiuramdisulfides leads to the corresponding (dialkylthiocarbamoyl)thioferrocenes [144]. For catalytic applications, sulfur and selenium derivatives with one [140, 141, 145] or two [146 – 148] chalcogen substituents have been prepared (see Sect. 4.5.3.1). The technique is essentially the same as for the chiral phosphines: lithiation of enantiomerically pure *N*,*N*-dimethyl-1-ferrocenylethylamine, followed either by the addition of a disulfide or a diselenide to the monosubstituted compounds, or by

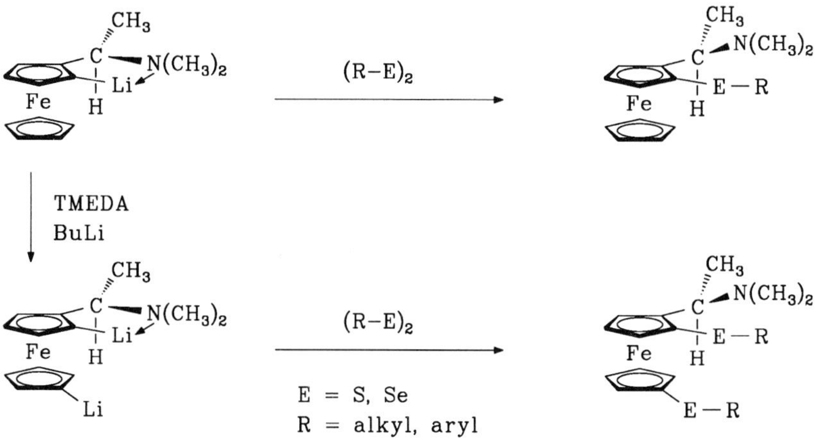

Fig. 4-25. Planar chiral sulfur and selenium derivatives.

adding a further equivalent of butyl lithium, together with TMEDA, to achieve lithiation in the second ring, and then addition of disulfide or diselenide (Fig. 4-25). Complexes of these ligands with palladium have been studied by X-ray crystallography [145, 148].

4.4.4 Silicon and Tin Compounds

Trimethylchlorosilane was among the first electrophiles that were introduced in lithiated ferrocenylalkyl amines [62]. The trimethylsilyl substituent is easily cleaved from the ferrocene ring with potassium *tert*-butoxide in DMSO [118, 129, 130], and its application for temporary blocking of reactive positions has already been mentioned (Sect. 4.4.2). Interestingly, the steric hindrance introduced by a trimethylsilyl substituent in the position adjacent to a dimethylaminoethyl group is so high that the formation of a cation by departure of the acylated dimethylamino group is possible only in the (*R*,*S*)-isomer, but not in the (*R*,*R*)-compound, due to strong steric repulsion in the cation that would be formed (see Fig. 4-26, top right) [130].

When 2-(*N*,*N*-dimethyl-1-aminoethyl)-1-trimethylsilylferrocene is treated with two equivalents of butyl lithium and TMEDA, lithiation occurs in the positions 3 and 1′. Trapping of the intermediate with trimethylchlorosilane leads to a ferrocene containing three trimethylsilyl groups (Fig. 4-26). According to the X-ray structure, the main product from this reaction is 2-(*N*,*N*-dimethyl-1-aminoethyl)-1,3,1′-tris(trimethylsilyl)ferrocene, a compound with severe steric hindrance [149]. It was expected that the rotation of the two cyclopentadienyl rings against one another might be hindered, a situation similar to that in heteroannular bridged ferrocenophanes (see Chapters 5 and 6), but no evidence for any hindered rotation was found by variable temperature NMR.

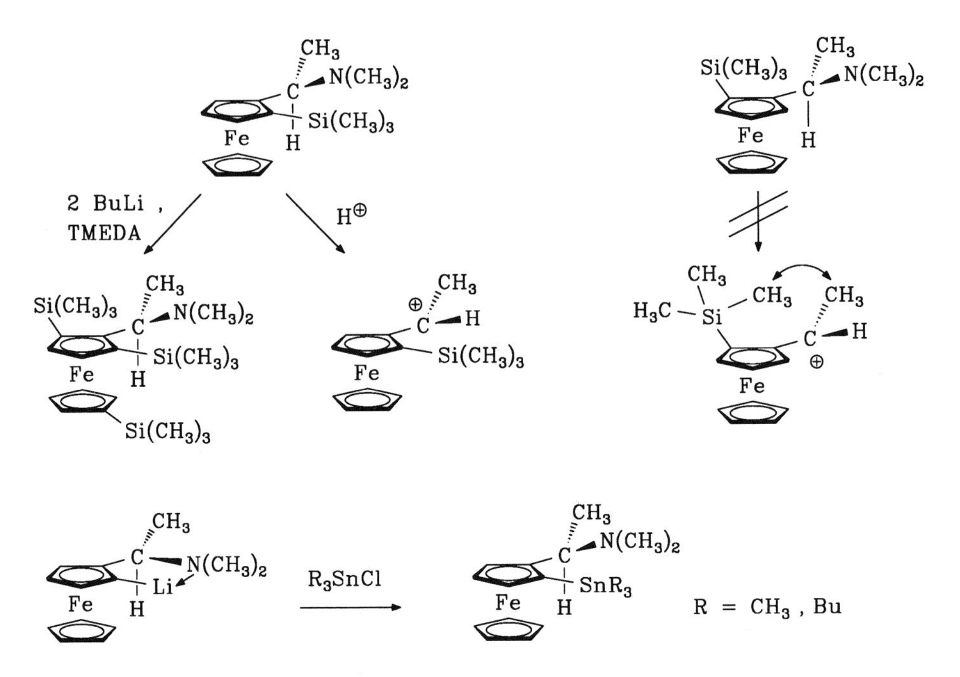

Fig. 4-26. Planar chiral silicon and tin derivatives.

In complete analogy to silicon, trialkyltin chlorides react with the lithiated ferrocenylalkyl amines to give the corresponding tin derivatives (Fig. 4-26, bottom); trimethyltin [129] and tributyltin substituents [106] have been introduced by this method. The tin compounds are more reactive towards transmetalation than the silicon analogues and are therefore useful intermediates in the preparation of other derivatives [129], e.g., by C–C coupling to give chiral biferrocenes [106]. Several racemic planar chiral derivatives of ferrocenyltriethylstannane have been prepared via lithiated *N*,*N*-dimethyl-ferrocenylmethylamine and similar compounds [150].

4.4.5 Other Derivatives

Many other electrophiles are able to react with metalated ferrocenylalkyl amines, e.g., trimethyl borate (Fig. 4-27a), which gives, after hydrolytic workup, compounds like (*S*,*S*)-1-(*N*,*N*-dimethyl-1-aminoethyl)ferrocene-2-boronic acid [106]. Important intermediates for further derivatization are the halogens. For the lithiation technique, I_2 (Fig. 4-27b) [151] and BrCN [106] lead to the desired compounds, but when BrCN is used, partial substitution of the dimethylamino group by cyanide occurs (see Sect. 4.3.3.2 and Fig. 4-17). For palladated amines, Br_2 is applicable [152]. (*R*,*S*)-1-Iodo-2-(*N*,*N*-dimethyl-1-aminoethyl)ferrocene is the starting material for catalytically active zinc compounds for the enantioselective addition of zinc alkyls to carbonyl compounds [151] (see Chapter 3 for this topic).

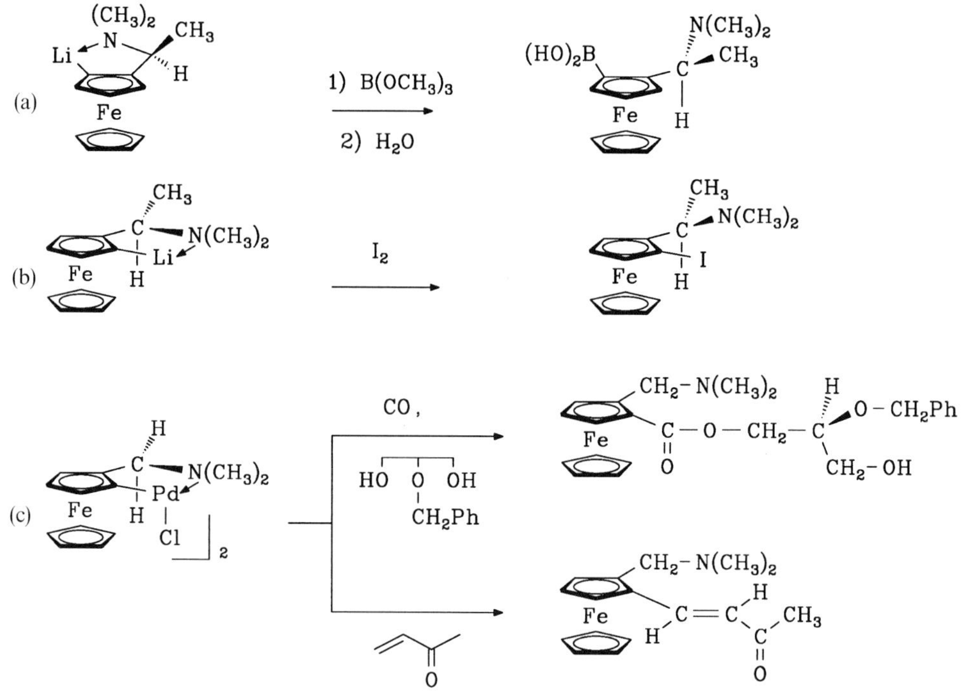

Fig. 4-27. Reaction of metalated chiral ferrocene derivatives with halogenes, boron and carbon electrophiles.

Among the carbon electrophiles, carbonyl compounds [113, 114] were first applied in the reaction with lithiated ferrocenylalkyl amines (Sect. 4.3.3.3 and Fig. 4-18). Analogously, carboxylic acids are obtained from CO_2 [153]. The reactivity pattern of palladated ferrocenylalkyl amines with carbon electrophiles is somewhat different. Carbon monoxide in alcohols leads to the formation of esters of substituted ferrocenecarboxylic acids [124]. With prochiral alcohols, a moderate asymmetric induction is observed [154]. α,β-Unsaturated ketones react with palladated ferrocenylalkyl amines not with addition to the carbonyl group, but with substitution of a hydrogen at the carbon–carbon double bond, allowing the introduction of longer side chains at the ferrocene ring (Fig. 4-27c) [124, 152].

Carbonyl complexes can be considered as a special kind of electrophilic carbonyl compound. Tungsten hexacarbonyl reacts with lithiated (*R*)-*N*,*N*-dimethyl-1-ferrocenyl-ethylamine to give products that depend on the reaction conditions (Fig. 4-28). *tert*-Butyl chloride acts as a proton source and protonates the intermediate acyl metalate at oxygen. In a secondary reaction, an aminocarbene complex is formed by a rearrangement in which the dimethylamino group and the hydroxy group exchange places. Triethyloxonium tetrafluoroborate, which smoothly alkylates most acyl metalates at oxygen, attacks here at nitrogen. The intermediate ammonium salt eliminates dimethylethylamine and a proton at higher temperature to give the vinyl carbene complex. At $\leq 20\,°C$, proton elimination does not occur, rather

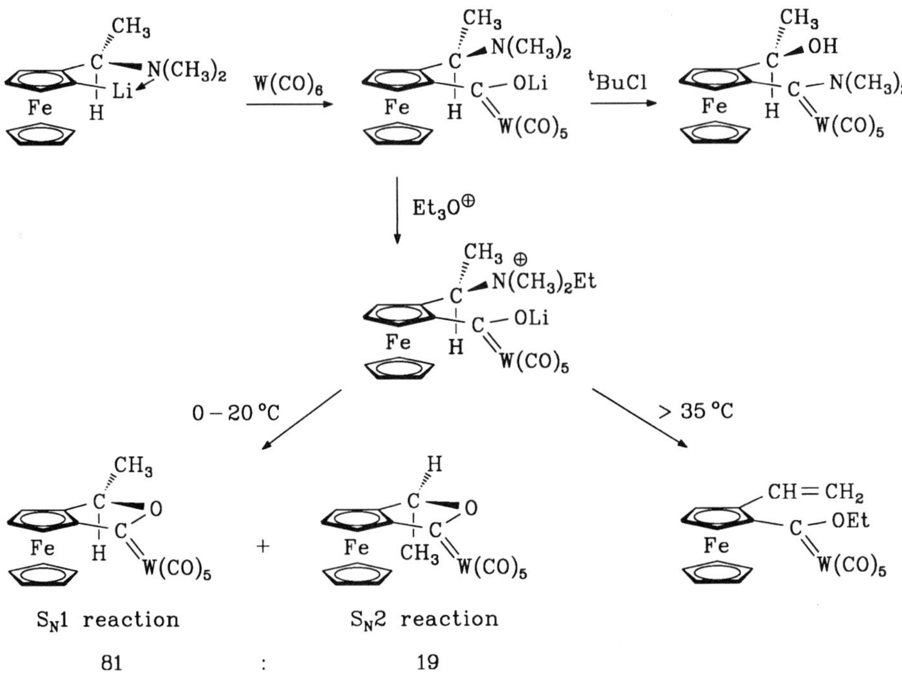

Fig. 4-28. Chiral tungsten carbene complexes.

nucleophilic substitution by the oxygen of the acyl metalate group, and two diastereoisomeric cyclic carbene complexes are obtained in the ratio 81 : 19. The main product has the same configuration at the aliphatic CH group as the starting *N*,*N*-dimethyl-1-ferrocenyl-ethylamine; its formation is due to the expected *retentive* S_N1 nucleophilic substitution. The minor isomer, however, has the opposite configuration; this seems to be the first (and, up to now, only) example of an S_N2 type nucleophilic displacement with *inversion* at a carbon adjacent to ferrocene. Of course, this is possible only because of the intramolecular character of the reaction in a rigid framework [155]. The complexes do not catalyze a simple metathesis reaction, as some chromium carbene complexes do, but other catalytic applications are under investigation.

4.5 Applications of Chiral Ferrocene Derivatives

This section is meant to give a short survey on the applications of chiral ferrocenes for the synthesis of (mostly chiral) organic compounds. Applications of phosphines will be treated very briefly, as they are the topic of a separate chapter of this book (Chapter 2). The same is true for the reaction of zinc alkyls with carbonyl compounds, in which chiral amino alcohols derived from ferrocene are active enantioselective catalysts. The ferrocenes can be used as chiral *starting materials*, *auxiliaries*, and *catalysts*.

4.5.1 As Chiral Starting Materials

As some synthetic effort is necessary to prepare chiral ferrocenes, they cannot be considered as cheap compounds and any application as chiral starting material (which finally leads to the destruction of the ferrocene moiety producing a purely organic product) is reasonable only when the product is a quite precious compound. So far, only prostaglandins and their analogues have been considered as target molecules in this sense. It seems a logical strategy to employ the special reactivity patterns of ferrocene to introduce side chains at the cyclopentadienyl ring and to cleave the iron later (see Fig. 4-29). The main problem with this technique is the high stability of the ferrocene system, which makes any decomplexation difficult and sensitive functions constructed at the beginning might be damaged under forcing conditions. The only decomplexation method not too destructive is catalytic hydrogenation in an acidic medium, typically with palladium in trifluoroacetic acid [156], which cleaves the ferrocene and reduces the double bonds of the cyclopenta-

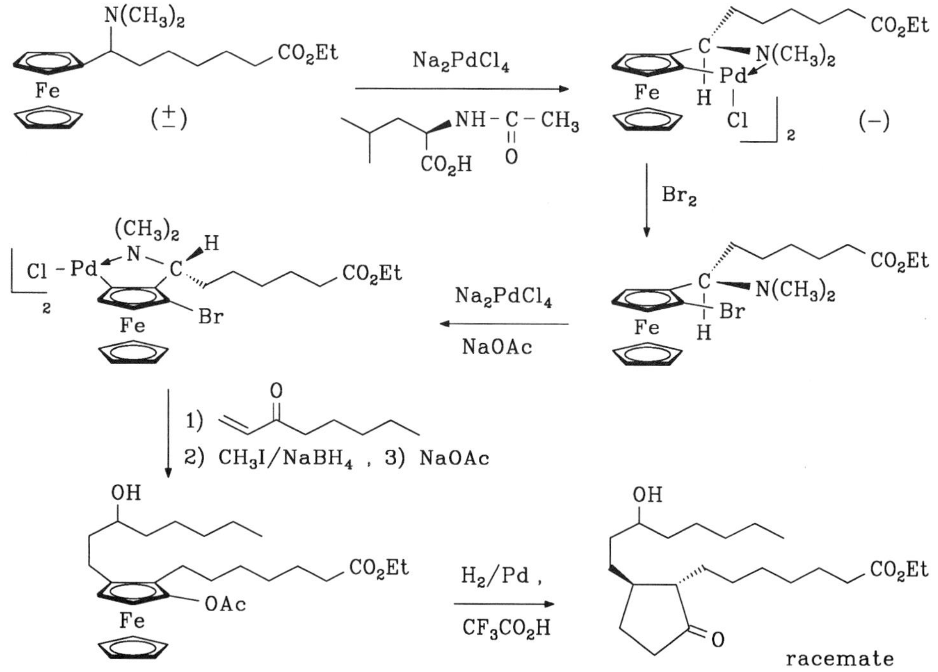

Fig. 4-29. Prostaglandin analogues from ferrocenes.

diene to the saturated cyclopentane derivative. When planar chiral ferrocenes are reduced, however, the chirality information is not transmitted to the newly formed chiral centres at the cyclopentane ring, i.e., racemic products are obtained. Only some diastereoselectivity towards the preferred formation of *trans*-cyclopentanes is observed (about 4:1 ratio) [157]. Nevertheless, several organometallic precursors of prostaglandins have been prepared by the enantioselective palladation method (Sect. 4.3.3.3), having up to three substituents at the ferrocene ring [152, 158]. Decomplexation by hydrogenolysis gave racemic prostaglandins, even when a chiral centre in the side chain was present [159]; it seems that no further research is being done in this field. There exists some hope that the organometallic precursors of the prostaglandins might be useful as drugs, provided that in vivo decomplexation occurs [6].

4.5.2 As Chiral Auxiliaries

A chiral auxiliary is a compound that is incorporated into a larger molecule, where it transmits its chirality information to newly formed chiral elements. It is cleaved from the target molecule in a later step of the reaction sequence and should in principle be recycled, to avoid loss of the precious chiral material. Among the ferrocene derivatives, chiral primary amines are typical auxiliaries; for their synthesis, see Section 4.3.3. All applications involve the intermediate formation of imines with carbonyl compounds.

4.5.2.1 Four Component Condensations

Mixing an amine, a carbonyl compound, an acid, and an isocyanide, may lead (under appropriate conditions) to a single product that is derived from the α-addition of the protonated imine and the anion of the acid to the isocyanide, followed by a rearrangement. Two amide bonds are formed during this sequence of events, and Ugi has coined the term *four component condensation* (*4CC*) (sometimes also called the Ugi reaction; Fig. 4-30). An aldehyde as carbonyl compound is thus transformed into an amino acid, the new asymmetric center of which is the carbon atom of the former carbonyl group. The idea of stereoselective four component condensation was then to make one of the components chiral and to determine whether the amino acids show some enantiomeric excess. Checking the possible candidates for chiral auxiliaries, chiral amines turned out to be the only compounds that gave a significant chiral induction. Chiral α-ferrocenylalkyl amines were chosen as auxiliaries for two reasons. First, they had the highest induction ability of all amines tested. Second, they could be readily cleaved from the reaction product, an *N*-alkylated peptide, by simple treatment with a comparatively weak acid, such as trifluoroacetic acid, as the α-ferrocenylalkyl cation. Adding 2-mercaptoacetic acid as scavenger for these cations removes them from the equilibrium and allows easy isolation of the free peptide [160]. As described in Section 4.3.3, the mercaptoacetic acid derivative

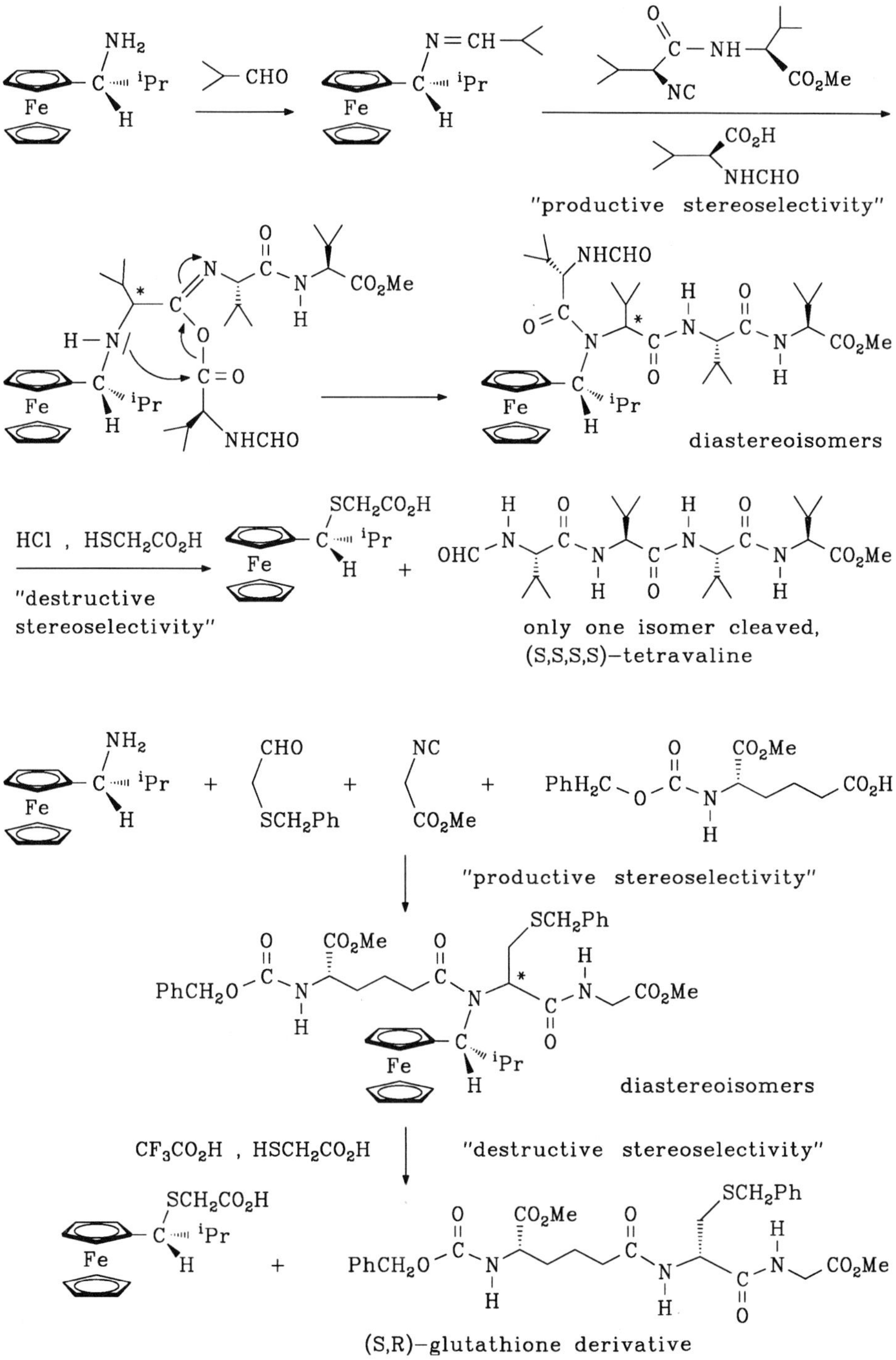

"productive stereoselectivity"

diastereoisomers

HCl, $HSCH_2CO_2H$

"destructive stereoselectivity"

only one isomer cleaved, (S,S,S,S)-tetravaline

"productive stereoselectivity"

diastereoisomers

CF_3CO_2H, $HSCH_2CO_2H$

"destructive stereoselectivity"

(S,R)-glutathione derivative

Fig. 4-30. Stereoselective four component condensation (4CC).

can be recycled to the enantiomerically pure ferrocenylalkyl amine by the $HgCl_2$ technique [46].

Initially, 1-ferrocenyl-2-methyl-propylamine was used almost exclusively as chiral auxiliary, as its induction ability was high and its synthesis not too difficult [90]. In the diastereoisomeric 4CC products, one isomer generally prevailed. The minor isomer had a higher reactivity towards cleavage with 2-mercaptoacetic acid and could therefore be removed from the mixture under thoroughly controlled reaction conditions [161]. This was called "destructive selectivity" (as opposed to "productive selectivity" in the 4CC step); in principle, the major isomer can be obtained by this technique in unlimited enantiomeric purity, but of course with decreasing yields along the reaction time, and the reaction has to be quenched at a suitable time when the content of the minor isomer falls below the detection limit. Remarkable products prepared in enantiomerically pure form by combined productive and destructive selectivity include tetravaline [161, 162] and glutathione derivatives [163].

In the meantime, the primary amines from the chiral pool (Sect. 4.3.1.3) and other amines have been tested for their induction ability, and found to give similar results as 1-ferrocenyl-2-methyl-propylamine [40, 75, 76]. Although there has been some optimism towards a broader range of applications [164], it seems that carbohydrate-derived primary amines have a superior induction power in stereoselective 4CC and will therefore become the reagents for the future.

4.5.2.2 Other Applications

In addition to four component condensation, several other applications of chiral primary ferrocenylalkyl amines have been published. Thus, an asymmetric synthesis of alanine was developed (Fig. 4-31 a), which forms an imine from 1-ferrocenylethyl amine and pyruvic acid, followed by catalytic reduction (Pd/C) to the amine. Cleavage of the auxiliary occurs readily by 2-mercaptoacetic acid, giving alanine in 61% *ee* and allowing for recycling of the chiral auxiliary from the sulfur derivative by the $HgCl_2$ technique [165]. Enantioselective reduction of imines is not limited to pyruvic acid, but has recently also been applied to the imine with acetophenone, although the diastereoisomeric ferrocenylalkyl derivatives of phenylethylamine were obtained only in a ratio of about 2:1 (Fig. 4-31 b). The enantioselective addition of methyl lithium to the imine with benzaldehyde was of the same low selectivity [57]. Recycling of the chiral auxiliary was possible by treatment of the secondary amines with acetic acid/formaldehyde mixture that cleaved the phenylethylamine from the cation and substituted it for acetate.

With respect to diastereoselectivity, the reaction of the imine of 1-ferrocenyl-2-methyl-propylamine and piperonal with a racemic cyclic anhydride gave a much better result (diastereomer ratio 81:10 of the isoindole derivatives, total yield 91%); the major isomer crystallized from the reaction mixture. This reaction was the key step in an enantioselective synthesis of the alkaloid (+)-corynoline (Fig. 4-31 c). The 1-ferrocenyl-2-methyl-propyl auxiliary could be cleaved by the 2-mercaptoacetic acid technique. The final elaboration of the intermediate to (+)-corynoline needed only five steps [166].

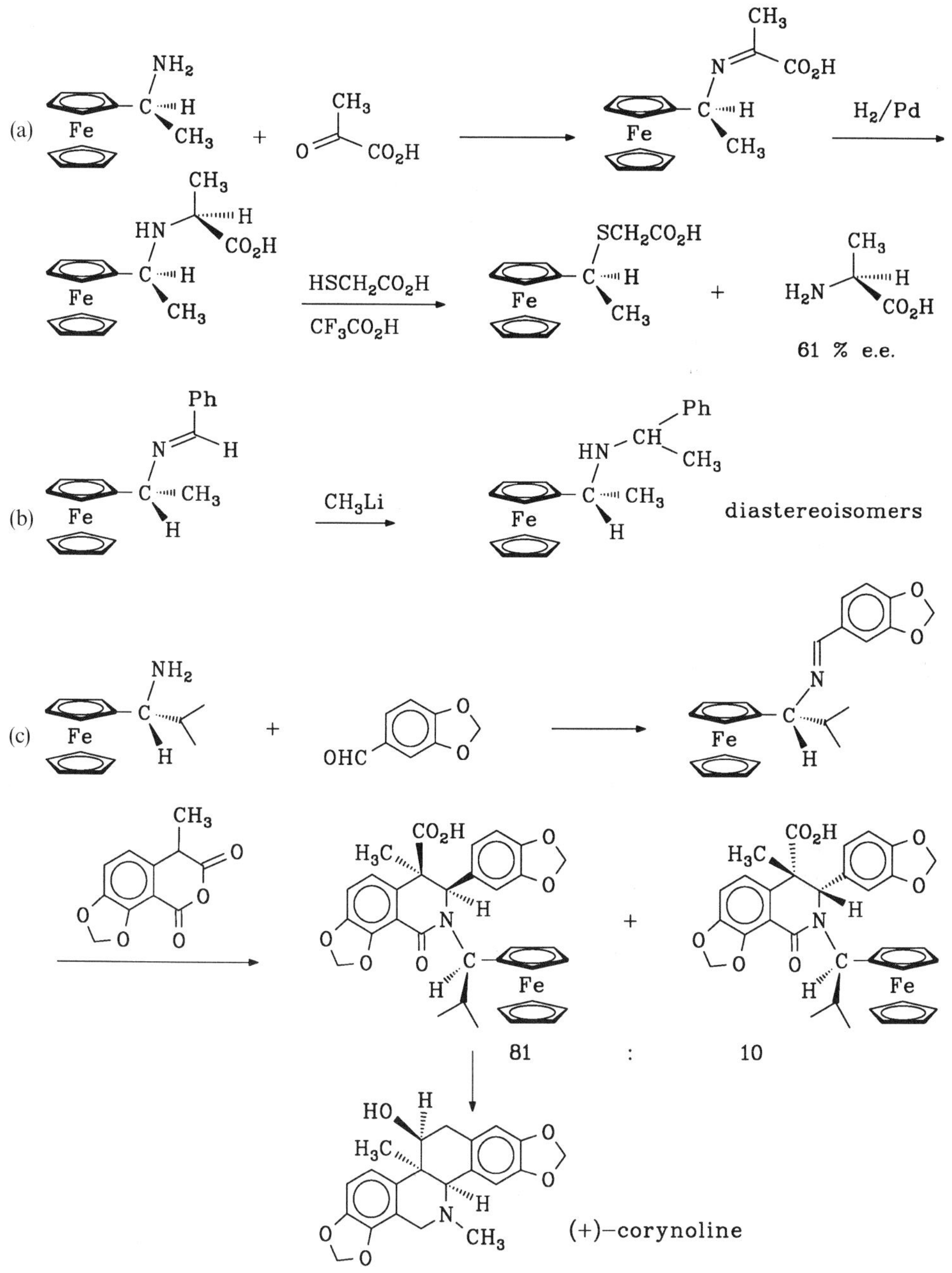

(a)
NH_2 C H CH_3 Fe + CH_3 O CO_2H → CH_3 N CO_2H C H Fe CH_3 H_2/Pd →

CH_3 HN H CO_2H C H Fe CH_3 $HSCH_2CO_2H$ / CF_3CO_2H → SCH_2CO_2H C H Fe CH_3 + CH_3 H_2N H CO_2H

61 % e.e.

(b)
Ph N H C CH_3 Fe H CH_3Li → Ph HN CH CH_3 C CH_3 Fe H

diastereoisomers

(c)
NH_2 C Fe H + OHC O O → O O N C Fe H

CH_3 O O O O O → H_3C CO_2H O O H N O O O C Fe H + H_3C CO_2H O O H N O O O C Fe H

81 : 10

HO H H_3C O O H N CH_3 O O

(+)–corynoline

Fig. 4-31. Ferrocenylalkyl amines as chiral auxiliaries.

4.5.3 As Chiral Catalysts

Catalytic reactions have the advantage over the methods discussed so far in that the chiral catalyst need not be added in stoichiometric amounts, but only in very small quantities, which is important if not only the metal (very often a precious one) but also the chiral ligand are expensive. Among the ferrocenes, phosphines are by far the most important catalysts for stereoselective reactions, and are covered in Chapter 2 of this book. We will therefore focus here mainly on the catalytic applications of chiral ferrocenes not containing phosphine groups. Only recently, some progress has been made with such compounds, mainly with sulfides and selenides, and with amino alcohols in the side chain (for this topic, see Chapter 3 on the addition of dialkyl zinc to aldehydes).

4.5.3.1 C – C Coupling, Hydrogenation, and Hydrosilylation

The first investigations on ferrocene derivatives as chiral ligands for transition metal-catalyzed asymmetric reactions were stimulated by the known power of chelating chiral phosphines, such as DIPAMP and DIOP, to induce chirality in rhodium-catalyzed hydrogenations of functionalized alkenes, e.g., in the synthesis of amino acids. Many chiral phosphine derivatives of ferrocene were prepared and tested for their ability in such hydrogenations. Although enantiomeric excesses up to 93% were achieved for phenylalanine [167], other chiral phosphines not containing organometallic residues may perform even better [168, 169]. For alkenes without polar functional groups, the hydrosilylation catalyzed by palladium and, for example (*R*,*S*)-2-(*N*,*N*-dimethyl-1-aminoethyl)-1-diphenylphosphino-ferrocene, leads to chiral silanes with reasonable *ee* (up to 64%). The same catalytic system allows the preparation of allylsilanes with up to 95% *ee* by asymmetric cross-coupling of silylalkyl-Grignard reagents with vinyl halides (Fig. 4-32a) [170]. The hydrosilylation of ketones to chiral secondary alcohols with rhodium was, however, less efficient (acetophenone gave 1-phenylethanol with 49% *ee*) [171]. Under high pressure of hydrogen, the direct reduction of the carbonyl group of pyruvic acid to lactic acid with up to 83% *ee* was possible [172]. The asymmetric Grignard cross coupling, catalyzed by nickel or palladium, has been studied in great detail. For the work done with ferrocene-derived phosphines, we will just mention as a highlight the synthesis of binaphthyl derivatives in over 95% *ee* by the reaction of substituted 1-bromonaphthalenes with the corresponding Grignard reagents under the influence of nickel complexes of chiral 1-diphenylphosphino-2-(1-methoxyethyl)ferrocene (Fig. 4-32b) [173, 174]. Recently, the cross-coupling between allyl magnesium chloride and 1-chloro-1-phenylethane has also been achieved with ligands containing thioethers and selenoethers instead of phosphines [7, 145, 148]. However, the enantioselectivity (highest *ee* 45%) cannot compete with that of the phosphine systems.

(a) $Et_3SiCH(CH_3)MgCl$ + PhCH=CHBr —$PdCl_2$, (R,S)–PPFA→ PhCH=CH–CH(SiEt$_3$)CH$_3$ 93 % e.e.

(b) 2-methyl-1-naphthylmagnesium bromide + 1-bromonaphthalene —$NiBr_2$, (R,S)–PPFOMe→ 2-methyl-1,1'-binaphthyl 80 % e.e.

(c) $PhCH_2NH_2$ + H_3C–CH=CH–CH$_2$–O–C(=O)–OCH$_3$ —Pd(II), (R,S)–BPPFNOO→ H_3C–CH(NHCH$_2$Ph)–CH=CH$_2$ 84 % e.e.

(d) $C_{13}H_{27}$–CH=CH–CHO + CN–CH$_2$–CO$_2$Me —Au(I), (R,S)–BPPFMor→ oxazoline ($C_{13}H_{27}$CH=CH, CO$_2$Me) → → $C_{13}H_{27}$–CH=CH–CH(OH)–CH(NH$_2$)–CH$_2$OH threo–sphingosine

(e) cis-$CH_3OC(O)O$–CH$_2$–CH=CH–CH$_2$–OC(O)OCH$_3$ + CH$_2$(CO$_2$Me)$_2$ —Pd(II), (R,S)–BPPFA→ 2-vinylcyclopropane-1,1-dicarboxylate (H, CO$_2$Me, CO$_2$Me)

(f) CH_2=C(CH$_2$SO$_2$Ph)(CH$_2$–O–C(=O)–OEt) + CH$_2$=CH–CO$_2$Me —Pd(II), (R,S)–BPPFNOO→ 3-methylene-cyclopentane (SO$_2$Ph, CO$_2$Me)

Ligand	X	Name
Fe, C(CH$_3$)(H)X, PPh$_2$	X = N(CH$_3$)$_2$	(R,S)–PPFA
	X = OCH$_3$	(R,S)–PPFOMe
Fe, C(CH$_3$)(H)X, PPh$_2$, PPh$_2$	X = N(CH$_3$)$_2$	(R,S)–BPPFA
	X = N(CH$_3$)–CH(CH$_2$OH)$_2$	(R,S)–BPPFNOO
	X = N(CH$_3$)–CH$_2$CH$_2$–N(morpholino)	(R,S)–BPPFMor

Fig. 4-32. Chiral ferrocene derivatives as chiral catalysts.

4.5.3.2 Allylations and Aldol-Type Reactions

The reactions of this section use stabilized carbanions formed from C–H-acidic compounds by deprotonation. As phosphines are the only successful ligands known up to now, these reactions have been discussed in detail in Chapter 2, and we will again restrict this section to a few highlights. Carbanions derived from 1,3-diketones react with allylic esters enantioselectively under palladium catalysis with more than 80% *ee* [175]. Benzylamine is allylated by the same catalytic system, leading to substituted allyl-benzylamines with up to 97% *ee* (Fig. 4-32c) [176].

The carbanion derived from an ester of isocyanoacetic acid reacts with carbonyl compounds under ring closure to oxazolines which may contain up to two chiral centers (Fig. 4-32d). Gold(I) is able to catalyze the reaction, so that there is no need to add strong base to generate the carbanion. Many ferrocene-derived chiral phosphines can coordinate to gold(I) and are able to induce some enantio- and diastereoselectivity in the product. The reaction has been studied thoroughly, with the industrial aspect of producing substituted chiral amino acids and amino alcohols. A fundamental investigation on many aspects of this reaction, particularly on the cooperativity of central and planar chirality, provided a model for the transition state geometry and allows the prediction of the stereoselectivity [130]. Modifications of the carbon side chain in the ferrocenylphosphines by introducing further donor atoms improved both enantio- and diastereoselectivity [177], and the state of the art today is quite impressive (as an example, the synthesis of sphingosines is given [178]).

4.5.4 Other Catalytic Applications

Considering the industrial importance of cyclopropanes in the pesticide field, it is not surprising that chiral ferrocenylphosphines have been applied as control ligands for the palladium-catalyzed enantioselective formation of cyclopropanes from the dicarbonate of 2-butene-1,4-diol and malonates, leading to 70% *ee* (Fig. 4-32e) [179]. Ferrocenylphosphines also induce chirality in the reaction of sulfonyl-substituted propenyl carbonates and acrylic esters to methylenecyclopentanes (up to 78% *ee* (Fig. 4-32f)) [180], with potential applications in natural product synthesis. These examples show that the synthetic potential of chiral ferrocene derivatives is not yet fully exploited, and one may look forward to new applications.

References

[1] K. Schlögl, M. Fried, *Monatsh. Chem.* **1964**, *95*, 558–575.
[2] L. Westman, K. L. Rinehart, Jr., *Acta Chem. Scand.* **1962**, *16*, 1199–1205.
[3] O. L. Carter, A. T. McPhail, G. A. Sim, *J. Chem. Soc. A* **1967**, 365–373.
[4] H. Falk, K. Schlögl, *Monatsh. Chem.* **1965**, *96*, 266–275.
[5] K. Schlögl, *Top. Stereochem.* **1967**, *1*, 39–91.

[6] V. I. Sokolov, *Chirality and Optical Activity in Organometallic Compounds*, Gordon and Breach Science Publishers, New York, **1990**, p. 83.
[7] H. M. Ali, C. H. Brubaker, Jr., *J. Mol. Catal.* **1990**, *60*, 331–342.
[8] R. S. Cahn, C. K. Ingold, V. Prelog, *Angew. Chem.* **1966**, *78*, 413–447; *Angew. Chem. Int. Ed. Engl.* **1966**, *5*, 385–415.
[9] H. Falk, K. Schlögl, *Monatsh. Chem.* **1965**, *96*, 1065–1080.
[10] H. Des Abbayes, R. Dabard, *Tetrahedron* **1975**, *31*, 2111–2116.
[11] B. Gautheron, R. Broussier, *Tetrahedron Lett.* **1971**, 513–516.
[12] T. Aratani, T. Gonda, H. Nozaki, *Tetrahedron Lett.* **1969**, 2265–2268.
[13] T. Aratani, T. Gonda, H. Nozaki, *Tetrahedron* **1970**, *26*, 5453–5464.
[14] Y. Yamazaki, K. Hosono, *Tetrahedron Lett.* **1988**, *29*, 5769–5770.
[15] Y. Yamazaki, M. Uebayasi, K. Hosono, *Europ. J. Biochem.* **1989**, *184*, 671–680.
[16] T. Izumi, T. Hino, K. Shoji, K. Sasaki, A. Kasahara, *Chem. Ind.* **1989**, 457.
[17] T. Izumi, T. Hino, A. Kasahara, *J. Chem. Technol. Biotechnol.* **1991**, *50*, 571–573.
[18] T. Izumi, T. Hino, *J. Chem. Technol. Biotechnol.* **1992**, *55*, 325–331.
[19] V. I. Sokolov, L. L. Troitskaya, B. Gautheron, G. Tainturier, *J. Organomet. Chem.* **1982**, *235*, 369–373.
[20] M. F. Brown, B. R. Cook, T. E. Sloan, *Inorg. Chem.* **1975**, *14*, 1273–1278.
[21] T. E. Sloan, *Top. Stereochem.* **1981**, *12*, 1–36.
[22] W. E. Watts, *J. Organomet. Chem. Libr.* **1979**, *7*, 399–459.
[23] R. Gleiter, R. Seeger, *Helv. Chim. Acta* **1971**, *54*, 1217–1220.
[24] S. Lupan, M. Kapon, M. Cais, F. H. Herbstein, *Angew. Chem.* **1972**, *84*, 1104–1106; *Angew. Chem. Int. Ed. Engl.* **1972**, *11*, 1025–1027.
[25] R. L. Sime, R. J. Sime, *J. Am. Chem. Soc.* **1974**, *96*, 892–896.
[26] U. Behrens, *J. Organomet. Chem.* **1979**, *182*, 89–98.
[27] T. D. Turbitt, W. E. Watts, *J. Chem. Soc., Chem. Commun.* **1973**, 182–183.
[28] T. D. Turbitt, W. E. Watts, *J. Chem. Soc., Perkin Trans. 2* **1974**, 177–184.
[29] S. G. Baxter, R. L. Collins, A. H. Cowley, S. F. Sena, *Inorg. Chem.* **1983**, *22*, 3475–3479.
[30] G. Gokel, P. Hoffmann, H. Klusacek, D. Marquarding, E. Ruch, I. Ugi, *Angew. Chem.* **1970**, *82*, 77–78; *Angew. Chem. Int. Ed. Engl.* **1970**, *9*, 64–65.
[31] G. Gokel, I. Ugi, *Angew. Chem.* **1971**, *83*, 178–179; *Angew. Chem. Int. Ed. Engl.* **1971**, *10*, 191–192.
[32] G. W. Gokel, D. Marquarding, I. K. Ugi, *J. Org. Chem.* **1972,** *37*, 3052–3058.
[33] P. Dixneuf, *Tetrahedron Lett.* **1971**, 1561–1563.
[34] P. Dixneuf, R. Dabard, *Bull. Soc. Chim. Fr.* **1972**, 2847–2854.
[35] C. A. Bunton, N. Carrasco, F. Davoudzadeh, W. E. Watts, *J. Chem. Soc., Perkin Trans. 2* **1980**, 1520–1528.
[36] C. A. Bunton, W. Crawford, W. E. Watts, *Tetrahedron Lett.* **1977**, 3755–3758.
[37] C. A. Bunton, W. Crawford, N. Cully, W. E. Watts, *J. Chem. Soc., Perkin Trans. 1* **1980**, 2213–2217.
[38] M. J. Nugent, R. Kummer, J. H. Richards, *J. Am. Chem. Soc.* **1969**, *91*, 6141–6145.
[39] M. J. Nugent, R. E. Carter, J. H. Richards, *J. Am. Chem. Soc.* **1969**, *91*, 6145–6151.
[40] R. Herrmann, G. Hübener, F. Siglmüller, I. Ugi, *Liebigs Ann. Chem.* **1986**, 251–268.
[41] K. Schlögl, *Pure Appl. Chem.* **1970**, *23*, 413–432.
[42] K. Schlögl, H. Falk, *Methodicum Chimicum* **1976**, Vol. 8, Chap. 30.
[43] H. Vogel, M. D. Rausch, H. Rosenberg, *J. Org. Chem.* **1957**, *22*, 1016–1018.
[44] F. S. Arimoto, A. C. Haven, *J. Am. Chem. Soc.* **1955**, *77*, 6295–6297.
[45] A. Ratajczak, B. Misterkiewicz, *J. Organomet. Chem.* **1975**, *91*, 73–80.
[46] G. Eberle, I. Ugi, *Angew. Chem.* **1976**, *88*, 509–510; *Angew. Chem. Int. Ed. Engl.* **1976**, *15*, 492–493.
[47] S. Stüber, I. Ugi, *Synthesis* **1973**, 309.
[48] S. Allenmark, *Tetrahedron Lett.* **1974**, 371–374.
[49] S. Allenmark, K. Kalen, A. Sandblom, *Chem. Scr.* **1975**, *7*, 97–101.
[50] R. Herrmann, I. Ugi, *Angew. Chem.* **1979**, *91*, 1023–1024; *Angew. Chem. Int. Ed. Engl.* **1979**, *18*, 956–957.

[51] R. Herrmann, I. Ugi, *Tetrahedron* **1981**, *37*, 1001–1009.
[52] B. Misterkiewicz, *J. Organomet. Chem.* **1982**, *224*, 43–47.
[53] V. I. Boev, A. V. Dombrovskii, *Zh. Obshch. Khim.* **1980**, *50*, 2520–2525.
[54] G. Gokel, I. Ugi, *J. Chem. Educ.* **1972**, *49*, 294–296.
[55] H. Martin, R. Herrmann, I. Ugi, *J. Organomet. Chem.* **1984**, *269*, 87–89.
[56] I. R. Butler, W. R. Cullen, *Can. J. Chem.* **1983**, *61*, 2354–2358.
[57] D. M. David, L. A. P. Kane-Maguire, S. G. Pyne, *J. Organomet. Chem.* **1990**, *390*, C6–C9.
[58] B. Misterkiewicz, R. Dabard, A. Darchen, H. Patin, *Compt. Rend. Acad. Sci., Ser. II* **1989**, *309*, 875–880.
[59] R. Dabard, B. Misterkiewicz, H. Patin, J. Wasielewski, *J. Organomet. Chem.* **1987**, *328*, 185–192.
[60] K. Schlögl, M. Widhalm, *Monatsh. Chem.* **1984**, *115*, 1113–1120.
[61] D. W. Armstrong, W. De Mond, B. P. Czech, *Anal. Chem.* **1985**, *57*, 481–484.
[62] D. Marquarding, H. Klusacek, G. Gokel, P. Hoffmann, I. Ugi, *J. Am. Chem. Soc.* **1970**, *92*, 5389–5393.
[63] H. Falk, K. Schlögl, *Monatsh. Chem.* **1971**, *102*, 33–36.
[64] N. W. Boaz, *Tetrahedron Lett.* **1989**, *30*, 2061–2064.
[65] S. I. Goldberg, W. D. Bailey, M. L. McGregor, *J. Org. Chem.* **1971**, *36*, 761–769.
[66] M. N. Nefedova, I. A. Mamedyarova, P. P. Petrovski, V. I. Sokolov, *J. Organomet. Chem.* **1992**, *425*, 125–130.
[67] A. Ratajczak, A. Palka, B. M. Misterkiewicz, *Bull. Acad. Pol. Sci., Ser. Sci. Chim.* **1982**, *28*, 593–598.
[68] R. Fornasier, F. Reniero, P. Scrimin, U. Tonellato, *J. Org. Chem.* **1985**, *50*, 3209–3211.
[69] V. I. Sokolov, V. L. Bondareva, *Izv. Akad. Nauk SSSR, Ser. Khim.* **1987**, 460.
[70] Y. Kawajiri, N. Motohashi, *J. Chem. Soc., Chem. Commun.* **1989**, 1336–1337.
[71] A. D. Ryabov, *Angew. Chem.* **1991**, *103*, 945–955; *Angew. Chem. Int. Ed. Engl.* **1991**, *30*, 931–941.
[72] A. Ratajczak, B. Misterkiewicz, *Chem. Ind.* **1976**, 902–903.
[73] V. I. Sokolov, L. L. Troitskaya, O. A. Reutov, *Dokl. Akad. Nauk SSSR, Ser. Khim.* **1977**, *237*, 1376–1379.
[74] Y. Yamazaki, K. Hosono, *Tetrahedron Lett.* **1989**, *30*, 5313–5314.
[75] F. Siglmüller, R. Herrmann, I. Ugi, *Liebigs Ann. Chem.* **1989**, 623–635.
[76] F. Siglmüller, R. Herrmann, I. Ugi, *Tetrahedron* **1986**, *42*, 5931–5941.
[77] A. Demharter, I. Ugi, *J. prakt. Chem.* **1993**, *335*, 244–254.
[78] A. Gieren, C.-P. Kaerlein, T. Hübner, R. Herrmann, F. Siglmüller, I. Ugi, *Tetrahedron* **1986**, *42*, 427–434.
[79] W. Himmele, H. Siegel, *Tetrahedron Lett.* **1976**, 907–910.
[80] W. Himmele, H. Siegel, *Tetrahedron Lett.* **1976**, 911–914.
[81] L. A. Paquette, M. Gugelchuk, M. L. McLaughlin, *J. Org. Chem.* **1987**, *52*, 4732–4740.
[82] R. L., Haltermann, K. P. C. Vollhardt, *Tetrahedron Lett.* **1986**, *26*, 1461–1464.
[83] M. L. H. Green, N. M. Walker, *J. Organomet. Chem.* **1988**, *344*, 379–382.
[84] J. C. Leblanc, C. Moise, *J. Organomet. Chem.* **1976**, *120*, 65–71.
[85] K.-L. Gibis, G. Helmchen, G. Huttner, L. Zsolnai, *J. Organomet. Chem.* **1993**, *445*, 181–186.
[86] T. R. Kelly, P. Meghani, *J. Org. Chem.* **1990**, *55*, 3684–3688.
[87] A. Sudhakar, T. J. Katz, *J. Am. Chem. Soc.* **1986**, *108*, 179–181.
[88] A. Sudhakar, T. J. Katz, *J. Am. Chem. Soc.* **1986**, *108*, 2790–2791.
[89] A. Ketter, R. Herrmann, *J. Organomet. Chem.* **1990**, *386*, 241–252.
[90] G. Eberle, I. Lagerlund, I. Ugi, R. Urban, *Tetrahedron* **1978**, *34*, 977–981.
[91] G. Ortaggi, R. Marčeč, *Gazz. Chim. Ital.* **1979**, *109*, 13–17.
[92] T. El-Shihi, F. Siglmüller, R. Herrmann, M. F. N. N. Carvalho, A. J. L. Pombeiro, *J. Organomet. Chem.* **1987**, *335*, 239–247.
[93] T. El-Shihi, F. Siglmüller, R. Herrmann, M. F. N. N. Carvalho, A. J. L. Pombeiro, *Port. Electrochim. Acta* **1987**, *5*, 179–185.
[94] K. Schlögl, M. Fried, H. Falk, *Monatsh. Chem.* **1964**, *95*, 576–597.
[95] R. Eberhardt, C. Glotzmann, H. Lehner, K. Schlögl, *Tetrahedron Lett.* **1974**, *15*, 4365–4368.
[96] P. Reich-Rohrwig, K. Schlögl, *Monatsh. Chem.* **1968**, *99*, 1752–1763.

[97] V. I. Boev, A. V. Dombrovskii, *Zh. Obshch. Khim.* **1984**, *54*, 1863–1873.
[98] A. Togni, R. Häusel, *Synlett* **1990**, 633–635.
[99] A. Togni, G. Rihs, R. E. Blumer, *Organometallics* **1992**, *11*, 613–621.
[100] H. Patin, G. Mignani, C. Mahe, J.-Y. Le Marouille, A. Benoit, D. Grandjean, *J. Organomet. Chem.* **1980**, *193*, 93–103.
[101] V. I. Boev, A. V. Dombrovskii, *Zh. Obshch. Khim.* **1982**, *52*, 1693–1694.
[102] V. I. Boev, M. S. Lyubich, S. M. Larina, *Zh. Org. Khim.* **1985**, *21*, 2195–2200.
[103] A. N. Nesmeyanov, E. G. Perevalova, M. D. Reshetova, *Izv. Akad. Nauk SSSR, Ser. Khim.* **1966**, 335–337.
[104] V. P. Tverdokhlebov, A. V. Sachivko, I. V. Tselinskii, *Zh. Org. Khim.* **1982**, *18*, 1958–1961.
[105] G. Marr, T. M. White, *J. Chem. Soc., Perkin Trans. 1* **1973**, 1955–1958.
[106] N. Deus, R. Herrmann, unpublished.
[107] V. I. Boev, *Zh. Obshch. Khim.* **1978**, *48*, 1594–1601.
[108] E. A. Hill, *J. Org. Chem.* **1963**, *28*, 3586–3588.
[109] A. N. Nesmeyanov, E. G. Perevalova, R. V. Golovnya, O. A. Nesmeyanova, *Dokl. Akad. Nauk SSSR* **1954**, *97*, 459–461.
[110] R. A. Benkeser, D. Goggin, G. Scholl, *J. Am. Chem. Soc.* **1954**, *76*, 4025–4026.
[111] D. W. Slocum, T. R. Engelmann, C. Ernst, C. A. Jennings, W. Jones, B. Koonsvitsky, J. Lewis, P. Shenkin, *J. Chem. Educ.* **1969**, *46*, 144–150.
[112] G. Gokel, P. Hoffmann, H. Kleimann, H. Klusacek, D. Marquarding, I. Ugi, *Tetrahedron Lett.* **1970**, 1771–1774.
[113] L. F. Battelle, R. Bau. G. W. Gokel, R. T. Oyakawa, I. Ugi, *Angew. Chem.* **1972**, *84*, 164–165; *Angew. Chem. Int. Ed. Engl.* **1972**, *11*, 138–140.
[114] L. F. Battelle, R. Bau, G. W. Gokel, R. T. Oyakawa, I. Ugi, *J. Am. Chem. Soc.* **1973**, *95*, 482–486.
[115] P. B. Valkovich, G. W. Gokel, I. Ugi, *Tetrahedron Lett.* **1973**, 2947–2950.
[116] I. R. Butler, W. R. Cullen, J. Reglinski, S. J. Rettig, *J. Organomet. Chem.* **1983**, *249*, 183–194.
[117] I. R. Butler, W. R. Cullen, F. G. Herring, N. R. Jagannathan, *Can. J. Chem.* **1986**, *64*, 667–669.
[118] D. Marquarding, H. Burghard, I. Ugi, R. Urban, H. Klusacek, *J. Chem. Res. (S)* **1977**, 82–83; *J. Chem. Res. (M)* **1977**, 915–958.
[119] L. G. Kuz'min, Y. T. Struchkov, L. L. Troitskaya, V. I. Sokolov, O. A. Reutov, *Izv. Akad. Nauk SSSR, Ser. Khim.* **1979**, 1528–1534.
[120] L. L. Troitskaya, A. I. Grandberg, V. I. Sokolov, O. A. Reutov, *Dokl. Akad. Nauk SSSR* **1976**, *228*, 367–370.
[121] V. I. Sokolov, L. L. Troitskaya, O. A. Reutov, *J. Organomet. Chem.* **1977**, *133*, C28–C30.
[122] V. V. Bashilov, V. I. Sokolov, O. A. Reutov, *Izv. Akad. Nauk SSSR, Ser. Khim.* **1982**, 2069–2089.
[123] V. I. Sokolov, L. L. Troitskaya, T. I. Rozhkova, *Gazz. Chim. Ital.* **1987**, *117*, 525–527.
[124] V. I. Sokolov, L. L. Troitskaya, O. A. Reutov, *J. Organomet. Chem.* **1979**, *182*, 537–546.
[125] V. I. Sokolov, K. S. Nechaeva, O. A. Reutov, *Zh. Org. Khim.* **1983**, *19*, 1103–1105.
[126] T. Komatsu, M. Nonoyama, J. Fujita, *Bull. Chem. Soc. Jap.* **1981**, *54*, 186–189.
[127] V. I. Sokolov, L. L. Troitskaya, O. A. Reutov, *J. Organomet. Chem.* **1980**, *202*, C58–C60.
[128] T. Hayashi, K. Yamamoto, M. Kumada, *Tetrahedron Lett.* **1974**, 4405–4408.
[129] T. Hayashi, T. Mise, M. Fukushima, M. Kagotani, N. Nagashima, Y. Hamada, A. Matsumoto, S. Kawakami, M. Konishi, K. Yamamoto, M. Kumada, *Bull. Chem. Soc. Jap.* **1980**, *53*, 1138–1151.
[130] A. Togni, S. Pastor, *J. Org. Chem.* **1990**, *55*, 1649–1654.
[131] N. Deus, G. Hübener, R. Herrmann, *J. Organomet. Chem.* **1990**, *384*, 155–163.
[132] G. Hübener, R. Herrmann, unpublished.
[133] R. Breslow, G. Trainor, A. Ueno, *J. Am. Chem. Soc.* **1983**, *105*, 2739–2744.
[134] D. Robles, N. Deus, R. Herrmann, *J. Organomet. Chem.* **1990**, *386*, 253–260.
[135] A. Ratajczak, H. Zmuda, *Bull. Acad. Pol., Ser. Sci. Chim.* **1974**, *22*, 261–266.
[136] A. Ratajczak, H. Zmuda, *Bull. Acad. Pol., Ser. Sci. Chim.* **1977**, *25*, 35–38.
[137] D. W. Slocum, S. P. Tucker, T. R. Engelmann, *Tetrahedron Lett.* **1970**, 621–624.
[138] H. Falk, H. Lehner, J. Paul, U. Wagner, *J. Organomet. Chem.* **1971**, *28*, 115–124.

[139] I. R. Butler, W. R. Cullen, F. W. B. Einstein, S. J. Rettig, A. J. Willis, *Organometallics* **1983**, *2*, 128–135.
[140] R. Herrmann, G. Hübener, I. Ugi, *Tetrahedron* **1985**, *41*, 941–947.
[141] R. V. Honeychuk, M. O. Okoroafor, L.-H. Shen, C. H. Brubaker, Jr., *Organometallics* **1986**, *5*, 482–490.
[142] G. Glahsl, R. Herrmann, *J. Chem. Soc., Perkin Trans. 1* **1988**, 1753–1757.
[143] Y. Yamazaki, K. Hosono, *Biotechnol. Lett.* **1989**, *11*, 627–628.
[144] L.-H. Shen, M. O. Okoroafor, C. H. Brubaker, Jr., *Organometallics* **1988**, *7*, 825–829.
[145] M. O. Okoroafor, D. L. Ward, C. H. Brubaker, Jr., *Organometallics* **1988**, *7*, 1504–1511.
[146] S. Pastor, *Tetrahedron* **1988**, *44*, 2883–2886.
[147] A. A. Naiini, C.-K. Lai, C. H. Brubaker, Jr., *Inorg. Chim. Acta* **1989**, *160*, 241–244.
[148] A. A. Nainii, C.-K. Lai, D. L. Ward, C. H. Brubaker, Jr., *J. Organomet. Chem.* **1990**, *390*, 73–90.
[149] I. R. Butler, W. R. Cullen, S. J. Rettig, *Organometallics* **1986**, *5*, 1320–1328.
[150] C. Krüger, K.-H. Thille, M. Dargatz, *Z. Anorg. Allg. Chem.* **1989**, *569*, 97–105.
[151] M. Watanabe, S. Araki, Y. Butsugan, M. Uemura, *J. Org. Chem.* **1991**, *56*, 2218–2224.
[152] V. I. Sokolov, L. L. Troitskaya, N. S. Krushchova, *J. Organomet. Chem.* **1983**, *250*, 439–446.
[153] W. R. Cullen, E. B. Wickenheiser, *Can. J. Chem.* **1990**, *68*, 705–707.
[154] L. L. Troitskaya, V. I. Sokolov, *J. Organomet. Chem.* **1985**, *285*, 389–393.
[155] R. Herrmann, I. Ugi, *Angew. Chem.* **1982**, *94*, 798–799; *Angew. Chem. Int. Ed. Engl.* **1982**, *21*, 788–789; *Angew. Chem. Suppl.* **1982**, 1630–1642.
[156] V. I. Sokolov, L. L. Troitskaya, N. S. Krushchova, O. A. Reutov, *Dokl. Akad. Nauk SSSR* **1985**, *281*, 861–863.
[157] V. I. Sokolov, T. M. Filippova, N. S. Krushchova, L. L. Troitskaya, *Izv. Akad. Nauk SSSR, Ser. Khim.* **1986**, 2600–2603.
[158] V. I. Sokolov, L. L. Troitskaya, N. S. Krushchova, O. A. Reutov, *Dokl. Akad. Nauk SSSR* **1985**, *274*, 342–347.
[159] V. I. Sokolov, L. L. Troitskaya, N. S. Krushchova, O. A. Reutov, *Izv. Akad. Nauk SSSR, Ser. Khim.* **1987**, 2387–2388.
[160] G. Gokel, P. Hoffmann, H. Kleimann, H. Klusacek, G. Lüdke, D. Marquarding, I. Ugi, in *Isonitrile Chemistry* (Ed.: I. Ugi), Academic Press, New York **1971**, Chap. 9.
[161] R. Urban, D. Marquarding, I. Ugi, *Hoppe-Seyler's Z. Physiol. Chem.* **1978**, *359*, 1541–1552.
[162] R. Urban, G. Eberle, D. Marquarding, D. Rehn, H. Rehn, I. Ugi, *Angew. Chem.* **1976**, *88*, 644–646; *Angew. Chem. Int. Ed. Engl.* **1976**, *15*, 627–628.
[163] R. Urban, *Tetrahedron* **1979**, *35*, 1841–1843.
[164] I. Ugi, M. Baumeister, C. Fleck, R. Herrmann, R. Obrecht, F. Siglmüller, J.-H. Youn, in *Proc. Eur. Pept. Symp. 19th, 1986* (Ed.: D. Theodoropoulos), de Gruyter, Berlin, **1987**, p. 103.
[165] A. Ratajczak, A. Czech, *Bull. Acad. Pol., Ser. Sci. Chim.* **1979**, *27*, 661–664.
[166] M. Cushman, J.-K. Chen, *J. Org. Chem.* **1987**, *52*, 1517–1521.
[167] T. Hayashi, T. Mise, S. Mitachi, K. Yamamoto, M. Kumada, *Tetrahedron Lett.* **1976**, 1133–1134.
[168] B. Bosnich, M. D. Fryzuk, *Top. Stereochem.* **1981**, *12*, 119–154.
[169] T. Hayashi, M. Kumada, *Acc. Chem. Res.* **1982**, *15*, 395–401.
[170] T. Hayashi, *Chem. Scr.* **1985**, *25*, 61–70.
[171] T. Hayashi, K. Yamamoto, *Tetrahedron Lett.* **1974**, 4405–4408.
[172] T. Hayashi, T. Mise, M. Kumada, *Tetrahedron Lett.* **1976**, 4351–4354.
[173] T. Hayashi, K. Hayashizaki, T. Kiyoi, Y. Ito, *J. Am. Chem. Soc.* **1988**, *110*, 8153–8156.
[174] T. Hayashi, K. H. Hayashizaki, Y. Ito, *Tetrahedron Lett.* **1989**, *30*, 215–218.
[175] T. Hayashi, K. Kanehira, T. Hagihara, M. Kumada, *J. Org. Chem.* **1988**, *53*, 113–120.
[176] T. Hayashi, A. Yamamoto, Y. Ito, E. Nishioka, H. Miura, K. Yanagi, *J. Am. Chem. Soc.* **1989**, *111*, 6301–6311.
[177] T. Hayashi, M. Sawamura, Y. Ito, *Tetrahedron* **1992**, *48*, 1999–2012.
[178] Y. Ito, M. Sawamura, T. Hayashi, *Tetrahedron Lett.* **1988**, *29*, 239–240.
[179] T. Hayashi, A. Yamamoto, Y. Ito, *Tetrahedron Lett.* **1988**, *29*, 669–672.
[180] A. Yamamoto, Y. Ito, T. Hayashi, *Tetrahedron Lett.* **1989**, *30*, 375–378.

Received: August 18, 1993

5 Ferrocene Compounds Containing Heteroelements

Max Herberhold

5.1 Introduction

In several respects, the cylindrical molecule ferrocene (di(η^5-cyclopentadienyl)iron, $FeCp_2$) is analogous to the planar molecule benzene (C_6H_6). Both ferrocene and benzene are electron-rich aromatic systems that undergo electrophilic substitution. Ferrocene reacts about 10^6 times faster than benzene in Friedel–Crafts acetylation and about 10^9 times faster in mercuration with $Hg(OAc)_2$.

However, in contrast to benzene, ferrocene is sensitive to oxidation, and the ferrocenium cation, $FeCp_2^+$, a paramagnetic 17-electron species, is readily formed in the presence of various oxidants. The ferrocenium cation is reluctant to undergo electrophilic substitution, and therefore reactions such as halogenation and nitration, which are important routes to substituted benzene derivatives, cannot be used for the synthesis of substituted ferrocenes. Only electrophilic substitution under non-oxidizing conditions (e.g., Friedel–Crafts acylation, Mannich reaction, borylation, lithiation or mercuration), and radical substitution are available as an entry into the chemistry of substituted ferrocenes.

The most frequently studied ferrocene derivatives are the monosubstituted and the 1,1′-disubstituted ferrocenes, for which the abbreviations Fc-X and fcX_2 (X = substituent) will be used throughout the following discussion (Fig. 5-1).

In general, the ferrocenyl compounds, Fc-X, are easily converted to the 1,1′-ferrocenediyl compounds, fcX_2, and isolation of the monosubstituted product Fc-X may become difficult even if an excess of ferrocene (FcH) is used.

The present survey describes selected ferrocene derivatives in which heteroatoms are directly attached to a cyclopentadienyl ring, with particular emphasis on ferrocenyl and 1,1′-ferrocenylene compounds of transition metals and of chalcogens

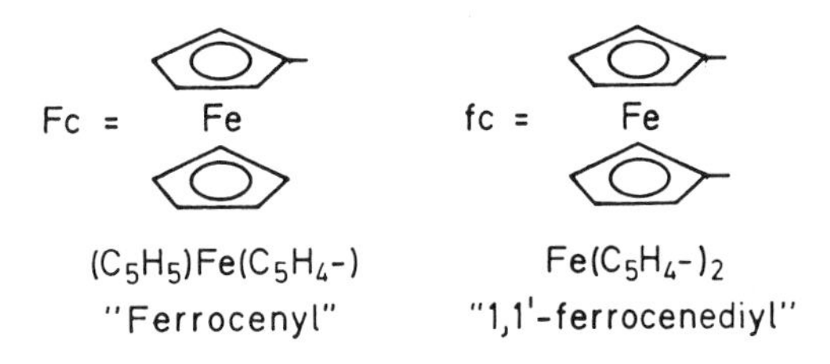

Fig. 5-1. Ferrocenyl (Fc) and 1,1′-ferrocenediyl (fc).

(E = S, Se, Te). An abundant number of reviews on ferrocene compounds is available, storehouses in particular are the series of Gmelin volumes [1] and the Annual Surveys published (with some delay) by the Journal of Organometallic Chemistry [2].

5.2 Synthesis of Ferrocene Derivatives Bearing Heteroelement Substituents

Three routes have been preferentially used to prepare ferrocene compounds that contain heteroelements directly attached to the metallocene unit: borylation, mercuration, and lithiation.

5.2.1 Borylation

Electrophilic substitution of ferrocene using BBr_3 or BI_3 in boiling CS_2 solution [3–5] can be conducted in a controlled manner to give either Fc-BX_2 or fc$(BX_2)_2$ (X = Br, I) as the main products (Scheme 5-1). The corresponding reaction with BCl_3 [3, 4, 6, 7] in CS_2 or cyclohexane is only a suitable method to prepare Fc-BCl_2, whereas fc$(BCl_2)_2$ is better obtained by halogen exchange [3, 5] using $AsCl_3$ in pentane, starting from fc$(BX_2)_2$ (X = Br, I). The difluoroboryl compounds Fc-BF_2 [3–5] and fc$(BF_2)_2$ [3, 5] are similarly accessible from Fc-BCl_2 or fc$(BX_2)_2$ (X = Br,I) with AsF_3 in pentane solution (Scheme 5-1).

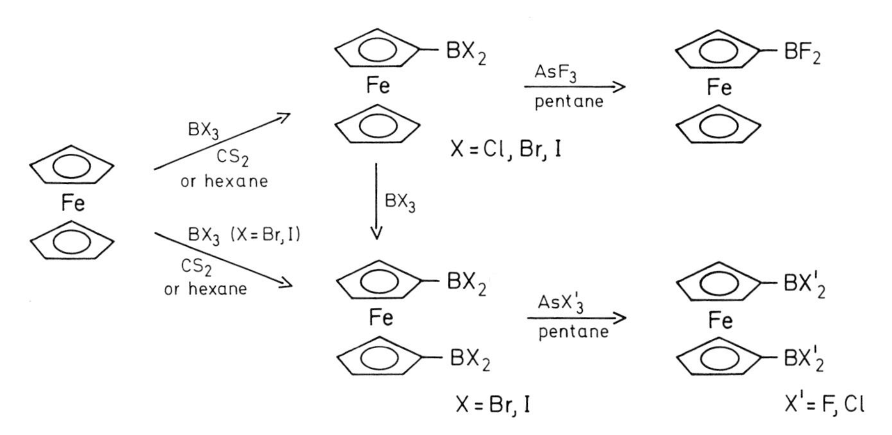

Scheme 5-1. Synthesis of dihaloboryl ferrocenes.

The dihaloboryl-substituted ferrocenes can be used to prepare various other boryl derivatives; thus, Fc-BX_2 compounds (X = Cl, Br or I) have been converted into Fc-BR_2 (R = NMe_2 [4, 7], NEt_2, OEt, SMe, Me [4, 5]). As an example, the synthesis

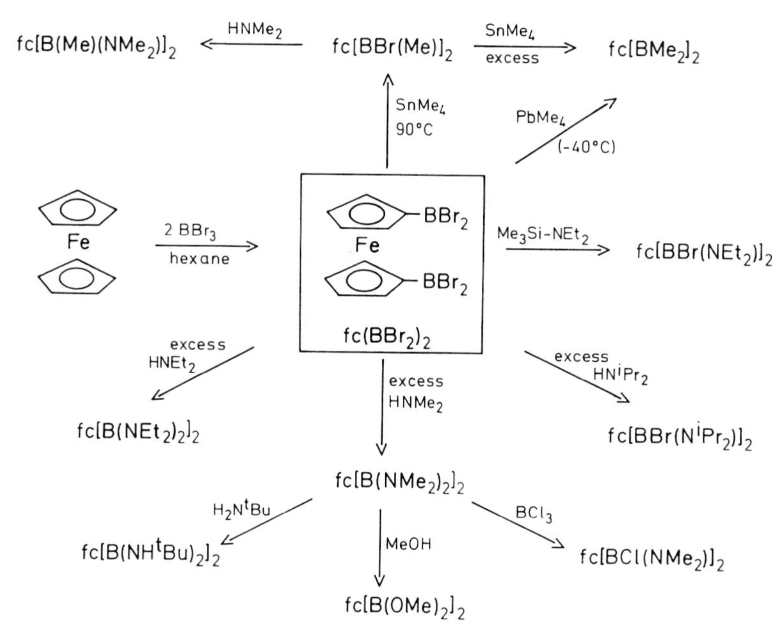

Scheme 5-2. Reactions of 1,1′-bis(dibromoboryl)ferrocene, fc$(BBr_2)_2$.

of a series of 1,1′-bis(aminoboryl) ferrocenes [8] starting from 1,1′-bis(dibromoboryl) ferrocene, fc$(BBr_2)_2$, is given in Scheme 5-2.

It has been shown that borylation of ferrocene with BBr_3 or BI_3 in hexane solution proceeds in a stepwise manner [9]. Excess BX_3 leads to mixtures containing essentially 1,3,1′-tris- and 1,3,1′,3′-tetrakis(dihalogenoboryl) ferrocene; the intermediates are Fc-BX_2 and fc$(BX_2)_2$, although small amounts of 1,3-bis(diahalogenoboryl) ferrocene have also been detected in the predominant 1,1′ isomer, fc$(BX_2)_2$ (X = Br, I).

5.2.2 Metallation

The metallation of ferrocene, for example lithiation and mercuration, has been the historically important route to the halogenated derivatives, Fc-X and fcX_2. Scheme 5-3 summarizes the basic compounds that are useful intermediates for the synthesis of many ferrocenes bearing heteroatoms.

The mercuration of ferrocene to chloromercuri-ferrocenes – first described by Nesmeyanov and coworkers [10] – is generally carried out in a one-pot reaction by using first mercury(II) acetate, $Hg(OAc)_2$, and then adding a chloride such as KCl [10] or LiCl [11]. The mixture of Fc-HgCl and fc$(HgCl)_2$ thus formed [11–13] can be separated by Soxhlet extraction of both Fc-HgCl and the sublimable starting ferrocene with dichloromethane. The yellow-brown 1,1′-bis(chloromercuri)ferrocene, fc$(HgCl)_2$, is essentially insoluble in all common organic solvents except hot dimethylformamide (DMF) [13].

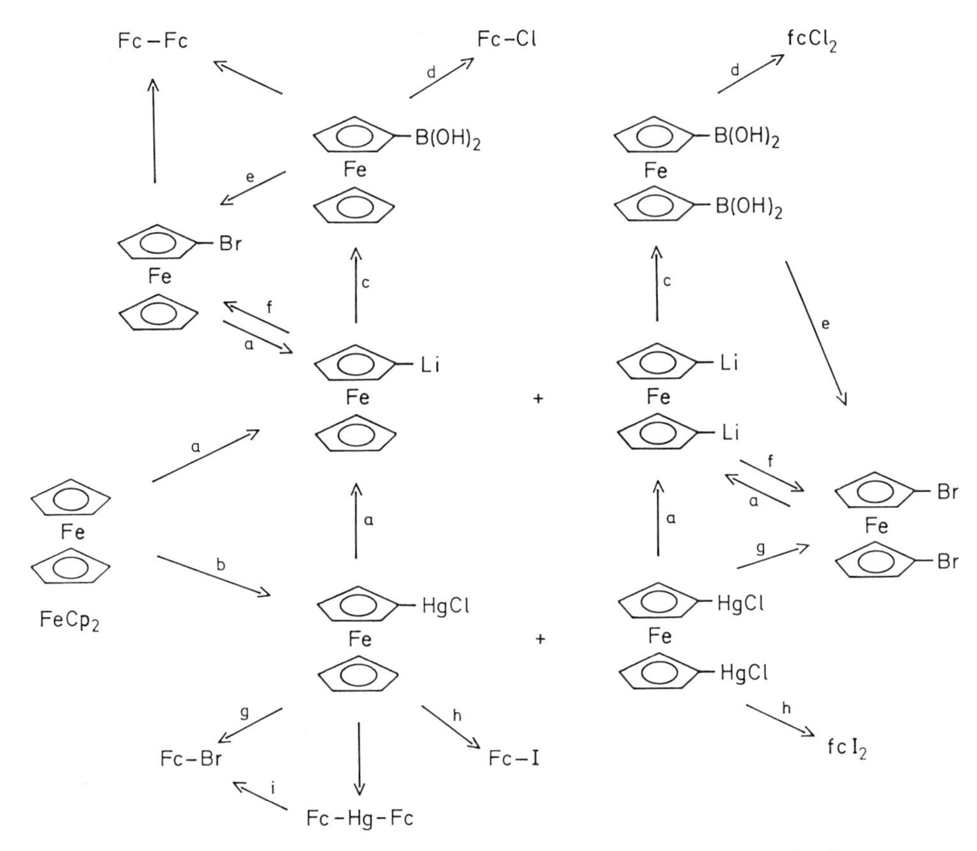

Scheme 5-3. Useful intermediates for the synthesis of ferrocene derivatives containing heteroatoms: (a) Lithiation (generally with nBuLi); (b) mercuration with $Hg(OAc)_2$, followed by reaction with LiCl or KCl; (c) borylation with $B(OnBu)_3$, followed by hydrolysis; (d) $CuCl_2$; (e) $CuBr_2$; (f) compounds containing reactive bromo substituents, e.g., $BrCX_2-CX_2Br$ (X = F, Cl) or tosyl bromide; (g) *N*-bromo succinimide; (h) I_2; (i) Br_2.

Transmetallation with alkyl-lithium, LiR, can be used to convert the chloromercury(II) compound into the corresponding lithioferrocenes Fc-Li and $fcLi_2$ [13, 14]. A mixture of these two lithioferrocenes is also easily obtained by direct metallation of ferrocene using the commercially available 1.6*M* solution of *n*-butyllithium in hexane. However, the reaction cannot be stopped at the monosubstituted stage, Fc-Li. An early method to separate mono- and 1,1′-disubstituted compounds involved reacting the mixture of lithioferrocenes with $B(OnBu)_3$ in ether or ether-containing solvent mixtures at low temperature, and then hydrolysing the product containing Fc-$B(OnBu)_2$ and Fc$[B(OnBu)_2]_2$ with either 10% H_2SO_4 solution [15–17] or 10% NaOH solution [18–20]. The monoboric acid, Fc-$B(OH)_2$, can be extracted from the solid product mixture with ether in a Soxhlet apparatus [18]; on the other hand, fractional extraction of an ethereal solution of the Fc-$B(OH)_2$/fc$[B(OH)_2]_2$ mixture with 10% KOH solution gives initially a fraction of almost pure disubstitution product, fc$[B(OH)_2]_2$ [15–17]. The di(hydroxy)boryl

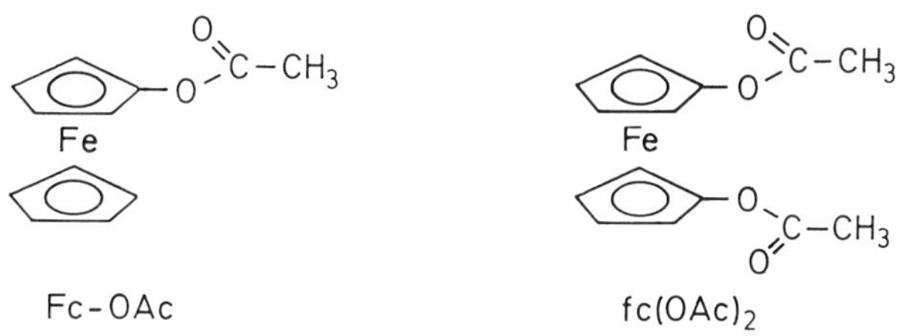

Fc-OAc

$fc(OAc)_2$

Fig. 5-2. Ferrocenyl acetate and 1,1′-ferrocenediyl diacetate.

ferrocenes, Fc-$B(OH)_2$ and fc$[B(OH)_2]_2$, are versatile precursors of various ferrocene derivatives; e.g., reaction with CuX_2 (X = Cl, Br, OAc) in hot water produces the halogeno-ferrocenes, Fc-X and fcX_2 (X = Cl, Br) [15, 16], and the acetates, Fc-OAc [16, 21] and $fc(OAc)_2$ (Fig. 5-2) [16, 22].

The lithiation of ferrocene by *n*-butyl-lithium in solution – which has been used [10, 23] since 1954 – generally leads to a mixture of mono- and 1,1′-dimetallated derivatives. The preparation of pure monolithio-ferrocene, Fc-Li, therefore requires the synthesis of a monosubstituted ferrocene precursor such as Fc-HgCl [14], Fc-Br, or Fc-I [24], which may then react with *n*BuLi or other organo-lithium reagents (Scheme 5-4). On the other hand, dilithioferrocene, $fcLi_2$, is almost exclusively formed if ferrocene is lithiated with *n*BuLi in the presence of *N*,*N*,*N*′,*N*′-tetramethyl-ethylenediamine (tmeda) in hexane solution [25–32]. The orange adduct of $fcLi_2$ with tmeda precipitates from the hexane solution; the adduct is pyrophoric but can be stored under inert gas at room temperature [29]. An X-ray structure determination of the red crystals obtained from $Fe(C_5H_5)_2$/*n*BuLi/tmeda (1 : 2 : 1) in ether/hexane revealed the trinuclear composition $[fcLi_2]_3(tmeda)_2$ [33]. However, the orange

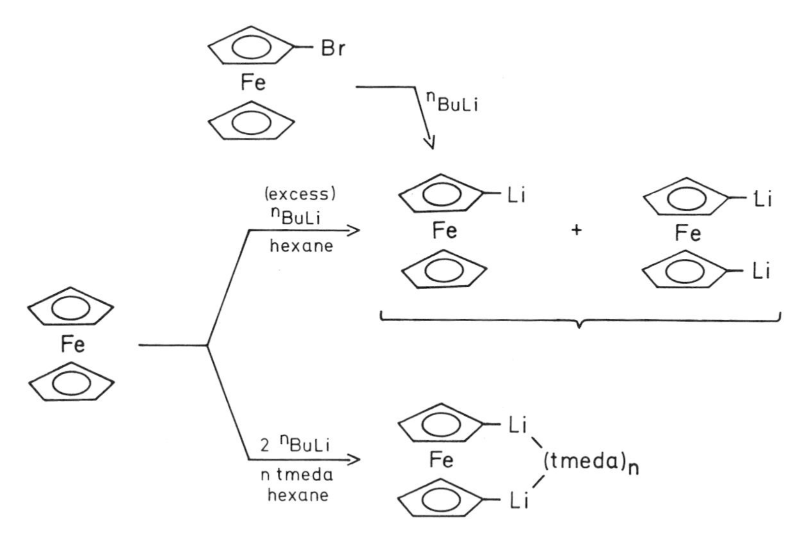

Scheme 5-4. Preparation of lithioferrocenes using *n*-butyl-lithium.

needles formed in the reaction between ferrocene and a three-fold excess of the (1:1) adduct of *n*BuLi with *N,N,N',N',N''*-pentamethyl-diethylenetriamine, *n*BuLi · (pmdta), in hexane consisted of binuclear units of $[fcLi_2 \cdot pmdta]_2$ [34].

In principle, more than two lithium atoms can be introduced into the ferrocene molecule with excess *n*BuLi [35]; the polylithiated species $FeC_{10}H_{10-n}Li_n$ ($n = 1-8$) were identified via reaction with either Me_3SiCl or D_2O. Highly chlorinated ferrocenes including decachloroferrocene, $Fe(C_5Cl_5)_2$, have been synthesized by repetitive metallation with *n*BuLi followed by exchange halogenation using hexachloroethane, C_2Cl_6 [36]. The persubstituted complex, $Fe(C_5Cl_5)_2$, is thermally quite stable (dec. 245 °C) and very resistant to oxidation [36]; lithiation with *n*BuLi in THF gives only the dilithio species $Fe(C_5Cl_4Li)_2$, from which octachloroferrocene, $Fe(C_5Cl_4H)_2$ (by hydrolysis) or other octachloroferrocenes of the type $Fe(C_5Cl_4R)_2$ (R = I, COOH, OCH_3, OC_2H_5) can be obtained.

Exhaustive mercuration of ferrocene with mercury(II) trifluoroacetate, $Hg(OCOCF_3)_2$, is another possibility to synthesize polyhalogenated ferrocenes [37]. The reaction of deca(trifluoroacetato-mercuri)ferrocene, $Fe[C_5(Hg\text{-}OCOCF_3)_5]_2$, with the trihalides KBr_3 and KI_3 can be used to prepare $Fe(C_5Br_5)_2$ and $Fe(C_5I_5)_2$, respectively; monosubstituted ferrocenes, Fc-R, react in an analogous manner (Scheme 5-5):

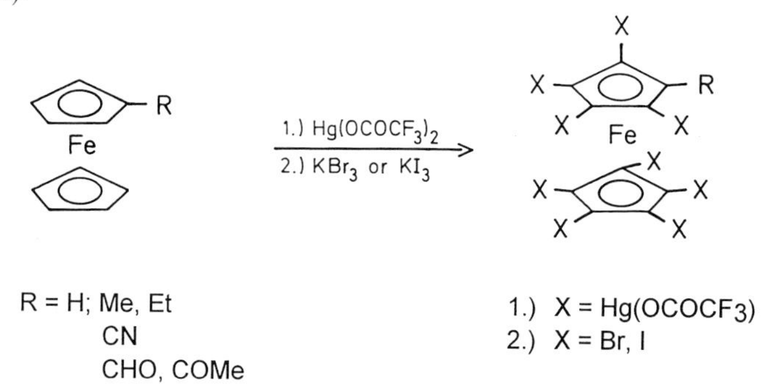

Scheme 5-5. Formation of highly substituted ferrocenyl derivatives.

Although *n*BuLi metallates ferrocene to give a mixture of both Fc-Li and $fcLi_2$ under almost any conditions (see above), the isomeric *t*BuLi gives exclusively and cleanly monolithio-ferrocene, Fc-Li, in Et_2O solution [38, 282] (see Appendix).

In addition to the lithioferrocenes Fc-Li and $fcLi_2 \cdot$ (tmeda), the haloferrocenes Fc-X and fcX_2 (X = Br, I) are the most versatile precursor compounds for ferrocenes bearing heteroelements. As investigated by M. Sato and coworkers [39–43], the halogeno substituent in Fc-X (X = Cl, but mainly Br and I) may be exchanged with other anionic groups in a kind of nucleophilic substitution in the presence of copper(I) salts and polar solvents such as pyridine. Scheme 5-6 summarizes some typical reactions of ferrocenyl bromide, Fc-Br, which can be described as copper-assisted substitution processes [39, 41]. For example, Fc-Cl [40] and Fc-CN [16, 39, 44] can be synthesized by this method, as well as ferrocene derivatives bearing nitrogen, oxygen or sulfur at the metallocene nucleus such as azido [45], diphenylamino [46], phthalimido [16, 44], acetato [16, 21, 41, 42, 44], phenolato [47, 214] and thiophenolato

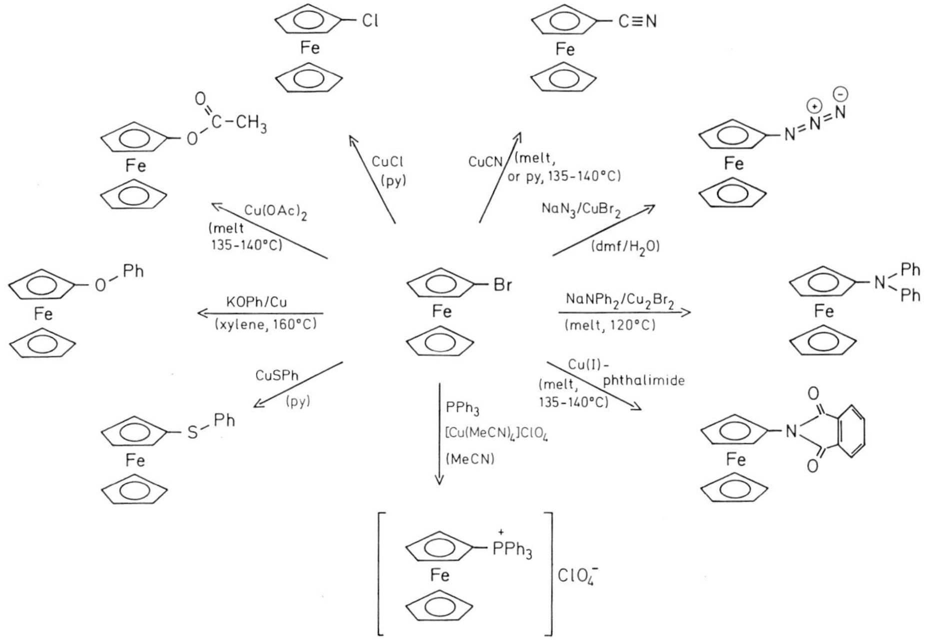

Scheme 5-6. Copper-assisted substitution reactions of bromoferrocene, Fc-Br.

[41, 48] ferrocene. A ferrocenyl phosphonium salt, [Fc-PPh_3]ClO_4, has also been described [43].

In an analogous manner, 1,1′-ferrocenylene dibromide, $fcBr_2$ [279] (cf. Scheme 5-3), has been used to prepare 1,1′-disubstituted ferrocene derivatives such as $fc(N_3)_2$ [45], $fc(OAc)_2$ [16] and $fc(OPh)_2$ [47] by copper-assisted substitution. Lithiation of $fcBr_2$ with *n*BuLi in THF/hexane or Et_2O solution leads to pure $fcLi_2$ [24, 49], while magnesium gives the bifunctional Grignard compound $fc(MgBr)_2$ [18].

5.3 Selected Ferrocene Derivatives with Reactive Functional Substituents

The ferrocenyl analogues of the phenyl compounds aniline, phenol and thiophenol are air-sensitive, and are generally generated from stable precursor compounds prior to use.

5.3.1 Aminoferrocenes and Related Compounds

Although aminoferrocene, Fc-NH_2, can be obtained from various precursors in high yield, the overall yields based on ferrocene are always mediocre [50]. The original preparation by Nesmeyanov et al. [51], who treated Fc-Li with the *O*-benzyl ether of hydroxylamine, $H_2N\text{-}OCH_2Ph$, is still an acceptable possibility (Kochetkov reaction, 25%). A good starting material is the azide of the ferrocenyl carboxylic acid, Fc-C(O)N_3, which may be converted by Curtius degradation to ferrocenyl amine derivatives such as the urethane Fc-NH-C(O)OCH_2Ph [52, 55] or the acetamide Fc-NH-C(O)Me [53, 54] (Scheme 5-7).

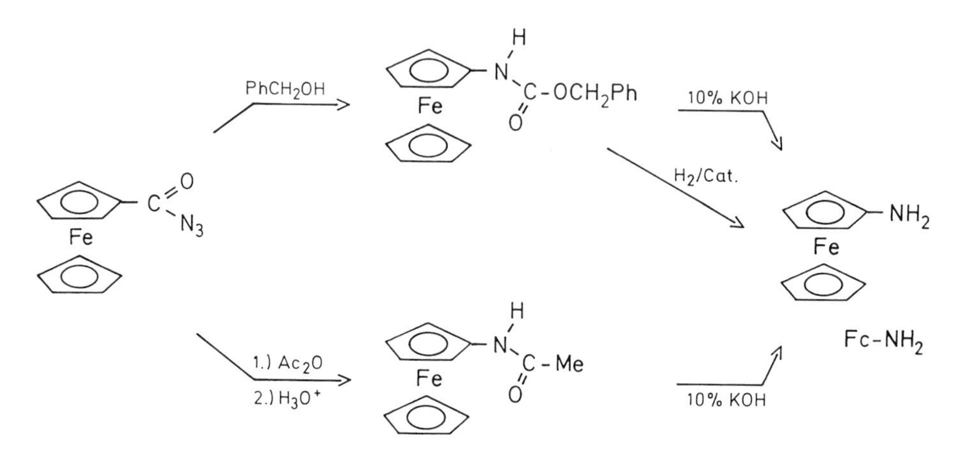

Scheme 5-7. Synthesis of aminoferrocene, Fc-NH_2.

Subsequent hydrolysis with degassed 10% KOH under N_2 leads to Fc-NH_2 in high yield; the orange product can be extracted into diethyl ether and purified via the crystalline hydrochloride, [Fc-NH_3^+]Cl^-, which is formed by introduction of gaseous HCl into the solution [54]. The urethane Fc-NH-C(O)OCH_2Ph has also been hydrogenated over Raney nickel [52] or palladium on carbon [55] to give Fc-NH_2 in high yield.

Other aminoferrocene precursors are *N*-ferrocenyl phthalimide, which can be converted to Fc-NH_2 by $N_2H_4 \cdot H_2O$ in boiling ethanol (82% yield) [16, 44], and ferrocenyl azide, Fc-N_3, which has been reduced with $LiAlH_4$ (72% yield) (cf. Scheme 5-6) [45]. Ferrocenyl amine, Fc-NH_2, is generally formed among other products in the thermal or photochemical decomposition of Fc-N_3 [56] and in the thermolysis of ferrocenyl isocyanate, Fc-NCO [57], in solvents such as cyclohexane, cyclohexene and benzene, probably by reaction of the intermediate nitrene, [Fc-N], with the solvent [56, 57]. The reduction of nitroferrocene, Fc-NO_2, provides another route to aminoferrocene, Fc-NH_2 [58–61]; however, Fc-NO_2 is not an easily accessible starting material.

Finally, ferrocenyl diazenes (azo compounds such as Fc-N=N-Fc) [62, 63] and *N*-ferrocenyl azomethines (Fig. 5-3) [64] can be used for the synthesis of Fc-NH_2 by hydrolytic cleavage of the N=N or N=C double bond, respectively.

Ferrocenyl amine, Fc-NH_2, is a somewhat stronger base than aniline [51, 60, 65]. Its chemistry has been studied in detail; alkylation and acylation – but not diazotation – proceed as expected. Fc-NH_2 is the parent compound for the

N–R
N
Fe
R = Fc, Ph
CHR
N
Fe
R = Ph

Fig. 5-3. Ferrocenyl diazenes and *N*-ferrocenyl azomethines.

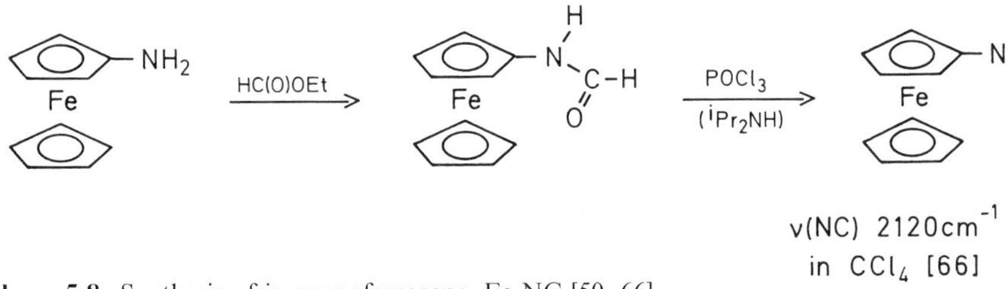

NH2
Fe
HC(O)OEt
H
N
C–H
O
Fe
POCl3
(iPr2NH)
N≡C
Fe
ν(NC) 2120cm-1
in CCl4 [66]

Scheme 5-8. Synthesis of isocyanoferrocene, Fc-NC [50, 66].

preparation of isocyanoferrocene, Fc-N≡C [50, 66], via ferrocenyl formamide (Scheme 5-8).

Isocyanoferrocene has been used as a ligand in carbonyl metal complexes such as $Cr(CO)_5$(CN-Fc) [66], *cis*- and *trans*-$Mo(CO)_4(CN\text{-}Fc)_2$ [50] and $Fe(CO)_4$(CN-Fc) [50]. With cyclo-S_8 or CS_2, sulfurization of Fc-NC to isothiocyanatoferrocene, Fc-NCS, takes place [50], which is also directly accessible from Fc-NH_2.

$$\text{Fc-NH}_2 + CS_2 + NH_3 \rightarrow [\text{Fc-NHCS}_2^-]NH_4^+ \xrightarrow{COCl_2} \text{Fc-NCS} + COS + NH_4Cl + HCl$$

The related isocyanatoferrocene, Fc-NCO (Fig. 5-4), is also best prepared from Fc-NH_2 (i.e., Fc-$NH_3^+Cl^-$), using the reaction with $COCl_2$ in boiling toluene (80%) although the thermolysis of Fc-C(O)N_3 in toluene (57%) is also an acceptable method [67]. Oxidation of Fc-NH_2 by air in the presence of CuX (X = Cl, Br) in benzene leads to azoferrocene, Fc-N=N-Fc, in almost quantitative yield [46],

N=C=O
Fe
N=C=S
Fe

Fc-NCO [67]
orange, mp 36-39°C

Fc-NCS [50]
orange, mp 58°C

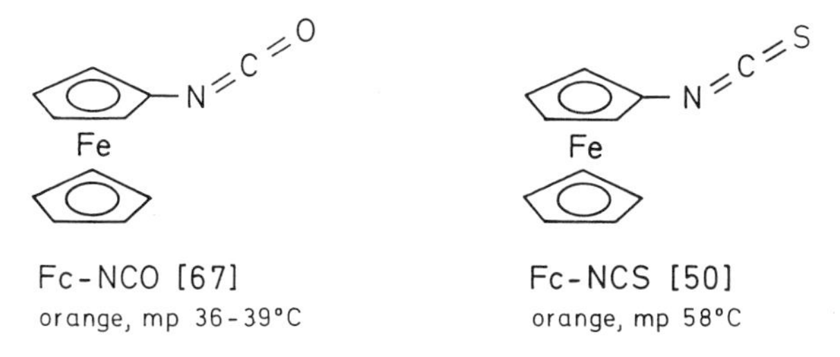

Fig. 5-4. Isocyanato- and isothiocyanatoferrocene.

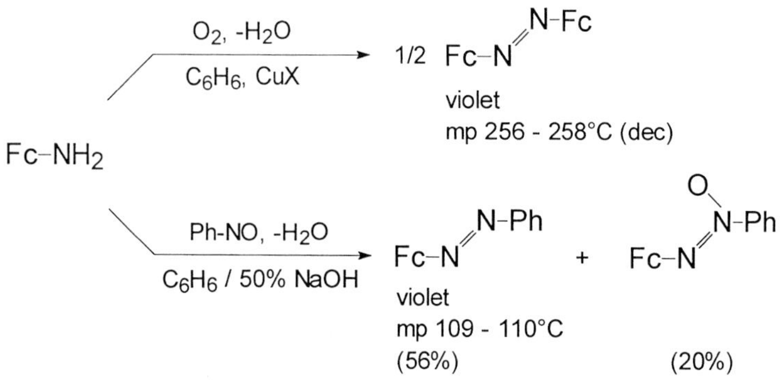

Scheme 5-9. Reactions of aminoferrocene, Fc-NH_2.

whereas condensation of Fc-NH_2 with nitrosobenzene gives *N*-ferrocenyl-*N*′-phenyl-diazene, Fc-N = N-Ph (Scheme 5-9) [63]. (Other methods to prepare Fc-N = N-Ph are available [71, 115].)

The reactions of Fc-NH_2 with thionyl chloride or sulfur dichloride lead to ferrocenyl thionylimide, Fc-NSO, and diferrocenyl sulfur diimide, Fc(NSN)Fc, respectively (Scheme 5-10) [68].

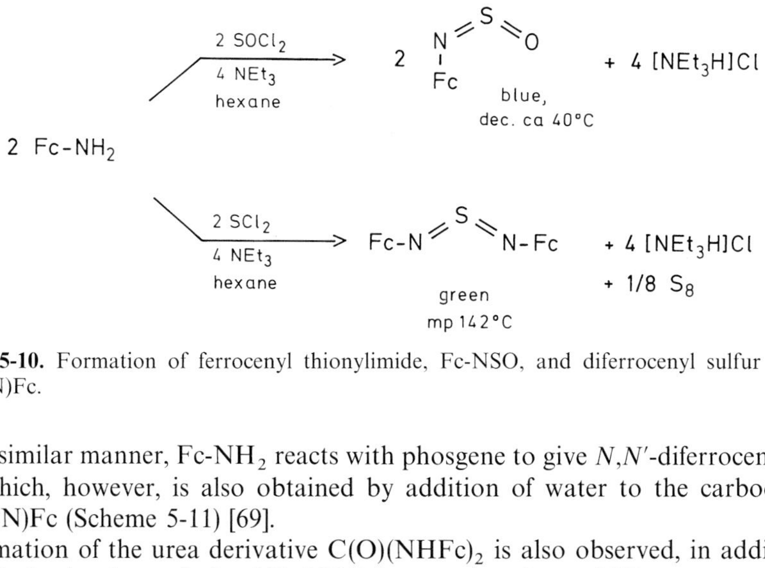

Scheme 5-10. Formation of ferrocenyl thionylimide, Fc-NSO, and diferrocenyl sulfur diimide, Fc(NSN)Fc.

In a similar manner, Fc-NH_2 reacts with phosgene to give *N*,*N*′-diferrocenyl urea [55], which, however, is also obtained by addition of water to the carbodiimide Fc(NCN)Fc (Scheme 5-11) [69].

Formation of the urea derivative $C(O)(NHFc)_2$ is also observed, in addition to Fc-NH_2, in the thermolysis of Fc-NCO in organic solvents [57].

The bifunctional 1,1′-diaminoferrocene, $fc(NH_2)_2$, has been synthesized either by catalytic hydrogenation of the bis(organodiazeno) ferrocenes, $fc(N{=}N\text{-}R)_2$ (R = Me [71] or Ph [70, 71]), over PtO_2 in glacial acetic acid, or by hydrogenation of $fc(N_3)_2$ with $LiAlH_4$ in ether [45]. The yields are reasonable, but the yellow solid $fc(NH_2)_2$ is highly air-sensitive, and only a few derivatives such as $fc[NH\text{-}C(O)OR]_2$ (R = Me [45, 71], Et and CH_2Ph [71]) have been described.

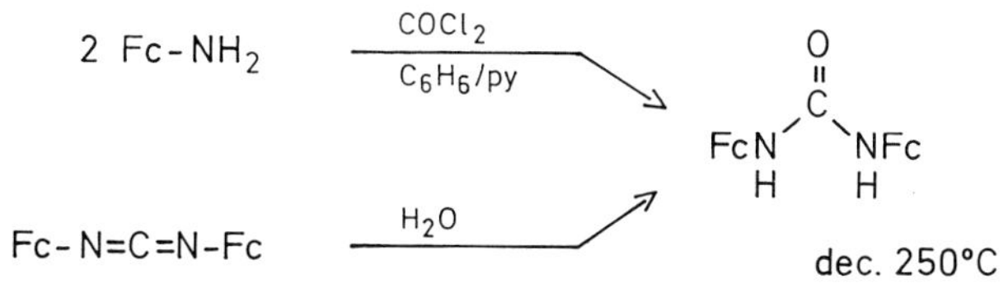

2 Fc-NH$_2$ —($COCl_2$, C_6H_6/py)→ FcN(H)–C(=O)–N(H)Fc (dec. 250°C)

Fc-N=C=N-Fc —(H_2O)→

Scheme 5-11. Formation of *N,N'*-diferrocenyl urea.

The instability of fc$(NH_2)_2$ towards oxidation is certainly the reason that it cannot be used as a building-block for the synthesis of ferrocene-containing polymers [72].

1,1'-Di(isocyano)ferrocene, fc$(NC)_2$, has not been described so far.

5.3.2 Hydroxyferrocenes and Related Compounds

Both hydroxyferrocene, Fc-OH [16, 21, 22], and 1,1'-di(hydroxy)ferrocene, fc$(OH)_2$ [16, 22], were first studied in 1959–60 by Nesmeyanov and coworkers. Both are yellow, air-sensitive compounds that can be easily generated from the acetates Fc-OAc and fc$(OAc)_2$ by alkaline hydrolysis with KOH. Thus, Fc-B$(OH)_2$ reacts with Cu$(OAc)_2 \cdot 2\,H_2O$ in hot water to give Fc-OAc [16, 21, 22], which is hydrolyzed by KOH in boiling ethanol–water to Fc-OH in 88% yield [16, 21]; Fc-OAc is also accessible from Fc-Br and Cu$(OAc)_2$, e.g., in the melt at 135–140 °C (Scheme 5-6).

Hydroxyferrocene, Fc-OH, is a slightly weaker acid than phenol, Ph-OH [16, 22]. Scheme 5-12 summarizes some typical reactions that have been carried out with Fc-OH, including methylation with Me_2SO_4 to methoxyferrocene, Fc-OMe [16], alkylation with (activated) halides R-X (such as Cl-CH_2COOH and Br-$CH_2CH{=}CH_2$) to give ferrocenyl ethers, Fc-OR [16, 22], and the Schotten-Baumann reaction with Ph-C(O)Cl to give the ferrocenyl ester of benzoic acid, Fc-OC(O)Ph [16, 21, 22]. Similar acylations using ferrocenoyl chloride or malonyl dichloride have been investigated [73]. Fc-OH has been added to 2,3-dihydropyrane [74], and also reacted with non-metal halides such as Me_3SiCl and Me_3SnCl to give the ferrocenoxy derivatives Fc-OSiMe_3 and Fc-OSnMe_3, respectively [73]. A further example is the synthesis of di(ferrocenyl) carbonate, C(O)$(OFc)_2$ [73] from Fc-OH and the trimeric precursor of phosgene, C(O)$(OCCl_3)_2$.

As expected, derivatives of the bifunctional 1,1'-dihydroxyferrocene, fc$(OH)_2$, such as fc[O-CH_2COOH]$_2$, fc[O-C(O)Ph]$_2$ and fc[O-SO_2Ph]$_2$, can be obtained without difficulty from fc$(OH)_2$ and the corresponding chlorides [16, 22]. Polyethers of the type fc[O$(CH_2CH_2O)_n$Me]$_2$ (n = 0, 1, 2, 3) have been synthesized in a one-pot reaction starting from fc$(OAc)_2$ and the corresponding halide (Scheme 5-13) [75].

In addition, fc$(OH)_2$ [76] or its potassium salt generated in situ [77, 78] may also be used to prepare crown-ether type polyoxa ferrocenophanes [76–78], which show interesting complex forming ability for metal cations. Two examples of simple [3]ferrocenophanes are Fe$(C_5H_4\text{-}O)_2$Pd(PPh_3) [79] and Fe$(C_5H_4\text{-}O)_2$C=O ("ferrocenylene carbonate") (Scheme 5-14; cf. Sect. 5.7.1 [80]).

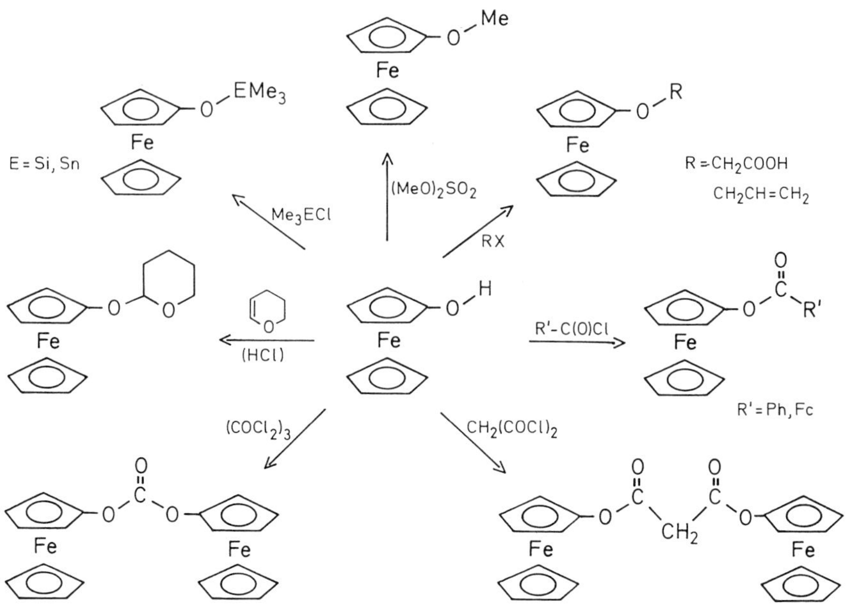

Me
O
Fe
EMe3
O
Fe
E = Si, Sn
R
O
Fe
R = CH_2COOH
$CH_2CH=CH_2$
Me_3ECl
$(MeO)_2SO_2$
RX
O
O
O
(HCl)
H
O
Fe
R'-C(O)Cl
O
C
R'
O
Fe
R' = Ph, Fc
$(COCl_2)_3$
$CH_2(COCl)_2$
O
C
O
O
Fe
Fe
O
O
C
C
O
O
CH_2
Fe
Fe

Scheme 5-12. Reactions of hydroxyferrocene, Fc-OH.

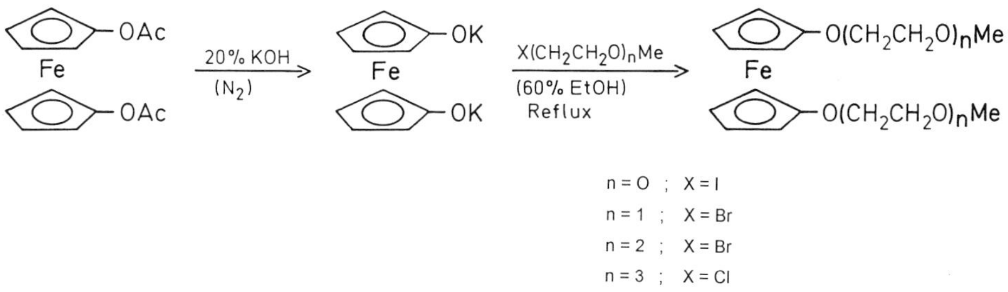

OAc
Fe
OAc
20% KOH
(N_2)
OK
Fe
OK
$X(CH_2CH_2O)_nMe$
(60% EtOH)
Reflux
$O(CH_2CH_2O)_nMe$
Fe
$O(CH_2CH_2O)_nMe$

n = 0 ; X = I
n = 1 ; X = Br
n = 2 ; X = Br
n = 3 ; X = Cl

Scheme 5-13. Synthesis of polyethers, fc[O(CH_2CH_2O)$_n$Me]$_2$ [75].

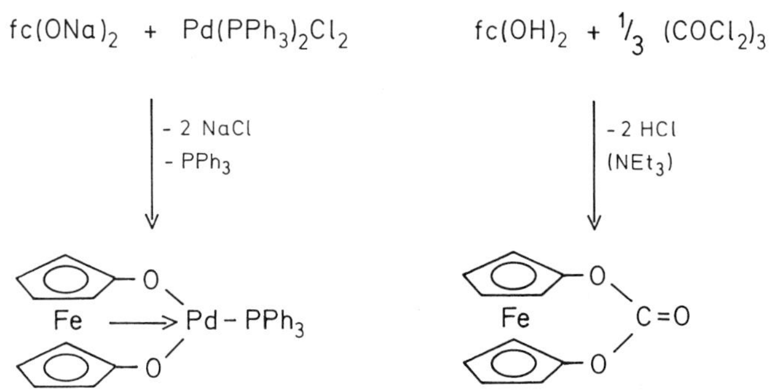

$fc(ONa)_2$ + $Pd(PPh_3)_2Cl_2$
- 2 NaCl
- PPh_3
O
Fe ⟶ Pd - PPh_3
O

$fc(OH)_2$ + 1/3 $(COCl_2)_3$
- 2 HCl
(NEt_3)
O
Fe C=O
O

Scheme 5-14. Formation of 1,3-dioxa-[3]ferrocenophanes.

5.3.3 Mercaptoferrocenes and Related Compounds

The introduction of sulfur into the ferrocene molecule may be effectively realized by two methods (Scheme 5-15): electrophilic sulfonation of $FeCp_2$ to give Fc-SO_3H, and insertion of sulfur into the carbon-lithium bond of lithioferrocene to give Fc-SLi. Mercaptoferrocene, Fc-SH, can then be prepared by hydrogenation of either Fc-SO_2Cl [81–83] or Fc-SS-Fc [81, 82, 84] with $LiAlH_4$. The original procedure [81, 82] via the sulfonic acid is improved if the ammonium salt, [Fc-SO_3^-]NH_4^+, is used as an intermediate that can be purified by recrystallization and then directly treated with PCl_3 to give the sulfonyl chloride, Fc-SO_2Cl; under these conditions the product of the $LiAlH_4$ reduction, Fc-SH, is isolated as an orange-brown solid. The alternative route, insertion of sulfur into Fc-Li in tetrahydofuran (THF) solution leads to the THF adduct, Fc-SLi(thf), which is a light-yellow solid [84]. Hydrolysis of the lithium ferrocenyl thiolate adduct, Fc-SLi(thf), to Fc-SH is possible in principle, but, in the presence of traces of oxygen, mixtures of Fc-SH and Fc-SS-Fc are always obtained. It is therefore advisable to convert the thiolate completely to the

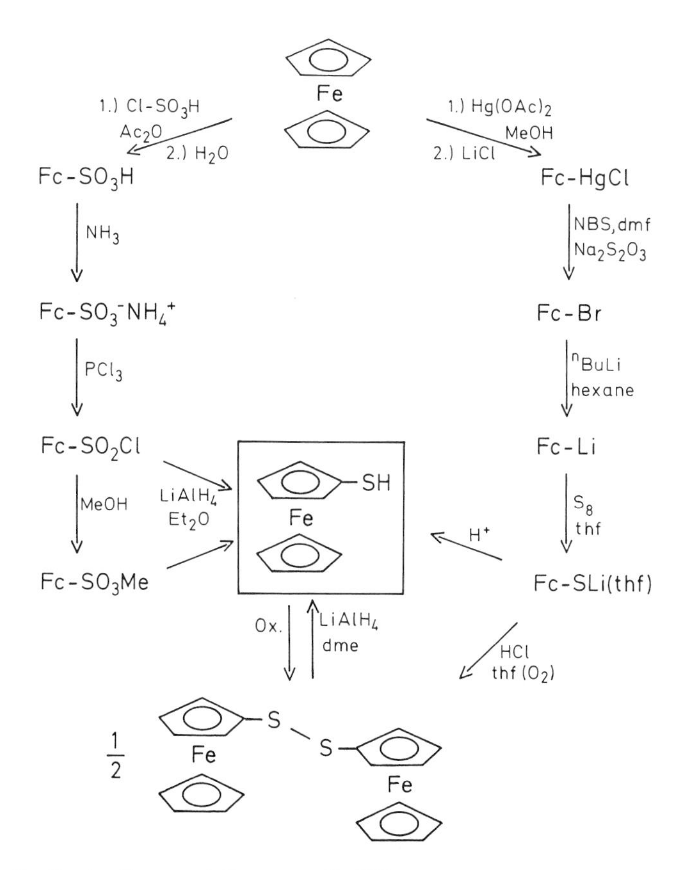

Scheme 5-15. Synthesis of mercaptoferrocene, Fc-SH.

diferrocenyl disulfide, Fc-SS-Fc, and to generate Fc-SH by reaction with $LiAlH_4$ in boiling dimethoxyethane (DME) as it is required [84]. The compound Fc-SH is then obtained as a foul-smelling red oil [81, 84], which solidifies at lower temperature to an orange-red solid.

Diferrocenyl disulfide, Fc-SS-Fc, can be considered as the most versatile precursor for mercaptoferrocene, Fc-SH. It is obtained in high yield (>90%) by air-oxidation of Fc-SH [81, 82] and from Fc-SLi [84] in various solvents; it has also been produced in high yield by hydrolysis of Fc-SCN in boiling methanolic NaOH [85]. It is always observed as a more or less important side-product in reactions in which ferrocenyl thiolate anions, $Fc\text{-}S^-$, or ferrocenyl thiyl radicals, Fc-S · [84], participate. Thus, reaction of the ferrocenyl thiolate complex (Fc-S)Au(PPh_3) (Fig. 5-5) with either HBF_4 in ether [86] or with Br_2 in CH_2Cl_2 solution [87] again produces high yields of Fc-SS-Fc.

Mercaptoferrocene, Fc-SH, is a highly reactive compound that may add to activated olefins such as acryl- and methacryl acetate [83], norbornene and norbornadiene [84], methyl vinyl ketone, methyl vinyl sulfone, benzylidene and di(benzylidene) acetone [88]. It reacts with acid chlorides such as Me-C(O)Cl [81], Ph-C(O)Cl [82] and Fc-C(O)Cl [89] to give ferrocenylthio esters, FcS-C(O)R (R = Me, Ph, Fc); however, these acyl derivatives are more conveniently obtained using THF solutions of lithium ferrocenyl thiolate, Fc-SLi, in situ [89] instead of the malodorous thiol itself. The intermediate formation of alkali metal ferrocenyl thiolates is similarly applied in the reactions of Fc-SH in aqueous alkaline solution with, for example, allyl bromide to give the allyl ether, $Fc\text{-}S\text{-}CH_2CH{=}CH_2$ [81], and with (Ph_3P)AuCl to give the ferrocenyl thiolate complex, (Fc-S)Au(PPh_3) [86] (Fig. 5-5). (Coordination compounds containing ferrocenyl thiolate ligands are described in Sect. 5.6.)

Just as mercaptoferrocene, Fc-SH, is most conveniently prepared from diferrocenyl disulfide, Fc-SS-Fc, the corresponding 1,1′-dimercaptoferrocene, $fc(SH)_2$, can be easily obtained from the stable precursor 1,2,3-trithia-[3]ferrocenophane, $fc(S_3)$, by hydrogenation using $LiAlH_4$ [29, 90, 91]. The standard route (Scheme 5-16) to $fc(SH)_2$ starts from ferrocene, which is dilithiated in the presence of tmeda to give $fcLi_2 \cdot (tmeda)_n$ (see above, Sect. 5.1.2) [25, 29]; subsequent stoichiometric insertion of 2/8 S_8 in THF solution produces dilithium 1,1′-ferrocenylene dithiolate as a yellow adduct $fc(SLi \cdot thf)_2$, which may be hydrolyzed to give the desired product $fc(SH)_2$ [91]. It is generally more convenient, however, to treat the dilithio intermediate, $fcLi_2 \cdot (tmeda)_n$, with excess sulfur and to prepare the precursor $fc(S_3)$ [29, 92].

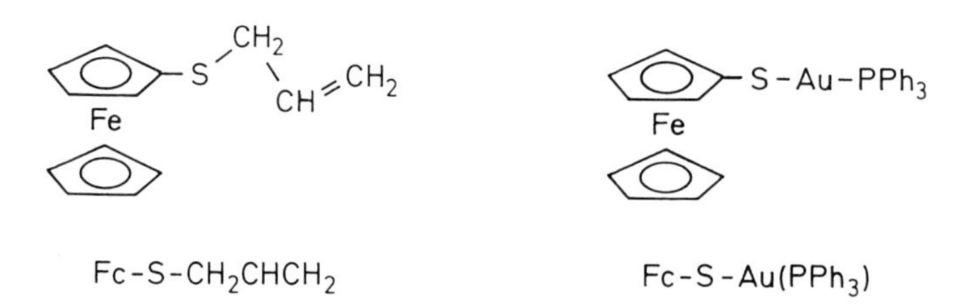

Fig. 5-5. Derivatives of mercaptoferrocene, Fc-SH.

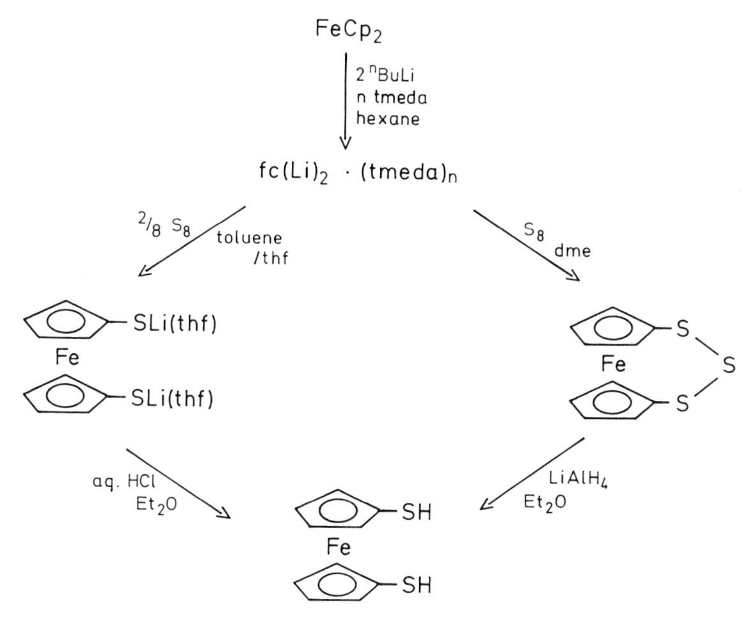

Scheme 5-16. Synthesis of 1,1′-dimercaptoferrocene, $fc(SH)_2$.

1,1′-Dimercaptoferrocene, $fc(SH)_2$, is a yellow, volatile, moderately air-sensitive solid [29, 90]. It reacts with activated olefins to give 1:2 adducts, for example, with styrene, methyl vinyl ketone, methyl vinyl sulfone, methyl methacrylate, and benzylidene acetone (PhCH=CHAc) under basic conditions (catalysis by NEt_3), or with vinyl ethers, such as isobutyl vinyl ether, under acidic conditions (catalysis by HCl) (Scheme 5-17) [88].

Under radical conditions (catalysis by AIBN), the formation of disulfido links predominates; thus, the low-molecular mass product of the reaction of $fc(SH)_2$ with

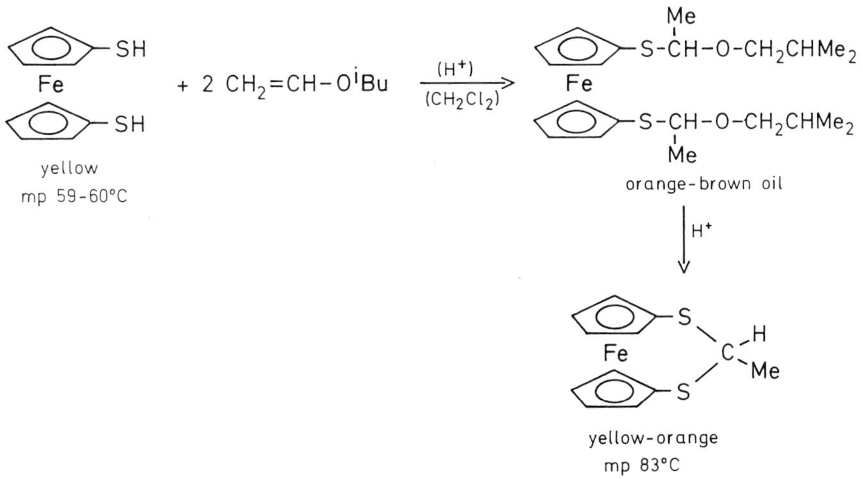

Scheme 5-17. Reaction of 1,1′-dimercaptoferrocene, $fc(SH)_2$, with isobutyl vinyl ether.

norbornene is $[(C_7H_{11}\text{-}S)fc\text{-}S]_2$ [84]. A comparable disulfide is obtained from the reaction between $fc(SH)_2$ and the oxonium salt $[(Ph_3PAu)_3O]BF_4$ (3:1), which leads to $[(Ph_3PAu\text{-}S)fc\text{-}S]_2$ (Scheme 5-18) [90].

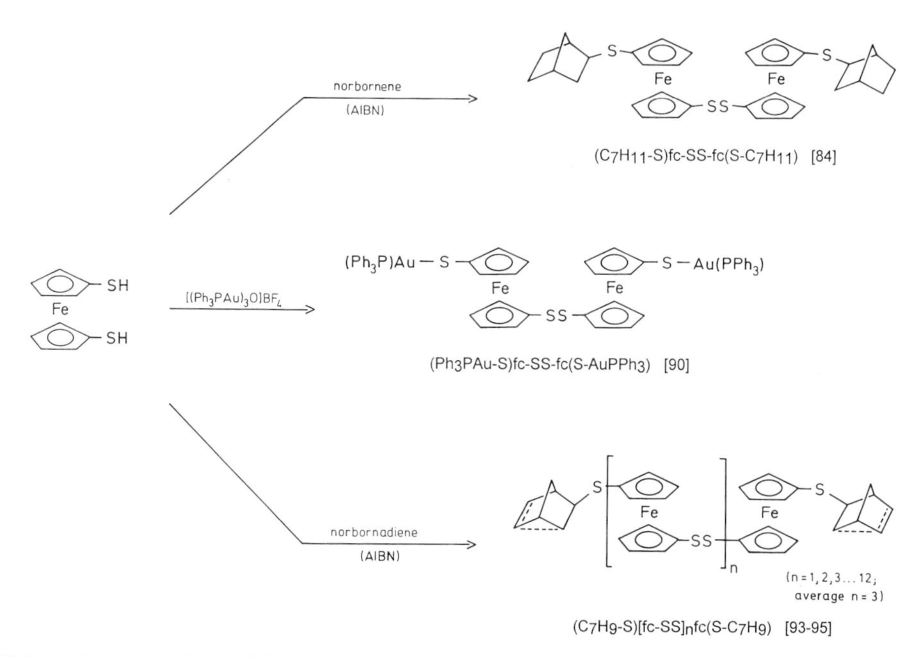

Scheme 5-18. Reactions of $fc(SH)_2$ leading to disulfide linkages.

The bifunctional ferrocene derivative $fc(SH)_2$ has been tested as a component in the synthesis of macromolecules [84, 93, 94]. However, the reaction (1:1) of $fc(SH)_2$ with norbornadiene (in the presence of AIBN) leads to oligomers containing only disulfide-bridged ferrocene units in the main chain [93, 95]; the diolefin appears solely in the form of norbornenyl or nortricyclyl terminations (Scheme 5-18). The preferred formation of disulfide linkages can be explained by the stability of the ferrocenyl thiyl radicals (Fc-S ·), which live long enough to terminate the chain growth by radical combination [94].

Alternating copolymers containing [S(fc)S] building blocks have been obtained [88] from the base-catalysed polyaddition of $fc(SH)_2$ to activated diolefins such as 1,4-butandiol-dimethacrylate and divinyl sulfone (Fig. 5-6); the oily brown products possess olefinic terminal groups.

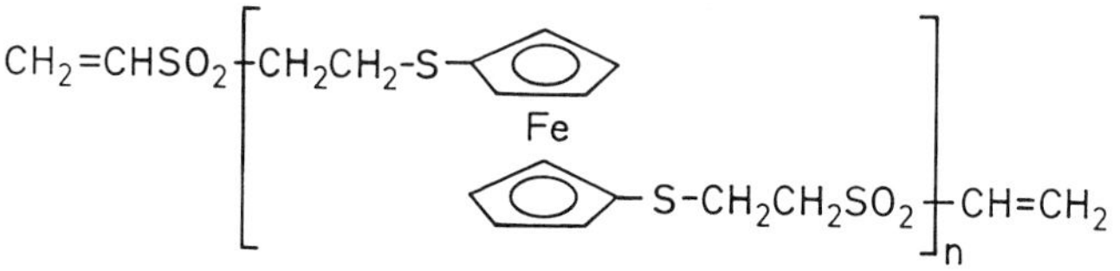

$$CH_2{=}CHSO_2\left[CH_2CH_2\text{-}S\text{-}C_5H_4\text{-}Fe\text{-}C_5H_4\text{-}S\text{-}CH_2CH_2SO_2\right]_n CH{=}CH_2$$

Fig. 5-6. Copolymers containing [Sfc)S] building blocks.

A polypropylene sulfide polymer with just a single ferrocene unit [88] is formed in the ring-opening polymerisation of propylene sulfide (i.e., methyl thiirane) initiated by $fc(SNa)_2$ (Scheme 5-19).

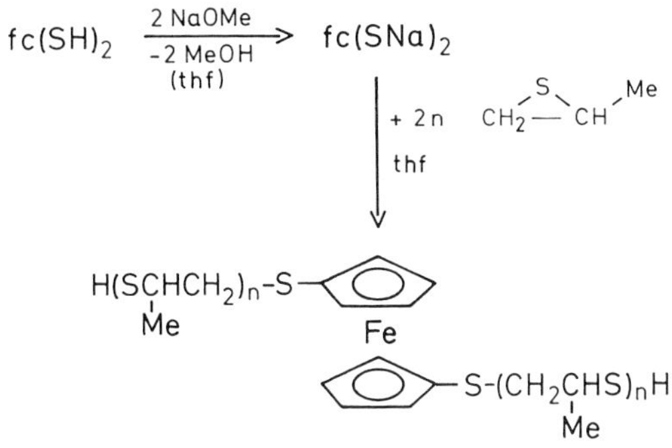

$fc(SH)_2$ $\xrightarrow[-2\,MeOH\ (thf)]{2\,NaOMe}$ $fc(SNa)_2$ $\xrightarrow{+2n\ CH_2(S)CHMe,\ thf}$ $H(SCH(Me)CH_2)_n$-S-C_5H_4FeC$_5H_4$-S-$(CH_2CH(Me)S)_nH$

Scheme 5-19. Polypropylene sulfide containing a single ferrocene unit.

If, however, $fc(SH)_2$ is first treated with 2 equivalents of propylene sulfide to give 1,1′-bis(2-mercapto-propylthio)ferrocene, fc[S-$CH_2CH(Me)SH]_2$, and this new dithiol is then added to divinyl sulfone (1 : 1), a regular alternating polymer is obtained (Scheme 5-20) [94].

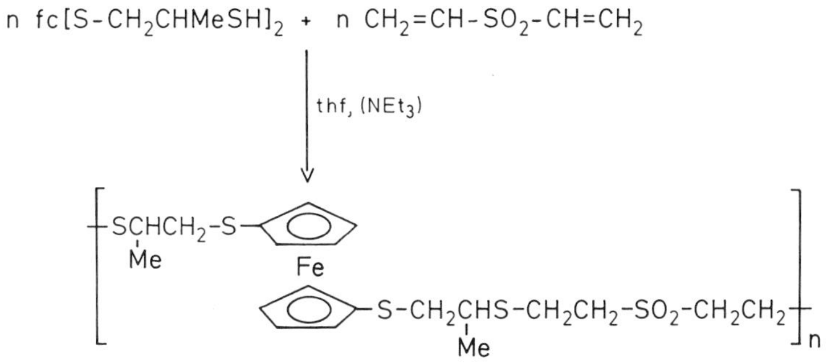

n fc[S-$CH_2CHMeSH]_2$ + n CH_2=CH-SO_2-CH=CH_2 $\xrightarrow{thf,\ (NEt_3)}$ [-SCH(Me)CH_2-S-C_5H_4FeC$_5H_4$-S-CH_2CH(Me)S-CH_2CH_2-SO_2-CH_2CH_2-]$_n$

Scheme 5-20. Copolymer from 1,1′-bis(2-mercapto-propylthio)ferrocene and divinyl sulfone [94].

1,1′-Dimercaptoferrocene, $fc(SH)_2$, may also be used in polycondensation reactions with bifunctional dicarboxylic acid chlorides, Cl(O)C$(CH_2)_n$C(O)Cl (e.g., $n = 2$, 8) to give polythioesters [88]. Aromatic dicarboxylic acid chlorides derived from terephthalic or isophthalic acid lead to cyclic products; thus, an [8.8]ferrocenophane (Scheme 5-21) is isolated using terephthaloyl chloride [96].

A large number of sulfur-containing [*n*]ferrocenophanes [96 – 100] has been synthesized starting either from $fc(SH)_2$ itself or its alkali metal derivatives $fc(SLi)_2$ and $fc(SNa)_2$; among them, the 1,3-dithia-[3]ferrocenophanes are particularly common. These compounds are considered together with the corresponding selenium and tellurium analogues in Section 5.6. A remarkable example is the tetrathio-substituted ethylene, $[fcS_2]C{=}C[S_2fc]$ (Scheme 5-22) [101].

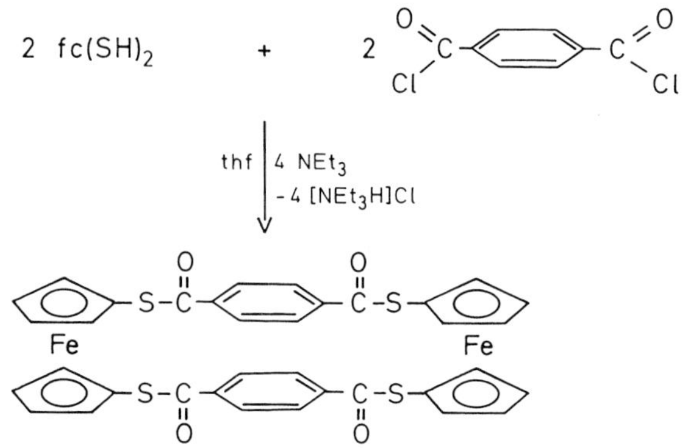

Scheme 5-21. Formation of the [8.8]ferrocenophane fc[S-C(O)-C_6H_4-C(O)-S]$_2$fc [96].

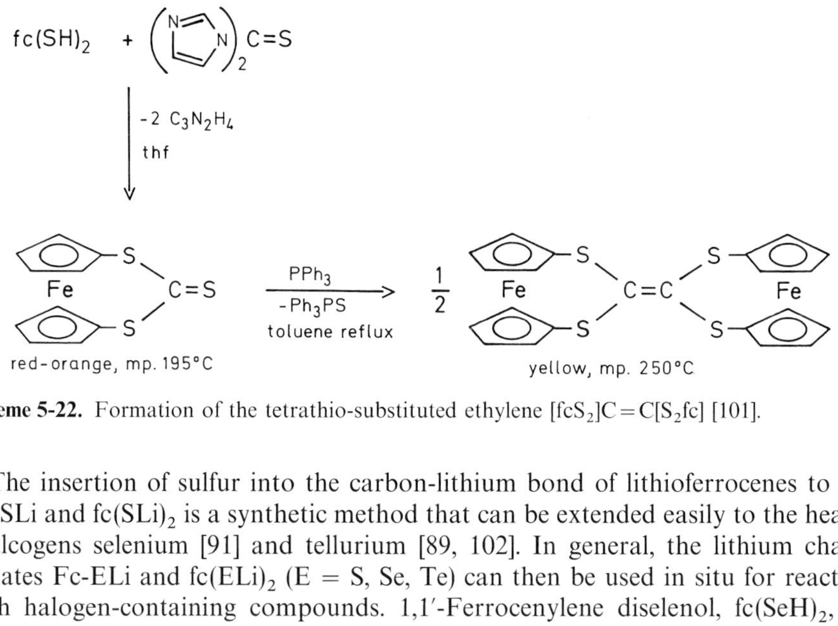

Scheme 5-22. Formation of the tetrathio-substituted ethylene [fcS_2]C=C[S_2fc] [101].

The insertion of sulfur into the carbon-lithium bond of lithioferrocenes to give Fc-SLi and fc(SLi)$_2$ is a synthetic method that can be extended easily to the heavier chalcogens selenium [91] and tellurium [89, 102]. In general, the lithium chalcogenates Fc-ELi and fc(ELi)$_2$ (E = S, Se, Te) can then be used in situ for reactions with halogen-containing compounds. 1,1′-Ferrocenylene diselenol, fc(SeH)$_2$, has been isolated as a volatile orange crystalline material (*m.p.* 36–39 °C) [103] both by protonation of the dilithium diselenolate, fc(SeLi · thf)$_2$ [91], and by $LiAlH_4$ reduction of the 1,2,3-triselena-[3]ferrocenophane, fc(Se_3) [103].

5.4 Homoleptic Ferrocenyl Derivatives of the Elements

The system of the homoleptic ("binary") element–ferrocenyl compounds is more limited than that of the corresponding element–phenyl compounds. Whereas stable homoleptic σ-phenyl derivatives of almost all main group elements (Groups 1, 2 and 13–17) have been characterized, the corresponding Periodic Table of the main

Table 5-1. Homoleptic ferrocenyl derivatives of the main group elements

Group 1	13	14	15	16	17
FcLi [10, 14, 24, 26, 38]	BFc_3 [108, 109]	CFc_4 [284]	NFc_3 [54, 129]		Fc-F [149, 167]
FcNa [14, 104]		$SiFc_4$ [113]	PFc_3 [132–134]	SFc_2 [48, 145, 146]	Fc-Cl [15, 16, 39, 40]
FcK [105]		$GeFc_4$ [113]		$SeFc_2$ [111, 147, 148]	Fc-Br [11, 16, 39, 47, 49, 150]
		$SnFc_4$ [113]	$SbFc_3$ [144]	$TeFc_2$ [102]	Fc-I [11, 18, 39, 150]
	$TlFc_3$ [112]	$PbFc_4$ [113]	$BiFc_3$ [144]		

group element σ-ferrocenyl derivatives (Table 5-1) contains several blanks. Thus, complexes of the alkaline earth metals, MFc_2, are completely absent, although the Grignard compounds Fc-MgX (X = Cl, Br, I) [18, 106, 107], in particular Fc-MgBr [106, 107], are available for synthetic purposes. In Group 13, homoleptic triferrocenyl complexes of Al, Ga, and In are unknown so far. In the case of boron, in addition to the expected BFc_3 [108, 109], which forms red crystals, a black tetraferrocenyl complex, BFc_4, is obtained in the reaction of either $B\mathit{n}Bu_3$ or BF_3 with excess lithioferrocene and subsequent oxidation of the BFc_4^- anion [110].

$$BF_3 + 4\,\text{Fc-Li} \xrightarrow{(-3\,\text{LiF})} \text{Li}[BFc_4] \xrightarrow{(\text{Oxid.})} Fc^+BFc_3^-$$

On the basis of an X-ray structure analysis [110], cyclic voltammetry [110] and Mößbauer data [111], BFc_4 is best described as a zwitterion containing one cationic ferrocenyl substituent and a negative charge at boron [110].

Tetraferrocenyl derivatives of the Group 14 elements Si, Ge, Sn, and Pb have appeared in the patent literature [113, 114]; the MFc_4 complexes were prepared by reaction between MCl_4 and Fc-Li (ca. 1:5). In the cases of Sn and Pb, the dihalides MCl_2 (M = Sn, Pb) were also used for the reactions with Fc-Li (ca. 1:2.4); the products $SnFc_4$ and $PbFc_4$ are reported to be formed by thermal decomposition of the intermediates $Fc_3M\text{-}MFc_3$ (M = Sn, Pb) [113].

Tetraferrocenylmethane, CFc_4, has recently been described [284]. Tetraferrocenylethylene [116] ("$Fc_2C{=}CFc_2$") and hexaferrocenylbenzene ("C_6Fc_6") are unknown; it is remarkable, however, that several tetrasubstituted ethylenes of the type Fc(R)C=C(R)Fc, (R = Me, *t*Bu, Ph, C(O)Fc) have been described, and that tetraferrocenylethane, $Fc_2CH\text{-}CHFc_2$, which is a chiral molecule resembling a four-bladed propeller [117], has also been repeatedly investigated [116, 117]. Ferrocenyl-substituted acetylenes such as Fc-C≡C-Fc [118–123], Fc-C≡C-C≡C-Fc [120, 124–128], $\text{Fc-(C≡C)}_n\text{-Fc}$ (n = 2, 4, 6, 8) [127], and fc(C≡C-Fc)_2 [120] are favored models for the study of linear conjugated systems and of mixed-valence cations.

All elements of Group 15 except arsenic can be found as the central atom in triferrocenyl complexes. Triferrocenyl amine, NFc_3, was synthesized starting from

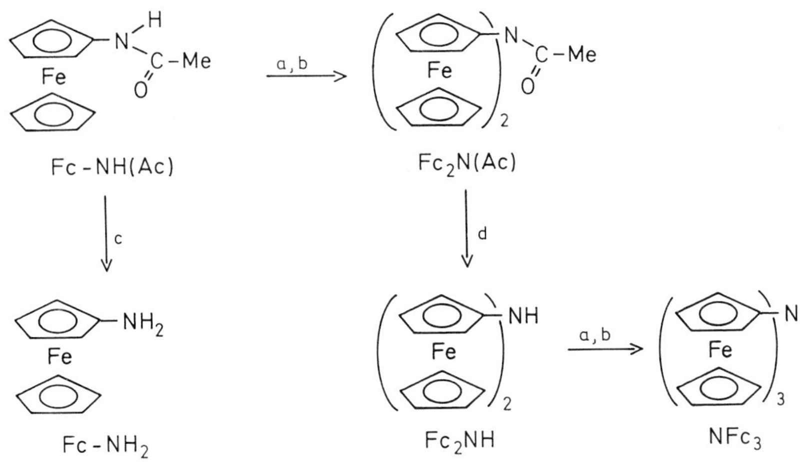

Scheme 5-23. Synthesis of the ferrocenyl amines Fc-NH_2, Fc_2NH and NFc_3 [54]. (a) Metallation ($NaNH_2$ in toluene); (b) ferrocenylation (Fc-Br/CuBr/py) in toluene, 110 °C; (c) hydrolysis (10% aq. KOH, reflux); (d) hydrolysis (KOH/EtOH), reflux).

acetamido ferrocene, Fc-NH(Ac), by stepwise introduction of ferrocenyl groups using ferrocenyl bromide, Fc-Br, in the presence of CuBr/pyridine in boiling toluene (Scheme 5-23) [54, 129]. Diferrocenyl amine, Fc_2NH [130], is the precursor of triferrocenyl amine, NFc_3.

According to the X-ray structure analysis [129], NFc_3 contains the nitrogen atom in an almost, although not completely, planar environment. The strong interaction of the nitrogen lone pair with the ferrocenyl substituents, which is reflected in the lack of basicity of NFc_3, appears to be an important factor in the "anomalous" electrochemistry of the ferrocenyl amines [131]. In contrast to NFc_3, the homologous PFc_3 [132–137] is a good two-electron ligand, in particular in carbonyl metal complexes [134, 137–139].

Whereas tetra(ferrocenyl)hydrazine ("N_2Fc_4") has not been described so far, the corresponding diphosphane P_2Fc_4 [140] has been obtained by reductive dehalogenation of Fc_2PCl. The analogue of azoferrocene (Fc-N=N-Fc) [46], di(ferrocenyl)diphosphene (Fc-P=P-Fc) [141], has recently been prepared by reductive dehalogenation of Fc-PCl_2. Inorganic ring systems containing ferrocenyl substituents, such as P_4Fc_4 and As_3Fc_3, have also been characterized [142]; in general, mixtures of rings, $[PFc]_n$ [142] or $[AsFc]_n$ [143], are expected to be formed on reductive dehalogenation of Fc-PCl_2 and Fc-$AsCl_2$, respectively [142].

Neither diferrocenyl ether ("Fc-O-Fc") nor diferrocenyl peroxide ("Fc-OO-Fc") are known so far. However, the corresponding compounds of sulfur, selenium and tellurium are well characterized (see Sect. 5.8). The dichalcogenides Fc-EE-Fc (E = S, Se, Te) are easily obtained by "oxidative dimerisation" of either the lithium ferrocenyl chalcogenate, Fc-ELi, or the chalcogenols, Fc-EH, for which they can be used as stable precursors (Sect. 5.3.3) [91, 95]. Reaction of the dichalcogenides Fc-EE-Fc with Fc-Li is a possible route to the monochalcogenides, Fc-E-Fc, but the yields are low [102]. A diferrocenyl complex containing a pentasulfide chain, $Fc(S_5)Fc$, has been detected among the products of the reaction of ferrocene with S_8 (1 : 1) in the presence of $Fe_3(CO)_{12}$ in boiling benzene [145].

A few homoleptic ferrocenyl compounds of transition metals have been described; a review on "ferrocene-containing metal complexes" is available (see also Chap. 7) [151]. The tetraferrocenyls MFc_4 (M = Ti, Zr, Hf) have been mentioned briefly in a patent [113]; studies on $TiFc_4$ [171] and $HfFc_4$ [172] have appeared recently. The ferrocenyls of the coinage metals copper and silver, CuFc [152] and AgFc [153], are obviously polymeric; they were prepared by the reactions of Fc-Li (from Fc-Br) with either $[CuBr\text{-}PPh_3]_4$ or $AgNO_3 \cdot 3\,PPh_3$, respectively. The orange ferrocenyl copper, CuFc, is a valuable intermediate for the acylation [152, 154], arylation [152, 155], or ferrocenylation [155] of ferrocene [cf. 170]; for synthetic purposes, the use of the dimethylsulfide-stabilised complex, $CuFc \cdot SMe_2$ (from Fc-Li and $CuBr \cdot SMe_2$), has been recommended [281, 282]. In a similar manner, 1,1′-ferrocenylene dicopper, $fcCu_2$ [283], can be directly applied in the form of $fc(Cu \cdot SMe_2)_2$ [281, 282].

The oldest [10] and most extensively studied diferrocenyl metal compound is the orange mercury derivative, $HgFc_2$ (Fig. 5-7), which is obtained readily from "chloromercuriferrocene", Fc-HgCl, with various "reductive agents" such as, for example, aq. $Na_2S_2O_3$ solution [10, 156], Na dispersion in nonane/benzene, NaI in ethanol or $SnCl_2 \cdot 2\,H_2O$ in aq. NaOH [12], $[Re(CO)_5]^-$ in THF solution [157], and cystein in DMF/H_2O at pH 7 [158].

The X-ray structure analysis of the monohydrate, $HgFc_2 \cdot H_2O$, has confirmed that the cyclopentadienyl rings in the ferrocenyl substituents possess an eclipsed conformation [158]. The mercurial $HgFc_2$ can be used as ferrocenyl transfer reagent; thus, rare-earth dihalides react with $HgFc_2$ to give new diferrocenyl metal compounds such as $YbFc_2$ [159, 160], $SmFc_2$ and $EuFc_2$ [160].

With regard to the rare homoleptic compounds containing the 1,1′-ferrocenylene unit, mention should be made of the red spiro complex bis(1,1′-ferrocenediyl)silane, $Si(fc)_2$, obtained from the reaction of $SiCl_4$ with a suspension of $fcLi_2 \cdot$ (tmeda) in hexane [28, 161]. As in the strained sila-[1]ferrocenophanes of the type $fc(SiR_2)$ (R = Cl [162, 163], Me [164, 165], Ph [28, 161, 164, 166]), the five-membered rings in the 1,1′-ferrocenediyl units of $Si(fc)_2$ can no longer be planar but are tilted towards silicon; the tilt angle is 20.8(5)° in $fc(SiMe_2)$ [165] and 19.2° in $fc(SiPh_2)$ [166]. Both $Si(fc)_2$ [168] and the corresponding $Ge(fc)_2$ [169] have been used for the derivatization of surfaces in photoelectrochemical studies on n-type silicon electrodes.

The 1,1′-ferrocenylene derivatives of the heavier chalcogens, the 1,2,3-trichalcogena-[3]ferrocenophanes, $fc(E_3)$ (E = S, Se, Te), are described in Section 5.7.

Fig. 5-7. Diferrocenylmercury, $HgFc_2$, and bis(1,1′-ferrocenediyl)silane, $Si(fc)_2$.

5.5 Transition Metal Complexes with Ferrocenyl and 1,1′-Ferrocenediyl Ligands

Coordination compounds containing the electron-donating ferrocenyl group (Fc) directly attached to a transition metal center are useful models in which the localisation of electric charges and intramolecular interactions can be studied by way of cyclic voltammetry and Mößbauer spectroscopy at the ferrocene iron, which acts as a probe.

The standard method of synthesis involves the reaction of suitable transition metal halides with Fc-Li [49, 113, 171–180, 189, 190], although $HgFc_2$ has also been used as ferrocenyl transfer reagent [181–183]. A different approach to ferrocenyl complexes proceeds via thermal or photo-induced decarbonylation of the corresponding ferrocenoyl compounds [178, 184]. All three routes have been applied to prepare the half-sandwich/sandwich complex $CpFe(CO)_2$-Fc, which contains two different iron centers (Scheme 5-24). It is surprising that Fc-$SnMe_3$ has rarely been used [185] for ferrocenyl transfer.

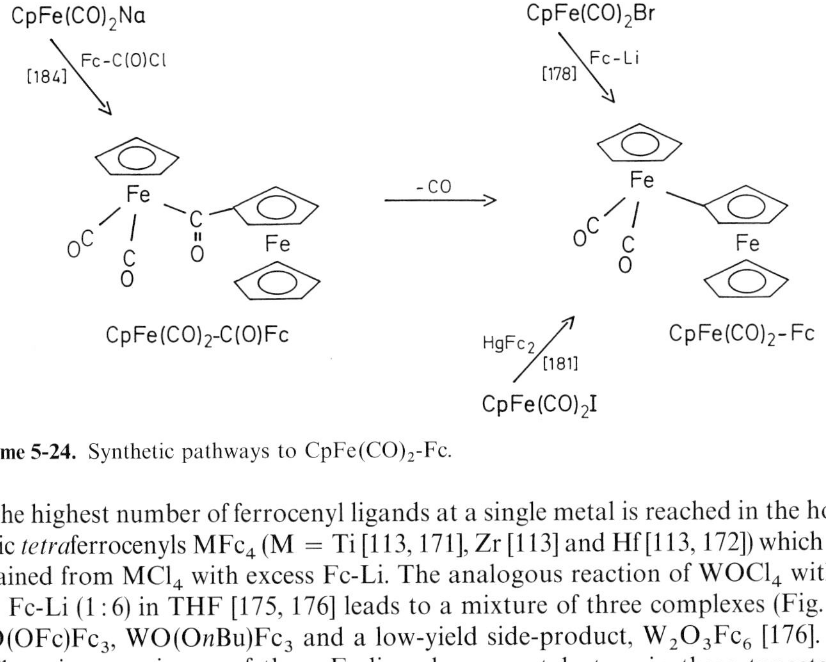

Scheme 5-24. Synthetic pathways to $CpFe(CO)_2$-Fc.

The highest number of ferrocenyl ligands at a single metal is reached in the homoleptic *tetra*ferrocenyls MFc_4 (M = Ti [113, 171], Zr [113] and Hf [113, 172]) which were obtained from MCl_4 with excess Fc-Li. The analogous reaction of $WOCl_4$ with excess Fc-Li (1:6) in THF [175, 176] leads to a mixture of three complexes (Fig. 5-8): $WO(OFc)Fc_3$, $WO(OnBu)Fc_3$ and a low-yield side-product, $W_2O_3Fc_6$ [176].

There is a maximum of three Fc ligands per metal atom in these tungsten(VI) oxo complexes; the fourth Fc substituent in $WO(OFc)Fc_3$ is bound via an oxo bridge. According to the X-ray crystallographic structure [175], $WO(OFc)Fc_3$ possesses a trigonal-pyramidal geometry with three ferrocenyl ligands occupying the equatorial positions and an axial ferrocenoxy ligand trans to the oxo ligand (Fig. 5-9).

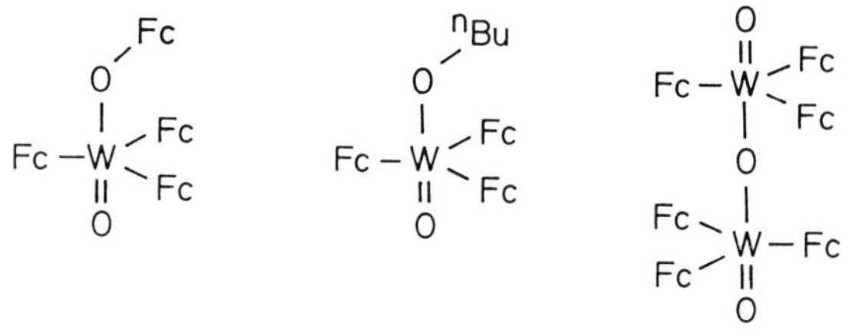

$WO(OFc)Fc_3$ dark-red | $WO(O^nBu)Fc_3$ light-red | $W_2O_3Fc_6$ violet

Fig. 5-8. Ferrocenyl-tungsten complexes [176].

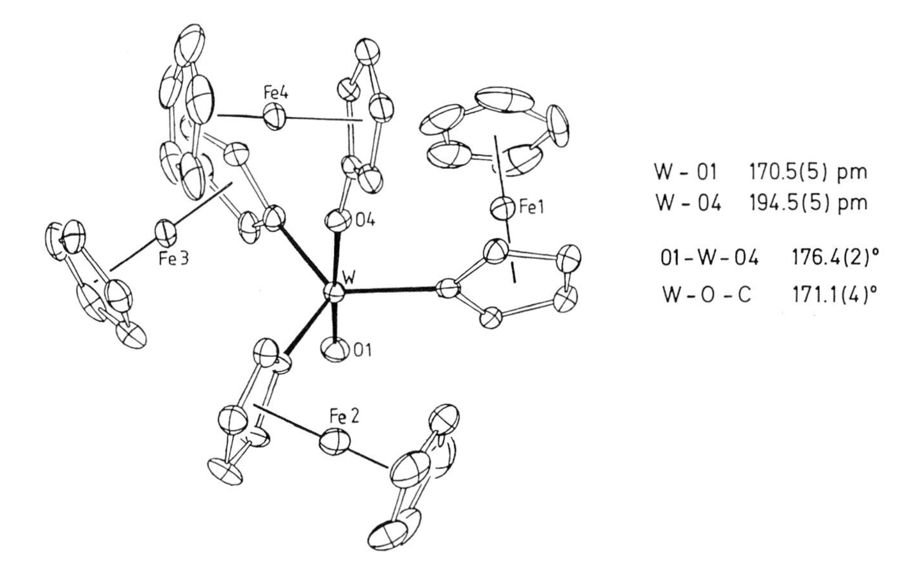

Fig. 5-9. Molecular geometry of $WO(OFc)Fc_3$ [175].

The reaction of $WOCl_4$ with a limited amount of Fc-Li (1:3) in THF solution produces the light-sensitive complex $WO(Cl)Fc_3$ [175, 176]; the reactive chloro ligand can be substituted by nucleophiles such as alkoxides (Scheme 5-25).

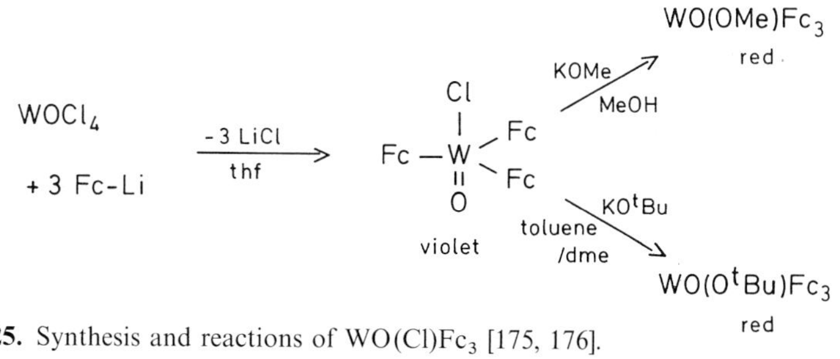

$WOCl_4$ + 3 Fc-Li —(-3 LiCl, thf)→ $WO(Cl)Fc_3$ (violet)
—(KOMe, MeOH)→ $WO(OMe)Fc_3$ (red)
—(KO^{t}Bu, toluene/dme)→ $WO(O^tBu)Fc_3$ (red)

Scheme 5-25. Synthesis and reactions of $WO(Cl)Fc_3$ [175, 176].

The violet fluoro derivative $WO(F)Fc_3$ is formed from $WO(X)Fc_3$ (X = Cl, OFc, O*n*Bu) in the presence of complex fluoride salts such as $Me_3O^+BF_4^-$ in CH_2Cl_2; it is almost insoluble in organic solvents [176]. The ^1H and ^{13}C NMR spectra of the diamagnetic oxotungsten(VI) complexes are temperature-dependent, suggesting a hindered rotation of the ferrocenyl ligands around the W-C(ferrocenyl) bond [175, 176]; in the case of $WO(OFc)Fc_3$ the free activation enthalpy, $\Delta G^{\ddagger}(T_c)$, was determined to be 62.5 ± 0.5 kJ/mol [175].

The ferrocenyl ligand is able to tolerate both high and low oxidation states of the central metal, as may be seen from the series of half-sandwich tungsten-ferrocenyl complexes $CpW(CO)_3$-Fc [178, 184], $CpW(NO)_2$-Fc [177], and $CpWO_2$-Fc [177] (Scheme 5-26).

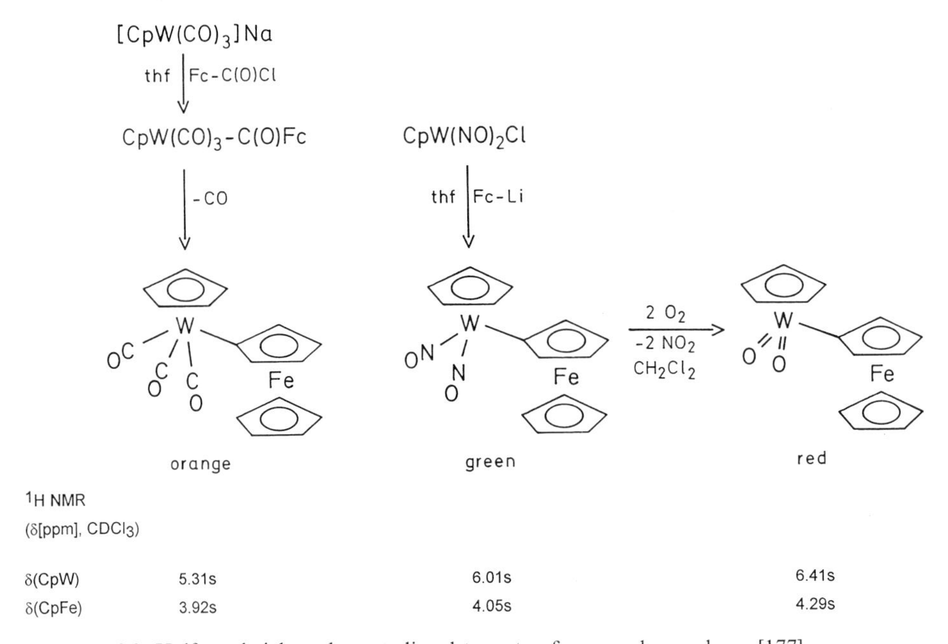

Scheme 5-26. Half-sandwich cyclopentadienyl tungsten-ferrocenyl complexes [177].

The 18-electron dinitrosyl compound $CpW(NO)_2$-Fc is spontaneously oxidized in air to the 16-electron dioxo complex $CpWO_2$-Fc [177]. The gradual increase in the formal oxidation number of tungsten from +II in $CpW(CO)_3$-Fc to +VI in $CpWO_2$-Fc is reflected in the ^1H and ^{13}C NMR spectra, in which the characteristic signals of both the cyclopentadienyl ring and the ferrocenyl group are gradually shifted to lower field as the shielding of the ^1H and ^{13}C nuclei is reduced due to decreasing charge density around the metals W and Fe.

*Tri*ferrocenyl complexes of transition metals are still rare. In addition to the triferrocenyl-tungsten(VI) compounds [175, 176] derived from $WOCl_4$, a few intensely colored half-sandwich *tri*ferrocenyl compounds of the group 4 metals have been obtained [172] by treating $CpMCl_3$ (M = Ti, Zr) and Cp^*HfCl_3 with three equivalents of Fc-Li in toluene solution (Table 5-2). In contrast to $CpTiCl_3$, the ring-

Table 5-2. Cyclopentadienyl-containing metal(VI) complexes of the group 4 and 5 transition elements with σ-ferrocenyl (Fc) ligands

Complex	Color	Ref.
$CpTiFc_3$	black-blue	[172]
$Cp^*Ti(Cl)Fc_2$	red-violet	[172]
$CpZrFc_3$	red-brown	[172]
Cp^*HfFc_3	orange	[172]
Cp_2TiFc_2	green	[173, 186][a]
$(MeC_5H_4)_2TiFc_2$	dark green	[172]
Cp_2ZrFc_2	red	[173]
$Cp^*_2Zr(Cl)Fc$	dark red	[172][a]
$(tBuC_5H_4)_2Zr(OH)Fc$	orange	[187, 188]
$(tBuC_5H_4)_2Zr(X)Fc$ (X = OMe, OCH_2Ph, $OCH_2C_6H_3$-3,5-$(OMe)_2$, OC_6H_3-2,6-Me_2 and OC_6H_3-3,5-Me_2, OC(O)Ph)	orange to red	[188]
Cp_2HfFc_2	red	[173]
Cp_2VFc_2	dark brown	[172]
Cp_2NbFc_2		[174]

a X-Ray structure analysis.

permethylated analogue Cp^*TiCl_3 reacts only with two equivalents Fc-Li to give the diferrocenyl-chloro derivative $Cp^*Ti(Cl)Fc_2$ [172].

A series of *di*ferrocenyl complexes can be prepared starting from the dicyclopentadienyl dichlorides of the early (groups 4 and 5) transition metals (Table 5-2); the central metal always possesses the formal oxidation number +IV. Similar Cp_2M complexes of the group 4 metals containing substituted ferrocenyl ligands (see below) have also been obtained [189, 190]. Some diferrocenyl derivatives of gold(III) are known; examples are $Fc_2AuCl(PMe_3)$ [191, 192] and the dialkyldithiocarbamate complexes $Fc_2Au[SC(S)\text{-}NR_2]$ (R = Me [193, 194], Et [194]), which are prepared by oxidative addition of tetraalkyl thiuram disulfides, $[R_2N\text{-}C(S)S\text{-}]_2$, to the ferrocenyl-gold(I) complex Fc-$Au(PPh_3)$.

The formation of coordination compounds with a single σ-ferrocenyl ligand is frequently observed if the central metal is in a formally low oxidation state and neighboring ligands with acceptor properties are present. Prominent examples (Fig. 5-10) are the pentacarbonyls $M(CO)_5$-Fc (M = Mn [179], Re [182]), the 1,5-cyclooctadiene platinum(II) complex $(C_8H_{12})Pt(Cl)Fc$ and its triphenylphosphine derivatives, *cis*- and *trans*-$[(PPh_3)_2Pt(Cl)Fc]$, which were studied by ^{31}P NMR spectroscopy in solution [185], and the triphenylphosphine-gold(I) complexes Fc-$Au(PPh_3)$ [192, 195, 196, 200] and $[Fc(AuPPh_3)_2]^+$ [183, 192, 196, 197, 200, 210]. The first actinoide complex of this type was Fc-UCp_3 [211].

Half-sandwich complexes of the group 6 and 8 transition metals carrying one σ-ferrocenyl ligand are presented in Table 5-3.

Chemical or electrochemical oxidation of σ-ferrocenyl complexes generally leads to the corresponding σ-ferrocenium cations in a reversible electron transfer process. Oxidation of $Mn(CO)_5$-Fc by iodine produces a black salt $[Mn(CO)_5\text{-}Fc^+]_2(I_8^{2-})$,

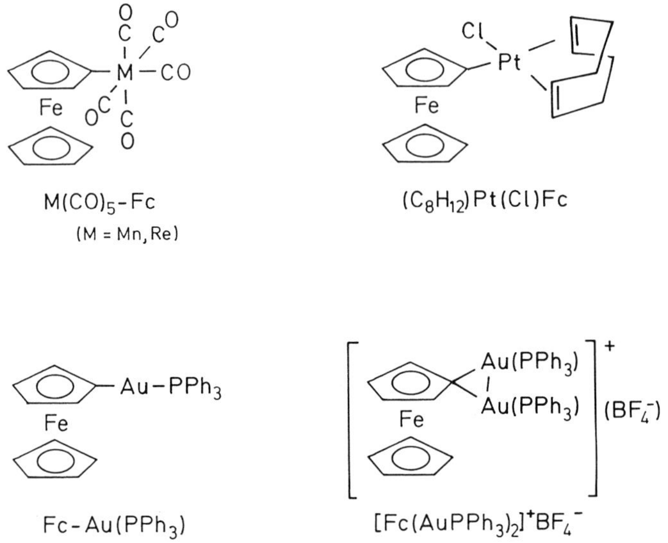

Fig. 5-10. Transition metal complexes containing one σ-ferrocenyl ligand.

Table 5-3. Half-sandwich complexes of group 6 and 8 transition elements with a σ-ferrocenyl (Fc) ligand

Complex	Color	Ref.
$CpMo(CO)_3Fc$		[184]
$CpW(CO)_3Fc$	orange	[178, 184]
$CpW(NO)_2Fc$	green	[177][a]
$CpWO_2Fc$	red	[177]
$CpFe(CO)_2Fc$	orange	[178, 181, 184]
$CpFe(CO)(PPh_3)Fc$	red	[181]
$CpRu(CO)_2Fc$	orange	[180]
$CpRu(CO)(L)Fc$	orange	[180]
(L = PMe_3, PPh_3, CN*t*Bu)		
$CpRu(CNtBu)_2Fc$	orange	[180]
$Cp^*Ru(CO)_2Fc$	orange	[180]
$Cp^*Ru(CO)(L)Fc$		[180]
(L = PMe_3, PPh_3, CN*t*Bu)		
$Cp^*Ru(CNtBu)_2Fc$	orange	[180]
$Cp^*Ru(PMe_3)_2Fc$	red	[198][a]
$(C_6Me_6)Ru(Cl)(CO)Fc$	red	[180]
$(C_6Me_6)Ru(Cl)(R)Fc$	orange-yellow	[180]
(R = *n*Bu, *p*-tolyl)		
$[CpRu(CO)_2Fc]^+BF_4^-$	black-red	[180]
$[Cp^*Ru(CO)_2Fc]^+BF_4^-$	black-violet	[180]
$[(C_6Me_6)Ru(Cl)(CO)Fc]^+BF_4^-$	black-violet	[180]

a X-Ray structure analysis.

which contains $[I_3(I_2)I_3]^{2-}$ dianions [179]. Oxidants such as $NO^+BF_4^-$ (in CH_2Cl_2) and $Ag^+BF_4^-$ (in THF) have been used to prepare the paramagnetic tetrafluoroborate salts $[Mn(CO)_5\text{-}Fc^+]BF_4^-$ [179], $[(C_5R_5)Ru(CO)_2\text{-}Fc^+]BF_4^-$ (R = H, Me) and $[(C_6R_6)Ru(Cl)(CO)Fc^+]BF_4^-$ (R = Me) (Table 5-3) [180]; reduction of the BF_4^- salts by $Na_2S_2O_3$ in THF/H_2O solution regenerates the neutral ferrocenyl complexes. Various other chemical (e.g., by *p*-quinone [182]) and electrochemical oxidations can be applied. Cyclovoltammetric studies involving the Cp and Cp* ruthenium carbonyl complexes (given in Table 5-3) have indicated that the first (reversible) oxidation potentials, $E_{1/2}$, correlate in a first approximation with the stretching frequency, $\nu(CO)$, of the carbonyl ligands [180], i.e., both $E_{1/2}$ and $\nu(CO)$ reflect the charge density in the half-sandwich ruthenium compounds.

It is interesting to note that an increasing number of ring-substituted ferrocenyl ligands are being used in transition metal complexes. The substituents may simply reduce the electron-donating power of the electron-rich σ-ferrocenyl unit, as in 1'-chloro- [199, 200], 2-chloro- [183], and 2,1'-dichloro- [190] ferrocenyl compounds (Scheme 5-27).

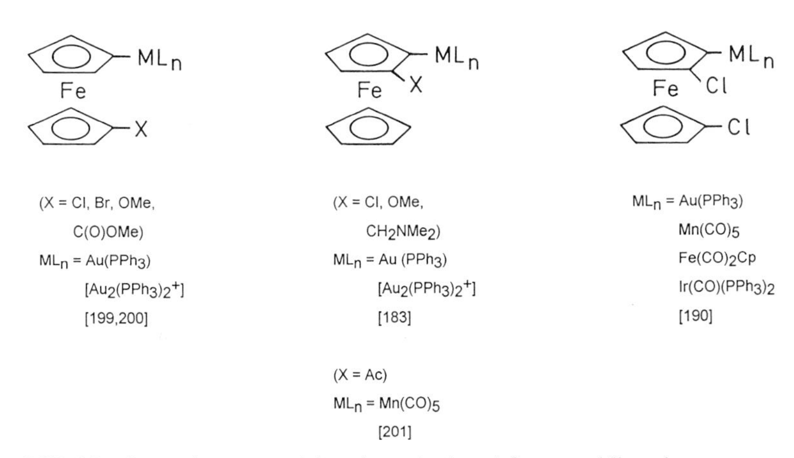

Scheme 5-27. Metal complexes containing ring-substituted ferrocenyl ligands.

However, if the additional substituent is able to coordinate to the metal, the substituted σ-ferrocenyl group (Fc') may become a potential bidentate-chelating ligand. Thus, 2-acetyl-ferrocenyl can behave both as a monodentate (in $Mn(CO)_5\text{-}Fc'$) and as a bidentate-chelating ligand (in $M(CO)_4(Fc')$, M = Mn, Re) [201]. The metal-carbon σ-bond is clearly stronger than the additional donor bond from the carbonyl oxygen of the acetyl group (Scheme 5-28).

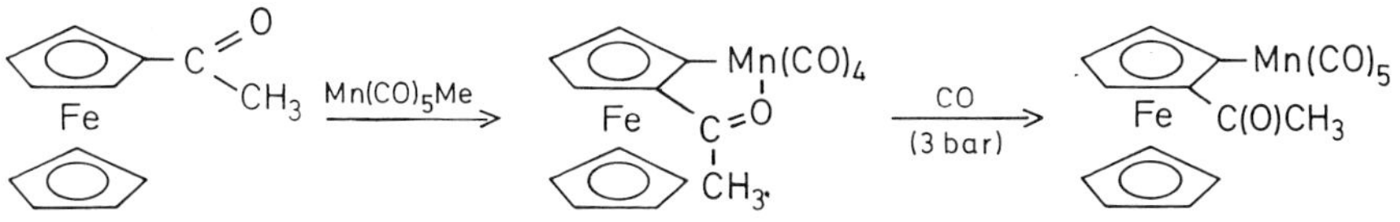

Fe C(=O)CH3 —Mn(CO)5Me→ Fe Mn(CO)4 C=O CH3 —CO (3 bar)→ Fe Mn(CO)5 C(O)CH3

Scheme 5-28. Carbonylmanganese complexes containing the 2-acetyl-ferrocenyl ligand [201].

In a similar manner, 2-dimethylaminomethyl-ferrocenyl (Fc′) may function both as a monodentate σ-ferrocenyl ligand (as in $Cp_2M(Fc')_2$ (M = Ti, Zr, Hf [189], $Cp^*_2Ti(Cl)(Fc')$ [189], or $(Fc')\text{-}Au(PPh_3)$ [183]) and, more frequently, as a bidentate-chelating ligand (as in e.g., $Re(CO)_4(Fc')$ [202], $[(Fc')PdCl]_2$ and its derivatives [203], $Pd(Fc')_2$, $(Fc')Pt(Cl)(PPh_3)$ [204] and $M(Fc')_2$ (M = Zn, Cd, Hg) [189]). Other potential bidentate chelate ligands are 1′-diphenylphosphino-ferrocenyl [205] and 1′-diphenylarsino-ferrocenyl [206], e.g., Scheme 5-29.

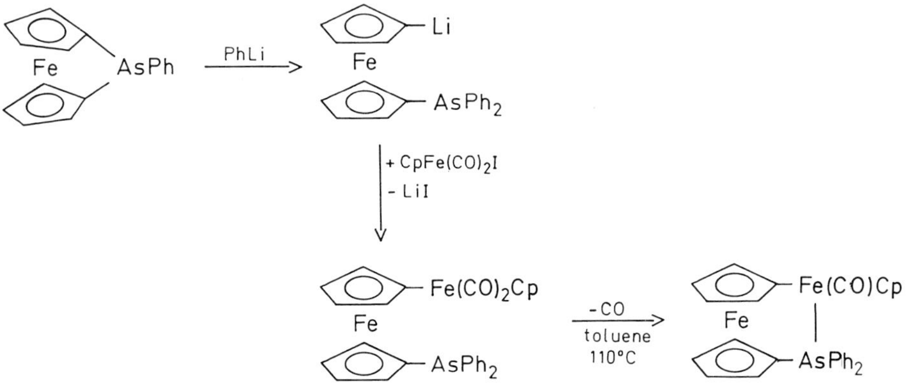

Fe AsPh —PhLi→ Li / Fe / $AsPh_2$

+ $CpFe(CO)_2I$
- LiI

$Fe(CO)_2Cp$ / Fe / $AsPh_2$ —(-CO, toluene, 110°C)→ Fe(CO)Cp / Fe / $AsPh_2$

Scheme 5-29. Cyclopentadienyl carbonyliron complexes containing the 1′-diphenylarsino-ferrocenyl ligand [206].

Coordination compounds containing transition metals directly attached to a 1,1′-ferrocenediyl unit are also well known. In general, the ferrocene sandwich acts as a bridge; examples are given in Scheme 5-30 (see Appendix).

ML_n	X		ML_n	
$Cr(CO)_5^-$	PPh_2Me^+	[207]	$Ti(NEt_2)_3$	[49,208]
$W(CO)_5^-$	PPh_2Me^+	[207]	VCp_2	[209]
$Fe(CO)_2Cp$	PPh_2	[205]	$Ru(CO)_2Cp$	[180]
$Fe(CO)_2Cp$	$AsPh_2$	[206]	$Ru(CO)_2Cp^*$	[180]
$Au(PPh_3)$	Cl, Br, OMe C(O)OMe	[199,200]	$Au(PPh_3)$	[210]
$[Au_2(PPh_3)_2]^+$	Cl, Br, OMe C(O)OMe	[199,200]	$[Au_2(PPh_3)_2{}^+]$	[210]
			$PtCl(C_8H_{12})$	[185,cf.285]
			UCp_3	[211]

Scheme 5-30. 1,1′-Ferrocenediyl compounds containing transition metal complex fragments.

On the other hand, transition metal compounds with chelating 1,1′-ferrocenylene, i.e., 1-metalla-[1]ferrocenophanes, are rare, the best-known representatives being the group 4 metal complexes (Fig. 5-11) derived from the di(cyclopentadienyl) metal dichlorides, fc[MCp_2] and fc[$M(C_5H_4tBu)_2$] (M = Ti, Zr, Hf) [188, 212].

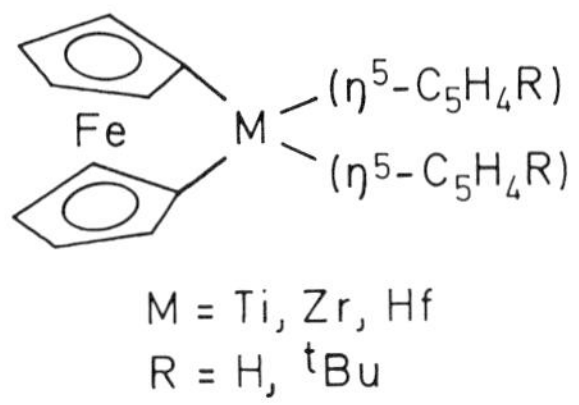

Fe M (η^5-C_5H_4R) (η^5-C_5H_4R)

M = Ti, Zr, Hf
R = H, ^{t}Bu

Fig. 5-11. 1-Metalla-[1]ferrocenophanes [188, 212].

5.6 Compounds with Ferrocenyl Chalcogenate and 1,1′-Ferrocenylene Dichalcogenate Groups

The analogy between phenyl thiolate ($Ph\text{-}S^-$) and ferrocenyl thiolate ($Fc\text{-}S^-$) anions leads to the expectation that many compounds containing ferrocenyl thiolate units as either substituents or ligands should exist. This is indeed the case and, although with fewer examples, is also true for ferrocenyl selenolate and tellurolate. Similarly, 1,1′-ferrocenylene dithiolate compounds (which correspond to *o*-phenylene dithiolates) are known, together with the analogous 1,1′-ferrocenylene diselenolates and ditellurolates.

The synthesis generally involves reaction of either the ferrocenyl chalcogenols and 1,1′-ferrocenylene dichalcogenols, Fc-EH and $fc(EH)_2$, or (preferably) the corresponding lithium derivatives, Fc-ELi and $fc(ELi)_2$, with halides such as acid chlorides, non-metal chlorides and chloro metal complexes (E = S, Se, Te).

Thus, all possible acyl derivatives of the three ferrocenyl chalcogenols and the three 1,1′-ferrocenylene dichalcogenols (E = S, Se, Te) have been obtained [89] by treating the lithio compounds with the carboxylic acid chlorides benzoyl, 2-thenoyl, and ferrocenoyl chloride (Scheme 5-31).

The chalcogeno esters (E = S, Se, Te) presented in Scheme 5-31 are yellow to red-brown compounds that are air-stable in the solid state [89]; the benzoic acid derivatives Fc-S-C(O)Ph [82] and $fc(E\text{-}C(O)Ph)_2$ (E = S, Se) [91] had been described earlier. The orange derivatives of the ferrocenyl carboxylic acid contain two or three metallocene units connected by ester links (Fig. 5-12). The corresponding esters of the alcohols Fc-OH and $fc(OH)_2$ (E = O) can be obtained [73] by direct ferrocen*o*ylation using Fc-C(O)Cl.

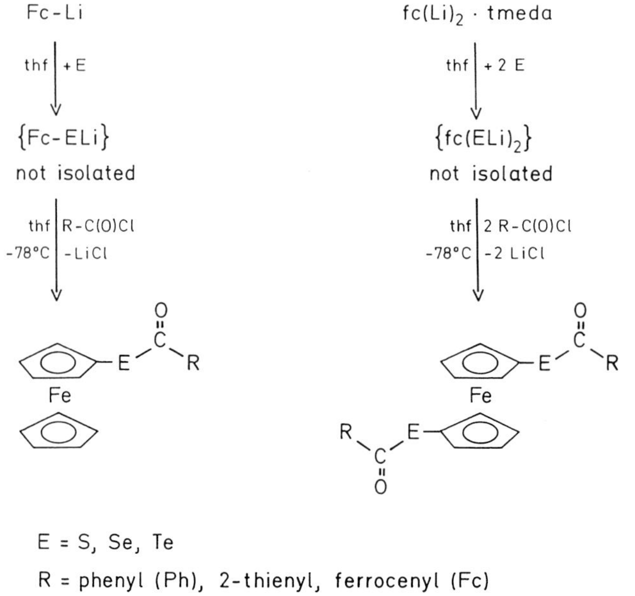

Fc-Li fc(Li)$_2$ · tmeda

thf | +E thf | +2 E

{Fc-ELi} not isolated {fc(ELi)$_2$} not isolated

thf, -78°C | R-C(O)Cl, -LiCl thf, -78°C | 2 R-C(O)Cl, -2 LiCl

E = S, Se, Te

R = phenyl (Ph), 2-thienyl, ferrocenyl (Fc)

Scheme 5-31. Synthesis of thio-, seleno- and telluroesters of carboxylic acids [89].

Scheme 5-32 shows a series of ferrocenyl compounds that can be formally considered as derivatives of the ferrocenyl chalcogenols, Fc-EH (E = O, S, Se, Te), including ferrocenyl ethers, thioethers, selenoethers and telluroethers. The analogous 1,1′-ferrocenediyl compounds, formally derived from fc(EH)$_2$ (E = O, S, Se, Te), are presented in Scheme 5-33 (see Appendix).

Among the 1,1′-disubstituted ferrocene derivatives, the bifunctional thio- and selenoethers, fc(SMe)$_2$ and fc(SeMe)$_2$, have found particular attention as chelating ligands (Scheme 5-34) [32, 226–230], being comparable in this respect with 1,1′-bis(diphenylphosphinyl) ferrocene, fc(PPh$_2$)$_2$ (cf. Chap. 1).

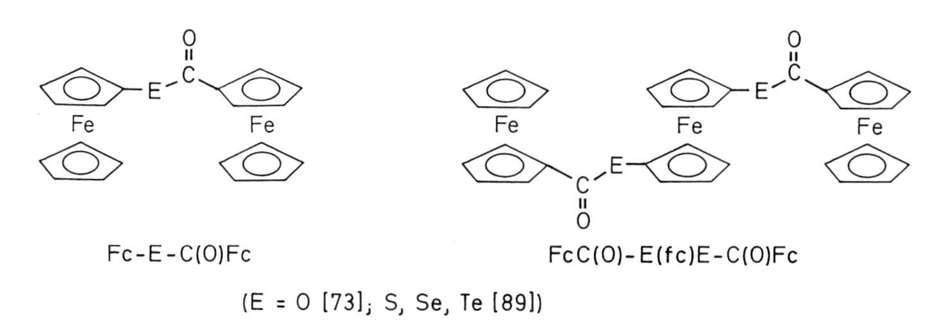

Fig. 5-12. Di- and trinuclear ferrocene derivatives containing ester links.

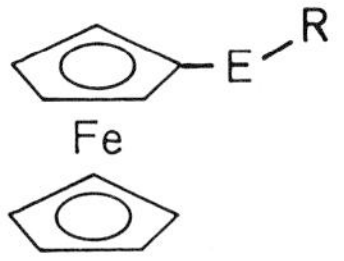

Fe

E = O; R = Me [16,21,213], Et [221], Ph [47,214]
$CH_2CH{=}CH_2$ [16,22]
$SiMe_3$ and $SnMe_3$ [73]

E = S; R = Me [47,81,85], Ph [41,48,225]
C_6H_4-Me-4 [48,215]
$C_6H_2(Me)_3$-2,4,6 [216]
CN [47,85,147]
C(S)-NMe_2 [31]
$SnMe_3$ [217]

E = Se; R = Me [218], ^nBu [148]
Ph [188]
CN [148,219], $SnMe_3$ [217]

E = Te; R = ^nBu [102]
C_6H_4-OMe-2 and -4 [102]
$SnMe_3$ [217]

Scheme 5-32. Compounds containing a ferrocenyl chalcogenate group, Fc-E (E = O, S, Se, Te).

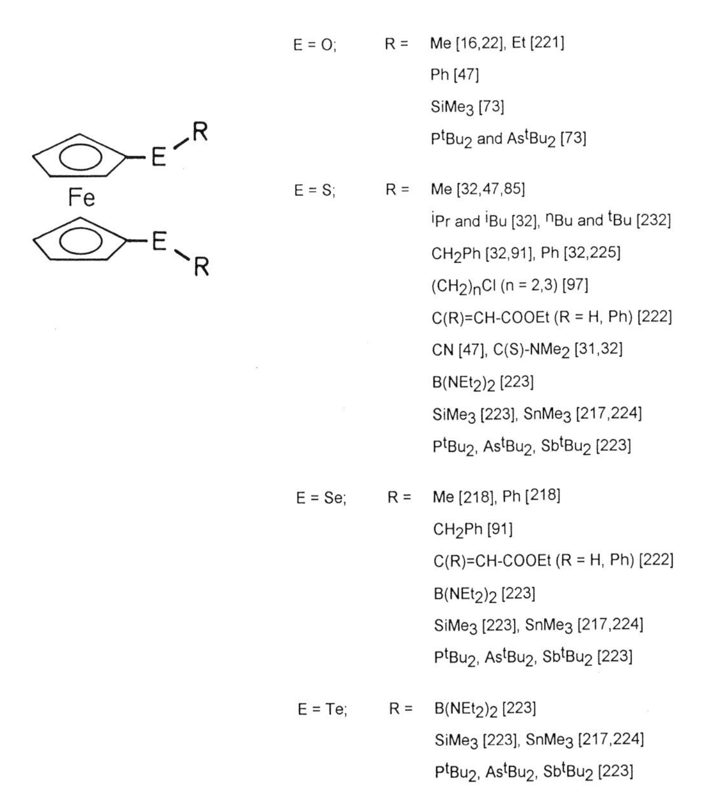

Scheme 5-33. Compounds containing a 1,1′-ferrocenylene dichalcogenate group, fcE_2 (E = O, S, Se, Te).

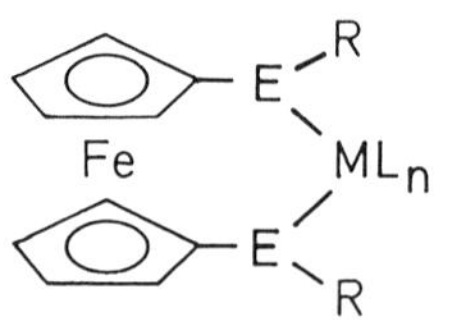

E = S and Se; R = Me

ML_n =		
$Cr(CO)_4$, $Mo(CO)_4$, $W(CO)_4$,		[226]
$Re(CO)_3X$ (X = Cl, Br, I)		[227]
$PtCl_2$		[228]
$Pt(Me)_3X$ (X = Cl, Br, I)		[229]

E = S; R = Me, ^{i}Pr, ^{i}Bu, Ph

ML_n = PdX_2 and PtX_2 (X = Cl, Br) [32,230]

Scheme 5-34. Metal complexes containing the chelate ligands 1,1′-bis(organylthiolato)- and 1,1′-bis(organylselenolato) ferrocene, $fc(SR)_2$ and $fc(SeR)_2$.

The chelate complexes (Scheme 5-34) are fluxional molecules in solution, as shown by variable-temperature NMR spectroscopy (^{1}H, ^{13}C, ^{77}Se, ^{195}Pt). The non-rigidity can be ascribed to two processes: pyramidal inversion at the chalcogen atoms (E = S, Se) at higher temperature, and bridge reversal involving the central metal M [32, 227, 229, 230]. The cationic palladium complexes $[fc(ER)_2Pd(PPh_3)]^{2+}$ (Scheme 5-35) are assumed to contain a weak iron-to-metal dative bond [231, 232] (see Appendix).

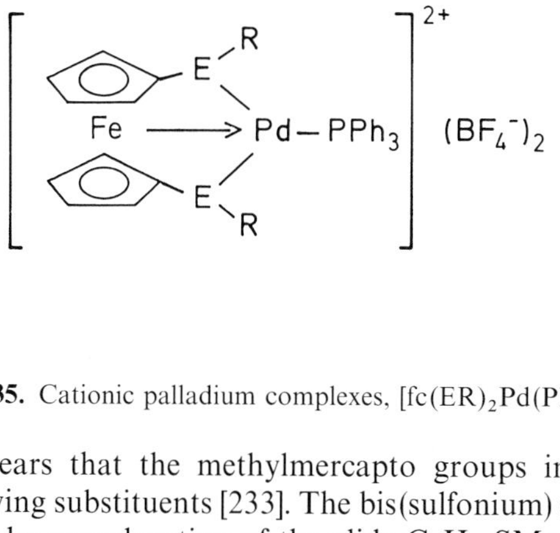

$[fc(ER)_2Pd(PPh_3)]^{2+}$ $(BF_4^-)_2$

E	R	Ref.
O:	R = Me	[232]
S:	R = Me, ^{i}Pr, ^{i}Bu, CH$_2$Ph, Ph	[231,232]
	^{n}Bu	[232]
Se:	R = Me, Ph	[232]

Scheme 5-35. Cationic palladium complexes, $[fc(ER)_2Pd(PPh_3)]^{2+}$.

It appears that the methylmercapto groups in $fc(SMe)_2$ behave as electron-withdrawing substituents [233]. The bis(sulfonium) salt $[fc(SMe_2^+)_2]FeCl_4^{2-}$ has been obtained by complexation of the ylide C_5H_4-SMe_2 with $FeCl_2$ in THF [234].

Main group element derivatives containing three or four ferrocenyl thiolate "ligands" are accessible from the corresponding element chlorides and Fc-SLi. Thus, the reaction of PCl_3 with a suspension of 3 equivalents Fc-SLi in toluene – THF (3:1) leads to tris(ferrocenyl thiolato) phosphine, $P(SFc)_3$; the oxide and sulfide, $P(O)(SFc)_3$ and $P(S)(SFc)_3$, are formed as side-products in addition to Fc-SS-Fc. The corresponding tris(ferrocenyl thiolato) element compounds $As(SFc)_3$ and $Sb(SFc)_3$ are also known; the antimony derivative decomposes under the influence of either air or light [235].

A comprehensive study of tin(IV) compounds containing ferrocenyl chalcogenate (FcE) and 1,1′-ferrocenylene dichalcogenate (fcE_2) ligands (E = S, Se, Te) has been carried out with particular emphasis on ^{1}H, ^{13}C, ^{119}Sn, ^{77}Se and ^{125}Te NMR spectroscopy [217]. In a formal consideration, all complexes are derived from tetramethylstannane, $SnMe_4$, by stepwise substitution of methyl groups by either FcE or fcE_2 ligands. Starting from lithioferrocene, the products are tetrasubstituted stannanes of the type $Me_{4-n}Sn(EFc)_n$ (Scheme 5-36).

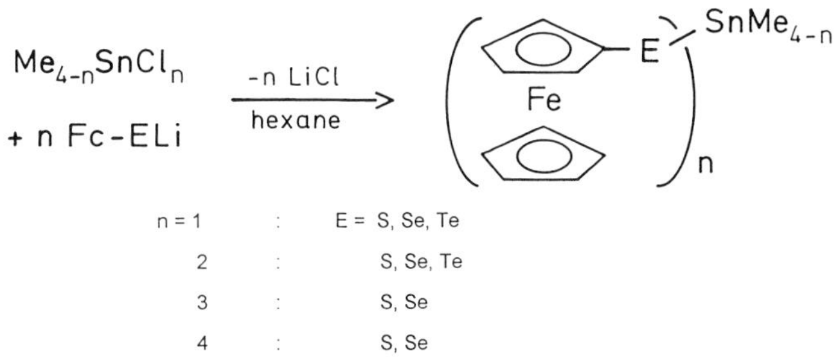

$Me_{4-n}SnCl_n$ + n Fc-ELi —(−n LiCl, hexane)→ $(FcE)_n SnMe_{4-n}$

n = 1 : E = S, Se, Te
2 : S, Se, Te
3 : S, Se
4 : S, Se

Scheme 5-36. Tin(IV) compounds with ferrocenyl chalcogenate ligands [217].

Starting from 1,1′-dilithioferrocene, the products can be 1,1′-disubstituted ferrocenes, $fc(E\text{-}SnMe_3)_2$ (E = S, Se, Te) (Scheme 5-33), but 1,3-dichalcogena-2-stanna-[3]ferrocenophane rings are formed whenever possible (Scheme 5-37).

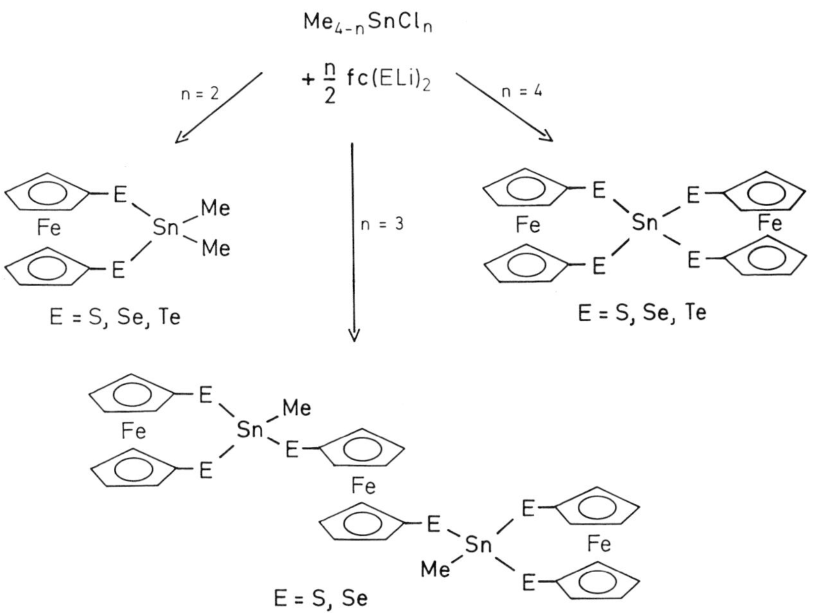

$Me_{4-n}SnCl_n$ + $\frac{n}{2}$ $fc(ELi)_2$

n = 2: E = S, Se, Te
n = 3: E = S, Se
n = 4: E = S, Se, Te

Scheme 5-37. Tin(IV) compounds with 1,1′-ferrocenylene dichalcogenate ligands [217].

Whereas all (eight) possible sulfur and (eight) selenium complexes have been prepared and unequivocally characterized [217], some of the tellurium-rich compounds could not be isolated due to the ready formation of either Fc-TeTe-Fc or $fc(Te_3)$.

It is surprising that transition metal complexes containing ferrocenyl thiolate ligands (FcS) have not been studied in detail so far. The only examples are gold complexes such as (Fc-S)Au(PPh_3) [86], [(FcS)Au_2(PPh_3)$_2$](BF_4) [86, 183], [(Ph_3PAu-S)fc-S-]$_2$ (Scheme 5-18) [90], and [(FcS)Au(nBu)$_2$]$_3$ [86].

On the other hand, the coordination chemistry of 1,1′-ferrocenylene dithiolate ligands (fcS_2) and their heavier analogs (fcE_2, E = Se, Te) has been the subject of numerous investigations. The reactions of 1,1′-dimercaptoferrocene, fc(SH)$_2$, with monochloro complexes of the type CpM(NO)$_2$Cl (M = Cr, Mo, W) and CpW(CO)$_3$Cl in the presence of triethylamine (NEt_3) proceed stepwise to give first the monothiol-thiolato and then the bis(thiolato) complexes, e.g., Scheme 5-38 [236].

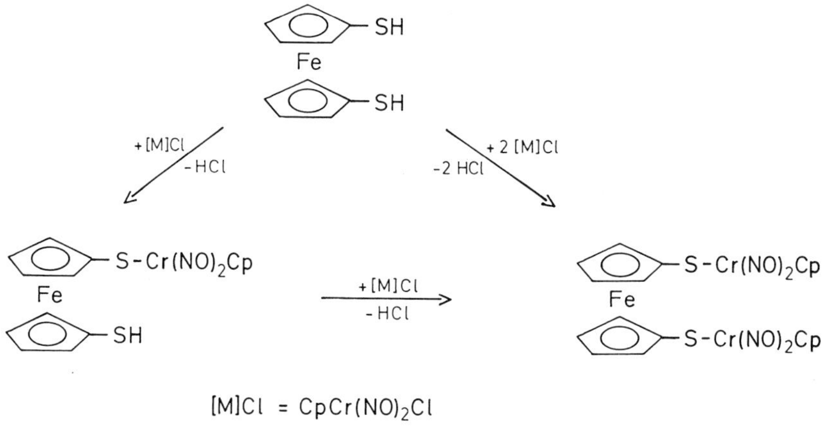

SH
Fe
SH
+ [M]Cl, − HCl
+ 2 [M]Cl, −2 HCl
S-Cr(NO)$_2$Cp
Fe
SH
+ [M]Cl, − HCl
S-Cr(NO)$_2$Cp
Fe
S-Cr(NO)$_2$Cp

[M]Cl = CpCr(NO)$_2$Cl

Scheme 5-38. Half-sandwich dinitrosylchromium complexes derived from 1,1′-dimercaptoferrocene [236].

Instead of CpW(CO)$_3$Cl, the more reactive salt [CpW(CO)$_3$(CH_2Cl_2)]PF_6 can be used for the reaction with fc(SH)$_2$. The carbonylmolybdenum complexes (obtained using CpMo(CO)$_3$F-PF_5) are only stable at low temperature and easily decompose to [CpMo(CO)$_3$]$_2$ under ambient conditions. Scheme 5-39 summarizes the products that have been characterized.

ML_n =	Fe(C$_5$H$_4$-S-ML_n)(C$_5$H$_4$-SH)	Fe(C$_5$H$_4$-S-ML_n)$_2$	
M(NO)$_2$Cp	M = Cr, Mo	M = Cr, Mo, W	[236]
M(CO)$_3$Cp	M = Mo, W	M = Mo, W	[236]
M(Cl)(NO)[HB(pz*)$_3$]		M = Mo	[237]
M(CO)$_2$Cp*		M = Ru	[238]

Scheme 5-39. Ferrocenyl thiolate and 1,1′-ferrocenylene dithiolate derivatives of half-sandwich transition metal complexes.

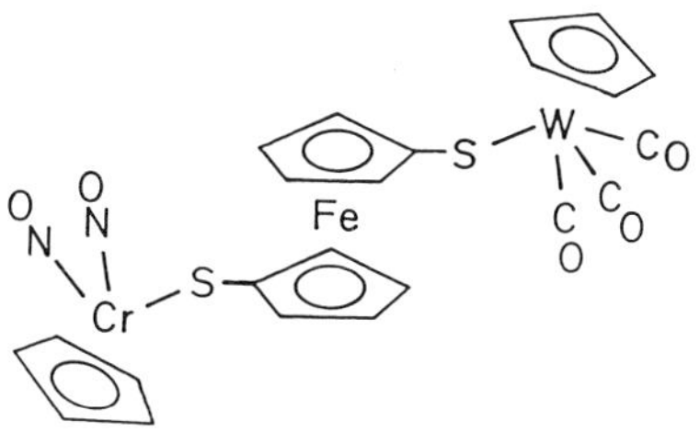

Fig. 5-13. The trimetallic complex $Cp(NO)_2Cr-[S(fc)S]-W(CO)_3Cp$ [236].

It is remarkable that the voluminous half-sandwich complex fragments $[M(NO)_2Cp]$ (M = Cr, Mo) and $[M(CO)_3Cp]$ (M = Mo, W) are able to protect the otherwise oxidation-sensitive SH ligand in the 1′ position [236], whereas $[Au(PPh_3)]$ is unable to avoid formation of a disulfide link [90]. By stepwise introduction of different half-sandwich units, a complex containing three metals (Cr, Fe, W) has been prepared (Fig. 5-13) [236].

The [S(fc)S] unit functions as a bridge between the two half-sandwich moieties. As expected, the ferrocene-derived thiolato ligands fcS_2 and (HS)fcS are stronger electron-donors than phenyl- and methyl thiolato ligands [236].

A different type of a 1,1′-ferrocenylene dithiolato bridge is observed in the hexacarbonyl dimetal complexes $(fcS_2)Fe_2(CO)_6$ [239], $(fcS_2)Ru_2(CO)_6$ [240], and $(fcS_2)Os_2(CO)_6(\mu\text{-S})$ [240], which are formed in the reaction of 1,2,3-trithia-[3]ferrocenophane, $fc(S_3)$, with the trinuclear carbonylmetal complexes $M_3(CO)_{12}$ (M = Fe, Ru, Os) (Scheme 5-40). In these cases, the sulfur atoms become part of an M_2S_2 tetrahedrane structure. The ruthenium complex has been characterized by an X-ray structure analysis [240], which indicated an Ru-Ru single bond (268 pm) and nearly parallel cyclopentadienyl rings with only a small tilt of 5.38°. The sulfur atoms are slightly displaced outwards from the corresponding C_5H_4 ring planes, and the rings assume an eclipsed conformation.

M = Fe [239], Ru [240]

[240]

Scheme 5-40. Hexacarbonyl dimetal complexes containing a 1,1′-ferrocenylene dithiolate bridge.

Oligonuclear clusters of osmium, such as $(fcS_2)Os_2(CO)_8$, $(fcS_2)Os_3(CO)_{10}$, $(fcS_2)Os_4(CO)_{10}(\mu_3\text{-}S)_2$ and $(fcS_2)Os_4(CO)_{11}(\mu_3\text{-}S)$, are slowly formed in the reaction of $fc(S_3)$ with $Os_3(CO)_{12}$ in refluxing toluene [240].

A tetrahedrane-type core geometry can also be ascribed to the violet rhodium complexes $Cp^*_2Rh_2(E_2fc)$ (E = S, Se, Te) obtained from the reaction of $fc(ELi)_2$ with $Cp^*_2Rh_2Cl_4$ in THF [267]. The binuclear fragment $[Cp^*_2Rh_2]$ is isoelectronic and isolobal with $[Ru_2(CO)_6]$.

5.7 [3]Ferrocenophanes Containing Chalcogens

The smallest rings that are able to incorporate a ferrocene building block without deformation of the sandwich structure (with its two planar and parallel cyclopentadienyl rings) are [3]ferrocenophanes. This becomes evident in the series of sila-ferrocenophanes $fc[(SiMe_2)_n]$ (n = 1, 2, 3) (Scheme 5-41) [243].

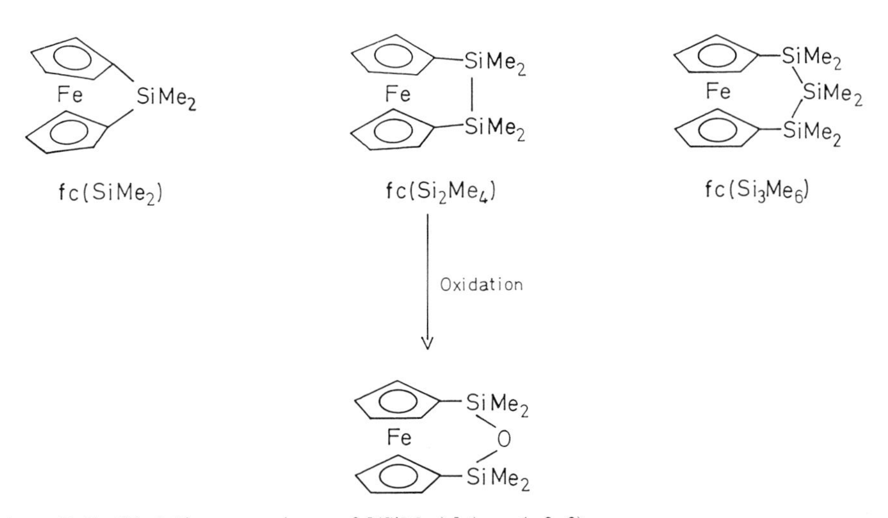

Scheme 5-41. Sila-[n]ferrocenophanes, $fc[(SiMe_2)_n]$ (n = 1, 2, 3).

According to the X-ray crystallographic structure determinations, the two cyclopentadienyl rings in 1-dimethylsila-[1]ferrocenophane, $fc(SiMe_2)$ [164, 165], are inclined towards the $SiMe_2$ group to include an angle of 20.8(5)°, whereas in 1,2-bis(dimethylsila)-[2]ferrocenophane, $fc(Si_2Me_4)$ [241, 242], the corresponding tilt angle is reduced to almost 4° (cf. 4.19(2)° [165], 4.3° [243], respectively); however, the carbon analog $fc(C_2Me_4)$ still holds an angle of 23° [244]. In contrast to the [1]ferrocenophane $fc(SiMe_2)$, which readily polymerizes to give sila-ferrocenylene polymers (e.g., in the melt above 130 °C [165]), the [2]ferrocenophane $fc(Si_2Me_4)$ requires a temperature above 350 °C to undergo ring-opening polymerization [165, 243]. It is easy to oxidize $fc(Si_2Me_4)$ to 1,3-bis(dimethylsila)-2-oxa-[3]ferrocenophane, fc[$(SiMe_2)_2O$], e.g., by photooxidation in methanol under O_2 in quartz tubes [243] or by acid-catalyzed alcoholysis (EtOH/HCl) [245]. The cyclovoltammetric (CV) oxidation is irreversible in the case of fc($SiMe_2$) and reversible for only a few cycles in the case of fc(Si_2Me_4), although fc(Si_3Me_6) [241] is perfectly stable on reversible CV redox cycling [243].

The most intensely studied mononuclear ferrocenophane rings are the [3]ferrocenophanes, fc(E_3), which contain three chalcogens E in the bridge (E = S, Se, Te). [1]Ferrocenophanes with a single chalcogen atom are unknown. [2]Ferrocenophanes containing a chalcogen chain link are rare, two examples that can be mentioned (Scheme 5-42) are the ferrocenylene thiazine 1,1-dioxide [246, 247], and the products [248] that are formed by insertion of chalcogen (E = S, Se, Te) into the [1]ferrocenophane fc[$Zr(C_5H_4tBu)_2$] (cf. Fig. 5-11).

The tilt angle between the two cyclopentadienyl rings in fc[$S(O)_2NH$] is as large as 23° [247].

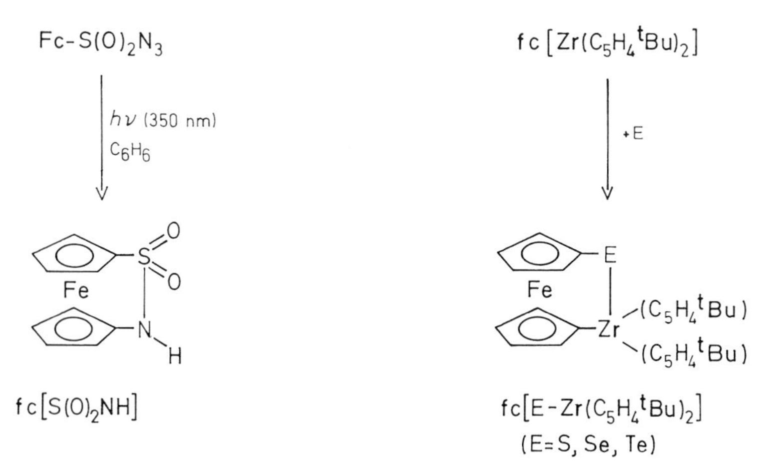

Scheme 5-42. [2]Ferrocenophanes containing a chalcogen chain link.

5.7.1 Compounds with One 1,3-Dichalcogena-[3]ferrocenophane Ring

The first 1,2,3-trichalcogena-[3]ferrocenophane, fc(S_3), was described [92] in 1969; it is widely used as the stable precursor for 1,1′-dimercaptoferrocene and its derivatives (cf. Sect. 5.3.3). The corresponding [3]ferrocenophanes of selenium and tellurium, fc(Se_3) [249] and fc(Te_3) [250], were announced in 1979 and 1990, respectively. The general method of preparation involves insertion of the chalcogen E into the carbon-lithium bonds of fcLi_2, and subsequent oxidation of the dilithium 1,1′-ferrocenylene dichalcogenate, fc$(ELi)_2$, by the chalcogen E (E = S, Se, Te) [29, 91, 251]. Today, the complete series of all nine 1,2,3-trichalcogena-[3]ferrocenophanes, fc(E_2E') (E, E′ = S, Se, Te) is available [251, 258] (cf. Fig. 5-15), although the oxa complexes "(fc(O_2E')", E′ = O, S, Se, Te) are still unknown. (The

Fig. 5-14. 1,2,3-Tritellura-[3]ferrocenophane, fc(Te_3) (two representations).

related 1,2,3-tri(phenylphospha)-[3]ferrocenophane, fc[$(PPh)_3$], has recently been prepared [220].) In the solid state, the 1,2,3-trichalcogena-[3]ferrocenophanes form a lattice of molecules that contain two almost coplanar and eclipsed cyclopentadienyl rings; two representations of fc(Te_3) [250, 251] are given in Fig. 5-14.

According to the X-ray structure determinations carried out for fc(S_3) [252], fc(S_2Se) [103], fc(Te_3) [250], and for the analogous [3]ruthenocenophane rc(Se_3) [253] and [3]osmocenophane oc(S_3) [280], the displacement of the ring-attached chalcogens E (in 1,1′ positions) from their cyclopentadienyl ring plane is small (the maximum being 22 pm in fc(Te_3) [250]), and the deviation of the rings from an exact parallel arrangement ("ring canting") is also minute, being largest in the cases of fc(Te_3) (4.6 – 5.0° [250]) and oc(S_3) (4.74° [280]).

In solution, the 1,2,3-trichalcogena-[3]ferrocenophanes are non-rigid molecules [92, 254 – 258]. A systematic study of the temperature-dependent ^1H and ^{13}C NMR spectra by Abel and coworkers [256 – 258] has confirmed a hindered intramolecular motion that is generally interpreted in a simplified manner as an inversion of the trichalcogen bridge. However, recent calculations have indicated that a transition state with a staggered conformation of the cyclopentadienyl rings might be

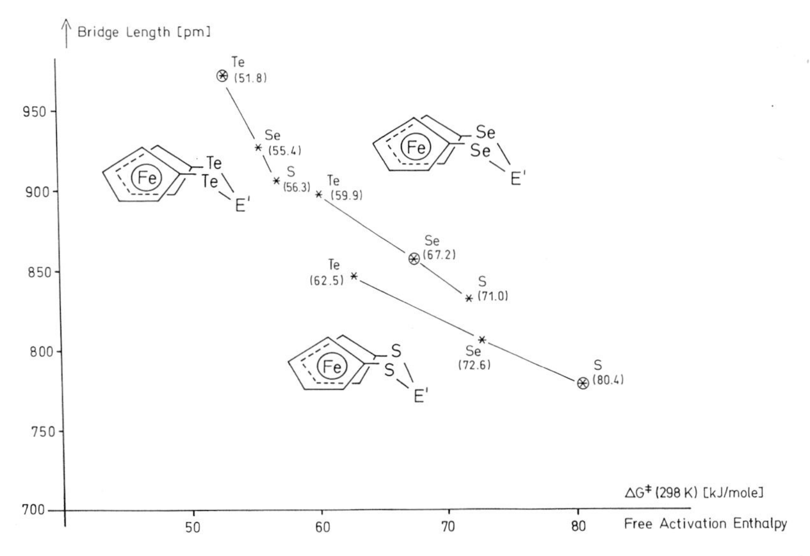

Fig. 5-15. Free activation enthalpy, ΔG^{+} (298 K), of the bridge-inversion rearrangement of 1,2,3-trichalcogena-[3]ferrocenophanes as a function of the bridge length [256, 258].

energetically preferable to a transition state in which the EE′E bridge becomes coplanar with the iron atom [258]. The free activation enthalpy, ΔG^{+}, of the "bridge-inversion" rearrangement decreases as the size of the chalcogen atoms in the bridge increases, e.g., in the order fc(S_3) (ΔG^{+} (298 K) = 80.4 ± 0.2) [256], fc(Se_3) (67.2 ± 0.1) [256]), fc(Te_3) (51.8 ± 0.2 kJ/mol) [258]; the mixed complexes fc(E_2E') assume intermediate values. The ΔG^{+} (298 K) values are summarized in Fig. 5-15 with respect to the bridge length, as expressed by the sum of the covalent radii.

A large number of 1,3-dichalcogena-[3]ferrocenophanes, fc(E_2E'), has been synthesized in which the central position of the triatomic bridge (E′) may vary within broad limits. In general, the 1,1′-dilithioferrocene intermediate, fcLi_2 · (tmeda), is simply treated with the appropriate dihalide, $E'X_2$, although the 1,1′-ferrocenylene dichalcogenols, fc$(EH)_2$, or the 1,1′-bis(trimethylstannyl-chalcogenato)ferrocenes, fc$(E\text{-}SnMe_3)_2$ [217] may have advantages as educts. All three routes can be used to prepare the nine 1,3-dichalcogena-2-(*tert*-butyl)pnicogena-[3]ferrocenophanes, fc(E_2Y*t*Bu) (E = S, Se, Te; E′ = Y*t*Bu = P*t*Bu, As*t*Bu, Sb*t*Bu) (Scheme 5-43) [223, 224]; the analogous three 1,3-dioxa compounds (E = O) were obtained from 1,3-dihydroxyferrocene, fc$(OH)_2$ [80].

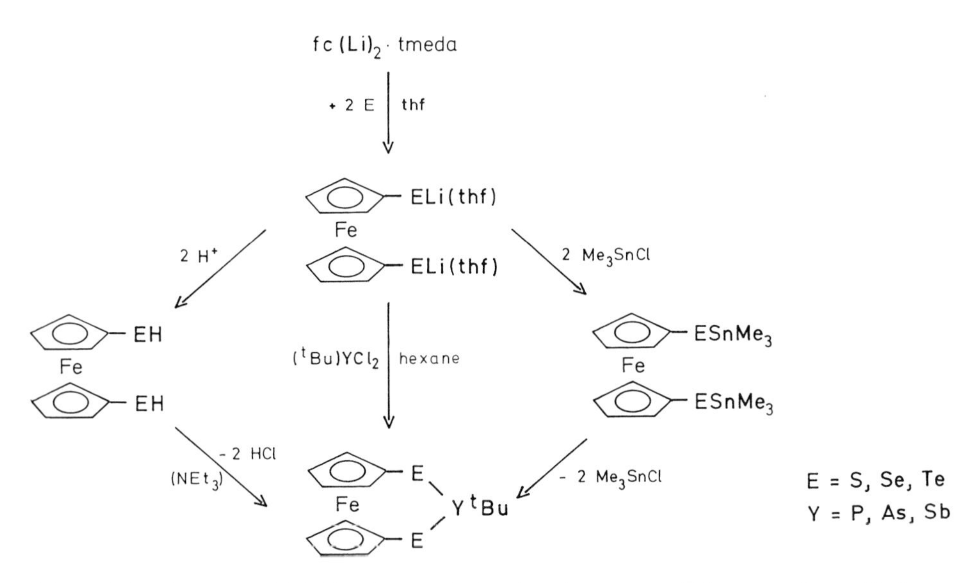

Scheme 5-43. Synthesis of 1,3-dichalcogena-2-(*tert*-butyl)pnicogena-[3]ferrocenophanes [224].

Table 5-4 summarizes the known 1,3-dichalcogena-[3]ferrocenophanes that contain an element of the main groups 13, 14 or 15 between the two ring-attached chalcogen atoms. Rapid configurational interchange appears to take place in solution at room temperature in the case of the methylene compounds, fc(E_2CH_2) (E = S [254, 257], Se [260], Te [251]); the activation barrier could only be determined for the sulfur-containing complex, fc(S_2CH_2) (ΔG^{+} (298 K) = 47.90 ± 0.23 kJ/mol) [257]. In a similar manner the triatomic bridge undergoes rapid inversion in fc(S_2E')

Table 5-4. 1,3-Dichalcogena-[3]ferrocenophanes, fc(E_2E'), containing elements of main groups 13, 14 and 15

E′	E			
BPh	O [80]			
BR (R = Me, Et, *n*Bu, C_6H_{13}, mesityl, N*i*Pr_2, N*i*PrEt)		S [259]		
B(NEt_2)		S [259]	Se [223, 224]	Te [223, 224]
CH_2		S [91, 251, 254, 257]	Se [91, 251, 260, 261]	Te [251]
CHMe		S [88][c]		
CHPh		S [254]		
CMe_2		S [254]		
CPh_2		S [254]	Se [260, 261]	
C=O	O [80][b]	S [80][a]		
C=S		S [101][d]		
Si(Cl)Me		S [223]	Se [223]	
SiMePh		S [260]	Se [260, 261]	
$SiMe_2$		S [254]	Se [260]	Te [224]
$GeCl_2$		S [262]	Se [262][a]	
$GeMe_2$		S [254]		
$SnCl_2$		S [262]	Se [262]	
$SnMe_2$		S [217, 254]	Se [217, 260]	Te [217]
$SnPh_2$			Se [260]	
Pb		S [260]	Se [260]	
PCl		S [223, 224]	Se [223, 224]	
PPh		S [263]	Se [261, 263]	
P(S)Ph		S [261, 263]	Se [261, 263]	
P*t*Bu	O [80]	S [224]	Se [224]	Te [224]
AsPh		S [263][a]	Se [263]	
As*t*Bu	O [80]	S [224]	Se [224]	Te [224]
Sb*t*Bu	O [80]	S [224]	Se [224]	Te [224]

a X-Ray structure analysis. b Scheme 5-14. c Scheme 5-17. d Scheme 5-22.

compounds, where E′ = CHMe [88], CMe_2 and CPh_2 [254], $SiMe_2$, $GeMe_2$ and $SnMe_2$ [254]). On the other hand, the ^{1}H and ^{13}C NMR data for the *tert*-butylpnicogena complexes, fc(E_2Y*t*Bu) (E = O, S, Se, Te; Y = P, As) are indicative of essentially rigid structures at room temperature in solution [80, 224].

With respect to intramolecular metal-metal interactions, a series of 1,3-dichalcogena-[3]ferrocenophanes has been investigated that contain a transition metal complex fragment in the center between the two ring-attached chalcogens (Table 5-5). Hetero-bimetallic complexes of this kind can be considered as either 1,3-dichal-

Table 5-5. 1,3-Dichalcogena-[3]ferrocenophanes, fc(E_2M), containing a transition metal complex fragment (M)

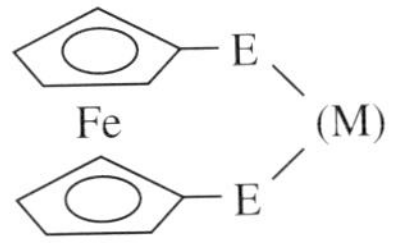

M	E	Ref.
$ZrCp_2$	Se	[264, 265]
$Hf(C_5H_4tBu)_2$	Se	[264, 265]
V(O)Cp	S, Se	[99]
V(O)Cp*	S, Se[a]	[99]
$Mo(NO)[HB(pz^*)_3]$	S[a]	[266]
Mo(NO)Cp*	S, Se	[267]
W(NO)Cp*	S, Se	[267]
Re(O)Cp*	S, Se	[267]
$Ru(CO)(C_6Me_6)$	S, Se	[238]
$Rh(PMe_3)Cp^*$	S[a], Se, Te	[100]
Ir(L)Cp*	S, Se, Te	[100]
(L = PMe_3, PPh_3, CNtBu)		
$Pd(PPh_3)$	O	[79]
$Pd(PPh_3)$	S[a]	[268, 269]
$Pt(PPh_3)$	S[a]	[269]
$Pt(PPh_3)_2$	S, Se	[269]

a X-Ray structure analysis.

cogena-2-metalla-[3]ferrocenophanes or as coordination compounds containing a dianionic 1,1′-ferrocenylene dichalcogenate, which formally behaves as a chelating 4-electron ligand (Fig. 5-16).

It is difficult to find direct bonding interaction between the ferrocene iron and the second transition metal in the hetero-bimetallic complexes. Only the mono-triphenylphosphine palladium and platinum compounds fc[$E_2M(PPh_3)$] (E = O, M = Pd (Scheme 5-14) [79]; E = S, M = Pd [268, 269], and Pt [269]) appear to

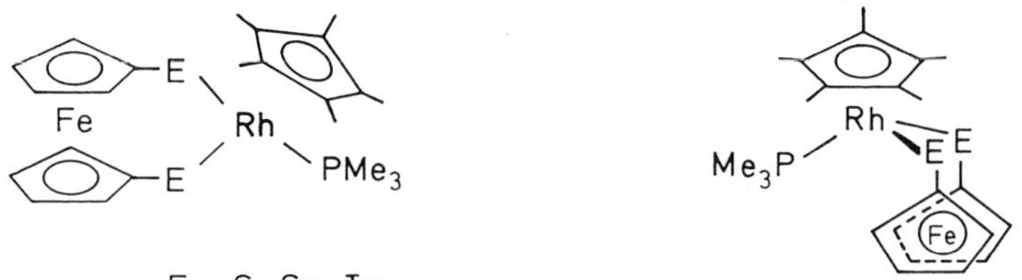

E = S, Se, Te

Fig. 5-16. 1,3-Dichalcogena-2-(pentamethylcyclopentadienyl-trimethylphosphine-rhoda)-[3]ferrocenophanes, fc[$E_2Rh(PMe_3)Cp^*$] [100], (two representations).

contain a weak metal–metal interaction, as may be concluded from the X-ray structure determinations carried out for the [3]ferrocenophanes fc[S_2M(PPh_3)] (Fe-Pd 287.8(1) pm [268] and Fe-Pt 293.5(2) pm [269]) and for the related [3]ruthenocenophanes rc[O_2Pd(PPh_3)] (Ru-Pd 269.2(1) pm) and rc[S_2Ni(PPh_3)] (Ru-Ni 286.4(1) pm) [79], which all revealed somewhat shortened metal–metal distances. In general, however, the iron remains in the cavity of the metallocene structure and tends to avoid external contacts. Thus, the large metal–metal distances in fc[S_2Rh(PMe_3)Cp*] (Fe-Rh 430.4(1) pm, Fig. 5-16) [100], fc[S_2Mo(NO)(HB(pz*)$_3$)] (Fe-Mo 414.7(2) pm) [266], fc[Se_2V(O)Cp*] (Fe-V 401.4(2) pm [99]) and fc[Se_2FeCl_2] (Fe-Fe 387 pm [224]) all clearly rule out any direct bonding interaction between the two transition metals, irrespective of whether the second transition metal is formally an 18-(Rh), 16-(Mo), 14-(V) or 12-(Fe) electron center.

In the case of the 1,3-dithia-[3]ferrocenophanes containing platinum in the bridging 2-position (Table 5-5), the iron-platinum bond in fc[S_2Pt(PPh_3)] is apparently formed as soon as one of the two PPh_3 ligands is lost from the precursor fc[S_2Pt(PPh_3)] [269]; the quantitative elimination of PPh_3 upon heating in solution has also been demonstrated for the corresponding [3]ruthenocenophanes, rc[E_2Pt(PPh_3)$_2$] (E = S, Se) [269]. However, attempts to induce the generation of a hetero-bimetallic bond in the 2-irida-[3]ferrocenophanes fc[E_2Ir(PPh_3)Cp*] (E = S, Se) by way of PPh_3 abstraction has led to dimerization via the 1,1′-ferrocenylene dichalcogenate ligand, which forms a bridge between two iridium centers (Scheme 5-44) [100].

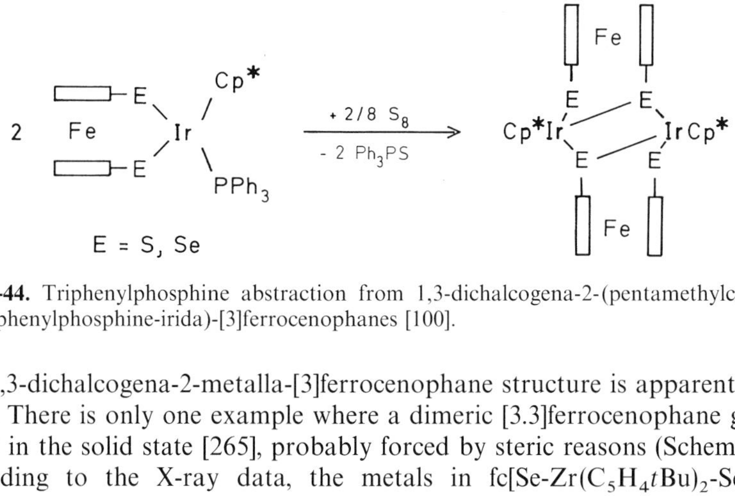

Scheme 5-44. Triphenylphosphine abstraction from 1,3-dichalcogena-2-(pentamethylcyclopentadienyl-triphenylphosphine-irida)-[3]ferrocenophanes [100].

The 1,3-dichalcogena-2-metalla-[3]ferrocenophane structure is apparently highly favored. There is only one example where a dimeric [3.3]ferrocenophane geometry is found in the solid state [265], probably forced by steric reasons (Scheme 5-45).

According to the X-ray data, the metals in fc[Se-Zr(C_5H_4tBu)$_2$-Se]$_2$fc are widely separated (Zr-Zr 936.3(3), Fe-Fe 600.0(3), Fe-Zr 558.8(3) and 555.3(3) pm) [265]. Remarkably, the related complexes fc[Se_2ZrCp_2] and fc[Se_2Hf(C_5H_4tBu)$_2$] (cf. Table 5-5) are monomeric [3]ferrocenophanes [265], as confirmed by cryoscopic determination of the molecular mass in benzene.

In an extension of the chemistry of 1,2,3-trithia-[3]ferrocenophane, fc(S_3), the corresponding bis(1,2,3-trithia) complex, i.e., 1,1′,2,2′-bis-(1,2,3-trithia-[3])ferrocenophane, [Fe(C_5H_3)$_2$](S_3)$_2$, has recently been prepared [271, 272] by tetralithiation of fc(SH)$_2$ with *n*BuLi. Following the initial deprotonation of fc(SH)$_2$, metallation

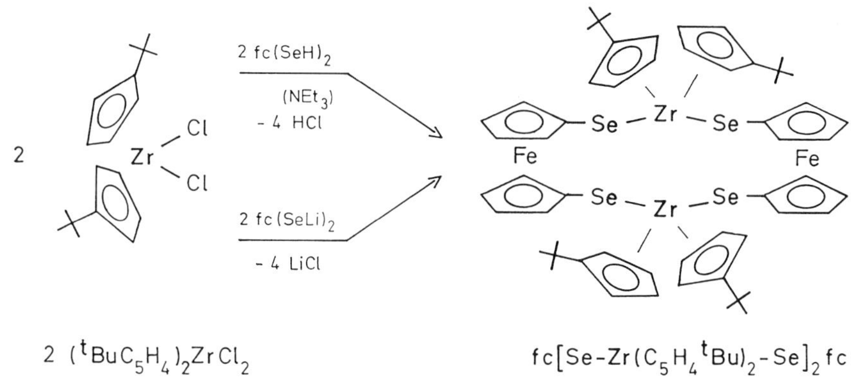

Scheme 5-45. Formation of a [3.3]ferrocenophane [265].

of the *ortho*-positions occurs regiospecifically and the tetralithio intermediate $Fe(C_5H_3(SLi)Li)_2$ is then treated with excess S_8 to give $[Fe(C_5H_3)_2](S_3)_2$. The ^1H NMR spectra indicate the presence of an equilibrium mixture of the (favored) chair-chair and the (less-favored) chair-boat isomers in solution at room temperature [271, 292], although the crystal contained only the chair-chair conformer. Similar results were obtained for the *tert*-butyl-substituted analog, $[Fe(tBuC_5H_2)(C_5H_3)](S_3)_2$ (Scheme 5-46) [272] (see Appendix).

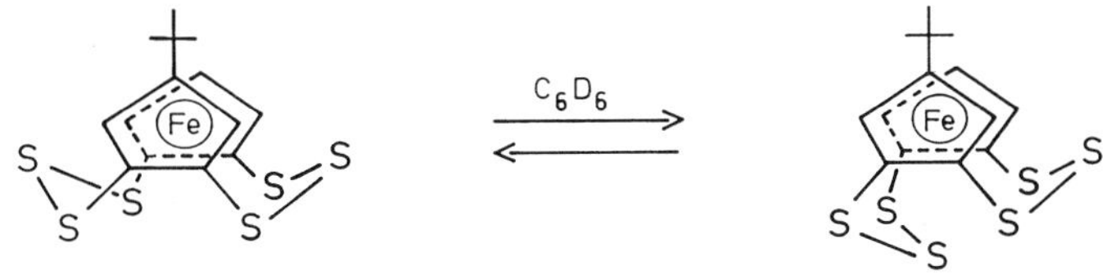

Scheme 5-46. Equilibrium mixture of the two isomers of 3-*tert*-butyl-1,1′,2,2′-bis(1,2,3-trithia-[3])ferrocenophane (K_{eq} 0.93 in C_6D_6, 19 °C) [272].

Desulfurization with tri(*n*-butyl)phosphine converts both $fc(S_3)$ (or its *tert*-butyl-substituted derivatives) [273, 274] and $[Fe(C_5H_3)_2](S_3)_2$ (or its *tert*-butyl-substituted analog) [272] to polymers in which the ferrocene units are linked through disulfide bridges (cf. Sect. 5.8).

5.7.2 Compounds with Two or Three 1,3-Dichalcogena-[3]ferrocenophane Rings

The group 4 elements Si, Ge and Sn are known to form spiro compounds in which two [3]ferrocenophane rings share the bridge atom in position 2. These complexes (Scheme 5-47) are generally obtained by the reaction of the tetrachlorides (ZCl_4) with either $fc(EH)_2$ [254, 260, 262] or $fc(ELi)_2$ [217, 224, 251], although Si_2Cl_6 [223] and $SnCl_2 \cdot 2\,H_2O$ [254] have also been used as starting halides.

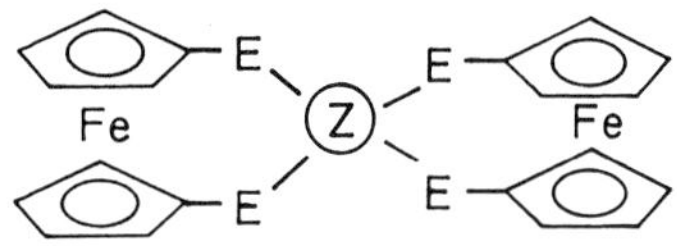

Z =						
Si:	E = S	[224]	Se	[224,261]	Te	[251]
Ge:	S	[262]	Se	[262]		
Sn:	S	[217,254]	Se	[217,260,261]	Te	[217]

Scheme 5-47. Spiro compounds of the type $Z(E_2fc)_2$.

The color deepens as the elements E and Z become heavier: the sila spiro complexes $Si(E_2fc)_2$ (E = S, Se) [223, 224] are yellow, the germa spiro complexes $Ge(E_2fc)_2$ [262] are red, and in the series of stanna spiro complexes $Sn(E_2fc)_2$ the color changes from red-golden (E = S) through dark-red (Se) to violet (Te) [217]. With the exception of the tellurium compounds the spiro complexes $Z(E_2fc)_2$ are thermally stable, decomposing only above 200 °C (Z = Si) or 250 °C (Z = Ge, Sn). However, $Si(Te_2fc)_2$ tends to decompose even at room temperature; products identified in the EI mass spectrum of $Si(Te_2fc)_2$ include $Si(fc)_2$ and $fc(Te_3)$ (Figs. 5-7 and 5-14, respectively) [251].

The molecular structure of $Sn(Se_2fc)_2$ [260] is based on a pseudo-tetrahedral tin(IV) center (angle Se-Sn-Se 108.6(1)°); the cyclopentadienyl rings of the $[fcSe_2]$ ligands are eclipsed with a mean twist angle of 5.7°.

The only transition metal complex known so far with two chelating feS_2 ligands is the anion of the diamagnetic brown rhenate(V) salt $(PPh_4)[ReO(S_2fc)_2]$, which was obtained by the reaction of $fc(SH)_2$ with $ReOCl_3(PPh_3)$ in the presence of Et_2NH in boiling ethanol [275]. The basic coordination geometry is assumed to be a tetragonal pyramid with the oxo ligand at the apex.

Diphosphine and diarsine derivatives such as $[fcE_2]P\text{-}P[E_2fc]$ and $[fcE_2]As\text{-}As[E_2fc]$ (E = S, Se) are formed as side-products in the reactions of $fc(ELi)_2$ with PCl_3 and $AsCl_3$, respectively [223]. They are related to the olefin $[fcS_2]C{=}C[S_2fc]$ (Scheme 5-22) [101].

Bifunctional bis(dichlorophosphinyl) and bis(dichloroarsinyl) compounds have also been used [223] to synthesize products in which two [3]ferrocenophane rings are kept at a defined distance, e.g., in molecules such as $[fcE_2]P(CH_2)_2P[E_2fc]$ and $[fcE_2]As(CH_2)As[E_2fc]$ (E = S, Se) (Fig. 5-17).

The study of these compounds is hampered by their low solubility in all common solvents.

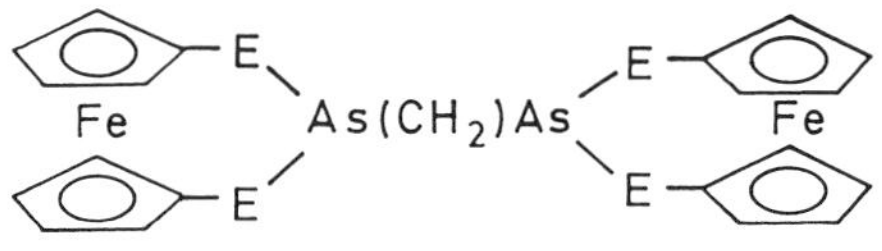

E = S, Se

Fig. 5-17. As,As′-Bis(1,3-dichalcogena-2-arsa-[3]ferrocenophanyl)methane.

If trihalides YCl_3 (Y = P, As, Sb) are allowed to react with 1,1'-bis(trimethylstannylchalcogenato)ferrocene, $fc(ESnMe_3)_2$, in the ratio 2:3, trinuclear products $fc[E\text{-}(YE_2fc)]_2$ (E = S, Se, Te) are formed in good yield (Scheme 5-48) [224].

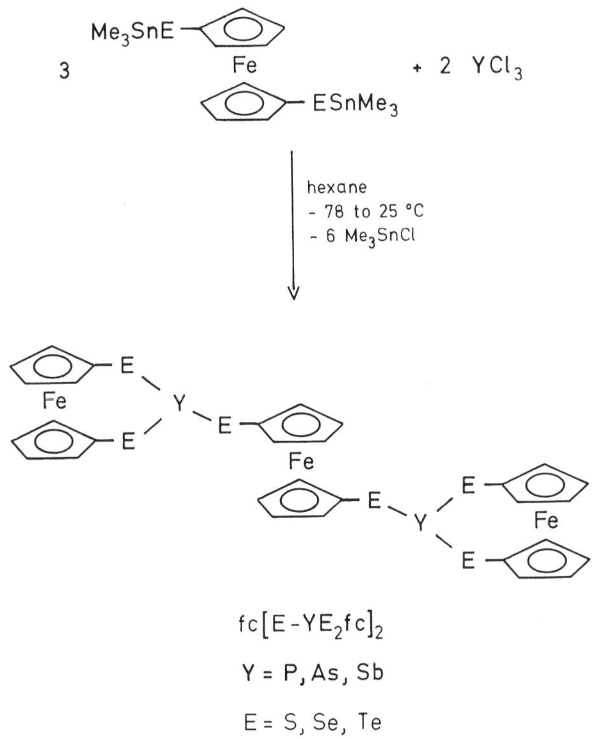

$$3\ fc(ESnMe_3)_2 + 2\ YCl_3 \xrightarrow[\text{- 78 to 25 °C; - 6 } Me_3SnCl]{\text{hexane}} fc[E\text{-}YE_2fc]_2$$

Y = P, As, Sb

E = S, Se, Te

Scheme 5-48. Synthesis of trinuclear ferrocene derivatives containing two terminal [3]ferrocenophane units [224].

The trinuclear complexes contain two [3]ferrocenophane rings linked through a 1,1'-ferrocenylene dichalcogenate bridge. The same trinuclear products are also formed preferentially, if the dilithium salts, $fc(ELi)_2$, are used as starting materials for the reactions with YCl_3 (Y = P, As, Sb [223]; MeSn [217], cf. Scheme 5-37). In the case of PCl_3, mononuclear $fc(E_2PCl)$ and binuclear $[fcE_2]P\text{-}P[E_2fc]$ are then obtained as side-products [224]. The trinuclear compounds $fc[E\text{-}YE_2fc]_2$ (Y = P; E = S, Se) have been oxidized at both phosphorus atoms with 3-chloroperbenzoic acid to give $fc[E\text{-}P(O)E_2fc]_2$. According to the ^1H and ^{13}C NMR spectra, the trinuclear compounds $fc[E\text{-}YE_2fc]_2$ (Y = P, As, Sb; E = Se, Te) possess rigid structures at room temperature in C_6D_6 solution [223]. In the Mössbauer spectra, only one type of iron can be observed; this might suggest charge delocalization within the trinuclear molecules.

The trimeric phosphazene dichloride, hexachloro-cyclotriphosphazene ($N_3P_3Cl_6$), reacts with dilithium 1,1'-ferrocenylene dichalcogenates, fc(ELi), under stepwise replacement of the three pairs of geminal chloro substituents to give spiro-cyclotri-

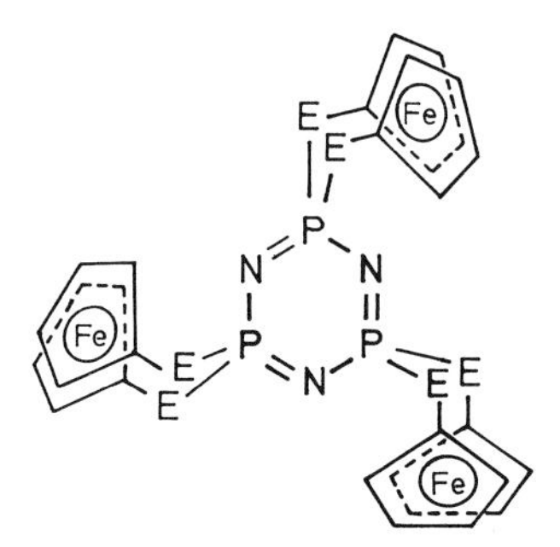

Fig. 5-18. P,P′,P″-Tri(ferrocenylene dichalcogenato)-cyclotriphosphazene, $N_3P_3(E_2fc)_3$ (E = S, Se).

phosphazenes, $N_3P_3Cl_{6-2n}[E_2fc]_n$ (n = 1, 2, 3; E=S, Se) [276]. The fully substituted products, $N_3P_3[E_2fc]_3$, are practically insoluble even in polar solvents; they were obtained from suspensions of $fc(ELi)_2$ in concentrated THF solutions of $N_3P_3Cl_6$. In analogy to the molecular structure determined for $N_3P_3Cl_2[S_2fc]_2$ [276], a molecular bowl containing three spiro phosphorus centers can be assumed (Fig. 5-18).

5.8 Oligo- and Polynuclear Ferrocenes with Chalcogen and Dichalcogen Bridges

The binuclear prototypes of the chalcogen-bridges ferrocenes, Fc-E-Fc and Fc-EE-Fc (E = S, Se, Te), are well-established compounds (Scheme 5-49) [95, 102].

The monosulfide, Fc-S-Fc, was first prepared in 1961 by Rausch [48] by reaction of the thiolate Fc-SNa with iodoferrocene, Fc-I, in the presence of copper bronze at 150 °C. The disulfide, Fc-SS-Fc, had been obtained independently by Knox and Pauson [81] and by Nesmeyanov et al. [82] at an even earlier date (1958) by aerial oxidation of mercaptoferrocene, Fc-SH; hydrolytic cleavage of ferrocenyl thiocya-

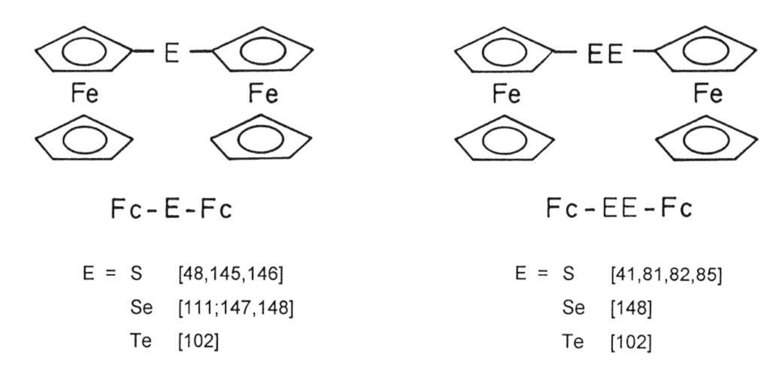

Scheme 5-49. Di(ferrocenyl) chalcogenides and dichalcogenides.

nate, Fc-SCN, is another reasonable route to Fc-SS-Fc [41, 85]. The monoselenide, Fc-Se-Fc, first prepared in 1958 by Nesmeyanov et al. [147] by the reaction of $HgFc_2$ with $SeCl_2$, is formed quantitatively from Fc-HgCl and Fc-SeCN [148]. Both Fc-Se-Fe and Fc-SeSe-Fc can be conveniently synthesized by the reaction of Fc-HgCl with copper selenocyanate, $Cu(SeCN)_2$, in acetonitrile [148]. The monosulfide and the monoselenide, Fc-E-Fc (E = S [146], Se [148]), have been studied with respect to mixed-valence chemistry by Cowan and coworkers. A cyclovoltammetric comparison of Fc-H, Fc-Fc and Fc-S-Fc has indicated that the ferrocenyl substituent (Fc) is electron-donating whereas the ferrocenyl thiyl substituent (Fc-S) is essentially electron-withdrawing [146]. Oxidation of $SeFc_2$ with iodine leads to a salt $[SeFc_2]I_3 \cdot 0.5\ CH_2Cl_2$, which has been extensively inestigated [111, 277]; its mixed-valent cation, $SeFc_2^+$, contains the positive charge localized in one of the two ferrocenyl units.

The tellurides, Fc-Te-Fc and Fc-TeTe-Fc, where only obtained in 1987 via the solvated lithium ferrocenyl tellurolate, Fc-TeLi(thf) [102]. Oxidation by air produces Te_2Fc_2 which, in turn, may be cleaved by Fc-Li to give eventually $TeFc_2$.

The trinuclear ferrocene chalcogenides, Fc-E-fc-E-Fc, are accessible from a general reaction [32] that involves the cleavage of dichalcogenides, E_2R_2, by dilithioferrocene (Scheme 5-50).

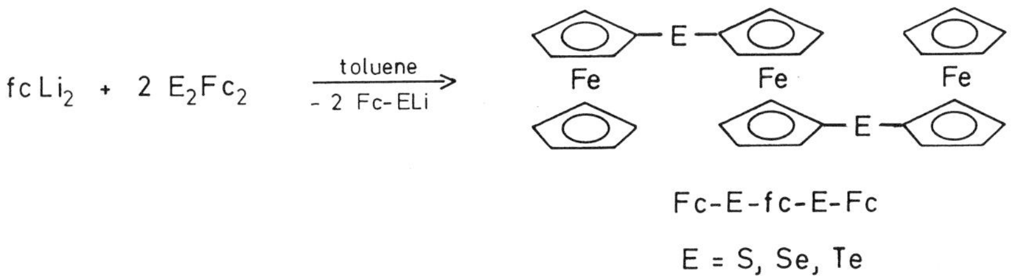

$$fcLi_2 + 2\ E_2Fc_2 \xrightarrow[-2\ Fc\text{-}ELi]{toluene} \text{Fc-E-fc-E-Fc}$$

E = S, Se, Te

Scheme 5-50. Synthesis of trinuclear ferrocene derivatives $fc(E\text{-}Fc)_2$ [95].

The color of the trinuclear chalcogenides $fc(E\text{-}Fc)_2$ intensifies as the two chalcogen atoms become heavier, i.e., in the order E = S (yellow), Se (yellow-orange), Te (orange) [95].

Oligonuclear ferrocene derivatives containing dichalcogenide bridges are so far only known with sulfur: examples are $fc(SS\text{-}Fc)_2$ and $[FcS\text{-}fc\text{-}S\text{-}]_2$ (Fig. 5-19) [95].

Disulfide bridges are generally formed as soon as thiol substituents – as in Fc-SH and $fc(SH)_2$ – come into contact with radical initiators such as AIBN. Thus, the radical-induced reaction of $fc(SH)_2$ with Fc-SH (1:2) gives some Fc-SS-Fc and a small amount of $fc(SS\text{-}Fc)_2$, but mainly polynuclear disulfide-bridged products that are insoluble in organic solvents [95].

Ferrocene derivatives containing a single disulfide link are occasionally obtained from reactions of $fc(SH)_2$, e.g., with norbornene [84] or with the oxonium salt $[(Ph_3PAu)_3O]BF_4$ (Scheme 5-18) [90]. Splitting of 1,2,3-trithia-[3]ferrocenophane, $fc(S_3)$, by organolithium reagents followed by air oxidation is another pathway to disulfides (Scheme 5-51) [95].

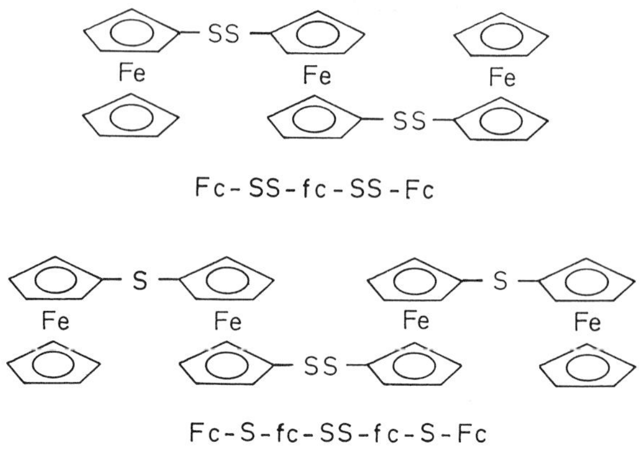

Fc-SS-fc-SS-Fc

Fc-S-fc-SS-fc-S-Fc

Fig. 5-19. Oligonuclear ferrocene derivatives containing disulfide bridges [95].

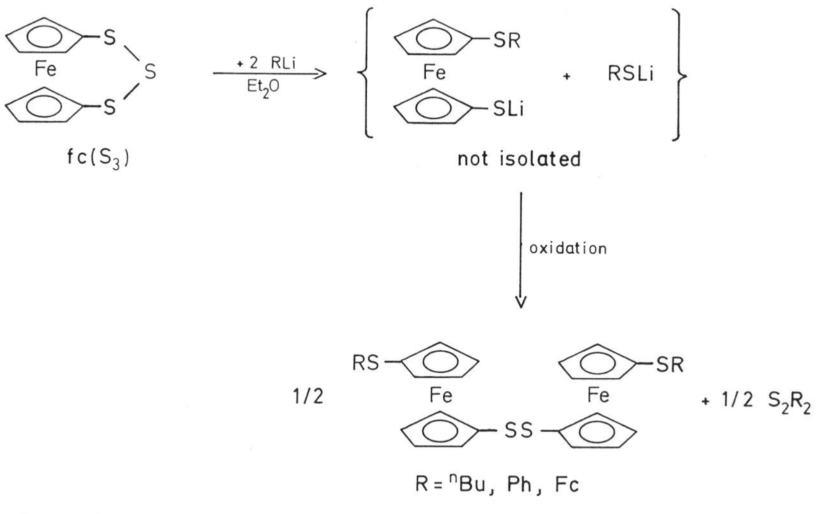

$fc(S_3)$ + 2 RLi, Et_2O → { [Fe(C5H4SR)(C5H4SLi)] + RSLi } not isolated → oxidation → 1/2 [RS-fc-SS-fc-SR] + 1/2 S_2R_2

$R = {}^nBu$, Ph, Fc

Scheme 5-51. Disulfide-bridged ferrocene derivatives by ring-opening in 1,2,3-trithia-[3]ferrocenophane, $fc(S_3)$ [95].

As a general rule, polymers containing ferrocene units connected by sulfide or disulfide bridges are insoluble in all common organic solvents. Thus, insoluble yellow poly(ferrocenylene sulfide), $[fcS]_n$, was obtained by reaction of 1,1′-diiodoferrocene, fcI_2, with the copper salt of 1,1′-dimercaptoferrocene, $fc(SCu)_2$, in boiling pyridine [278]. An insoluble bright-orange precipitate of poly(ferrocenylene disulfide), $[fcS_2]_n$, was formed when $fc(S_3)$ was desulfurized by tri(*n*-butyl)phosphine to undergo ring-opening polymerization [273]. However, as shown by Rauchfuss and coworkers [272–274], soluble polymers are obtained starting from *n*-butyl- [273] or *tert*-butyl- [274] substituted 1,2,3-trithia-[3]ferrocenophane monomers (Scheme 5-52).

The polymerization process (Scheme 5-52) is highly dependent on the solvent [272–274]. Whereas $[Fe(tBuC_5H_3)(C_5H_4)](S_3)$ produces a one-dimensional polymer

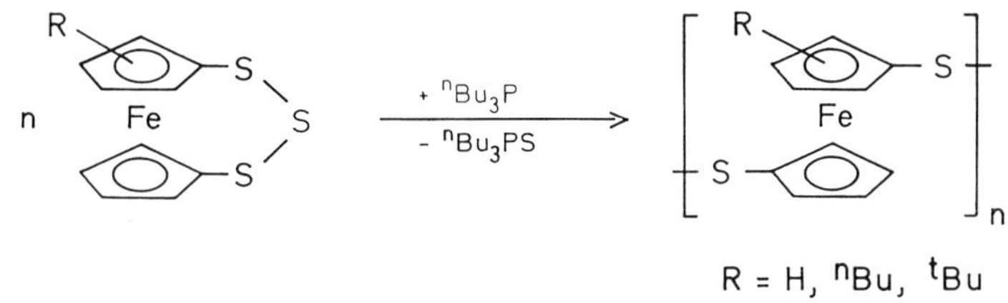

Scheme 5-52. Ring-opening polymerisation of 1,2,3-trithia-[3]ferrocenophanes [272–274].

with low molar mass ($M_n < 6000$), the 4-*tert*-butyl-1,1′,2,2′-bis(1,2,3-trithia-[3])ferrocenophane, $[Fe(tBuC_5H_2)(C_5H_3)](S_3)_2$ (Scheme 5-46), leads to a three-dimensional polymer with a broad distribution of molar mass (up to $M_n \approx 850000$); after removal of the fractions with lower molar mass, a giant network with molar mass $>10^6$ Da was isolated (Scheme 5-53) [272].

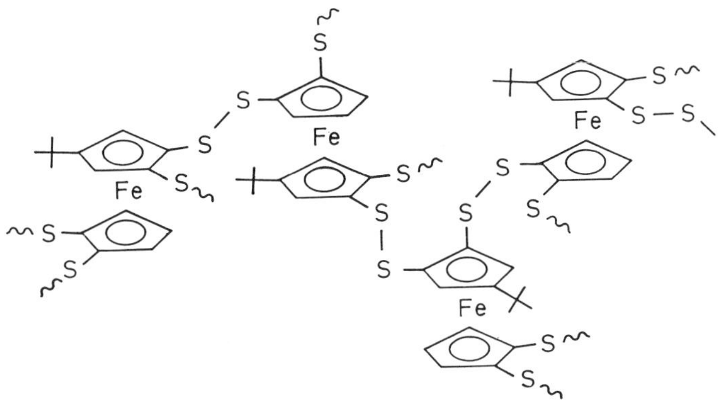

Scheme 5-53. Disulfide links in three-dimensional polymers derived from $[Fe(tBuC_5H_2)(C_5H_3)](S_3)_2$ (schematic representation) [272].

The disulfide links in the organometallic ferrocene polymers can be reductively cleaved, as in other disulfide-connected structures such as proteins, and the iron centers can be oxidized electrochemically to give ferrocenium units [273]; both these redox processes are reversible in principle.

5.9 Conclusions

While the majority of all ferrocene derivatives described so far [1, 2] contain the sandwich unit in an organic environment, i.e., attached to carbon, nearly all elements of the Periodic Table (except the rare gases) are apparently able to bind directly to the cyclopentadienyl rings of ferrocene. In particular, homoleptic ferrocenyl compounds of almost all non-metal elements are known (Table 5-1), two remarkable missing links being "OFc_2" and "$AsFc_3$". Furthermore, the number of homoleptic

1,1′-ferrocendiyl compounds of the non-metals is growing, including spiro compounds such as $Si(fc)_2$ and $Ge(fc)_2$, [3]ferrocenophanes such as $fc(E_3)$ (E = S, Se, Te) (Sect. 5.7.1) and macromolecules of the type $[fcS]_n$ and $[S(fc)S]_n$ (Sect. 5.8).

The ferrocenyl group (Fc) is also a versatile ligand system in transition metal complexes (Sect. 5.4) and can coordinate to metals in both high (e.g., W(VI) in $WO(X)Fc_3$ (X = Cl, F, OFc) or Ti(IV) in $TiFc_4$) and low oxidation state (e.g., Au(I) in Fc-$Au(PPh_3)$ or Mn(I) and Re(I) in $M(CO)_5$-Fc). Up to four ferrocenyl ligands have been attached to a single metal atom, although the number of ferrocenyl ligands remains restricted to two or one in most cases, due to the steric requirements of the bulky sandwich unit (Tables 5-2 and 5-3).

Ferrocenyl and ferrocenyl thiolate derivatives of the metals are heterobimetallic compounds, one component of which is iron. In principle, complex fragments of different metals may be combined with a ferrocene unit, as in the trimetallic 1,1′-ferrocenylene dithiolate, $Cp(NO)_2Cr$-[S(fc)S]-$W(CO)_3Cp$. Another versatile method to synthesize heterobimetallic complexes makes use of chelating ligands such as 1,1′-bis(diphenylphosphinyl) ferrocene, $fc(PPh_2)_2$ (see Chapter 1), and 1,1′-bis(methylchalcogenyl) ferrocene, $fc(EMe)_2$ (E = S, Se), (Sect. 5.4).

In 1,3-dichalcogena-[3]ferrocenophanes of the type $fc(E_2M)$ (E = O, S, Se, Te), a transition metal complex fragment (M) is held firmly between the two chalcogen atoms (E) (Table 5-5). Although both the ferrocene iron and the transition metal are essential parts of the cyclic structure, the two metals generally remain isolated from each other. Only a few examples have been found where an intramolecular, transannular interaction between Fe and M can be taken into consideration, as in the palladium(II) and platinum(II) complexes $fc[E_2M(PPh_3)]$ (E = O, M = Pd; E = S, M = Pd and Pt) (Sect. 5.7.1).

A large number of ferrocene compounds containing heteroelements is now available that can be used as model systems to study the steric and electronic influence of a sandwich ferrocene unit on the rest of the molecule. This is important with regard to the fact that ferrocene building blocks are frequently incorporated into macrocycles and polymers either as structure-determining spacer groups or as redox-active probes. In addition to X-ray crystallographic determinations of the molecular structure, investigations at the ferrocene iron center by cyclic voltammetry and Mößbauer spectroscopy are particularly informative.

Appendix (Notes added in proof, August 1994)

The lithiation of ferrocene and ruthenocene by *t*BuLi and *n*BuLi can be selectively conducted to give predominantly either the mono- or the 1,1′-dilithio-metallocenes, which may then be converted into the corresponding carbaldehydes using *N*,*N*-dimethylformamide [286].

Starting from diferrocenyl ditelluride, Fc-TeTe-Fc [102], a series of 20 ferrocenyl organyl tellurides, Fc-Te-R, has been prepared (R = long-chain alkyl, alkenyl or alkynyl) and their reactivity towards oxidation has been studied [287]. Several 1,1′-ferrocenylene di(organotellurides), $fc(TeR)_2$, were synthesised from $fcLi_2$ and R-TeTe-R; the corresponding ferrocenyl organyl tellurides, Fc-Te-R (R = Me, *n*Bu, Ph, 4-MeO-C_6H_4 and 4-EtO-C_6H_4), were also obtained [288]. The molecular structure of $fc(TePh)_2$ is now available [288].

Ferrocenophanes containing transition metals directly attached to the cyclopentadienyl rings are attractive hetero-bimetallic model compounds. Two remarkable new examples include the dark-red 1,2-ditungsta-[2]ferrocenophane, $fc[W_2(NMe_2)_4]$, (with a formal W≡W triple bond, 228.8 pm) [289], and the green, paramagnetic 1,3-diplatina-2-hydroxa-[3]ferrocenophanes, $fc[PtCl(L)]_2OH$ (L = PPh_3, $P(C_6H_4\text{-}OMe\text{-}4)_3$) [290] which are formed in the reactions of 1,1′-ferrocenylene-bis(cyclooctadiene-platinumchloride), $fc[Pt(C_8H_{12})Cl]_2$ [185], and molecular oxygen in the presence of triarylphosphines L [290].

A trimetallic product $fc[Pt(C{\equiv}CPh)_3]_2Cu_3$ has been obtained from either $fc[Pt(C_8H_{12})Cl]_2$ [185] or the phenylacetylido analog $fc[Pt(C_8H_{12})C{\equiv}CPh]_2$ with excess PhC≡CCu; the three copper atoms are each coordinated to two triple bonds [291].

In a dynamic ^1H NMR study of the doubly trithia-bridged 1,1,2,2′,-bis(1,2,3-trithia-[3])ferrocenophane, $[Fe(C_5H_3)_2](S_3)_2$, the activation barriers of the two possible bridge-reversal processes, chair-boat to chair-chair ($\Delta G^{\neq}$ (298 K) 93.9 $\pm$ 0.8 kJ/mol) and chair-boat to boat-chair (89.0 $\pm$ 1 kJ/mol), were found to be higher [292] than in the singly-bridged $fc(S_3)$ (80.4 $\pm$ 0.2 [256]). The relative abundance of the two diastereomers depends on the polarity of the solvent, the less-favored chair-boat isomer being favored in polar solvents [292]. Mixed thia/selena-bridged species, $[Fe(C_5H_3)_2](S_3)(SSeS)$ and $[Fe(C_5H_3)_2](SSeS)_2$, were also investigated.

The molecular structure of the cationic 1,1′-bis(methylthiolato) ferrocene complex $[fc(SMe)_2Pt(PPh_3)](BF_4)_2$ suggests a weak Fe-Pt interaction (285.1 (2) pm) [293].

References

[1] Gmelin Handbuch der Anorganischen Chemie, 8th ed. Organoiron Compounds, Part A, Ferrocene, **1974–1991** Vols. *1* (1974), *2* (1977), *3* (1978), *4* (1980), *5* (1981), *6* (1977), *7* (1980), *8* (1986), *9* (1989), *10* (1991), *11* (in preparation). Starting with Vol. *8*, the volumes are published in English. Springer Verlag, Berlin–Heidelberg–New York–London– Paris–Tokyo–Hongkong–Barcelona.

[2] B. W. Rockett, G. Marr, *J. Organomet. Chem.* **1991**, *416*, 327–398 (covering the year 1989); ibid. **1990**, *392*, 93–160 (covering 1988); ibid. **1988**, *357*, 247–318 (covering 1987); ibid. **1988**, *343*, 79–146 (covering 1986); ibid. **1987**, *318*, 231–297 (covering 1985) and earlier articles in the series.

[3] W. Ruf, M. Fueller, W. Siebert, *J. Organomet. Chem.* **1974**, *64*, C45–C47.

[4] T. Renk, W. Ruf, W. Siebert, *J. Organomet. Chem.* **1976**, *120*, 1–25.

[5] W. Ruf, T. Renk, W. Siebert, *Z. Naturforsch.* **1977**, *31b*, 1028–1034.

[6] J. C. Kotz, E. W. Post, *J. Am. Chem. Soc.* **1968**, *90*, 4503–4504.

[7] J. C. Kotz, E. W. Post, *Inorg. Chem.* **1970**, *9*, 1661–1669; cf. E. W. Post, R. G. Cookes, J. C. Kotz, *Inorg. Chem.* **1970**, *9*, 1670–1677.

[8] U. Dörfler, B. Wrackmeyer, M. Herberhold, to be published; cf. Diplomarbeit U. Dörfler, Univ. Bayreuth, **1992**.

[9] B. Wrackmeyer, U. Dörfler, M. Herberhold, *Z. Naturforsch.* **1993**, *48b*, 121–123.

[10] A. N. Nesmeyanov, E. G. Perevalova, R. V. Golovnya, O. A. Nesmeyanova, *Dokl. Akad. Nauk SSSR* **1954**, *97*, 459–461; *Chem. Abstr.* **1955**, *49*, 9633.

[11] R. W. Fish, M. Rosenblum, *J. Org. Chem.* **1965**, *30*, 1253–1254.

[12] M. Rausch, M. Vogel, H. Rosenberg, *J. Org. Chem.* **1957**, *22*, 900–903.

[13] M. D. Rausch, L. P. Klemann, A. Siegel, R. F. Kovar, T. H. Gund, *Synth. Reactiv. Inorg. Metal-Org. Chem.* **1973**, *3*, 193–199.

[14] D. Seyferth, H. P. Hofmann, R. Burton, J. F. Helling, *Inorg. Chem.* **1962**, *1*, 227–231.

[15] A. N. Nesmeyanov, V. A. Sazonova, V. N. Drozd, *Dokl. Akad. Nauk SSSR* **1959**, *126*, 1004–1006; *Dokl. Chem. Proc. Acad. Sci. USSR* **1959**, *124–126*, 437–439; *Chem. Abstr.* **1960**, *54*, 6673h.

[16] A. N. Nesmeyanov, W. A. Ssazonowa, V. N. Drosd, *Chem. Ber.* **1960**, *93*, 2717–2729.

[17] A. N. Nesmeyanov, V. A. Sazonova, N. S. Sazonova, *Izv. Akad. Nauk SSSR Ser. Khim.* **1968**, 2371–2372; *Bull. Acad. Sci. USSR Div. Chem. Sci.* **1968**, 2240–2241; *Chem. Abstr.* **1969**, *71*, 3454c.

[18] H. Shechter, J. F. Helling, *J. Org. Chem.* **1961**, *26*, 1034–1037.

[19] J. E. Mulvaney, J. J. Bloomfield, C. S. Marvel, *J. Polym. Sci.* **1962**, *62*, 59–72.

[20] S. I. Goldberg, R. L. Matteson, *J. Org. Chem.* **1964**, *29*, 323–326.

[21] A. N. Nesmeyanov, V. A. Sazonova, V. N. Drozd, *Dokl. Akad. Nauk SSSR* **1959**, *129*, 1060–1063; *Dokl. Chem. Proc. Acad. Sci. USSR* **1959**, *124–129*, 1113–1116; cf. *Tetrahedron Letters* **1959**, No. 17, 13–15.

[22] A. N. Nesmeyanov, V. A. Sazonova, V. N. Drozd, L. A. Nikonova, *Dokl. Akad. Nauk SSSR* **1960**, *133*, 126–129; *Dokl. Chem. Proc. Acad. Sci. USSR* **1960**, *130–135*, 751–754; *Chem. Abstr.* **1960**, *54*, 24616h.

[23] R. A. Benkeser, D. Goggin, G. Schroll, *J. Am. Chem. Soc.* **1954**, *76*, 4025–4026.

[24] F. L. Hedberg, H. Rosenberg, *Tetrahedron Letters* **1969**, No. 46, 4011–4012.

[25] M. D. Rausch, D. J. Ciappenelli, *J. Organomet. Chem.* **1967**, *10*, 127–136.

[26] M. D. Rausch, G. A. Moser, C. F. Meade, *J. Organomet. Chem.* **1973**, *51*, 1–11.

[27] A. J. Lee Hanlan, R. C. Ugolick, J. G. Fulcher, S. Togashi, A. B. Bocarsly, J. A. Gladysz, *Inorg. Chem.* **1980**, *19*, 1543–1551.

[28] A. G. Osborne, R. H. Whiteley, R. E. Meads, *J. Organomet. Chem.* **1980**, *193*, 345–357.

[29] J. J. Bishop, A. Davison, M. L. Katcher, D. W. Lichtenberg, R. E. Merrill, J. C. Smart, *J. Organomet. Chem.* **1971**, *27*, 241–249.

[30] J. D. Unruh, J. R. Christenson, *J. Mol. Catal.* **1982**, *14*, 19–34.

[31] B. McCulloch, C. H. Brubaker, Jr., *Organometallics* **1984**, *3*, 1707–1711.

[32] B. McCulloch, D. L. Ward, J. D. Woollins, C. H. Brubaker, Jr., *Organometallics* **1985**, *4*, 1425–1532.

[33] I. R. Butler, W. R. Cullen, J. Ni, S. J. Rettig, *Organometallics* **1985**, *4*, 2196–2201.

[34] M. Walczak, K. Walczak, R. Mink, M. D. Rausch, G. Stucky, *J. Am. Chem. Soc.* **1978**, *100*, 6382–6388.

[35] A. F. Halasa, D. P. Tate, *J. Organomet. Chem.* **1970**, *24*, 769–773.

[36] F. L. Hedberg, H. Rosenberg, *J. Am. Chem. Soc.* **1970**, *92*, 3239–3240; ibid. **1973**, *95*, 870–875.

[37] V. I. Boev, A. V. Dombrovskii, *Zh. Obshch. Khim.* **1977**, *47*, 727–728; *J. Gen. Chem. USSR* **1977**, *47*, 663–664; *Izv. Vyssh. Uchebn. Zaved., Khim. Khim. Tekhnol.* **1977**, *20*, 1789–1793; *Chem. Abstr.* **1978**, *88*, 136762z.

[38] F. Rebière, O. Samuel, H. Kagan, *Tetrahedron Letters* **1990**, *31*, 3121–3124; cf. E. Moret, O. Desponds, M. Schlosser, *J. Organomet. Chem.* **1991**, *409*, 83–91.

[39] M. Sato, T. Ito, I. Motoyama, K. Watanabe, K. Hata, *Bull. Chem. Soc. Japan* **1969**, *42*, 1976–1981.

[40] M. Sato, I. Motoyama, K. Hata, *Bull. Chem. Soc. Japan* **1970**, *43*, 1860–1863.

[41] M. Sato, I. Motoyama, K. Hata, *Bull. Chem. Soc. Japan* **1970**, *43*, 2213–2217.

[42] M. Sato, Y. P. Lam, I. Motoyama, K. Hata, *Bull. Chem. Soc. Japan* **1971**, *44*, 808–812.

[43] M. Sato, I. Motoyama, K. Hata, *Bull. Chem. Soc. Japan* **1971**, *44*, 812–815.

[44] A. N. Nesmeyanov, V. A. Sazonova, V. N. Drozd, *Dokl. Akad. Nauk SSSR* **1960**, *130*, 1030–1032; *Dokl. Chem. Proc. Acad. Sci. USSR* **1960**, *130–135*, 153–155; *Chem. Abstr.* **1960**, *54*, 12089e.

[45] A. N. Nesmeyanov, V. N. Drozd, V. A. Sazonova, *Dokl. Akad. Nauk SSSR* **1963**, *150*, 321–324; *Dokl. Chem. Proc. Acad. Sci. USSR* **1963**, *148–153*, 416–419; *Chem. Abstr.* **1963**, *59*, 5196a.

[46] A. N. Nesmeyanov, V. A. Sazonova, V. I. Romanenko, *Dokl. Akad. Nauk SSSR* **1964**, *157*, 922–925; *Dokl. Chem. Proc. Acad. Sci. USSR* **1964**, *154–159*, 765–768; *Chem. Abstr.* **1964**, *61*, 13343h.

[47] V. A. Nefedov, M. N. Nefedova, *Zh. Obshch. Khim.* **966**, *36*, 122–126; *J. Gen. Chem. USSR* **1966**, *36*, 127–130; *Chem. Abstr.* **1966**, *64*, 14215a.

[48] M. D. Rausch, *J. Org. Chem.* **1961**, *26*, 3579–3580.

[49] H. Bürger, C. Kluess, *J. Organomet. Chem.* **1973**, *56*, 269–277.

[50] G. R. Knox, P. L. Pauson, D. Willison, E. Solčaniova, S. Toma, *Organometallics* **1990**, *9*, 301–306.

[51] A. N. Nesmeyanov, E. G. Perevalova, R. V. Golovnya, L. S. Shilovtseva, *Dokl. Akad. Nauk SSSR* **1955**, *102*, 535–538; *Chem. Abstr.* ***1956***, *50*, 4925g.

[52] F. S. Arimoto, A. C. Haven, *J. Am. Chem. Soc.* **1955**, *77*, 6295–6296.

[53] D. W. Hall, J. H. Richards, *J. Org. Chem.* **963**, *28*, 1549–1554.

[54] M. Herberhold, M. Ellinger, W. Kremnitz, *J. Organomet. Chem.* **1983**, *241*, 227–240.

[55] E. M. Acton, R. M. Silverstein, *J. Org. Chem.* **1959**, *24*, 1487–1490.

[56] R. A. Abramovitch, C. I. Azogu, R. G. Sutherland, *Chem. Commun.* **1971**, 134–135.

[57] R. A. Abramovitch, R. G. Sutherland, A. K. V. Unni, *Tetrahedron Letters* **1972**, No. 11, 1065–1068.

[58] J. F. Helling, H. Shechter, *Chem. Ind.* [London] **1959**, 1157; *Chem Abstr.* **1960**, *54*, 3360a.

[59] H. Grubert, K. L. Rinehart, Jr., *Tetrahedron Letters* **1959**, No. 12, 16–17.

[60] E. G. Perevalova, K. I. Grandberg, N. A. Zharikova, S. P. Gubin, A. N. Nesmeyanov, *Izv. Akad. Nauk SSSR Ser. Khim.* **1966**, 832–839; *Bull. Acad. Sci. USSR Div. Chem. Sci.* **1966**, 796–802; *Chem. Abstr.* **1966**, *65*, 5344e.

[61] L. N. Nekrasov, *Faraday Discussions Chem. Soc.* **1973**, *56*, 308–316; *Elektrokhimiya* **1975**, *11*, 851–859, *Soviet. Electrochem.* **1975**, *11*, 789–795.

[62] A. N. Nesmeyanov, E. G. Perevalova, T. V. Nikitina, *Dokl. Akad. Nauk SSSR* **1961**, *138*, 1118–1121; *Dokl. Chem. Proc. Akad. Sci. USSR* **1961**, *136–141*, 584–587; *Chem. Abstr.* **1961**, *55*, 24707c.

[63] A. N. Nesmeyanov, T. V. Nikitina, E. G. Perevalova, *Izv. Akad. Nauk SSSR Ser. Khim.* **1964**, 197–199; *Bull. Acad. Sci. USSR Div. Chem. Sci.* **1964**, 184–185; *Chem. Abstr.* **1964**, *60*, 9310c.
[64] K. Schlögl, H. Mechtler, *Monatsh. Chem.* **1966**, *97*, 150–167.
[65] A. N. Nesmeyanov, V. I. Romanenko, V. A. Sazonova, *Izv. Akad. Nauk SSSR Ser. Khim.* **1966**, 357–358; *Bull. Acad. Sci. USSR Div. Chem. Sci.* **1966**, 330–331; *Chem. Abstr.* **1966**, *64*, 18509b.
[66] T. El-Shishi, F. Siglmüller, R. Herrmann, M. F. N. N. Carvalho, A. J. L. Pombeiro, *J. Organomet. Chem.* **1987**, *335*, 239–247.
[67] K. Schlögl, H. Seiler, *Naturwissenschaften* **1958**, *45*, 337.
[68] M. Herberhold, B. Distler, in preparation.
[69] K. Schlögl, H. Mechtler, *Angew. Chem.* **1966**, *78*, 606–607; *Angew. Chem. Int. Ed. Engl.* **1966**, *5*, 596.
[70] G. R. Knox, *Proc. Chem. Soc.* **1959**, 56–57.
[71] G. R. Knox, P. L. Pauson, *J. Chem. Soc.* **1961**, 4615–4618.
[72] C. E. Carraher, Jr., *Interfacial Synth.* **1977**, 367–416.
[73] M. Herberhold, H.-D. Brendel, in preparation.
[74] R. Epton, G. Marr, G. K. Rogers, *J. Organomet. Chem.* **1976**, *110*, C42–C44.
[75] S. Akabori, M. Ohtomi, M. Sato, S. Ebine, *Bull. Chem. Soc. Japan* **1983**, *56*, 1455–1458.
[76] J. F. Biernat, T. Wilczewski, *Tetrahedron* **1980**, *36*, 2521–2523.
[77] S. Akabori, Y. Habata, Y. Sakamoto, M. Sato, S. Ebine, *Bull. Chem. Soc. Japan* **1983**, *56*, 537–541.
[78] S. Akabori, Y. Habata, M. Sato, S. Ebine, *Bull. Chem. Soc. Japan* **1983**, *56*, 1459–1461.
[79] S. Akabori, T. Kumagai, T. Shirahige, S. Sato, K. Kawazoe, C. Tamura, M. Sato, *Organometallics* **1987**, *6*, 2105–2109.
[80] M. Herberhold, H.-D. Brendel, *J. Organomet. Chem.* **1993**, *458*, 205–209.
[81] G. R. Knox, P. L. Pauson, *J. Chem. Soc.* **1958**, 692–696.
[82] A. N. Nesmeyanov, E. G. Perevalova, S. S. Churanov, O. A. Nesmeyanova, *Dokl. Akad. Nauk SSSR* **1958**, *119*, 949–952; *Dokl. Chem. Proc. Akad. Sci. USSR* **1958**, *118–123*, 281–284; *Chem. Abstr.* **1958**, *52*, 17225b.
[83] A. Ratajczak, M. Dominiak, *Rocnicki Chem.* **1974**, *48*, 175–176, *Chem. Abstr.* **1974**, *80*, 146269t.
[84] M. Herberhold, O. Nuyken, T. Pöhlmann, *J. Organomet. Chem.* **1991**, *405*, 217–227.
[85] G. R. Knox, I. G. Morrison, P. L. Pauson, *J. Chem. Soc.* **1967**, *C*, 1842–1847.
[86] E. G. Perevalova, D. A. Lemenovskii, K. I. Grandberg, A. N. Nesmeyanov, *Dokl. Akad. Nauk SSSR* **1972**, *203*, 1320–1323; *Dokl. Chem. Proc. Akad. Sci. USSR* **1972**, *202–207*, 375–378; *Chem. Abstr.* **1972**, *77*, 114528y.
[87] A. N. Nesmeyanov, K. I. Grandberg, D. A. Lemenovskii, O. B. Afanasova, E. G. Perevalova, *Izv. Akad. Nauk SSSR Ser. Khim.* **1973**, 887–890; *Bull. Acad. Sci. USSR Div. Chem. Sci.* **1973**, 856–859; *Chem. Abstr.* **1973**, *79*, 42655d.
[88] M. Herberhold, T. Pöhlmann, O. Nuyken, in preparation.
[89] M. Herberhold, P. Leitner, C. Dörnhöfer, J. Ott-Lastic, *J. Organomet. Chem.* **1989**, *377*, 281–289.
[90] E. G. Perevalova, T. V. Baukova, M. M. Sazonenko, K. I. Grandberg, *Izv. Akad. Nauk SSSR Ser. Khim.* **1985**, 1873–1876; *Bull. Acad. Sci. USSR Div. Chem. Sci.* **1985**, *34*, 1722–1726; *Chem. Abstr.* **1985**, *105*, 97597x.
[91] R. Broussier, A. Abdulla, B. Gautheron, *J. Organomet. Chem.* **1987**, *332*, 165–173.
[92] A. Davison, J. C. Smart, *J. Organomet. Chem.* **1969**, *19*, P7–P8.
[93] O. Nuyken, V. Burkhardt, T. Pöhlmann, M. Herberhold, *Makromol. Chem., Macromol. Symp.* **1991**, *44*, 195–206.
[94] O. Nuyken, T. Pöhlmann, M. Herberhold, *Macromolecular Reports* **1992**, *A29* (Suppl. 3), 211–220.
[95] M. Herberhold, H.-D. Brendel, O. Nuyken, T. Pöhlmann, *J. Organomet. Chem.* **1991**, *413*, 65–78.
[96] M. Herberhold, C. Dörnhöfer, H. I. Hayen, B. Wrackmeyer, *J. Organomet. Chem.* **1988**, *355*, 325–335.

[97] M. Sato, S. Tanaka, S. Ebine, S. Akabori, *Bull. Chem. Soc. Japan* **1984**, *57*, 1929–1934.
[98] M. Sato, K. Suzuki, S. Akabori, *Bull. Chem. Soc. Japan* **1986**, *59*, 3611–3615; cf. M. Sato, S. Tanaka, S. Akabori, Y. Habata, ibid. **1986**, *59*, 1515–1519.
[99] M. Herberhold, M. Schrepfermann, A. L. Rheingold, *J. Organomet. Chem.* **1990**, *394*, 113–120.
[100] M. Herberhold, G.-X. Jin, A. L. Rheingold, G. F. Sheats, *Z. Naturforsch.* **1992**, *47b*, 1091–1098.
[101] M. Sato, S. Akabori, *Bull. Chem. Soc. Japan* **1985**, *58*, 1615–1616.
[102] M. Herberhold, P. Leitner, *J. Organomet. Chem.* **1987**, *336*, 153–161.
[103] A. G. Osborne, R. E. Hollands, J. A. K. Howard, R. F. Bryan, *J. Organomet. Chem.* **1981**, *205*, 395–406.
[104] E. W. Post, T. F. Crimmins, *J. Organomet. Chem.* **1978**, *161*, C17–C19.
[105] A. G. Osborne, R. H. Whiteley, *J. Organomet. Chem.* **1978**, *162*, 79–81.
[106] W. Reeve, E. F. Group, Jr., *J. Org. Chem.* **1967**, *32*, 122–125.
[107] S. Kato, M. Wakamatsu, M. Mizuta, *J. Organomet. Chem.* **1974**, *78*, 405–414.
[108] E. W. Post, R. G. Cooks, J. C. Kotz, *Inorg. Chem.* **1970**, *9*, 1670–1677.
[109] T. Lopez, A. Campero, *J. Organomet. Chem.* **1989**, *378*, 91–98.
[110] D. O. Cowan, P. Shu, F. L. Hedberg, M. Rossi, T. J. Kistenmacher, *J. Am. Chem. Soc.* **1979**, *101*, 1304–1306.
[111] cf. M. J. Cohn, M. D. Tinken, D. N. Hendrickson, *J. Am. Chem. Soc.* **1984**, *106*, 6683–6689.
[112] A. N. Nesmeyanov, D. A. Lemenovskii, E. G. Perevalova, *Izv. Akad. Nauk SSSR Ser. Khim.* **1975**, 1667–1668; *Bull. Acad. Sci. USSR Div. Chem. Sci.* **1975**, 1558–1559; *Chem. Abstr.* **1975**, *83*, 179186d.
[113] H. Rosenberg, US 3410883 (1966); *Chem. Abstr.* **1969**, *71*, P 13149w.
[114] A. N. Nesmeyanov, N. S. Kochetkova, *Uspekhi Khim.* **1974**, *43*, 1513–1523; *Russ. Chem. Rev.* **1974**, *43*, 710–715.
[115] G. R. Knox, P. L. Pauson, D. Willison, *J. Organomet. Chem.* **1993**, *450*, 177–184.
[116] M. Sato, M. Asai, *J. Organomet. Chem.* **1992**, *430*, 105–110.
[117] H. Paulus, K. Schlögl, W. Weissensteiner, *Monatsh. Chem.* **1982**, *113*, 767–780.
[118] P. L. Pauson, W. E. Watts, *J. Chem. Soc.* **1963**, 2990–2996.
[119] M. D. Rausch, A. Siegel, L. P. Klemann, *J. Org. Chem.* **1966**, *31*, 2703–2704.
[120] M. Rosenblum, N. Brawn, J. Papenmeier, M. Applebaum, *J. Organomet. Chem.* **1966**, *6*, 173–180.
[121] M. Rosenblum, N. Brawn, B. King, *Tetrahedron Letters* **1967**, *45*, 4421–4425.
[122] M. D. Rausch, F. A. Higbie, G. F. Westover, A. Clearfield, R. Gopal, J. M. Troup, I. Bernal, *J. Organomet. Chem.* **1978**, *149*, 245–264.
[123] C. Lewanda, D. O. Cowan, C. Leitch, K. Beechgard, *J. Am. Chem. Soc.* **1974**, *96*, 6788–6789.
[124] cf. C. Lewanda, K. Beechgard, D. O. Cowan, *J. Org. Chem.* **1976**, *41*, 2700–2704.
[125] K. Schlögl, H. Egger, *Monatsh. Chem.* **1963**, *94*, 376–392.
[126] K. Schlögl, W. Steyrer, *Monatsh. Chem.* **1965**, *96*, 1520–1535.
[127] K. Schlögl, W. Steyrer, *J. Organomet. Chem.* **1966**, *6*, 399–411.
[128] T. S. Abram, W. E. Watts, *Synth. Reactiv. Inorg. Metal-Org. Chem.* **1976**, *6*, 31–53.
[129] M. Herberhold, M. Ellinger, U. Thewalt, F. Stollmeier, *Angew. Chem.* **1982**, *94*, 70; *Angew. Chem. Int. Ed. Engl.* **1982**, *21*, 74–75.
[130] A. N. Nesmeyanov, V. A. Sazonova, V. I. Romanenko, *Dokl. Akad. Nauk SSSR* **1965**, *161*, 1085–1088; *Dokl. Chem. Proc. Acad. Sci. USSR* **1965**, *161*, 343–346; *Chem. Abstr.* **1965**, *61*, 4331e.
[131] W. E. Britton, R. Kashyap, M. El-Hashash, M. El-Kady, M. Herberhold, *Organometallics* **1985**, *5*, 1029–1031.
[132] G. P. Sollott, E. Howard, Jr., *J. Org. Chem.* **1962**, *27*, 4034–4040.
[133] G. P. Sollott, W. R. Petersen, Jr., *J. Organomet. Chem.* **1965**, *4*, 491–493; ibid. **1969**, *19*, 143–159.
[134] A. N. Nesmeyanov, D. N. Kursanov, N. V. Setkina, V. D. Vil'chevskaya, N. K. Baranetskaya, A. I. Krylova, L. A. Glushchenko, *Dokl. Akad. Nauk SSSR* **1971**, *199*, 1336–1338; *Dokl. Chem. Proc. Acad. Sci. USSR* **1971**, *199*, 719–721; *Chem. Abstr.* **1972**, *76*, 3999n.

[135] G. V. Gridunova, V. E. Shklover, Yu. T. Struchkov, V. D. Vil'chevskaya, N. L. Podobedova, A. I. Krylova, *J. Organomet. Chem.* **1982**, *238*, 297–305.
[136] F. Delgado-Pena, D. R. Talham, D. O. Cowan, *J. Organomet. Chem.* **1983**, *253*, C43–C46.
[137] J. C. Kotz, C. L. Nivert, J. M. Lieber, R. C. Reed, *J. Organomet. Chem.* **1975**, *91*, 87–95.
[138] C. U. Pittman, Jr., G. O. Evans, *J. Organomet. Chem.* **1972**, *43*, 361–367.
[139] J. C. Kotz, C. L. Nivert, *J. Organomet. Chem.* **1973**, *52*, 387–406.
[140] S. G. Baxter, R. L. Collins, A. H. Cowley, S. F. Sena, *Inorg. Chem.* **1983**, *22*, 3475–3479.
[141] D. J. Woollins, personal communication, **1993**.
[142] C. Spang, F. T. Edelmann, M. Noltemeyer, H. W. Roesky, *Chem. Ber.* **1989**, *122*, 1247–1254.
[143] G. P. Sollott, W. R. Petersen, Jr., *J. Org. Chem.* **1965**, *30*, 389–393.
[144] A. E. Ermoshkin, N. P. Makarenko, K. J. Sakodynskii, *J. Chromatogr.* **1984**, *290*, 377–391; *Chem. Abstr.* **1984**, *101*, 103306n.
[145] N. S. Nametkin, V. D. Tyurin, S. A. Sleptsova, A. M. Krapivin, A. Ya. Sideridu, *Izv. Akad. Nauk SSSR Ser. Khim.* **1982**, 955; *Bull. Acad. Sci. USSR Div. Chem. Sci.* **1982**, *31*, 848; *Chem. Abstr.* **1982**, *97*, 110145m.
[146] D. C. O'Connor Salazar, D. O. Cowan, *J. Organomet. Chem.* **1991**, *408*, 227–231.
[147] A. N. Nesmeyanov, E. G. Perevalova, O. A. Nesmeyanova, *Dokl. Akad. Nauk SSSR* **1958**, *119*, 288–291; *Dokl. Chem. Proc. Acad. Sci. USSR* **1958**, *119*, 215–217; *Chem. Abstr.* **1958**, *52*, 14579d.
[148] P. Shu, K. Bechgaard, D. O. Cowan, *J. Org. Chem.* **1976**, *41*, 1849–1852.
[149] F. L. Hedberg, H. Rosenberg, *J. Organomet. Chem.* **1971**, *28*, C14–C16.
[150] A. N. Nesmeyanov, E. G. Perevalova, O. N. Nesmeyanova, *Dokl. Akad. Nauk SSSR* **1955**, *100*, 1099–1101; *Chem. Abstr.* **1956**, *50*, 2558a.
[151] W. R. Cullen, J. D. Woollins, *Coord. Chem. Rev.* **1982**, *39*, 1–30.
[152] A. N. Nesmeyanov, N. N. Sedova, V. A. Sazonova, S. K. Moiseev, *J. Organomet. Chem.* **1980**, *185*, C6–C8.
[153] A. N. Nesmeyanov, V. A. Sazonova, N. N. Sedova, S. K. Moiseev, *Dokl. Akad. Nauk SSSR* **1980**, *252*, 361–363; *Dokl. Chem. Proc. Acad. Sci. USSR* **1980**, *250–255*, 227–229; *Chem. Abstr.* **1980**, *93*, 220883a.
[154] S. K. Moiseev, N. N. Sedova, V. A. Sazonova, *Dokl. Akad. Nauk SSSR* **1982**, *267*, 1374–1378; *Chem. Abstr.* **1983**, *98*, 160894w.
[155] S. K. Moiseev, N. N. Meleshonkova, V. A. Sazonova, *Koord. Khim.* **1988**, *14*, 328–331; *Chem. Abstr.* **1989**, *110*, 135440d.
[156] A. N. Nesmeyanov, E. G. Perevalova, D. A. Lemonovskii, V. P. Alekseev, K. I. Grandberg, *Dokl. Akad. Nauk SSSR* **1971**, *198*, 1099–1101; *Dokl. Chem. Proc. Acad. Sci. USSR* **1971**, *198*, 505–507; *Chem. Abstr.* **1971**, *75*, 63937e.
[157] M. I. Bruce, P. W. Jolly, F. G. A. Stone, *J. Chem. Soc. A* **1966**, 1602–1606.
[158] L. Zhu, L. M. Daniels, L. M. Peerey, N. M. Kostic, *Acta Cryst., Sect. C (Cryst. Struct. Commun.)* **1988**, *C44*, 1727–1729.
[159] G. Z. Suleimanov, P. V. Petrovskii, Yu. S. Bogachev, I. L. Zhuravleva, E. I. Fedin, I. P. Beletskaya, *J. Organomet. Chem.* **1984**, *262*, C35–C37.
[160] G. Z. Suleimanov, Yu. S. Bogatchev, L. T. Abdullaeva, I. L. Zhuravleva, Kh. S. Khalilov, L. F. Rybakova, I. P. Beletskaya, *Polyhedron* **1985**, *4*, 29–31.
[161] A. G. Osborne, R. H. Whiteley, *J. Organomet. Chem.* **1975**, *101*, C27–C28.
[162] M. S. Wrighton, M. C. Palazzotto, A. B. Bocarsly, J. M. Bolts, A. B. Fischer, L. Nadjo, *J. Am. Chem. Soc.* **1978**, *100*, 7264–7271.
[163] I. R. Butler, W. R. Cullen, S. J. Rettig, *Can. J. Chem.* **1987**, *65*, 1452–1456.
[164] A. B. Fischer, J. B. Kinney, R. H. Staley, M. S. Wrighton, *J. Am. Chem. Soc.* **1979**, *101*, 6501–6506.
[165] W. Finckh, B.-Z. Tang, D. A. Foucher, D. B. Zamble, R. Ziembinski, A. Lough, I. Manners, *Organometallics* **1993**, *12*, 823–829.
[166] H. Stoeckli-Evans, A. G. Osborne, R. H. Whiteley, *Helv. Chim. Acta* **1976**, *59*, 2402–2406.
[167] V. I. Popov, M. Lib, A. Haas, *Ukr. Khim. Zh. (Russ. Ed.)* **1990**, *56*, 1115–1116; *Chem. Abstr.* **1991**, *114*, 185701t.
[168] F. R. Mayers, A. G. Osborne, D. R. Rosseinsky, *J. Chem. Soc. Dalton Trans.* **1990**, 3419–3425

[169] A. J. Blake, F. R. Mayers, A. G. Osborne, D. R. Rosseinsky, *J. Chem. Soc. Dalton Trans.* **1982**, 2379–2383.
[170] N. N. Meleshonkova, M. Se Phan, S. K. Moiseev, *Zh. Obshch. Khim.* **1987**, *57*, 2606–2612; *Chem. Abstr.* **1989**, *110*, 24056p.
[171] N. B. Patrikeeva, O. N. Savorova, G. A. Domrachev, *Metalloorg. Khim.* **1991**, *4*, 682–683; *Chem. Abstr.* **1991**, *115*, 92529y.
[172] M. Wedler, H. W. Roesky, F. T. Edelmann, U. Behrens, *Z. Naturforsch. B* **1988**, *43b*, 1461–1467.
[173] G. A. Razuvaev, G. A. Domrachev, V. V. Sharutin, O. N. Suvorova, *J. Organomet. Chem.* **1977**, *141*, 313–317.
[174] G. A. Razuvaev, V. V. Sharutin, G. A. Domrachev, O. N. Suvorova, *Izv. Akad. Nauk SSSR Ser. Khim.* **1978**, 2177–2178; *Chem. Abstr.* **1979**, *90*, 23208w.
[175] M. Herberhold, H. Kniesel, L. Haumaier, U. Thewalt, *J. Organomet. Chem.* **1986**, *301*, 355–367.
[176] M. Herberhold, H. Kniesel, *J. Organomet. Chem.* **1989**, *371*, 205–218.
[177] M. Herberhold, H. Kniesel, L. Haumaier, A. Gieren, C. Ruiz-Pérez, *Z. Naturforsch.* **1986**, *41b*, 1431–1436.
[178] A. N. Nesmeyanov, L. G. Makarova, V. N. Vinogradova, *Izv. Akad. Nauk SSSR, Ser. Khim.* **1973**, 2796–2798; *Bull. Acad. Sci. USSR Div. Chem. Sci.* **1973**, 2731–2733; *Chem. Abstr.* **1974**, *80*, 96133w.
[179] M. Herberhold, H. Kniesel, *J. Organomet. Chem.* **1987**, *334*, 347–358.
[180] M. Herberhold, W. Feger, U. Kölle, *J. Organomet. Chem.* **1992**, *436*, 333–350.
[181] A. N. Nesmeyanov, L. G. Makarova, V. N. Vinogradova, *Izv. Akad. Nauk SSSR Ser. Khim.* **1971**, 892 and **1972**, 1600–1604; *Bull. Akad. Sci. USSR Div. Chem. Sci.* **1971**, 818 and **1972**, 1541–1544; *Chem. Abstr.* **1971**, *75*, 49290g and **1972**, *77*, 140255b.
[182] T. M. Miller, K. J. Ahmed, M. S. Wrighton, *Inorg. Chem.* **1989**, *28*, 2347–2355.
[183] A. N. Nesmeyanov, E. G. Perevalova, K. I. Grandberg, D. A Lemenovskii, T. V. Baukova, O. B. Afanassova, *J. Organomet. Chem.* **1974**, *65*, 131–144.
[184] K. H. Pannell, J. B. Cassias, G. M. Crawford, A. Flores, *Inorg. Chem.* **1976**, *15*, 2671–2675.
[185] Z. Dawoodi, C. Eaborn, A. Pidcock, *J. Organomet. Chem.* **1979**, *170*, 95–104.
[186] L. N. Zakharov, Yu. T. Struchkov, V. V. Sharutin, O. N. Suvarova, *Cryst. Struct. Commun.* **1979**, *8*, 439–444.
[187] A. Da Rold, Y. Mugnier, R. Broussier, B. Gautheron, E. Laviron, *J. Organomet. Chem.* **1989**, *362*, C27–C30.
[188] R. Broussier, A. Da Rold, B. Gautheron, *J. Organomet. Chem.* **1992**, *427*, 231–244.
[189] K.-H. Thiele, C. Krüger, T. Bartik, M. Dargatz, *J. Organomet. Chem.* **1988**, *352*, 115–124.
[190] A. G. Osborne, R. H. Whiteley, *J. Organomet. Chem.* **1979**, *181*, 425–437.
[191] E. G. Perevalova, K. I. Grandberg, D. A. Lemonovskii, T. V. Baukova, *Izv. Akad. Nauk SSSR Ser. Khim.* **1971**, 2077–2078; *Bull. Acad. Sci. USSR Div. Chem. Sci.* **1971**, 1967–1969; *Chem. Abstr.* **1972**, *76*, 14673c.
[192] A. N. Nesmeyanov, E. G. Perevalova, K. I. Grandberg, D. A. Lemonovskii, *Izv. Akad. Nauk SSSR Ser. Khim.* **1974**, 1124–1137; *Bull. Acad. Sci. USSR Div. Chem. Sci.* **1974**, 1068–1078 (Review).
[193] E. G. Perevalova, K. I. Grandberg, V. P. Dyadchenko, O. N. Kalinina, *Koord. Khim.* **1988**, *14*, 1145; *Chem. Abstr.* **1989**, *111*, 39486y.
[194] E. G. Perevalova, K. I. Grandberg, V. P. Dyadchenko, O. N. Kalinina, *J. Organomet. Chem.* **1988**, *352*, C37–C41.
[195] A. N. Nesmeyanov, E. G. Perevalova, D. A. Lemonovskii, A. N. Kosina, K. I. Grandberg, *Izv. Akad. Nauk SSSR Ser. Khim.* **1969**, 2030–2031; *Bull. Acad. Sci. USSR Div. Chem. Sci.* **1969**, 1876–1878; *Chem. Abstr.* **1970**, *72*, 21768h.
[196] E. G. Perevalova, D. A. Lemonovskii, K. I. Grandberg, A. N. Nesmeyanov, *Dokl. Akad. Nauk SSSR* **1972**, *202*, 93–96; *Dokl. Chem. Proc. Acad. Sci. USSR* **1972**, *202–207*, 11–14; *Chem. Abstr.* **1972**, *76*, 99798v.
[197] V. G. Andrianov, Yu. T. Struchkov, E. R. Rossinskaya, *J. Chem. Soc. Chem. Commun.* **1973**, 338–339; *Zh. Strukt. Khim.* **1974**, *15*, 74–82; *J. Struct. Chem. [USSR]* **1974**, *15*, 65–72.

[198] H. Lehmkuhl, R. Schwickardi, C. Krüger, G. Raabe, *Z. Anorg. Allg. Chem.* **1990**, *581*, 41–47; cf. H. Lehmkuhl, *Pure Appl. Chem.* **1990**, *62*, 731–740.

[199] A. N. Nesmeyanov, E. G. Perevalova, O. B. Afanasova, M. N. Elinson, K. I. Grandberg, *Izv. Akad. Nauk SSSR Ser. Khim.* **1975**, 477–478; *Bull. Acad. Sci. USSR Div. Chem. Sci.* **1975**, 408–409; *Chem. Abstr.* **1975**, *83*, 10347w.

[200] A. N. Nesmeyanov, E. G. Perevalova, O. B. Afanasova, M. V. Tolstaya, K. I. Grandberg, *Izv. Akad. Nauk SSSR Ser. Khim.* **1978**, 1118–1122; *Bull. Acad. Sci. USSR Div. Chem. Sci.* **1978**, 969–972; *Chem. Abstr.* **1978**, *89*, 109836r.

[201] S. Schreiber Crawford, H. D. Kaesz, *Inorg. Chem.* **1977**, *16*, 3193–3201.

[202] S. Schreiber Crawford, G. Firestein, H. D. Kaesz, *J. Organomet. Chem.* **1975**, *91*, C57–C60.

[203] J. C. Gaunt, B. L. Shaw, *J. Organomet. Chem.* **1975**, *102*, 511–516; M. Pfeffer, M. A. Rotteveel, J.-P. Sutter, A. De Cian, J. Fischer, *J. Organomet. Chem.* **1989**, *317*, C21–C25.

[204] L. L. Troitskaya, V. I. Sokolov, *J. Organomet. Chem.* **1987**, *328*, 169–172.

[205] I. R. Butler, W. R. Cullen, *Organometallics* **1984**, *3*, 1846–1851.

[206] I. R. Butler, W. R. Cullen, S. J. Rettig, *Organometallics* **1987**, *6*, 872–880.

[207] I. R. Butler, W. R. Cullen, F. W. B. Einstein, A. C. Willis, *Organometallics* **1985**, *4*, 603–604.

[208] U. Thewalt, D. Schomburg, *Z. Naturforsch.* **1975**, *30b*, 636.

[209] F. H. Köhler, W. A. Geike, P. Hofmann, U. Schubert, P. Stauffert, *Chem. Ber.* **1984**, *117*, 904–914.

[210] E. G. Perevalova, T. V. Baukova, M. M. Sazonenko, K. I. Grandberg, *Izv. Akad. Nauk SSSR Ser. Khim.* **1985**, 1877–1881; *Bull. Acad. Sci. USSR Div. Chem. Sci.* **1985**, *34*, 1727–1730; *Chem. Abstr.* **1985**, *105*, 95633f.

[211] M. Tsutsui, N. Ely, *J. Am. Chem. Soc.* **1974**, *96*, 3560–3561; M. Tsutsui, N. Ely, A. Gebala, *Inorg. Chem.* **1974**, *14*, 78–81.

[212] R. Broussier, A. Da Rold, B. Gautheron, Y. Dromzee, Y. Jeannin, *Inorg. Chem.* **1990**, *29*, 1817–1822.

[213] I. G. Morrison, P. L. Pauson, *Proc. Chem. Soc.* **1962**, 177; S. McVay, I. G. Morrison, P. L. Pauson, *J. Chem. Soc.* **1967C**, 1847–1850.

[214] M. D. Rausch, *J. Org. Chem.* **1961**, *26*, 1802–1805.

[215] V. A. Nefedov, L. K. Tarygina, *Zh. Org. Khim.* **1976**, *12*, 2012–2019; *J. Org. Chem. [USSR]* **1976**, *12*, 1960–1965; *Chem. Abstr.* **1977**, *86*, 16722q.

[216] V. N. Drozd, V. A. Sazonova, A. N. Nesmeyanov, *Dokl. Akad. Nauk SSSR* **1964**, *159*, 591–594; *Dokl. Chem. Proc. Acad. Sci. USSR* **1964**, *154–159*, 1213–1216; *Chem. Abstr.* **1965**, *62*, 7794g.

[217] M. Herberhold, M. Hübner, B. Wrackmeyer, *Z. Naturforsch.* **1993**, *48b*, 940–950.

[218] R. V. Honeychuck, M. O. Okoroafor, L. H. Shen, C. H. Brubaker, Jr., *Organometallics* **1986**, *5*, 482–490.

[219] V. A. Nefedov, *Zh. Obshch. Khim.* **1968**, *38*, 2191–2193; *J. Gen. Chem. [USSR]* **1968**, *38*, 2122–2123.

[220] A. G. Osborne, H. M. Pain, M. B. Hursthouse, M. A. Mazid, *J. Organomet. Chem.* **1993**, *453*, 117–120.

[221] A. Eisenstadt, G. Scharf, B. Fuchs, *Tetrahedron Letters* **1971**, No. 8, 679–682.

[222] R. Broussier, M. El Mjidi, B. Gautheron, *J. Organomet. Chem.* **1991**, *408*, 381–393.

[223] M. Herberhold, C. Dörnhöfer, A. Scholz, in preparation.

[224] M. Herberhold, C. Dörnhöfer, A. Scholz, G.-X. Jin, *Phosphorus, Sulfur and Silicon* **1992**, *64*, 161–168.

[225] J. A. Adeleke, Yu-W. Chen, L.-K. Liu, *Organometallics* **1992**, *11*, 2543–2550.

[226] E. W. Abel, N. J. Long, K. G. Orrell, A. G. Osborne, V. Šik, P. A. Bates, M. B. Hursthouse, *J. Organomet. Chem.* **1989**, *367*, 275–289.

[227] E. W. Abel, N. J. Long, K. G. Orrell, A. G. Osborne, V. Šik, P. A. Bates, M. B. Hursthouse, *J. Organomet. Chem.* **1990**, *383*, 253–269.

[228] E. W. Abel, N. J. Long, K. G. Orrell, A. G. Osborne, V. Šik, *J. Organomet. Chem.* **1991**, *405*, 375–382.

[229] E. W. Abel, N. J. Long, K. G. Orrell, A. G. Osborne, V. Šik, *J. Organomet. Chem.* **1989**, *378*, 473–483.

[230] K. G. Orrell, V. Šik, C. H. Brubaker, Jr., B. McCulloch, *J. Organomet. Chem.* **1984**, *276*, 267–279.

[231] M. Sato, M. Sekino, S. Akabori, *J. Organomet. Chem.* **1988**, *344*, C31–C34.

[232] M. Sato, M. Sekino, M. Katada, S. Akabori, *J. Organomet. Chem.* **1989**, *377*, 327–337.

[233] T. Vondrák, M. Sato, *J. Organomet. Chem.* **1989**, *364*, 207–215.

[234] N. L. Holy, T. Nalesnik, L. Warfield, M. Mojesky, *J. Coord. Chem.* **1983**, *12*, 157–162.

[235] M. Herberhold, Cl. Breutel, to be published.

[236] M. Herberhold, J. Ott, *Z. Naturforsch.* **1988**, *43b*, 682–686.

[237] P. D. Beer, S. M. Charlsley, C. J. Jones, J. A. McCleverty, *J. Organomet. Chem.* **1986**, *307*, C19–C22.

[238] M. Herberhold, G.-X. Jin, I. Trukenbrod, W. Milius, in preparation.

[239] D. Seyferth, B. W. Hames, *Inorg. Chim. Acta* **1983**, *77*, L1–L2.

[240] W. R. Cullen, A. Talaba, S. J. Rettig, *Organometallics* **1992**, *11*, 3152–3156.

[241] M. Kumada, T. Kondo, K. Mimura, M. Ishikawa, K. Yamamoto, S. Ikeda, M. Kondo, *J. Organomet. Chem.* **1972**, *43*, 293–305.

[242] W. Finckh, B.-Z. Tang, A. Lough, I. Manners, *Organometallics* **1992**, *11*, 2904–2911.

[243] V. V. Dement'ev, F. Cervantes-Lee, L. Parkanyi, H. Sharma, K. H. Pannell, M. T. Nguyen, A. Diaz, *Organometallics* **1993**, *12*, 1983–1987.

[244] M. Burke Laing, K. N. Trueblood, *Acta Cryst.* **1965**, *19*, 373–381; see also T. H. Barr, W. E. Watts, *Tetrahedron* **1968**, *24*, 6111–6118.

[245] M. Kumuda, T. Kondo, K. Mimura, K. Yamamoto, M. Ishikawa, *J. Organomet. Chem.* **1972**, *43*, 307–314.

[246] R. A. Abramovitch, C. I. Azogu, R. G. Sutherland, *Chem. Commun.* **1969**, 1439–1440.

[247] R. A. Abramovitch, J. L. Atwood, M. L. Good, B. A. Lampert, *Inorg. Chem.* **1975**, *14*, 3085–3089.

[248] R. Broussier, personal communication (1989).

[249] R. E. Hollands, A. G. Osborne, I. Townsend, *Inorg. Chim. Acta* **1979**, *37*, L541.

[250] M. Herberhold, P. Leitner, U. Thewalt, *Z. Naturforsch.* **1990**, *45b*, 1503–1507.

[251] M. Herberhold, P. Leitner, *J. Organomet. Chem.* **1991**, *411*, 233–237.

[252] B. R. Davis, I. Bernal, *J. Cryst. Mol. Struct.* **1972**, *2*, 107–114.

[253] A. J. Blake, R. O. Gould, A. G. Osborne, *J. Organomet. Chem.* **1986**, *308*, 297–302.

[254] A. Davison, J. C. Smart, *J. Organomet. Chem.* **1979**, *174*, 321–334.

[255] E. W. Abel, M. Booth, K. G. Orrell, *J. Organomet. Chem.* **1980**, *186*, C37–C41.

[256] E. W. Abel, M. Booth, K. G. Orrell, *J. Organomet. Chem.* **1981**, *208*, 213–224.

[257] E. W. Abel, M. Booth, C. A. Brown, K. G. Orrell, R. L. Woodford, *J. Organomet. Chem.* **1981**, *214*, 93–105.

[258] E. W. Abel, K. G. Orrell, A. G. Osborne, V. Šik, W. Guoxiong, *J. Organomet. Chem.* **1991**, *411*, 239–249.

[259] D. Fest, C. D. Habben, *J. Organomet. Chem.* **1990**, *390*, 339–344.

[260] A. G. Osborne, R. E. Hollands, R. F. Bryan, S. Lockhart, *J. Organomet. Chem.* **1982**, *226*, 129–142.

[261] A. G. Osborne, R. E. Hollands, A. G. Nagy, *J. Organomet. Chem.* **1989**, *373*, 229–234.

[262] A. G. Osborne, A. J. Blake, R. E. Hollands, R. F. Bryan, S. Lockhart, *J. Organomet. Chem.* **1985**, *287*, 39–47.

[263] A. G. Osborne, R. E. Hollands, R. F. Bryan, S. Lockhart, *J. Organomet. Chem.* **1985**, *288*, 207–217.

[264] B. Gautheron, G. Teinturier, *J. Organomet. Chem.* **1984**, *262*, C30–C34

[265] R. Broussier, Y. Gobet, R. Amardeil, A. Da Rold, M. M. Kubicki, B. Gautheron, *J. Organomet. Chem.* **1993**, *445*, C4–C5.

[266] R. P. Sidebotham, P. D. Beer, T. A. Hamor, C. J. Jones, J. A. McCleverty, *J. Organomet. Chem.* **1989**, *371*, C31–C34.

[267] M. Herberhold, G.-X. Jin, to be published.

[268] D. Seyferth, B. W. Hames, T. G. Rucker, M. Cowie, R. S. Dickson, *Organometallics* **1983**, *2*, 472–474; M. Cowie, R. S. Dickson, *J. Organomet. Chem.* **1987**, *326*, 269–280.

[269] S. Akabori, T. Kumagai, T. Shirahige, S. Sato, K. Kawazoe, C. Tamura, M. Sato, *Organometallics* **1987**, *6*, 526–531.
[270] B. Czech, A. Piorko, R. Annunziata, *J. Organomet. Chem.* **1983**, *255*, 365–369.
[271] N. J. Long, S. J. Sharkey, M. B. Hursthouse, M. A. Mazid, *J. Chem. Soc. Dalton* **1993**, 23–26.
[272] C. P. Galloway, T. B. Rauchfuss, *Angew. Chemie* **1983**, *105*, 1407–1409; *Angew. Chem. Int. Ed. Engl.* **1993**, *32*, 1319–1321.
[273] P. F. Brandt, T. B. Rauchfuss, *J. Am. Chem. Soc.* **1992**, *114*, 1926–1927.
[274] D. L. Compton, T. B. Rauchfuss, *ACS Polym. Prepr.* **1993**, *34*, 351–354.
[275] J. R. Dilworth, S. K. Ibrahim, *Trans. Met. Chem.* **1991**, *16*, 239–240.
[276] M. Herberhold, C. Dörnhöfer, U. Thewalt, *Z. Naturforsch.* **1990**, *45b*, 741–746.
[277] J. A. Kramer, F. H. Herbstein, D. N. Hendrickson, *J. Am. Chem. Soc.* **1980**, *102*, 2293–2301.
[278] J. C. W. Chien, R. O. Gooding, C. P. Lillya, *Polym. Mater. Sci. Eng.*, **1983**, *49*, 107–111; *Chem. Abstr.* **1984**, *100*, 149187g.
[279] R. F. Kovar, M. D. Rausch, H. Rosenberg, *Organomet. Chem. Synth.* **1970**, *1*, 173–181; *Chem. Abstr.* **1971**, *74*, 112178y.
[280] E. W. Abel, N. J. Long, A. G. Osborne, M. B. Hursthouse, M. A. Mazid, *J. Organomet. Chem.* **1992**, *430*, 117–122.
[281] M. Buchmeiser, H. Schottenberger, *J. Organomet. Chem.* **1992**, *436*, 223–230.
[282] H. Schottenberger, M. Buchmeiser, J. Polin, K.-E. Schwarzhans, *Z. Naturforsch.* **1993**, *48b*, 1524–1532.
[283] N. N. Sedova, S. K. Moiseev, V. A. Sazonova, *J. Organomet. Chem.* **1982**, *224*, C53–C56.
[284] T. Kinoshita, S. Tsuji, T. Otake, K. Takeuchi, *J. Organomet. Chem.* **1993**, *458*, 187–191.
[285] K. Sonogashira, T. Yoshida, K. Onitsuka in *J. Organomet. Chem. Conference* (Ed.: W. A. Herrmann), Elsevier, Amsterdam, **1993**, p. 121.
[286] U. T. Mueller-Westerhoff, Z. Zang, G. Ingram, *J. Organomet. Chem.* **1993**, *463*, 163–167.
[287] Y. Nishibayashi, T. Chiba, J. D. Singh, S. Uemura, S. Fukuzawa, *J. Organomet. Chem.* **1994**, *473*, 205–213.
[288] H. B. Singh, A. Regina V, J. P. Jasinski, E. S. Paight, R. J. Butcher, *J. Organomet. Chem.* **1994**, *464*, 87–94.
[289] H. Schulz, K. Folting, J. C. Huffman, W. E. Streib, M. H. Chisholm, *Inorg. Chem.* **1993**, *32*, 6056–66.
[290] K. Sonogashira, personal communication (1994).
[291] S. Tanaka, T. Yoshida, T. Adachi, T. Yoshida, K. Onitsuka, K. Sonogashira, *Chemistry Letters* **1994**, 877–880.
[292] E. W. Abel, N. J. Long, K. G. Orrell, V. Šik, G. N. Ward, *J. Organomet. Chem.* **1993**, *462*, 287–293.
[293] M. Sato, K. Suzuki, H. Asano, M. Sekino, Y. Kawata, Y. Habata, S. Akabori, *J. Organomet. Chem.* **1994**, *470*, 263–269.

Received: December 1, 1993

6 Macrocycles and Cryptands Containing the Ferrocene Unit

C. Dennis Hall

6.1 Introduction

There are thousands of discoveries in molecular science reported every year but very few of these are destined to promote a new generation of research activity. The serendipitous preparation of di-benzo-18-crown-6 **1** by Pedersen in 1967 [1] and the subsequent discovery [1, 2] that **1** and other crown ethers selectively complex biologically relevant alkali and alkaline earth cations was, however, the catalyst for a huge explosion of activity in the field of host–guest or "supramolecular" chemistry. The resulting inspired and innovative work by Lehn [3, 4] on, in particular, the 3-dimensional bicyclic cryptands (e.g., **2**) and by Cram [5] on chiral crown ethers and rigid spherands (e.g., **3**) was recognised by the award to Pedersen [3], Lehn [4] and Cram [5] of the 1987 Nobel Prize for Chemistry. Some excellent reviews and monographs are available on the general subject [6–10], which are recommended as supplementary reading.

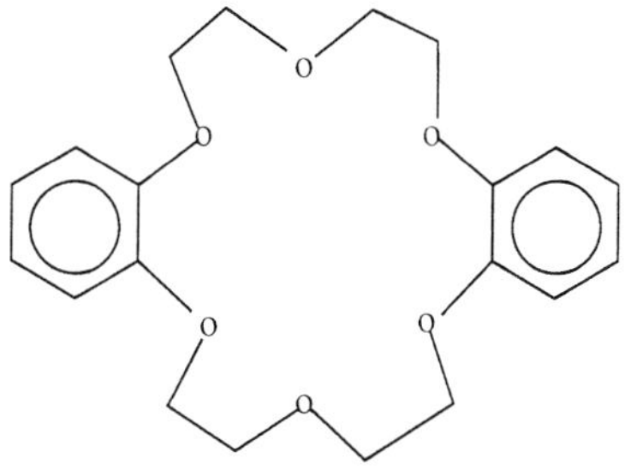

(1) 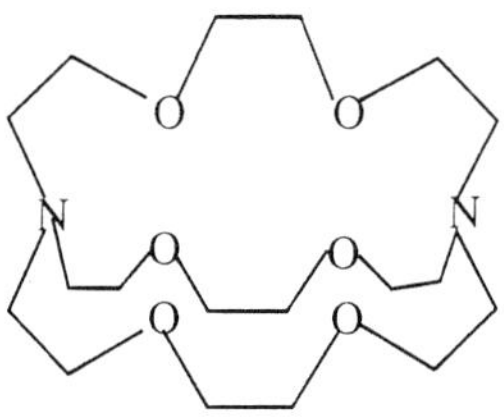 (2)

Stimulated by the concept of creating models for reactions that play a fundamental role in biological processes [11], such as ion transport across membranes and oxidative phosphorylation, recent attention has focused on a new generation of abiotic host molecules that contain responsive functions either attached to, or an integral part of, a macrocyclic framework. The signal associated with complex formation may originate from pH-responsive [12], photochemically responsive [13], or thermally responsive [14] receptors, whose binding strength and selectivity may be influenced by the appropriate chemical or physical stimulus. This chapter, however, will focus on macrocycles and cryptands that contain an electrochemical

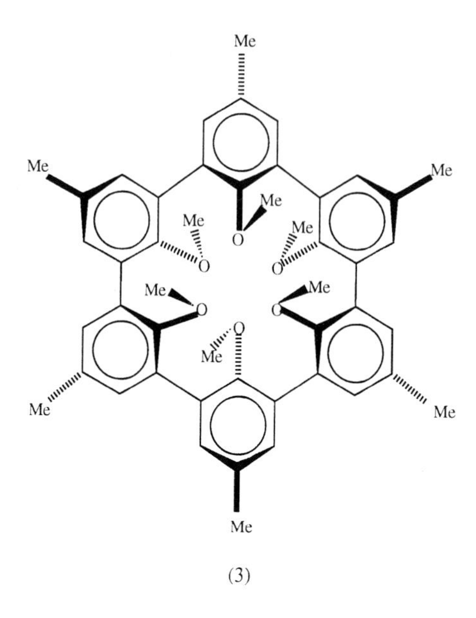

(3)

response function (a redox-active centre) – in particular, the ferrocene unit. By suitable variation of the size and structure of the host cavity these molecules can be made to bind cationic, anionic or even neutral guest species by means of electrostatic interactions either through space or, in appropriate cases, through bonds between the binding sites and the metallocene unit. Selective binding of guest species produces supramolecular systems that have potential applications in the fields of chemical sensors [15], cation transport across membranes [16], molecular electronics [17], and as catalysts either in chemical processes (e.g., polymerisation) or as mimics for metallo-enzymes.

In general, there are two types of macrocycles or cryptands containing the ferrocene unit. The first is that in which the ferrocene is appended to either a macrocycle – represented schematically by Fig. 6-1, or to a cryptand, spherand or cavitand (Fig. 6-2). The second class of compounds has ferrocene incorporated

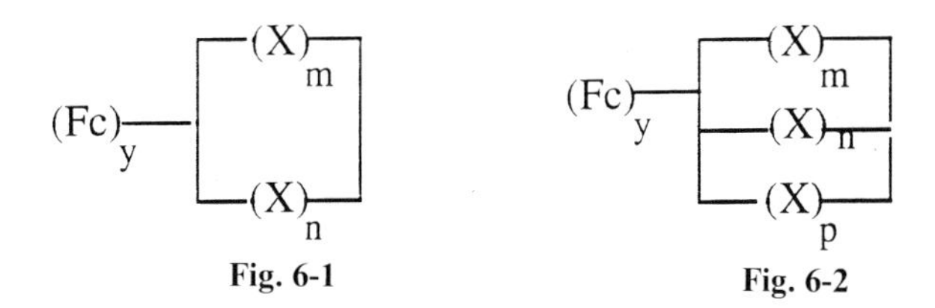

Fig. 6-1 **Fig. 6-2**

Fig. 6-1. Schematic representation of a ferrocene-containing macrocycle with the ferrocene moiety appended to the macrocycle. Fc = 1-substituted or 1,1′-disubstituted ferrocene; x = O, S, or NR; m, n, p = 1, 2, or 3; y = number of ferrocene units attached (usually between 1 and 4).

Fig. 6-2. Schematic representation of a ferrocene-containing macrocycle in which the ferrocene moiety is attached to a cryptand, spherand or cavitand. Fc = 1-substituted or 1,1′-disubstituted ferrocene; x = O, S, or NR; m, n, p = 1, 2, or 3; y = number of ferrocene units attached, usually 1 to 4.

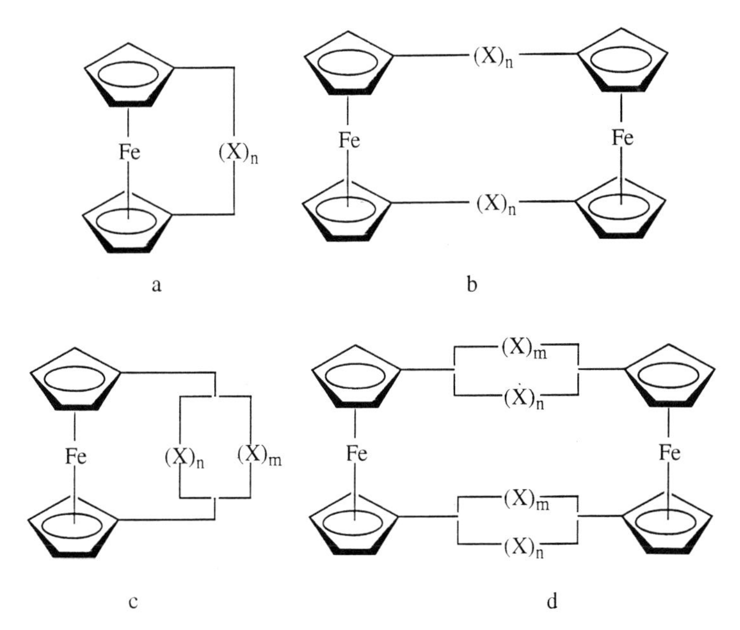

Fig. 6-3. Schematic representation of a ferrocene-containing macrocycle where the ferrocene is incorporated into the macrocyclic frame (ferrocenophane); x = O, S, or NR; m, n = 1, 2 or 3.

within the macrocycle (Fig. 6-3a, b) or cryptand (Fig. 6-3c, d). The permutation of ring size, type, and number of heteroatoms and, of course, the metallocene metal, clearly gives rise to a large range of compounds that have been classified as ferrocenophanes and our discussion will begin with this category of ferrocene-containing molecules. It should be noted that the material will not be treated in a chronological order but instead will be discussed in terms of the various structural types represented by Fig. 6-3a–d.

6.2 The Ferrocenophanes

Although a number of ferrocene crown ethers **5a**–**d** were first reported by Biernat and Wilczewski [18], the range was extended and the yields improved by a method developed by Akabori et al. starting from **4** [19]. The reaction is readily adaptable to the synthesis of ferrocenophanes containing other metallic units (e.g., ruthenocene) [20], but this is outside the scope of the present review. Pedersen's method [21] was used to test the extraction of alkali and alkaline earth metal cations from the aqueous phase into organic media (usually dichloromethane), which was generally found to be poor (see Table 6-1), except for Tl^+, which was extracted at a level of about 80% with **5a**. A size effect is evident here (e.g., extraction of K^+, **5a** > **5b**) and, interestingly, the use of Ag^+ ion (diameter = 2.52 Å) led to oxidative decomposition indicating an interaction between the complexed silver and the

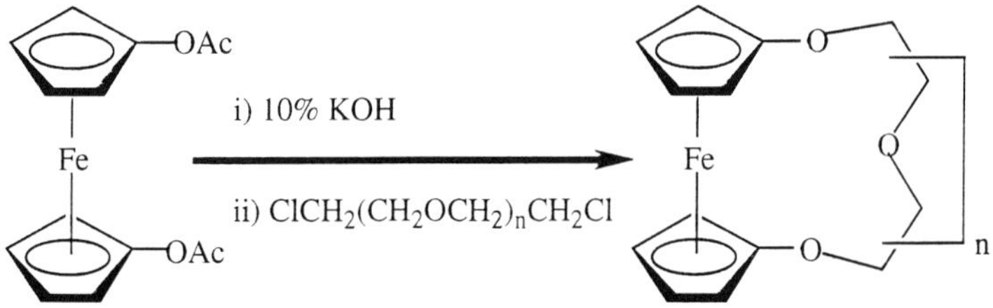

i) 10% KOH

ii) $ClCH_2(CH_2OCH_2)_nCH_2Cl$

(4)

(5a-e)

a) n = 4, 60%
b) n = 3, 58%
c) n = 2, 50%
d) n = 1, 32%
e) n = 0, 26%

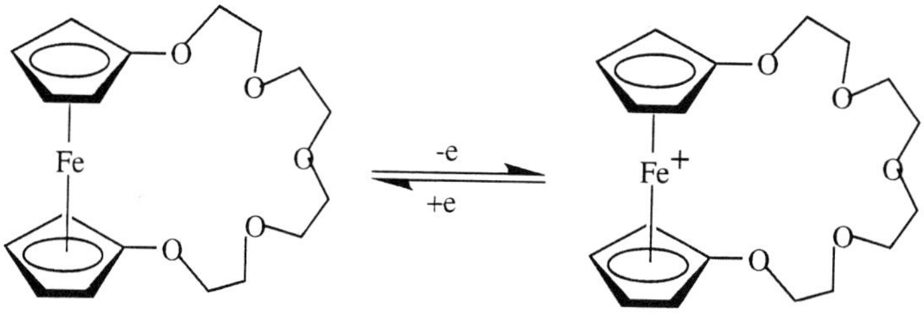

-e

+e

(5b)

$(5b)^+$

macrocycle. Macrocycle–cation interaction was confirmed by Saji, who first reported anodic shifts in the oxidation potential of **5b** resulting from addition of alkali metal salts [22]. Two distinct cyclic voltammetry (CV) waves corresponding to complexed and uncomplexed **5b** were observed for both Na^+ and Li^+ guest cations (Fig. 6-4). The oxidation of free and complexed macrocycle may be represented by Scheme 6-1, for which K_2 would be expected to be less than K_1, due to electrostatic repulsion. A digital simulation led Kaifer et al. [23] to conclude that two distinct waves will be observed only when K_1 is large (i.e., $>10^4$). When this is the case, however, quantitative values of binding *enhancements* may be

Table 6-1. Extraction of metal picrates from aqueous to organic phase (%)[a]

Compound	Metal cation (diameter, Å)				
	Na^+ (1.90)	K^+ (2.66)	Rb^+ (2.96)	Cs^+ (3.34)	Tl^+ (2.99)
5a	1.62	22.7	29.1	18.4	79.7
5b	1.49	4.5	4.6	5.6	25.2
B-18-C-6	22.3	94.7	–	–	97.3
B-15-C-5	23.9	50.1	–	–	56.0

a Solvent: H_2O/CH_2Cl_2, 1:1; [Macrocycle] $= 7\times10^{-4}$ *M* [picric acid] $= 7\times10^{-5}$ *M*, $[MNO_3]$ $= 0.1$ *M*.

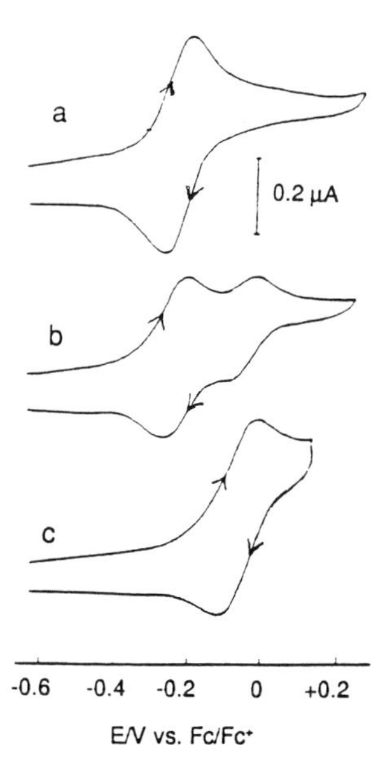

Fig. 6-4. Cyclic voltammograms of **5b** in CH_2Cl_2 in the presence of $NaClO_4$: (a) 0.0 equiv., (b) 0.5 equiv., (c) 1.0 equiv. of $NaClO_4$.

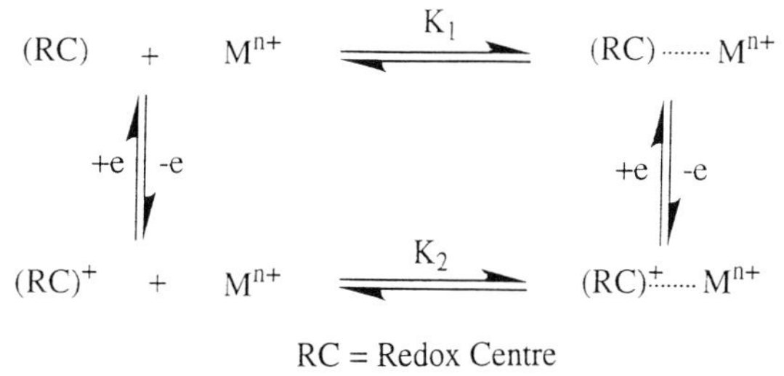

$$\begin{array}{ccccc} (RC) & + & M^{n+} & \underset{}{\overset{K_1}{\rightleftharpoons}} & (RC)\cdots M^{n+} \\ {\scriptstyle +e}\uparrow\downarrow{\scriptstyle -e} & & & & {\scriptstyle +e}\uparrow\downarrow{\scriptstyle -e} \\ (RC)^{+} & + & M^{n+} & \underset{}{\overset{K_2}{\rightleftharpoons}} & (RC)^{+}\cdots M^{n+} \end{array}$$

RC = Redox Centre

Scheme 6-1

calculated from the differences of formal redox potentials for the free ligand E_f^{free} or the metal complex $E_f^{complex}$ using Equation 6-1, where ΔE_f is the shift in reduction potential caused by the complexation with metal cations and n represents the number of electrons involved in the redox process.

$$\Delta E_f = (E_f^{complex} - E_f^{free}) = -RT/nF \cdot \ln (K_1/K_2) \quad (6\text{-}1)$$

Thus, the quantitative decrease in metal binding capacity on oxidation may be calculated as 40 and 742 for Na^+ and Li^+, respectively. The facility to switch cation binding on and off electrochemically has been used by Saji et al. [24] to transport alkali metal cations across a membrane using **5b** as the carrier. Scheme 6-2 shows the mechanism of electrochemical redox-driven metal ion transport. Initially at the left aqueous-dichloromethane interface, **5b** extracts a metal cation to form the complex **5b** · M^+. Diffusion across the membrane followed by electrochemical

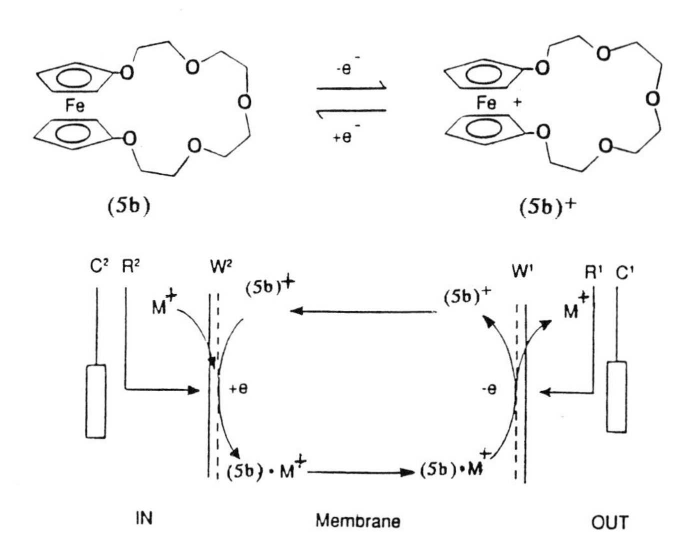

Scheme 6-2. Electrochemical ion transport utilising a redox active crown ether. W^1, W^2: minigrid platinum working electrodes. C^1, C^2: platinum plate counter electrodes. R^1, R^2: saturated calomel reference electrodes.

oxidation by electrode W^1 at the right hand phase interface releases the metal ion to the aqueous phase. The ferrocenium derivative $\mathbf{5b}^+$ then diffuses back to the left interface, where it is electrochemically reduced at electrode W^2 back to **5b**.

The polyoxaferrocenophanes **5a** – **e** have been extended to include sulfur and nitrogen as the heteroatoms in the macrocyclic ring system. Thus compounds of type **7** containing one sulfur atom were prepared in modest yields by using sodium sulfide in dimethylformamide (DMF) to cyclize 1,1′-bis-(ω-chloropolyethyleneoxy)-ferrocene **6a** – **c** [25].

Analogous compounds **8a** – **c** with more than one sulfur atom in the ring have also been prepared by Akabori [26] and the synthesis led to ring-expanded products **9** and **10**, probably by the route illustrated in Scheme 6-3. All the polyoxathiaferrocenophanes showed little or no ability to extract hard alkali or alkaline earth cations

Fe
Cl
Cl
Na_2S
DMF
Fe
S

(6a-c) a , n = 3
b , n = 2
c , n = 1

(7a-c)

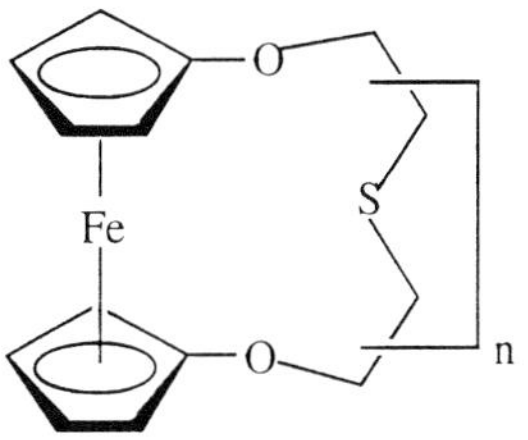

a, n = 4
b, n = 3
c, n = 2

(8 a-c)

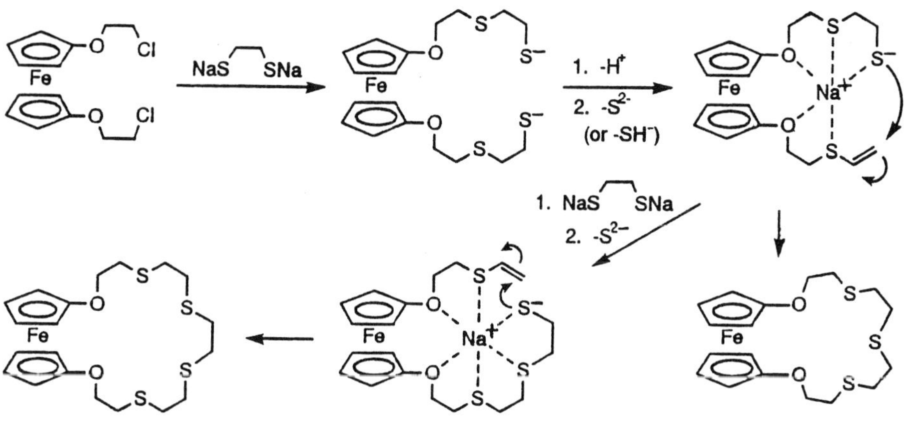

(9)

(10)

Scheme 6-3

from water but the softer Ag^+ ion was extracted in the order **8b** (76%) > **7c** (30%) > **5b** (12%), that is, in the same order as the decreasing number of sulfur atoms in the ring. Silver ion again decomposed the ferrocenophanes, but Cu^{2+} was extracted strongly by **8a** and **8b** [26]. A second group of oxathiaferrocenophanes in which the sulfur atoms are attached directly to the ferrocene rings is represented by **11a–d** obtained in moderate yield by the reaction of disodium-1,1′-ferrocenedithiolate and the dibromide of the appropriate polyethyleneglycol [27]. Dimers of type **12** were also obtained in various amounts during the synthesis and complex formation between both series of compounds and a variety of metal cations was examined by extraction and by spectroscopic methods. Alkali metals were not extracted by **11** or **12**, but both sets of compounds showed a great affinity for Ag^+ (50–94%, depending on the macrocyclic structure) and interestingly, in this case, there was no evidence for decomposition of the host molecules. Hypsochromic shifts and a decrease in intensity of the d–d absorption of ferrocene at $\lambda = 440$ nm were observed for **11** and **12** on complexation with silver nitrate and a stable 1 : 1 complex of **11b** with $AgNO_3$ was isolated as yellow crystals from acetonitrile.

The syntheses of dithiaoxa-[n]-ferrocenophanes of types **13** and **14** have also been reported [28a]. These compounds showed some affinity for alkali metal cations but

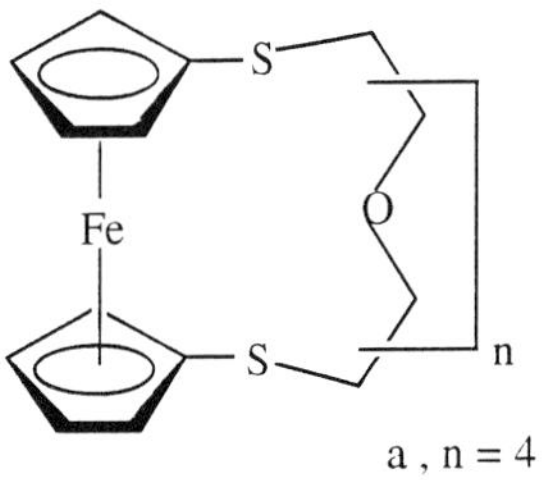

a, n = 4
b, n = 3
c, n = 2
d, n = 1

(11 a-d)

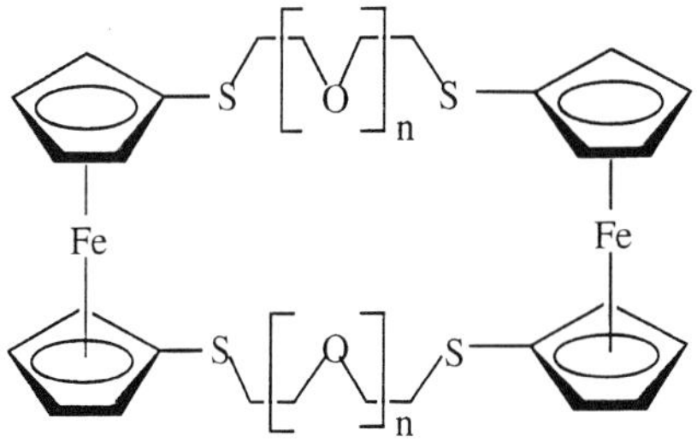

(12)

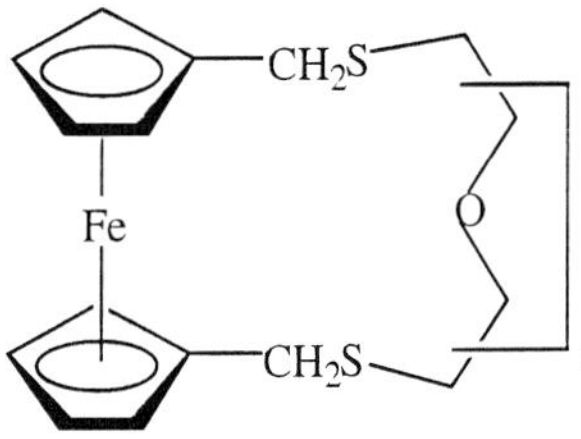

(13)

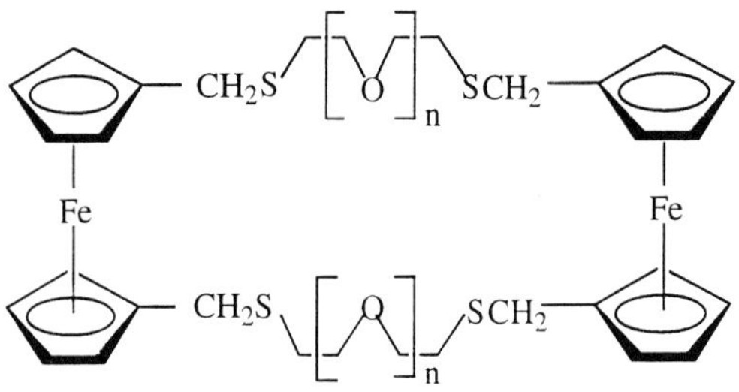

(14)

there was no evidence of interaction between the ferrocene units and the complexed cation, presumably because of the distance involved and the insulation provided by the methylene groups to through-bond interaction. Single crystal X-ray structures have been reported for macrocycles of this type [29] and a modification of the synthetic method has been used to prepare some all-oxygen analogues of **13** [28b].

It is well known that sulfide ligands form stable complexes with soft metal cations [30] and numerous complexes of macrocyclic thioethers with various metal salts have been prepared and characterized [31]. The iron atom of ferrocene is also basic and therefore affords a potential coordination site as evidenced by derivatives of ferrocene that are known to form complexes with certain metal salts [32]. Thus,

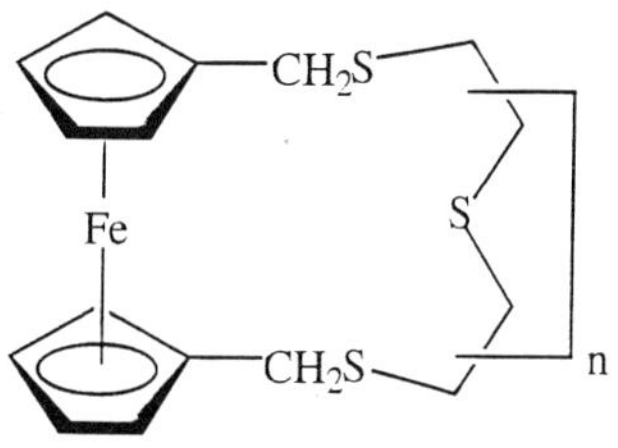

(15) 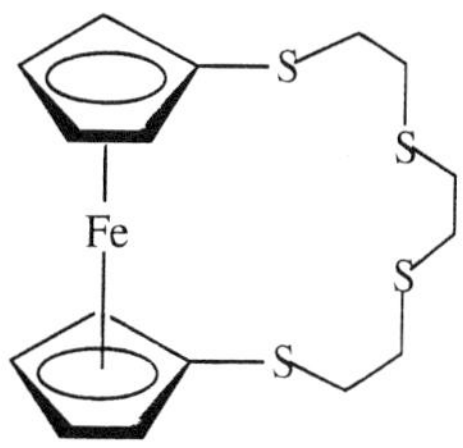 (16)

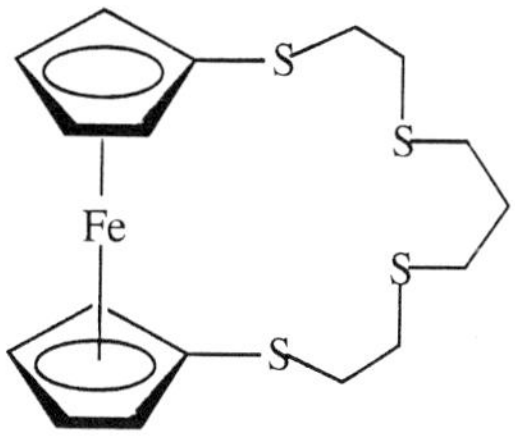

(17) 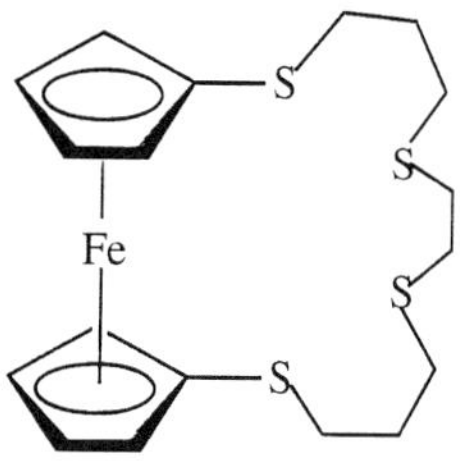 (18)

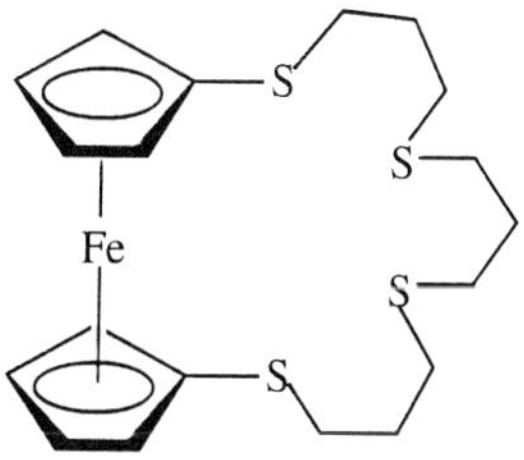

(19) 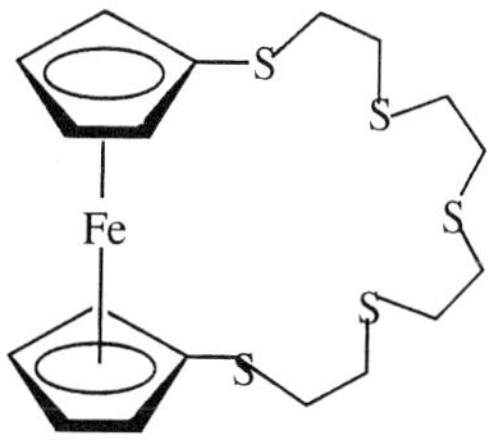 (20)

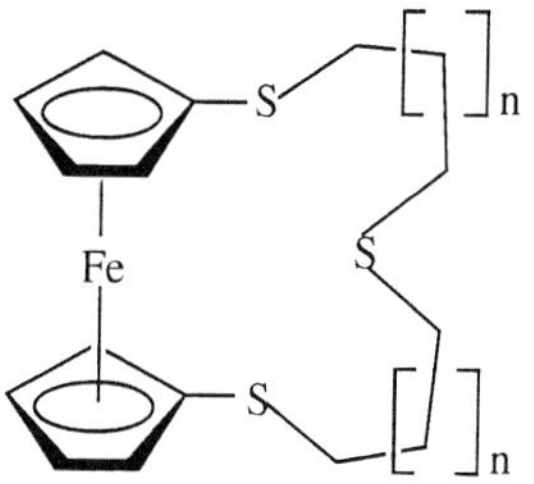

(21) a, n=0; b, n=1 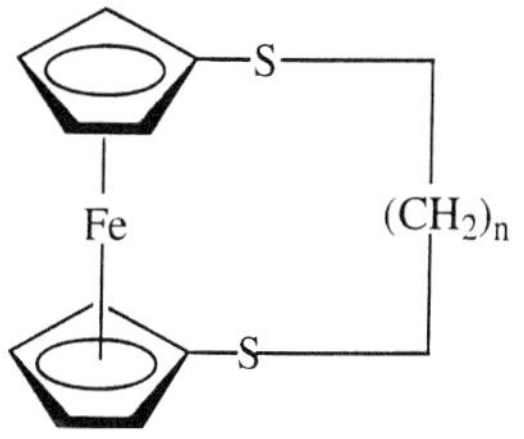 (22) n=2-8

polythiamacrocyclic compounds containing metallocene units afford organometallic ligands that may be useful for the generation of heterobimetallic complexes in which metal-metal interactions may be possible. The first polythia-[*n*]-ferrocenophane **15** reported in 1980 [28] contained methylene bridges between the *cpd* rings and the sulfur atoms of the macrocyclic chain. Polythia-[*n*]-ferrocenophanes in which the sulfur atoms were connected directly to the ferrocene nucleus (e.g., **16**–**19**) were subsequently prepared by Sato et al. [33, 34] and by analogy with the oxathiamacrocycles, ring expanded products typified by **20** were also obtained. Trithiaferrocenophanes **21** have also been synthesized [35] and show "abnormal" bathochromic shifts ($\Delta\lambda \approx 20$ nm) of the d–d transition of the ferrocene unit compared with their larger macrocyclic analogues. This result prompted an examination of a series of dithia-[n]-ferrocenophanes **22** ($n = 2-8$), which showed the largest shift in electronic and ^{13}C NMR spectra and the largest decrease in the redox potential of the ferrocene nucleus for $n = 7$ [36]. These data were interpreted as a new type of stereoelectronic dπ–pπ interaction between the sulfur atoms and the ferrocene nucleus, a conclusion that was substantiated by photoelectron spectroscopy [37]. Similar spectroscopic and electrochemical abnormalities were observed for the heteroaromatic analogues **23** and **24** [36].

On addition of $AgNO_3$ to a solution of **16**, **17** or **18** in acetonitrile, the d–d transition at *ca.* 440 nm showed a hypsochromic shift and a decrease in absorbance indicative of 1:1 complex formation [38]. An X-ray crystal structure of the 1:1 silver perchlorate complex of **17** revealed a five-coordinate Ag^+ ion bound within a distorted square pyramid to four sulfur atoms and a perchlorate ligand [39].

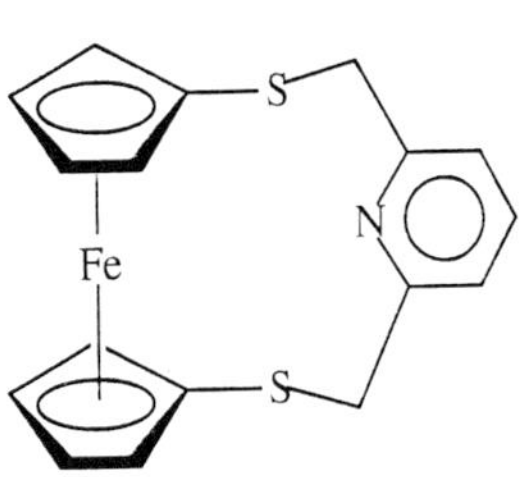

(23)

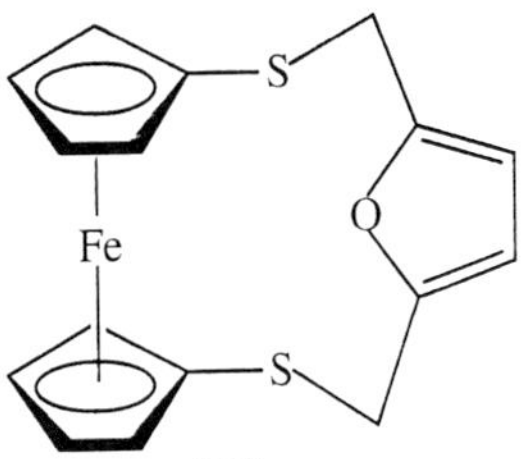

(24)

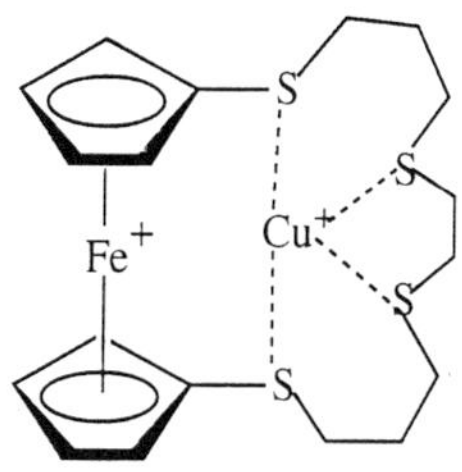

(25)

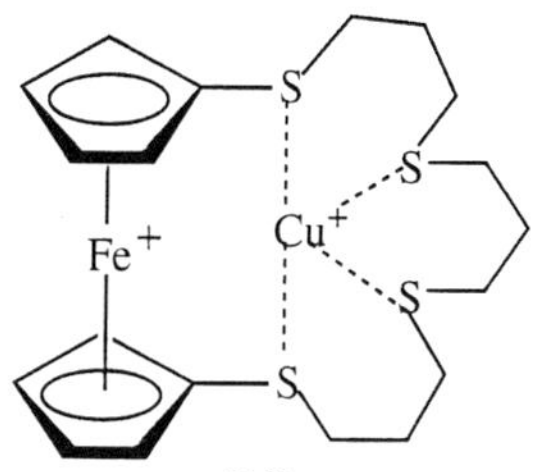

(26)

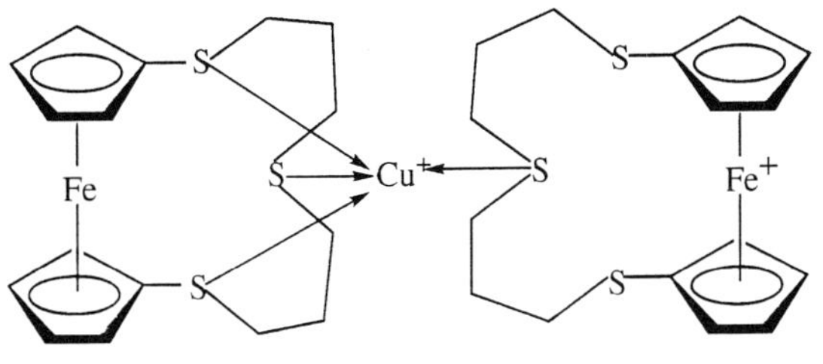

(27)

Similar complexes were formed with Cu^+, Hg^{2+} and Pd^{2+}, but no clear evidence was found for interaction between the iron atom and the bound cations [38]. On the other hand, electronic spectra and NMR data of complexes formed between **16**–**18** and $Pd(BF_4)_2$ [40] or $Pt(BF_4)_2$ [41] suggested that there was some form of charge interaction between the ferrocene moiety and the bound cation which was controlled by the size of the macrocycle thiacrown cavity. Definite interaction was, however, observed with $Cu(BF_4)_2 \cdot n\,H_2O$ in nitromethane to give 1:1 complexes **25** and **26** as black crystals with IR and electronic spectra very similar to those of the ferrocenium ions of **18** and **19** [42]. The quadrupole splitting of the Mössbauer spectra and the magnetic moments of these complexes showed a good linear correlation with the Prins parameter of low-symmetry perturbation obtained from ESR measurements [43]. Trithiamacrocycle **21** ($n = 1$) also reacts with $Cu(BF_4)_2$ in ethanol to give a 2:1 complex **27** as black plates. Quadrupole splittings (QS) of 2.38 and 0.70 mm s^{-1} were observed indicating two kinds of iron atom within the molecule, the latter being similar to that of the ferrocenium ion of **21** (QS = 0.65 mm s^{-1}). This result suggests that the copper is reduced to the Cu^+ state and that *one* of the iron atoms is oxidised in an intramolecular redox reaction to give a complex in which the neutral ligand is tricoordinated and the oxidised ligand monocoordinated to Cu^+ [40, 43]. Compounds **21a** and **21b** also react with $Pd(BF_4)_2 \cdot 4\,CH_3CN$ in acetone to give **28a** and **28b**, respectively, each containing one molecule of acetone in the crystal, which may be replaced by acetonitrile [44, 45]. Downfield shifts of the ^{13}C NMR signals of the methylene groups adjacent to sulfur in the complexes, relative to the free ligands, indicated coordination of all three sulfur atoms to the Pd atom in **28a**,. Dramatic shifts in the ^1H NMR signals of the *cpd* protons of **28a**, were also observed and these data, coupled with shifts in the out-of-plane bending vibration of the CH bonds of the ferrocene rings, suggested a large perturbation of the ferrocene nucleus on complexation. The central Pd atom in **28a**, seems to need electron donation from the iron atom since it is coordinated to only three sulfur atoms giving an unsaturated 14-electron system around Pd. Some support for this hypothesis was provided by the X-ray crystal structure of the selenium analogue **29** in which the pentacoordinated Pd atom is situated in a distorted *tbp* configuration involving coordination to acetonitrile and an interatomic distance of 3.09 Å between the Pd^{2+} ion and iron atom. This distance, is however, somewhat larger than the sum (2.52 Å) of the covalent radii of both metal atoms [46].

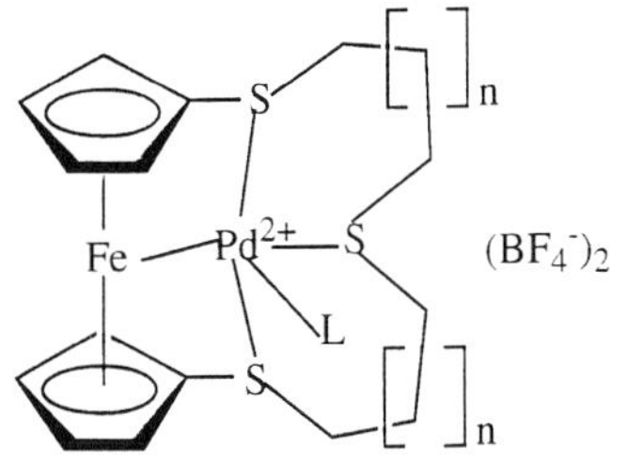

(28) a, n = 0; b, n = 1

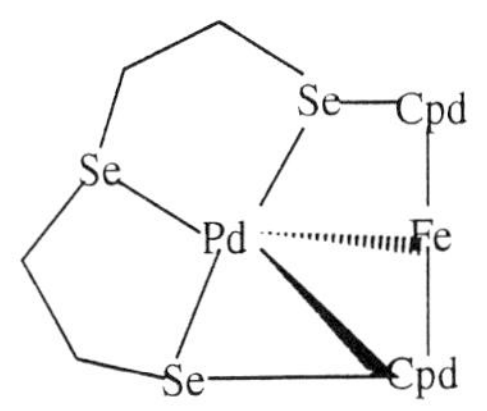

(29)

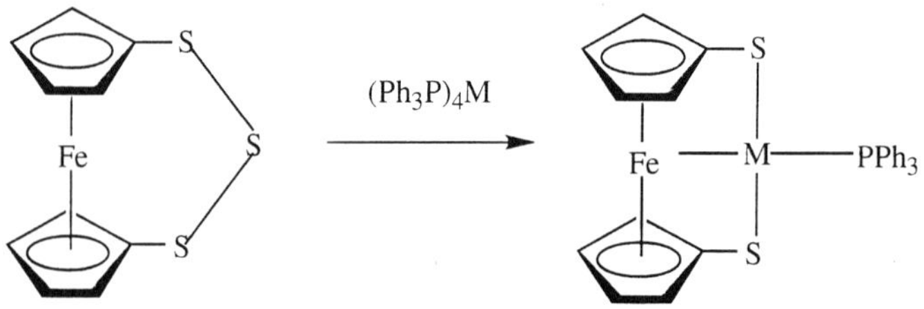

$(Ph_3P)_4M$

(30)

(31) M=Pd

(32) M=Pt

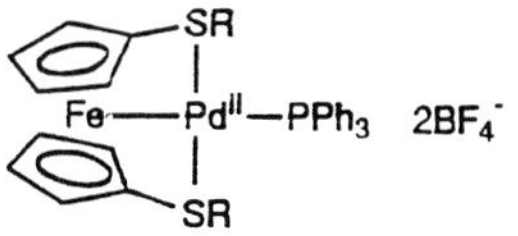

(33)

Several years ago, Seyferth [47] prepared complex **31** by reaction of **30** with $Pd(PPh_3)_4$ and subsequently Akabori prepared the Pt^{2+} analogue **32** and the dioxygen-palladium analogue [48]. In a similar manner, 1,1′-bis-(alkylthio)-ferrocenes react with $Pd(BF_4)_2 \cdot (MeCN)_4$ in the presence of triphenylphosphine to give **33** [49, 50], results which reveal that formation of a dative metal-metal bond is not dependent on a macrocylic unit as part of the coordination sphere. The final set of

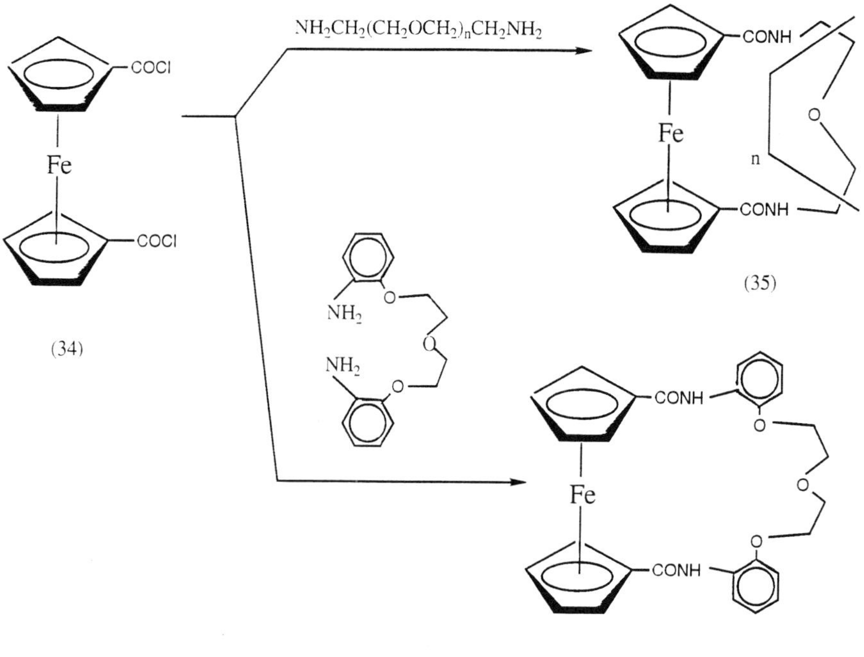

$NH_2CH_2(CH_2OCH_2)_nCH_2NH_2$

(34)

(35)

(36)

compounds that may be categorised as ferrocenophanes are synthesized from ferrocene bis-acid chloride **34** and they utilise the ester or more frequently, the amide function as the cyclization link. Thus, bis-amides such as **35** and **36** were prepared by reaction of **34** with the appropriate diamine in the presence of a tertiary amine [51, 52]. The cation-binding capacity of these ferrocene-containing macrocycles is, to a large extent, associated with the coordination by the amide carbonyl functions, but this topic will be discussed in more detail in the next section dealing with cryptands containing the ferrocene unit.

6.3 Cryptands Containing the Ferrocene Unit

The first synthesis of molecules of this kind was reported independently by Vögtle [52] and Hall [51] and both involved condensation of **34** with a range of diazamacrocycles **37** under high dilution conditions to form the corresponding amide cryptands **38** together with dimeric analogues **39** and some polymers. The compounds were easily separated by column chromatography on alumina and were fully characterised by mass spectrometry, elemental analysis, and multinuclear NMR

(34) + HN NH O O m n

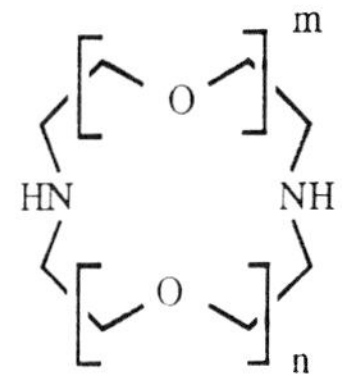

(37) m, n = 1, 2 or 3

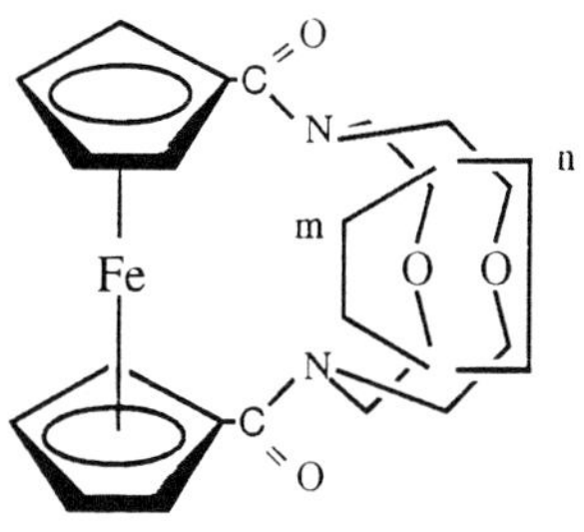

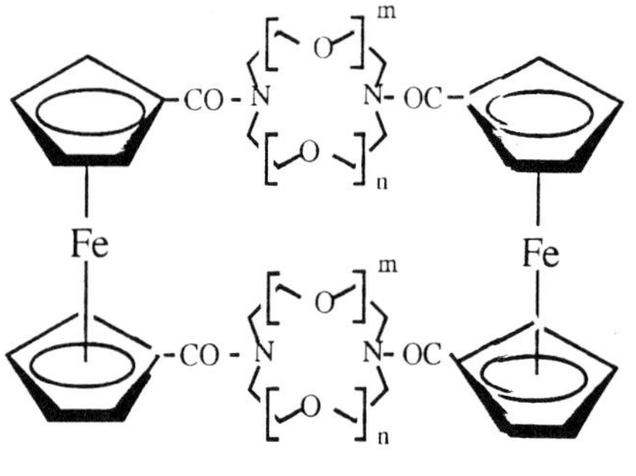

(38) + (39)

spectroscopy, including COSY and heteronuclear $^{1}H/^{13}C$ correlation spectra [53–55]. The NMR data showed clearly that compounds of type **38** adopted the *trans* configuration of the carbonyl groups in solution since **38**, with $m = 1$ and $n = 2$, gave a total of twenty two ^{13}C NMR signals consisting of two carbonyl carbons, ten ferrocene carbons, four NCH_2 carbon signals, and six OCH_2, carbons denoting the absence of an element of symmetry within the molecule. The configuration was confirmed by X-ray crystal structures **38** ($m = 1$, $n = 2$) [56], **38** ($m, n = 2$) [57, 58] and the ruthenium analogue of the latter [59]. The energy barriers for rotation about the amide bonds in **38** were also determined by dynamic NMR revealing $t_{1/2}$ (298 K) $\approx 1 \times 10^{-2}$ s for **38** ($m = 1, n = 2$) [55]. Thus, although the amide rotation is slow on the NMR time scale ($\Delta G^{\ddagger} = 16-18$ kcal mol^{-1}, depending upon the size of the macrocyclic unit), the molecular process actually occurs about 100 times per second at ambient temperature. This molecular mobility allows divalent (or trivalent) cations to be complexed by the two carbonyl groups switching to a *cis*-configuration in order to form a powerful chelating function. This is shown by the fact that the coordination of **38** ($m = 1$, $n = 2$) with, for example, Ba^{2+}, caused the 22 carbon NMR signals of the host macrocycle to collapse to 11 (1 CO, 5 cpd, 2 NCH_2 and 3 OCH_2), consistent with the plane of symmetry expected for the *cis*-configuration of the host [60]. Similar results were obtained with Be^{2+}, Mg^{2+}, Ca^{2+} and Sr^{2+} and a convincing argument was presented on the basis of the NMR data for complexation of the cation on the short side of the macrocyclic chain. Careful studies with **38** ($m, n = 2$) and a variety of divalent cations revealed that both 1:1 and 2:1 (ligand:cation) complexes could be formed in ratios dependent on the conditions of concentration and temperature [61, 62]. In all cases it was evident from NMR and IR data that the amide carbonyl groups were the principal coordinating functions for the cations with the macrocyclic units playing a minor role in the binding process. Complexation of cations by cryptands of this type also resulted in a *bathochromic* shift (and *increase* of intensity) of the d–d transition of the ferrocene unit at ca. 440 nm, a feature that was then used to determine the stoichiometry and stability constants of the resultant complexes [63]. Analysis of the data showed that although **38** ($m, n = 2$) formed 1 : 1 complexes with Be^{2+}, Mg^{2+}, Ca^{2+}, Sr^{2+} and Ba^{2+}, an equilibrated mixture of 1:1 and 2:1 (ligand:cation) complexes was formed with **40**, **41**, **42** and **43**. In every case, the association constants in acetonitrile for a given stoichiometry were of the same order of magnitude at 10^4 for the 1:1 complexes and $10^6 - 10^7$ for the 2:1 complexes. This result illustrates beyond doubt that the amide carbonyl functions are crucial to complex formation, at least for di- and trivalent cations with the macrocyclic heteroatoms playing, at best, a secondary role. The conclusion was reinforced by the observation that replacement of the amide functions of **40** by thioamide groups, as in **44**, removed coordination with the hard alkaline earth or lanthanide cations although complex formation was detected with Ag^{+} and Hg^{2+}. On the other hand, the reaction of **38** ($m, n = 2$) with $LiClO_4$ in methanol under reflux gave a crystalline, 1:1 complex that was isolated and characterised by elemental analysis, IR [$\nu(C{=}O) = 1600$ cm^{-1}] and $^{1}H/^{13}C$ NMR [58]. The data suggested that the small Li^{+} cation was bound *within* the cavity by the ethyleneoxy bridges with no evidence for involvement of the ferrocene moiety or the carbonyl groups. The Mössbauer

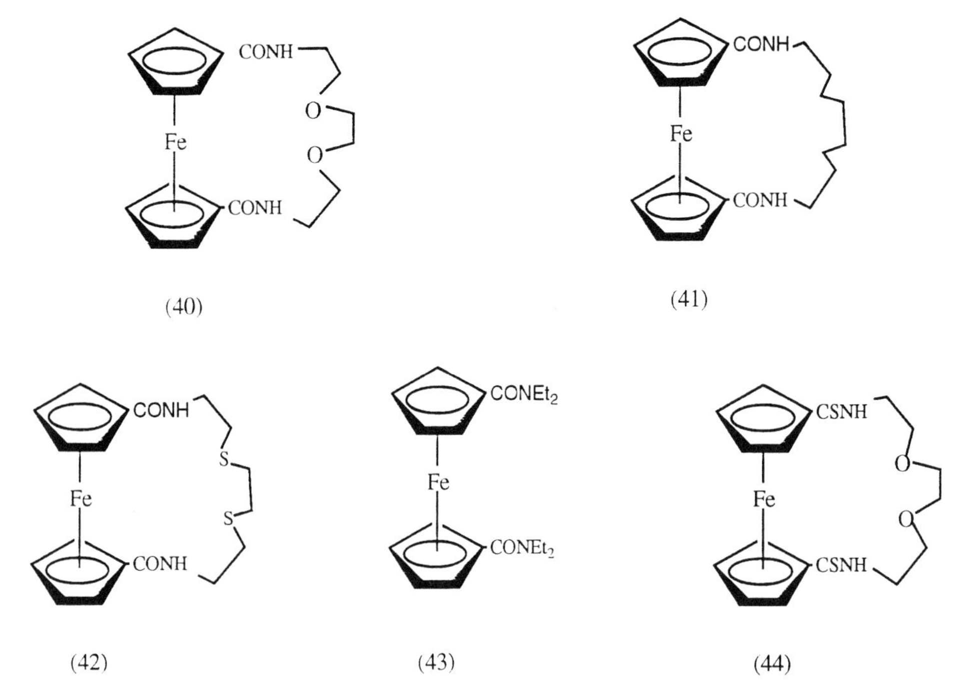

spectrum of this complex also suggested that there was no interaction between the Li^+ cation and the iron atom. Similar Mössbauer results were obtained with a series of divalent cations and **39** ($m, n = 2$), where the carbonyl groups are definitely involved in the coordination process [64], and it seems, therefore, that in complexes formed between amide cryptands and hard (alkali or alkaline earth) cations, the electron density at iron is not sufficiently perturbed to achieve dramatic changes in the Mössbauer spectrum.

The first confirmation of *cis*-binding of cations by the amide functions came from an X-ray crystal structure of the complex between **38** ($m, n = 2$) and yttrium perchlorate [65]. The stoichiometry was 1:2:1 (Y^{3+}:ligand:H_2O) and the seven-coordinate yttrium was situated at the centre of a capped *tbp* with the oxygens of the four amide groups occupying four corners of the prism and the water molecule situated at the cap (Fig. 6-5). The last two coordination sites were occupied by the oxygen atoms of one of the macrocyclic rings. The ^{89}Y and ^{13}C NMR spectra were also found to be consistent with the stoichiometry revealed by the X-ray structure. Multinuclear (^1H, ^{13}C and ^{89}Y) NMR has also been used to study complex formation between (**38**) ($m = 2, n = 1$) and $Y(ClO_4)_3$. The results showed that in acetonitrile complexes of 1:1 and 2:1 stoichiometry were formed, both of which involved coordination of the cation by a *cis*-configuration of the amide groups. The carbonyl coordination was also characterised by an unusual two-bond coupling between ^{89}Y and the carbonyl carbons [66]. The NMR data suggested that the coordination of the cation occurred on the short (one oxygen) side of the macrocyclic ring, but unfortunately no suitable crystals could be grown to verify this by X-ray crystallography. An analogous system, however, was reported recently involving a pyridine unit

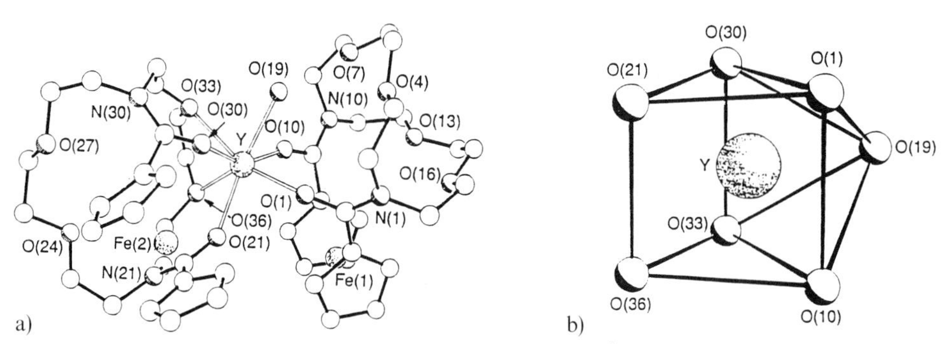

Fig. 6-5. The X-ray crystal structure of the complex between Y^{3+} and **38**; (a) molecular structure, (b) coordination geometry around Y^{3+}.

within the macrocyclic structure **46**, synthesized from **34**, and the aza-oxamacrocycle **45** [67]. The $^{1}H/^{13}C$ NMR of **46** again showed that the amide groups adopted a *trans* configuration in solution (27 carbon signals including 2 CO groups), which became *cis* on coordination with divalent or trivalent cations in either a 2:1 or 1:1 (ligand:cation) stoichiometry with 14 carbon signals from each complex. Crystals suitable for X-ray analysis were obtained from the Ca^{2+} complex **47** (stoichiometry 2:1, ligand:cation), which revealed a seven coordinate calcium cation at the centre of a capped trigonal biprism, the seventh coordination site being occupied by a molecule of the solvent, acetonitrile (Fig. 6-6). Clearly, as predicted by the NMR data for both **38** ($m = 2, n = 1$) and **47**, the cation is bound on the short, pyridine side of the macrocyclic ring, thus allowing the amide carbonyl groups to bend inwards towards the cation, with the larger, more flexible arm of the bridge expanding to accommodate a molecular pincer movement.

(34)

\+

NH O O O N NH

(45)

Fe CON CON O O O N

(46)

$Ca^{2+} (ClO_4^-)_2$

$2\ (46).\ Ca^{2+}.CH_3CN\ (ClO_4^-)$

(47)

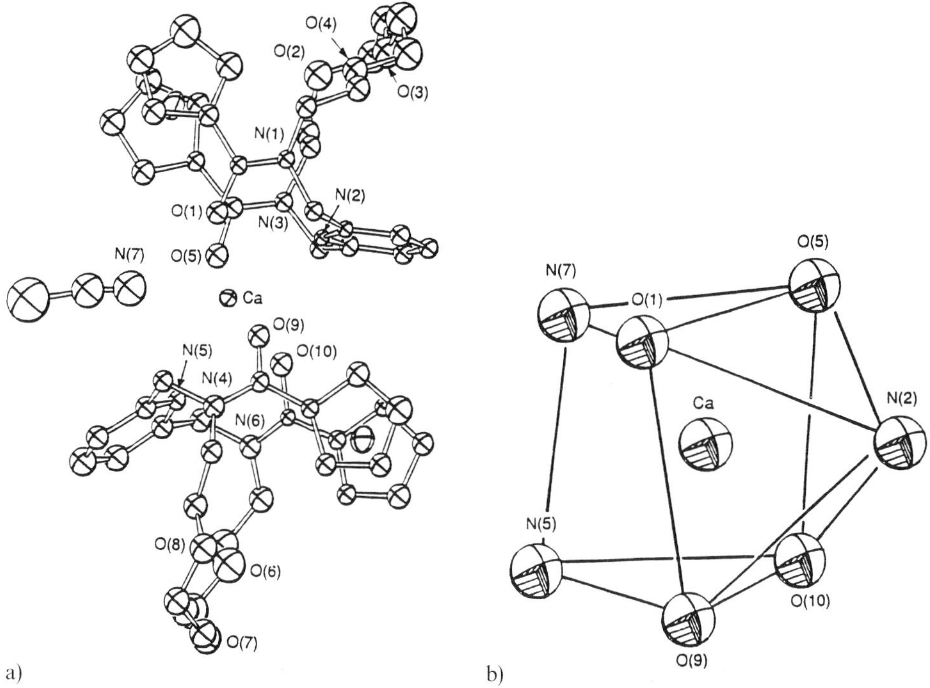

Fig. 6-6. The X-ray crystal structure of the complex between Ca^{2+} and **46**; (a) molecular structure, (b) coordination geometry around Ca^{2+}. Bond lengths to Ca (Å) O(1), 2.290(13); O(5), 2.309(13); O(9), 2.299(14); O(10), 2.277(14); N(2), 2.795(14); N(5), 2.751(15); N(7), 2.517(24).

The electrochemistry associated with these redox-active cryptands is quite intriguing. As pointed out earlier, anodic shifts of the ferrocene redox potential may be used via the Nernst equation to estimate the decrease in binding capacity (K_2/K_1) on coordination with a cation. Beyond this, however, if K_1 is determined independently as is the case for **38** ($m, n = 2$) [63] then K_2 may be calculated and correlated with the ratio of cationic radius/charge (Fig. 6-7) – data that reveal that increasing charge density on the cation destabilizes the complex between the oxidised cryptand and the cation, presumably by charge repulsion [68]. Alkali metal cations gave only small (<20 mV) anodic shifts with this cryptand.

Laser flash photolysis has also been used to study the photophysics and photochemistry of metallocene-containing cryptands and their complexes with rare earth cations [69]. The metallocene moiety was shown to act as an efficient centre for the radiationless deactivation of the lanthanide excited state. Detailed time-resolved studies permitted the characterisation of the coordination chemistry about Dy^{3+} within the cryptate and showed, once again, that the functions within the host cryptand primarily responsible for coordination of the guest cation were the amide carbonyl groups.

Reduction of the carbonyl groups of **38** to **48** had to await the development of an elegant method by Gokel using $LiAlH_4$ in a mixture of CH_2Cl_2 and THF [70].

(38) $\xrightarrow[CH_2Cl_2\ /\ THF]{LiAlH_4}$ Fe CH_2N CH_2N O O m n

(48)

The cyclic voltammogram of **48** ($m = n = 2$) in acetonitrile exhibited two-wave behavior in the presence of sub-stoichiometric amounts of Na^+ analogous to the results reported by Saji for **5b** [22] and suggesting a large association constant for Na^+. Similar results were found with K^+ and Ca^+, with the latter affording the largest anodic shift in accordance with its high charge density. The X-ray crystal structures of **48** $\cdot$ 2 H_2O and its complexes with $NaClO_4$ and $AgClO_4$ showed that the iron atom is a donor for Ag^+ but not for Na^+, the Fe-Ag^+ distance being about 1 Å shorter than that for Fe-Na^+ [71]. The same paper reported stability constants in CH_3OH or CH_3CN for complexes between **48** and Li^+, Na^+, K^+, Ca^+ and Ag^+, together with an assessment of complexation by mass spectrometry. The evidence suggests that, in contrast to the amide ligands **38**, complexation by

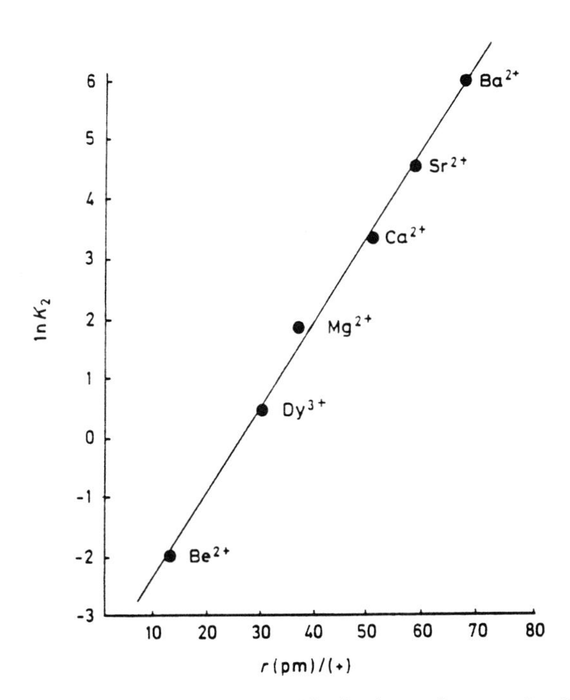

Fig. 6-7. Plot of ln K_2 *vs.* ionic radius r (in pm)/ionic charge for complex formation of **38** with divalent and trivalent cations.

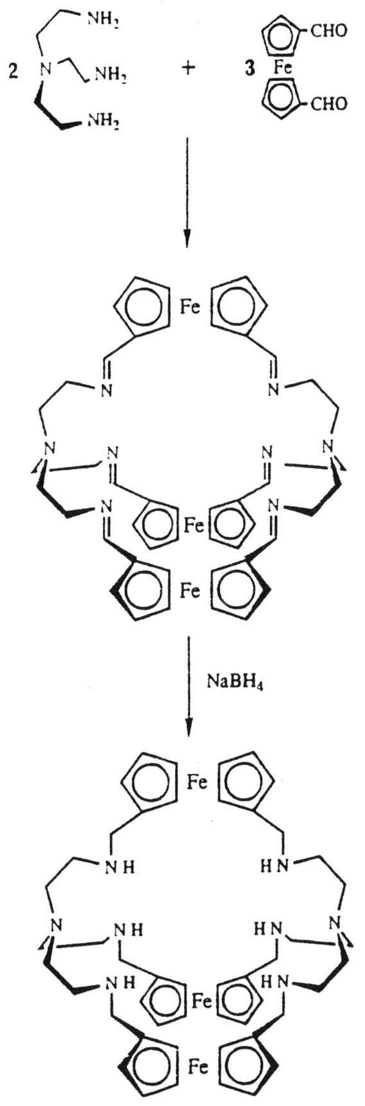

(49)

48 involves cation binding within the cavity of the cryptand. In fact, the ferrocene subunit has been discussed as an essential feature in the design and synthesis of a family of molecular receptors containing clefts that vary in size, shape and electron density [72].

The cryptand **49** has also been prepared and although electrochemically insensitive to group 1A cations, the cryptand recognises Zn^{2+} electrochemically and forms an isolable 2 Zn^{2+} · **49** bimetallic complex [73].

As mentioned earlier, the synthesis of molecules of the type shown in Fig. 6-3c, e.g., **38** is usually accompanied by the formation of the dimeric analogues (Fig. 6-3d), e.g., **39**, compounds that in principle have molecular clefts large enough to bind more than one cation. X-ray crystallographic studies (e.g., of **39**, $m = n = 2$) revealed a centrosymmetric structure with all the amide groups mutually *trans* to each other [74]. In solution, however, the molecules are fluxional with two energetically distinct rotations about the amide links ($\Delta G^{\ddagger} \approx 67\ \mathrm{kJ\ mol^{-1}}$) and the ferrocene-carbonyl bond ($\Delta G^{\ddagger} = 50\ \mathrm{kJ\ mol^{-1}}$) [51, 53, 54]. Attempts to synthesise the thio-analogues of **38** – compound **50** for example, actually resulted in the formation of **51**. An X-ray crystal structure of **51** (Fig. 6-8) revealed that in this case, the amide carbonyl groups were *cis* (crown)/*trans* (ferrocene)/c/t to each other with the *cpd* rings eclipsed rather than staggered as in **39** [75]. In solution, multinuclear NMR studies showed that the molecule is fluxional with a simple 7 signal ^{13}C NMR spectrum at 360 K

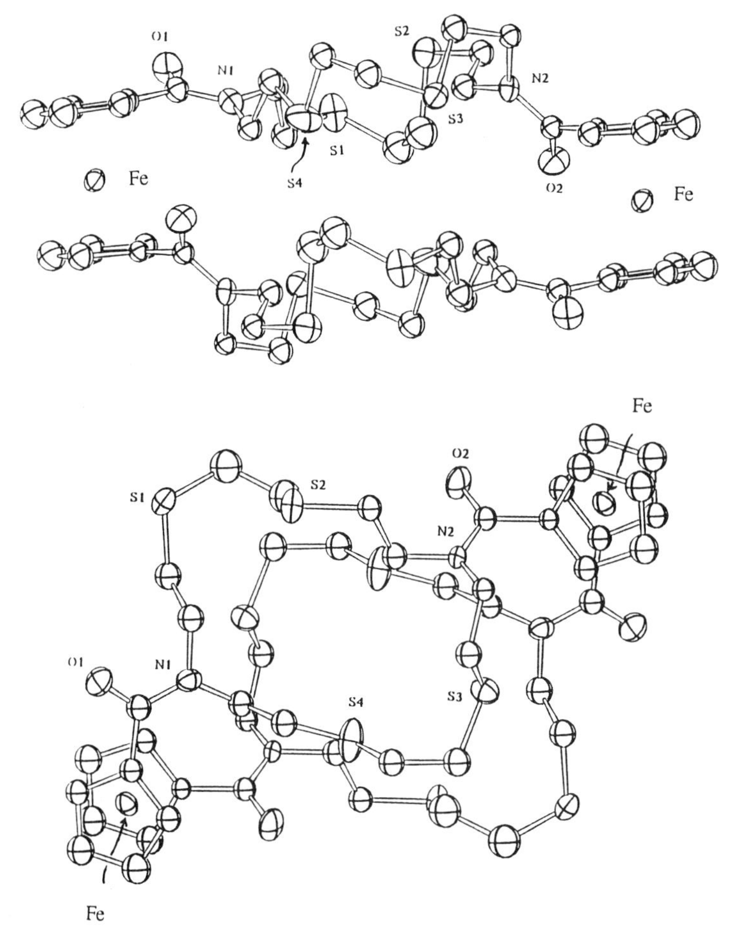

Fig. 6-8. The X-ray crystal structure of **51** showing the configuration of the carbonyl groups and the eclipsed ferrocene rings.

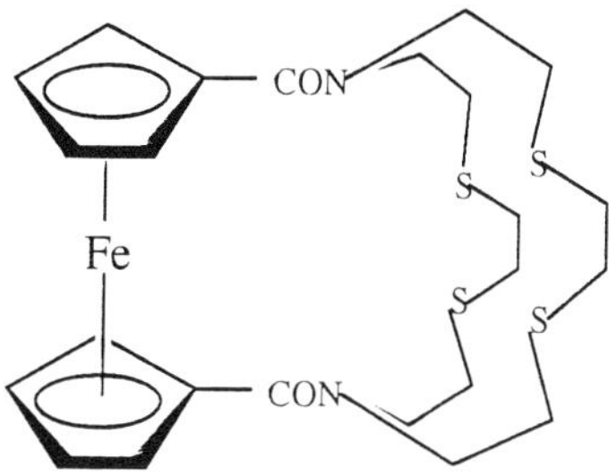

(50)

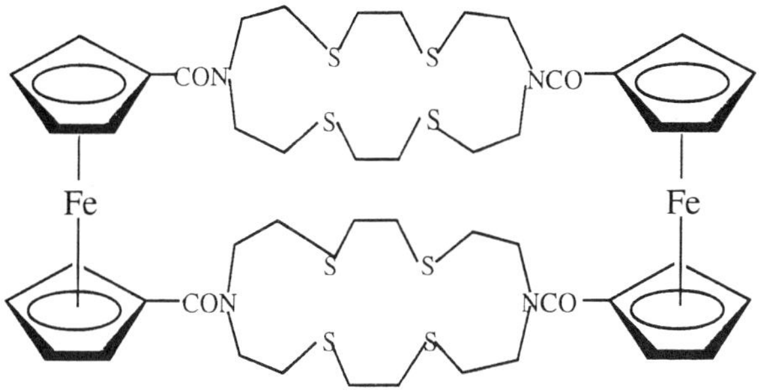

(51)

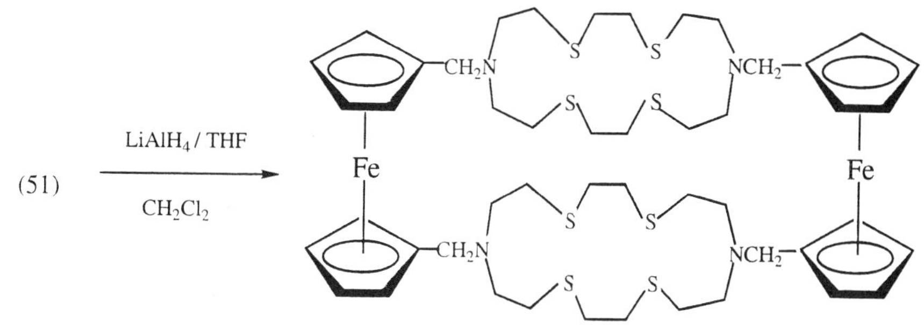

(51) $\xrightarrow[CH_2Cl_2]{LiAlH_4 / THF}$

(52)

($C_2D_2Cl_4$) but at 260 K (CD_2Cl_2), a complex 24-line ^{13}C NMR spectrum emerged representing a 60:40 mixture of two diastereomeric forms of **51** associated in each case with slow rotation about the amide function. Reduction of **51** using Gokel's method gave a highly fluxional thioamino cryptand **52** with a simple 7-line ^{13}C NMR spectrum and a capacity to bind transition metal cations. The amide dimer **39** ($m = n = 2$) may also be reduced by this method to produce **53**, which is also highly fluxional [71, 76]. Complexation with monovalent (e.g., Na^+), divalent (e.g., Ca^{2+}) and trivalent (e.g., Y^{3+}) cations occurred in a ratio of 1:2 (host:guest) with shifts in the ferrocene protons and the NCH_2 protons but *not* the central OCH_2 protons, indicating that the cations were complexed within the cavity close to the

(39) $\xrightarrow[CH_2Cl_2]{LiAlH_4 / THF}$ (53)

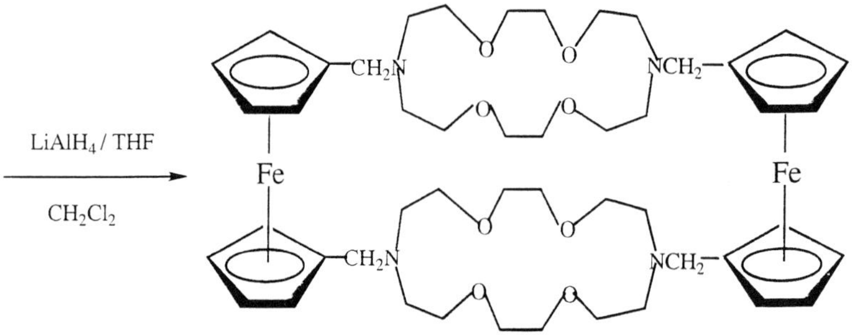

iron atoms [76]. This concept was proposed earlier for complex formation between **53** and Na^+ or Ag^+ on the basis of FAB MS data and cyclic voltammetry [71]. Thus a rich tapestry of host : guest chemistry appears to be developing around these redox-active cryptands and it remains to be seen whether or not their potential utility will be fulfilled.

6.4 Macrocyclic Cryptand and Cavitand Derivatives of Ferrocene

6.4.1 Ferrocene Bound to Macrocycles

Amido-linked ferrocene bis-crown ethers (e.g., **55**) are obvious analogues of the ferrocenophanes and among the earliest examples of macrocyclic derivatives of ferrocene [77–81].

Multinuclear NMR experiments suggest 1 : 1 ligand : M^+ intramolecular sandwich complexes with Na^+, K^+ and Cs^+. The crystal structure with KPF_6 as the guest (Fig. 6-9) confirmed this finding [79] with all ten oxygen atoms of the benzocrown units coordinated to K^+. The sodium cation formed 1:2 (ligand : M^+) complexes

(34) + 2 H_2N–(54) ⟶ (55)

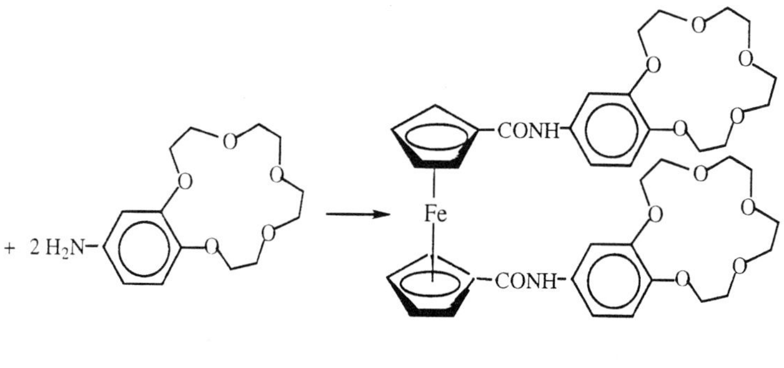

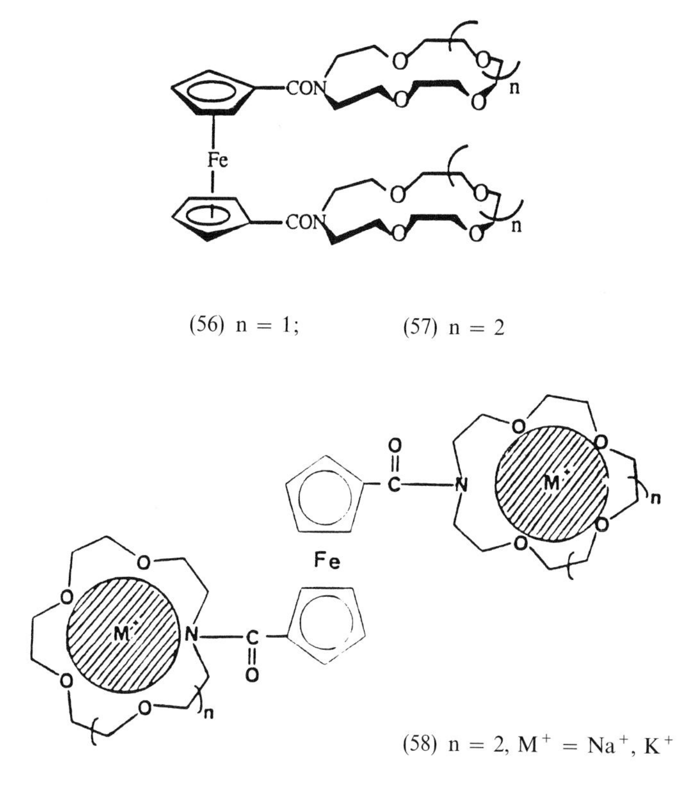

with **56** and **57**, whereas the K^+ ion formed a 1:1 intramolecular sandwich complex with the former (analogous to that in Fig. 6-9) and a 1:2 complex **58** with the latter [81]. The FAB MS technique was also used to test the selectivity of ferrocene bis-crown ethers towards cations and it was found that in competition experiments with Li^+, Na^+, K^+ and Cs^+, receptor **55** exhibited exclusive selectivity for K^+ whereas **56** and **57** displayed no preference for any alkali metal cation [77, 79]. Electrochemical complexation studies of **55** were disappointing since the reversible ferrocene oxidation wave was not perturbed on addition of either Na^+

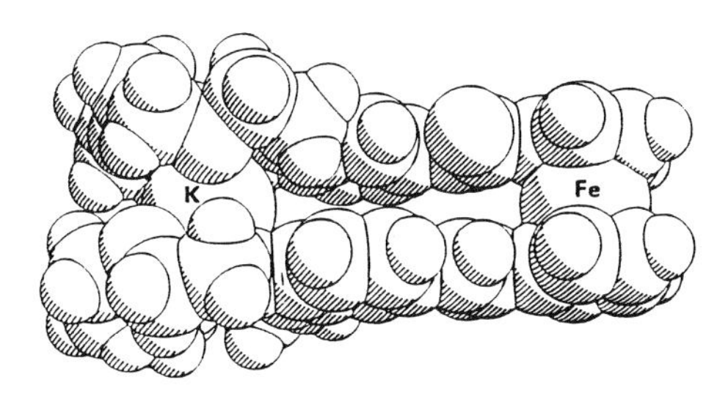

Fig. 6-9. Space-filling representation of the solid state structure of the 1:1 complex between **55** and the potassium cation.

Table 6-2. Electrochemical data for ferrocene amide bis-azacrown ethers and their group IA metal complexes compared with $Fc(CONMe_2)_2$, **59**

Compound	**56**	**57**	**59**
$E_{1/2}$ (V)[a]	0.67	0.67	0.68
ΔE (Na^+, mV)[b]	40	35	<10
ΔE (K^+, mV)[b]	20	20	<10
ΔE (Li^+, mV)[c]	70	75	360

a Obtained in CH_3CN solutions containing 0.2 *M* $[Bu_4N]BF_4$ as supporting electrolyte with a standard calomel electrode (SCE) as reference and with solutions 2×10^{-3} *M* in complex.
b One wave shift in oxidation potential produced by 4 equivalents of metal cation perchlorate.
c Two wave situation; ΔE is the difference between E_f^{free} and position of new redox couple.

or K^+, implying that the complexed cation was too far away to influence electron density at the iron atom via inductive, conjugative or through space effects. On the other hand, significant one-wave anodic shifts were observed with **56** and **57**, as summarised in Table 6-2. The magnitude of ΔE is dependent on the polarising power of the cation guest as discussed earlier. It was suggested that the apparently anomalous (two-wave) shifts observed with Li^+ were due to cation coordination with the amide carbonyl oxygens rather than the macrocyclic portion of **56** or **57** [81, 82].

In order to elucidate the primary electrostatic "through bond" mode of electrochemical communication between a ferrocene redox centre and the heteroatoms of the crown ionophore, a variety of conjugated ferrocene benzocrown ether systems such as **60** and **61** [83, 84], **62** and **63** [85, 86], and **64** or **65** [87, 88] were synthesized and their complex formation with mono- and divalent cations studied by multinuclear NMR, X-ray crystallography and cyclic voltammetry.

The sodium cation forms 1 : 1 complexes with **60** – **64** and the potassium ion forms 1 : 2 intermolecular sandwich complex with the same ferrocenyl ionophores. Electro-

(60) a, cis
b, trans

(61) a, cis
b, trans

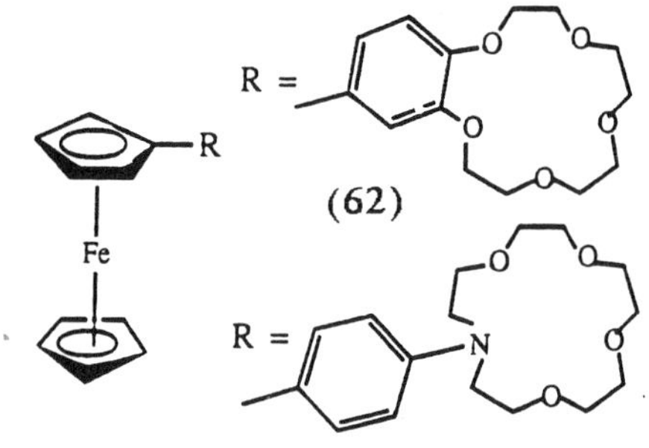

chemical studies revealed that the addition of Li^+ or Mg^{2+} to acetonitrile solutions of **60** gave a two-wave CV with considerable shifts to more anodic potentials, whereas addition of Na^+ or K^+ gave more modest one-wave shift behavior. With compounds **61**–**65**, one-wave anodic shifts of the ferrocene redox couple were observed with Na^+, K^+ and Mg^{2+}, with the magnitudes of the shifts again increasing with the polarizing power of the cation (Table 6-3).

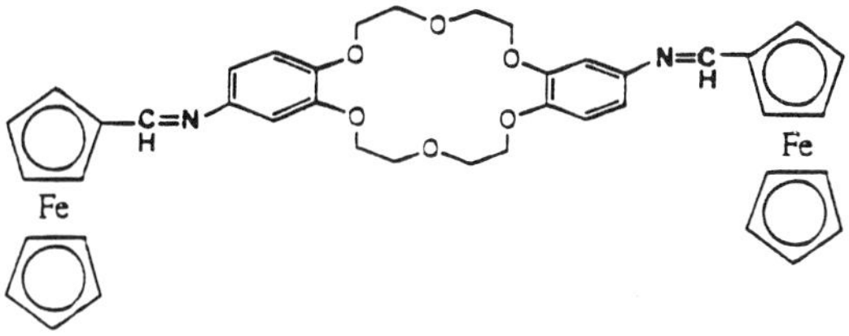

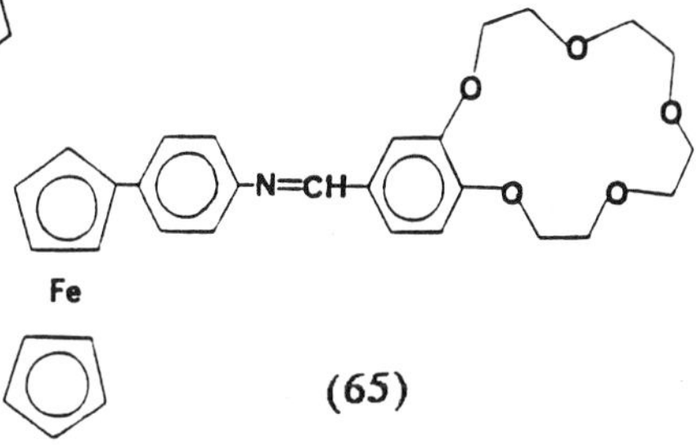

Table 6-3. Electrochemical data for crown ether derivatives of ferrocene and some metal ion complexes

Compound	**61a**	**61b**	**60a**	**60b**	**62**	**63**	**64**	**65**
E_f (V)[a]	0.43	0.40	0.33	0.34	0.54	0.58	0.44	0.50
ΔE (Na^+, mV)[b]	30	30	50	65	60	90	30	40
ΔE (K^+, mV)[b]	20	20	20	20	20	40	20	<10
ΔE (Mg^{2+}, mV)	70[b]	60[b]	100[c]	100[c]	70[b]	110[b]	70[b]	70[b]
ΔE (Li^+, mV)	–	–	110[c]	110[c]	–	–	–	–

a Obtained in CH_3CN solutions containing 0.2 *M* $[Bu_4N]BF_4$ as supporting electrolyte with SCE as reference and with solutions 2×10^{-3} *M* in complex.
b One wave shift in oxidation potential produced by 4 equivalents of metal cation perchlorate.
c Two wave situation; ΔE as difference between E_f^{free} and position of new redox couple.

The redox potentials for the complexes of **60** with Li^+ and Mg^{2+} are both close to that observed for the methiodide **66** suggesting that cations of high charge density induce a positive charge on the nitrogen atom of the macrocycle that is close to unity.

Hydrogenation of **60** gave **67** and $NaBH_4$ reduction of **64** gave **68** in which the methylene bridge was found to serve as an insulator since both compounds were electrochemically insensitive to the presence of alkali metal cation. One concludes, therefore, that in these ligand systems, anodic perturbations of the ferrocene oxidation wave are observed if a conjugated π-electron system (C=C or C=N) links the heteroatoms of the crown to the redox centre and that the polarizing power of the cation determines the magnitude and type (one or two-wave) of the observed redox shift.

Schiff base **69** and sulfur-linked **70** ferrocene bis-crown ether ligands have also been prepared and both were shown to be potassium ion selective, especially the S-linked macrocycle **70** [89]. Compound **69** did not behave well electrochemically and somewhat unexpectedly addition of K^+ to **70** resulted in a *cathodic* shift of the ferrocene redox couple, possibly due to conformational steric effects involving the electron lone pairs on sulfur, but certainly reminiscent of Sato's thia-ferrocenophanes discussed earlier.

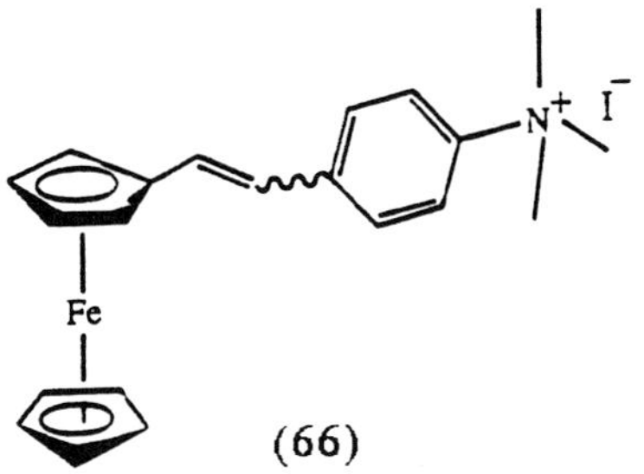

(66)

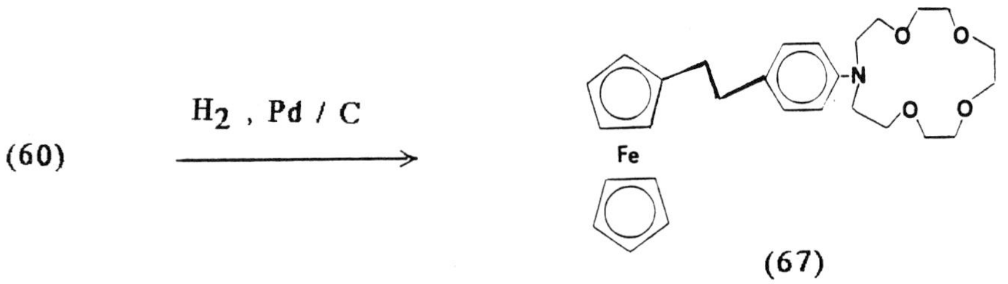

(60) $\xrightarrow{H_2,\ Pd/C}$ (67)

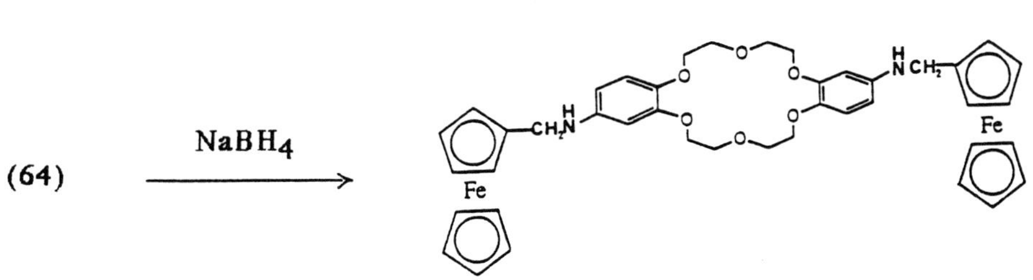

(64) $\xrightarrow{NaBH_4}$ (68)

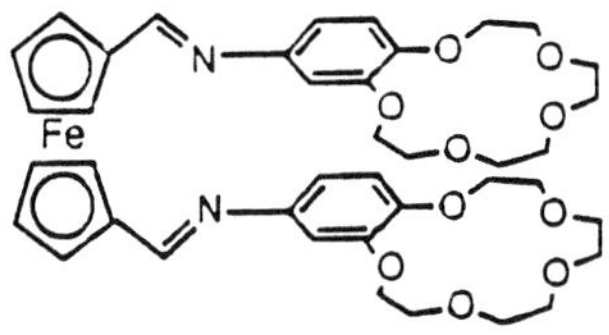

(69)

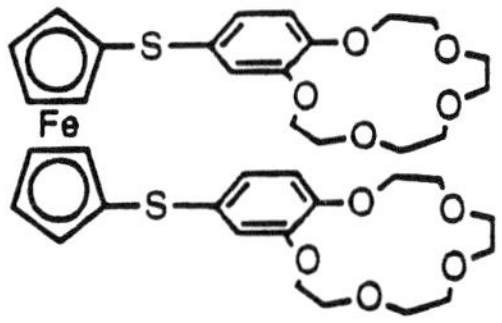

(70)

Mixed aza-thiamacrocyclic derivatives of ferrocene **71** and **72** have also been prepared and the X-ray crystal structures of **72** and its copper complex reported [90]. Electrochemical studies revealed that the redox couples of **71** and **72** were perturbed to more positive potentials by 60 and 40 mV, respectively, on coordination with Cu^{2+}, even though the X-ray structure of the complex with **72** showed square planar coordination of the Cu^{2+} by the four sulfur atoms rather than by nitrogen. The cyclam ligand **73** derived from ferrocene has also been prepared and shown to recognise Ni^{2+} and Zn^{2+} cations electrochemically [90].

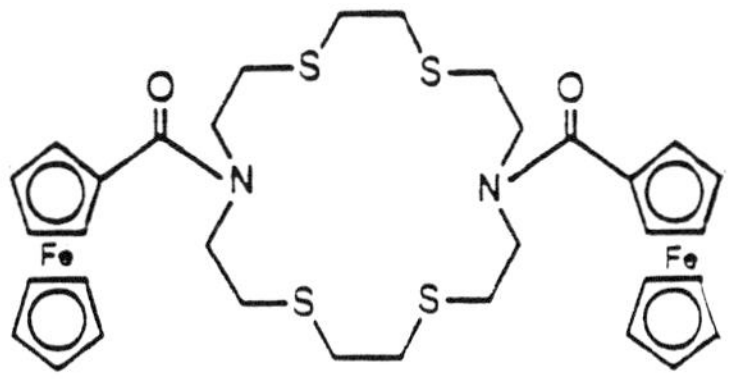

(71)

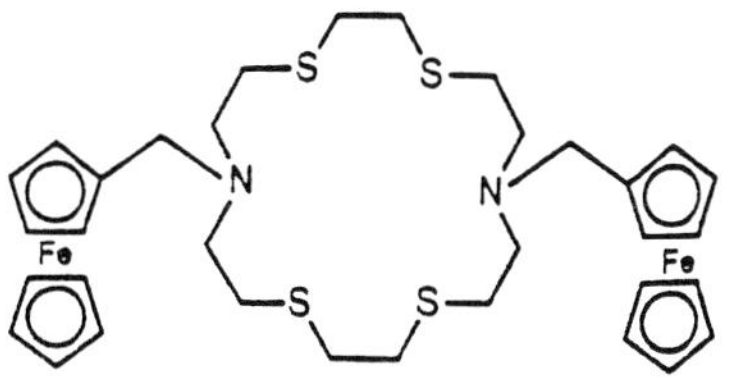

(72)

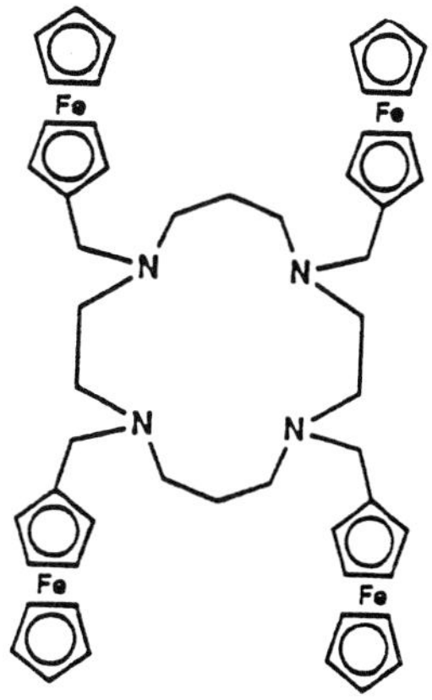

(73)

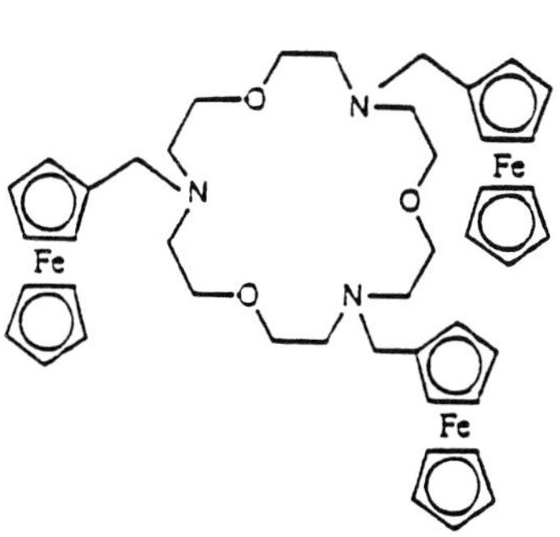

(74)

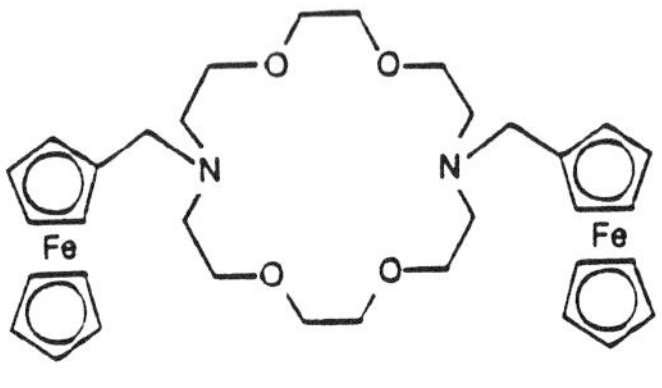

(75)

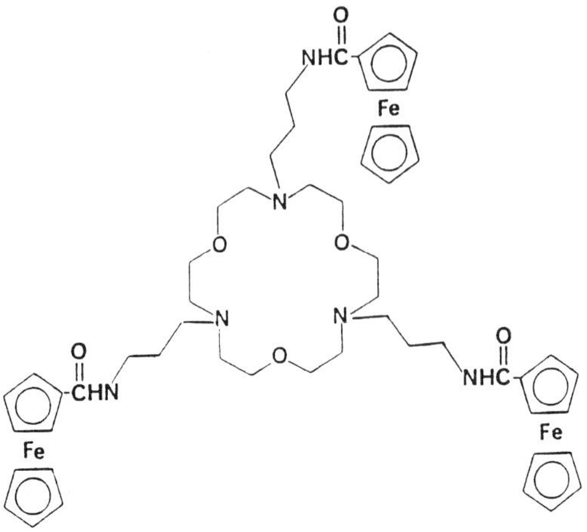

(76)

The complexation of ammonium ions by redox-active macrocycles has attracted considerable attention recently and the development of aza-macrocycles bound to ferrocene units has created the appropriate molecular systems to achieve such host-guest interactions. Significant one-wave anodic shifts of the ferrocene redox couple were produced on addition of NH_4^+ to solutions of **74**–**77**, but no response was observed with the amide analogue **78** showing that the amine nitrogen donor atoms were essential for successful NH_4^+ binding [91]. The largest anodic shift of 220 mV was observed with **74** probably resulting from a combination of through space electrostatic effects and N^+-H ... O=C-hydrogen bonds as illustrated in

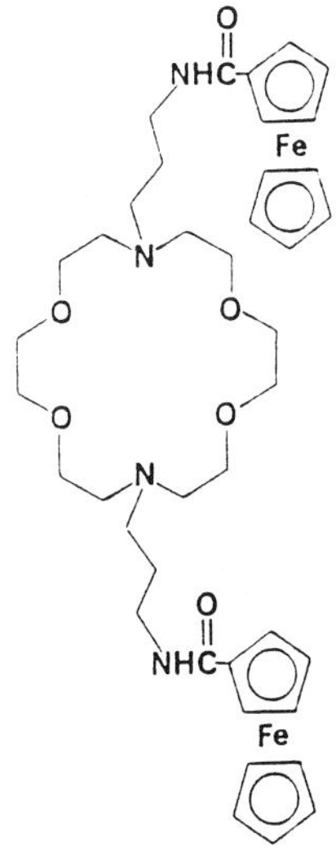

(77)

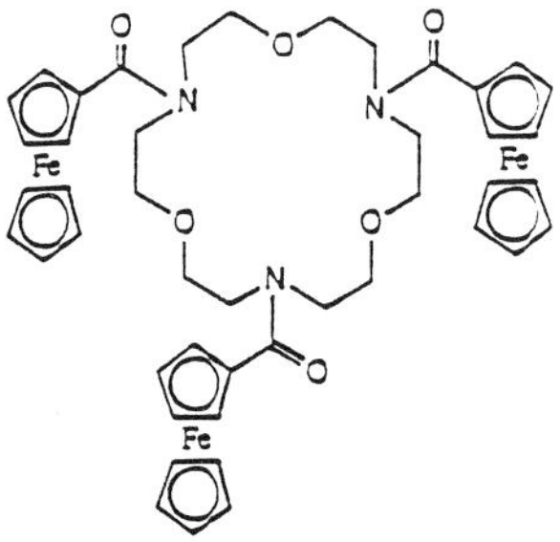

(78)

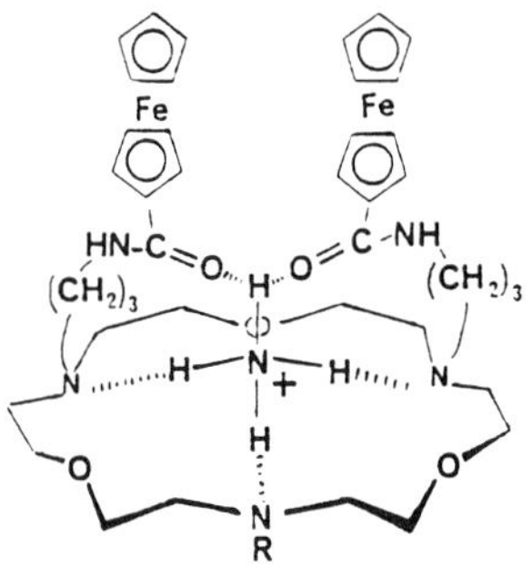

(79) 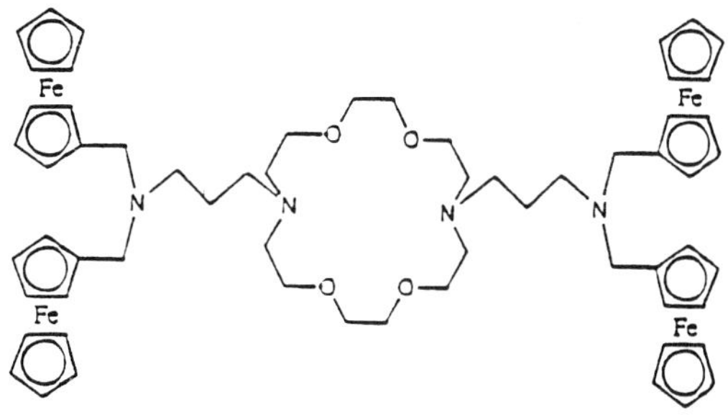 (80)

79 and similar to those reported by Gokel for quinone [92, 93] and nitroaromatic [94, 95] lariat crown ethers. Support for this concept was provided by the related system **80**, which, although it contained no amide carbonyl groups, was found to be electrochemically unresponsive to NH_4^+ [93]. Finally, it should be noted that ferrocene has also been attached to the porphyrin ring system through the ester link to form compounds such as **81** [96] and a porphyrin–ferrocene–quinone linked molecular system **82** [97].

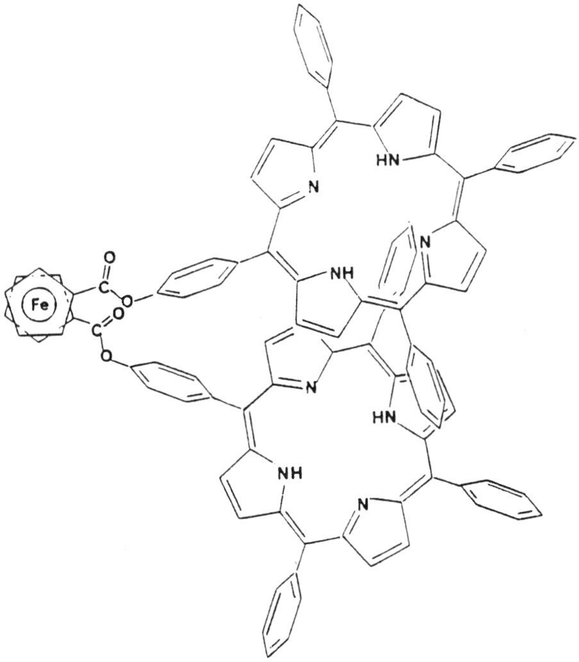

(81)

(82)

6.4.2 Ferrocene Bound to Cavitand Molecules

In an attempt to mimic enzymes that selectively bind organic guest substrates as part of the process of catalysis, several research groups have designed and synthesized a variety of cyclophane host molecules [98–101]. Receptors of this kind include the cavitands [98, 101], which contain enforced, *rigid*, hydrophobic cavities with dimensions large enough to encapsulate simple *neutral* organic guests. These abiotic hosts have been modified to incorporate redox-active units (e.g., ferrocene) either adjacent to, or as part of, the hydrophobic cavities with a view to creating sensory devices capable of detecting the inclusion of a neutral guest electronically (Fig. 6-10) and with the potential to catalyse reactions at the guest substrate [102–107].

The reaction of ferrocene carboxaldehyde **83** with resorcinol **84** gave the macrocycle **85**, which, on treatment with bromochloromethane, gave the first redox-active cavitand **86** [106, 107], the X-ray crystal structure of which revealed the inclusion of CH_2Cl_2 within the cavity. The same paper reported the synthesis of two related cavitands **87** and **88**, but electrochemical investigations failed to produce significant changes in the CV when CH_2Cl_2 was added to, or CO_2 bubbled through, DMF solutions of **86** or **87**.

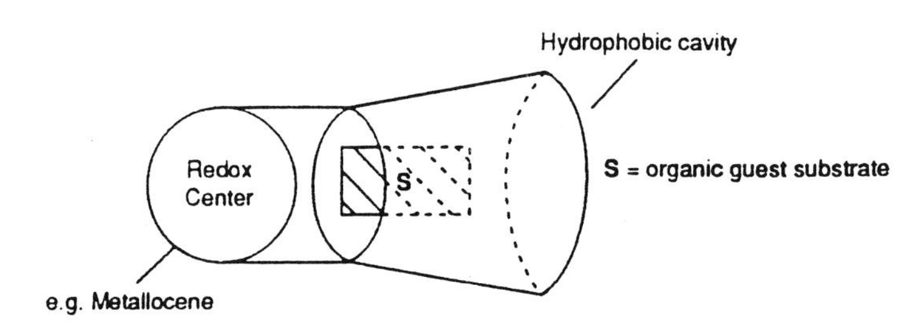

Fig. 6-10. Diagrammatic representation of the binding of an organic guest (S) in a hydrophobic cavity in close proximity to a redox centre.

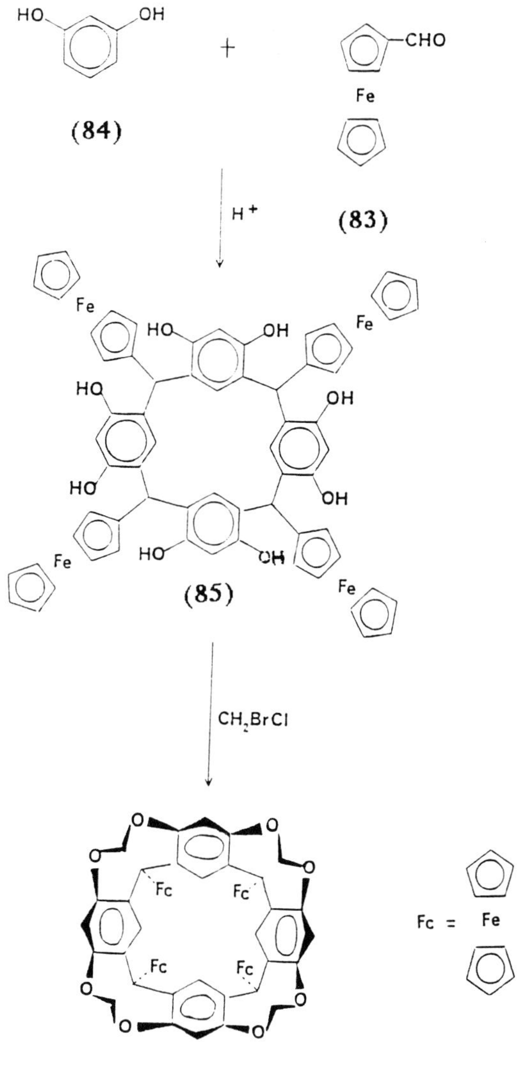

(84) (83) (85) (86)

Further examples of related multi redox-active macrocycles **89**–**91** have also been synthesised and their electrochemistry examined [108–110]. Calix-4-arenes have also been derived from ferrocene to form monomeric macrocycles of type **92** or **93** together with dimeric compounds of type **94** [102] and crystal structures of the latter two compounds were reported. Solution ^{1}H NMR and electrochemical studies in CH_3CN, $CHCl_3$, or CH_2Cl_2 showed, however, that there was no complex formation between these calixarenes and either aliphatic or aromatic amines.

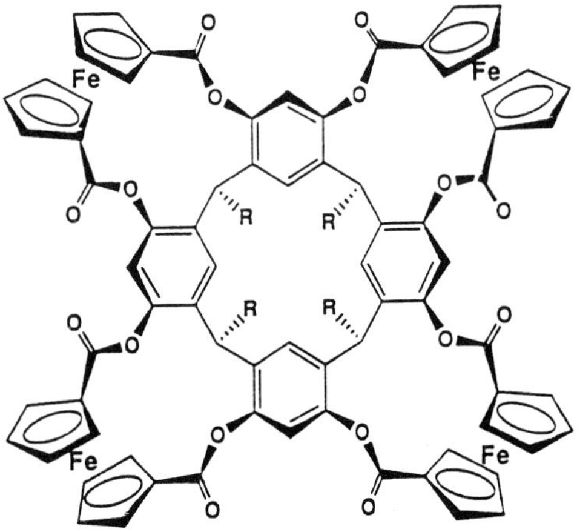

(87) R=Me

(88) R= ferrocenyl

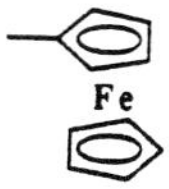

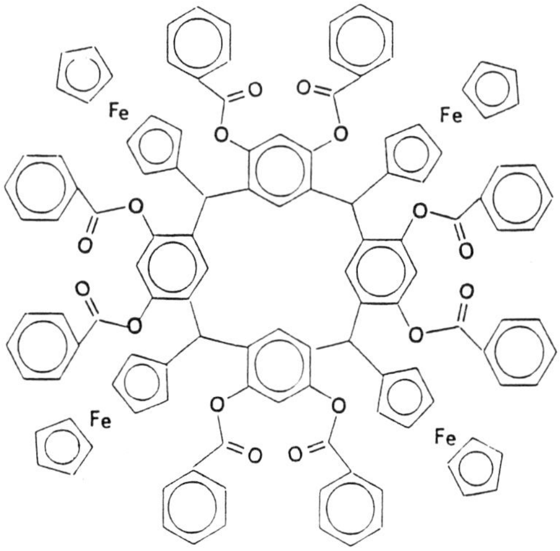

(89)

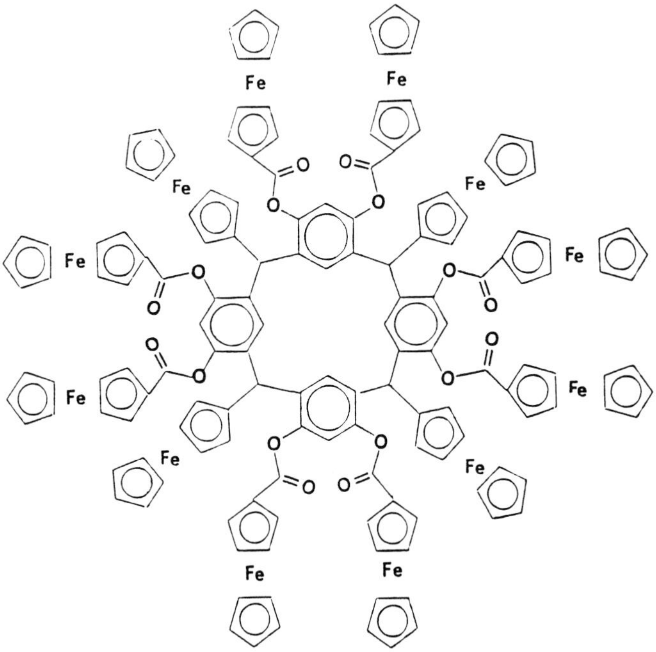

(90)

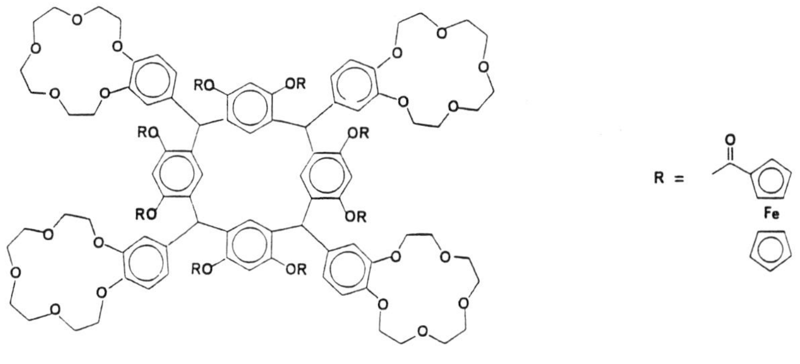

R =

Fe

(91)

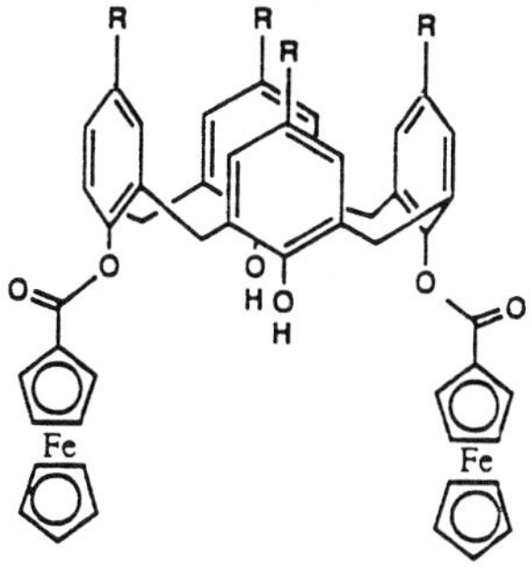

R R R R

O O
H O
H

Fe Fe

(92) R = Bu^t (93) R = H

O O
O M O
OH HO
OH HO
O M O
O O

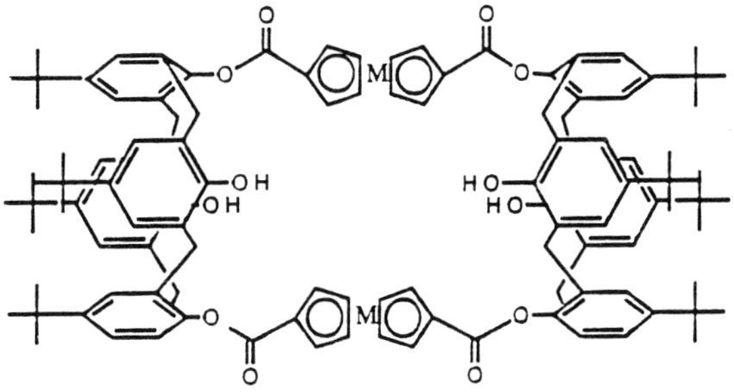

(94) M = Fe

6.5 Summary

It is clear from this review that a wide variety of macrocycles, cryptands, and cavitands containing the ferrocene unit have been synthesized and characterized. In the majority of cases, complexation by these ligands affects the electrochemistry of the redox centre, particularly when the guest species is either in close proximity to the iron atom or coordinated by functional groups that are conjugated with the ferrocene system. Conversely, a large distance between the guest and the redox centre or insulation by saturated bonds between the guest and the ferrocene unit effectively eliminates the electrochemical perturbation.

Ferrocene is not the only redox active molecule to be incorporated in systems of this kind. Macrocycles have also been derived from cobalticene in attempts to detect anions and quinones attached to or incorporated within macrocycles provide another series of molecules with pronounced electrochemical activity [112]. These electrochemical features occasionally occur together in one molecular system, as evidenced by a recent communication describing the self-assembly of molecular devices containing a ferrocene, a porphyrin, and a quinone in a triple macrocyclic architecture **95** [113]. Ferrocene has also been incorporated as the redox-active center in molecular systems that are designed to fluoresce through energy transfer from the host ligand to guest lanthanide cations [114]. Clearly the molecular systems containing ferrocene are becoming more exotic and further examples of this are provided by the encapsulation of ferrocene by carcerands [115] and the hydrosilylation of vinyl ferrocene to form ferrocenyl-substituted octasilsesquioxanes (e.g., **96**), which have been used as electrochemically active coatings for Pt electrodes [116]. It seems likely,

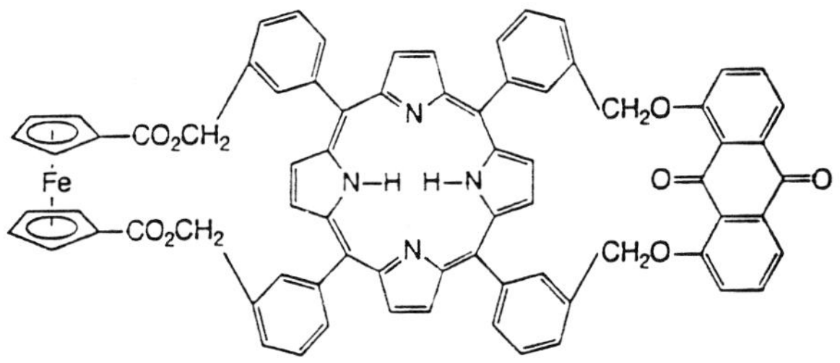

(95)

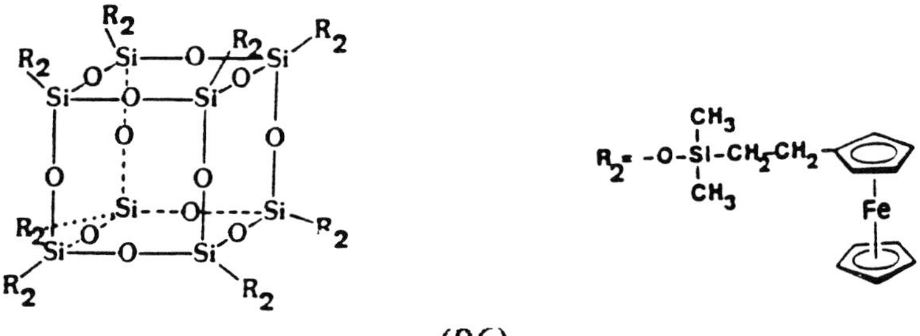

(96)

therefore, that some fascinating chemistry and valuable applications will develop as the multidisciplinary nature of this field of molecular science continues to attract an abundance of high quality research.

Acknowledgement. I should like to acknowledge the contributions of all my co-workers to the chemistry reported from our group and especially the assistance of Sunny Chu, Ti Khim Truong and Niki Sachsinger in the preparation of the manuscript.

References

[1] C. J. Pedersen, *J. Am. Chem. Soc.* **1967**, *89*, 7017–7036.

[2] C. J. Pedersen, Nobel Lecture, *Angew. Chem. Int. Ed. Engl.* **1988**, *27*, 1021–1027.

[3] J. M. Lehn, *Structure and Bonding*, Springer Verlag, Berlin, **1973**.

[4] J. M. Lehn, Nobel Lecture, *Angew. Chem. Int. Ed. Engl.* **1988**, *27*, 89–112.

[5] D. J. Cram, Nobel Lecture, *Angew. Chem. Int. Ed. Engl.* **1988**, *27*, 1009–1020.

[6] "Synthesis of Macrocycles", *Progress in Macrocyclic Chemistry*, Vol. 3 (Eds. R. M. Izatt and J. J. Christensen), J. Wiley, New York, **1987**.

[7] a) "Host-Guest Chemistry I", *Topics Curr. Chem.* Springer-Verlag, Berlin, **1980**, *98*, 1–190; b) "Supramolecular Chemistry I", *Topics Curr. Chem.* Springer-Verlag, Berlin, **1993**, *165*, 1–315.

[8] *Synthetic Multidentate Macrocyclic Compounds* (Eds. R. M. Izatt and J. J. Christensen), Academic Press, New York, **1978**.

[9] B. Dietrich, P. Viout and J. M. Lehn, *Macrocyclic Chemistry*, VCH, Weinheim, **1993**.

[10] G. W. Gokel, in *Crown Ethers and Cryptands*, Monographs in Supramolecular Chemistry (Ed. J. F. Stoddart), RSC, Cambridge, **1991**.

[11] L. Stryer, *Biochemistry*, 2nd ed., Freeman, San Francisco, **1981**.

[12] L. A. Frederick, T. M. Fyles, V. A. Malik-Diemer and D. M. Whitfield, *J. Chem. Soc., Chem. Commun.* **1980**, 1211–1212.

[13] a) S. Shinkai and O. Manabe, *Top. Curr. Chem.* **1984**, *121*, 67–104; b) H.-G. Lohr and F. Vögtle, *Acc. Chem. Res.* **1985**, *18*, 65–72; c) S. Misumi, *Pure Appl. Chem.* **1990**, *62*, 493–498; d) A. Prasanna de Silva and K. R. A. Samenkumara Sandanayake, *Angew. Chem. Int. Ed. Engl.* **1990**, *29*, 1173–1175.

[14] A. Warshawsky and N. Kahana, *J. Am. Chem. Soc.* **1982**, *104*, 2663–2664.

[15] T. E. Edmonds, in *Chemical Sensors* (Ed. T. E. Edmonds), Chap. 8, Blackie, Glasgow and London, **1988**.

[16] R. L. Bruening, R. M. Izatt and J. S. Bradshaw, in *Cation Binding by Macrocycles* (Eds. Y. Inoue and G. W. Gokel), Marcel Dekker, New York, **1990**.

[17] F. Vögtle, *Supramolecular Chemistry*, Chaps. 10 and 11, Wiley, Chichester, England, **1991**.

[18] J. F. Biernat and T. Wilczewski, *Tetrahedron* **1980**, *36*, 2521–2523.

[19] S. Akabori, Y. Habata, Y. Sakamoto, M. Sato and S. Ebine, *Bull. Chem. Soc. Jpn.* **1983**, *56*, 537–541.

[20] S. Akabori, Y. Habata and M. Sato, *Bull. Chem. Soc. Jpn.* **1985**, *58*, 3540–3546.

[21] a) C. J. Pedersen, *J. Am. Chem. Soc.* **1967**, *89*, 2495–2496; b) C. J. Pedersen, *J. Org. Chem.* **1971**, *36*, 254–257; c) C. J. Pedersen, *Fed. Proc.* **1968**, *27*, 1305.

[22] T. Saji, *Chem. Lett.* **1986**, 275–276.

[23] S. R. Miller, D. A. Gustowski, Z. C. Chen, G. W. Gokel, L. Echegoyen and A. E. Kaifer, *Anal. Chem.* **1988**, *60*, 2021–2024.

[24] T. Saji and J. Kinoshita, *J. Chem. Soc., Chem. Commun.* **1986**, 716–717.

[25] A. Akabori, Y. Habata, M. Sato and S. Ebine, *Bull. Chem. Soc. Jpn.* **1983**, *56*, 1455–1461.

[26] A. Akabori, S. Shibahara, Y. Habata and M. Sato, *Bull. Chem. Soc. Jpn.* **1984**, *57*, 63–67.
[27] M. Sato, M. Kubo, S. Ebine and S. Akabori, *Bull. Chem. Soc. Jpn.* **1984**, *57*, 421–425.
[28] a) B. Czech and A. Rarajczak, *Polish J. Chem.* **1980**, *54*, 767–776; b) S. J. Rao, C. I. Milberg and R. C. Petter, *Tet. Lett.* **1991**, *31*, 3775–3778.
[29] a) I. Bernal, E. Raube, G. M. Reisner, R. A. Bartsch, R. A. Holwerda and B. P. Czech, *Organometallics* **1988**, *7*, 247–252; b) I. Bernal, G. M. Reisner, R. A. Bartsch, R. A. Holwerda and B. P. Czech, *Organometallics* **1988**, *7*, 255–258.
[30] S. G. Murray and R. R. Hartley, *Chem. Rev.* **1981**, 365–414.
[31] S. R. Cooper, *Acc. Chem. Res.* **1988**, *21*, 141–146.
[32] a) M. Watanabe, H. Ichikawa, I. Motoyama and H. Sano, *Bull. Chem. Soc. Jpn.* **1983**, *56*, 3291–3293; b) M. Watanabe, H. Ichikawa, I. Motoyama and H. Sano, *Chem. Lett.* **1983**, 1009–1012.
[33] M. Sato, H. Watanabe, S. Ebine and S. Akabori, *Chem. Lett.* **1982**, 1753–1756.
[34] M. Sato, S. Tanaka, S. Ebine and S. Akabori, *Bull. Chem. Soc. Jpn.* **1984**, *57*, 1929–1934.
[35] M. Sato, K. Suzuki and S. Akabori, *Bull. Chem. Soc. Jpn.* **1986**, *59*, 3611–3615.
[36] a) M. Sato, S. Tanaka, S. Ebine, K. Morinaga and S. Akabori, *J. Organomet. Chem.* **1985**, *282*, 247–253; b) M. Sato, S. Tanaka, S. Ebine, K. Morinaga and S. Akabori, *J. Organomet. Chem.* **1985**, *289*, 91–95.
[37] T. Vandrak and M. Sato, *J. Organomet. Chem.* **1989**, *364*, 207–215.
[38] M. Sato, S. Tanaka and S. Akabori, *Bull. Chem. Soc. Jpn.* **1986**, *59*, 1515–1519.
[39] S. Akabori, S. Sato, M. Sato and Y. Takunohoshi, *Bull. Chem. Soc. Jpn.* **1989**, *62*, 1582–1586.
[40] M. Sato, K. Suzuki and S. Akabori, *Chem. Lett.* **1987**, 2239–2242.
[41] M. Sato, H. Asano and S. Akabori, *J. Organomet. Chem.* **1991**, *401*, 363–370.
[42] M. Sato, S. Akabori, M. Katuda, I. Motoyama and H. Sano, *Chem. Lett.* **1987**, 1847–1850.
[43] M. Sato, M. Katuda, S. Nakashima, H. Sano and S. Akabori, *J. Chem. Soc. Dalton Trans.* **1990**, 1979–1984.
[44] M. Sato, H. Asano, K. Suzuki, M. Katada and S. Akabori, *Bull. Chem. Soc. Jpn.* **1989**, *62*, 3828–3834.
[45] Y. Yokomori and Y. Ohushi, *39th Symp. on Coordination Chem. Jpn.* **1989**, p. 769.
[46] S. Akabori, Y. Takunohoshi and S. Takagi, *Synth. Commun.* **1990**, *20*, 3187–3191.
[47] D. Seyforth, B. W. Homes, T. G. Rucker, M. Cowie and R. S. Dickson, *Organometallics* **1983**, *2*, (3), 472–474.
[48] a) S. Akabori, T. Kumagai, T. Shirahige, S. Sato, K. Kawazoe, C. Tamura and M. Sato, *Organometallics* **1987**, *6*, 526–531; b) S. Akabori, T. Kumagai, T. Shirahige, S. Sato, K. Kawazoe, C. Tamura and M. Sato, *Organometallics* **1987**, *6*, 2105–2109.
[49] M. Sato, M. Sekino and S. Akabori, *J. Organomet. Chem.* **1988**, *344*, C31–C34.
[50] M. Sato, M. Sekino, M. Katuda and S. Akabori, *J. Organomet. Chem.* **1989**, *377*, 327–337.
[51] a) A. P. Bell and C. D. Hall, *J. Chem. Soc. Chem. Commun.* **1980**, 163–165.
b) P. J. Hammond, P. D. Beer, C. Dudman, I. P. Danks, C. D. Hall, J. Knychala and M. C. Grossel, *J. Organomet. Chem.* **1986**, *306*, 367–373.
[52] G. Voepen and F. Vögtle, *Liebigs Ann. Chem.* **1979**, 1094–1101.
[53] P. J. Hammond, A. P. Bell and C. D. Hall, *J. Chem. Soc., Perkin Trans. 1* **1983**, 707–715.
[54] P. D. Beer, J. Elliot, P. J. Hammond, C. Dudman and C. D. Hall, *J. Organomet. Chem.* **1984**, *263*, C37–C42.
[55] C. D. Hall and N. W. Sharpe, *Organometallics* **1990**, *9*, 952–959.
[56] C. D. Hall, I. P. Danks, S. C. Nyburg, A. W. Parkins and N. W. Sharpe, *Organometallics* **1990**, *9*, 1602–1607.
[57] P. D. Beer, C. D. Bush and T. A. Harmor, *J. Organomet. Chem.* **1988**, *338*, 133–138.
[58] M. C. Grossel, M. R. Goldspink, J. A. Hriljac and S. C. Weston, *Organometallics* **1991**, *10*, 851–860.
[59] C. D. Hall, A. W. Parkins, S. C. Nyburg and N. W. Sharpe, *J. Organomet. Chem.* **1991**, *407*, 107–113.
[60] C. D. Hall, J. H. R. Tucker and N. W. Sharpe, *Organometallics* **1991**, *10*, 1727–1731.
[61] C. D. Hall, I. P. Danks, M. C. Lubienski and N. W. Sharpe, *J. Organomet. Chem.* **1990**, *384*, 139–146.
[62] C. D. Hall and N. W. Sharpe, *J. Organomet. Chem.* **1991**, 365–373.
[63] C. D. Hall, I. P. Danks and N. W. Sharpe, *J. Organomet. Chem.* **1990**, *390*, 227–235.

[64] C. D. Hall, I. P. Danks, P. J. Hammond, N. W. Sharpe and M. J. K. Thomas, *J. Organomet. Chem.* **1990**, *388*, 301–306.
[65] C. D. Hall, J. H. R. Tucker, A. Sheridan and D. J. Williams, *J. Chem. Soc., Dalton Trans.* **1992**, 3133–3136.
[66] C. D. Hall, J. H. R. Tucker and S. Y. F. Chu, *J. Organomet. Chem.* **1993**, *448*, 175–179.
[67] C. D. Hall, J. H. R. Tucker, S. Y. F. Chu, A. W. Parkins and S. C. Nyburg, *J. Chem. Soc., Chem. Commun.* **1993**, 1505–1507.
[68] C. D. Hall, N. W. Sharpe, I. P. Danks and Y. P. Sang, *J. Chem. Soc., Chem. Commun.* **1989**, 419–421.
[69] C. D. Hall and N. W. Sharpe, *J. Photochem. and Photobiol. A; Chem.* **1991**, *56*, 255–265.
[70] J. C. Medina, T. T. Goodnow, S. Bott, A. L. Atwood, A. E. Kaifer and G. W. Gokel, *J. Chem. Soc., Chem. Commun.* **1991**, 290–292.
[71] J. C. Medina, T. T. Goodnow, M. J. Rojas, J. L. Atwood, B. C. Lynn, A. E. Kaifer and G. W. Gokel, *J. Am. Chem. Soc.* **1992**, *114*, 10,583–10,595.
[72] G. W. Gokel, J. C. Medina and C. Li, *Synlett.* **1991**, 677–683.
[73] P. D. Beer, O. Kocian, R. J. Mortimer and P. Spencer, *J. Chem. Soc., Chem. Commun.* **1992**, 602–603.
[74] M. C. Grossel, M. R. Goldspink, J. P. Knychala, A. K. Cheetham and A. J. Hriljac, *J. Organomet. Chem.* **1988**, *532*, C13–C16.
[75] C. D. Hall, I. P. Danks, P. D. Beer, S. Y. F. Chu and S. C. Nyburg, *J. Organomet. Chem.* **1994**, *468*, 193–198.
[76] C. D. Hall, J. H. R. Tucker and S. Y. F. Chu, *Pure Appl. Chem.* **1993**, *65*, (3), 591–594.
[77] P. D. Beer, *J. Chem. Soc., Chem. Commun.* **1985**, 1115–1116.
[78] P. D. Beer, *J. Organomet. Chem.* **1985**, *297*, 313–317.
[79] P. D. Beer, H. Sikanyika, A. M. Z. Slawin and D. J. Williams, *Polyhedron* **1989**, *8*, 879–886.
[80] P. D. Beer and A. D. Keefe, *J. Organomet. Chem.* **1986**, *306*, C10–C12.
[81] P. D. Beer, A. D. Keefe, H. Sikanyika, C. Blackburn and J. F. McAleer, *J. Chem. Soc., Dalton Trans.* **1990**, 3289–3300.
[82] P. D. Beer, H. Sikanyika, C. Blackburn and J. F. McAleer, *J. Organomet. Chem.* **1988**, *350*, C15–C19.
[83] P. D. Beer, C. Blackburn, J. F. McAleer and H. Sikanyika, *Inorg. Chem.* **1990**, *29*, 378–381.
[84] M. P. Andrews, C. Blackburn, J. F. McAleer and V. D. Patel, *J. Chem. Soc., Chem. Commun.* **1987**, 1122–1124.
[85] P. D. Beer, H. Sikanyika, C. Blackburn, J. F. McAleer and M. G. B. Drew, *J. Organomet. Chem.* **1988**, *350*, C15–C19.
[86] P. D. Beer, H. Sikanyika, C. Blackburn, J. F. McAleer and M. G. B. Drew, *J. Chem. Soc., Dalton Trans.* **1990**, 3295–3300.
[87] P. D. Beer and H. Sikanyika, *Polyhedron* **1990**, *9*, 1091–1094.
[88] P. D. Beer and K. Wild, *Proc. 16th Int. Macrocylic Conf.* **1991**, p. 106.
[89] P. D. Beer, J. P. Danks, D. Hesek and J. F. McAleer, *J. Chem. Soc., Chem. Commun.* **1993**, 1735–1737.
[90] P. D. Beer, J. E. Martin, S. L. W. McWhinnie, M. E. Harman, M. B. Hursthouse, M. I. Ogden and A. H. White, *J. Chem. Soc., Dalton Trans.* **1991**, 2485–2492.
[91] P. D. Beer, D. B. Crowe and B. Main, *J. Organomet. Chem.* **1989**, *375*, C35–C39.
[92] L. Echegoyen, G. A. Gustowski, V. J. Gatto and G. W. Gokel, *J. Chem. Soc., Chem. Commun.* **1986**, 220–223.
[93] D. A. Gustowski, M. Delgado, V. J. Gatto, L. Echegoyen and G. W. Gokel, *J. Am. Chem. Soc.* **1986**, *108*, 7553–7560.
[94] A. Kaifer, L. Echegoyen, D. A. Gustowski, D. M. Goli and G. W. Gokel, *J. Am. Chem. Soc.* **1983**, *105*, 7168–7169.
[95] A. Kaifer, D. A. Gustowski, L. Echegoyen, V. J. Gatto, R. A. Schultz, T. P. Cleary, C. R. Morgan, D. M. Goli, A. M. Rios and G. W. Gokel, *J. Am. Chem. Soc.* **1985**, *107*, 1958–1965.
[96] P. D. Beer and S. S. Kurek, *J. Organomet. Chem.* **1987**, *336*, C17–C21.
[97] P. D. Beer and S. S. Kurek, *J. Organomet. Chem.* **1989**, *366*, C6–C8.
[98] D. J. Cram, *Science* **1983**, *219*, 1177–1183.
[99] F. Diederich, *Angew. Chem. Int. Ed. Engl.* **1988**, *27*, 362–386.

[100] C. D. Gutsche, *Monographs – Supramolecular Chemistry Series* (Ed. J. F. Stoddart), RSC, Cambridge, England, **1989**.
[101] D. J. Cram, *Angew. Chem. Int. Ed. Engl.* **1986**, *25*, 1039–1057.
[102] P. D. Beer and A. D. Keefe, *J. Inclusion Phenom.* **1987**, *5*, 499.
[103] P. D. Beer, M. G. B. Drew and A. D. Keefe, *J. Organomet. Chem.* **1988**, *353*, C10–C12.
[104] P. D. Beer, M. G. B. Drew and A. D. Keefe, *J. Organomet. Chem.* **1989**, *378*, 437–447.
[105] P. D. Beer and E. L. Tite, *Tetrahedron Lett.* **1988**, 2349–2352.
[106] P. D. Beer, M. G. B. Drew, A. Ibbotson and E. L. Tite, *J. Chem. Soc., Chem. Commun.* **1988**, 1498–1500.
[107] P. D. Beer, E. L. Tite, M. G. B. Drew and A. Ibbotson, *J. Chem. Soc., Dalton Trans.* **1990**, 2343–2350.
[108] P. D. Beer, E. L. Tite and A. Ibbotson, *J. Chem. Soc., Chem. Commun.* **1989**, 1874–1876.
[109] P. D. Beer, E. L. Tite and A. Ibbotson, *J. Chem. Soc., Dalton Trans.* **1991**, 1691–1698.
[110] P. D. Beer, A. D. Keefe, V. Böhmer, H. Goldmann, W. Vogt, S. Lerocq and M. Perrin, *J. Organomet. Chem.* **1991**, *421*, 265–273.
[111] P. D. Beer, A. D. Keefe, A. M. Z. Slaivin and D. J. Williams, *J. Chem. Soc., Dalton Trans.* **1990**, 3675–3682.
[112] P. D. Beer, *Adv. Inorg. Chem.* **1992**, *39*, 79–157.
[113] R. W. Wagner, P. A. Brown, T. E. Johnson and J. S. Lindsay, *J. Chem. Soc., Chem. Commun.* **1993**, 1463–1466.
[114] a) J.-M. Lehn, *Angew. Chem. Int. Ed. Engl.* **1987**, *26*, 266–267; b) V. Balzani, F. Barigalletti and L. De Cola, *Top. Curr. Chem.* **1990**, *158*, 31–71; c) J. H. R. Tucker and C. D. Hall, unpublished results.
[115] D. J. Cram, M. E. Tanner and R. Thomas, *Angew. Chem. Int. Ed. Engl.* **1991**, *30*, 1024–1027.
[116] M. Moran, C. M. Casado and I. Cuadrado, *Organometallics* **1993**, *12*, 4327–4333.

Received: January 27, 1994

7 Electrochemical and X-ray Structural Aspects of Transition Metal Complexes Containing Redox-Active Ferrocene Ligands

Piero Zanello

7.1 Introduction

One of the best known properties of ferrocene molecules is their ability to lose one electron at potentials that are a function of the electron-donating ability of the substituents attached to the cyclopentadienyl rings [1]. Such electron removal commonly does not involve fragmentation of the original molecular framework. As a typical example, Table 7-1 summarizes the redox potentials and the main structural changes accompanying the one-electron oxidation of some ferrocene molecules.

Table 7-1. Formal electrode potentials (*vs.* SCE) and main structural changes observed in the one-electron oxidation of some ferrocene molecules

Complex	$E^{\circ\prime}$,[a] V	Fe-C, Å	Fe-Cp, Å	C-C, Å	Reference
$Fe(C_5H_5)_2$	+0.45	2.03	1.66	1.39	[2]
$[Fe(C_5H_5)_2]^{+}$ [b]		2.05	1.70	1.34	[3]
$Fe(C_5Me_5)_2$	−0.10	2.05	1.66	1.42	[4]
$[Fe(C_5Me_5)_2]^{+}$ [c]		2.10	1.71	1.42	[5]
$Fe\{C_5(CH_2Ph)_5\}_2$	+0.38	2.06	1.66	1.43	[6, 7]
$[Fe\{C_5(CH_2Ph)_5\}_2]^{+}$ [d]		2.10	1.71	1.42	[8]

a Measured in $CH_2Cl_2/[NBu_4][ClO_4]$ 0.2 M solution.
b $[PF_6]^-$ counteranion.
c $[Br_3]^-$ counteranion.
d $[BF_4]^-$ counteranion.

Beyond the well known interest in the study of mixed-valent complexes [9], the incorporation of redox-actice ferrocene molecules into transition metal fragments is now opening promising fields of research. Heterometallic complexes able to undergo sequential electron-transfer steps have been shown to display the following:

1. to exhibit unusual catalytic activity [10], likely because of the availability of different, but cooperatively conjugated, metal sites;
2. to offer the unique opportunity to control the chemical, physico-chemical and structural variations in the coordination sphere of the central transition metal fragment through change of the oxidation state of the remote ferrocene subunit(s) [11];

3. to exhibit non-linear optical properties, which are of great relevancy in molecular materials for optical data transmission in the telecommunications industry [12];
4. to add synergistically the antitumor activity of platinum [13] or gold [14] complexes to that of metallocene [15] derivatives.

A simple citation will be made for those complexes of which the crystal structure has been solved, but for which the electrochemistry has not been studied.

In part, the present paper updates a previous review by Cullen and Woollins [16].

Unless otherwise specified, the potential values $E^{\circ\prime}$ here quoted are referred to the saturated calomel electrode (SCE).

7.2 Ferrocene Complexes of Homometallic Metal Fragments

7.2.1 Group 4 Metal Complexes

7.2.1.1 Titanium Complexes

Figure 7-1 shows the crystal structure of the diferrocenyl-titanium complex $(\eta^5\text{-}C_5H_5)_2Ti[(\eta^5\text{-}C_5H_4)Fe(\eta^5\text{-}C_5H_5)]_2$ [17]. The titanium atom possesses a pseudo-tetrahedral geometry. The cyclopentadienyl rings of the two ferrocenyl units are parallel and eclipsed.

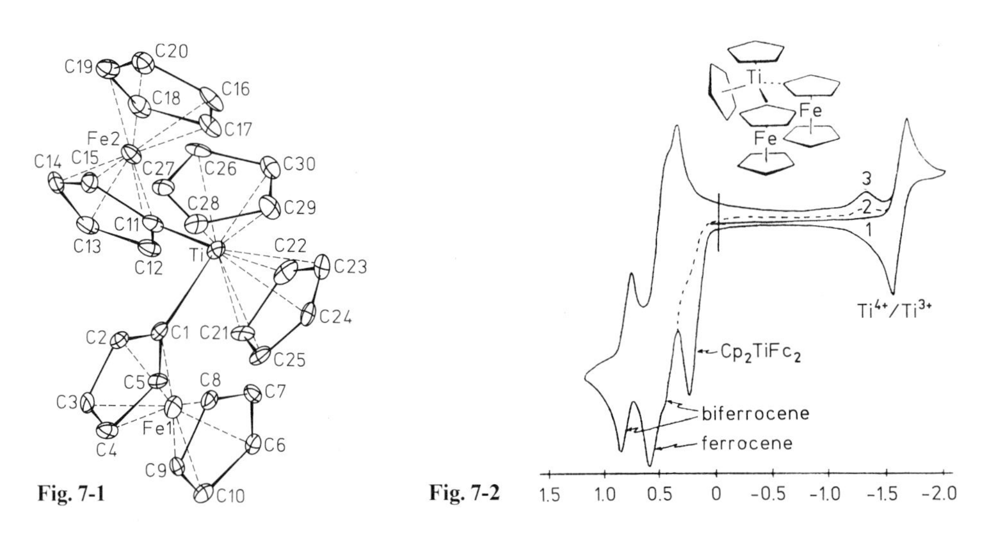

Fig. 7-1. X-ray structure of $(\eta^5\text{-}C_5H_5)_2Ti[(\eta^5\text{-}C_5H_4)Fe(\eta^5\text{-}C_5H_5)]_2$. Ti-C1 = Ti-C11 = 2.19 Å; Ti-C (Cp rings), average 2.41 Å (reproduced by permission of the International Union of Crystallography).

Fig. 7-2. Cyclic voltammogram recorded at a platinum electrode on a CH_2Cl_2 solution of $(C_5H_5)_2Ti[(C_5H_4)Fe(C_5H_5)]_2$. Scan rate 0.2 V s^{-1} (reproduced by permission of the American Chemical Society).

As illustrated in Fig. 7-2, in dichloromethane solution the complex undergoes a rather complicated redox pathway [18]. The first, irreversible anodic step has been tentatively attributed to the oxidation of one ferrocene ligand, at an electrode potential E_p of 0.25 V, according to the following reaction:

$$Cp_2Ti^{IV}(Fc^{II})_2 \xrightarrow[-e]{E_p\ =\ +0.25\ V} [Cp_2Ti^{IV}(Fc^{III})(Fc^{II})]^+$$

The subsequent anodic processes, which possess features of chemical reversibility, have been assigned to the oxidation of biferrocene and ferrocene, respectively; both molecules should derive from the fast degradation of the instantaneously electrogenerated monocation $[Cp_2Ti(Fc)_2]^+$. Finally, the reversible cathodic step is attributed to the reduction of the central titanium ion, according to the following reaction.

$$E^{\circ\prime} = -1.66\ V$$

$$Cp_2Ti^{IV}(Fc^{II})_2 \underset{-e}{\overset{+e}{\rightleftharpoons}} [Cp_2Ti^{III}(Fc^{II})_2]^-$$

It must be recognized that the high instability of the present compound in solution, even under a nitrogen atmosphere, makes an accurate electrochemical investigation difficult. The complete irreversibility of the first anodic step seems rather unusual for a ferrocenyl ligand. Indeed, we examined the redox behavior of the related methylated derivative, $(\eta^5\text{-}C_5H_4Me)_2Ti[(\eta^5\text{-}C_5H_4)Fe(\eta^5\text{-}C_5H_5)]_2$, and we reached somewhat different conclusions [19]. As shown in Fig. 7-3, this compound exhibits an overall cyclic voltammetry (CV) profile quite similar to that of the unmethylated species. However, the first anodic step, although complicated by subsequent chemical reactions, has features of transient chemical reversibility. In addition, controlled potential coulometric tests suggested that it consumes two electrons/molecule (the near closeness of the successive anodic steps prevents us performing exhaustive electrolysis; nevertheless, consumption of about 1.5 electrons/molecule was reached when the current decayed to 10% of that of the starting value). Thus, we attribute the first anodic process to the concomitant oxidation of the two ferrocenyl ligands, which, based on the occurrence of a single two-electron step, are likely to be electronically non-communicating [20]. In summary, the following redox processes occur.

$$E^{\circ\prime} = +0.90\ V$$

$$(MeCp)_2Ti(Fc)_2 \underset{+2e}{\overset{-2e}{\rightleftharpoons}} [(MeCp)_2Ti(Fc)_2]^{2+} \xrightarrow{\text{relatively fast}} \text{likely decomposition to biferrocene and ferrocene}$$

$$E^{\circ\prime} = -1.73\ V$$

$$(MeCp)_2Ti(Fc)_2 \underset{-e}{\overset{+e}{\rightleftharpoons}} [(MeCp)_2Ti(Fc)_2]^- \xrightarrow{\text{relatively slow}} \text{decomposition}$$

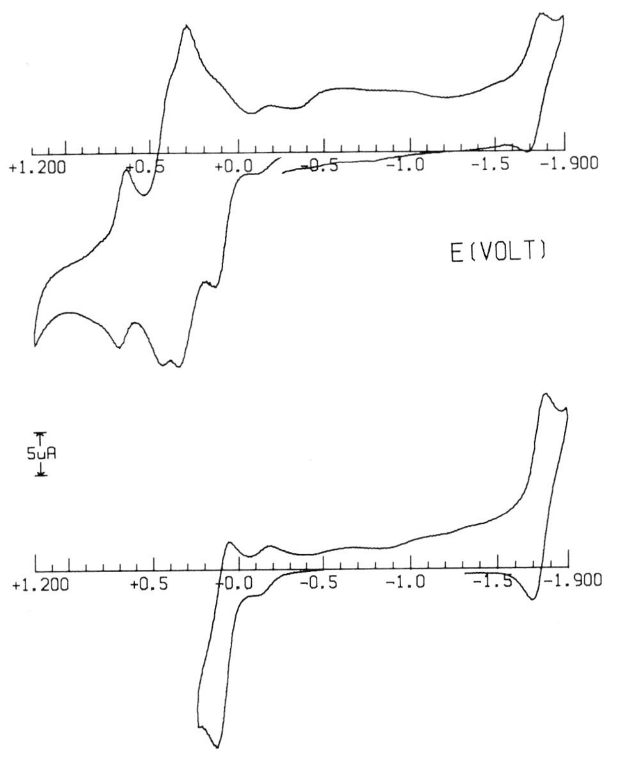

Fig. 7-3. Cyclic voltammograms recorded at a platinum electrode on a CH_2Cl_2 solution of $(C_5H_4Me)_2Ti[(C_5H_4)Fe(C_5H_5)]_2$. Scan rate 0.2 V s^{-1}.

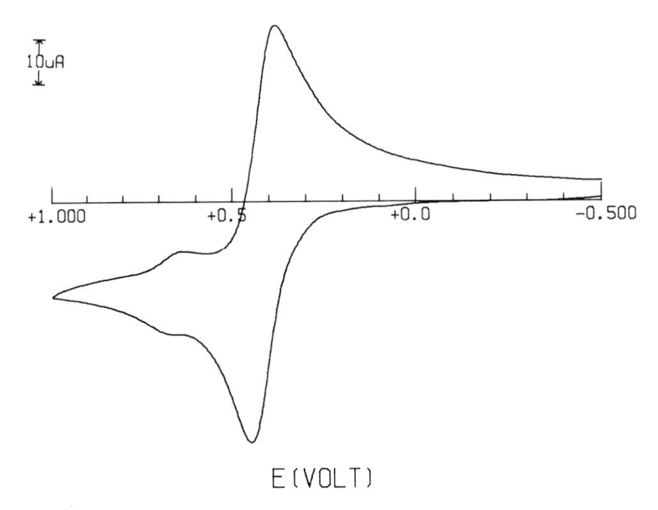

Fig. 7-4. Cyclic voltammogram recorded at a platinum electrode on a CH_2Cl_2 solution of $(C_5H_5)Ti[(C_5H_4)Fe(C_5H_5)]_3$. Scan rate 0.1 V s^{-1}.

The redox behavior exhibited by the triferrocenyl complex $(\eta^5\text{-}C_5H_5)Ti([\eta^5\text{-}C_5H_4)Fe(\eta^5\text{-}C_5H_5)]_3$ is less complicated [19]. As illustrated in Fig. 7-4, in dichloromethane solution it exhibits a single oxidation process, reversible in character ($E^{\circ\prime} = +0.41$ V). Controlled potential coulometry showed that this process involves three electrons/molecule.

Also in this case, the occurrence of a single-step three-electron process indicated that the three ferrocenyl ligands, from which the electrons are concomitantly removed, are non-communicating. In addition, if one considers that, under the same experimental conditions, ferrocene undergoes oxidation at $E^{\circ\prime} = +0.45$ V, it is evident that the ferrocenyl ligands are only slightly electronically perturbed by complex formation with the CpTi fragment.

7.2.1.2 Zirconium Complexes

The crystal structure of $(\eta^5\text{-}C_5H_4\text{-}t\text{Bu})_2Zr[(\eta^5C_5H_4)_2Fe]$ is shown in Fig. 7-5 [21]. The zirconium atom has a pseudotetrahedral geometry. The cyclopentadienyl rings of the ferrocene unit are eclipsed and deviate somewhat from the parallel disposition (dihedral angle 6°).

As illustrated in Fig. 7-6, in tetrahydrofuran (THF) solution the complex undergoes a one-electron oxidation to the corresponding stable monocation ($E^{\circ\prime}(A_1'/A_1) = +0.52$ V), which, in the presence of traces of water, evolves to a more easily oxidizable hydroxo species ($E^{\circ\prime}(A_2'/A_2) = +0.25$ V) [22].

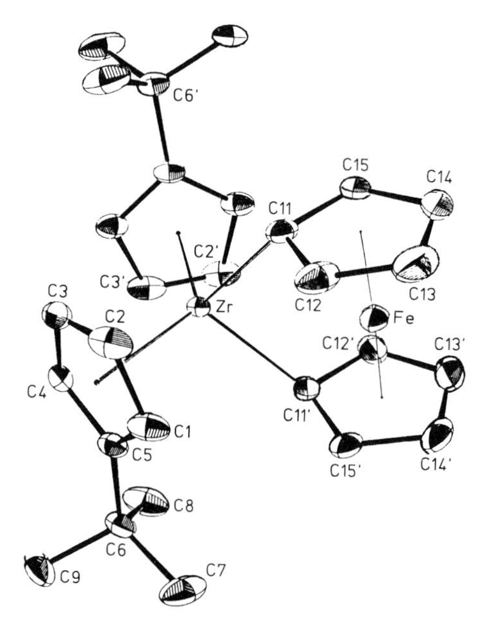

Fig. 7-5. Perspective view of $(\eta^5\text{-}C_5H_4\text{-}t\text{Bu})_2Zr[(\eta^5\text{-}C_5H_4)_2Fe]$. Zr-C11, 2.28 Å; Zr-C (*t*-butyl-Cp rings), average 2.55 Å. Zr ··· Fe, 2.96 Å (reproduced by permission of the American Chemical Society).

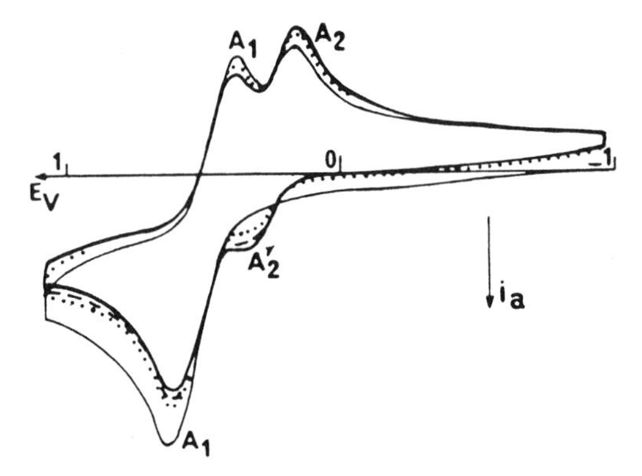

Fig. 7-6. Cyclic voltammogram recorded at a platinum electrode on a THF solution of $(C_5H_4\text{-}t\text{Bu})_2Zr[(C_5H_4)_2Fe]$. Scan rate 0.5 V s^{-1} (reproduced by permission of Elsevier Sequoia S.A.).

The crystal structure of $(\eta^5\text{-}C_5H_4\text{-}t\text{Bu})_2Zr[(\eta^5\text{-}C_5H_4Se)_2Fe]$ has been reported recently [23].

7.2.1.3 Hafnium Complexes

The triferrocenyl-hafnium complex $(\eta^5\text{-}C_5Me_5)Hf[(\eta^5\text{-}C_5H_4)Fe(\eta^5\text{-}C_5H_5)]_3$ exhibits a redox behavior quite similar to that illustrated for the analogous titanium complex $(\eta^5\text{-}C_5H_5)Ti[(\eta^5\text{-}C_5H_4)Fe(\eta^5\text{-}C_5H_5)]_3$ [19]. In fact, in dichloromethane solution it undergoes a reversible, single-step, three-electron oxidation according to the following equation.

$$E^{\circ\prime} = +0.43\ \text{V}$$

$$(\text{Cp}^*)\text{Hf}(\text{Fc})_3 \underset{+3e}{\overset{-3e}{\rightleftharpoons}} [(\text{Cp}^*)\text{Hf}(\text{Fc})_3]^{3+}$$

The fact that the oxidation process occurs at a potential value slightly more positive than that exhibited by $(\text{Cp})\text{Ti}(\text{Fc})_3$, despite the presence of the more electron-donating pentamethylcyclopentadienyl ligand, suggests that the hafnium atom is more electron-withdrawing than the titanium atom.

7.2.2 Group 5 Metal Complexes: Vanadium

To our knowledge, the onyl vanadium-ferrocene complexes structurally characterized are $Fe[\eta^5\text{-}C_5H_4\text{-}V(\eta^5\text{-}C_5H_4Me)_2]_2$ [24] and $[Fe(\eta^5\text{-}C_5H_4Se)_2]V(O)(\eta^5\text{-}C_5Me_5)$ [25]. Their redox behavior is not known.

7.2.3 Group 6 Metal Complexes

The heterometallic ferrocene complexes of Group 6 metals are those that have been most widely investigated.

7.2.3.1 Chromium Complexes

Scheme 7-1 shows a series of ferrocenyl-benzenetricarbonylchromium(I) complexes that have been studied by electrochemistry, in which the chromium atom assumes a pseudotetrahedral geometry.

These complexes undergo two separate oxidations, both in acetonitrile (MeCN) [26, 27] and *N*,*N*-dimethylformamide (DMF) [28] solution. As shown in Table 7-2,

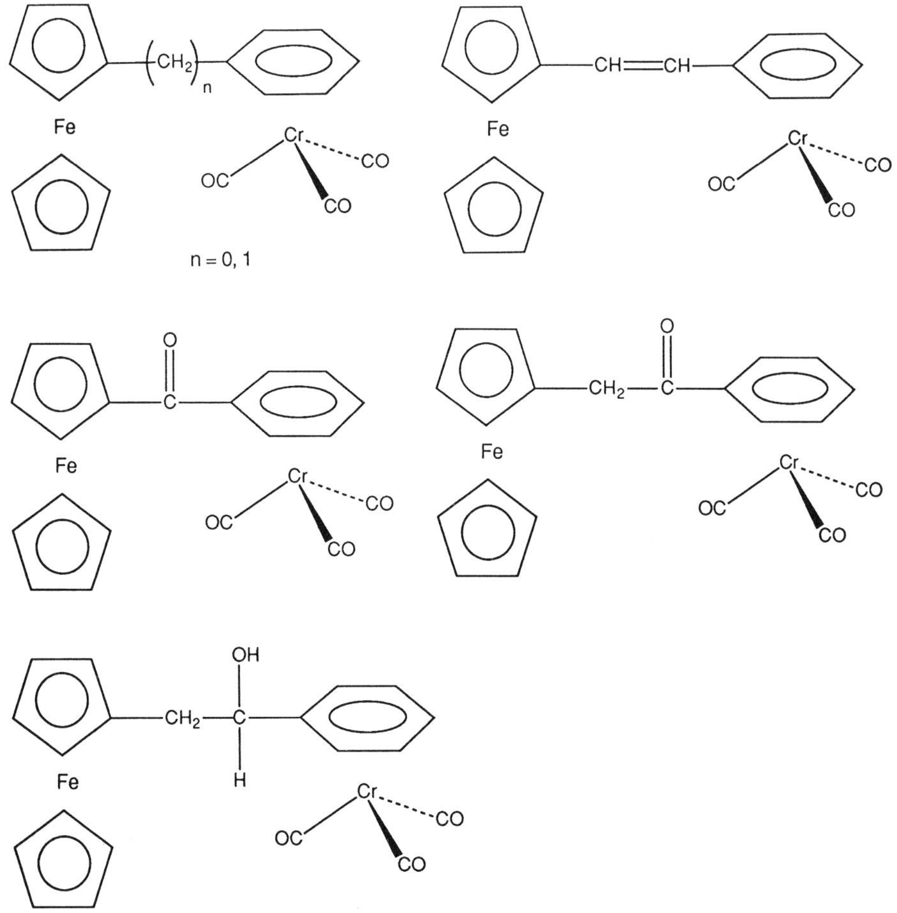

CH_2 n Fe Cr OC CO CO n = 0, 1

CH=CH Fe Cr OC CO CO

O C Fe Cr OC CO CO

O CH_2 C Fe Cr OC CO CO

OH CH_2 C H Fe Cr OC CO CO

Scheme 7-1

Table 7-2. Half-wave potentials (*vs.* SCE) for the sequential oxidation of the chromium-ferrocenyl complexes $(\eta^5\text{-}C_5H_5)Fe[(\eta^5C_5H_4\text{-}\{R\}(\eta^6\text{-}C_6H_5)Cr(CO)_3]$ shown in Scheme 7-1

Complex	$E_{1/2}(0/+)$, V	$E_{1/2}(+/2+)$, V	Solvent	Reference
R = CH_2				
$n = 0$	+0.55	+0.96	MeCN	[26]
	+0.48	+0.84	MeCN	[27]
	+0.49	+0.84	DMF	[28]
$n = 1$	+0.49	+0.96	MeCN	[26]
	+0.41	+0.80	MeCN	[27]
	+0.44	+0.83	DMF	[28]
R = CH=CH	+0.42	+0.77	MeCN	[27]
	+0.46	+0.80	DMF	[28]
R = C=O	+0.63	+0.95	MeCN	[27]
	+0.68	+1.0	DMF	[28]
R = CH_2CO	+0.62	+0.62	MeCN	[27]
	+0.66	+0.66	DMF	[28]
R = $CH_2CH(OH)$	+0.40	+0.80	MeCN	[27]
	+0.40	+0.76	DMF	[28]
$Fe(C_5H_5)_2$	+0.39		MeCN	a
	+0.45		DMF	[29]
$(C_6H_6)Cr(CO)_3$	+0.74		MeCN	[26]
	+0.72		MeCN	[27]
	+0.82		DMF	[28]

a Unpublished results.

comparison with the redox potentials of the two metal-fragment precursors suggests that the first electron removal is centered on the ferrocenyl fragment, whereas the second oxidation step is centered on the phenylchromiumtricarbonyl unit. It has long been known that in acetonitrile solution the two electron removals involve reversible, one-electron transfers [26]. Indeed, more recent investigations support the view that the two oxidation processes are more complicated in both acetonitrile [27] and in DMF [27] solution. They are found to involve multielectron processes that are primed by intramolecular oxidation of the benzenechromiumtricarbonyl (Bct) unit from the instantaneously electrogenerated ferrocenium (Fc^+) moiety. This regenerates the ferrocene (Fc) fragment, which in turn undergoes reoxidation at the electrode, according to the following mechanism.

$$\text{Bct-Fc} \underset{+e}{\overset{-e}{\rightleftharpoons}} \text{Bct-Fc}^+$$

$$\text{Bct-Fc}^+ \overset{K}{\rightleftharpoons} \text{Bct}^+\text{-Fc}$$

$$\text{Bct}^+\text{-Fc} \xrightarrow{-e} \text{decomposition}$$

Only in the case of the complex in which the ferrocene and the benzenechromium-tricarbonyl fragments are separated from each other by an aliphatic bridge

R= $-CH_2C(H)(OH)-$ precluding electronic interaction, are both oxidation steps (the first on the ferrocene center and the second on the chromium center) chemically reversible. It should be noted that removal of both of the two electrons becomes more difficult on complexation of the two metal fragments.

We have recently investigated the redox properties of the series of ferrocenylsilanechromium complexes shown in Scheme 7-2 [30].

Me
(-Si-)n
Me
Fe
Cr
OC CO CO

Scheme 7-2 n = 1 - 3

As an example, Fig.7-7 illustrates the X-ray structure of one member of this series, $(\eta^5\text{-}C_5H_5)Fe[\eta^5\text{-}C_5H_4\text{-}(SiMe_2)_2\text{-}(\eta^6\text{-}C_6H_5)Cr(CO)_3]$. The cyclopentadienyl rings of the ferrocenyl unit assume an eclipsed configuration.

As shown in Fig. 7-8, which refers to $(C_5H_5)Fe[C_5H_4\text{-}(SiMe_2)_2\text{-}(C_6H_5)Cr(CO)_3]$, these complexes undergo two successive one-electron oxidations in dichloromethane solution. The first one-electron removal is chemically reversible and is centered on the ferrocenyl fragment; the second, which has features of an electron transfer complicated by following chemical reactions (particularly when tetrabutylammonium perchlorate is used as supporting electrolyte) [31], is centered on the benzenechromiumtricarbonyl fragment.

Table 7-3 summarizes the redox potentials of the two oxidation steps, also in comparison with those of the corresponding free units.

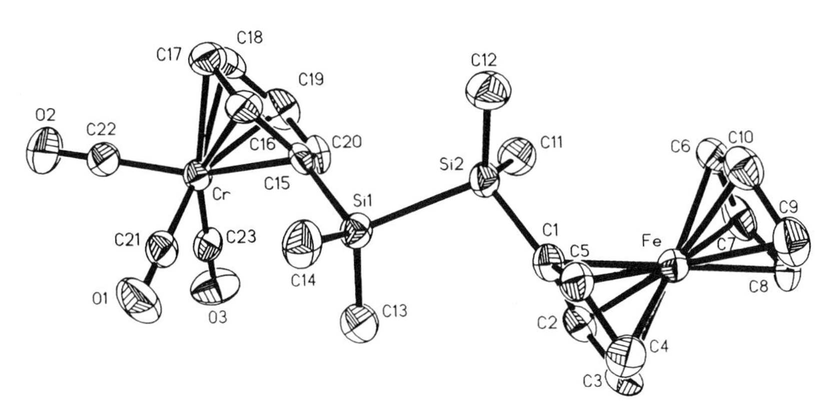

Fig. 7-7. X-ray crystal structure of $(\eta^5\text{-}C_5H_5)Fe[\eta^5\text{-}C_5H_4\text{-}(SiMe_2)_2\text{-}(\eta^6\text{-}C_6H_5)Cr(CO)_3]$. Average bond lengths: Cr-C (cyclopentadienyl), 2.22 Å; Cr-C (carbonyl), 1.82 Å; Fe-C (cyclopentadienyl), 2.40 Å. C15-Si1, 1.90 Å; Si1-Si2, 2.34 Å; Si2-C1, 1.85 Å.

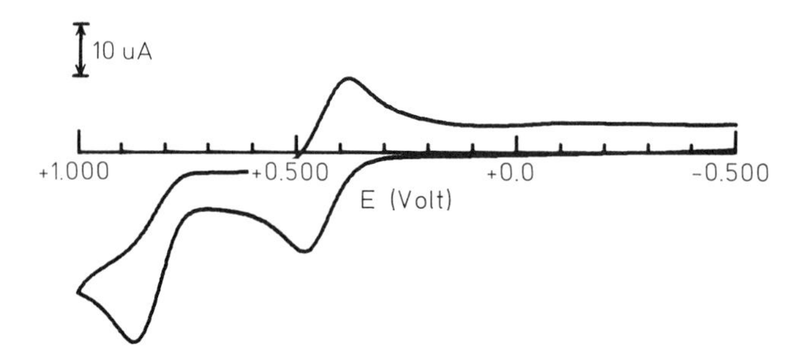

Fig. 7-8. Cyclic voltammogram recorded at a platinum electrode on a CH_2Cl_2 solution of $(C_5H_5)Fe[C_5H_4\text{-}(SiMe_2)_2\text{-}(C_6H_5)Cr(CO)_3]$. Scan rate 0.2 V s^{-1}.

Table 7-3. Formal electrode potentials (*vs.* SCE) for the sequential oxidation of the ferrocenylsilane-chromium complexes $(\eta^5\text{-}C_5H_5)Fe[(\eta^5\text{-}C_5H_4)\text{-}\{SiMe_2\}_n(\eta\text{-}C_6H_5)Cr(CO)_3]$ shown in Scheme 7-2. Dichloromethane solution. Fc = $(C_5H_5)Fe(C_5H_4)$

Complex	$E^{\circ\prime}$ (Fc/Fc$^+$), V	$E_p{}^a$ (Cr^0/Cr^I), V
Fc-$SiMe_2(C_6H_5)Cr(CO)_3$	+0.48	+0.89
Fc-$\{SiMe_2\}_2(C_6H_5)Cr(CO)_3$	+0.44	+0.87
Fc-$\{SiMe_2\}_3(C_6H_5)Cr(CO)_3$	+0.42	+0.84
$(C_6H_6)Cr(CO)_3$		+0.83
Fc-$SiMe_2(C_6H_5)$	+0.44	
Fc-$\{SiMe_2\}_2(C_6H_5)$	+0.42	
Fc-$\{SiMe_2\}_3(C_6H_5)$	+0.41	

a Measured at 0.2 V s^{-1}.

It can be seen that, on complexation, either the ferrocene or the chromium oxidations become slightly more difficult, but the difference in redox potential decreases with increasing number of $SiMe_2$ units. This suggests that they tend to prevent interactions between the two metal centers.

The ferrocenylcarbene complexes shown in Scheme 7-3 were the first octahedral chromium fragments studied from an electrochemical viewpoint [32].

M = Cr

Scheme 7-3

In spite of the presence of the two metal centers, these complexes exhibit a single, reversible, one-electron oxidation. Despite the well known propensity of ferrocene to undergo reversible one-electron oxidation, it has been assumed that both metal centers contribute to the molecular orbital (HOMO) from which the electron is removed. Table 7-4 summarizes the relevant redox potentials.

Replacement of the *O*-heteroatomic group bound to the carbene-carbon atom by the *N*-heteroatomic group makes removal of an electron significantly easier.

Furthermore, the ferrocenylcarbyne complexes illustrated in Scheme 7-4 are related to the carbene complexes discussed above.

Figure 7-9 illustrates the molecular structure of $(\eta^5\text{-}C_5H_5)Fe[\eta^5\text{-}C_5H_4{\equiv}Cr(CO)_4Br]$ [34], showing that the chromium atom possesses octahedral coordination; the

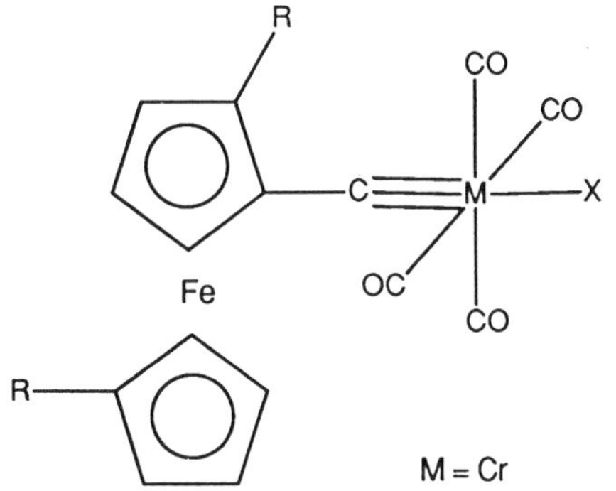

Scheme 7-4

Table 7-4. Formal electrode potentials (*vs.* SCE) for the one-electron oxidation of the ferrocenylcarbene complexes $(1\text{-R-}C_5H_4)Fe\{C_5H_3(2'\text{-R})[1'\text{-C(R')}{=}Cr(CO)_5]\}$

Complex		$E^{\circ\prime}$, V	Solvent	Reference
R = H	R′ = OEt	+0.70	CH_2Cl_2	[32, 33]
		+0.68	DME[a]	[34, 36]
R = Me	R′ = OEt	+0.71	DME[a]	[36]
R = H	R′ = OMe	+0.69	CH_2Cl_2	[32]
R = H	R′ = Pyrrolidin-1-yl	+0.51	CH_2Cl_2	[32]
$Fe(C_5H_5)_2$		+0.39	CH_2Cl_2	[33]
		+0.49	DME[a]	[35]

a DME = dimethoxyethane.

Table 7-5. Formal electrode potentials (*vs.* SCE) for the one-electron oxidation of the ferrocenylcarbyne complexes $(1\text{-R-}C_5H_4)Fe\{C_5H_3(2'\text{-R})[1'\text{-C}{\equiv}Cr(CO)_4X]\}$ in dimethoxyethane solution [34]

R	X	$E^{\circ\prime}$, V
H	Cl	+0.82
H	Br	+0.82
H	I	+0.84

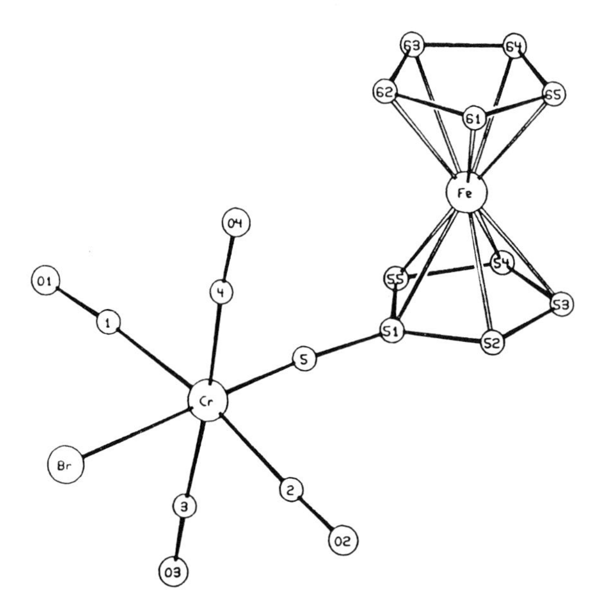

Fig. 7-9. X-Ray molecular structure of $(\eta^5\text{-}C_5H_5)Fe[\eta^5\text{-}C_5H_4\text{-}C{\equiv}Cr(CO)_4Br]$. Cr-Br, 2.58 Å; Cr-C1, Cr-C2, Cr-C3, Cr-C4, average1,94 Å; Cr-C5, 1.71 Å; C5-C51, 1.40 Å; Fe-C (Cp rings), average 2.07 Å; C-C (Cp rings), average 1.46 Å (reproduced by permission of VCH Publishers).

Cr-C (carbyne) distance is about 0.2 Å shorter than those of Cr-C (CO); the bond distances in the ferrocenyl fragment basically parallel those in the unsubstituted ferrocene [2], with the cyclopentadienyl rings being parallel and nearly eclipsed.

These carbyne complexes also undergo a single, electrochemically reversible, one-electron oxidation. The electrochemical reversibility of the oxidation process suggests that, independent of the localization of the electron removal, it does not involve important stereochemical reorganizations with respect to the original geometry [37]. The potentials of such oxidations are reported in Table 7-5.

Another series of octahedral chromium complexes, the electrochemical behavior of which has been studied, is represented by the isocyanides shown in Scheme 7-5.

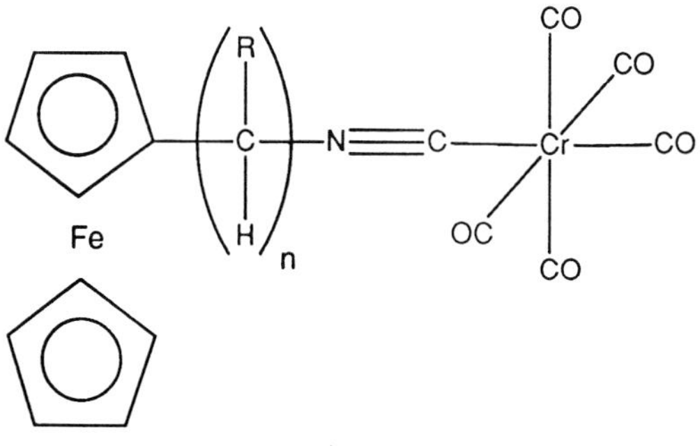

Scheme 7-5

Table 7-6. Formal electrode potentials[a] (*vs.* SCE) for the two one-electron oxidations exhibited by the ferrocenyl-isocyanide complexes $(C_5H_5)Fe[C_5H_4\text{-}(CHR)_nNCCr(CO)_5]$ in acetonitrile solution [38, 39]

Complex	$E^{\circ\prime}$ (Fc^{II}/Fc^{III}), V	(Cr^0/Cr^I), V
$n = 0$ –	+0.70	+1.16
$n = 1$ $R = H$	+0.52	+1.10
$n = 1$ $R = Men$[b]	+0.50	+1.10
$(C_5H_5)Fe(C_5H_4\text{-}NC)$	+0.70	
$(C_5H_5)Fe(C_5H_4\text{-}CH_2\text{-}NC)$	+0.45	
$(C_5H_5)Fe(C_5H_4\text{-}CH(Men)\text{-}NC]$	+0.47	
$(CO)_5Cr(MeCN)$		+0.90[c]

a Under the actual conditions: $[FcH]^{0/+} = +0.40$ V.
b Men = [(1*R*,2*S*,5*R*)-2-isopropyl-5-methylcyclohexyl]-(*R*)-methyl.
c In CH_2Cl_2 solution [32].

These complexes undergo two subsequent one-electron oxidations, which are reversible. Table 7-6 summarizes the relevant redox potentials.

Comparison with the redox potentials of the precursors of the two metal fragments suggests that the first electron removal is localized on the ferrocene moiety, whereas the second is localized on the chromium center [38, 39].

As illustrated in Fig. 7-10, quite the opposite trend has been assigned to the sequential oxidations exhibited by the chromium complex of 1-[(dimethylamino)methyl]-2-(diphenylphosphino)ferrocene (FcCNP), $(FcCNP)Cr(CO)_4$ [40].

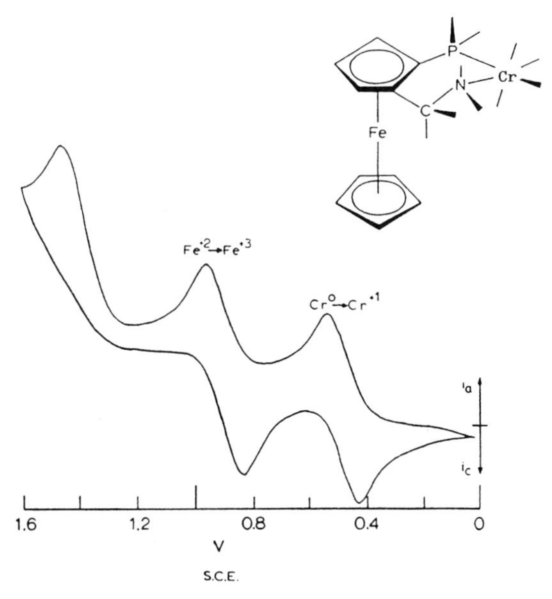

Fig. 7-10. Cyclic voltammogram recorded at a platinum electrode on a CH_2Cl_2 solution of $(FcCNP)Cr(CO)_4$ (reproduced by permission of Elsevier Sequoia S.A.).

Based on spectroscopic measurements on the gradually electrogenerated monocation and dication species, oxidation of the chromium fragment ($E^{\circ\prime} = +0.48$ V) now precedes that of the ferrocene moiety ($E^{\circ\prime} = +0.90$ V).

To conclude the present Section, we would like to remark that a few ferrocene-chromium complexes have been characterized by X-ray analysis, but that their redox properties are still unknown: $(\eta^5\text{-}C_5H_5)Fe\{(\eta^5\text{-}C_5H_4)\text{-}C(O)[(\eta^5\text{-}C_5H_4)Cr(CO)_2(NO)]\}$ [41]; $\{[(CO)_2(NO)Cr(\eta^5\text{-}C_5H_4)]CH_2(\eta^5\text{-}C_5H_4)\}Fe\{(\eta^5\text{-}C_5H_4)C(O)[(\eta^5\text{-}C_5H_4)Cr(CO)_2(NO)]\}$ [42]; $[(CO)_5Cr(\eta^5\text{-}C_5H_4PPh_2)]Fe[(\eta^5\text{-}C_5H_4PPh_2)Cr(CO)_5]$ [43]; $(\eta^5\text{-}C_5H_5)Fe\{(\eta^5\text{-}C_5H_3)[1\text{-}CH_2(\eta^5\text{-}C_5H_4)Cr(CO)_2(NO)][2\text{-}C(O)(\eta^5\text{-}C_5H_4)Cr(CO)_2(NO)\}$ [44]; $(\eta^5\text{-}C_5H_5)Fe[\eta^5\text{-}C_5H_4C(Ph){=}S]Cr(CO)_5]$ [45]; and $[(\eta^5\text{-}C_5H_4\text{-}PPh_2Me)^+]Fe[\{\eta^5\text{-}C_5H_4\text{-}Cr(CO)_5\}^-]$ [46].

7.2.3.2 Molybdenum Complexes

Like the corresponding chromium complex shown in Scheme 7-3, the molybdenum-carbene complex $(\eta^5\text{-}C_5H_4Me)Fe\{\eta^5\text{-}C_5H_3[1\text{-}(OEt)C{=}Mo(CO)_5](2\text{-}Me)\}$ undergoes a single one-electron oxidation in DME solution. The redox potential ($E^{\circ\prime} = +0.73$ V) is nearly coincident with that of the chromium complex (see Table 7-4) [36].

In the ferrocenyl-carbene complex shown in Scheme 7-6, the molybdenum fragment possesses a tetrahedral geometry. It has been reported that this complex onyl undergoes a ferrocene-centred oxidation at +0.52 V in THF solution [47].

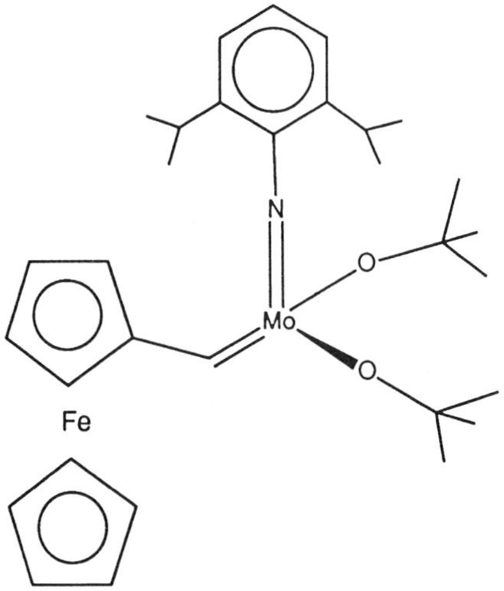

N
O
Mo
O
Fe

Scheme 7-6

Finally, the octahedral molybdenum-carbyne complex $(\eta^5\text{-}C_5H_5)Fe[\eta^5\text{-}C_5H_4\text{-}C{\equiv}Mo(CO)_4Br]$ undergoes a one-electron oxidation at $E^{\circ\prime} = +0.79$ V in DME [34], which is very near to that of the corresponding chromium complex illustrated in Scheme 7-4 and Table 7-5.

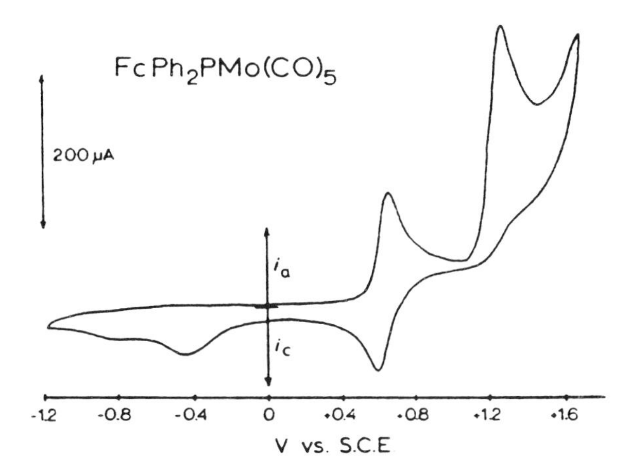

Fig. 7-11. Cyclic voltammetric response exhibited by $(C_5H_5)Fe[(C_5H_4PPh_2)Mo(CO)_5]$ in MeCN solution. Platinum working electrode (reproduced by permission of Elsevier Sequoia S.A.).

Diphenylphosphinoferrocene $(\eta^5\text{-}C_5H_5)Fe(\eta^5\text{-}C_5H_4PPh_2)$, $FcPPh_2$, and 1,1′-bis-(diphenylphosphino)ferrocene $Fe(\eta^5\text{-}C_5H_4PPh_2)_2$, $Fc(PPh_2{}')_2$, are particularly apt to coordinate metal fragments (see Chap. 1). The redox propensity of a series of complexes formed by molybdenum pentacarbonyl with ferrocenyl-phosphines has been studied.

As shown in Fig. 7-11, in acetonitrile solution $(\eta^5\text{-}C_5H_5)Fe[(\eta^5\text{-}C_5H_4PPh_2)\text{-}Mo(CO)_5]$ undergoes a first reversible one-electron oxidation, followed by a second irreversible multielectron oxidation [48]. It is thought that the first step is centered on the ferrocene fragment, whereas the second one on the molybdenum fragment.

The X-ray structure of the complex is not available, but its geometry is likely to be derived from that of the bis(diphenylphosphino) complex $(\eta^5\text{-}C_5H_4PPh_2)\text{-}Fe[(\eta^5\text{-}C_5H_4PPh_2)Mo(CO)_5]$, shown in Fig. 7-12 [49]. The molybdenum fragment possesses octahedral geometry.

The gradual substitution of the phenyl groups in the phosphine ligand for ferrocenyl subunits affords the diferrocenyl- and triferrocenyl-phosphinepentacarbonyl-molybdenum complexes, respectively. With respect to the redox pathway shown in Fig. 7-11, each added ferrocenyl ligand involves the appearance of a further one-electron oxidation [50]. The relevant redox potentials are given in Table 7-7.

It is evident that, with respect to the corresponding free ferrocenylphosphines, delicate (and difficult to predict) electronic effects govern the redox aptitude of these complexes. The presence of the molybdenum fragment affects appreciably the ability of the ferrocenyl ligands to lose one electron, thus indicating that some electronic conjugation between the two metals must exist.

Figure 7-13 shows the crystal structure of the bis(diphenylphosphino)ferrocene-molybdenum complex $[Fe(\eta^5\text{-}C_5H_4PPh_2)_2]Mo(CO)_4$ [51].

Once again the molybdenum atom has octahedral coordination. At variance with the previously described complexes, the cyclopentadienyl rings here are slightly tilted from the usual parallel disposition (2.2°) and assume a nearly staggered

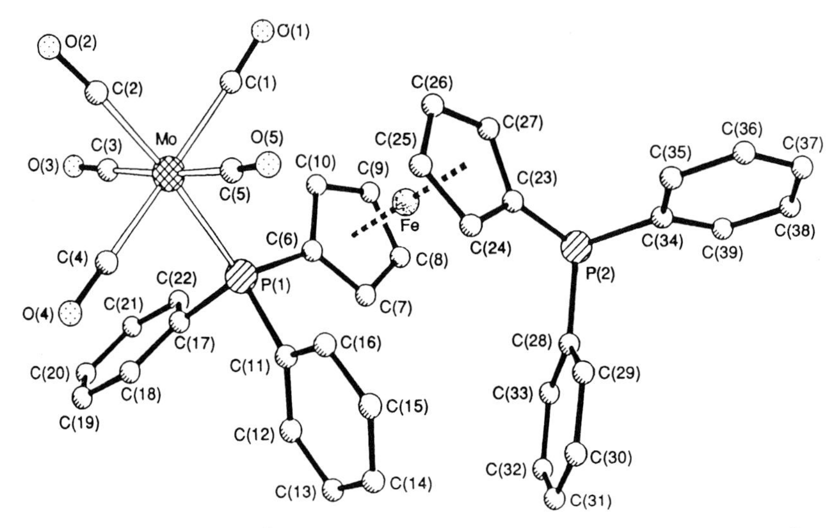

Fig. 7-12. X-Ray structure of $(\eta^5\text{-}C_5H_4PPh_2)Fe[(\eta^5\text{-}C_5H_4PPh_2)Mo(CO)_5]$. Mo-P, 2.56 Å, Mo-C2, 1.99 Å; Mo-C (1, 3, 5), average 2.04 Å; P1-C6, 1.81 Å; Fe-Cp (centroid), 1.64 Å reproduced by permission of the Royal Society of Chemistry).

conformation. This minor distortion is possibly due to the constraints imposed by coordination of the two phosphine substituents to the molybdenum atom.

In THF, $[Fe(C_5H_4PPh_2)_2]Mo(CO)_4$ undergoes a ferrocene-centered oxidation, irreversible because of fast degradation of the instantaneously electrogenerated monocation [52]. This result is not at all unexpected, because the one-electron oxidation of the free disubstituted ferrocenylphosphine $Fe(C_5H_4PPh_2)_2$ is followed by chemical complications [52, 53]. Furthermore, theoretical arguments on the

Table 7-7. Comparison between the formal electrode potentials (*vs.* SCE) for the oxidation processes exhibited by the pentacarbonylmolybdenum complexes of ferrocenylphosphines with those of the parent ferrocenylphosphines

Complex	Ferrocene-centred oxidations			Molybdenum-centred oxidation	Solvent	Ref.
	$E^{\circ\prime}$ (0/+), V	$E^{\circ\prime}$ (+/2+), V	$E^{\circ\prime}$ (2+/3+), V	E_p, V		
$(FcPPh_2)Mo(CO)_5$	+0.62			+1.3	MeCN	[48]
	+0.73			+1.56	CH_2Cl_2	[50]
$(Fc_2PPh)Mo(CO)_5$	+0.59	+0.78		+1.35	CH_2Cl_2	[50]
$(Fc_3P)Mo(CO)_5$	+0.61	+0.79	+0.84	+2.05	CH_2Cl_2	[50]
$FcPPh_2$	+0.48				MeCN	[48]
	+0.57				CH_2Cl_2	[50]
Fc_2PPh	+0.69	+0.85			CH_2Cl_2	[50]
Fc_3P	+0.65	+0.81	+0.81		CH_2Cl_2	[50]

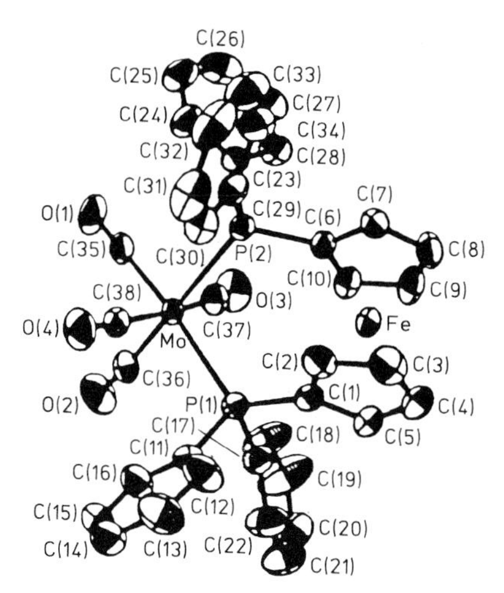

Fig. 7-13. X-Ray structure of $[Fe(\eta^5\text{-}C_5H_4PPh_2)_2]Mo(CO)_4$. Mo-P, 2.56 Å; Mo-C (*trans* to P), 1.97 Å; Mo-C (*cis* to P), 2.04 Å (reproduced by permission of the American Chemical Society).

instability of 1,1′-bis(diphenylphosphino)ferrocenium species have been put forward [54]. In contrast, the related complex with two monosubstituted ferrocenylphosphine ligands, $[(\eta^5\text{-}C_5H_5)Fe(\eta^5\text{-}C_5H_4PPh_2)]_2Mo(CO)_4$, undergoes a reversible oxidation in CH_2Cl_2, which involves two overlapping one-electron removals (one for each ferrocenyl ligand). This result is congruent with the chemical reversibility of the oxidation process of the free monosubstituted ligand diphenylphosphinoferrocene. Irreversible oxidation of the molybdenum centre occurs at higher potential [50].

More contrasting behavior is displayed by the somewhat related molybdenum complex of 1-[(dimethylamino)methyl]-2-(diphenylphosphino)ferrocene, (FcCNP)-$Mo(CO)_4$, in which the octahedral geometry around the molybdenum atom is achievd by coordination of one P atom and one N atom of the adjacent substituents attached to a single cyclopentadienyl ring (see the sketch in Fig. 7-10). In CH_2Cl_2, the irreversible oxidation of the molybdenum fragment precedes the reversible oxidation of the ferrocene moiety [40].

Table 7-8 summarizes the redox potentials of the oxidation processes of these ferrocenylphosphino-$Mo(CO)_4$ complexes.

Table 7-8. Formal electrode potentials (*vs.* SCE) for the oxidation processes exhibited by different tetracarbonylmolybdenum-ferrocenylphosphine complexes [Fc = $(C_5H_5)Fe(C_5H_4)$ or $(C_5H_5)Fe(C_5H_3)$]

Complex	$E^{\circ\prime}$ (Fc/Fc$^+$), V	$E^{\circ\prime}$ (Fc/Fc$^+$), V	E_p (Mo/Mo$^+$), V	Solvent	Ref.
$[Fe(C_5H_4PPh_2)_2]Mo(CO)_4$	+0.73[a]	–	–	THF	[52]
$(FcPPh_2)_2Mo(CO)_4$	+0.63	+0.63	+1.16	CH_2Cl_2	[50]
$(FcCNP)Mo(CO)_4$	+0.92	–	+.076	CH_2Cl_2	[40]

a Irreversible process.

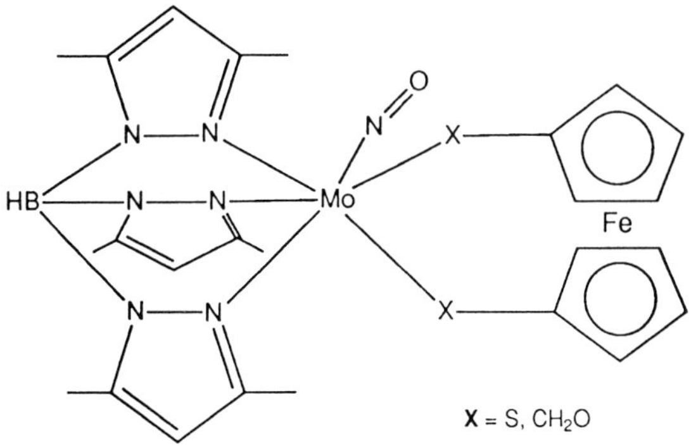

HB N N Mo N O X X Fe

X = S, CH_2O

1

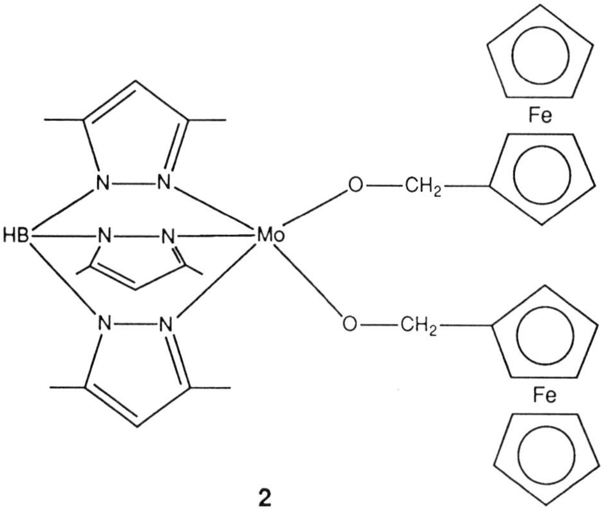

HB N N Mo O CH_2 Fe

2

HB N N Mo N O Y Z R N N n Fe

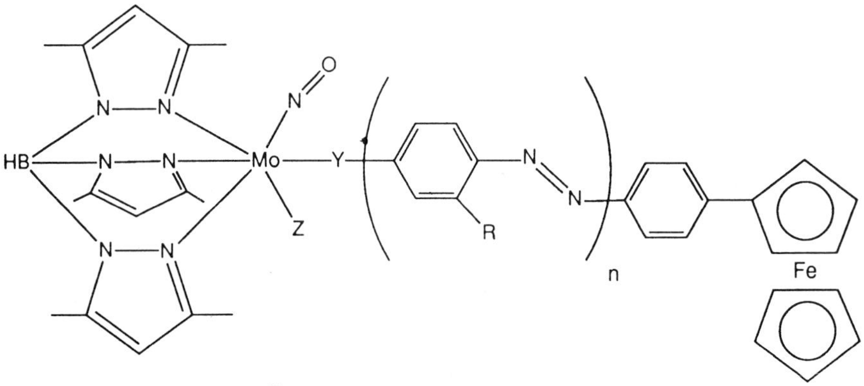

3

Scheme 7-7 (part 1)

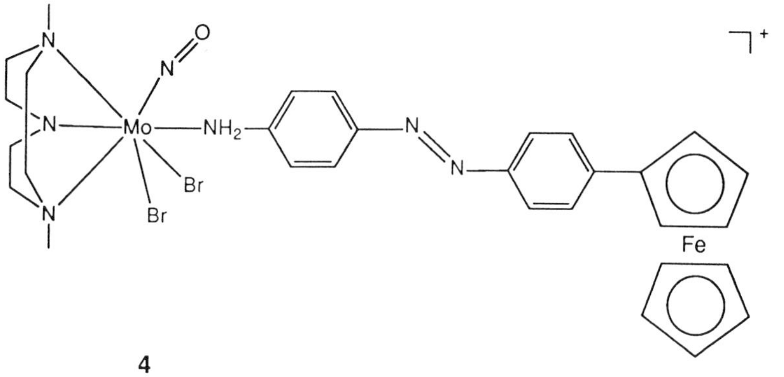

4

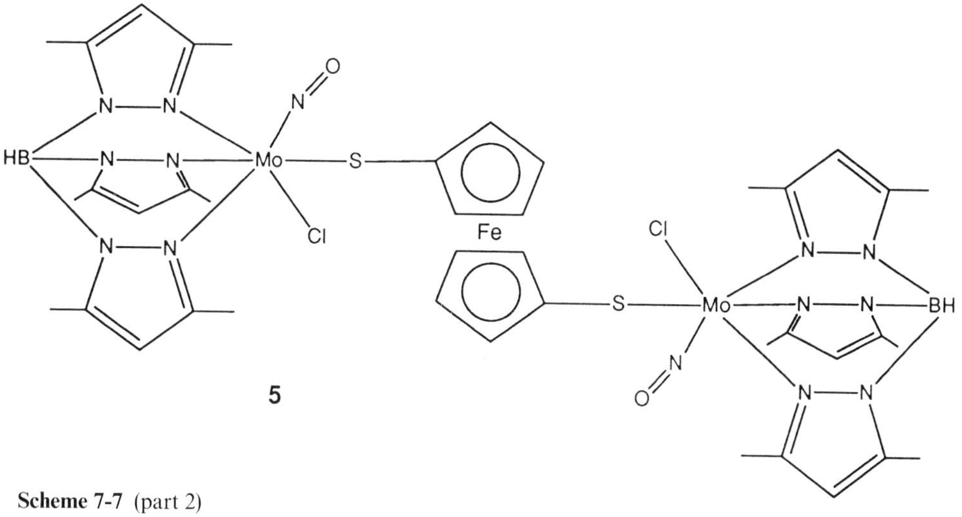

5

Scheme 7-7 (part 2)

A wide series of ferrocenyl-molybdenum complexes of known redox properties bear the tris(3,5-dimethylpyrazolyl)borato ligand or the *N,N′,N″*-trimethyl-1,4,7-triazacyclononane ligand. They are illustrated in Scheme 7-7 [55–59].

In the 3,5-dimethylpyrazolylborato complexes, the molybdenum atom possesses octahedral coordination, similar to that of $Mo[HB(3,5\text{-}Me_2C_3N_2H)_3(NO)(I)\text{-}(OCH_2CH_2Br)]$ [60] or that of $Fe(\eta^5\text{-}C_5H_4S)_2Mo(NO)\{HB(3,5\text{-}Me_2C_3N_2H)_3\}$, which is illustrated in Fig. 7-14 [59]. The cyclopentadienyl rings of the ferrocene fragment are parallel and eclipsed.

In the *N,N′,N″*-trimethyl-1,4,7-triazacyclononane complex, the molybdenum atom probably assumes a 4:3 piano stool hepta-coordination, similar to that found in the X-ray structure of $[Mo([9]aneN_3)(CO)_3Br]^+$ [61].

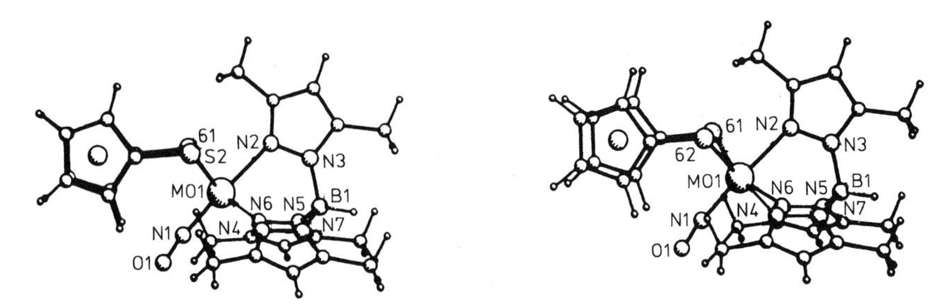

Fig. 7-14. X-Ray structure of $[Fe(\eta^5\text{-}C_5H_4S)_2]Mo(NO)\{HB(3,5\text{-}Me_2C_3N_2H)_3\}$. Mo-S, average 2.36 Å; Mo-N1, 1.84 Å; Mo-N2, 2.26 Å; Mo-N4, Mo-N6, average 2.20 Å; Fe···Mo, 4.15 Å (reproduced by permission of Elsevier Sequoia S.A.).

All these derivatives display both the one-electron oxidation of the ferrocenyl center and the one-electron reduction of the molybdenum fragment [55–59]. Interestingly, the two molybdenum centers in **5** communicate to some extent, in that they exhibit two separate one-electron reductions (see Fig. 7-15) [57]; in contrast, the two ferrocenyl units in **2** are electronically noninteracting, in that a single-step two-electron oxidation is observed [58]. It would appear that, the methylene-bridging groups do not permit electron flow along the molecular framework, in contrast to the sulfur-bridging atoms.

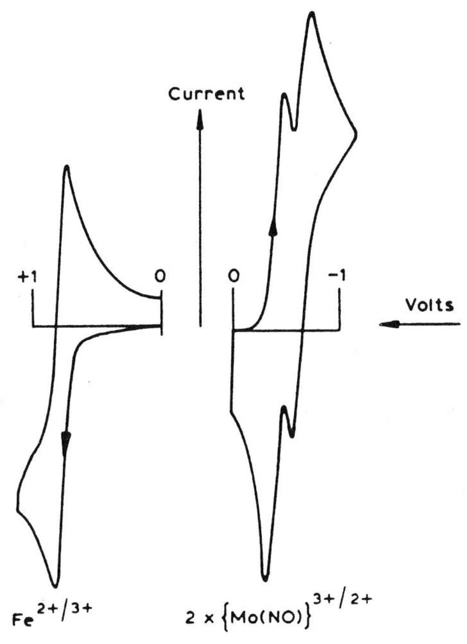

Fig. 7-15. Cyclic voltammogram exhibited by $Fe(C_5H_4S\{Mo(NO)Cl[HB(3,5\text{-}Me_2C_3N_2H]_3\})_2$ **5** in dichloromethane solution. Platinum working electrode (reproduced by permission of Elsevier Sequoia S.A.).

Table 7-9. Formal electrode potentials (*vs.* SCE) for the redox processes exhibited by the ferrocenyl-molybdenum complexes shown in Scheme 7-7 and related mononuclear species

Complex				$E^{\circ\prime}$ (ferrocenyl oxidation), V	$E^{\circ\prime}$ (molybdenum reduction), V	Solvent	Ref.
1							
X = S				+0.54	−0.60	CH_2Cl_2	[59]
X = CH_2O				+0.39	−1.39	CH_2Cl_2	[58]
2				+0.38[a]	−1.49	CH_2Cl_2	[58]
3							
Z	Y	R	n				
Cl	NH	H	1	+0.63	−0.57	CH_2Cl_2	[55]
Cl	NH	Me	1	+0.60	−0.58	CH_2Cl_2	[55]
Cl	O	H	1	+0.61	−0.21	CH_2Cl_2	[55]
I	NH	Me	1	+0.60	−0.54	CH_2Cl_2	[55]
I	O	H	1	+0.60	−0.15	CH_2Cl_2	[55]
I	NH	–	0	+0.59	−0.78	CH_2Cl_2	[55]
I	O	–	0	+0.57	−0.35	CH_2Cl_2	[55]
4				+0.51	−1.53	MeCN	[56]
5				+0.78	−0.41[b]		
					−0.66[c]	CH_2Cl_2	[57]
	$Fe(C_5H_5)_2$			+0.56		CH_2Cl_2	[55]
				+0.42		MeCN	[56]
$(C_5H_5)Fe(C_5H_4C_6H_4$-*p*-N=N-$C_6H_5)$				+0.39		MeCN	[62]
$(C_5H_5)Fe(C_5H_4C_6H_4$-*p*-OH)				+0.29		MeCN	[62]
$(C_5H_5)Fe(C_5H_4C_6H_4$-*p*-$NH_2)$				+0.24		MeCN	[62]
Mo[HB$(Me_2pz)_3$(NO)(I)(NH-C_6H_4-*p*-N=N-C_6H_5)]					−0.57	MeCN	[60]
Mo[HB$(Me_2pz)_3$(NO)(I)(O-C_6H_4-*p*-N=N-C_6H_5)]					−0.14	MeCN	[60]
Mo[HB$(Me_2pz)_3$(NO)(I)(NH-C_6H_5)]					−0.83	MeCN	[60]

a Two-electron oxidation.
b First molybdenum-centred reduction.
c Second-molybdenum centred reduction.

The redox potentials for all these complexes are compiled in Table 7-9.

It can be seen that in the heterodimetallic complexes, the oxidation of the ferrocenyl ligand is generally more difficult than in the related mononuclear species, whereas the reduction of the molybdenum center seems to be less affected.

Finally, the complexes $[\eta^5\text{-}C_5H_4C(O)Me]Fe[\eta^5\text{-}C_5H_4C(O)CH_2(\eta^6\text{-}C_7H_7)\text{-}Mo(CO)_3]$ [63], $[(CO)_5Mo(\eta^5\text{-}C_5H_4PPh_2)]Fe[(\eta^5\text{-}C_5H_4PPh_2)Mo(CO)_5]$ [43, 64], $[\{(\eta^5\text{-}C_5H_5)Fe(\eta^5\text{-}C_5H_4CHC{\equiv}CCH_2CH_2CH_3)\}\{\eta^5\text{-}C_5H_5)_2Mo_2(CO)_4\}][BF_4]$ [65] have been characterized by X-ray analysis, but no redox properties have been reported.

7.2.3.3 Tungsten Complexes

Like the chromium- and molybdenum-carbene complexes of the type illustrated in Scheme 7-3, the corresponding octahedral tungsten complexes exhibit only a one-electron oxidation, which is thought to be centered on both the ferrocenyl and the tungsten fragments. The redox potentials are summarized in Table 7-10.

Table 7-10. Formal electrode potentials (*vs.* SCE) for the one-electron oxidation of the ferrocenyl-carbene complexes $(C_5H_4R)Fe\{C_5H_3[1\text{-}C(R')=W(CO)_5](2\text{-}R)\}$

R	R′	$E^{\circ\prime}$, V	Solvent	Reference
H	OEt	+0.68	CH_2Cl_2	[32, 33]
		+0.80	DME[a]	[34, 36]
Me	OEt	+0.73	DME[a]	[36]
H	$ONMe_4$	−0.04	CH_2Cl_2	[32]
H	Pyrrolidin-1-yl	+0.50	CH_2Cl_2	[32]

a DME = dimethoxyethane.

There is no appreciable difference between the chromium, molybdenum, and tungsten homologs. The strong electron-donating power of the $ONMe_4$ group can be pointed out.

Analogous redox behavior is displayed by the octahedral carbyne complexes of the type shown in Scheme 7-4.

Figure 7-16 shows nicely the notably high electron withdrawing ability exerted on the ferrocenyl fragment by the carbyne-tungsten group. The oxidation process is made more difficult by about 0.28 V with respect to that of unsubstituted ferrocene [35].

The potential values of the single oxidation step for the ferrocenyl carbyne-tungsten complexes are reported in Table 7-11.

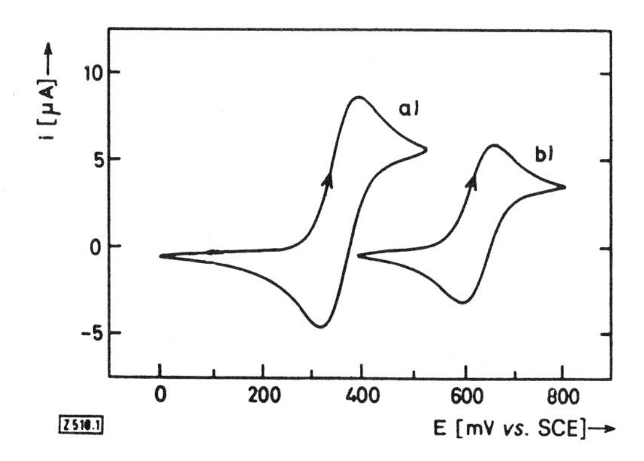

Fig. 7-16. Cyclic voltammograms recorded at a platinum electrode on a 1,2-dimethoxyethane solution of: (a) $(C_5H_5)Fe(C_5H_5)$; (b) $(C_5H_5)Fe\{C_5H_4\text{-}C\equiv W(CO)_4Br\}$. Scan rate 0.1 V s^{-1}. $T = -20$ °C (reproduced by permission of VCH Publishers).

Table 7-11. Formal electrode potentials (*vs.* SCE) for the one-electron oxidation of the ferrocenylcarbyne complexes $(1\text{-R-C}_5H_4)Fe\{C_5H_3\text{-}(2'\text{-R})[1'\text{-C}{\equiv}W(CO)_4X]\}$ in dimethoxyethane solution

Complex		$E^{\circ\prime}$, V	Reference
R = H	X = Cl	+0.62	[34]
R = H	X = Br	+0.64	[34, 35, 36]
R = Me	X = Br	+0.55	[36]
R = H	X = I	+0.55	[34]
$Fe(C_5H_5)_2$		+0.36	[35]

The tungsten complexes are slightly easier to oxidize than the corresponding chromium and molybdenum complexes.

The tetrahedrally coordinated tungsten-carbyne complex $(\eta^5\text{-C}_5H_5)Fe[\eta^5\text{-C}_5H_4\text{-C}{\equiv}W(CO)_2(\eta^5\text{-C}_5H_5)]$ also undergoes one-electron oxidation ($E^{\circ\prime}$ = +0.64 V) in DME solution [34].

Like the molybdenum complexes, a series of monoferrocenyl-, diferrocenyl- and triferrocenyl-phosphine pentacarbonyltungsten complexes have been prepared. Each ferrocenyl ligand undergoes reversible one-electron oxidation, whereas the tungsten moiety undergoes irreversible oxidation. The relevant potential values are reported in Table 7-12.

With respect to the corresponding molybdenum complexes, the tungsten derivatives are slightly more difficult to oxidize.

Interestingly, the related complex $(Fc_2PPh)W(CO)_4(PhPFc_2)$ undergoes two separate two-electron oxidations (at $E^{\circ\prime}$ = +0.64 V and +0.80 V, respectively), each attributable to the simultaneous oxidation of two crossing ferrocenyl fragments, followed by irreversible oxidation of the central tungsten unit (E_p = +1.32 V) [50].

Like the case of the molybdenum analog, in the 1-[(dimethylamino)methyl-2-(diphenylphosphino)ferrocene complex $(FcCNP)W(CO)_4$ the one-electron oxidation of the tungsten fragment ($E^{\circ\prime}$ = +0.78 V) precedes that of the ferrocenyl fragment ($E^{\circ\prime}$ = +0.92 V). Both steps are reversible in dichloromethane solution and occur at potential values similar to that of the corresponding molybdenum complex [40].

Table 7-12. Formal electrode potentials (*vs.* SCE) for the oxidation processes exhibited by ferrocenylphosphinepentacarbonyltungsten complexes in dichloromethane solution [50]

Complex	Ferrocene-centered oxidation			Tungsten-centered oxidation
	$E^{\circ\prime}$ (0/+), V	$E^{\circ\prime}$ (+/2+), V	$E^{\circ\prime}$ (2+/3+), V	E_p, V
$(FcPPh_2)W(CO)_5$	+0.75			+1.54
$(Fc_2PPh)W(CO)_5$	+0.71	+0.89		+1.52
$(Fc_3P)W(CO)_5$	+0.65	+0.82	+0.90	+1.92

We have conducted a preliminary investigation of the redox behavior of the complex shown in Scheme 7-8.

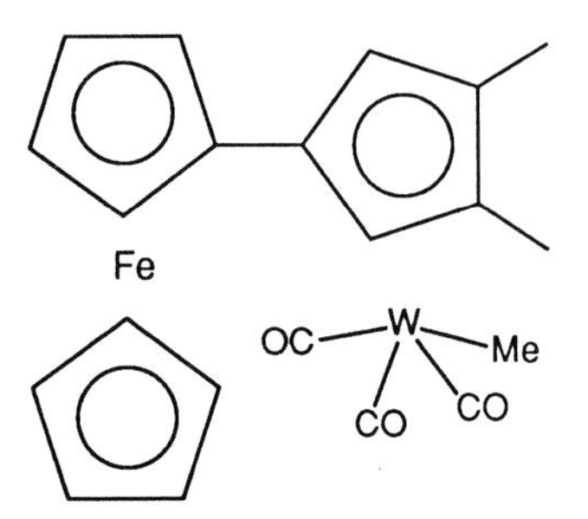

Scheme 7-8

As illustrated in Fig. 7-17, in CH_2Cl_2 this complex undergoes a first one-electron oxidation ($E^{\circ\prime} = +0.47$ V) followed by a second two-electron step ($E^{\circ\prime} = +0.81$ V).

In spite of the apparent chemical reversibility of the first anodic step, likely centered on the ferrocenyl fragment, in the long times of macroelectrolysis the monocation $[(C_5H_5)Fe\{C_5H_4[C_5H_3(3,4\text{-Me})_2]W(CO)_3(Me)\}]^+$ undergoes slow decomposition. The same happens for the subsequent, tungsten-based, oxidation product [66].

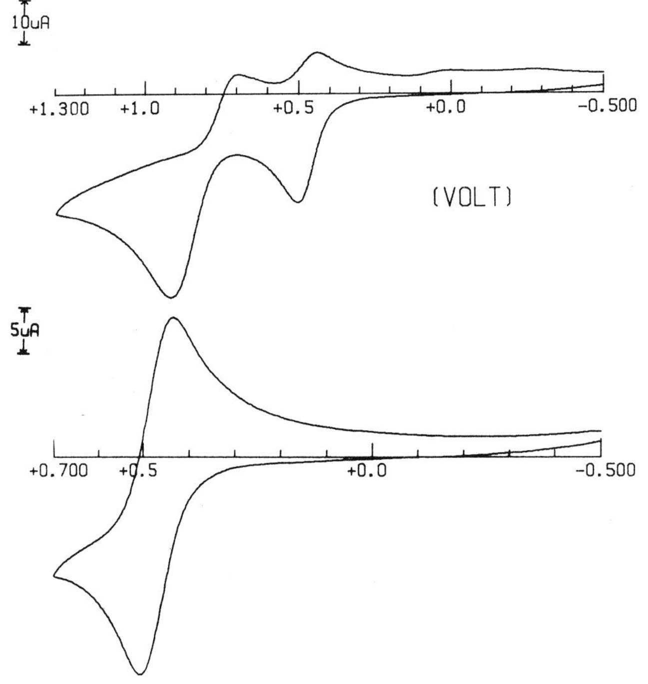

Fig. 7-17. Cyclic voltammograms recorded at a platinum electrode on a CH_2Cl_2 solution of $(C_5H_5)Fe\{C_5H_4[C_5H_3(3,4\text{-Me})_2]W(CO)_3(Me)\}$. Scan rate 0.1 V s^{-1}.

Table 7-13. Formal electrode potentials (*vs.* SCE) for the redox processes exhibited by the 3,5-dimethylpyrazolylboratoferrocenyl-tungsten complexes analogues of the molybdenum complexes shown in Scheme 7-7. Dichloromethane solution [55]

Complex				$E^{\circ\prime}$ (ferrocenyl oxidation), V	$E^{\circ\prime}$ (molybdenum reduction), V
X	Y	R	*n*		
Cl	NH	Me	1	+0.62	−1.01
Cl	NH	–	0	+0.56	−1.23
Cl	O	–	0	+0.56	−0.86

The last series of tungsten complexes that have been electrochemically studied belong to the 3,5-dimethylpyrazolyl-borato complexes of the type shown in Scheme 7-7. Also in this case, they exhibit the one-electron oxidation of the ferrocenyl centre and the one-electron reduction of the molybdenum fragment [55]. The relevant redox potentials are summarized in Table 7-13.

With respect to the molybdenum congeners, only the reduction of the tungsten fragment seems affected to a significant extent, in that it occurs at potential values notably more negative than that of the molybdenum fragment. In contrast, there is no significant difference in the oxidation potentials of the ferrocenyl moieties.

The X-ray structure of the following complexes has been reported, but their redox ability is unknown: $(\eta^5\text{-}C_5H_5)Fe\{\eta^5\text{-}C_5H_4C(O)CH_2[W(CO)_3(\eta^5\text{-}C_5H_5)]\}$ [67]; $(\eta^5\text{-}C_5H_5)Fe\{\eta^5\text{-}C_5H_4[W(NO)_2(\eta^5\text{-}C_5H_5)]\}$ [68]; $[(\eta^5\text{-}C_5H_5)Fe(\eta^5\text{-}C_5H_4)]_3W(O)\text{-}[(\eta^5\text{-}C_5H_4O)Fe(\eta^5\text{-}C_5H_5)]$ [69]; and $[\{\eta^5\text{-}C_5H_4C(OMe)\}Fe(\eta^5\text{-}C_5H_4PPh_2)]\text{-}W(CO)_4$ [46].

7.2.4 Group 7 Metal Complexes

7.2.4.1 Manganese Complexes

The only manganese-ferrocene complex that has been the subject of electrochemical studies is the carbene complex $(\eta^5\text{-}C_5H_5)Fe[\eta^5\text{-}C_5H_4C(OMe){=}Mn(CO)_2(\eta^5\text{-}C_5H_4Me)]$ [33]. In contrast to the similar carbene complexes of Group 6 metals, here both metal centers undergo one-electron oxidation ($E^{\circ\prime}$ $(0/+) = +0.26$ V; $E^{\circ\prime}$ $(+/2+) = +0.99$ V) in CH_2Cl_2. ESR spectra seem to indicate that the first electron removal is centered on the manganese fragment [33].

X-Ray studies have been performed on the following complexes: $(\eta_5\text{-}C_5H_5)Fe[(\eta^5\text{-}C_5H_4CH_2N(Me)CH_2Mn(CO)_4]$ [70]; $[Fe(\eta^5\text{-}C_5H_4PPh_2)_2](CO)Mn(\eta^5\text{-}C_5H_4Me)$ [71–73]; $(\eta^5\text{-}C_5H_4PPh_2)Fe[(\eta^5\text{-}C_5H_4PPh_2)Mn(CO)_2(\eta^5\text{-}C_5H_4Me)$ [72]; $[ClMn(CO)_4(\eta^5\text{-}C_5H_4PPh_2)]Fe[(\eta^5\text{-}C_5H_4PPh_2)(CO)_4MnCl]$ [72, 74]; and $\{[(CO)_9Mn_2]\text{-}(\eta^5\text{-}C_5H_4PPh_2)\}Fe\{(\eta^5\text{-}C_5H_4PPh_2)[Mn_2(CO)_9]\}$ [75, 76].

7.2.4.2 Rhenium Complexes

Scheme 7-9 illustrates a series of ferrocenylphosphine-rhenium complexes, the electrochemical behaviour of which has been studied.

Figure 7-18 shows the molecular structure of the two diphenylferrocenylphosphine-rhenium complexes $[Fe(\eta^5\text{-}C_5H_4PPh_2)_2]Re(CO)_3Cl$ and $[(\eta^5\text{-}C_5H_5)Fe(\eta^5\text{-}C_5H_4PPh_2)]_2Re(CO)_3Cl$ [11].

In both cases, the rhenium atom possesses octahedral geometry with the carbonyl groups assuming a facial arrangement. In $[Fe(C_5H_4PPh_2)_2]Re(CO)_3Cl$ the cyclopentadienyl rings of the ferrocene ligand are staggered, whereas in $[(C_5H_5)Fe(C_5H_4PPh_2)]_2Re(CO)_3Cl$ the cyclopentadienyl rings of the two ferrocenyl ligands assume an eclipsed and a staggered conformation, respectively.

The cyclic voltammetric profiles exhibited by such rhenium complexes are shown in Fig. 7-19 [11]. $[Fe(C_5H_4PPh_2)_2]Re(CO)_3Cl$ exhibits a reversible one-electron oxidation, whereas $[(C_5H_5)Fe(C_5H_4PPh_2)]_2Re(CO)_3Cl$ exhibits two closely spaced one-electron oxidations. In both cases the oxidation processes are centered on the ferrocene fragments. The occurrence of two separate waves for the diferrocenyl complex indicates electronic interaction between the two redox centres. In contrast, the corresponding bis(diferrocenylpyridine) complex $[(\eta^5\text{-}C_5H_5)Fe(\eta^5\text{-}C_5H_4C_5H_4N)]_2\text{-}Re(CO)_3Cl$ exhibits a single-step two-electron oxidation, indicating that no coupling exists between the two ferrocene fragments.

Scheme 7-10 shows a series of ketophosphine-rhenium-ferrocenyl complexes of known redox behavior.

Figure 7-20 illustrates the molecular structure of the complex $(\eta_5\text{-}C_5H_5)Fe[(\eta^5\text{-}C_5H_4C(O)CH_2PPh_2)Re(CO)_3Br]$ [77]. In this case also, the rhenium atom has octahedral coordination with the carbonyl group in a facial arrangement. The cyclopentadienyl rings of the pendant ferrocenyl group are nearly staggered.

These ketophosphine complexes all exhibit reversible one-electron oxidation of the ferrocene fragment. In addition, as shown in Fig. 7-21, $[(\eta^5\text{-}C_5H_5)Fe\{(\eta^5\text{-}C_5H_4C(O)CHPPh_2)Re(CO)_3Br\}][NMe_4]$ also displays a subsequent two-electron oxidation. Access to such a rhenium-centered process is likely to be electrostatically favored because of the anionic nature of the complex [77].

Table 7-14 reports the redox potentials of one-electron oxidation of the ferrocenyl ligands in all the discussed rhenium complexes.

The main difference between ferrocenylphosphine and ferrocenylketophosphine complexes lies in the fact that the free ferrocenylketophosphine is more difficult to oxidize than the free ferrocenylphosphines. On the other hand, there is no simple unequivocal trend on coordination to the rhenium fragment. With respect to free ferrocenyl ligands, in some cases (e.g., $FcRe(CO)_5$ and $[FcC(O)CHPPh_2\text{-}Re(CO)_3Br]^-$) oxidation is easier, in some cases (e.g., $Fc(PPh_2)Re(CO)_4Cl$, $[Fc(PPh_2)]_2Re(CO)_3Cl$, $(FcPy)_2Re(CO)_3Cl$ and $FcC(O)CH_2PPh_2Re(CO)_4Br$) no substantial difference exists, whereas in other cases (e.g., $[Fc(PPh_2)_2]Re(CO)_3Cl$ and $FcC(O)CH_2PPh_2Re(CO)_3Br$) oxidation is more difficult.

Ph Ph Cl CO P Re CO Fe P CO Ph Ph

Cl Ph Ph CO P Re CO Fe OC CO

CO CO Re CO Fe OC CO

Fe Ph Ph Cl P CO Re CO P CO Ph Ph Fe

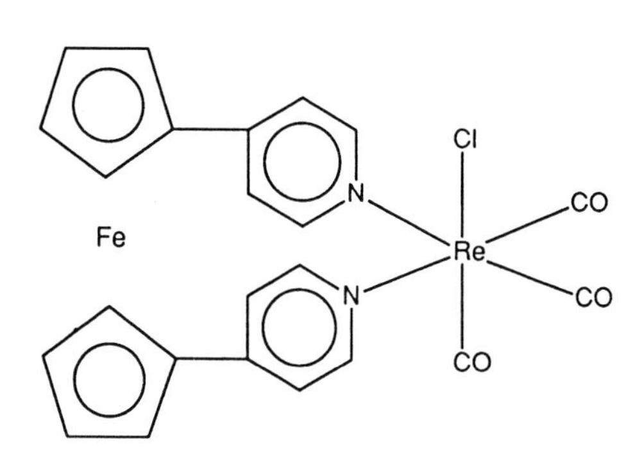

Scheme 7-9

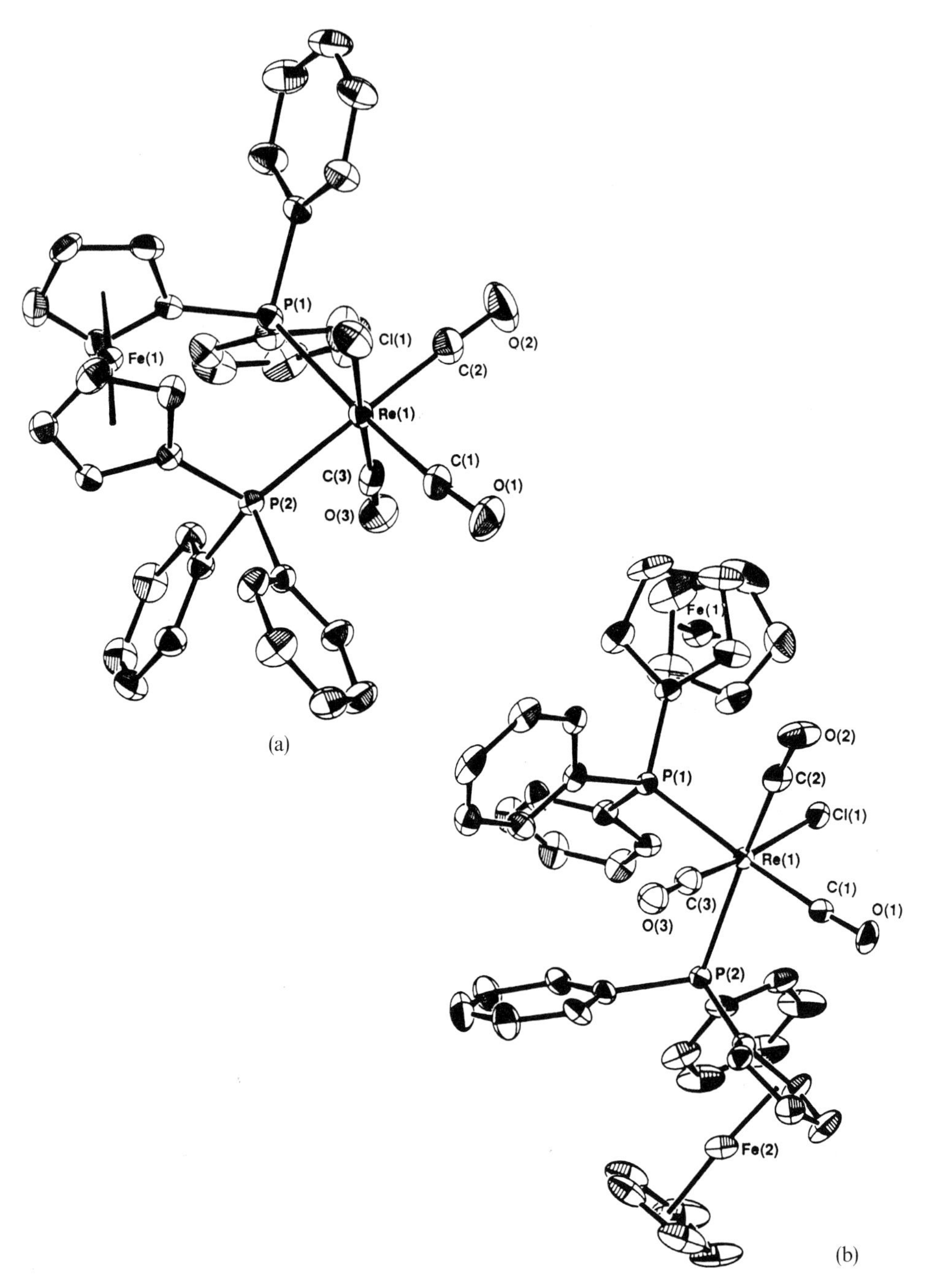

Fig. 7-18. X-Ray structures of: (a) $[Fe(\eta^5\text{-}C_5H_4PPh_2)_2]Re(CO)_3Cl$. Re-P, 2.50 Å; Re-C (*trans* to P), 1.93 Å; Re-C (*cis* to P), 1.94 Å; Re-Cl, 2.49 Å; (b) $[(\eta^5\text{-}C_5H_5)Fe(\eta^5\text{-}C_5H_4PPh_2)]_2Re(CO)_3Cl$. Re-P, 2.51 Å; Re-C (*trans* to P), 1.93 Å; Re-C (*cis* to P), 1.84 Å; Re-Cl, 2.48 Å (reproduced by permission of the American Chemical Society).

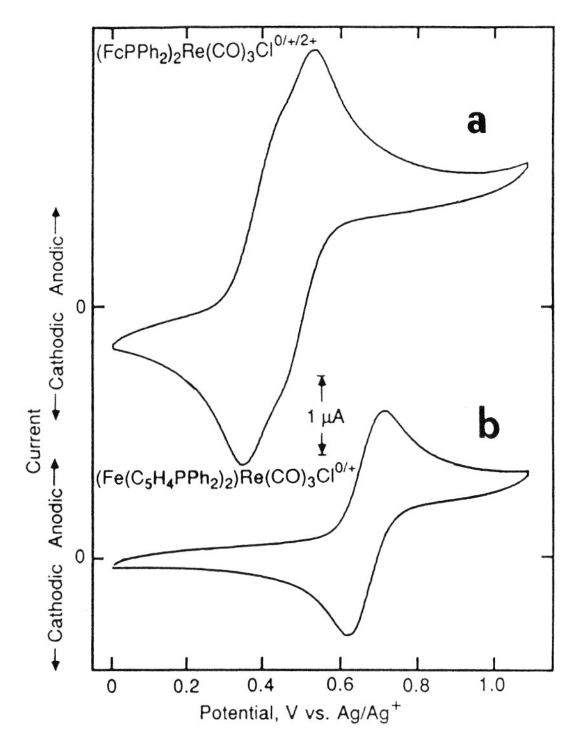

Fig. 7-19. Cyclic voltammograms recorded at a platinum electrode on a CH_2Cl_2 solution containing: (a) $[(C_5H_5)Fe(C_5H_4PPh_2)]_2Re(CO)_3Cl$; (b) $[Fe(C_5H_4PPh_2)_2]Re(CO)_3Cl$. Scan rate $0.1\ V\ s^{-1}$ (reproduced by permission of the American Chemical Society).

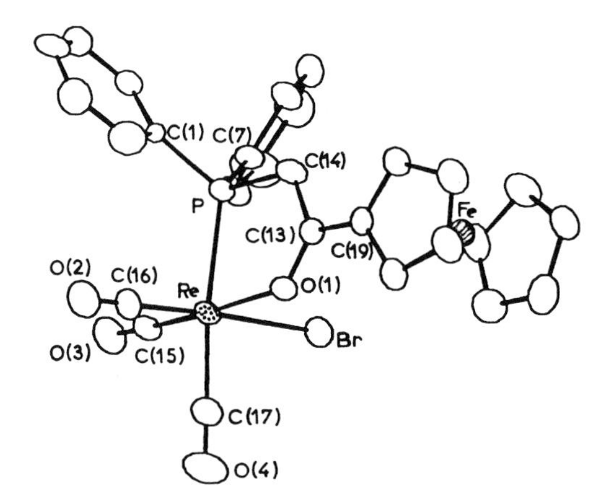

Fig. 7-20. Crystal structure of $(\eta^5\text{-}C_5H_5)Fe[\eta^5\text{-}C_5H_4C(O)CH_2PPh_2)Re(CO)_3Br]$. Re-P, 2.46 Å; Re-C (*trans* to P); 1.97 Å; Re-C (*cis* to P), average 1.91 Å; Re-Br, 2.62 Å (reproduced by permission of CNRS-Gauthier-Villars).

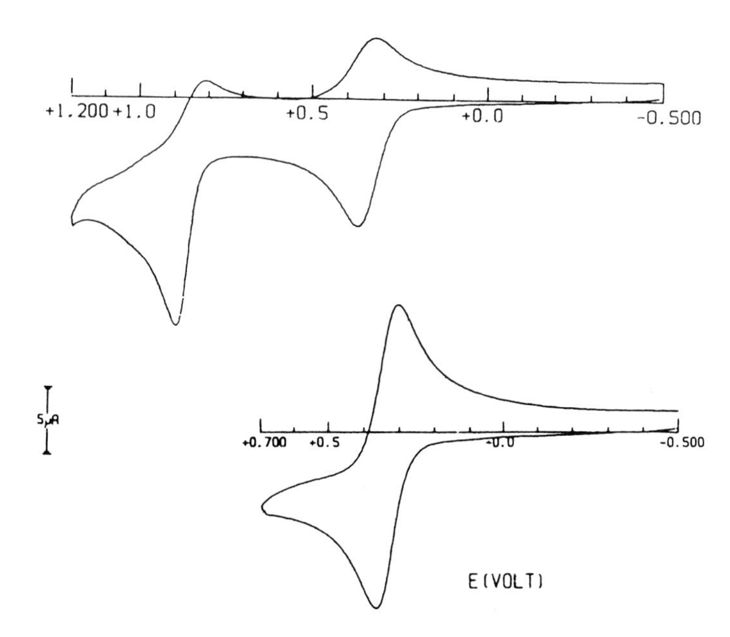

Fig. 7-21. Cyclic voltammograms recorded at a platinum electrode on a CH_2Cl_2 solution of $[(C_5H_5)Fe(C_5H_4C(O)CHPPh_2)Re(CO)_3Br]^-$. Scan rate $0.2\ V\ s^{-1}$ (reproduced by permission of CNSR-Gauthier-Villars).

Table 7-14. Formal electrode potentials (*vs.* S.C.E.) for the one-electron oxidation of the ferrocene ligands in the complexes illustrated in Schemes 7-9 and 7-10 [Fc = $(C_5H_5)Fe(C_5H_4)$] or $(C_5H_4)Fe(C_5H_4)$]

Complex	$E^{\circ\prime}$ (Fc/Fc^+), V	$E^{\circ\prime}$ (Fc/Fc^+), V	Solvent	Reference
$[Fc(PPh_2)_2]Re(CO)_3Cl$	+0.86		CH_2Cl_2	[11]
$Fc(PPh_2)Re(CO)_4Cl$	+0.62		CH_2Cl_2	[11]
$FcRe(CO)_5$	+0.20		CH_2Cl_2	[11]
$[Fc(PPh_2)]_2Re(CO)_3Cl$	+0.61	+0.71	CH_2Cl_2	[11]
$(FcPy)_2Re(CO)_3Cl$	+0.57	+0.57	MeCN	[11]
$FcC(O)CH_2PPh_2Re(CO)_4Br$	+0.70		CH_2Cl_2	[77]
$FcC(O)CH_2PPh_2Re(CO)_3Br$	+0.85		CH_2Cl_2	[77]
$[FcC(O)CHPPh_2Re(CO)_3Br^-$	+0.34		CH_2Cl_2	[77]
FcH	+0.45		CH_2Cl_2	[11, 77]
$FcPPh_2$	+0.59		CH_2Cl_2	[11]
$Fc(PPh_2)_2$	+0.72[a]		CH_2Cl_2	[11]
FcPy	+0.55		CH_2Cl_2	[11]
$FcC(O)CH_2PPh_2$	+0.72		CH_2Cl_2	[77]

a Peak potential value for irreversible process.

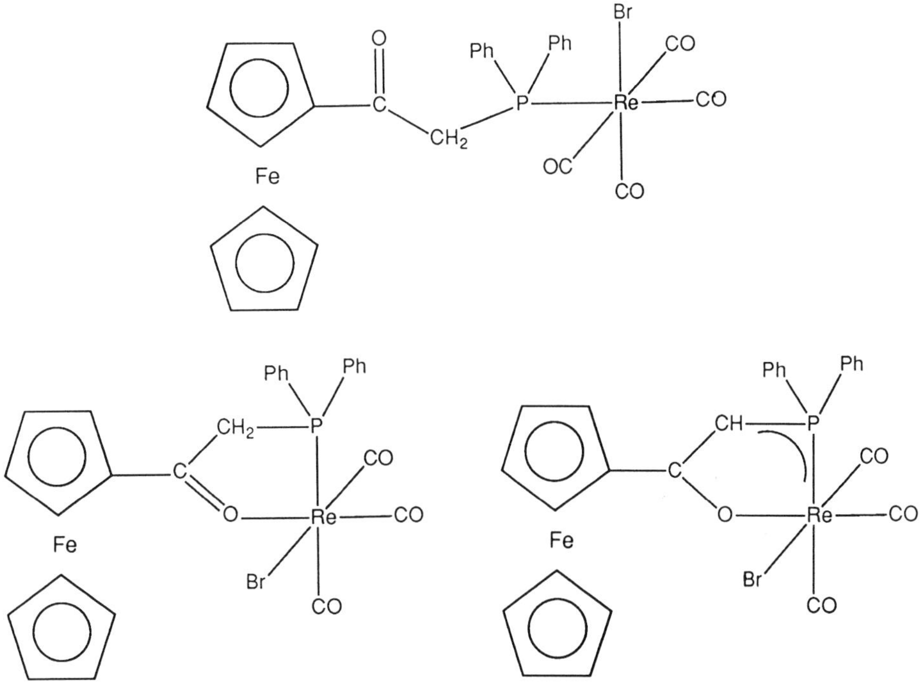

Scheme 7-10

The X-ray structure of the following complexes has been solved: $(\eta^5$-$C_5H_5)$-$Fe\{\eta^5$-$C_5H_4C(O)[Re(CO)_5]\}$ [63]; $\{[(CO)_9Re_2](\eta_5$-$C_5H_4PPh_2)]\}Fe\{(\eta^5$-$C_5H_4$-$PPh_2)[Re_2(CO)_9]\}$; and $(\eta^5$-$C_5H_4PPh_2O)Fe\{(\eta^5$-$C_5H_4PPh_2)[Re_2(CO)_9]\}$ [76].

7.2.5 Group 8 Metal Complexes

7.2.5.1 Iron Complexes

Even if the field of bi- and polyferrocenes deserves a review paper by itself [78], we would like to recall here the most simple compound, i.e. biferrocenyl, the crystal structure of which is shown in Fig. 7-22 [79].

The two ferrocenyl units assume a *trans* disposition. In each ferrocenyl fragment, the cyclopentadienyl rings are neither eclipsed nor staggered and nearly parallel. The two linked rings are coplanar.

As shown in Fig. 7-23, in CH_2Cl_2-MeCN solution biferrocenyl undergoes two distinct one-electron steps, which are reversible in character.

Removal of the first electron occurs at $E^{\circ\prime} = +0.31$ V, which is lower than that of ferrocene ($E^{\circ\prime} = +0.40$ V); removal of the second electron occurs at $E^{\circ\prime} = +0.65$ V [80]. As expected, the ferrocenyl substituent is more electron-donating than the hydrogen atom.

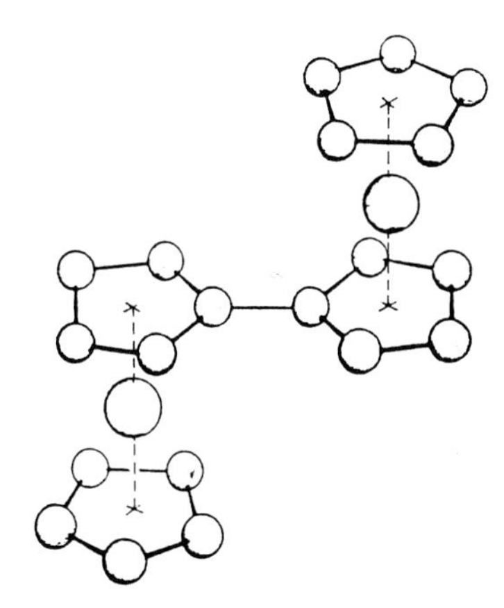

Fig. 7-22. Crystal structure of $[(\eta^5\text{-}C_5H_5)Fe(\eta^5\text{-}C_5H_4)]\text{-}[(\eta^5\text{-}C_5H_4)Fe(\eta^5C_5H_5)]$. Fe-C, average 2.04 Å; C-C (Cp rings), average 1.40 Å; C-C (between rings), 1.48 Å (reproduced by permission of the International Union of Crystallography).

In MeCN solution $(\eta^5\text{-}C_5H_5)Fe\{\eta^5\text{-}C_5H_4CH_2[Fe(CO)_2(\eta^5\text{-}C_5H_5)]\}$ exhibits an initial irreversible oxidation centered on the iron-carbonyl fragment ($E_p \approx +0.3$ V) and a second reversible anodic process centered on the ferrocenyl fragment ($E^{\circ\prime} \approx +0.5$ V) [81]. Qualitatively similar behavior is displayed by $(\eta^5\text{-}C_5H_5)Fe\{\eta^5\text{-}C_5H_4C(O)[Fe(CO)_2(\eta^5\text{-}C_5H_5)]\}$ and $(\eta^5\text{-}C_5H_5)Fe\{\eta^5\text{-}C_5H_4[Fe(CO)_2(C_5H_5)]\}$ [81].

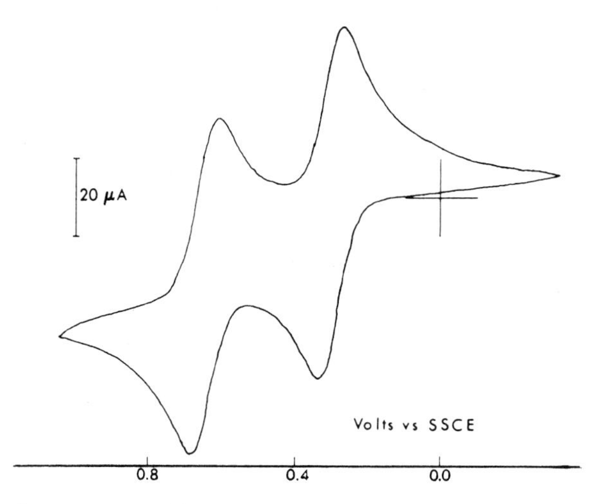

Fig. 7-23. Cyclic voltammogram exhibited at a platinum electrode by biferrocenyl in 1:1 (v/v) CH_2Cl_2-MeCN solution. Scan rate 0.2 V s^{-1} (reproduced by permission of the American Chemical Society).

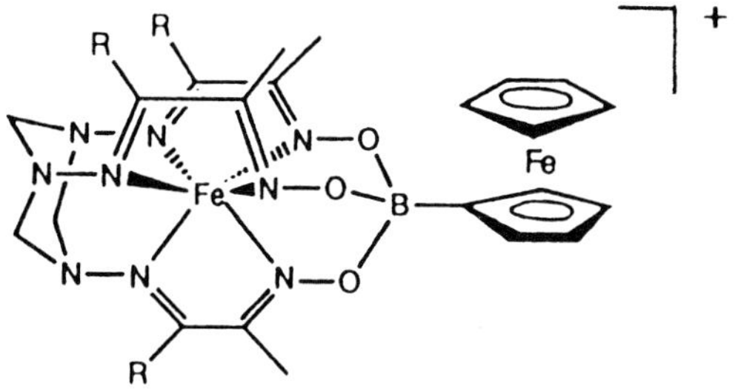

R = $-CH_3$: $[Fe(BMXY)_3(CH_2)_3\{B(cp)_2Fe\}]PF_6$

$-C_6H_5$: $[Fe(PMXY)_3(CH_2)_3\{B(cp)_2Fe\}]PF_6$

$-H$: $[Fe(PAXY)_3(CH_2)_3\{B(cp)_2Fe\}]PF_6$

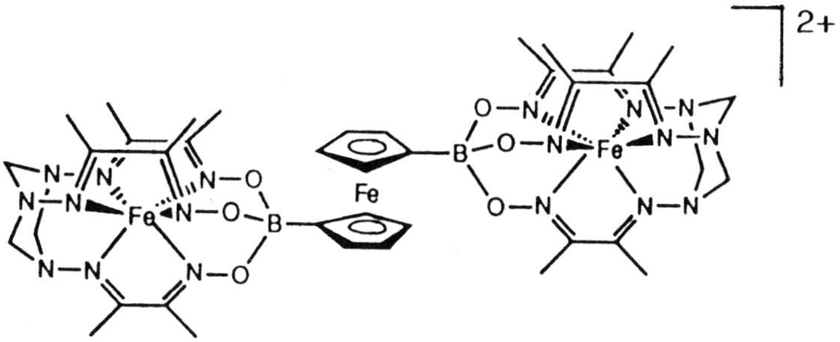

Scheme 7-11 $[\{Fe(BMXY)_3(CH_2)_3\}_2\{B_2(cp)_2Fe\}](PF_6)_2$

Scheme 7-11 shows the molecular assemblies of a few dinuclear and trinuclear iron(II) clathrochelate complexes linked to a ferrocenylboronic or a 1,1′-ferrocenyldiboronic fragment, which have been recently prepared [82].

As illustrated in Fig. 7-24 for $[Fe(BMXY)_3(CH_2)_3\{B(cp)_2Fe\}][PF_6]$ and $[\{Fe(BMXY)_3(CH_2)_3\}_2\{B(cp)_2Fe\}][PF_6]_2$, respectively, in acetonitrile solution these complexes undergo one reduction and two oxidation processes.

Removal of the first electron, which is chemically reversible, is thought to be centered on the ferrocene fragment. The two remaining redox changes, which are coupled to chemical complications, are assigned to the encapsulated iron atom. Easy confirmation comes from the fact that in the trinuclear complex, the redox processes of the encapsulated iron(II) centers have peak heights that are twice the size of those in the dinuclear complex [82]. This also indicates that all the iron centers are non-communicating. Table 7-15 summarizes the relevant redox potentials.

With respect to free ferrocenylboronic acid, oxidation of the ferrocenylboronic fragment is made significantly easier by complexation to the clathrochelate core(s),

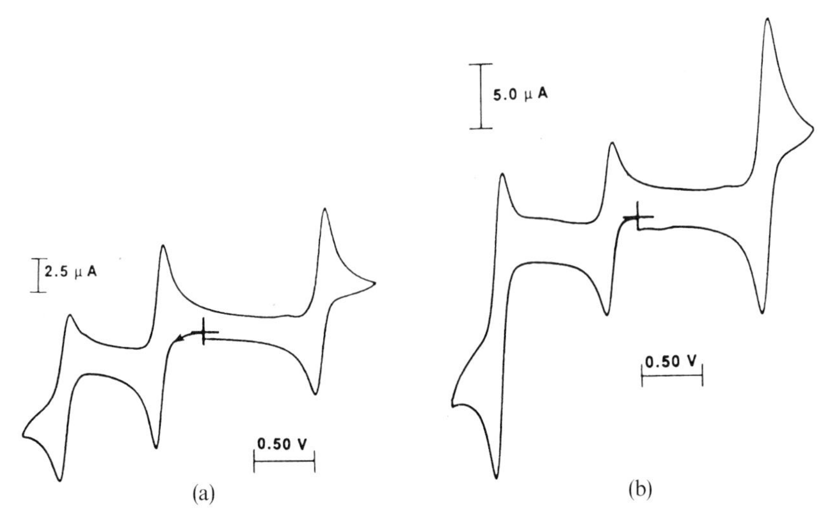

Fig. 7-24. Cyclic voltammograms recorded at a platinum electrode on a MeCN solution of: (a) $[Fe(BMXY)_3(CH_2)_3\{B(cp)_2Fe\}][PF_6]$; (b) $[\{Fe(BMXY)_3(CH_2)_3\}_2\{B(cp)_2Fe\}][PF_6]_2$ (reproduced by permission of the American Chemical Society).

the redox changes of which in turn occur at nearly identical potentials of the non-ferrocenylic analogue $[Fe(BMXY)_3(CH_2)_3\{B(C_6H_5)\}]^+$ [82]. This indicates that the clathrochelate fragments donate appreciable electron density to the ferrocene centre.

Crystallographically characterized iron-ferrocenyl complexes include the following: $[Fe(\eta^5\text{-}C_5H_4PPh_2)_2]Fe(CO)_3$ [83, 84]; $(\eta^5\text{-}C_5H_4PPh_2)Fe\{\eta^5\text{-}C_5H_4PPh_2[Fe(CO)_4]\}$; $\{[(CO)_4Fe](\eta^5\text{-}C_5H_4PPh_2)\}Fe\{\eta^5\text{-}C_5H_4PPh_2[Fe(CO)_4]\}$ and $[Fe(\eta^5\text{-}C_5H_4PPh_2)_2]Fe_3(CO)_{10}$ [85]; $\{Fe[(\eta^5\text{-}C_5H_4)_2]P(C_6H_5)\}Fe(H)(\eta^5\text{-}C_5H_5)(CO)$, and $\{[(\eta^5\text{-}C_5H_5)Fe(\eta^5\text{-}C_5H_4)]P(C_6H_5)[(\eta^5\text{-}C_5H_5)Fe(\eta^5\text{-}C_5H_3)(CO)]\}Fe(CO)(\eta^5\text{-}C_5H_5)$ [86].

Table 7-15. Formal electrode potentials (*vs.* SCE) for the redox changes exhibited in acetonitrile solution by the clathrochelate-ferrocene complexes illustrated in Scheme 7-11 [82]

Complex	Ferrocene-centered oxidation	Iron(II) clathrochelate centre	
		Oxidation	Reduction
	$E^{\circ\prime}$, V	$E^{\circ\prime}$, V	$E^{\circ\prime}$, V
$[Fe(BMXY)_3(CH_2)_3\{B(cp)_2Fe\}]^+$	+0.32	+1.15	−1.05
$[Fe(PMXY)_3(CH_2)_3\{B(cp)_2Fe\}]^+$	+0.33	+1.38[a]	−0.90
$[Fe(PAXY)_3(CH_2)_3\{B(cp)_2Fe\}]^+$	+0.33	+1.6[a]	−1.05
$[\{Fe(BMXY)_3(CH_2)_3\}_2\{B(cp)_2Fe\}]^{2+}$	+0.23	+1.16	−1.08
Ferrocenylboronic acid	+0.45		

a Irreversible process.

7.2.5.2 Ruthenium Complexes

Also in the case of ruthenium-ferrocene complexes, we would like to start with the bis-sandwich complexes illustrated in Scheme 7-12.

The X-ray structure of complex **3** is shown in Fig. 7-25 [87].

In each metallocene unit, the cyclopentadienyl rings are parallel and staggered with respect to each other, the rings are twisted at 17.7°, thereby relieving the steric repulsions of the hydrogen atoms in the bridging methylene groups [87].

The crystal structure of the somewhat related complex $(C_5H_5)Fe[C_5H_4\text{-}C(O)\text{-}C_5H_4]Ru(C_5H_5)$ has been reported [88], but its redox properties are unknown.

Figure 7-26 shows the cyclic voltammetric behavior of ferrocenylruthenocene **1** in acetonitrile solution [89]. It undergoes an initial, reversible, one-electron oxidation at a potential that is almost coincident with that of ferrocene, followed by two closely spaced, one-electron steps, with features of transient chemical reversibility. If one considers that ruthenocene undergoes an irreversible, single-step two electron oxidation, it appears that conjugation with ferrocene tends to stabilize the ruthenocenium cation fragments. This could be attributed to the steric hindrance afforded by the ferrocenyl fragment, which prevents rapid dimerization (followed by

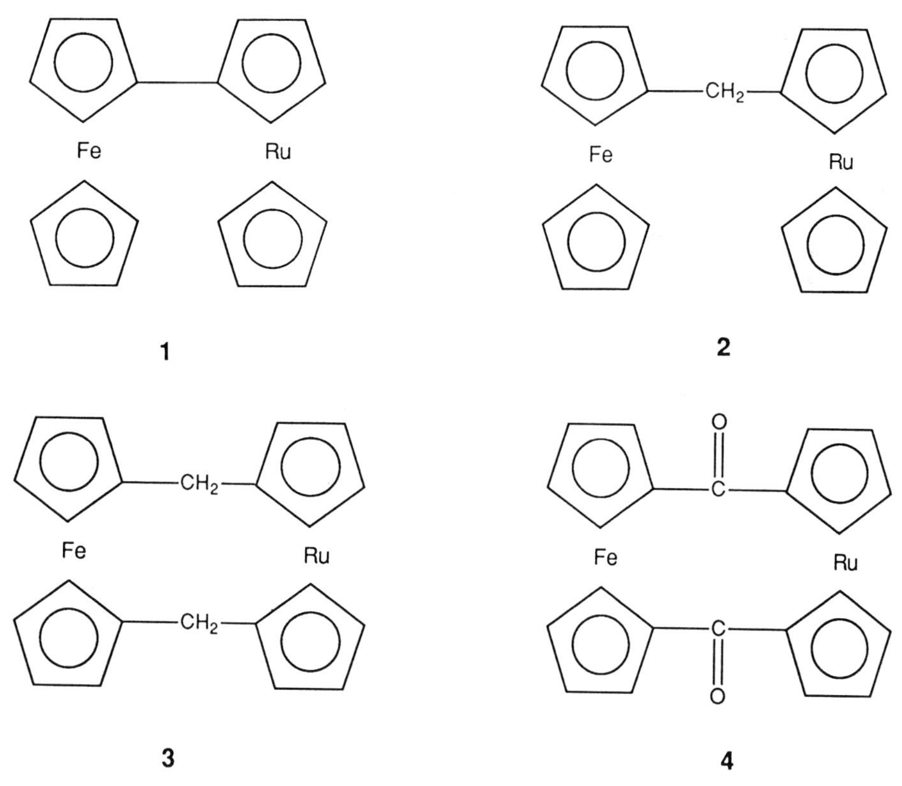

Scheme 7-12

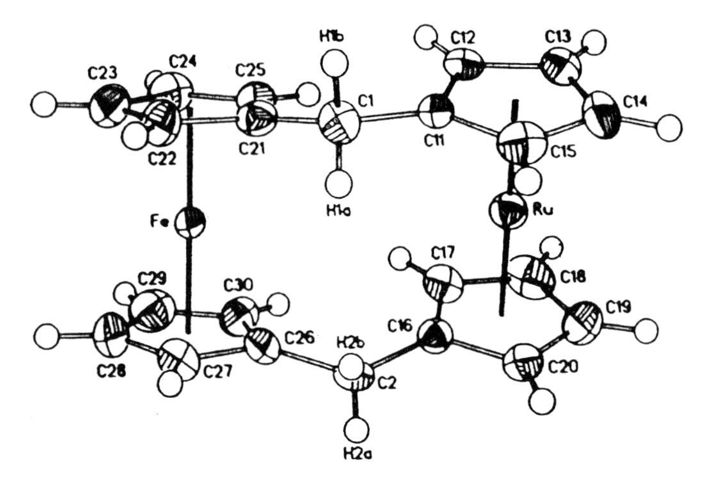

Fig. 7-25. Perspective view of [1.1]ferrocenoruthenocenophane **3**. Fe-C, average 2.05 Å; Ru-C, average 2.15 Å; Fe···Ru, 4.79 Å (reproduced by permission of the American Chemical Society).

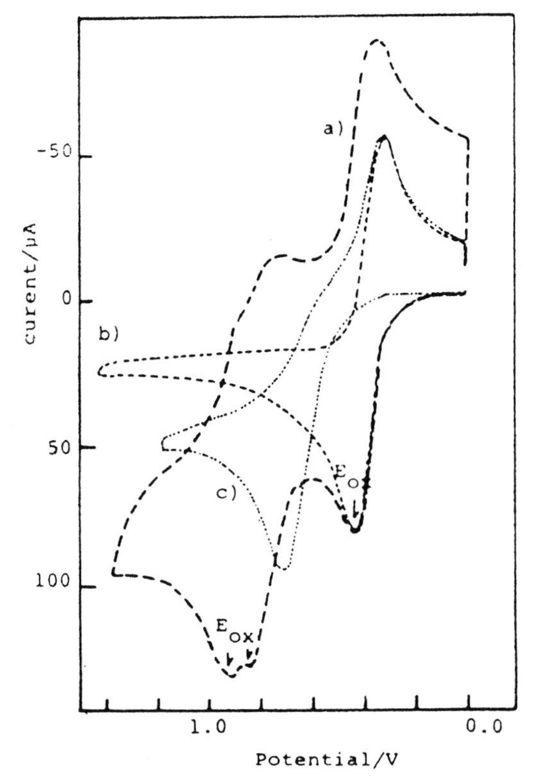

Fig. 7-26. Cyclic voltammograms recorded at a platinum electrode on MeCN solutions containing: (a) **1**; (b) ferrocene; (c) ruthenocene. Scan rate 0.3 V s^{-1}. Potential values refer to the Ag/AgCl reference electrode (reproduced by permission of the Chemical Society of Japan).

Table 7-16. Formal electrode potentials (*vs.* SCE) for the oxidation processes exhibited by the bis-sandwich complexes illustrated in Scheme 7-12

Complex	Ferrocene-centered oxidation	Ruthenocene-centered oxidation[a]		Solvent	Reference
	$E^{\circ\prime}$, V	$E^{\circ\prime}$, V	$E^{\circ\prime}$, V		
1	+0.41	+0.81	+0.88	MeCN	[89]
2	+0.35	+0.83	+0.83	MeCN	[91]
3	+0.35	+0.89	+0.89	MeCN	[89]
	+0.40	+0.94	+0.94	BzCN	[92]
4	+0.88	+1.17	+1.17	MeCN	[89]
Ferrocene	+0.39			MeCN	[89, 91]
	+0.50			BzCN	[92]
Ruthenocene		+0.68	+0.68	MeCN	[89]
		+0.81	+0.81	MeCN	[91]
		+0.92	+0.92	BzCN	[92]

a Complicated by subsequent chemical decomposition.

disproportionation) of the instantaneously electrogenerated ruthenocenium monocation [90].

The same behavior essentially holds for the remaining complexes in Scheme 7-12. Table 7-16 summarizes the relevant redox potentials. It should be noted how removal of electrons becomes more difficult in the case of [1]ferroceno[1]ruthenocenophane-1,13-dione **4**.

Scheme 7-13 shows a series of ruthenium complexes with cyanoferrocene ligands.

Figure 7-27 illustrates the redox fingerprints of all the ruthenium-ferrocenyl complexes in acetonitrile solution [9, 93].

It is clearly apparent that, while the complexes of cyanoferrocene and acrylonitrileferrocene undergo two separate and reversible one-electron oxidations, the two complexes of 1,1′-dicyanoferrocene also undergo two separate oxidation steps, but only the first is reversible. In all cases, the first oxidation step is assigned to the Ru(II)/Ru(III) couple. In particular, in the case of the trinuclear tetracationic species, the two ruthenium centers undergo one-electron oxidation at coincident potentials, thereby indicating the lack of interaction. The second oxidation process is thought to be ferrocene-based. The relevant redox potentials are summarized in Table 7-17, together with those of related mononuclear species.

On complex formation, oxidation of the ruthenium center seems to become easier, while that of the ferrocene moiety becomes more diffcult.

Another series of ruthenium complexes containing a ferrocenyl fragment is shown in Scheme 7-14 [94].

The X-ray structure of the somewhat similar complex $[(\eta^5\text{-}C_5H_5)Fe(\eta^5\text{-}C_5H_4)]Ru(PMe_3)_2(C_5Me_5)$ has been reported [95].

As can be deduced from Table 7-18, in CH_2Cl_2 such derivatives undergo a first reversible one-electron oxidation, centered on the ferrocenyl fragment, followed by an irreversible oxidation, centered on the (pseudo)tetrahedral ruthenium center [94].

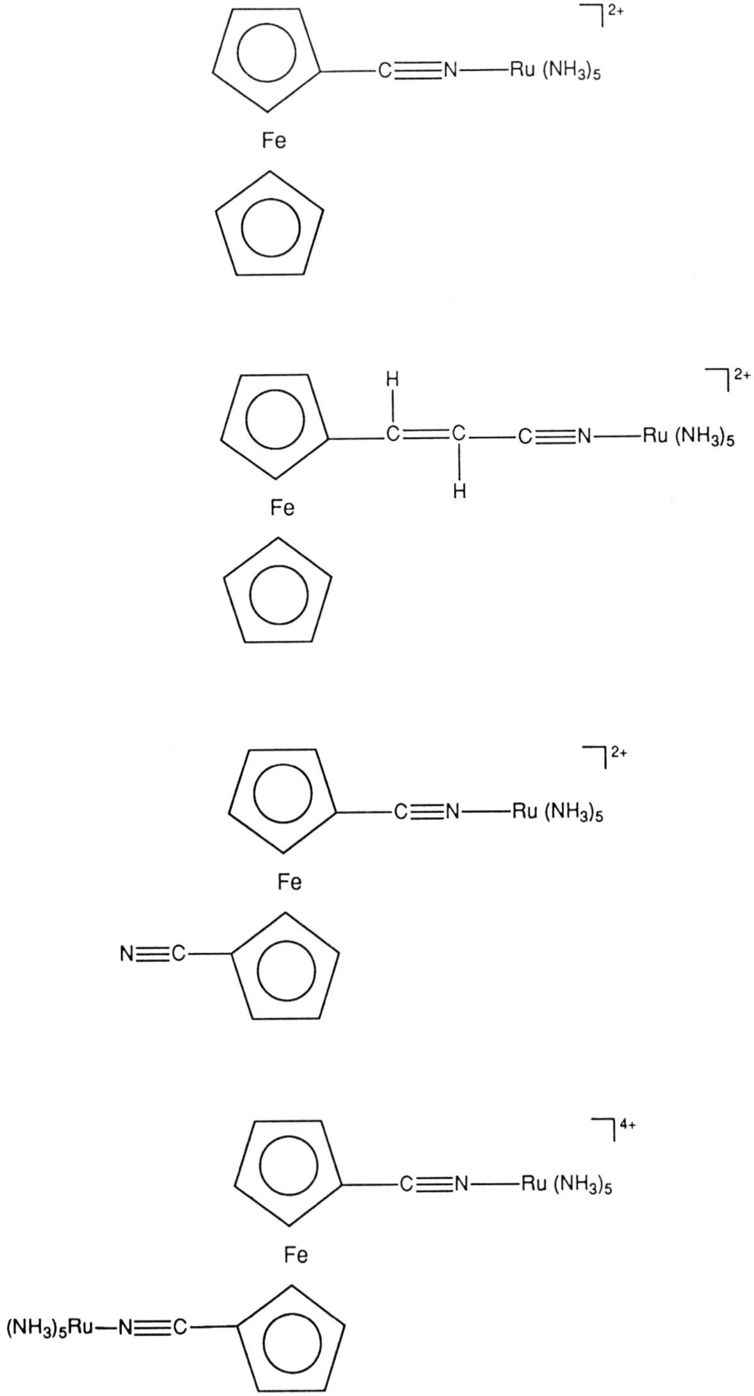

$[CpFe(C_5H_4-C\equiv N-Ru(NH_3)_5)]^{2+}$

$[CpFe(C_5H_4-CH=CH-C\equiv N-Ru(NH_3)_5)]^{2+}$

$[(N\equiv C-C_5H_4)Fe(C_5H_4-C\equiv N-Ru(NH_3)_5)]^{2+}$

$[((NH_3)_5Ru-N\equiv C-C_5H_4)Fe(C_5H_4-C\equiv N-Ru(NH_3)_5)]^{4+}$

Scheme 7-13

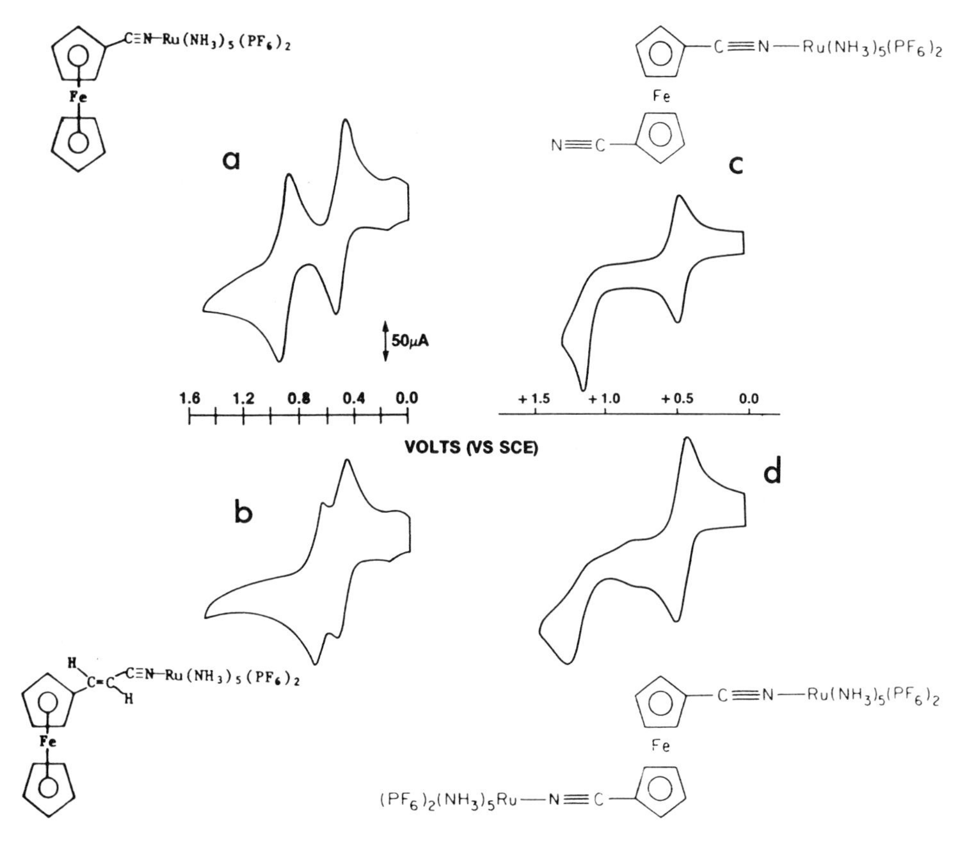

Fig. 7-27. Cyclic voltammograms recorded at a platinum electrode on MeCN solutions of: (a) $[(C_5H_5)Fe\{C_5H_4C{\equiv}NRu(NH_3)_5\}][PF_6]_2$; (b) $[(C_5H_5)Fe\{C_5H_4CH{=}CHC{\equiv}NRu(NH_3)_5\}][PF_6]_2$; (c) $[(C_5H_4C{\equiv}N)Fe\{C_5H_4C{\equiv}NRu(NH_3)_5\}][PF_6]_2$; (d) $[\{(NH_3)_5RuN{\equiv}CC_5H_4\}Fe\{C_5H_4C{\equiv}NRu(NH_3)_5\}][PF_6]_4$. Scan rate 0.2 V s^{-1} (reproduced by permission of the American Chemical Society).

Table 7-17. Formal electrode potentials (*vs.* SCE) for the oxidation processes exhibited by the complexes illustrated in Scheme 7-13 in acetonitrile solution [9, 93]

Complex	$E^{\circ\prime}$ (Ru^{II}/Ru^{III}), V	$E^{\circ\prime}$ (Fe^{II}/Fe^{III}), V
$[(C_5H_5)Fe\{C_5H_4CNRu(NH_3)_5\}]^{2+}$	+0.44	+0.85
$[(C_5H_5)Fe\{C_5H_4CHCHCNRu(NH_3)_5\}]^{2+}$	+0.44	+0.66
$[(C_5H_4CN)Fe\{C_5H_4CNRu(NH_3)_5\}]^{2+}$	+0.47	+1.19[a]
$[Fe\{C_5H_4CNRu(NH_3)_5\}_2]^{4+}$	+0.42[b]	+1.28[a]
$(NH_3)_5RuNCPh$	+0.50	
$(C_5H_5)Fe(C_5H_4CN)$		+0.77
$(C_5H_5)Fe(C_5H_4CHCHCN)$		+0.56
$Fe(C_5H_4CN)_2$		+1.09

a Peak potential value for irreversible processes.
b Two-electron step.

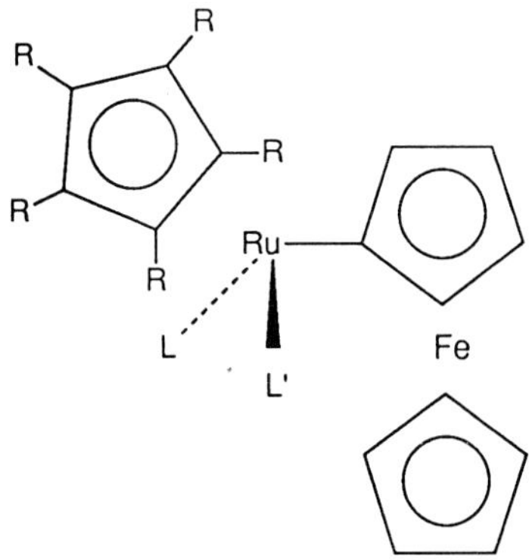

Scheme 7-14

Table 7-18. Formal electrode potentials (*vs.* SCE) for the oxidation processes exhibited by the complexes $(\eta^5\text{-}C_5H_5)Fe[\eta^5\text{-}C_5H_4Ru(L)(L')(C_5R_5)]$ in dichloromethane solution [94]

Complex			Ferrocenyl-centered oxidation	Ruthenium-centered oxidation
R	L	L′	$E^{\circ\prime}$, V	E_p,[a] V
H	CO	CO	+0.13	+1.3
	CO	CN*t*Bu	−0.01	+1.0
	CO	PPh_3	−0.10	+0.9
	CO	PMe_3	−0.08	+0.7
	CN*t*Bu	CN*t*Bu	−0.20	+0.5
Me	CO	CO	+0.07	+1.5
	CO	CN*t*Bu	−0.06	+1.1
	CO	PPh_3	−0.17	+1.0
	CO	PMe_3	−0.17	+0.8
	CN*t*Bu	CN*t*Bu	−0.24	+0.5

a Peak potential value for irreversible processes.

If one considers that ferrocene undergoes oxidation at $E^{\circ\prime} = +0.45$ V, the strong electron-donating ability of the $(C_5R_5)Ru(L)(L')$ assembly is evident. The electronic interaction between the ruthenium atom and the ferrocenyl fragment is illustrated by the fact that the inductive effects of the substituents directly linked to the ruthenium atom appreciably affect the redox potential of the ferrocenyl group. Accordingly, a linear correlation holds between the $\nu_{(CO)}$ stretching frequencies and the relevant redox potentials [94].

A rich redox chemistry is displayed by the ruthenium complex shown in Scheme 7-15.

In fact, in acetonitrile solution this compex exhibits (i) a reversible one-electron cathodic response ($E^{\circ\prime} = -0.80$ V), assigned to the Ru(III)/Ru(II) reduction; (ii) a first reversible one-electron oxidation ($E^{\circ\prime} = +0.55$ V), assigned to the ferrocene-centered Fe(II)/Fe(III) process, (iii) a second reversible one-electron oxidation ($E^{\circ\prime} = +1.01$ V), assigned to the Ru(III)/Ru(IV) redox change.

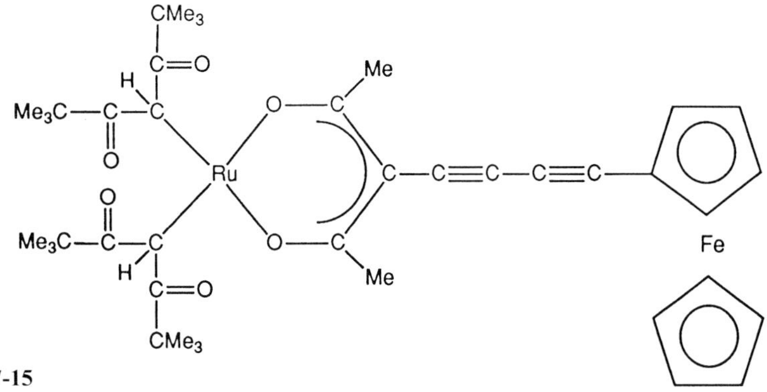

Scheme 7-15

Finally, Scheme 7-16 shows the octahedral structure assigned to two ruthenium complexes, the electrochemistry of which has been recently reported [97].

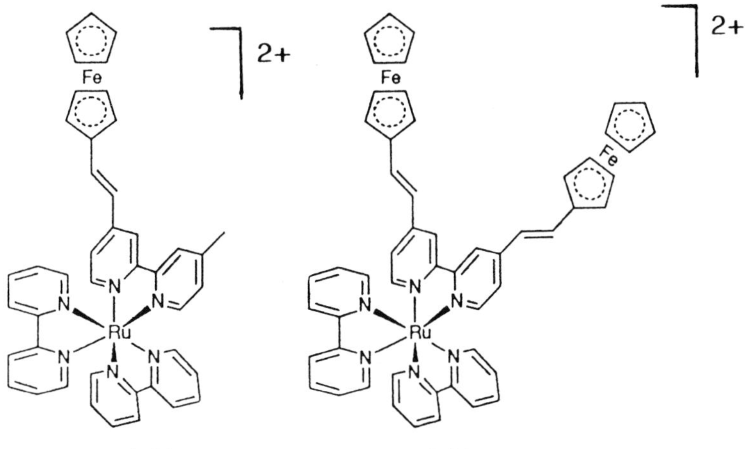

$[Ru(bipy)_2(L^1)]^{2+}$ $[Ru(bipy)_2(L^2)]^{2+}$

Scheme 7-16

Both complexes exhibit two oxidation processes, centered on the ferrocenyl fragment(s) and the central ruthenium atom, respectively. In addition, they undergo three sequential cathodic processes centered on the bipyridyl ligands. Table 7-19 summarizes the redox potentials of the only anodic processes that are pertinent to the present review paper.

It is evident that, on complex formation, both the metal fragments of the two species become slightly easier to oxidize than the corresponding free molecules.

The X-ray structures of $[Fe(\eta^5\text{-}C_5H_4PPh_2)_2]Ru(\eta^5\text{-}C_5H_5)(H)$ [98]; $\{(\eta^5\text{-}C_5H_5)\text{-}Fe[\eta^5\text{-}C_5H_3(CHMeNMe_2)(P\text{-}i Pr)^2\}RuCl_2(PPh_3)$; $\{(\eta^5\text{-}C_5H_5)Fe[\eta^5\text{-}C_5H_3(CHMe\text{-}NMe_2)(P\text{-}i Pr)_2]\}RuH(Cl)(PPh_3)$ and $\{(\eta^5\text{-}C_5H_5)Fe[\eta^5\text{-}C_5H_3(CHMeNMe_2)(P\text{-}i Pr)_2]\}_2(\eta^2\text{-}H_2)Ru(\mu\text{-}Cl)_2Ru(PPh_3)_2(H)$ [99]; $[Fe(\eta^5\text{-}C_5H_4S)_2]Ru_2(CO)_6$ [100]; $\{[(\eta^5\text{-}C_5H_5)Fe(\eta^5\text{-}C_5H_4)]_2C_4H_2\}Ru_2(CO)_6$ [101]; $[\mu\text{-}PPh(\eta^5\text{-}C_5H_4)Fe(\eta^5\text{-}C_5H_5)]_2(\mu_3\text{-}\eta^2\text{-}$

Table 7-19. Formal electrode potentials (*vs.* SCE) for the oxidation processes exhibited by the complexes illustrated in Scheme 7-16 [116]

Complex	$E^{\circ\prime}$ (Fc/Fc$^+$), V	$E^{\circ\prime}$ (Ru^{II}/Ru^{III}), V	Solvent
$[Ru(bipy)_2(L^1)]^{2+}$	+0.51	–	DMF
	+0.40	+1.25	MeCN
$[Ru(bipy)_2(L^2)]^{2+}$	+0.50[a]	–	DMF
$[Ru(bipy)_3]^{2+}$		+1.24	DMF
		+1.32	MeCN
L^1	+0.52		DMF
L^2	+0.53[a]		DMF
$Fe(C_5H_5)_2$	+0.49		DMF

a Single stepped two-electron wave.

$C_6H_4)Ru_3(CO)_7$ [102]; $[Fe(\eta^5\text{-}C_5H_4PPh_2)_2]Ru_3(CO)_{10}$ and $[(\eta^5\text{-}C_5H_5)Fe(\eta^5\text{-}C_5H_4PPh_2)]Ru_3(CO)_{11}$ [103]; $[(Ph_2P\text{-}\eta^5\text{-}C_5H_4)Fe\{\eta^5\text{-}C_5H_4PPh(C_6H_4)\}]Ru_3(CO)_8\text{-}(H)$; $[\mu\text{-}P(C_6H_4)(\eta^5\text{-}C_5H_4)Fe(\eta^5\text{-}C_5H_4PPh_2)]Ru_3(CO)_9$; $[\mu\text{-}PPh(\eta^5\text{-}C_5H_4)Fe(\eta^5\text{-}C_5H_5)](\mu\text{-}PPh_2)(\mu_3\text{-}C_6H_4)Ru_3(CO)_7$, $[\mu_3\text{-}PPh_2(\eta^1,\eta^5\text{-}C_5H_3)Fe(\eta^5\text{-}C_5H_4PPh_2)]Ru_3\text{-}(CO)_7(H)$ and $\{\mu_4\text{-}P[(\eta^5\text{-}C_5H_4)Fe(\eta^5\text{-}C_5H_5)]\}(\mu_4\text{-}C_6H_4)Ru_4(CO)_{11}$ [104]; $[\mu_3\text{-}P(\eta^5\text{-}C_5H_4)Fe(\eta^5\text{-}C_5H_4P\text{-}i\text{Pr}_2)]Ru_3(CO)_8(H)_2$ and $[(P\text{-}i\text{Pr}_2\text{-}\eta^5\text{-}C_5H_4)_2Fe](Cl)(OH)\text{-}Ru_3(CO)_8$ [105]; and $[Fe(\eta^5\text{-}C_5H_4PPh_2)_2]Au_2HRu_4(CO)_{12}B$ [106] have been reported.

7.2.5.3 Osmium Complexes

Figure 7-28 shows the crystal structure of the osmium-ferrocenyl cluster $(\eta^5\text{-}C_5H_5)Fe\{(\eta^5\text{-}C_5H_4OC)[Os_3H(CO)_{10}]\}$ [107].

The complex is obtained by reaction of $Os_3(CO)_{10}(MeCN)_2$ with $(C_5H_5)\text{-}Fe(C_5H_4CHO)$. The ferrocenyl group binds to the $Os_3H(CO)_{10}$ cluster fragment through the CO vector of the formyl group. The cyclopentadienyl rings of the ferrocenyl substituent are eclipsed.

In CH_2Cl_2 solution the present derivative exhibits a reversible one-electron oxidation ($E^{\circ\prime} = +0.75$ V), centered on the ferrocenyl group (under the same experimental conditions, $(C_5H_5)Fe(C_5H_4CHO)$ undergoes reversible oxidation at +0.73 V), and an irreversible two-electron reduction ($E_p = -1.68$ V), centered on the triosmium cluster [107]. Spectroelectrochemical measurements seem to indicate that, on removal of one-electron, no important stereochemical changes of the original molecular structure occur [107].

The crystal structures of $[(\eta^5\text{-}C_5H_4)Fe(\eta^5\text{-}C_5H_3\text{-}PPh_2)]Os_3(CO)_8(H)_2$, $[(\eta^5\text{-}C_5H_4)Fe\{\eta^5\text{-}C_5H_3\text{-}PPh(\eta^5\text{-}C_5H_4)Fe(\eta^5\text{-}C_5H_5)\}]Os_3(CO)_8(H)_2$ and $[(\eta^5\text{-}C_5H_4)\text{-}Fe(\eta^5\text{-}C_5H_4\text{-}P\text{-}i\text{Pr}_2)]Os_3(CO)_7(H)(P\text{-}i\text{Pr})$ [108], $[\mu\text{-}(\eta^5\text{-}C_5H_4P\text{-}i\text{Pr}_2)Fe(\eta^5\text{-}C_5H_3\text{-}P\text{-}i\text{Pr})]Os_3(CO)_8(\mu\text{-}H)_2$, $[\mu\text{-}(\eta^5\text{-}C_5H_4P\text{-}i\text{Pr}_2)Fe(\eta^5\text{-}C_5H_4\text{-}P\text{-}i\text{PrCHMeCH}_2CO)]Os_3\text{-}(CO)_8(\mu\text{-}H)$, $[\mu\text{-}(\eta^5\text{-}C_5H_4P\text{-}i\text{Pr}_2)Fe(\eta^5\text{-}C_5H_4\text{-}P\text{-}i\text{Pr})]Os_3(CO)_9(\mu\text{-}H)_2$ [109], $[\mu_3\text{-}(\eta^5\text{-}C_5H_3)Fe(\eta^5\text{-}C_5H_5)][\mu_3\text{-}P(\eta^5\text{-}C_5H_4)Fe(\eta^5\text{-}C_5H_5)]Os_3(CO)_9$, $[Os_3(H)_2(CO)_8(P\text{-}i\text{Pr}_2\text{-}$

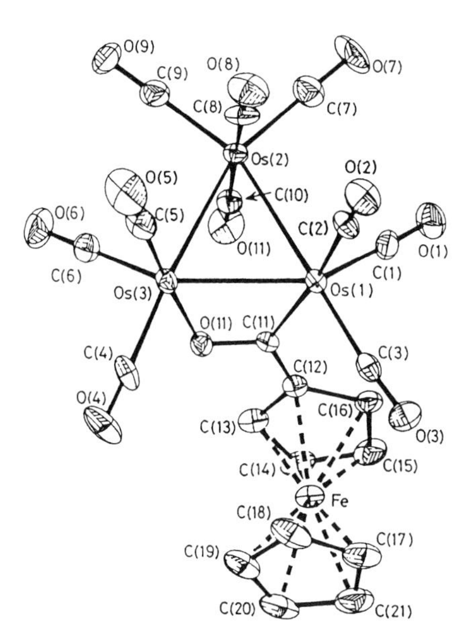

Fig. 7-28. X-Ray structure of $(\eta^5\text{-}C_5H_5)Fe[(\eta^5\text{-}C_5H_4CO)Os_3H(CO)_{10}]$. Os-Os, average 2.87 Å; C11-C12, 1.48 Å; Os1-C11, 2.07 Å; Os3-O11, 2.09 Å (reproduced by permission of the Royal Society of Chemistry).

$\eta^5\text{-}C_5H_2)]Fe[(\eta^5\text{-}C_5H_2\text{-}P\text{-}i\text{Pr}_2)Os_3(H)_2(CO)_8]$, $[\mu_3\text{-}P(\eta^5\text{-}C_5H_4)Fe(\eta^5\text{-}C_5H_5)](\mu_3\text{-}C_6\text{-}H_4)Os_3(CO)_9$ [110], $[\mu_3\text{-}(\eta^5\text{-}C_5H_4)Fe(\eta^5\text{-}C_5H_4PPh)]Os_3(CO)_9$, $[\mu_3\text{-}(\eta^5\text{-}C_5H_4)Fe(\eta^5\text{-}C_5H_4\text{-}P\text{-}\eta^5\text{-}C_5H_4)Fe(\eta^5\text{-}C_5H_5)]Os_3(CO)_9$ [111], $\{(\eta^5\text{-}C_5H_5)Fe[\eta^5\text{-}C_5H_4\text{-}P(Ph)(C_6H_4)]\}Os_3(CO)_9(H)$ and $\{(\eta^5\text{-}C_5H_5)Fe[\eta^5\text{-}C_5H_4P(Ph)(C_6H_4)]\}Os_3(CO)_8\text{-}(H)(\eta^5\text{-}C_5H_5)Fe[\eta^5\text{-}C_5H_4P(Ph)_2]$ [112], $[\mu_3\text{-}(\eta^5\text{-}C_5H_4)Fe(\eta^5\text{-}C_5H_3P\text{-}i\text{Pr}_2)]Os_3(CO)_8(H)_2$, $\{\mu_3\text{-}(\eta^5\text{-}C_5H_4)Fe(\eta^5\text{-}C_5H_3P(Et)[(\eta^5\text{-}C_5H_4)Fe(\eta^5\text{-}C_5H_5)]\}Os_3(CO)_8(H)_2$ [113], $\{[\eta^5\text{-}C_5H_5)Fe(\eta^5\text{-}C_5H_4)]_2P(Ph)\}1,2\text{-}Os_3(CO)_{10}$ [114] and $[Fe(\eta^5\text{-}C_5H_4S)_2]\text{-}(\mu_3\text{-}S)Os_4(CO)_{11}$ [100] have been reported.

7.2.6 Group 9 Metal Complexes

7.2.6.1 Cobalt Complexes

Scheme 7-17 shows a series of cyclobutadiene-cobalt complexes bearing ferrocenyl substituents.

The molecular structure of the *trans*-diferrocenyl-cobalt isomer is illustrated in Fig. 7-29 [115].

The cyclobutadiene ring is distorted square planar, in that one carbon is out of the plane formed by the remaining three carbon atoms by 0.035 Å. The cyclopentadienyl rings of the two ferrocenyl groups are eclipsed.

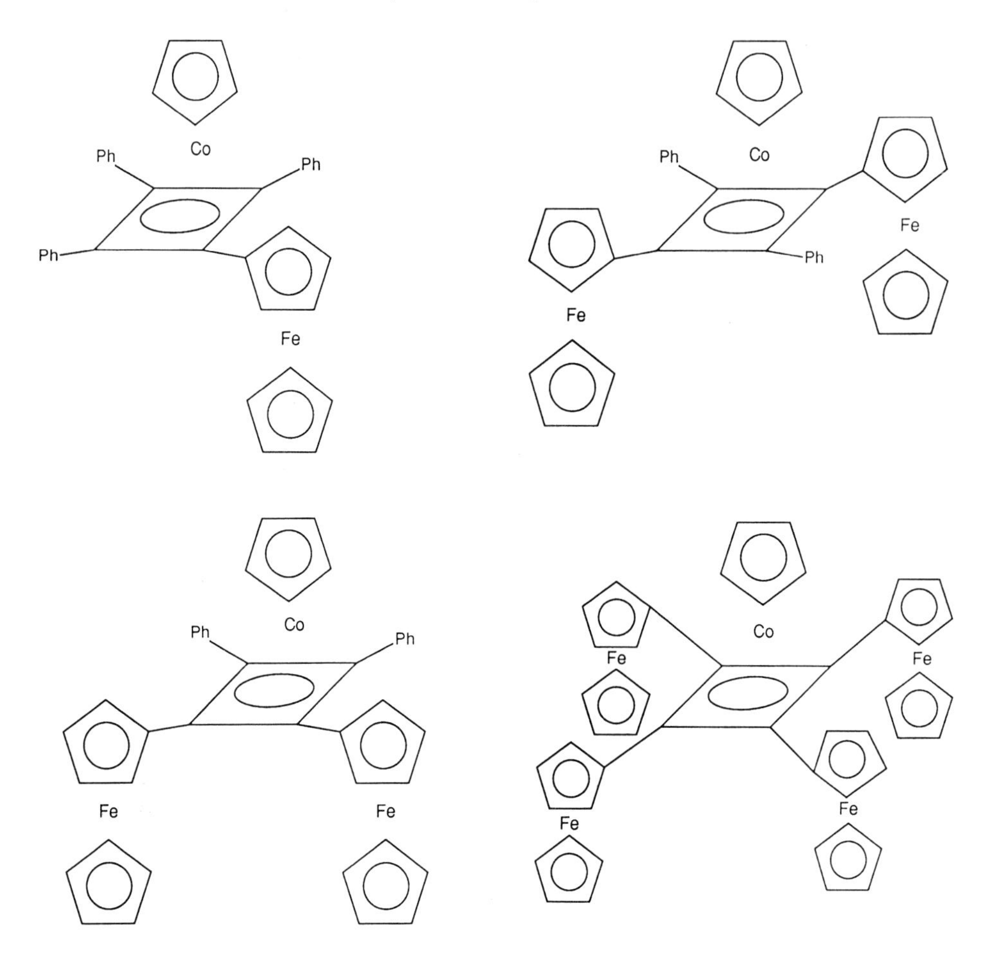

Scheme 7-17

Figure 7-30 shows the cyclic voltammetric response exhibited by *trans*-$CpCo[Fc_2Ph_2C_4]$ [116]. Two reversible one-electron oxidations, which are probably centered on the two ferrocenyl substituents, precede the irreversible oxidation of the cobalt center.

In the case of the tetraferrocenyl compound, the first one-electron oxidation is well shaped, whereas the successive oxidation waves of the three remaining ferrocenyl groups merge in a single broad peak.

Table 7-20 reports the redox potentials of the oxidation steps exhibited by the present cyclobutadiene-cobalt complexes.

There are no significant differences among the redox potentials of the complexes. It has only to be noted that the appearance of discrete oxidation steps for the ferrocenyl groups indicates that the cyclobutadiene ring allows electronic interaction between its substituents.

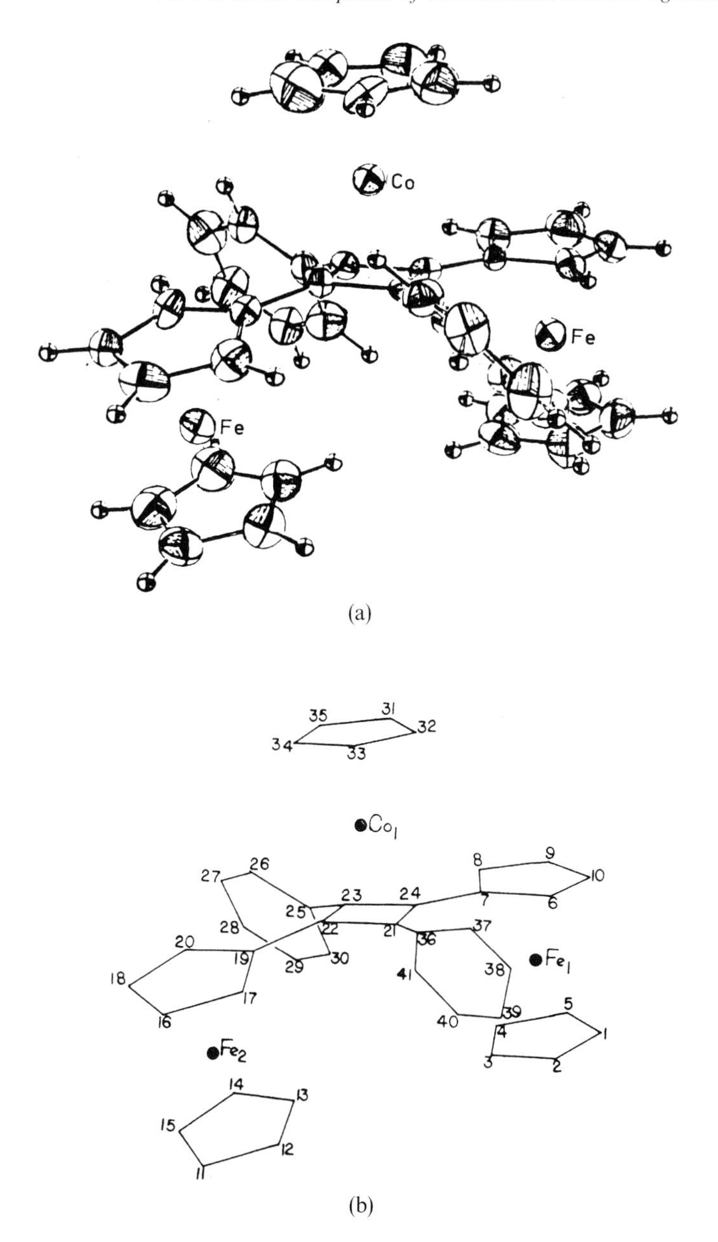

Fig. 7-29. Crystal structure of *trans*-$CpCo[Fc_2Ph_2C_4]$. (a) ORTEP drawing; (b) numbering scheme. C-C (cyclobutadiene ring), 1.47 Å; C7-C24, 1.45 Å; C19-C22, 1.46 Å; C25-C23, 1.47 Å; C21-C36, 1.46 Å; Co-centroid C_5 ring, 1.68 Å; Co-centroid C_4 ring, 1.69 Å, Fe-centroid C_5 ring, average 1.65 Å (reproduced by permission of Elsevier Sequoia S.A.).

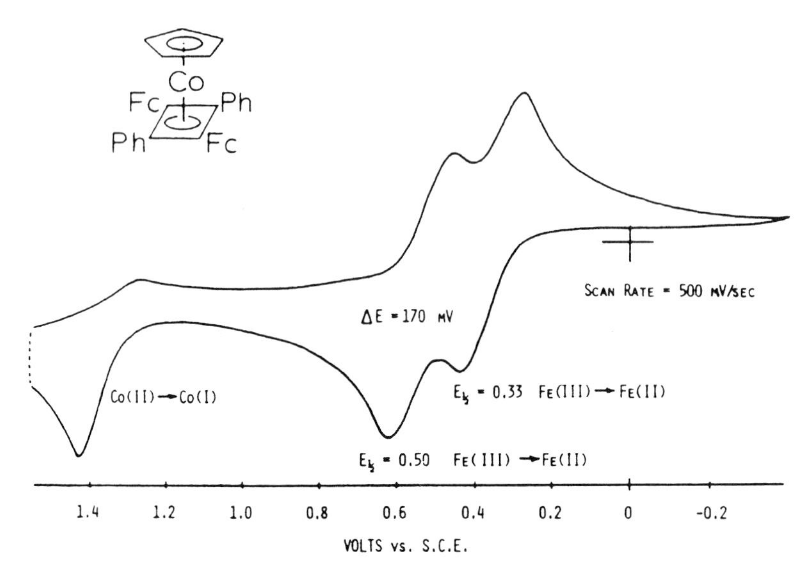

Fig. 7-30. Cyclic voltammogram exhibited by *trans*-$CpCo[Fc_2Ph_2C_4]$ in CH_2Cl_2 solution. Platinum working electrode (reproduced by permission of the American Chemical Society).

Table 7-20. Formal electrode potentials (*vs.* SCE) for the oxidation processes exhibited by the complexes illustrated in Scheme 7-17 in dichloromethane solution [116] [Fc = $(C_5H_5)Fe(C_5H_4)$]

Complex	Ferrocenyl-centered oxidation				Cobalt(I) oxidation
	$E^{\circ\prime}$, V	$E^{\circ\prime}$, V	$E^{\circ\prime}$, V	$E^{\circ\prime}$, V	E_p, V
$CpCo(FcPh_3C_4)$	+0.35	–	–	–	–
cis-$CpCo(Fc_2Ph_2C_4)$	+0.33	+0.52	–	–	+1.38
trans-$CpCo(Fc_2Ph_2C_4)$	+0.33	+0.50	–	–	+1.44
$CpCo(Fc_4C_4)$	+0.29	+0.4	+0.4	+0.4	–

A somewhat related family of cobalt-ferrocenyl complexes is shown in Scheme 7-18 [117].

As illustrated in Fig. 7-31 [117], the two isomeric complexes undergo (in dichloromethane solution) three separate oxidation steps, only the first two ($E^{\circ\prime}$ 0/+ = −0.20 V, $E^{\circ\prime}$ +/2+ = +0.19 V, *vs.* Ag/Ag^+, respectively) having features of full chemical reversibility.

The first one-electron removal is thought to be centered on the cobaltcyclopentadiene moiety, whereas the second is assigned to the oxidation of one ferrocenyl fragment. Finally, the third step might be centered on the second ferrocenyl unit, as well as on the central cobalt site.

It should be pointed out that the redox fingerprint of complex **2** only holds at very low temperature (−64 °C), whereas at room temperature the first oxidation process

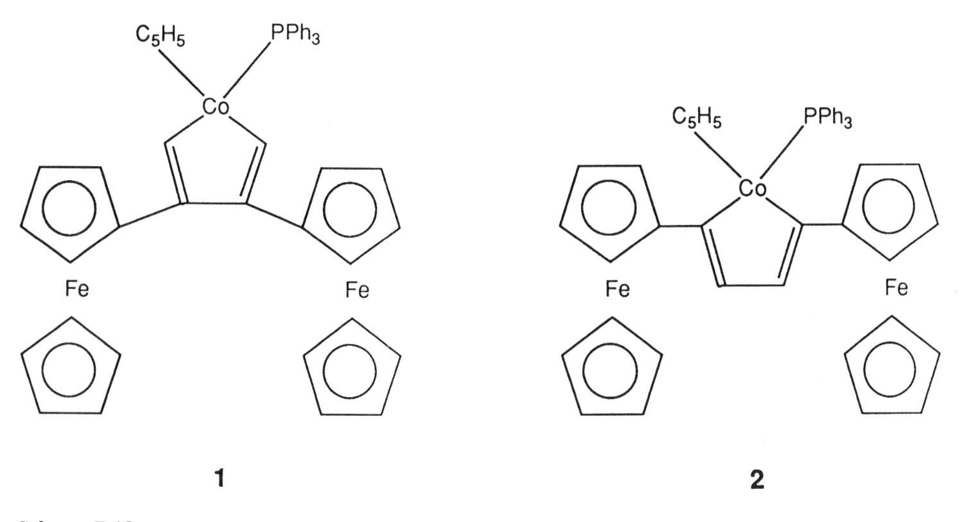

Scheme 7-18

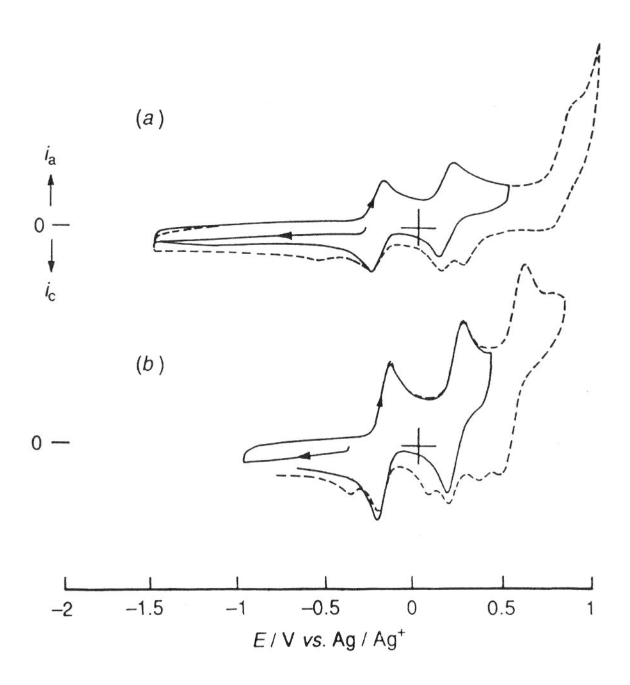

Fig. 7-31. Cyclic voltammograms recorded at a glassy-carbon electrode on dichloromethane solutions of: (a) **1**, at room temperature; (b) **2**, at $-64\,°C$. Scan rate 0.1 V s^{-1} (reproduced by permission of the Royal Society of Chemistry).

is complicated by destruction of the metallacycle with release of unsaturated diferrocenylphosphines [117].

Scheme 7-19 shows the probable structure of two ferrocene-capped cobalt clathrochelates [118].

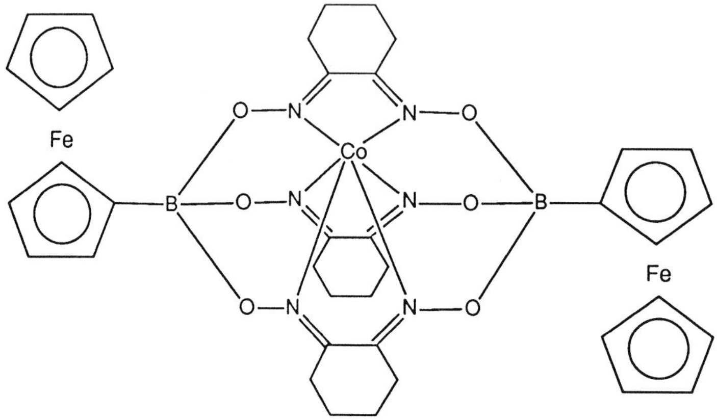

$Co(nox)_3[B(Cp)Fe(Cp)]_2$

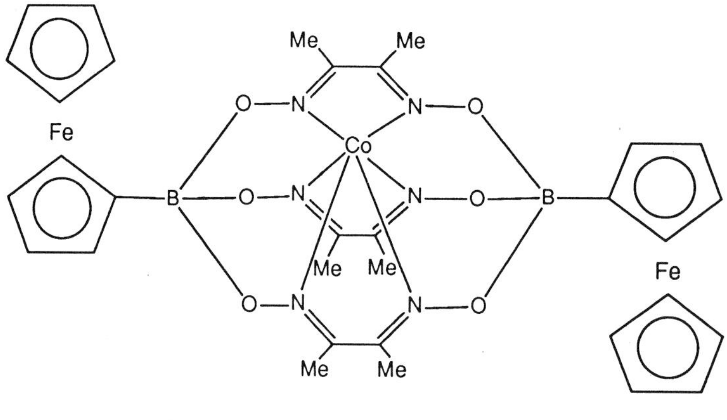

Scheme 7-19 $Co(dmg)_3[B(Cp)Fe(Cp)]_2$

In dichloromethane solution they undergo a first, cobalt-based, one-electron oxidation ($Co^{II/III}$: $Co(nox)_3[B(Cp)Fe(Cp)]_2$, $E_p = +0.21$ V; $Co(dmg)_3[B(Cp)$-$Fe(Cp)]_2$, $E_p = +0.19$ V), followed by a second, ferrocene-based, two-electron oxidation ($Fe^{II/III}$: $Co(nox)_3[B(Cp)Fe(Cp)]_2$, $E_p = +0.36$ V; $Co(dmg)_3[B(Cp)$-$Fe(Cp)]_2$, $E_p = +0.38$ V). Both oxidation processes appear to be complicated by subsequent chemical reactions [118]. The single-step two-electron removals are evidence for the lack of electronic conjugation between the two ferrocene-capping units.

Another cobalt-ferrocenyl complex, whose electrochemistry has been studied, is shown in Scheme 7-20 [119].

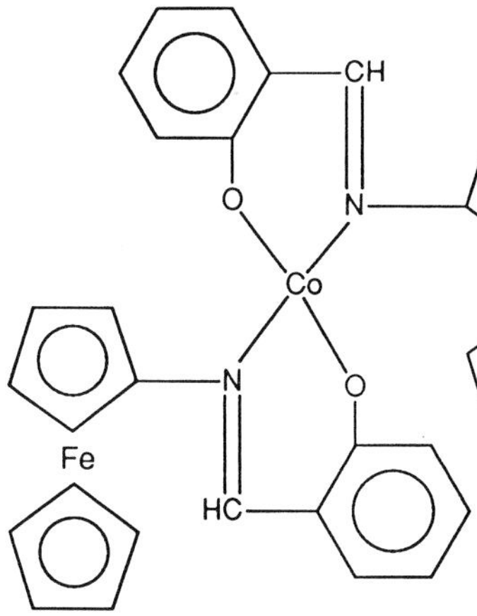

Scheme 7-20

It has been briefly reported that in DMF solution this complex undergoes a ferrocene-based, single-stepped two-electron oxidation ($E^{\circ\prime} = +0.58$ V), which once again indicates that no electronic contact exists between the two ferrocenyl units. The free *N*-(*o*-hydroxybenzylidene)ferroceneamine ligand undergoes reversible one-electron oxidation at a slightly lower potential ($E^{\circ\prime} = +0.51$ V) [119].

Scheme 7-21 shows the proposed structure for a further tetracoordinated cobalt(II) complex [120].

As in the preceding complex, this species undergoes a two-electron oxidation (in MeCN) centered on the two non-interacting ferrocenyl fragments, at $E^{\circ\prime} = +0.46$ V. This does not significantly differ from that of the parent ligand formylferrocenethiosemicarbazone, which undergoes oxidation at $E^{\circ\prime} = +0.45$ V. It also exhibits two

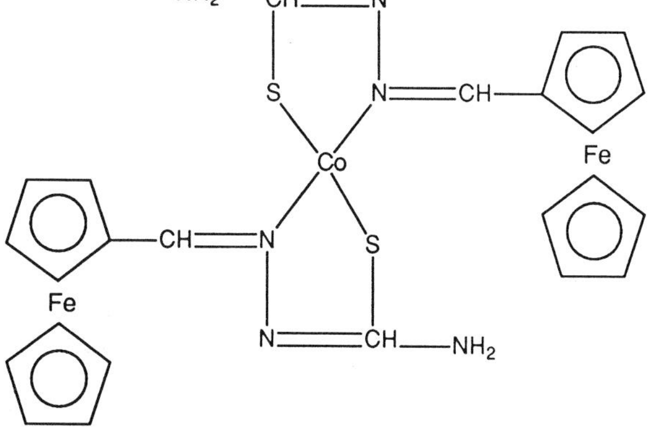

Scheme 7-21

separate one-electron reductions in DMF, only the first being reversible, at −1.21 V and −1.50 V, respectively. They are attributed to the sequential reduction (Co(II)/Co(I)/Co(0) of the central cobalt fragment; the relevant redox potentials are significantly more negative than the irreversible two-electron reduction at −1.10 V exhibited by free Co(II) ions [120].

Scheme 7-22 shows the octahedral coordination of the cobalt(III) center in two substituted enolate complexes [121].

As illustrated in Fig. 7.32, in CH_2Cl_2 the two complexes display quite different redox propensities. In fact, the phenyl-substituted complex undergoes reduction to the transient cobalt(II) congener ($E^{\circ\prime} = -1.13$ V), as well as a few closely-spaced irreversible oxidation steps at potential values higher than +0.9 V, probably ligand centered. In contrast, the ferrocenyl substituted species simply undergoes a single-stepped three-electron oxidation ($E^{\circ\prime} = +0.68$ V), centered on the ferrocenyl

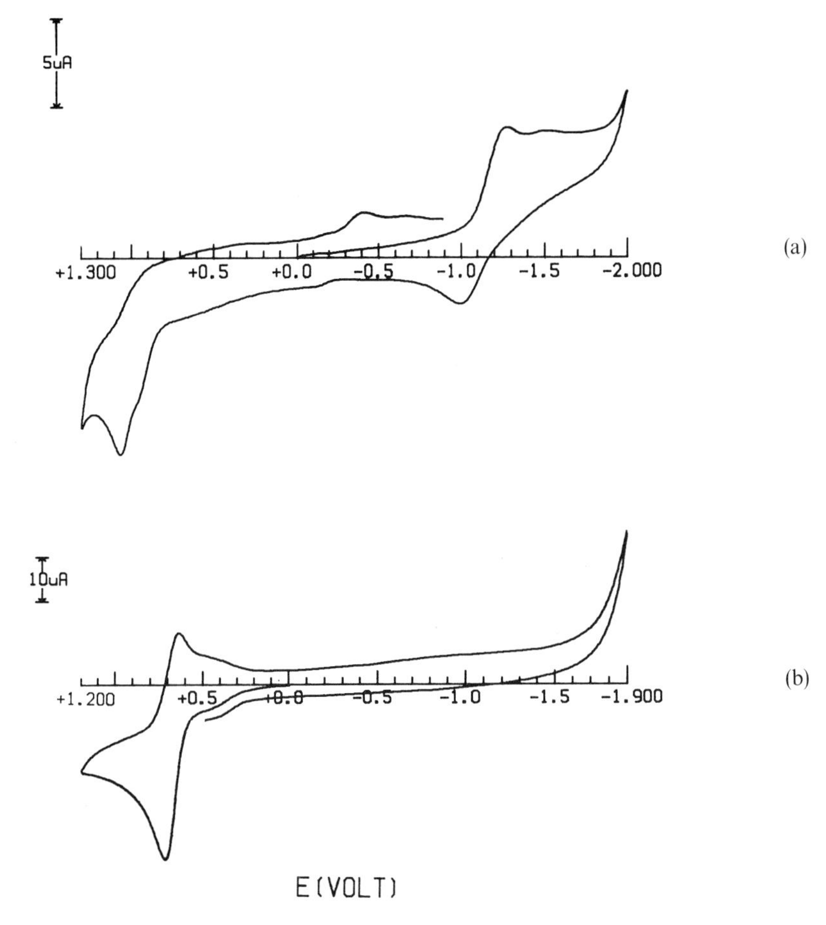

Fig. 7-32. Cyclic voltammograms recorded at a platinum electrode on dichloromethane solutions of: (a) $[PPh_2CHC(Ph)O]_3Co$; (b) $\{PPh_2CHC[(C_5H_5)Fe(C_5H_4)]O\}_3Co$.

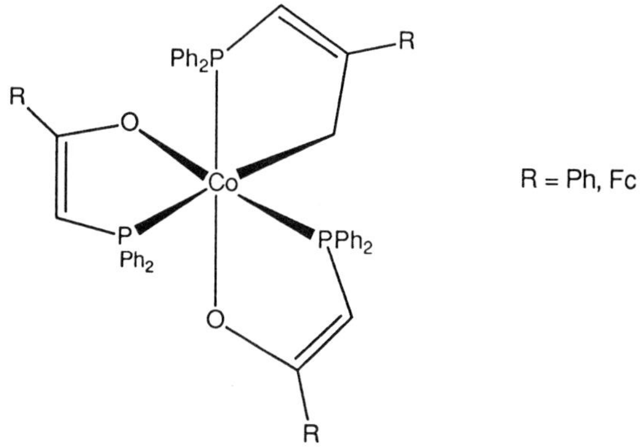

R = Ph, Fc

Scheme 7-22

fragments. These results are not only evidene for the non-interaction of the ferrocenyl units, but also provide support for the strong electronic perturbation brought about by the introduction of the redox-active ferrocenyl fragments, which hinder access to the cobalt-based LUMO of the complex [122].

The phosphino-ferrocene complexes illustrated in Scheme 7-23 also possess octahedral coordination around the cobalt atom [123].

Figure 7-33 illustrates the molecular structure of the dimethylphosphinoferrocene derivatives $[Co(acac)_2(dmpf)]^+$ and $[Co(dtc)_2(dmpf)]^+$ [123].

It is apparent that the coordination geometry at the cobalt atom is distorted octahedral. The cyclopentadienyl rings of the phosphinoferrocene fragment are staggered and tilted towards the cobalt atom (by 4.4° in the acetylacetonate complex and by 5.0° in the dithiocarbamate complex).

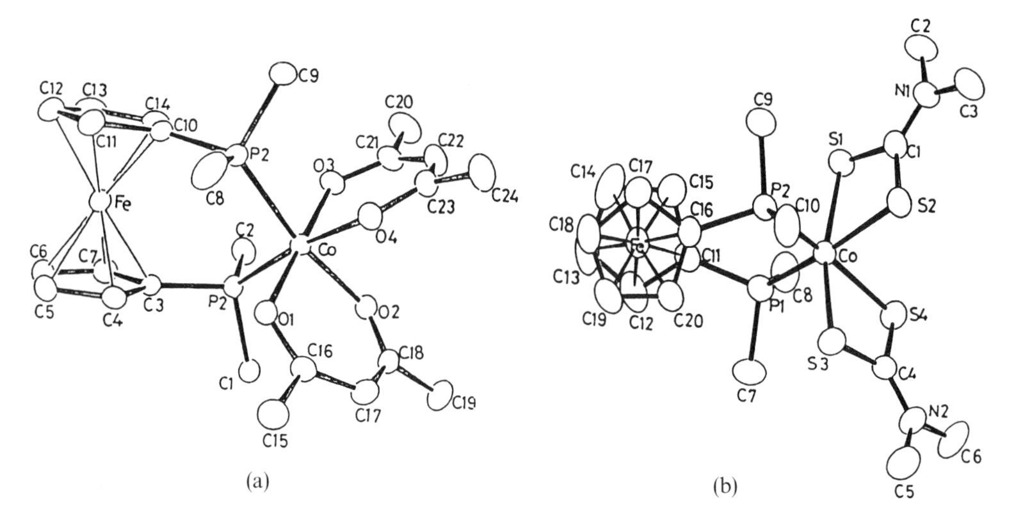

Fig. 7-33. Perspective views: (a) $[Co(acac)_2(dmpf)]^+$. Average bond lengths: Co-P, 2.25 Å, Co-O (*trans* to O), 1.90 Å; Co-O (*trans* to P), 1.94 Å. (b) $[Co(dtc)_2(dmpf)]^+$. Average bond lengths: Co-P, 2.25 Å; Co-S (*trans* to S), 2.28 Å; Co-S (*trans* to P), 2.30 Å (reproduced by permission of the Chemical Society of Japan).

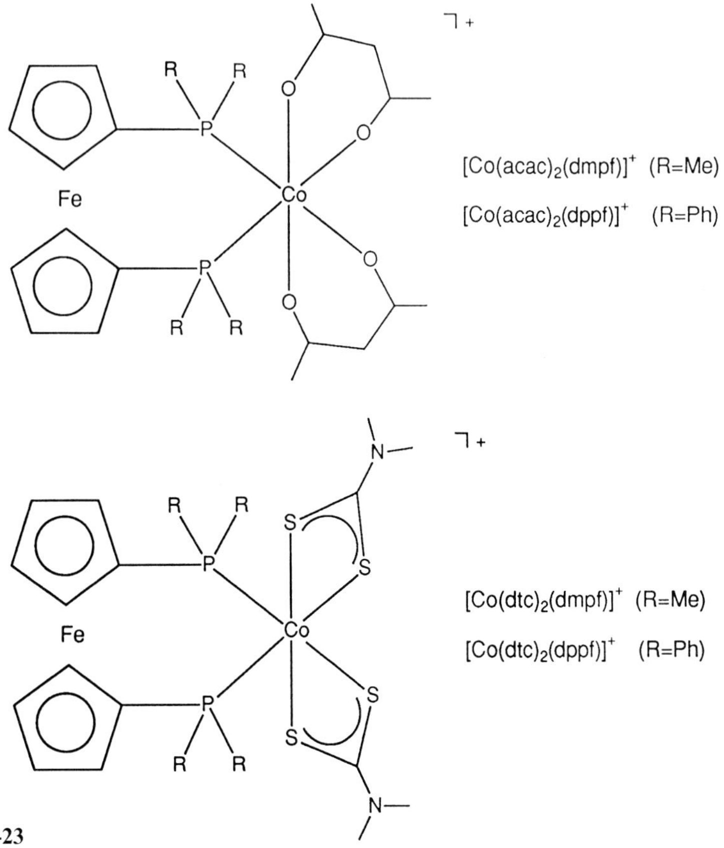

$[Co(acac)_2(dmpf)]^+$ (R=Me)

$[Co(acac)_2(dppf)]^+$ (R=Ph)

$[Co(dtc)_2(dmpf)]^+$ (R=Me)

$[Co(dtc)_2(dppf)]^+$ (R=Ph)

Scheme 7-23

In acetonitrile solution these complexes undergo both a Co(III)/Co(II) reduction and a Fe(II)/Fe(III) oxidation, which in some cases (indicated in Table 7-21) are coupled to chemical complications [123]. As shown in Table 7-21, the electron-donating ability of the various ligands is mainly directed towards the cobalt center. Substitution of either diphenylphosphinoferrocene for dimethylphosphinoferrocene

Table 7-21. Formal electrode potentials (*vs.* SCE) for the redox processes exhibited in acetonitrile solution by the phosphinoferrocene complexes illustrated in Scheme 7-23 [123]

Complex	$E^{\circ\prime}$ (Co^{III}/Co^{II}), V	$E^{\circ\prime}$ (Fe^{II}/Fe^{III}), V
$[Co(acac)_2(dppf)]^+$	−0.91[a]	+0.98
$[Co(acac)_2(dmpf)]^+$	−1.33[a]	+0.97[a]
$[Co(dtc)_2(dppf)]^+$	−1.23	+0.91
$[Co(dtc)_2(dmpf)]^+$	−1.57	+0.86[a]
dppf		+0.58

a Irreversible process.

or propanedionate for dimethyldithiocarbamate makes, as expected, the Co(III)/Co(II) redox change more difficult, while leaving the ferrocene oxidation almost unperturbed. Nevertheless, there are no doubts that complexation to cobalt makes the one-electron removal from the ferrocene fragment significantly more difficult, with respect to that for the free phosphinoferrocene molecule.

Scheme 7-24 shows a series of ferrocenyl-cobalt complexes, the electrochemical behavior of which has been accurately studied.

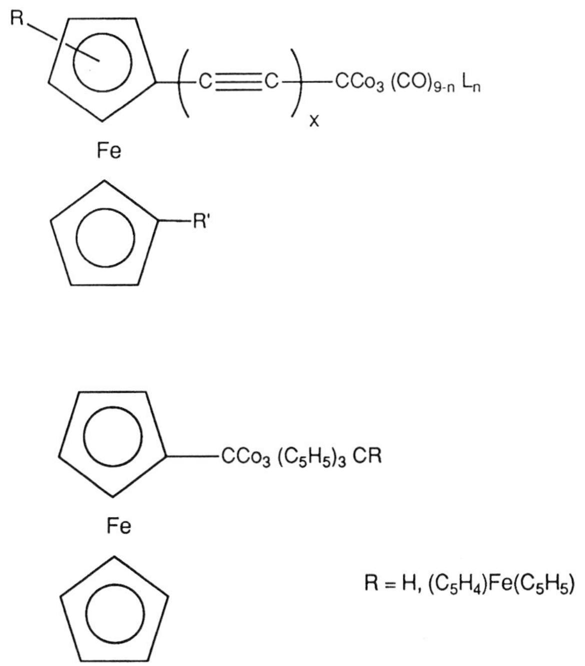

R

$-(C{\equiv}C)_x-CCo_3(CO)_{9-n}L_n$

Fe

R'

$CCo_3(C_5H_5)_3CR$

Fe

R = H, $(C_5H_4)Fe(C_5H_5)$

Scheme 7-24

Figure 7-34 shows a perspective view of the two species $(\eta^5\text{-}C_5H_5)Fe\{(\eta^5\text{-}C_5H_4)[CCo_3(CO)_9]\}$ [124] and $(\eta^5\text{-}C_5H_5)Fe\{(\eta^5\text{-}C_5H_4\text{-}C{\equiv}C)[CCo_3(CO)_9]\}$ [125].

In $(\eta^5\text{-}C_5H_5)Fe\{(\eta^5\text{-}C_5H_4)[CCo_3(CO)_9]\}$ the ferrocenyl group is directly linked to the pyramidal tricobalt-carbonyl cluster fragment through the capping carbyne atom. The cyclopentadienyl rings adopt an eclipsed configuration, with a slight tilting (3.6°) from the parallel disposition. Unexpectedly, the ferrocenyl unit is bent away from the threefold axis of the CCo_3 cluster core, a phenomenon that has been attributed to the geometrical requirement for the maximum pπ overlap between the orbitals of the apical carbon atom of the tricobalt cluster fragment and those of the cyclopentadienyl ring [124].

In contrast, in $(\eta^5\text{-}C_5H_5)Fe\{(\eta^5\text{-}C_5H_4\text{-}C{\equiv}C)[CCo_3(CO)_9]\}$ the cyclopentadienyl rings of the ferrocenyl unit are almost perpendicular to the Co_3 basal plane of the cluster, indicating the lack of hyperconjugation between the two building blocks [125]. With respect to each other, the cyclopentadienyl rings are less tilted (2.2°) and slightly rotated from the eclipsed conformation.

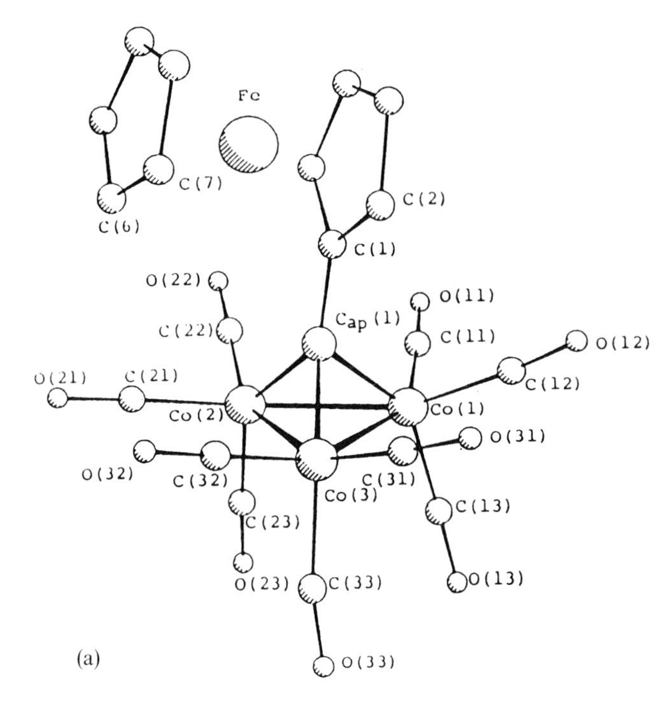

Fig. 7-34. X-Ray structures: (a) $(\eta^5\text{-}C_5H_5)Fe\{(\eta^5\text{-}C_5H_4)[CCo_3(CO)_9]\}$. Co-Co, average 2.47 Å; Co-C_{ap}, 1.92 Å; C_{ap}-C1, 1.45 Å (reproduced by permission of Elsevier Sequoia S.A.); (b) $(\eta^5\text{-}C_5H_5)Fe\{(\eta^5\text{-}C_5H_4\text{-}C{\equiv}C)[CCo_3(CO)_9]\}$. Co-Co, average 2.47 Å; Co-C1, 1.91 Å; C1-C2, 1.37 Å; C2-C3, 1.37 Å; C3-C41, 1.39 Å (reproduced by permission of the American Chemical Society).

As illustrated in Fig. 7-35, the complexes $Fc(C{\equiv}C)_xCCo_3(CO)_{9-n}L_n$ generally undergo a reversible one-electron oxidation at the ferrocene center and a one-electron reduction assigned to the tricobalt fragment. This last process is sometimes complicated by subsequent fast chemical decomposition. In addition, multiple substitution of carbonyl ligands for phosphorus-containing Lewis bases also makes oxidation of the carbon-capped tricobalt group possible [126–129].

The redox potentials of the actual processes are summarized in Table 7-22.

With respect to the free ferrocene and tricobalt-phenylcarbyne molecules, respectively, conjugation of the two redox centers makes the ferrocenyl ligand more difficult to oxidize and the cobalt cluster less reducible. It is also evident that the progressive introduction of Lewis bases make electron removal easier and electron addition more difficult.

As far as the second series of compounds containing a carbyne-bicapped tricobalt cluster is concerned, Fig. 7-36 illustrates the molecular structure of $(\eta^5\text{-}C_5H_5)\text{-}Fe\{(\eta^5\text{-}C_5H_4)[CCo_3CH(\eta^5\text{-}C_5H_5)_3]\}$ [124].

Also in this case, the cyclopentadienyl rings of the ferrocenyl fragment adopt an eclipsed configuration with a slight tilting (1.9°) from the parallel arrangement.

The most interesting redox behavior of the two complexes $FcCCo_3(Cp)_3CH$ and $FcCCo_3(Cp)_3CFc$ is shown in Fig. 7-37 [130].

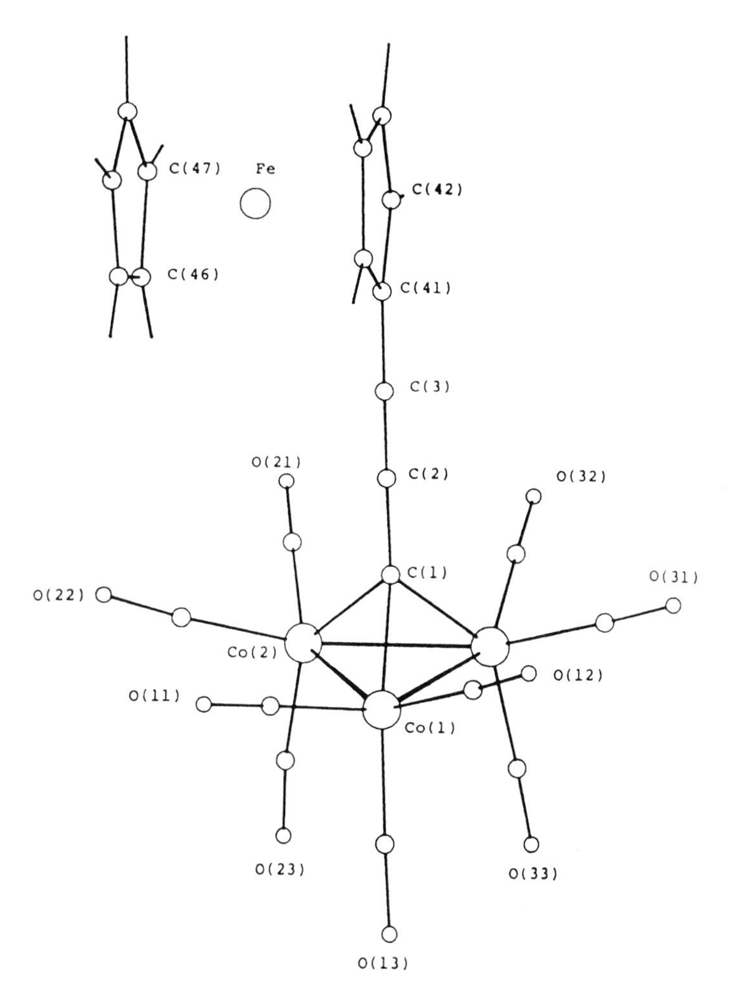

Fig. 7-34b

In both cases, the reversible one-electron oxidation of the $Co_3(Cp)_3$ fragment precedes that of the ferrocenyl ligand(s). Further processes, not shown in Fig. 7-37, concern the irreversible one-electron reduction of the tricobalt cluster as well as its two-electron oxidation. The relevant redox potentials are reported in Table 7-23.

In confirmation of the chemical reversibility of the oxidation process, the oxidized congeners have been characterized spetroscopically [130].

The occurrence of discrete oxidation steps for each ferrocenyl substituent in the diferrocene complex indicates that the biscarbyne cobalt core CCo_3C does not constitute an impassable barrier to electronic interaction between the two redox centers.

Figure 7-38 shows the X-ray structure of another carbon-capped tricobalt complex, namely $PhCCo_3(CO)_7[Fe(\eta^5\text{-}C_5H_4PPh_2)_2]$, in which the bis(diphenylphosphino)-ferrocene molecule, acting as a bidentate ligand, substitutes two axial CO groups of the precursor $PhCCo_3(CO)_9$, rather than the capping organic substituent [131].

Table 7-22. Formal electrode potentials (*vs.* SCE) for the redox processes exhibited by the complexes $[R'\text{-}C_5H_4]Fe[(C_5H_3)(R)(C{\equiv}C)_xCCo_3(CO)_{9-n}L_n]$

Complex R′	R	x	n	L	$E^{\circ\prime}$ (CCo_3/CCo_3^-), V	$E^{\circ\prime}$ (Fc/Fc^+), V	$E^{\circ\prime}$ (CCo_3/CCo_3^+), V	Solvent	Reference
H	H	0	0	–	−0.63	+0.63	–	Me_2CO	[127]
H	2-C(O)Me	0	0	–	−0.73	+0.73	–	Me_2CO	[127]
C(O)Me	H	0	0	–	−0.63	+0.75	–	Me_2CO	[127]
H	2-Me	0	0	–	−0.53	+0.70	–	Me_2CO	[127]
H	3-Me	0	0	–	−0.52	+0.66	–	Me_2CO	[127]
Me	H	0	0	–	−0.53	+0.69	–	Me_2cO	[127]
H	H	1	0	–	−0.70	+0.62	–	CH_2Cl_2	[128]
H	H	0	1	PPh_3	−0.84	+0.64	–	Me_2CO	[127]
H	H	0	1	$P(C_6H_{11})_3$	−1.04	+0.66	–	Me_2CO	[127]
H	H	0	1	$P(OMe)_3$	−0.99	+0.46	–	Me_2CO	[127]
H	H	0	2	$P(OMe)_3$	−1.32[a]	+0.42	+0.69[a]	Me_2CO	[127]
H	H	0	2	$P(OPh)_3$	−1.03[a]	+0.49	+1.04[a]	Me_2CO	[127]
H	H	0	3	$P(OMe)_3$	−1.55[a]	+0.30	+0.57	Me_2CO	[129]
H	H	0	3	$P(OPh)_3$	−1.22[a]	+0.41	+0.89	Me_2CO	[129]
$PhCCo_3(CO)_9$					−0.54			Me_2CO	[127]
$Fe(C_5H_5)_2$						+0.46[b]		Me_2CO	
						+0.45[b]		CH_2Cl_2	

a Peak potential value for irreversible processes.
b Unpublished results.

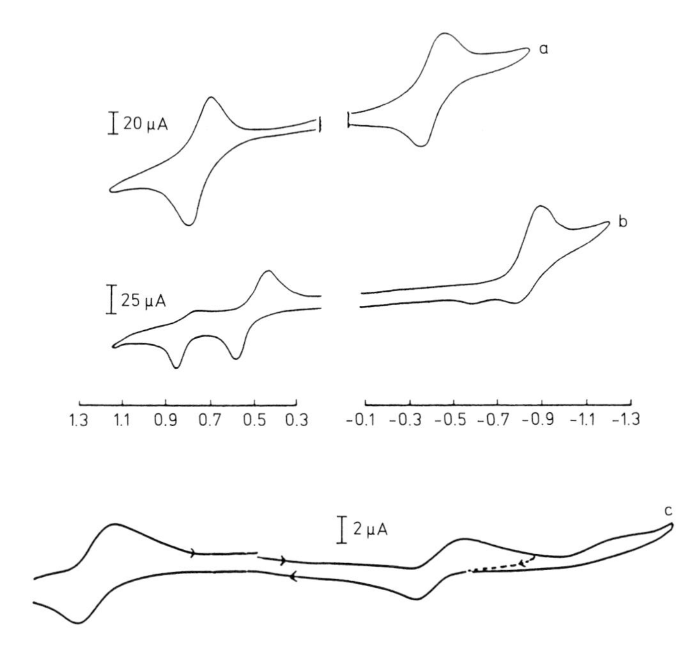

Fig. 7-35. Cyclic voltammograms recorded at a platinum electrode: (a) acetone solution of $FcCCO_3(CO)_9$; (b) acetone solution of $FcCCo_3(CO)_7[P(OMe)_3]_2$; (c) dichloromethane solution of $FcC{\equiv}CCCo_3(CO)_9$. Scan rate 0.2 V s^{-1} (reproduced by permission of the American Chemical Society).

In CH_2Cl_2 solution this cluster compound undergoes the expected (on the basis of the cited oxidation behavior of free bis(diphenylphosphino)ferrocene) irreversible one-electron oxidation of the ferrocene fragment (E_p 0/+ = +0.88 V), followed by the (once again expected on the basis of the multiple carbonyl substitution) one-electron oxidation of the Co_3 assembly (E_p +/2+ = +1.31 V). The reduction process (E_p 0/− = −1.16 V) typical of the CCo_3 assembly is also complicated by following chemical reactions [131].

Table 7-23. Formal electrode potentials (*vs.* SCE) for the redox processes exhibited by the complexes $(C_5H_5)Fe[C_5H_4\text{-}CCo_3(C_5H_5)_3CR]$, in acetone solution [130]

Complex	$E^{\circ\prime}$ (CCo_3/CCo_3^-), V	$E^{\circ\prime}$ (CCo_3/CCo_3^+), V	$E^{\circ\prime}$ (Fc/Fc^+), V	$E^{\circ\prime}$ (Fc/Fc^+), V	$E^{\circ\prime}$ (CCo_3^+/CCo_3^{2+}), V
R = H	−1.78[a]	+0.28	+0.60	−	+1.39[a]
R = $(C_5H_5)Fe(C_5H_4)$	−1.83[a]	+0.22	+0.46	+0.77	+1.53[a]
$Fe(C_5H_5)_2$			+0.46[b]		

a Peak potential value for irreversible processes.
b Unpublished results.

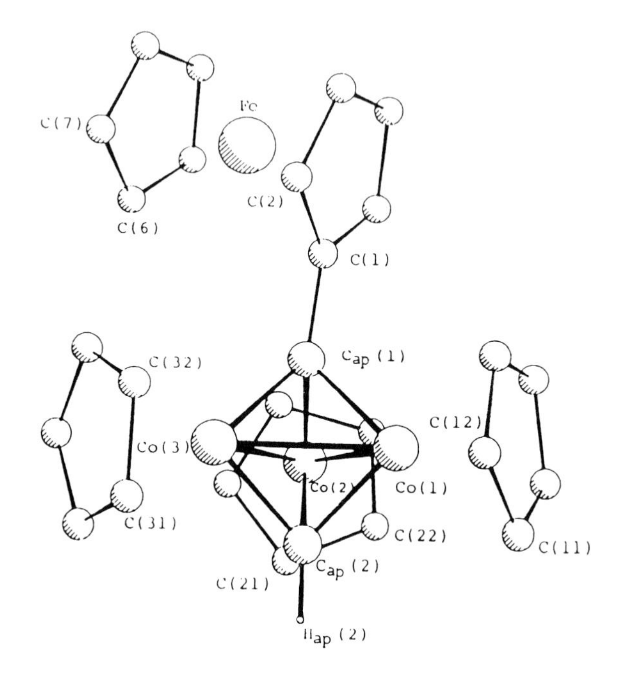

Fig. 7-36. X-Ray structure of $(\eta^5\text{-}C_5H_5)Fe\{(\eta^5\text{-}C_5H_4)[CCo_3CH(\eta^5\text{-}C_5H_5)_3]\}$. Co1-Co3 = Co2-Co3, 2.39 Å; Co1-Co2, 2.36 Å; Co-C_{ap}(1), 1.87 Å; Co-C_{ap}(2), 1.85 Å; Co-C (Cp ring), 2.08 Å; C_{ap}(1)-C1, 1.45 Å (reproduced by permission of Elsevier Sequoia S.A.).

A somewhat related series of complexes is illustrated in Scheme 7-25 [132].

The crystal structure of $[(CO)_9Co_3C(Me)_2Si\text{-}(\eta^5\text{-}C_5H_4)]Fe(\eta^5\text{-}C_5H_4)\text{-}Si(Me)_2\text{-}CCo_3(CO)_9]$ is shown in Fig. 7-39 [132]. The cyclopentadienyl rings assume a staggered conformation, with the silyl-bridged cluster fragments placed in a transoid orientation in order to minimize steric interaction.

As illustrated in Fig. 7-40, the complexes $(\eta^5\text{-}C_5H_5)Fe[\eta^5\text{-}C_5H_4\text{-}Si(Me)_2\text{-}CCo_3(CO)_9]$ and $[(CO)_9Co_3C(Me)_2Si\text{-}\eta^5\text{-}C_5H_4]Fe[\eta^5\text{-}C_5H_4\text{-}Si(Me)_2CCo_3(CO)_9]$ undergo a one-electron oxidation in CH_2Cl_2, centered on the ferrocenyl moiety, and a one-electron reduction per cobalt cluster unit. Electron removal from the central ferrocene group is always chemically reversible, whereas the one-electron addition to each cluster fragment tends to be complicated by decomposition reactions, particularly in the phosphine/phosphite substituted species [132].

It should be noted from Fig. 7-40b that the two-one-electron reductions of the capped tricobalt units occur in a single-step process, which provides evidence for the non-interaction of the two peripheral fragments.

Table 7-24 reports the redox potentials for all the ferrocenylsilyl-tricobaltcarbonyl complexes.

In contrast to the ability of ferrocenyl or ferrocenylsilyl ligands to link carbyne-capped tricobalt clusters via the capping atom, 1,1′-bis(diphenylphosphino)ferrocene

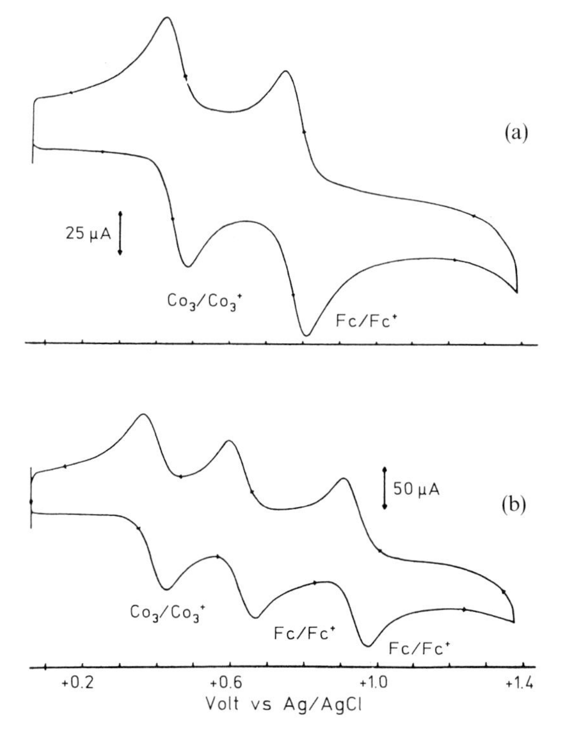

Fig. 7-37. Cyclic voltammograms recorded at a platinum electrode on acetone solutions containing: (a) $FcCCo_3(Cp)_3CH$; (b) $FcCCo_3(Cp)_3CFc$. Scan rate 0.2 V s^{-1} (reproduced by permission of the American Chemical Society).

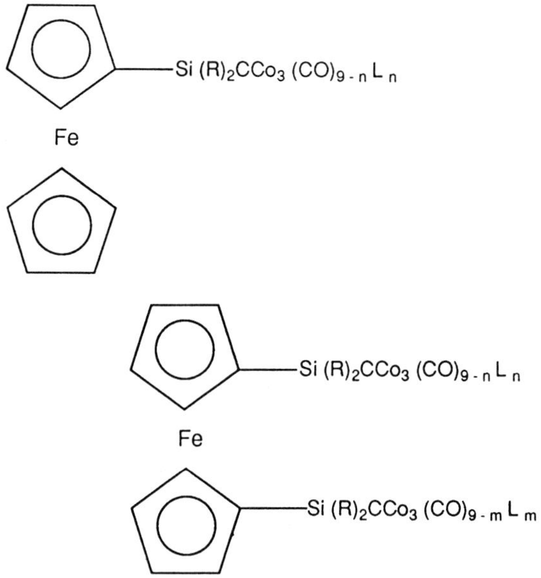

$-Si(R)_2CCo_3(CO)_{9-n}L_n$

Fe

$-Si(R)_2CCo_3(CO)_{9-n}L_n$

Fe

$-Si(R)_2CCo_3(CO)_{9-m}L_m$

Scheme 7-25

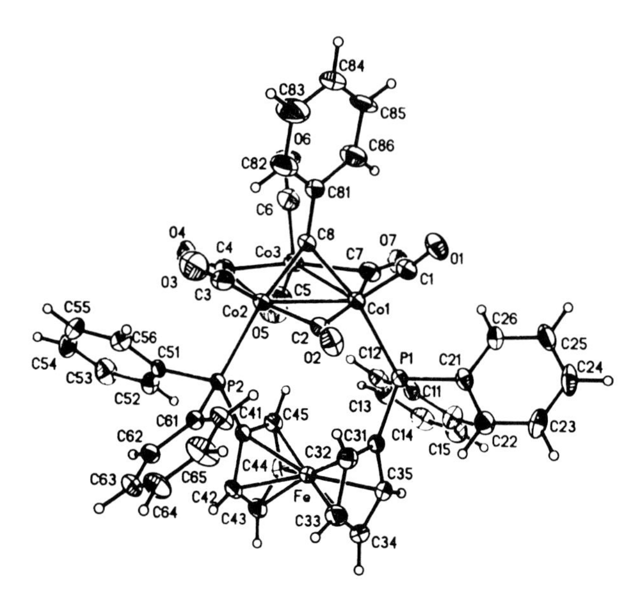

Fig. 7-38. Perspective view of $PhCCo_3(CO)_7[Fe(\eta^5\text{-}C_5H_4PPh_2)_2]$. Co1-Co2, 2.52 Å; Co1-Co3 = Co2-Co3, 2.48 Å; Co-C_{ap}, average 1.92 Å (reproduced by permission of Elsevier Sequoia S.A.).

Table 7-24. Formal electrode potentials (*vs.* S.C.E.) for the electron transfer processes exhibited in dichloromethane solution by the complexes $(C_5H_5)Fe[C_5H_4\text{-}Si(R)_2CCo_3(CO)_{9-n}L_n]$ and $[L_n(CO)_{9-n}Co_3C(R)_2Si\text{-}C_5H_4)]Fe[C_5H_4\text{-}Si(R)_2CCo_3(CO)_{9-m}L_m]$. Platinum working electrode [132]

Complex				$E^{\circ\prime}$ (+/0)[a], V	$E^{\circ\prime}$ (0/−)[a], V
$(C_5H_5)Fe[C_5H_4\text{-}Si(R)_2CCo_3(CO)_{9-n}L_n]$					
R = Me	$n = 0$			+0.72	−0.58
	$n = 1$		L = PPh_3	+0.69	−0.94[b]
	$n = 1$		L = PCy_3	+0.72	
	$n = 1$		L = $P(OMe)_3$	+0.67	−1.05[b]
	$n = 2$		L = $P(OMe)_3$	+0.69	−1.24[b]
R = Ph	$n = 0$			+0.77	−0.66
$[L_n(CO)_{9-n}Co_3C(R)_2Si\text{-}C_5H_4]Fe$ $[C_5H_4\text{-}Si(R)_2CCo_3(CO)_{9-m}L_m]$					
R = Me	$n = 0$	$m = 0$		+0.74	−0.70
	$n = 0$	$m = 1$	L = PPh_3	+0.75	−0.64
	$n = 1$	$m = 1$	L = PPh_3	+0.70	
	$n = 0$	$m = 2$	L = $P(OPh)_3$	+0.67	−0.62
	$n = 1$	$m = 2$	L = $P(OPh)_3$	+0.68	−0.67
R = Ph	$n = 0$	$m = 0$		+0.80	−0.67

a Converted to potential values *vs.* SCE.
b Peak potential values for processes complicated by subsequent decomposition.

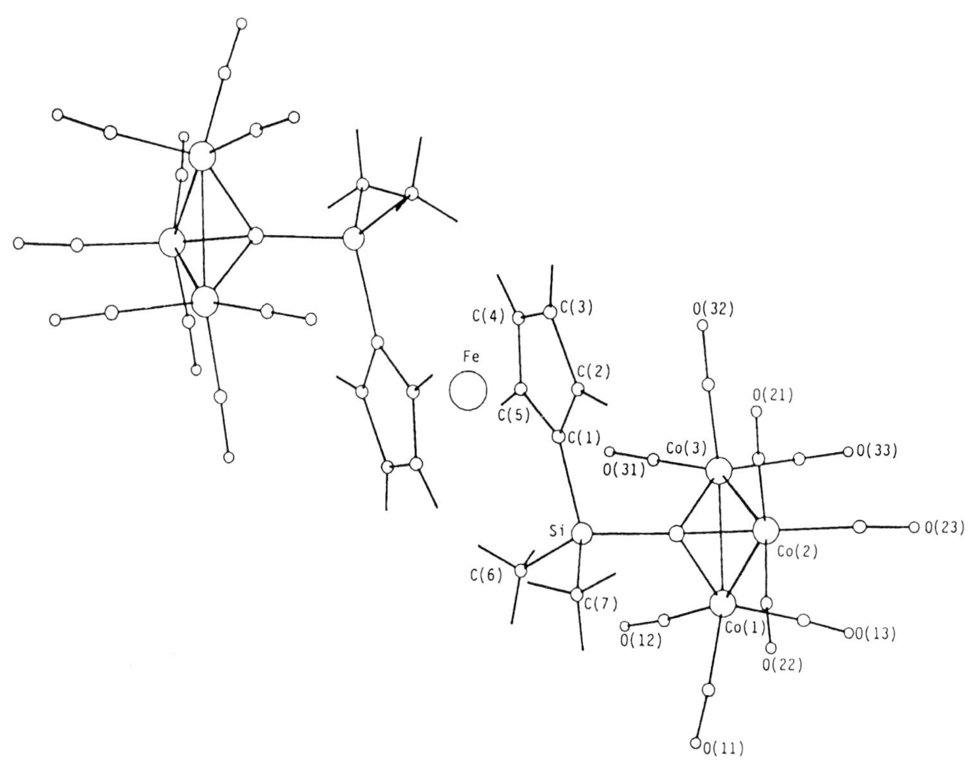

Fig. 7-39. Perspective view of $[(CO)_9Co_3C(Me)_2Si\text{-}\eta^5\text{-}C_5H_4]Fe[\eta^5\text{-}C_5H_4\text{-}Si(Me)_2CCo_3(CO)_9]$. Co1-Co3 = Co2-Co3, 2.47 Å; Co1-Co2, 2.46 Å; Co-C_{ap}, average 1.90 Å; Si-C1, 1.86 Å; Si-C_{ap}, 1.91 Å (reproduced by permission of Elsevier Sequoia S.A.).

links directly to two cobalt atoms by substitution of two carbonyl ligands, affording $[Fe(\eta^5\text{-}C_5H_4PPh_2)_2]Co_3(CO)_7Me$. The X-ray structure of this compound has been reported [72, 74], but its redox properties are still unknown.

Two other complexes for which the X-ray structure is known are $[(3\text{-}SiMe_3, 1\text{-}PPh_2\text{-}C_5H_3)_2Fe]CoCl_2$ [133] and $(\eta^5\text{-}C_5H_5)Fe[\eta^5\text{-}C_5H_4\text{-}CH(OH)\text{-}C_6H_5Co_4(CO)_9]$ [134].

7.2.6.2 Rhodium Complexes

A first class of heteronuclear ferrocene-rhodium complexes, the electrochemical behavior of which has been studied, is shown in Scheme 7-26.

The crystal structure of the sulfur-bridged complex $[Fe(\eta^5\text{-}C_5H_4S)_2]Rh(\eta^5\text{-}C_5Me_5)\text{-}(PMe_3)$ is illustrated in Fig. 7-41 [135].

The rhodium atom assumes a (distorted) tetrahedral coordination. The cyclopentadienyl rings of the ferrocene unit are nearly eclipsed and nearly parallel (tilting angle 0.7°).

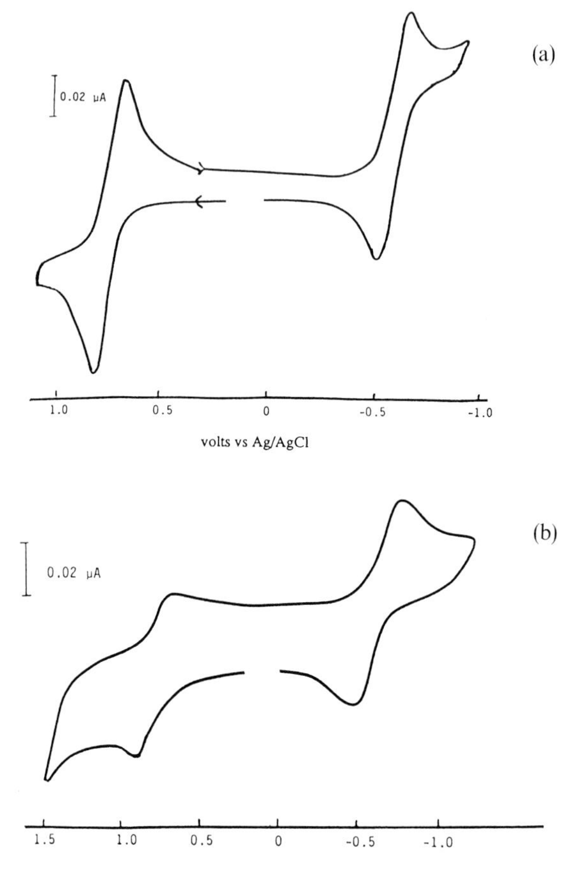

Fig. 7-40. Cyclic voltammograms recorded at a platinum electrode on dichloromethane solutions of: (a) $(C_5H_5)Fe[C_5H_4\text{-}Si(Me)_2CCo_3(CO)_9]$, scan rate $0.05\ V\ s^{-1}$; (b) $[(CO)_9Co_3C(Me)_2Si\text{-}C_5H_4]Fe[C_5H_4\text{-}Si(Me)_2CCo_3(CO)_9]$, scan rate $0.1\ V\ s^{-1}$ (reproduced by permission of Elsevier Sequoia S.A.).

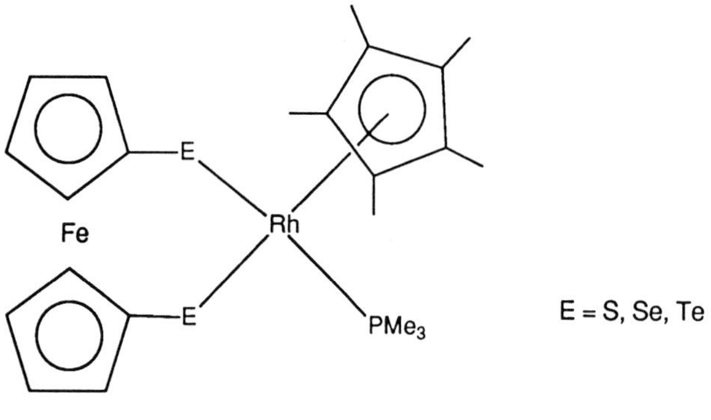

Fe E E Rh PMe_3

E = S, Se, Te

Scheme 7-26

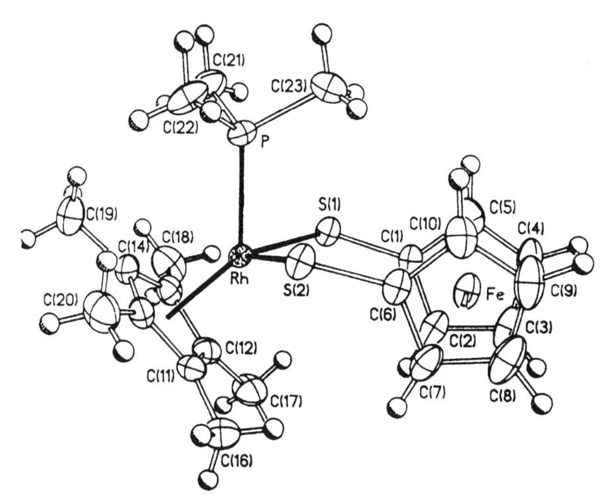

Fig. 7-41. Perspective view of $[Fe(\eta^5\text{-}C_5H_4S)_2]Rh(\eta^5\text{-}C_5Me_5)(PMe_3)$. Rh-P, 2.25 Å; Rh-S, average 2.36 Å, Rh-centroid(C_5Me_5), 1.88 Å; Rh···Fe 4.30 Å (reproduced by permission of Verlag der Zeitschrift für Naturforschung).

Figure 7-42 shows the redox behavior of such complex [136]. Two subsequent one-electron oxidations are displayed, only the first being reversible. The most naive expectation should be that such electron removals are centered on the iron and rhodium fragments, respectively; in particular, the first reversible step looks like the usual ferrocene-based response. Unexpectedly, the ESR spectrum of the electrogenerated monocation $[\{Fe(\eta^5\text{-}C_5H_4S)_2\}Rh(\eta^5\text{-}C_5Me_5)(PMe_3)]^+$ does not exhibit the features of a metal-centered radical. In addition, theoretical calculations indicate that the two electrons lost are mainly centered on the bridging sulfur atoms, even if the first SOMO receives a contribution of 15% from the rhodium atom *vs.* 4% from the iron atom, whereas the second SOMO receives a contribution of 25% from the iron atom *vs.* 16% from the rhodium atom [136].

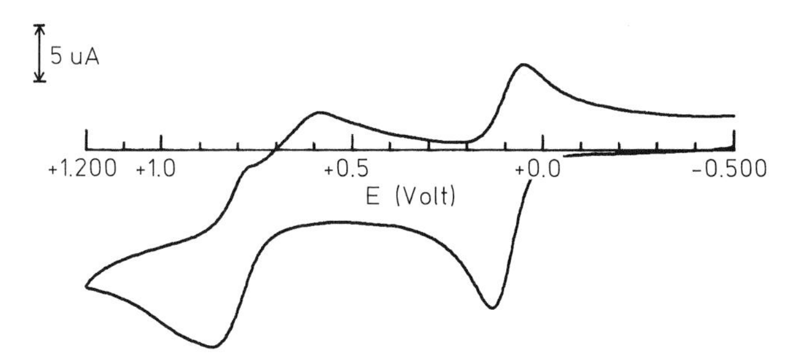

Fig. 7-42. Cyclic voltammetric response exhibited by $[Fe(C_5H_4S)_2]Rh(C_5Me_5)(PMe_3)$ in dichloromethane solution. Platinum working electrode. Scan rate 0.1 V s^{-1}.

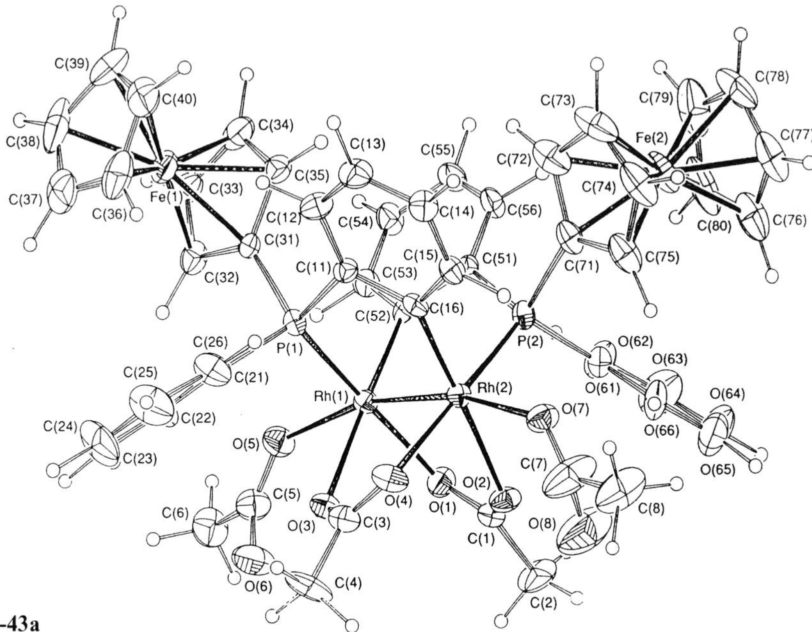

Fig. 7-43a

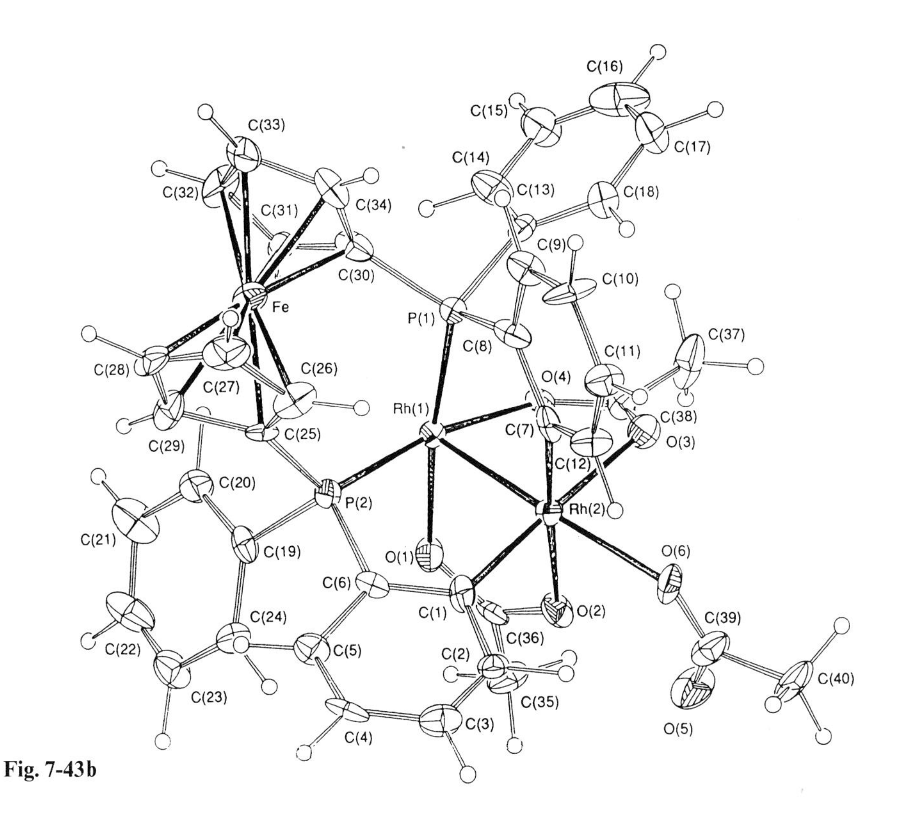

Fig. 7-43b

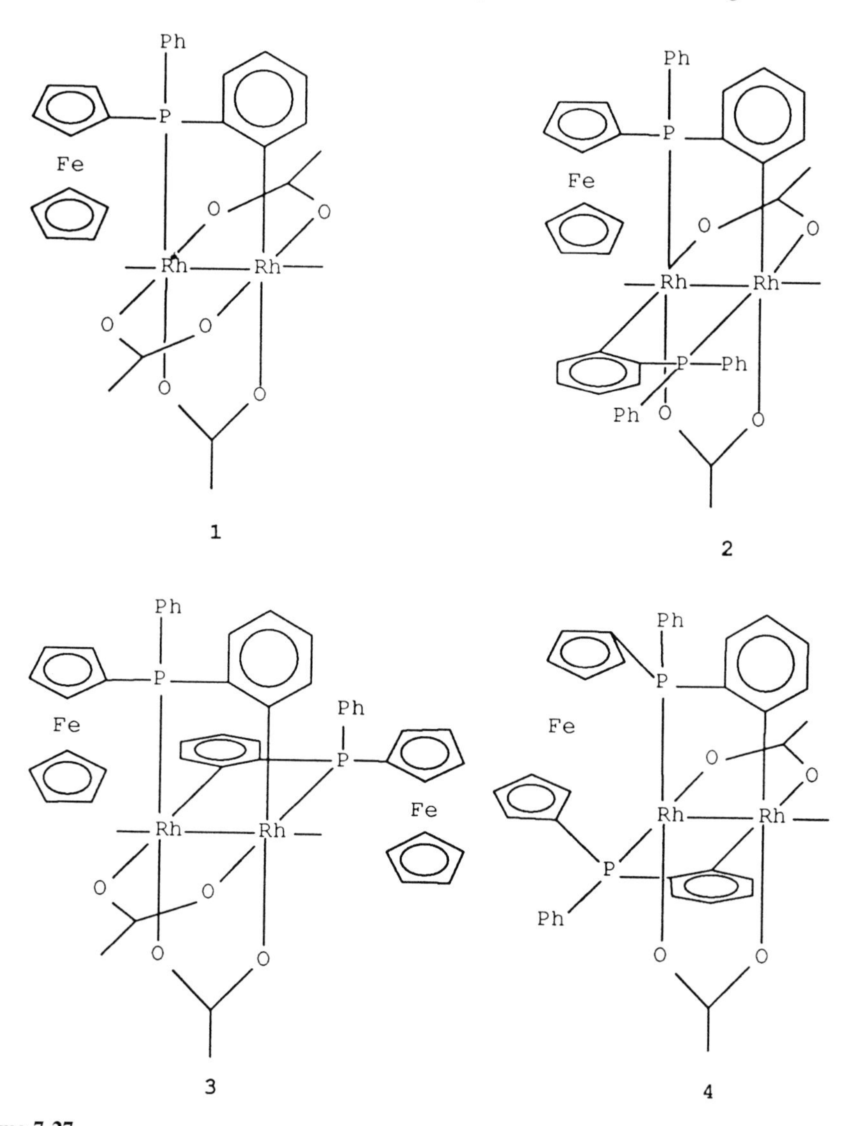

Scheme 7-27

◄

Fig. 7-43. X-Ray molecular structure of: (a) $[Rh_2(O_2CMe)_2\{[(C_6H_4)PhP(C_5H_4)]Fe(C_5H_5)\}_2$-$(HO_2CMe)_2]$. Rh-Rh, 2.50 Å; Rh-P, 2.20 Å. (b) $[Rh_2(O_2CMe)_2\{[(C_6H_4)PhP(C_5H_4)]_2Fe\}(HO_2CMe)]$. Rh-Rh, 2.51 Å, Rh-P, 2.22 Å (average) (reproduced by permission of the Royal Society of Chemistry).

Table 7-25. Formal electrode potentials (*vs.* SCE) for the oxidation processes exhibited by the complex $[Fe(C_5H_4E)_2]Rh(C_5Me_5)(PMe_3)$ in dichloromethane solution

Complex	$E^{\circ\prime}$ (0/+), V	$E_p{}^a$ (+/2+), V	Reference
E = S	+0.10	+0.86	[136]
E = Se	+0.12	+0.68	[136]
E = Te	+0.03[a]	+0.72	[136]
$Fe(C_5H_5)_2$	+0.46		[136]
$Fe(C_5H_5)_2$	+0.56		[59]
$Fe(C_5H_4SH)_2$	+0.65		[59]

a Peak potential value for an irreversible process.

Table 7-25 summarizes the redox potentials for the whole series of complexes.

As illustrated in Scheme 7-27, a second class of rhodium-ferrocene complexes of known redox behavior contains a central dirhodium acetate fragment bound, via bridging metallation, to either diphenylphosphinoferrocene or 1,1′-bis(diphenylphospino)ferrocene [137, 138].

Figure 7-43 shows the crystal structure of complexes **3** and **4** [137].

In complex **3**, both the rhodium atoms have octahedral coordination, whereas in **4** one rhodium atom has square pyramidal geometry and the other octahedral geometry.

From the electrochemical viewpoint, the complexes display a first oxidation process based on the ferrocene unit(s) followed by a second one-electron oxidation centered on the dirhodium moiety [137, 138]. Both the processes have features of chemical reversibility. The relevant redox potentials are summarized in Table 7-26, also in comparison with the corresponding metallated molecules containing triphenylphosphine instead of the ferrocenylphosphines.

The fact that the triphenylphosphino substituted analogues are more easily oxidized (by about 0.2 V) suggests that the introduction of the ferrocene molecules withdraws significant electron density from the central dirhodium fragment.

The crystal structure of a number of rhodium-phosphinoferrocene complexes has been reported: $\{(\eta^5\text{-}C_5H_5)Fe[\eta^5\text{-}C_5H_3(1\text{-}PPh_2)(2\text{-}CH(Me)NMe_2)]\}Rh(NBD)$ [139]; $\{Fe[\eta^5\text{-}C_5H_4P(CMe_3)_2]_2\}Rh(NBD)$ [140]; $\{[\eta^5\text{-}C_5H_4P(CMe_3)_2]Fe(\eta^5\text{-}C_5H_4PPh_2)\}$-

Table 7-26. Formal electrode potentials (*vs.* SCE) for the oxidation processes exhibited by the dirhodium complexes (illustrated in Scheme 7-27), in dichloromethane solution. Values in parentheses refer to the same dirhodium complexesd but metallated with PPh_3

Complex	$E^{\circ\prime}$ (Fc/Fc$^+$)	$E^{\circ\prime}$ (Rh_2^{4+}/Rh_2^{5+})	Reference
1	+0.68	+1.18 (+1.02)	[137]
2	+0.58	+1.04	[138]
3	+0.60[a]	+1.20 (+0.90)	[137]
4	+0.82	+1.12 (+0.84)	[137]

a Two-electron process.

Rh(NBD), {Fe(η^5-C_5H_4P[(CMe$_3$)(Ph)])$_2$}Rh(NBD), and [Fe(η^5-$C_5H_4PPh_2$)$_2$]-Rh(NBD) [141], [NBD = norbornadiene]; and (η^5-C_5H_5)Fe[η^5-C_5H_4-C(O)-CHC(O)CF_3]Rh(CO)(PPh_3) [142].

7.2.6.3 Iridium Complexes

The ferrocene-iridium complexes, analogs of the rhodium complexes illustrated in Scheme 7-26, are shown in Scheme 7-28.

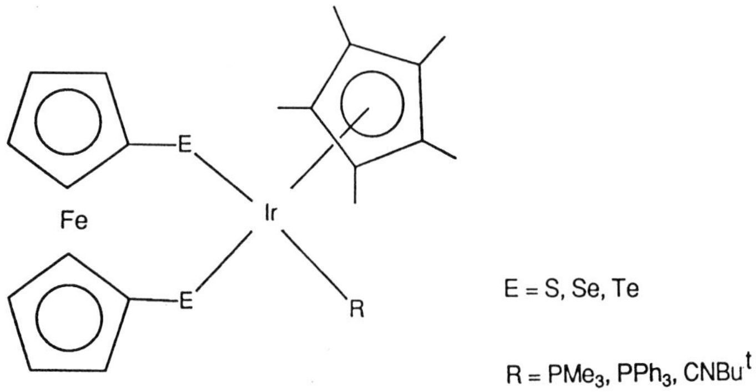

E
Fe
Ir
E
R
E = S, Se, Te
R = PMe3, PPh3, CNBut

Scheme 7-28

Like the corresponding rhodium complexes, the present iridium species undergo two successive one-electron oxidations. As illustrated in Fig. 7-44 for [Fe(η^5-C_5H_4S)$_2$]Ir(η^5-C_5Me_5)(PMe_3), in these complexes loss of the second electron is also generally reversible [136]. The redox potentials of these steps are reported in Table 7-27.

5 uA
+1.100
+0.5
+0.0
-0.500
E (Volt)

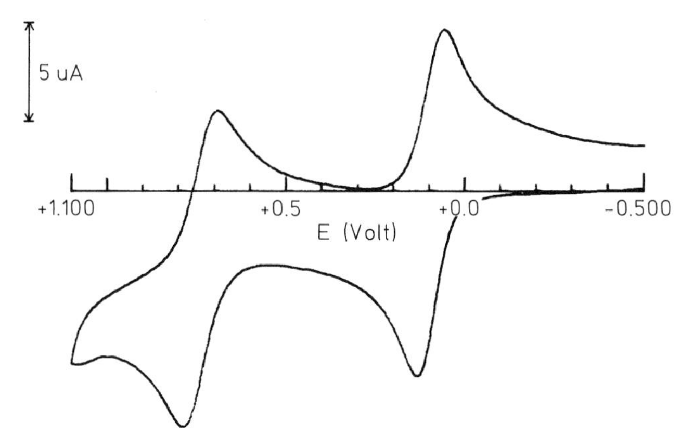

Fig. 7-44. Cyclic voltammogram recorded at a platinum electrode on a dichloromethane solution of [Fe(C_5H_4S)$_2$]Ir(C_5Me_5)(PMe_3). Scan rate 0.1 V s^{-1}.

Table 7-27. Formal electrode potentials (*vs.* SCE) for the oxidation processes exhibited by the complex $[Fe(C_5H_4E)_2]Ir(C_5Me_5)(R)$ in dichloromethane solution [136]

E	R	$E^{\circ\prime}$ (0/+), V	$E^{\circ\prime}$ (+/2+), V
S	PMe_3	+0.10	+0.74
	PPh_3	+0.09	+0.74
	CN*t*Bu	+0.05	+0.60
Se	PMe_3	+0.09	+0.64
	PPh_3	+0.12	+0.62
	CN*t*Bu	+0.04[a]	+0.56[a]
Te	PMe_3	+0.00[a]	+0.44[a]
	PPh_3	−0.02[a]	+0.69[a]

a Peak potential value for irreversible processes.

It is clear that only minor variations in redox potential are induced by changing the R ligand linked to the iridium atom, as well as going from iridium to rhodium. This would further confirm that the sites of the electron removals do not correspond with the metal centers.

The crystal structures of $[\{Fe(\eta^5\text{-}C_5H_4PPh_2)_2\}_2Ir][BPh_4]$ [143] and $[\{Fe(\eta^5\text{-}C_5H_4PPh_2)_2\}Ir(COD)][PF_6]$ (COD = 1,5-cyclooctadiene) [144] have been reported.

7.2.7 Group 10 Metal Complexes

7.2.7.1 Nickel Complexes

The open or locked polyamine-ferrocenyl complexes of nickel(II), illustrated in Scheme 7-29, have been prepared recently [145–147].

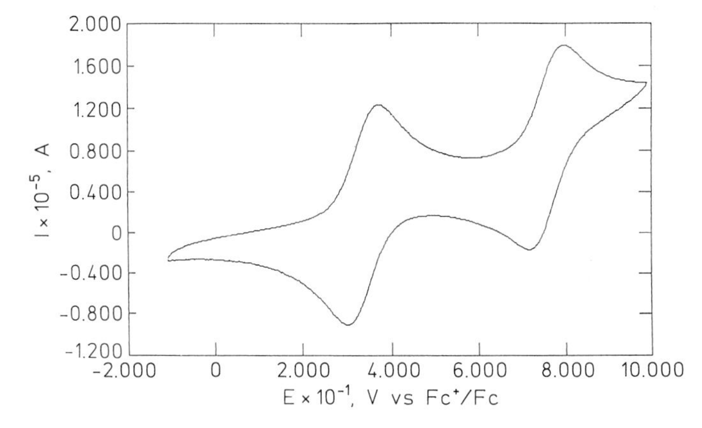

Fig. 7-45. Cyclic voltammogram exhibited at a platinum electrode by complex $[NiL^1]^{2+}$ in MeCN solution. Scan rate 0.1 V s^{-1} (reproduced by permission of Elsevier Sequoia S.A.).

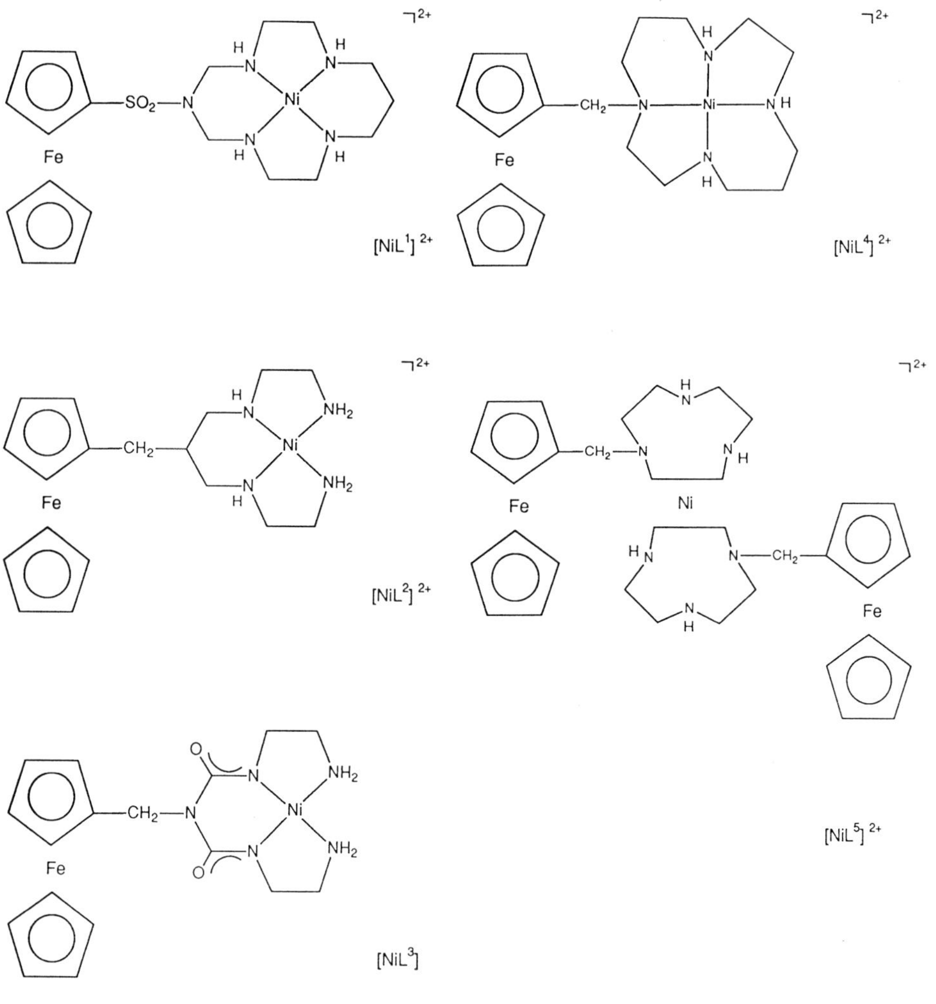

[NiL¹]²⁺ [NiL⁴]²⁺ [NiL²]²⁺ [NiL⁵]²⁺ [NiL³]

Scheme 7-29

Figure 7-45 shows the cyclic voltammetric response exhibited by the heterodinuclear complex $[NiL^1]^{2+}$ in acetonitrile solution [146].

As can be seen, the molecule undergoes two separate one-electron oxidations, reversible in character. The first has been assigned to electron loss from the ferrocenyl unit, the second to the $Ni^{II/III}$ redox change. As summarized in Table 7-28, coupling the ferrocenylsulfonamide fragment to the azacyclam-nickel(II) unit makes electron removal slightly more difficult with respect to the corresponding free molecules.

The same electronic effects hold as far as the complex $[NiL^2]^{2+}$ is concerned. In aqueous solution it undergoes two distinct one-electron oxidations, centered on the ferrocenyl and open polyamine-nickel(II) fragments, respectively [145]. As shown

Table 7-28. Formal electrode potentials (*vs.* SCE) for the redox processes exhibited by the nickel(II) complexes shown in Scheme 7-29 (Fc = $(C_5H_5)Fe(C_5H_4)$]

Complex	$E^{\circ\prime}$ (Fc/Fc$^+$), V	$E^{\circ\prime}$ (Ni$^{II/III}$), V	Solvent	Reference
$[NiL^1]^{2+}$	+0.75	+1.18	MeCN	[146]
$[NiL^2]^{2+}$	+0.51	+1.20	aqueous (pH 7.0)	[145]
$[NiL^3]$	+0.49		aqueous (pH 7.0)	[145]
$[NiL^4]^{2+}$	+0.44	+1.14	MeCN	[147]
$[NiL^5]^{2+}$	+0.45	+1.37	MeCN	[147]
L^2	+0.48		MeCN	[147]
L^3	+0.53		aqueous (pH 7.0)	[145]
L^4	+0.42		MeCN	[147]
$FcSO_2NH_2$	+0.68		MeCN	[146]
$FcSO_2NEt_2$	+0.59		MeCN	[147]
FcH	+0.13		aqueous (pH 7.0)	[148]
	+0.38		MeCN	a

a Unpublished results.

in Table 7-28, electron removal from the ferrocenyl unit is slightly more difficult than that of the uncomplexed ferrocenyl molecule.

As deducible from Table 7-28, oxidation of the ferrocenyl fragment in the nickel complex $[NiL^3]$ is unexpectedly easier than that of the uncomplexed ferrocenyl molecule [145]. This effect can be explained in terms of electrostatic arguments. Complexation of the ferrocenyl-dioxotetramino molecule by nickel(II) involves deprotonation. This leads to localization of two negative charges on the carbonyl fragments, which are not completely neutralized by the inclusion of nickel(II). The partially negative charge in the proximity of the ferrocene unit should favor loss of an electron [145].

Finally, Fig. 7-46 shows the redox behavior of the trinuclear complex $[NiL^5]^{2+}$ in acetonitrile solution [147].

A first single-step two-electron oxidation is followed by a second one-electron oxidation. The first step is assigned to the simultaneous removal of one electron from the two non-interacting ferrocenyl fragments, whereas the second step involves the Ni(II)/Ni(III) redox changes. The first process has features of chemical reversibility, whereas the second nickel-based step is complicated by subsequent decomposition reactions.

Figure 7-47 shows the molecular structure of the diphenylfosphinoferrocene-nickel complex $[Fe(\eta^5\text{-}C_5H_4PPh_2)_2]NiCl_2$ [150]. The nickel atom assumes an essentially tetrahedral coordination. The cyclopentadienyl rings of the ferrocene group are parallel and nearly eclipsed.

It has been briefly reported that in 1,2-dichloroethane solution $[Fe(C_5H_4PPh_2)_2]$-$NiCl_2$ undergoes an irreversible oxidation [151]. In this context, it must be kept in mind that the free ligand 1,1′-bis(diphenylphosphino)ferrocene itself undergoes a one-electron oxidation followed by fast chemical complications [53].

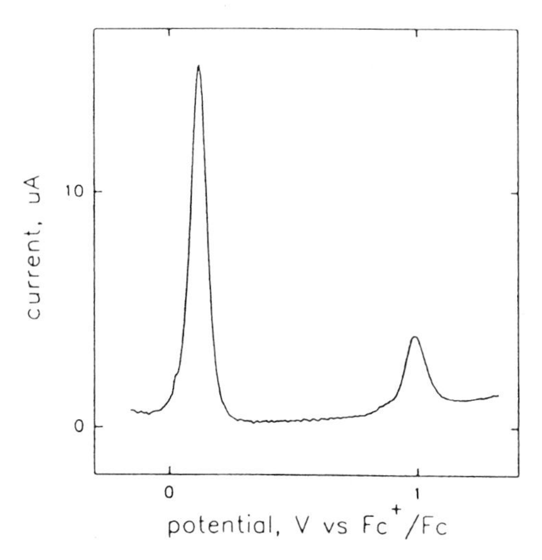

Fig. 7-46. Differential pulse voltammogram recorded at a platinum electrode on a MeCN solution of $[NiL^5]^{2+}$. Scan rate 0.005 V s^{-1}; pulse amplitude 10 mV (reproduced by permission of the American Chemical Society).

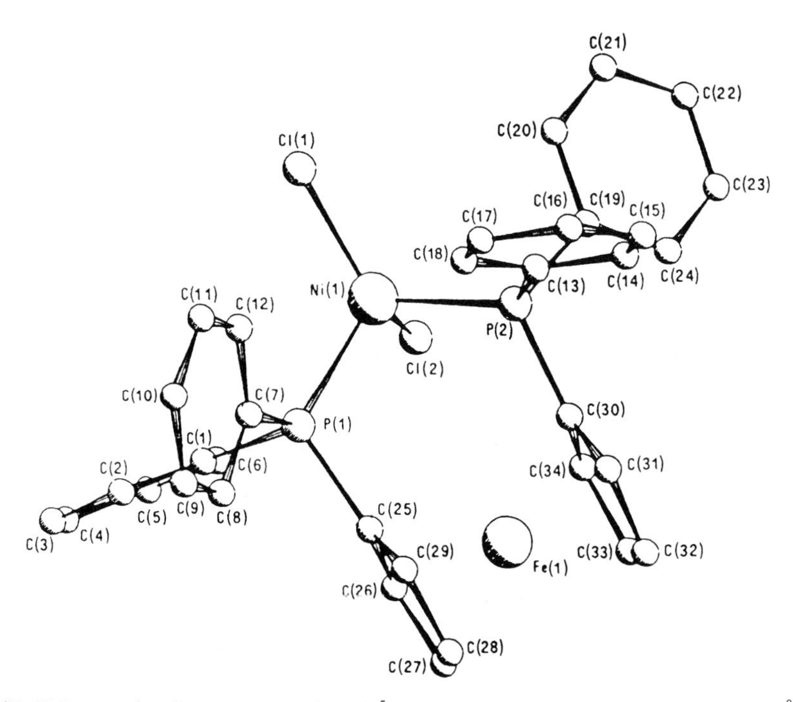

Fig. 7-47. X-Ray molecular structure of $[Fe(\eta^5\text{-}C_5H_4PPh_2)_2]NiCl_2$. Ni-Cl, average 2.22 Å; Ni-P, average 2.31 Å; P-C (Cp ring), average 1.83 Å (reproduced by permission of Plenum Publishing Corporation).

As illustrated in Fig. 7-48, in CH_2Cl_2 the nickel complex $(C_5H_5)Fe(C_5H_4$-$C(O){=}CHPPh_2)Ni(C_5Ph_5)$, shown in Scheme 7-30, undergoes either two reversible one-electron oxidations ($E^{\circ\prime}$ Fe(II)/Fe(III) = +0.46 V; $E^{\circ\prime}$ Ni(II)/Ni(III) = +0.68 V), or a one electron reduction ($E^{\circ\prime}$ Ni(II)/Ni(I) = −1.29 V) [152].

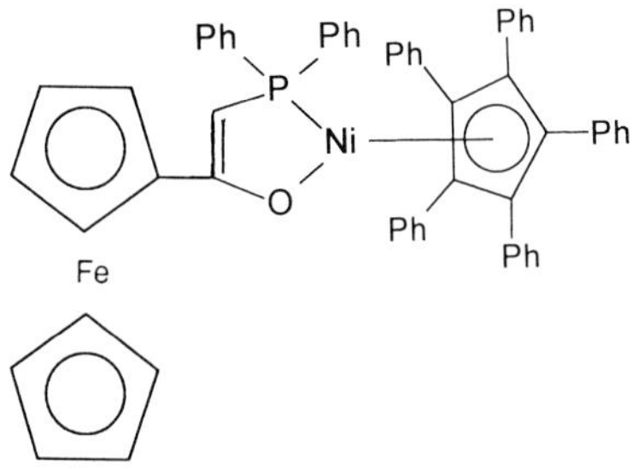

Scheme 7-30

The ferrocene/ferrocenium oxidation occurs at about the same potential as that of free ferrocene ($E^{\circ\prime}$ = +0.45 V), thus indicating that complexation by the nickel fragment does not involve important electronic variations on the ferrocenyl assembly.

In spite of the apparent chemical reversibility of these redox changes, chemical oxidation by one equivalent of $AgBF_4$ afforded the protonated monocation $[(C_5H_5)Fe(C_5H_4$-$C(O)CH_2PPh_2)Ni(C_5Ph_5)]^+$, the structure of which is shown in Fig. 7-49 [153]. It is thought that such a minor chemical complication arises from the fact that the first Fe(II)/Fe(III)-ferrocenyl oxidation is followed by an intramolecular Fe(III)/Fe(II) reduction operated by the chelating P, O system, which undergoes solvent-induced protonation during this electron exchange [153]. The nickel atom assumes a pseudo-square planar coordination through the P, O, C1 atoms and the centre of the C3-C4 bond. The cyclopentadienyl rings of the ferrocene appendix are eclipsed [153].

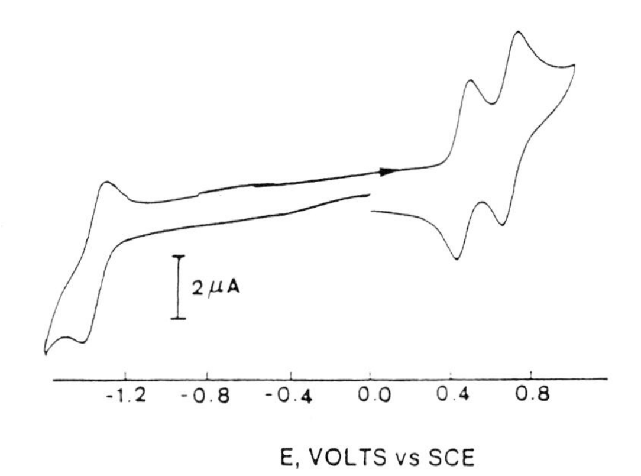

Fig. 7-48. Cyclic voltammogram recorded at a platinum electrode on a CH_2Cl_2 solution of $(C_5H_5)Fe(C_5H_4$-$C(O){=}CHPPh_2)Ni(C_5Ph_5)$. Scan rate 0.1 V s^{-1} (reproduced by permission of the American Chemical Society).

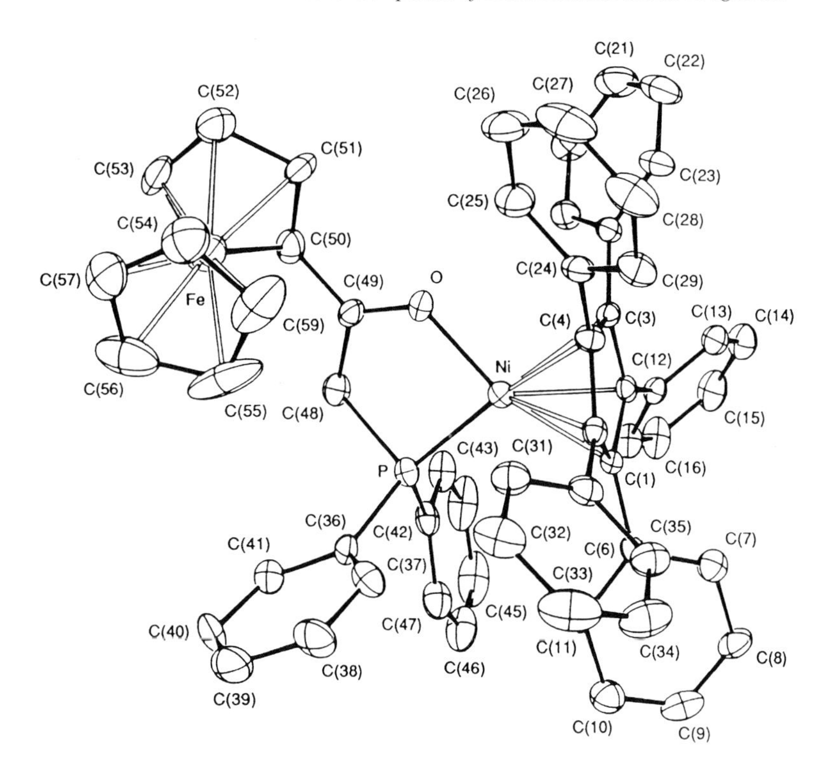

Fig. 7-49. Perspective view of $[(C_5H_5)Fe(C_5H_4\text{-}C(O)CH_2PPh_2)Ni(C_5Ph_5)]^+$. Ni-O, 1.88 Å; Ni-P, 2.17 Å; Ni-C1, 2.07 Å; Ni-C2, 2.18 Å; Ni-C3, 2.14 Å; Ni-C4, 2.16 Å; Ni-C5, 2.17 Å (reproduced by permission of the Royal Society of Chemistry).

In DMF the nickel complex $[(\eta^5\text{-}C_5H_5)Fe(\eta^5\text{-}C_5H_4N{=}CHC_6H_4O)]_2Ni$ undergoes a ferrocenyl-based, two-electron oxidation at $E^{\circ\prime} = +0.45$ V, which is slightly favored over that of the corresponding cobalt complex shown in Scheme 7-20 [119].

Analogously, the complex $[(\eta^5\text{-}C_5H_5)Fe(\eta^5\text{-}C_5H_4CHN(N){=}C(S)NH_2)]_2Ni$, an analog of the cobalt complex shown in Scheme 7-21, undergoes in acetonitrile solution a ferrocene-centered two-electron oxidation at +0.48 V. In DMF it also undergoes a nickel-centered, irreversible, two-electron reduction at −1.34 V. The same conclusions drawn for the corresponding cobalt complex hold [120].

Finally, Fig. 7-50 shows the crystal structure of bis[2-(ferrocenylmethyleneamino)-benzenethiolato]nickel(II) [149]. The square planar geometry of the nickel atom is slightly distorted by the bulkiness of the ligands. One of the ferrocenyl units (B) assumes an eclipsed conformation, whereas the other (D) slightly rotates its cyclopentadienyl rings.

In acetonitrile solution, such a complex displays two closely spaced reversible one-electron oxidations ($E^{\circ\prime} = +0.56$ V and +0.62 V, respectively) [149]. Based on the redox activity shown by bis[2-(phenylmethyleneamino)benzenethiolato-nickel(II), which oxidizes at +0.71 V, the authors attribute the two steps to loss of electrons from both the ferrocenyl and the nickel(II) centres. Indeed, the data reported do

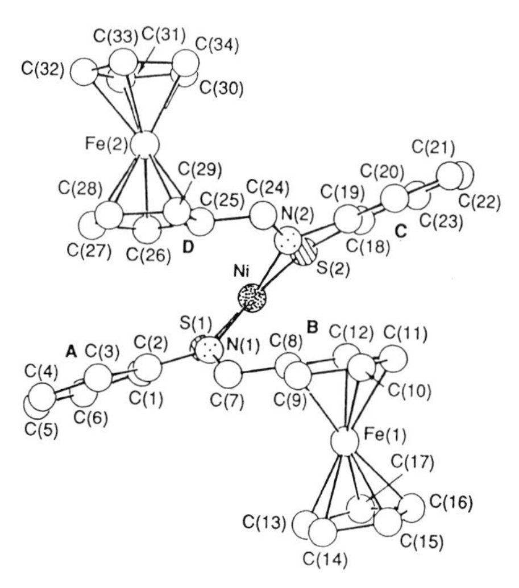

Fig. 7-50. Crystal structure of $\{(\eta^5\text{-}C_5H_5)Fe[\eta^5\text{-}C_5H_4\text{-}CH{=}N(C_6H_4)S]\}_2Ni$. Ni-S, average 2.16 Å; Ni-N, average 1.93 Å (reproduced by permission of the Royal Society of Chemistry).

not allow us to neglect the hypothesis that both oxidation steps are centered on the two (communicating) ferrocenyl units. It is, however, evident that the electron loss from the ferrocene fragments of the complex is more difficult than from free ferrocene.

The crystal structures of $[Fe(\eta^5\text{-}C_5H_4PPh_2)_2]NiBr_2$ [47] and $[Fe(\eta^5\text{-}C_5H_4AsMe_2)_2]\text{-}NiI_2(CO)$ [154] have been reported.

7.2.7.2 Palladium Complexes

Figure 7-51 shows the molecular structure of dichloro[1,1′-bis(diphenylphosphino)ferrocene]palladium(II) [51, 155]. The palladium atom possesses a square planar coordination. The cyclopentadienyl rings of the ferrocene group are nearly staggered and tilted from the parallel disposition by 6.2° [51, 155].

As illustrated in Fig. 7-52, in the related 1-[1,1′-bis(diphenylphosphino)ferrocene]-palladatetraborane complex, the palladium atom still possesses a pseudo-square-planar environment, two bonds being formed with the phosphorus atoms and the two remaining bonds with the triborane ligand. The cyclopentadienyl rings of the sandwich fragment are parallel and nearly staggered [156].

Such bis(diphenylphosphino)ferrocene-palladium complexes commonly undergo a reversible one-electron oxidation, centered on the ferrocene moiety, and an irreversible one-electron reduction, centered on the palladium fragment. The relevant redox potentials are reported in Table 7-29, together with those of related complexes. It must be noted that the ferrocene-based one-electron oxidation leads to ferrocenium-palladium complexes that are more stable than the free diphenylphosphinoferrocenium ion.

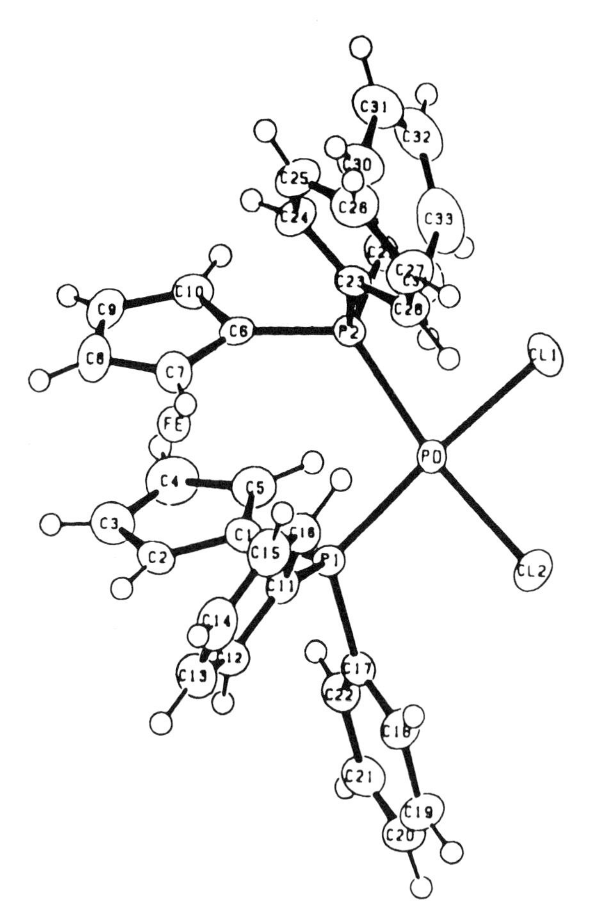

Fig. 7-51. Crystal structure of $[Fe(\eta^5\text{-}C_5H_4PPh_2)_2]PdCl_2$. Pd-Cl, average 2.35 Å; Pd-P, average 2.29 Å; P-C (Cp ring), average 1.81 Å (reproduced by permission of the American Chemical Society).

Substitution of the two diphenylphosphino substituents for thiolate groups affords a series of complexes, the structure of which is typically represented by that of dichloro[1,1′-bis(*i*butylsulfido)ferrocene]palladium(II) shown in Fig. 7-53 [10]. The assembly is similar to that of the preceding bis(diphenylphosphino) complexes, except for the eclipsed and less tilted (1.9°) conformation of the cyclopentadienyl rings of the ferrocene group.

These complexes also undergo both a one-electron oxidation and a one-electron reduction [10]. Table 7-30 summarizes the relevant redox potentials.

It is clearly evident that, on complexation to palladium, oxidation of the ferrocene group becomes significantly more difficult.

An interesting structural-electrochemical relationship arises from the two complexes shown in Scheme 7-31 [157].

Spectroscopic measurements have shown that a dative Fe-Pd bond is present in the 1,1′-bis(isobutylthio)ferrocene cation [158, 159]. This cation exhibits spec-

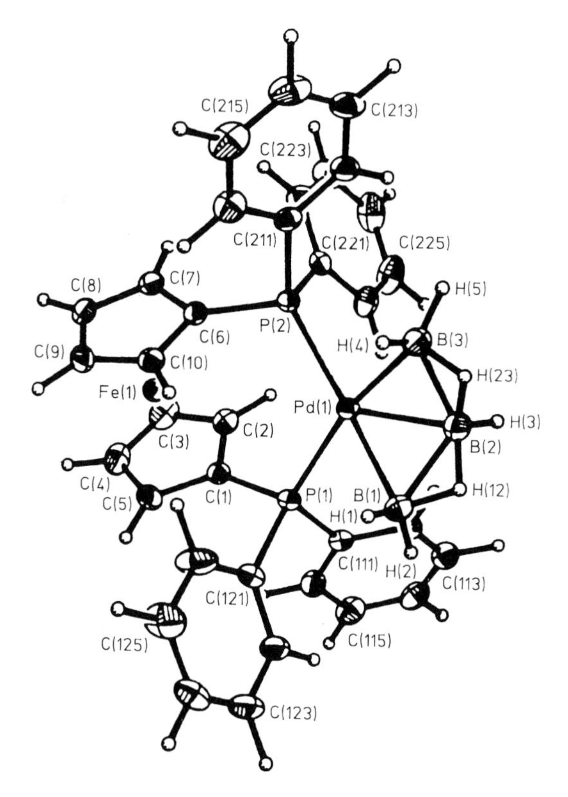

Fig. 7-52. Perspective view of 1-[$Fe(\eta^5\text{-}C_5H_4PPh_2)_2$]$PdB_3H_7$. Pd-P, average 2.37 Å; Pd-B1, Pd-B3, 2.19 Å; Pd-B2, 2.15 Å; Fe···Pd, 4.27 Å (reproduced by permission of the American Chemical Society).

Table 7-29. Formal electrode potentials (*vs.* SCE) for the redox processes exhibited by some bis(diphenylphosphino)ferrocene-palladium complexes

Complex	$E^{\circ\prime}$ (Fc/Fc^+), V	E_p (Pd^{II}/Pd^{I}), V	Solvent	Reference
[$Fe(C_5H_4PPh_2)_2$]$PdCl_2$	+1.03	—	$C_2H_4Cl_2$	[151]
	+1.03	−1.31	CH_2Cl_2	[10]
	+0.95	—	MeCN	[156]
[$Fe(C_5H_4PPh_2)_2$]PdB_3H_7	+1.07[a]	—	MeCN	[156]
$Fe(C_5H_5)_2$	+0.46		$C_2H_4Cl_2$	b
	+0.45		CH_2Cl_2	b
	+0.38		MeCN	b
$Fe(C_5H_4PPh_2)_2$	+0.64[a]		$C_2H_4Cl_2$	[151]
	+0.61[a]		CH_2Cl_2	b
	+0.58		MeCN	[156]
	+0.59[a,c]		MeCN	b

a Peak potential value for irreversible processes.
b Unpublished results.
c Poorly soluble.

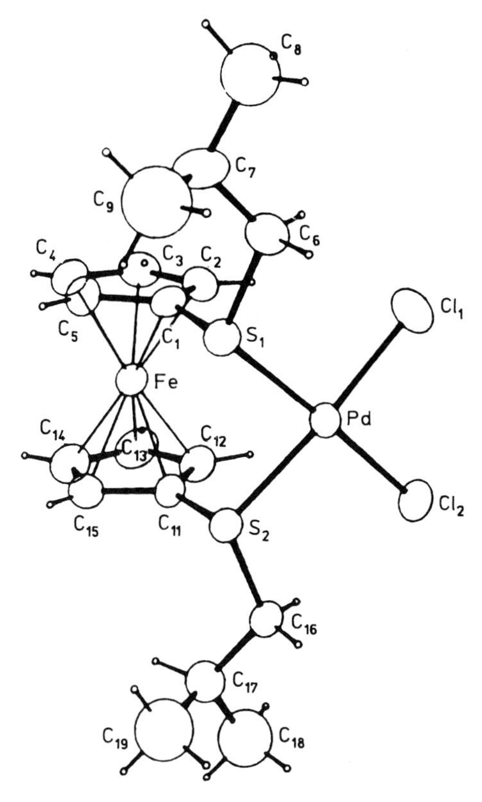

Fig. 7-53. X-Ray molecular structure of $[Fe(\eta^5\text{-}C_5H_4S\text{-}i\text{Bu})_2]PdCl_2$. Pd-Cl, 2.30 Å; Pd-S, average 2.32 Å; S-C (Cp ring), average 1.76 Å (reproduced by permission of the American Chemical Society).

Table 7-30. Formal electrode potentials (*vs.* SCE) for the redox processes exhibited by some ferrocenyl sulfide-palladium complexes in dichloromethane solution [10]

Complex	$E^{\circ\prime}$ (Fc/Fc$^+$), V	E_p (Pd^{II}/Pd^{I}), V
$[Fe(C_5H_4S\text{-}i\text{Bu})_2]PdCl_2$	+1.04	−0.90
$[Fe(C_5H_4S\text{-}Me)_2]PdCl_2$	+0.97	−0.77
$[Fe(C_5H_4S\text{-}Ph)_2]PdCl_2$	+0.96	−1.56
$Fe(C_5H_4S\text{-}i\text{Bu})_2$	+0.44	
$Fe(C_5H_4S\text{-}Me)_2$	+0.46	
$Fe(C_5H_4S\text{-}Ph)_2$	+0.63	

troscopic properties quite similar to those of the 1,1′-bis(diphenylphosphino)ferrocene complex $[\{Fe(C_5H_4\text{-}PPh_2)_2\}Pd(PPh_3)]^{2+}$ [160], the structure of which is illustrated in Fig. 7-54 in comparison with that of the 1,10-dithiapyridino-ferrocenophane cation. This last complex is dimeric in the solid state and shows a weak dative Fe-Pd interaction [157].

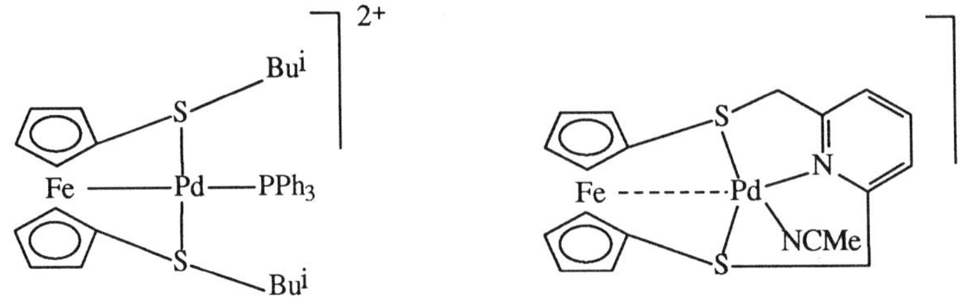

Scheme 7-31

It has been briefly reported that in acetonitrile the bis(isobutylthio)ferrocene complex undergoes the Fc/Fc^+ oxidation at $E_{pa} = +0.98$ V and the Pd(II)/Pd(I) reduction at $E_{pc} = -0.58$ V, whereas the dithiapyridinoferrocene complex is oxidized at $E_{pa} = +0.68$ V ($E^{\circ\prime} = +0.64$ V) and reduced at $E_{pc} = -0.49$ V [157]. As a result it has been argued that the strong dative Fe-Pd in the (isobutylthio)ferrocene dication removes more electron-density from the Fe(II) centre than in the dithiapyridinoferrocene dication, thus making its oxidation more difficult [157].

The electrochemical behavior of the two palladium complexes illustrated in Scheme 7-32 has been reported recently [161]. They contain the ferrocenyl-β-ketophosphine ligand previously mentioned in connection with the rhenium complexes illustrated in Scheme 7-10.

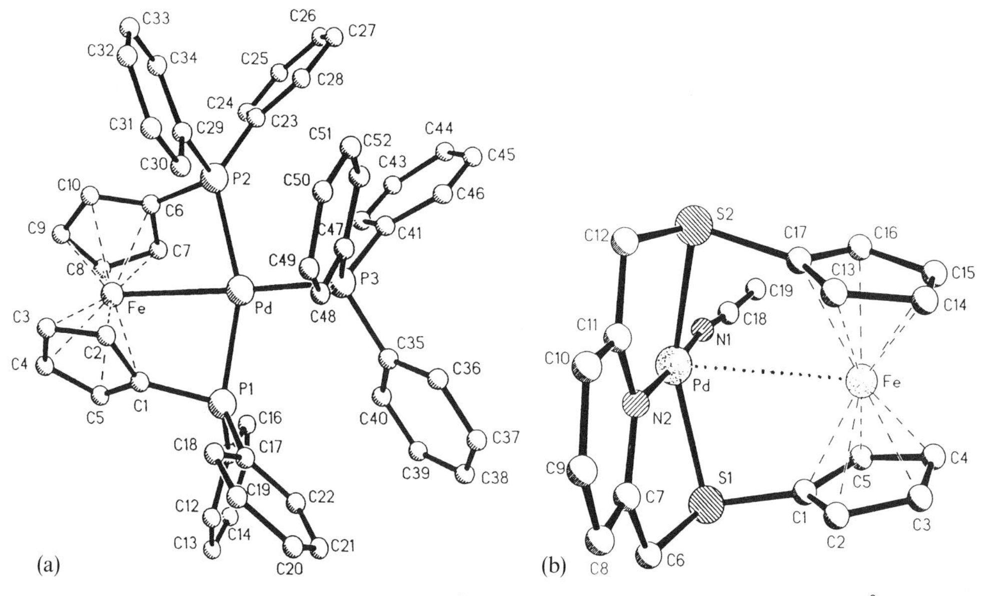

Fig. 7-54. (a) Molecular structure of $[\{Fe(\eta^5\text{-}C_5H_4\text{-}PPh_2)_2\}Pd(PPh_3)]^{2+}$. Fe-Pd, 2.88 Å; Pd-P1, 2.29 Å; Pd-P2, 2.32 Å; Pd-P3, 2.27 Å. (b) Side view of one molecule of the dimer (acetonitrile) {1.10-dithia[2](2,6)pyridino[2](1,1′)ferrocenophane}. (Pd-Pd, 3.28 Å). Fe-Pd, 3.23 Å; Pd-S, 2.28 Å and 2.32 Å; Pd-N (py), 1.91 Å; Pd-N (MeCN), 2.00 Å (reproduced by permission of Elsevier Sequoia S.A.).

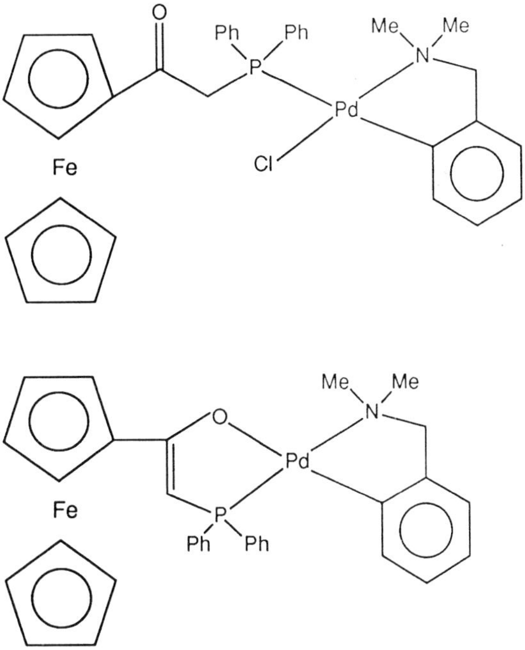

Scheme 7-32

Their redox properties are illustrated in Fig. 7-55. As can be seen, they undergo a first ferrocene-based one-electron oxidation, with features of chemical reversibility, followed by a second palladium-centered anodic process, which is coupled to subsequent chemical complications. Indeed, in the longer times of macroelectrolysis, the first ferrocene-centered oxidation steps are also complicated by subsequent reactions, which, in both cases, ultimately afford the chelate monocation shown in Scheme 7-33 [161]. The redox potentials are reported in Table 7-31.

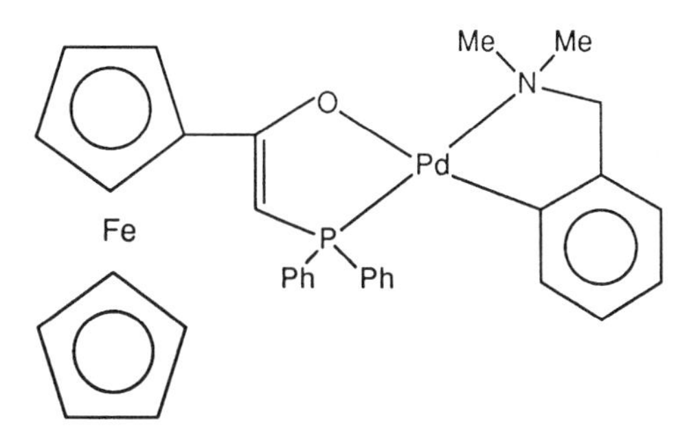

Scheme 7-33

In both cases, it is apparent that complexation to palladium increases the electron density on the ferrocenylketophosphine fragment.

Scheme 7-34 shows a series of palladium complexes of dimethylaminomethylferrocene, the electrochemical behavior of which has been reported [162].

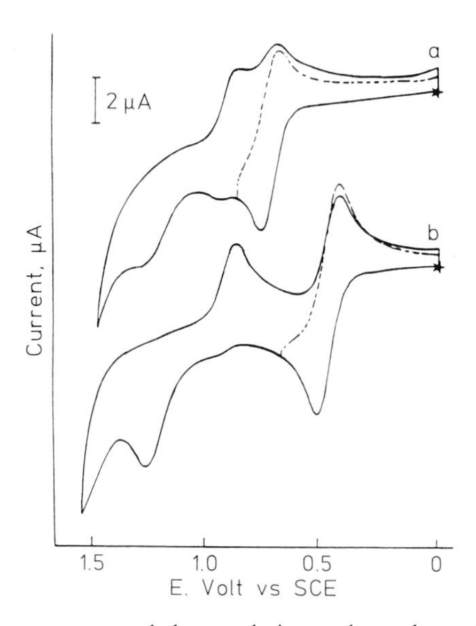

Fig. 7-55. Cyclic voltammograms recorded at a platinum electrode on dichloromethane solution of: (a) $(C_5H_5)Fe\{[C_5H_4\text{-}C(O)CH_2PPh_2]Pd(Cl)(o\text{-}C_6H_4CH_2NMe_2)\}$; (b) $(C_5H_5)Fe\{[C_5H_4\text{-}C(O){=}CHPPh_2]Pd(o\text{-}C_6H_4CH_2NMe_2)\}$. Scan rate 0.1 V s^{-1} (reproduced by permission of Elsevier Sequoia S.A.).

Also in this case, a first ferrocenyl-based one-electron oxidation, reversible in character, has been observed, followed by a more anodic irreversible oxidation, attributed to the oxidation of the palladium fragment [162]. The relevant redox potentials are summarized in Table 7-32.

Whereas the palladium complexation of dimethylaminomethyl-ferrocene renders the ferrocenyl oxidation slightly more difficult, it seems that the presence of electron-donating substituents X, L plays the major role in determining the accessibility of the ferrocenyl oxidation.

Figure 7-56 illustrates the molecular structure of *bis*[2-(ferrocenylmethyleneamino)benzenethiolato]palladium(II) [149].

Table 7-31. Formal electrode potentials (*vs.* SCE) for the redox processes exhibited by the palladium-ferrocenylketophosphine complexes illustrated in Scheme 7-31. Dichloromethane solution [161]

Complex	$E^{\circ\prime}$ (Fc/Fc^+), V	E_p (Pd^{II}/Pd^{III}), V
$(C_5H_5)Fe\{[C_5H_4\text{-}C(O)CH_2PPh_2]Pd(Cl)\cdot(o\text{-}C_6H_4CH_2NMe_2)\}$	+0.72	1.25
$(C_5H_5)Fe\{[C_5H_4\text{-}C(O){=}CHPPh_2]Pd\cdot(o\text{-}C_6H_4CH_2NMe_2)\}$	+0.47	1.25
$(C_5H_5)Fe[C_5H_4\text{-}C(O)CH_2PPh_2]$	+0.80[a]	

a Complicated by relatively fast reactions.

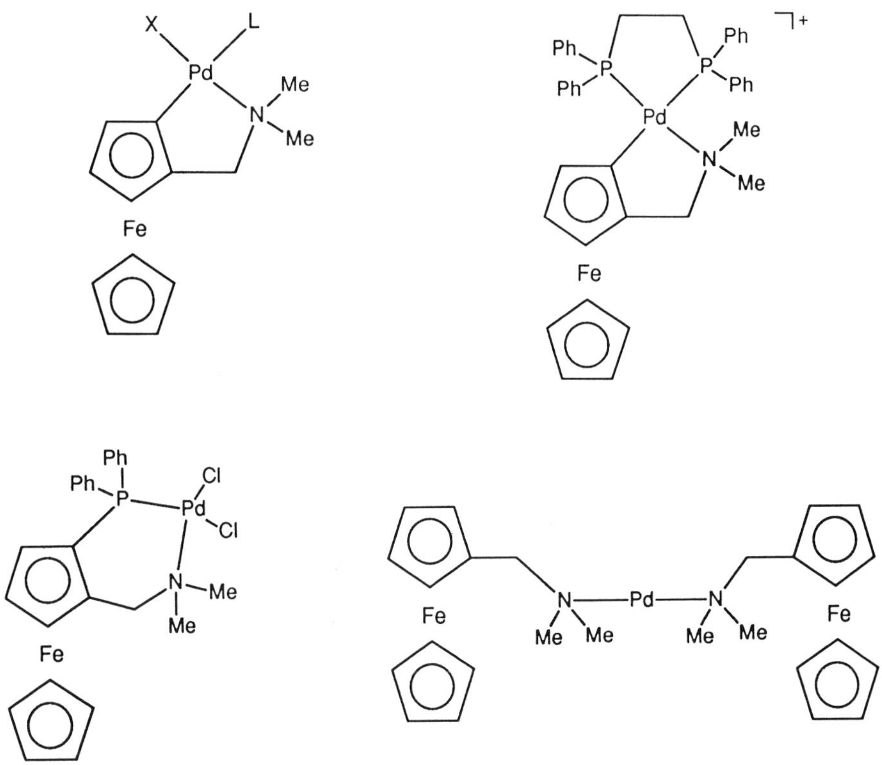

Scheme 7-34

Table 7-32. Formal electrode potentials (*vs.* SCE) for the first one-electron oxidation exhibited by the palladium-dimethylaminomethylferrocenyl complexes illustrated in Scheme 7-34. All the complexes show a further irreversible process at about +1.8 V. Dichloromethane solution [162]

Complex		$E^{\circ\prime}$ (Fc/Fc$^+$), V
$(C_5H_5)Fe[C_5H_3CH_2NMe_2]Pd(X)(L)$		
X = Cl	L = PPh_3	+0.23
	L = $AsPh_3$	+0.24
	L = $P(OPh)_3$	+0.27
	L = $P(OMe)_3$	+0.22
	L = $P(OEt)_3$	+0.23
	L = C_5H_5N	+0.26
X = I	L = PPh_3	+0.22
X = SCN	L = PPh_3	+0.23
$[(C_5H_5)Fe[C_5H_3CH_2NMe_2]Pd(Ph_2PCH_2CH_2PPh_2)]^+$		+0.33
$(C_5H_5)Fe[(C_5H_3)(1\text{-}CH_2NMe_2)(2\text{-}PPh_2)]Pd(Cl)_2$		+0.79
$[(C_5H_5)Fe(C_5H_4CH_2NMe_2)]_2Pd$		+0.49
$(C_5H_5)Fe(C_5H_4CH_2NMe_2)$		+0.43[a]

a Unpublished results.

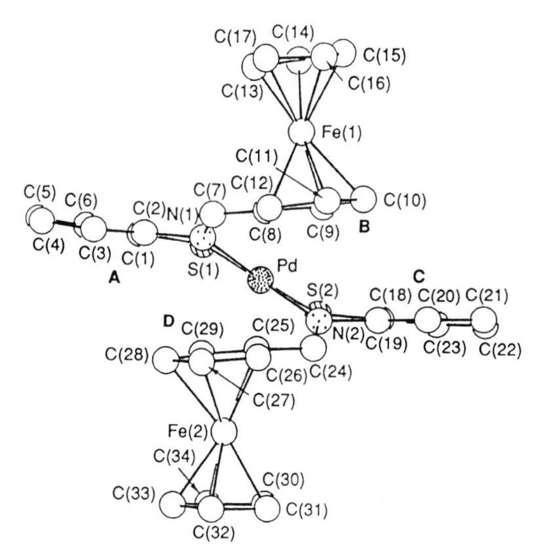

Fig. 7-56. Crystal structure of $\{(\eta^5\text{-}C_5H_5)Fe[\eta^5\text{-}C_5H_4\text{-}CH{=}N(C_6H_4)S]\}_2Pd$. Pd-S, average 2.26 Å; Pd-N, average 2.07 Å (reproduced by permission of the Royal Society of Chemistry).

Analogously to the corresponding nickel(II) complex previously discussed, the palladium atom lies in a square planar environment.

As shown in Fig. 7-57, like the nickel(II) analogue, in acetonitrile solution this complex undergoes two close oxidation steps, with features of chemical reversibility [149].

Both the two electron removals are thought to be centered on the electronically interacting ferrocenyl units ($E^{\circ\prime}$ = +0.60 V and +0.79 V, respectively) [149].

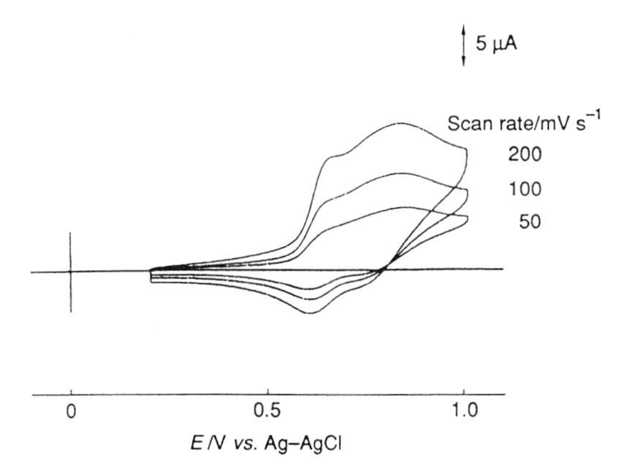

Fig. 7-57. Cyclic voltammetric responses recorded at a glassy-carbon electrode on a MeCN solution of $\{(C_5H_5)Fe[C_5H_4\text{-}CH{=}N(C_6H_4)S]\}_2Pd$ (reproduced by permission of the Royal Society of Chemistry).

Scheme 7-35 shows the probable structure of two complexes, the electrochemical behavior of which has been reported recently [163].

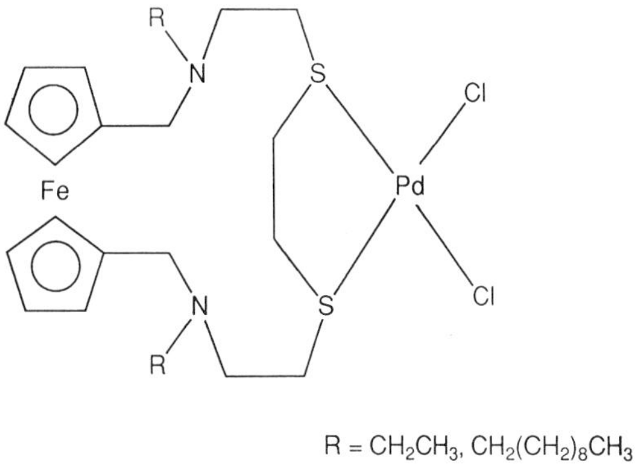

Scheme 7-35 $R = CH_2CH_3, CH_2(CH_2)_8CH_3$

In CH_2Cl_2 no appreciable difference exists between the redox potential of the ferrocene-centered one-electron oxidation of the *N*-decyl substituted complex ($E^{\circ\prime} = +0.20$ V) and the *N*-ethyl substituted complex ($E^{\circ\prime} = +0.21$ V). A rather significant role is played by palladium complexation, in that the one-electron removal for both of the two complexes becomes more difficult by about 80 mV than those of the corresponding free diazadithiaferrocenophane molecules [163].

The two dimeric complexes illustrated in Scheme 7-36 are related to the diphenylphosphino-palladium complexes above described.

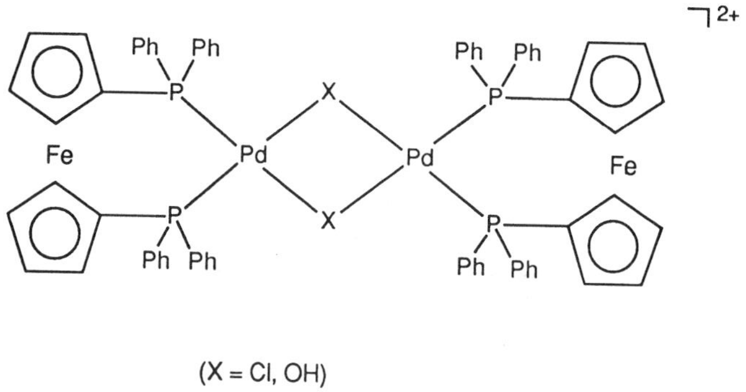

Scheme 7-36 (X = Cl, OH)

In 1,2-dichloroethane solution these complexes exhibit a single-step two-electron oxidation (X = Cl, $E^{\circ\prime}$ (2+/4+) = +1.19 V; X = OH, $E^{\circ\prime}$(2+/4+) = +1.13 V) [164]. The appearance of a single oxidation step and the similarity in redox potentials between the monoferrocene and the diferrocene complexes both suggest that the two ferrocene fragments are non-interacting.

The dimeric palladium complex of dimethylaminomethylferrocene, shown in Scheme 7-37, also displays a single-step two-electron oxidation of the ferrocenyl fragment in CH_2Cl_2 ($E^{\circ\prime} = +0.27$ V) [162].

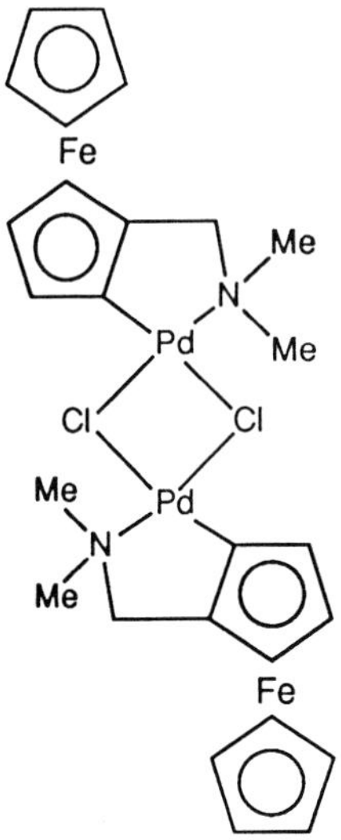

Scheme 7-37

The dipalladium complex of 1,1′-*bis*(3-methyl-5-oxo-2-aza-3-hexenyl)ferrocene (illustrated in Scheme 7-38) exhibits in acetonitrile solution a ferrocene-centered, reversible one-electron oxidation ($E^{\circ\prime} = +0.40$ V) [165]. The fact that the corresponding free ferrocenediyl molecule is oxidized at slightly higher potential ($E^{\circ\prime} = +0.45$ V) indicates that the palladium fragments donate electron density to the central ferrocene moiety.

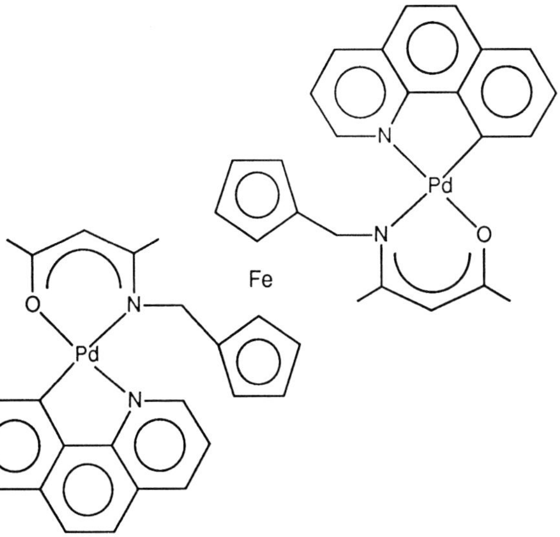

Scheme 7-38

A final ferrocenyl-dipalladium(0) complex studied from the electrochemical viewpoint is shown in Scheme 7-39 [166].

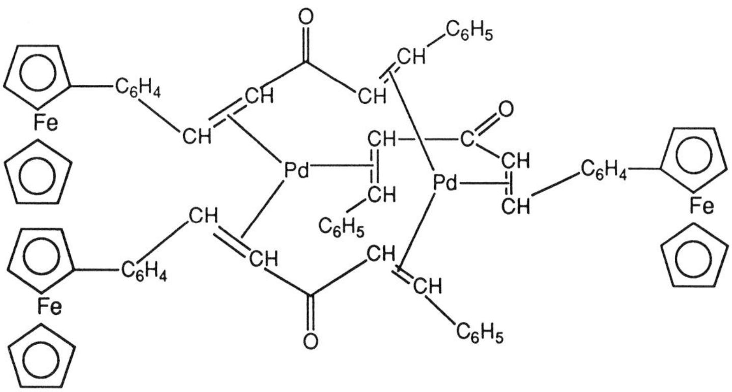

Scheme 7-39

The crystal structure of this complex is not known, but the structure of the tris(dibenzylideneacetone)dipalladium(0) precursor has been reported [167, 168]. The two palladium atoms have a trigonal planar geometry arising from bonding to the olefinic groups of the three ligands.

In dichloromethane the dipalladium complex exhibits a ferrocenyl-based reversible oxidation ($E^{\circ\prime} = +0.62$ V) and a palladium-centered irreversible oxidation ($E_p = +1.33$ V) [166]. Under the same experimental conditions, the 4-ferrocenyl-dibenzylideneacetone ligand displays reversible one-electron oxidation at $E^{\circ\prime} = +0.64$ V, thereby indicating that the complexation by palladium atoms does not appreciably perturb its electronic structure.

X-ray characterized ferrocene-palladium complexes include the following: [Fe(η^5-$C_5H_4S)_2$]Pd(PPh_3) [169, 170]; {(η^5-$C_5H_4PPh_2$)Fe[η^5-C_5H_3(1-CHMeNMe_2)(2-PPh_2)]}PdCl_2 [171]; {(η^5-C_5H_5)Fe[η^5-C_5H_3(CH=N$(CH_2)_2$Ph)]}PdCl(PEt_3) [172]; {Fe[η^5-C_5H_3(1-CHMeNMe_2)(2-PPh_2)]$_2$}PdCl_2 [173]; and [(η^5-C_5H_5)Fe(η^5-C_5H_4-C(O)CH_2PPh_2)]$_2$PdCl_2 [174].

7.2.7.3 Platinum Complexes

Figure 7-58 shows the molecular structure of the platinum-diphenylphosphinoferrocene complex [Fe(η^5-$C_5H_4PPh_2)_2$]PtCl_2. As in the case of the palladium analog, the platinum atom has a square-planar geometry and the cyclopentadienyl rings of the ferrocene group are staggered and significantly tilted (5.9°) from the parallel arrangement [175].

Figure 7-59 compares the anodic behavior of [Fe$(C_5H_4PPh_2)_2$]PtCl_2 with that of the free ferrocenylphosphine Fe$(C_5H_4PPh_2)_2$ [151]. As previously noted for the palladium analog, coordination to the platinum fragment kinetically stabilizes the oxidation product, but simultaneously making the one-electron oxidation of the ferrocene group thermodynamically more difficult.

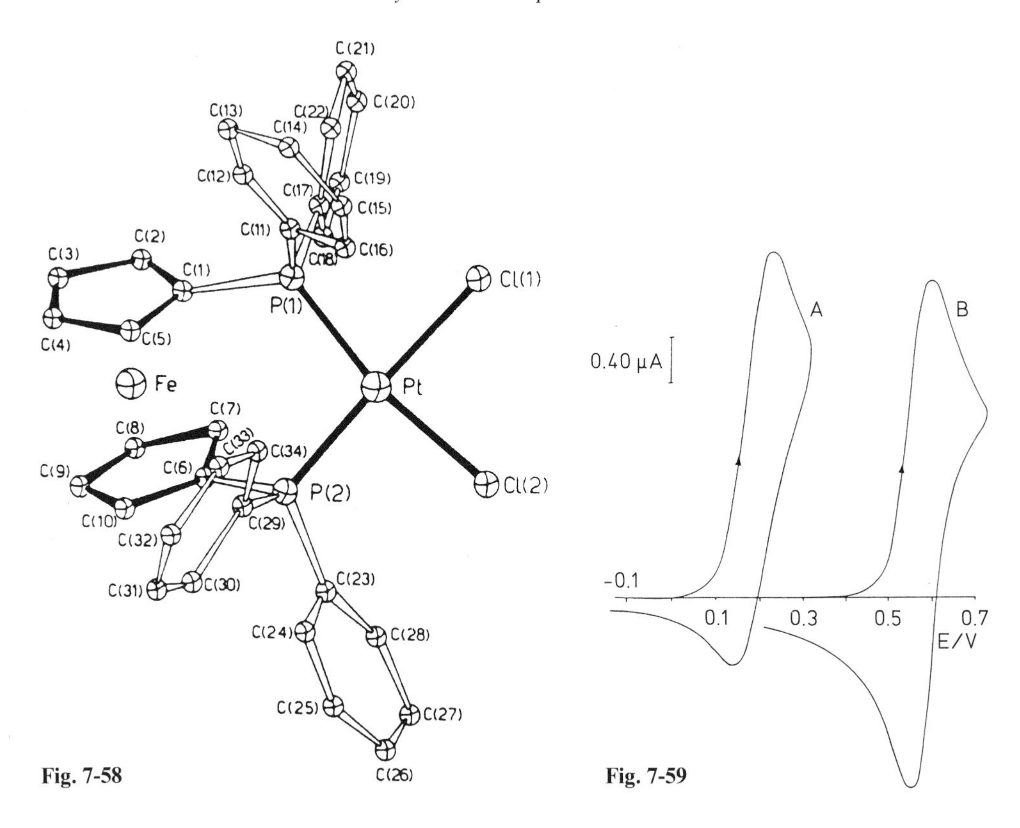

Fig. 7-58

Fig. 7-59

Fig. 7-58. X-Ray structure of $[Fe(\eta^5\text{-}C_5H_4PPh_2)_2]PtCl_2$. Pt-Cl, average 2.40 Å; Pt-P, average 2.26 Å (reproduced by permission of Elsevier Sequoia S.A.).

Fig. 7-59. Cyclic voltammograms recorded at a platinum electrode on 1,2-dichloroethane solutions of: (A) $Fe(C_5H_4PPh_2)_2$; (B) $[Fe(C_5H_4PPh_2)_2]PtCl_2$. Potential values are referred to the Fc/Fc^+ couple (reproduced by permission of Elsevier Sequoia S.A.).

Figure 7-60 shows the structure of the 3-ferrocenylpyridine-dichloroplatinum(II) complex, $[(\eta^5\text{-}C_5H_5)Fe(\eta^5\text{-}C_5H_4\text{-}m\text{-}C_5H_4N)]_2PtCl_2$ [176]. The coordination around the platinum atom is also square-planar in this case. In both ferrocenyl ligands, the cyclopentadienyl rings assume a nearly eclipsed conformation.

In nonaqueous solution such a diferrocenyl complex displays a single-step two-electron oxidation, which indicates that the two ferrocene fragments are electronically independent [176].

The redox potentials for the electron-transfer processes exhibited by the present platinum complexes, together with those of the alkylsulfidoferrocene-platinum derivatives, analog of the previously discussed palladium species, are reported in Table 7-33.

There is no significant variation with respect to the corresponding palladium complexes.

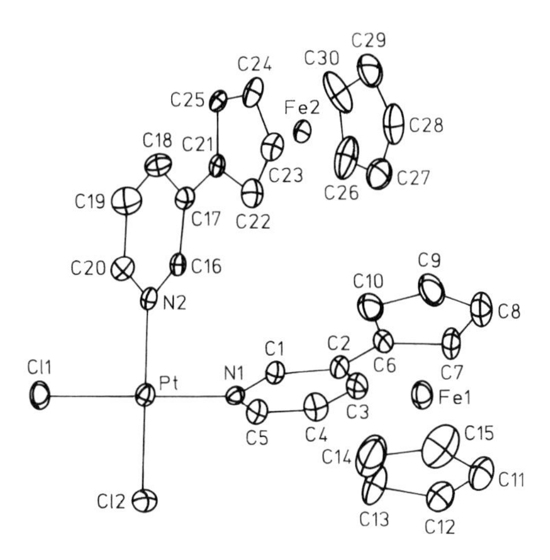

Fig. 7-60. Crystal structure of $[(\eta^5\text{-}C_5H_5)Fe(\eta^5\text{-}C_5H_4\text{-}m\text{-}C_5H_4N)]_2PtCl_2$. Pt-Cl, average 2.30 Å; Pt-N, average 2.02 Å (reproduced by permission of the American Chemical Society).

Table 7-33. Formal electrode potentials (*vs.* SCE) for the redox processes exhibited by some ferrocene-platinum complexes

Complex	$E^{\circ\prime}$ (Fc/Fc^+), V	E_p (Pt^{II}/Pt^{I}), V	Solvent	Reference
$[Fe(C_5H_4PPh_2)_2]PtCl_2$	+1.03	–	$C_2H_4Cl_2$	[151, 175]
	+1.03	–	CH_2Cl_2	[10]
$[(C_5H_5)Fe(C_5H_4\text{-}m\text{-}C_5H_4N)]_2PtCl_2$	+0.55[a]		MeCN	[176]
	+0.54[a]		DMSO	[176]
	+0.59[a]		Me_2CO	[176]
$[Fe(C_5H_4S\text{-}i\text{Bu})_2]PtCl_2$	+1.04	−1.75	CH_2Cl_2	[10]
$[Fe(C_5H_4S\text{-Me})_2]PtCl_2$	+1.04	−1.61	CH_2Cl_2	[10]
$[Fe(C_5H_4S\text{-Ph})_2]PtCl_2$	+1.13	−1.49	CH_2Cl_2	[10]
$Fe(C_5H_5)_2$	+0.46		$C_2H_4Cl_2$	[b]
	+0.45		CH_2Cl_2	[b]
	+0.38		MeCN	[b]
$(C_5H_5)Fe(C_5H_4\text{-}m\text{-}C_5H_4N)$	+0.50		MeCN	[176]
	+0.48		DMSO	[176]
	+0.53		Me_2CO	[176]
$Fe(C_5H_4PPh_2)_2$	+0.64		$C_2H_4Cl_2$	[151]
	+0.61		CH_2Cl_2	[b]
$Fe(C_5H_4S\text{-}i\text{Bu})_2$	+0.44		CH_2Cl_2	[10]
$Fe(C_5H_4S\text{-Me})_2$	+0.46		CH_2Cl_2	[10]
$Fe(C_5H_4S\text{-Ph})_2$	+0.63		CH_2Cl_2	[10]

a Two electron step.
b Unpublished results.

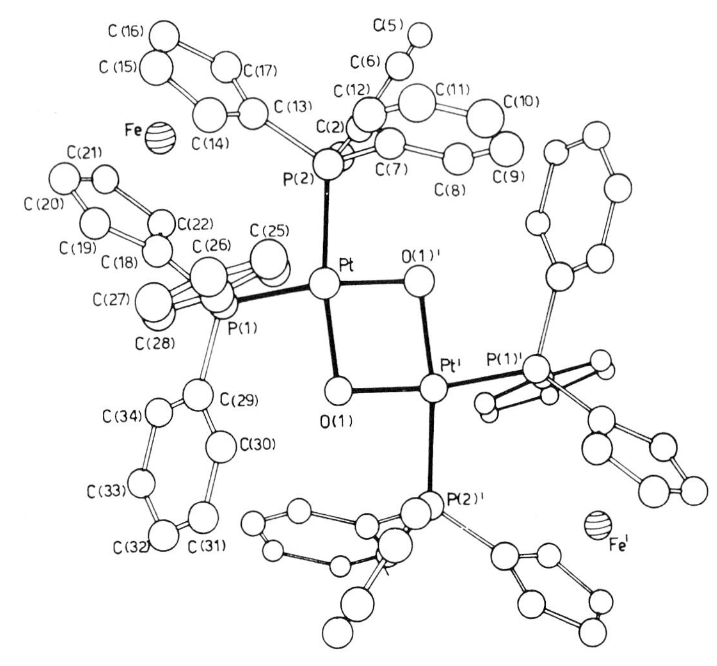

Fig. 7-61. X-ray structure of $[\{Fe(\eta^5\text{-}C_5H_4PPh_2)_2Pt(OH)\}_2]^{2+}$. Pt-O, 2.10 Å; Pt-P, average 2.24 Å; Pt···Pt, 3.23 Å (reproduced by permission of the American Chemical Society).

Figure 7-61 shows the crystal structure of the dimeric platinum complex $[\{Fe(\eta^5\text{-}C_5H_4PPh_2)_2Pt(OH)\}_2]^{2+}$ [164], congener of the palladium species shown in Scheme 7-36.

Each platinum atom maintains a square planar geometry and the whole $P_4Pt_2O_2$ core is also planar. The cyclopentadienyl rings of the two ferrocene units are staggered and deviate by 5.0° from the mutual parallel disposition.

The present oxo-bridged and the similar chloro-bridged platinum complexes both undergo a single-step two-electron oxidation in 1,2-dichloroethane solution at potential values that are coincident with those of the corresponding palladium complexes (X = OH, $E^{\circ\prime}$ $(2+/4+) = +1.13$ V; X = Cl, $E^{\circ\prime}$ $(2+/4+) = +1.18$ V) [164].

Finally, Fig. 7-62 illustrates the molecular structure of one of the two complexes shown in Scheme 7-40, namely the cyclometallated diplatinum complex [177]. Both platinum atoms have a square planar coordination. The cyclopentadienyl rings of the ferrocene fragment are eclipsed and somewhat tilted (4.9°) from the parallel arrangement.

The two complexes shown in Scheme 7-40 exhibit a ferrocene-centered reversible one-electron oxidation in acetone, which is notably easier for $[\eta^5\text{-}PtCl(DMSO)\text{-}Me_2NCH_2C_5H_3]_2Fe$ ($E^{\circ\prime} = -0.12$ V) than for $[\eta^5\text{-}trans\text{-}PtCl_2(DMSO)Me_2NCH_2\text{-}C_5H_4]_2Fe$ ($E^{\circ\prime} = +0.41$ V) [177].

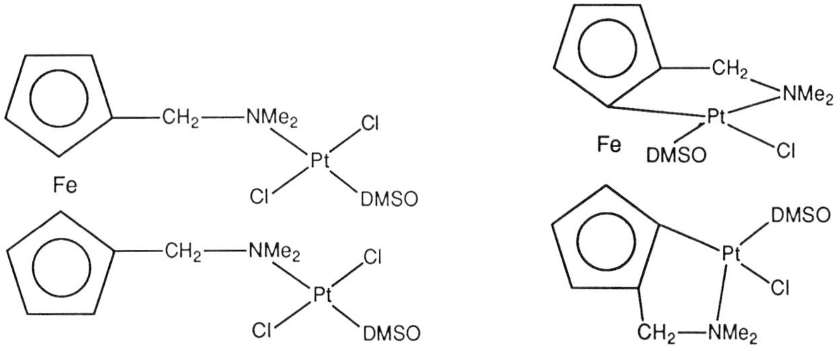

Scheme 7-40

Further platinum-ferrocene complexes that have been characterized by X-ray analysis, but that lack electrochemical studies, include the following: $[Fe(\eta^5\text{-}C_5H_4S)_2]Pt(PPh_3)$ [178]; $[(\eta^5\text{-}C_5H_5)Fe\{\eta^5\text{-}C_5H_3(1\text{-}CHMeNMe_2)[2\text{-}P(CHMe_2)_2]\}]\text{-}PtCl_2$ [179]; $\{Fe(\eta^5\text{-}C_5H_4PPh_2)_2\}PtS_2N_2$ [180]; $[Fe(\eta^5\text{-}C_5H_4Cl)(C_5H_3Cl)]\text{-}2\,Pt(COD)$ (COD = *cis, cis*-cycloocta-1,5-diene) [181]; $\{[Fe(\eta^5\text{-}C_5H_4PPh_2)_2]\text{-}Pt(\mu\text{-}H)(\mu\text{-}CO)Pt[Fe(\eta^5\text{-}C_5H_4PPh_2)_2]\}^+$ [182]; and $[\{Fe(\eta^5\text{-}C_5H_4PPh_2)_2\}_2\text{-}Pt_2H_3]^+$ [183].

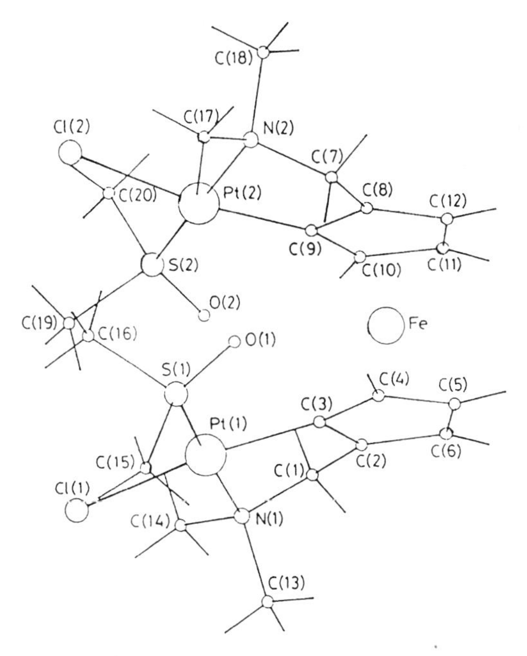

Fig. 7-62. Crystal structure of $[\eta^5\text{-}PtCl(DMSO)Me_2NCH_2C_5H_3]_2Fe$. Pt-Cl, average 2.42 Å; Pt-N, average 2.14 Å; Pt-S, average 2.20 Å; Pt-C (Cp ring), 2.01 Å; Pt···Pt, 4.28 Å (reproduced by permission of the Royal Society of Chemistry).

7.2.8 Group 11 Metal Complexes

7.2.8.1 Copper Complexes

Figure 7-63 shows the molecular structure of the two ferrocenecarboxylato-imidazolato-copper(II) complexes $[(\eta^5\text{-}C_5H_5)Fe(\eta^5\text{-}C_5H_4CO_2)]_2Cu(dmim)_2$ (dmim = 1,2-dimethylimidazole) and $[(\eta^5\text{-}C_5H_5)Fe(\eta^5\text{-}C_5H_4CO_2)]_2Cu(mim)_2$ (mim = *N*-methylimidazole) [184]. In both complexes, the copper atom possesses a CuN_2O_2 square planar coordination. The cyclopentadienyl rings of the ferrocenyl units are nearly eclipsed and deviate only slightly ($\approx 2.5-3°$) from the parallel configuration.

The dimethylimidazolate and the methylimidazolate complexes undergo a chemically reversible two-electron oxidation in dichloromethane at the same potential value ($E^{\circ\prime}$ (0/2+) = +0.56 V), which demonstrates the lack of interaction between the two ferrocenyl centers of these species. Nevertheless, if one considers that free

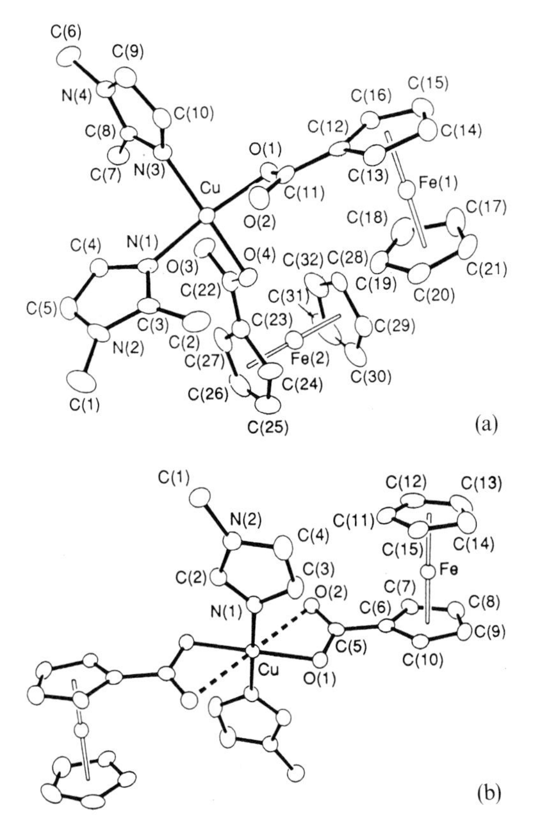

Fig. 7-63. X-ray structures of: (a) $[(\eta^5\text{-}C_5H_5)Fe(\eta^5\text{-}C_5H_4CO_2)]_2Cu(1,2\text{-dimethylimidazole})_2$. Cu-N, average 2.00 Å; Cu-O, average 1.97 Å. (b) $[(\eta^5\text{-}C_5H_5)Fe(\eta^5\text{-}C_5H_4CO_2)]_2Cu(N\text{-methylimidazole})_2$. Cu-N, 1.97 Å; Cu-O1, 1.98 Å; Cu ··· O2, 2.60 Å (reproduced by permission of the Royal Society of Chemistry).

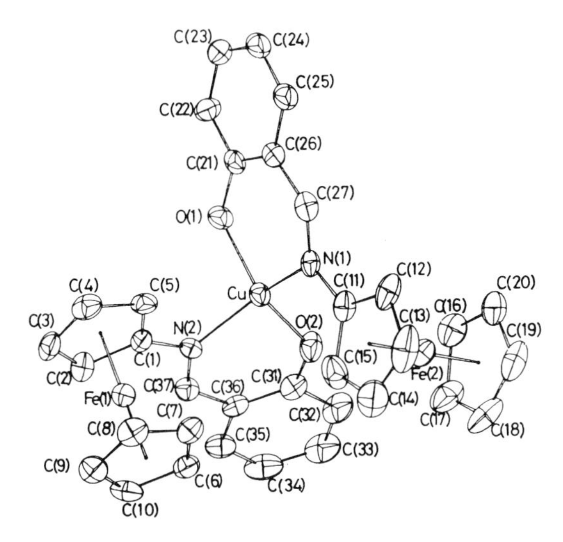

Fig. 7-64. Molecular structure of $[(\eta^5\text{-}C_5H_5)Fe(\eta^5\text{-}C_5H_4N{=}CHC_6H_4O)]_2Cu$. Cu-O, average 1.89 Å; Cu-N, average 1.97 Å; N-C (Cp ring), 1.41 Å (reproduced by permission of the Royal Society of Chemistry).

ferrocenecarboxylic acid undergoes oxidation at $E^{\circ\prime} = +0.67$ V [185], it can be deduced that the coordination to the copper fragment modifies the electronic level of the ferrocenyl moieties to some extent. Ill-defined reduction processes, centered on the copper moiety, are also present between -0.20 V and -0.50 V [184].

Figure 7-64 shows the crystal structure of the copper homolog of the cobalt compound represented in Scheme 7-20 [119]. The copper(II) center assumes a distorted tetrahedral coordination. The cyclopentadienyl rings of the two ferrocenyl fragments are eclipsed.

As in the case of the corresponding cobalt and nickel complexes, this copper complex undergoes a ferrocene-based two-electron oxidation in DMF in the same range of potential values ($E^{\circ\prime} = +0.53$ V) [119].

The ferrocenylthiosemicarbazonate-copper(II) complex $[(\eta^5\text{-}C_5H_5)Fe(\eta^5\text{-}C_5H_4\text{-}CHN(N){=}C(S)NH_2)]_2Cu$ behaves like the cobalt(II) and nickel(II) congeners (see Scheme 7-21). It undergoes a single, reversible, two-electron oxidation (in acetonitrile, $E^{\circ\prime} = +0.47$ V), and two one-electron reductions, only the first being reversible (in acetonitrile, $E^{\circ\prime}$ $Cu^{II/I} = -0.61$ V) [120].

Figure 7-65 shows the X-ray structure of the complex [7,16-bis(ferrocenylmethyl)-1,4,10,13-tetrathia-7,16-diaza-cyclooctadecane]copper(I), $[CuL']^+$ and that of the free ligand L′ [186]. The copper(I) centre assumes a distorted tetrahedral geometry. With respect to the macrocyclic conformation of the free ligand, which possesses two *gauche* and two *anti* C-S-C-C bonds and two *anti* N-C-C-S bonds, complexation with copper(I) maintains the two *gauche* and two *anti* C-S-C-C bonds, but the two N-C-C-S bonds become *gauche*.

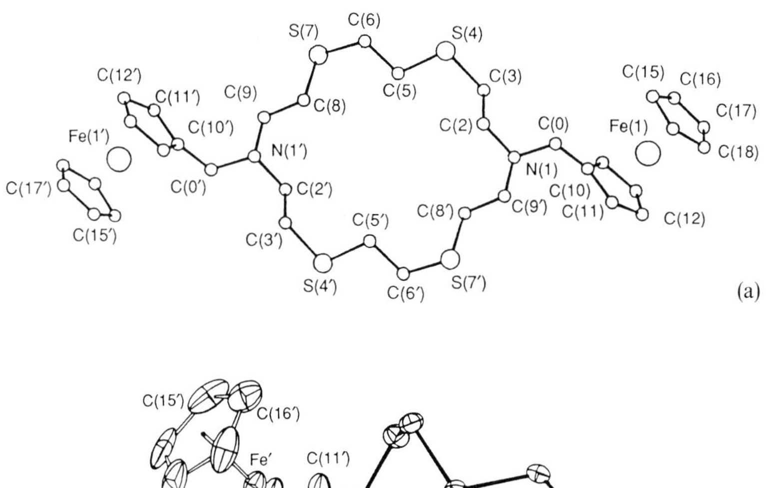

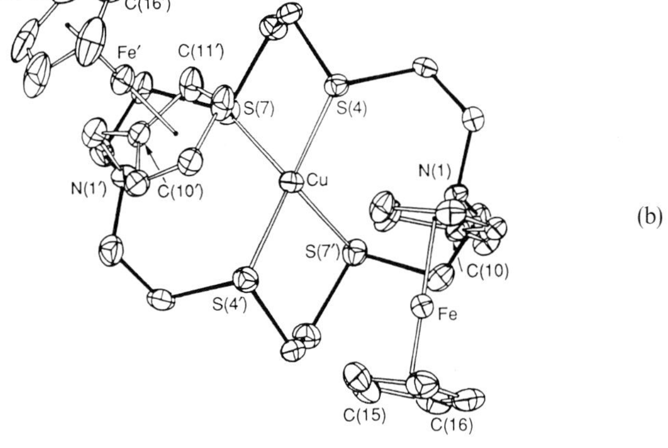

Fig. 7-65. Crystal structure of (a) the macrocyclic ligand 7,16-bis(ferrocenylmethyl)-1,4,10,13-tetrathia-7,16-diaza-cyclooctadecane; (b) one of the three independent molecules of its copper(I) complex (as PF_6 salt). Cu-S4, 2.27 Å; Cu-S7, 2.35 Å (reproduced by permission of the Royal Society of Chemistry).

The similar complex [7,16-bis(ferrocenylcarbonyl)-1,4,10,13-tetrathia-7,16-diaza-cyclooctadecane]copper(I), $[CuL'']^+$ has also been prepared [186].

Both complexes undergo a chemically reversible copper(I)/copper(II) oxidation in acetonitrile ($E^{\circ\prime}$ [CuL′] +/2+ = −0.28 V; $E^{\circ\prime}$ [CuL″] +/2+ = −0.18 V), followed by a subsequent ferrocene-centered oxidation ($E^{\circ\prime}$ [CuL′] 2+/4+ = +0.52 V; $E^{\circ\prime}$ [CuL″] 2+/4+ = +0.72 V). The formal electrode potentials for these ferrocene-based oxidations are marginally higher than those of the corresponding free ligands ($E^{\circ\prime}$ [L′] 0/2+ = +0.48 V; $E^{\circ\prime}$ [L″] 0/2+ = +0.66 V) [186].

Another copper(I) complex possessing tetrahedral coordination is shown in Fig. 7-66 [187]. The central copper(I) ion is coordinated both to one 1,1′-bis(diphenylphosphino)ferrocene molecule (dppf) and one 1,1′-bis(oxodiphenylphosphoranyl)-ferrocene molecule (odppf). All the cyclopentadienyl rings are planar and mutually staggered (A/C) or half-staggered (B/D).

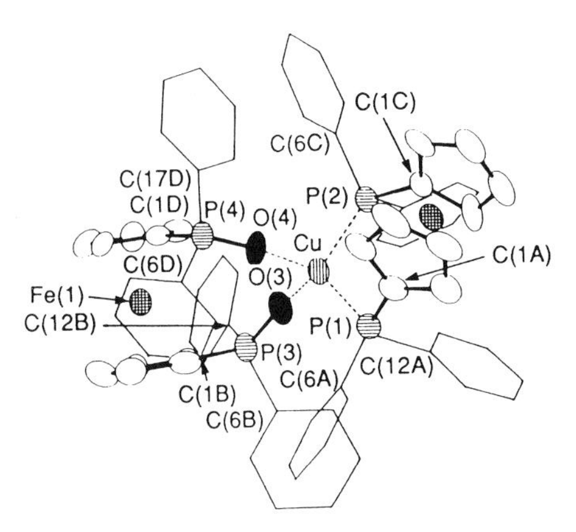

Fig. 7-66. X-ray structure of [Cu(dppf)(odppf)][BF_4]. Average Cu-P distance, 2.27 Å; average Cu-O distance, 2.11 Å (reproduced by permission of the Royal Society of Chemistry).

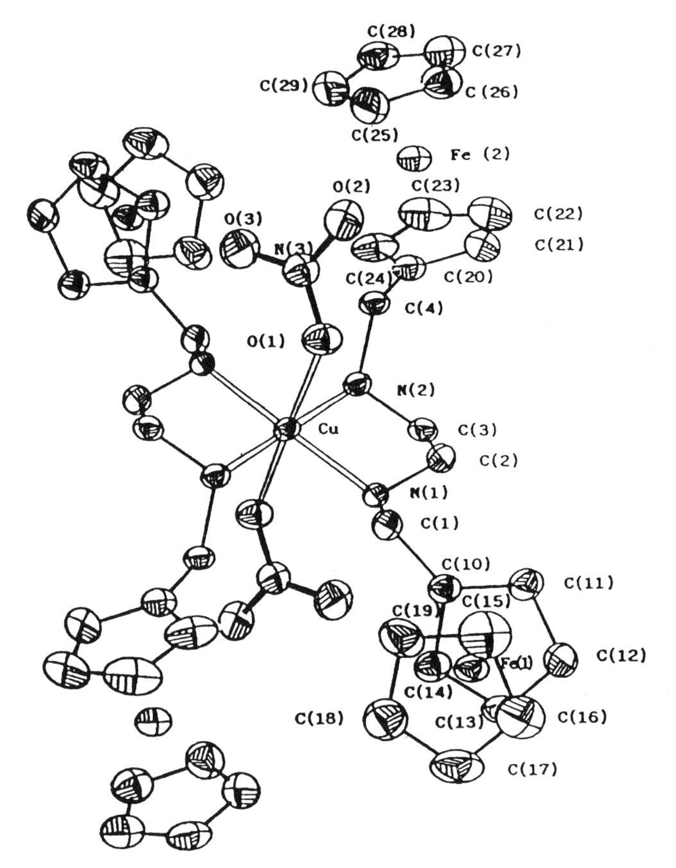

Fig. 7-67. Perspective view of $\{(C_5H_5)Fe[C_5H_4\text{-}CH_2\text{-}NH\text{-}(CH_2)_2\text{-}NH\text{-}CH_2\text{-}C_5H_4]Fe(C_5H_5)\}_2$-$Cu(NO_3)_2$. Cu-N, 2.04 Å; Cu-O, 2.51 Å (reproduced by permission of American Chemical Society).

In 1,2-dichloroethane solution the complex displays two subsequent one-electron oxidation steps, reversible in character, that are assigned to dppf/[dppf]$^+$ ($E^{\circ\prime}$ = +0.78 V) and odppf/[odppf]$^+$ ($E^{\circ\prime}$ = +1.16 V), respectively. In agreement with the strong bonding between the metal and the two ferrocene fragments, the free ferrocenylphosphines are significantly easier to oxidize [$E^{\circ\prime}$ (dppf/[dppf]$^+$) = +0.64 V; $E^{\circ\prime}$ (odppf/[odppf]$^+$) = +0.94 V) [187].

Figure 7-67 shows the crystal structure of $\{(C_5H_5)Fe[C_5H_4$-CH_2-NH-$(CH_2)_2$-NH-CH_2-$C_5H_4]Fe(C_5H_5)\}_2Cu(NO_3)_2$ [188]. Because of the axial coordination of the two nitrate anions, the central copper(II) ion assumes octahedral coordination. Both the ferrocenyl fragments are in a nearly eclipsed conformation.

In DMSO this compound undergoes a single-step four-electron process at $E^{\circ\prime}$ = +0.61 V, thus suggesting that the four ferrocene fragments are electronically isolated from each other. Such a potential value is significantly higher than that of the free *N,N*-bidentate diferrocene ligand ($E^{\circ\prime}$ = +0.47 V). The electrogenerated tetracation is unstable [188].

Figure 7-68 shows the molecular structure of the dicopper(I) cation $[\{Fe(\eta^5$-$C_5H_4PPh_2)_2\}Cu\{(\eta^5$-$C_5H_4PPh_2)Fe(\eta^5$-$C_5H_4PPh_2)\}Cu\{Fe(\eta^5$-$C_5H_4PPh_2)_2\}]^{2+}$ [189]. The copper(I) centres assume a nearly planar trigonal geometry, the displacement of the copper ion from the P_3 plane being 0.1 Å. In the central bridging diphenylphosphinoferrocene unit the cyclopentadienyl rings are staggered and parallel to each other. In contrast, in the lateral chelating diphenyl-phosphinoferrocene groups, the cyclopentadienyl rings are partially staggered and slightly deviated (by 1.6°) from the parallel disposition.

As illustrated in Fig. 7-69, the dication undergoes a single-step three-electron oxidation process at $E^{\circ\prime}$ = +0.92 V (*vs.* SCE) [190].

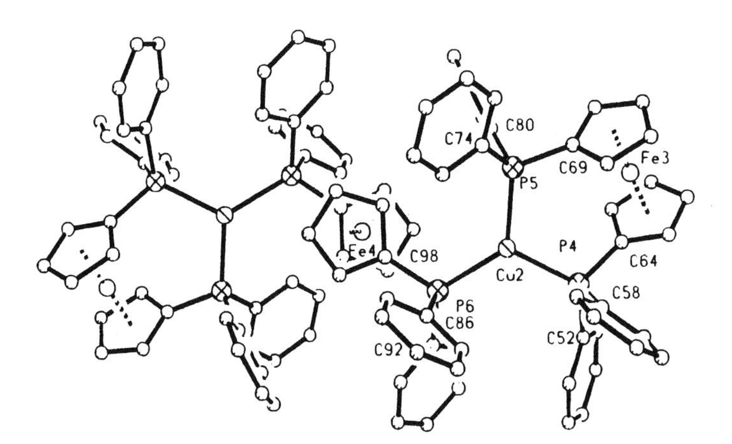

Fig. 7-68. X-ray structure of $[\{Fe(\eta^5$-$C_5H_4PPh_2)_2\}Cu\{(\eta^5$-$C_5H_4PPh_2)Fe(\eta^5$-$C_5H_4PPh_2)\}Cu\{Fe(\eta^5$-$C_5H_4PPh_2)_2\}]^{2+}$. Cu-P, average 2.28 Å; P1-C13, 1.81 Å; P2-C18, 1.83 Å; P3-C47, 1.79 Å; Cu1 ··· Fe1, 4.1 Å; Cu1 ··· Fe2, 4.7 Å (reproduced by permission of Plenum Publishing Corporation).

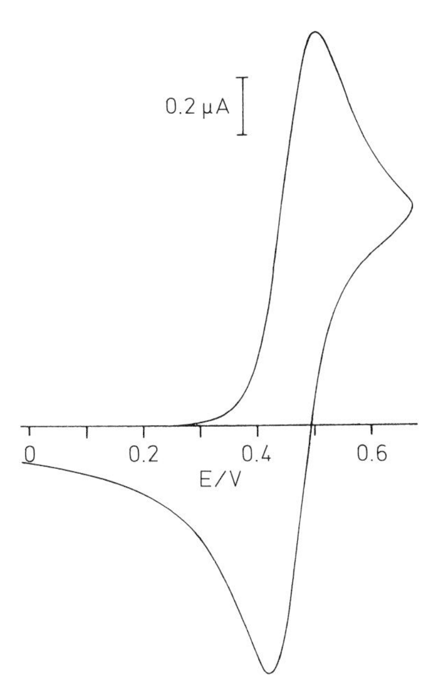

Fig. 7-69. Cyclic voltammogram recorded at a platinum electrode on a 1,1-dichloroethane solution of $[\{Fe(C_5H_4PPh_2)_2\}Cu\{Fe(C_5H_4PPh_2)_2\}_2]^{2+}$. Scan rate 0.2 V s^{-1}. Potential values are referred to $Ag/AgClO_4$ (0.1 *M*) (reproduced by permission of Elsevier Sequoia S.A.).

The simultaneous removal of one electron from each phosphinoferrocene unit apparently displays features of chemical reversibility. Indeed, in the longer times of macroelectrolysis the primarily generated pentacation decomposes to the monomeric copper(I)-diphenylphosphinoferrocenium dication $[Cu\{Fe(\eta^5\text{-}C_5H_4PPh_2)_2\}]^{2+}$ and bis(diphenylphosphino)ferrocenium radical ion $[Fe(\eta^5\text{-}C_5H_4PPh_2)_2]^+$.

Figure 7-70 shows the crystal structure of the tetrakis(ferrocenecarboxylato)-dicopper(II) THF-solvate complex [191]. The Cu_2O_{10} core has a lantern geometry in which each copper(II) atom assumes square-pyramidal coordination. The basal oxygen atoms are donated by the carboxylate groups, whereas the apical oxygen stems from the THF molecule. The copper atom is displaced from the basal plane (towards the apex) by 0.18 Å. The cyclopentadienyl rings of two ferrocene groups (Fe1, Fe1′) are eclipsed (tilting angle 3.6°), whereas the other two (Fe2, Fe2′) are staggered (tilting angle 3.0°).

It has been reported that this dicopper complex undergoes an irreversible copper-centered reduction at -1.0 V in DMSO and two oxidation steps at $E_p = +0.20$ V and $E^{\circ\prime} = +0.62$ V, respectively. The less anodic process, which is not fully reversible and possesses a return peak at $E_p = -0.1$ V, is difficult to assign, whereas the more anodic process, which is chemically reversible, has been assigned to electron removal processes from the ferrocenyl fragments [191]. Unfortunately, controlled potential coulometric tests have not been performed to determine the number of electrons involved in each electron-transfer step. We should like to think

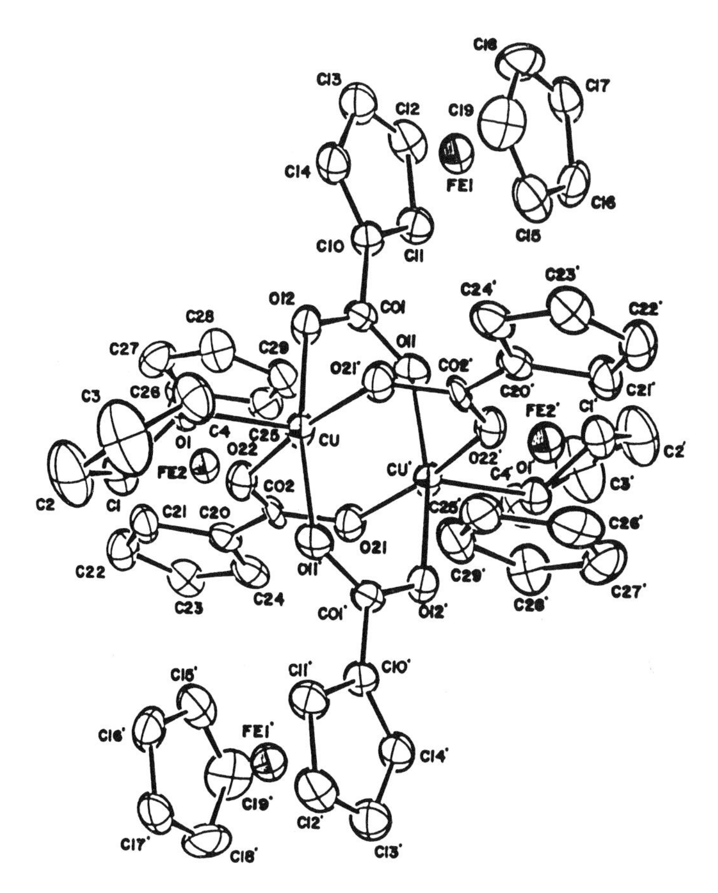

Fig. 7-70. X-ray structure of $[(\eta^5\text{-}C_5H_5)Fe(\eta^5\text{-}C_5H_4CO_2)]_4Cu_2(THF)_2$. Cu-O (carboxylate), average 1.96 Å; Cu-O (THF), 2.23 Å; Cu···Cu, 2.60 Å (reproduced by permission of the American Chemical Society).

that the nominally first anodic step is really the cathodic Cu(II)/Cu(I) process followed by the subsequent reduction Cu(I)/Cu(0) at more negative potential values.

Finally, Scheme 7-41 represents the probable structure of the two ferrocenyl-tetracopper complexes, $[LCuCl]_4O_2$ and $[LCuCl]_4(CO_3)_2$. Attempted crystallization of these complexes from dichloromethane afforded $(\mu_4\text{-}O)L_4Cu_4Cl_6$, the crystal structure of which has been solved [192].

In dichloromethane the two complexes undergo an oxidation step at potential values ($E^{\circ\prime}$ (oxo-species) = +0.43 V; $E^{\circ\prime}$ (carbonato-species) = +0.47 V) that are nearly coincident with that of the free ligand *N*,*N*-dimethylaminomethylferrocene ($E^{\circ\prime}$ = +0.43 V) [192].

The complexes $[\{(\eta^5\text{-}C_5H_5)Fe[\eta^5\text{-}C_5H_3(1\text{-}PPh_2)(2\text{-}CH(CH_3)S)]\}Cu]_3$ [193] and $\{(\eta^5\text{-}C_5H_5)Fe[(\eta^5\text{-}C_5H_3)(1\text{-}CH_2NMe_2)(2\text{-}Cu)]\}_4$ [194] have been characterized by X-ray analysis.

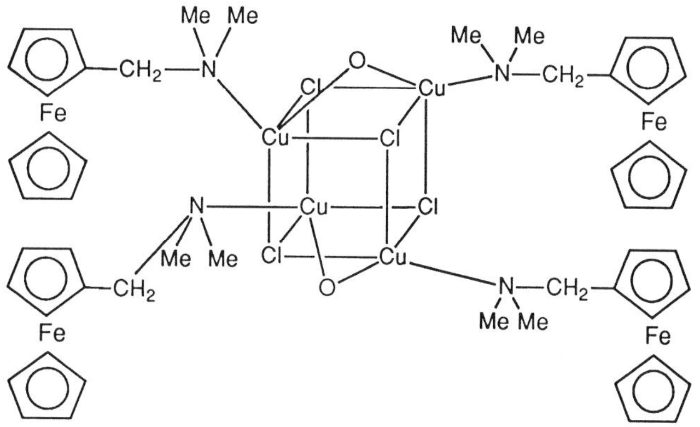

$[LCuCl]_4O_2$

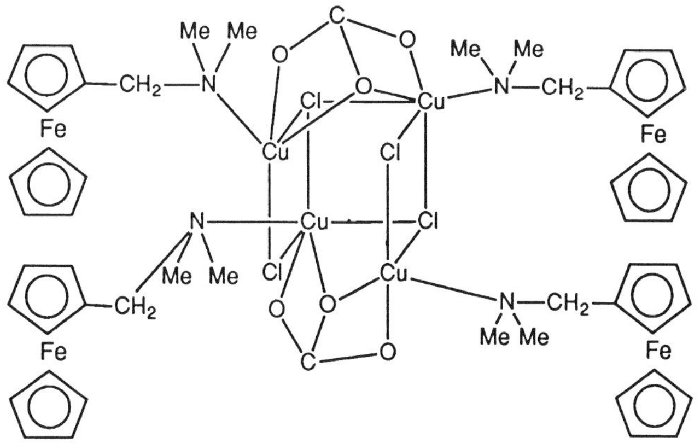

Scheme 7-41 $[LCuCl]_4(CO_3)_2$

7.2.8.2 Silver Complexes

Scheme 7-42 shows three ferrocene-crown derivatives that are able to incorporate silver(I) ions [195].

The X-ray structure of the monoferrocenyl-silver complex is shown in Fig. 7-71 [195]. The cryptand-like cavity of the peripheral tetraoxa-dizazadecane ring is able to accomodate one Ag^+ ion, but it is thought that a direct Fe-Ag interaction also occurs. The cyclopentadienyl rings of the ferrocene portion are tilted by 10°.

As illustrated in Fig. 7-72, complexation of silver ions in acetonitrile solution causes the one-electron oxidation of the ferrocene fragment to shift anodically by

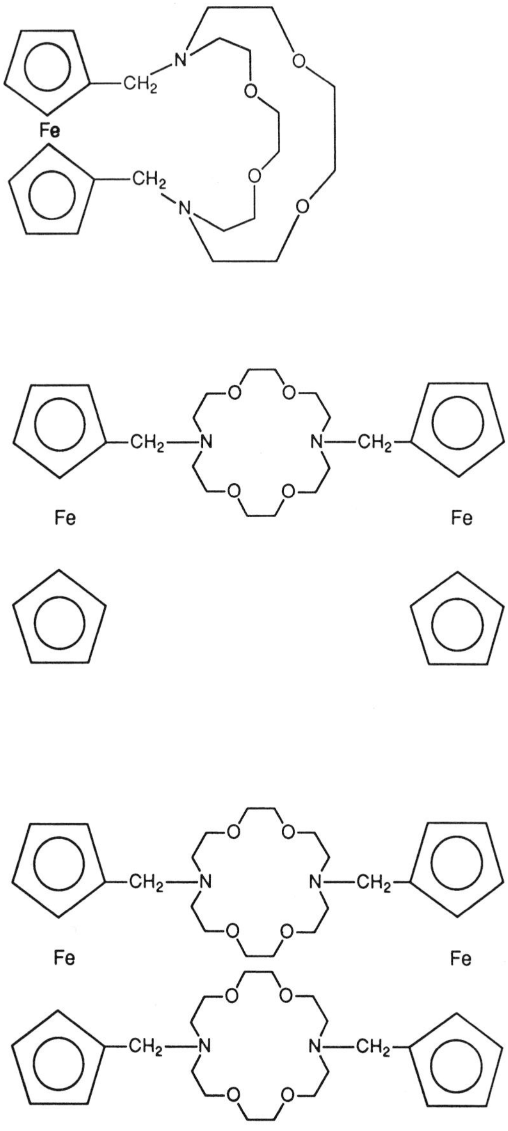

Scheme 7-42

282 mV with respect to that of the original ferrocene molecule ($E^{\circ\prime} = +0.22$ V). It is likely that such an increase in redox potential, which is one of the highest so far reported, may arise from the diminished electron density on the iron(II) center caused by the Fe $\rightarrow$ Ag^+ interaction [195].

A minor shift in redox potential is caused by silver ion complexation of the diferrocenyl-monocryptand complex. In fact, the silver-free ferrocene ligand undergoes a single-step two-electron oxidation at $E^{\circ\prime} = +0.39$ V in MeCN. On incorporation of silver ion, the two-electron redox change occurs at $E^{\circ\prime} = +0.50$ V.

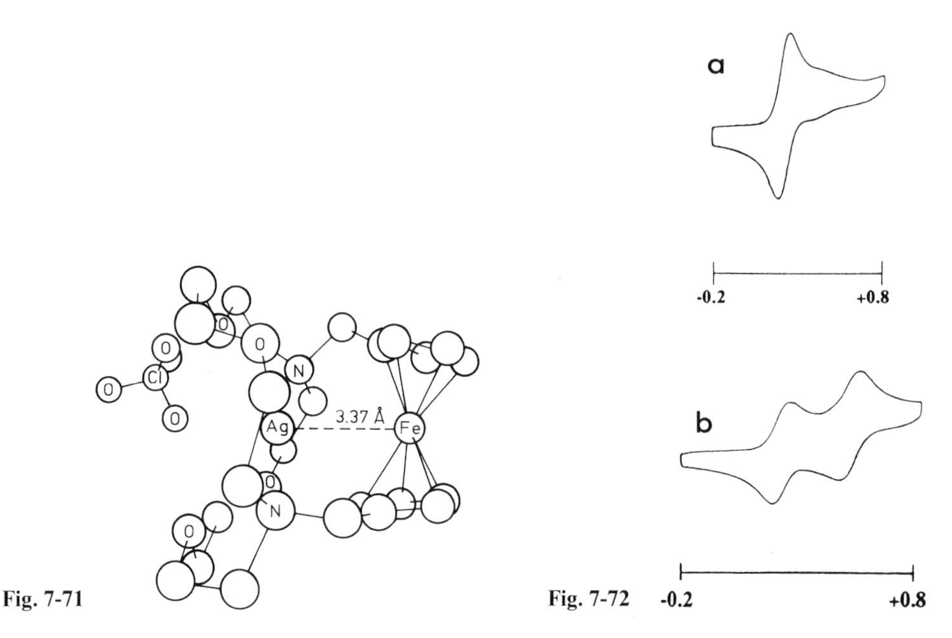

Fig. 7-71. Crystal structure of 1,1′-(1,4,10,13,-tetraoxa-7,16-diazacyclooctadecane-7,16-dimethyl)-ferrocene · $AgClO_4$ (reproduced by permission of the American Chemical Society).

Fig. 7-72. Cyclic voltammograms recorded at a glassy carbon electrode on an acetonitrile solution containing: (a) 1,1′-(1,4,10,13,-tetraoxa-7,16-diazacyclooctadecane-7,16-dimethyl)ferrocene; (b) after the addition of Ag^+ ions in a 2:1 ligand-to-metal ratio (reproduced by permission of the American Chemical Society).

Finally, complexation of two silver ions by the diferrocene-dicryptand molecule, according to the postulated structure shown in Scheme 7-43, again causes a notably high anodic shift of the redox potential (from +0.38 V to +0.58 V).

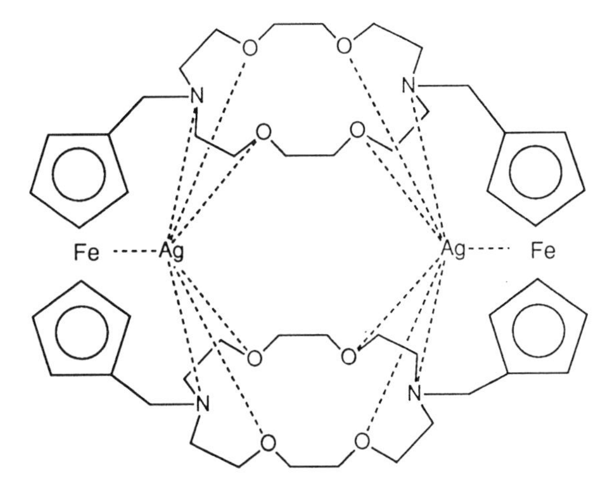

Scheme 7-43

Other X-ray characterized silver-ferrocenyl complexes include $\{(\eta^5\text{-}C_5H_5)\text{-}Fe[(\eta^5\text{-}C_5H_3)(1\text{-}CH_2NMe_2)(2\text{-}Ag)]\}_4$ [196], $[Fe(\eta^5\text{-}C_5H_4PPh_2)_2]Ag_2(C_6H_5CO_2)_2$, $[Fe(\eta^5\text{-}C_5H_4PPh_2)_2]_2Ag_2(NO_3)_2$, $[Fe(\eta^5\text{-}C_5H_4PPh_2)_2]_2Ag_4(CH_3CO_2)_4$, and $[Fe\text{-}(\eta^5\text{-}C_5H_4PPh_2)_2]_3Ag_2(HCO_2)_2$ [197].

7.2.8.3 Gold Complexes

To the best of our knowledge, the only electrochemical report on gold complexes containing ferrocene fragments states unexpectedly that $Fe(\eta^5\text{-}C_5H_4PPh_2AuCl)_2$ (the structure of which is shown in Fig. 7-73) [198], $[Fe(\eta^5\text{-}C_5H_4PPh_2AuCl)]_2$, $[\{Fe(\eta^5\text{-}C_5H_4PPh_2)\}_2Au]^+Cl^-$, and $\{Fe(\eta^5\text{-}C_5H_4PPh_2)_2\}AuCl$ do not exhibit redox activity in DMF or CH_2Cl_2 [199].

This seems to indicate that the AuCl fragments possess such a high electron-withdrawing ability to make the HOMO levels of the ferrocene moieties inaccessible.

Other X-ray structures concern $\{(\eta^5\text{-}C_5H_4PPh_2)(\eta^5\text{-}C_5H_3(PPh_2)CH(Me)N(Me)\text{-}CH_2CH_2NMe_2)Fe\}_2(AuCl)_3$ [200], and $[(\eta^5\text{-}C_5H_5)Fe\{(\eta^5\text{-}C_5H_4)Au_2(PPh_3)_2\}]^+$ [201, 202].

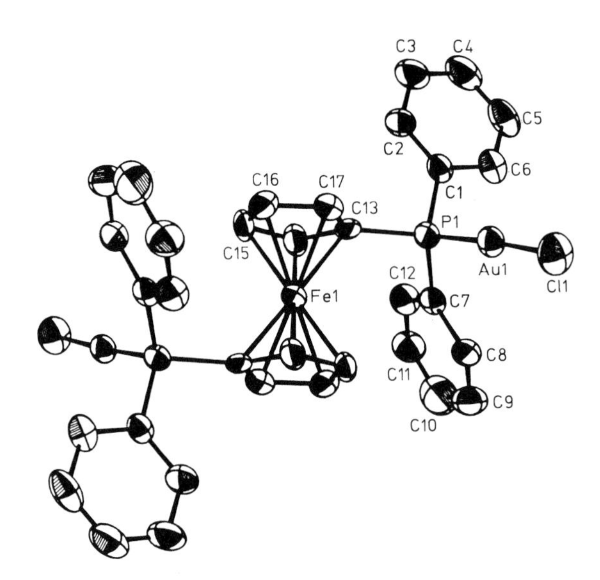

Fig. 7-73. Crystal structure of the ferrocen-diyl-staggered molecule of $Fe(\eta^5\text{-}C_5H_4PPh_2AuCl)_2$. C13-P1, 1.79 Å; P1-Au1, 2.24 Å; Au1-C11, 2.30 Å. Slight variations occur in the crystallographically independent, partially staggered, molecule (reproduced by permission of the American Chemical Society).

7.2.9 Group 12 Metal Complexes

7.2.9.1 Zinc Complexes

The 1,1′-bis(diphenylphosphino)ferrocene derivative $[Fe(\eta^5\text{-}C_5H_4PPh_2)_2]ZnCl_2$ has been assigned a tetrahedral geometry, similar to that of $[Fe(\eta^5\text{-}C_5H_4PPh_2)_2]NiCl_2$ [151]. It has been briefly reported that it exhibits irreversible one-electron oxidation in 1,2-dichloroethane solution, thus indicating the high instability of the ferrocenium cation $[\{Fe(C_5H_4PPh_2)_2\}ZnCl_2]^+$ [151].

As illustrated in Fig. 7-74, the complex $[(\eta^5\text{-}C_5H_5)Fe(\eta^5\text{-}C_5H_4N{=}CHC_6H_4O)]_2$-Zn, analog of the cobalt complex shown in Scheme 7-20, undergoes a two-electron oxidation step in DMF ($E^{\circ\prime} = +0.69$ V) [119]. Analogous behavior was previously illustrated for the corresponding cobalt, nickel, and copper complexes [119]. The instantaneously electrogenerated zinc dication is much less stable than the monocation of the corresponding free ligand.

In contrast, the complex $[(\eta^5\text{-}C_5H_5)Fe(\eta^5\text{-}C_5H_4CHN(N){=}C(S)NH_2)]_2Zn$, belonging to the series of the metal complexes illustrated in Scheme 7-21, undergoes a reversible two-electron oxidation in acetonitrile ($E^{\circ\prime} = +0.47$ V) [120].

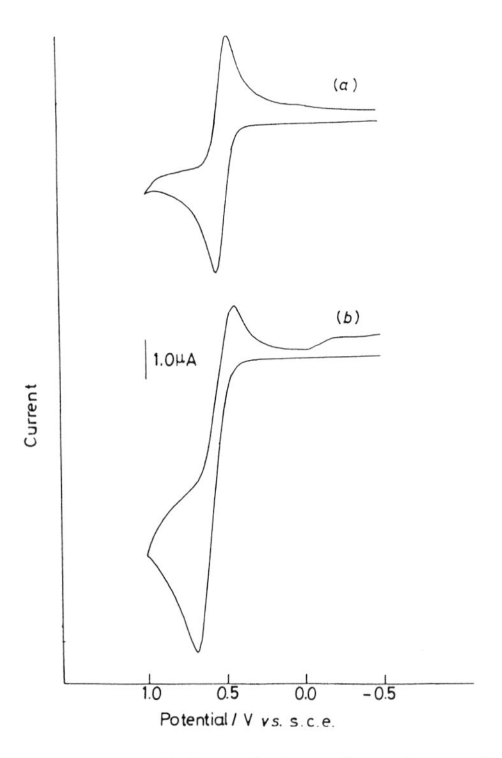

Fig. 7-74. Cyclic voltammograms recorded at a platinum electrode on a dimethylformamide solution of: (a) $(C_5H_5)Fe(C_5H_4NH{=}CHC_6H_4\text{-}m\text{-}OH)$; (b) $[(C_5H_5)Fe(C_5H_4N{=}CHC_6H_4O)]_2Zn$. Scan rate 0.1 V s^{-1} (reproduced by permission of the Royal Society of Chemistry).

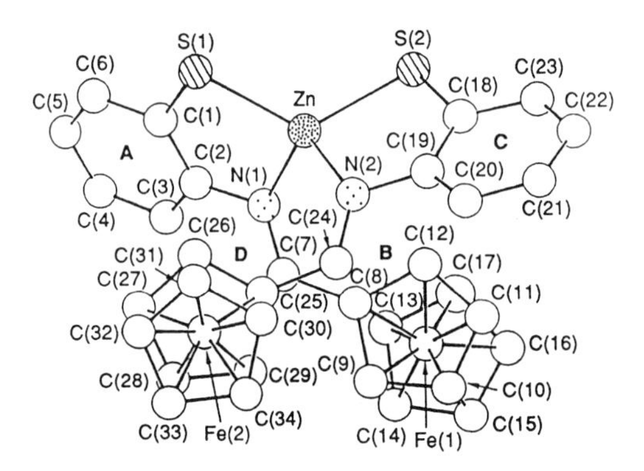

Fig. 7-75. Crystal structure of $\{(\eta^5\text{-}C_5H_5)Fe[\eta^5\text{-}C_5H_4\text{-}CH{=}N(C_6H_4)S]\}_2Zn$. Zn-S, average 2.27 Å; Zn-N, average 2.08 Å (reproduced by permission of the Royal Society of Chemistry).

Figure 7-75 shows the X-ray structure of bis[2-(ferrocenyl-methyleneamino)benzenethiolato]zinc(II) [149].

In contrast to the square planar geometry of the corresponding nickel(II) and palladium(II) complexes, the zinc atom displays a distorted tetrahedral coordination. Both of the two ferrocenyl molecules assume an eclipsed configuration.

In acetonitrile this complex undergoes one reversible oxidation process centered on the ferrocenyl fragments ($E^{\circ\prime} = +0.74$ V) [149].

The two complexes shown in Scheme 7-44 have recently been prepared [203].

It has been briefly reported that the monoferrocenyl dication undergoes reversible one-electron oxidation at a potential value that is 0.1 V more anodic than that of the oxidation of the triazaferrocenophane parent ligand. In contrast, in CH_2Cl_2 the related triferrocenyl tetracation undergoes a single three-electron oxidation at the same potential as that of the precursor ligand, this process being complicated by following chemical reactions. This not only suggests that no electronic communication exists between the peripheral ferrocene fragments, but also that the two zinc ions are non-interacting [203].

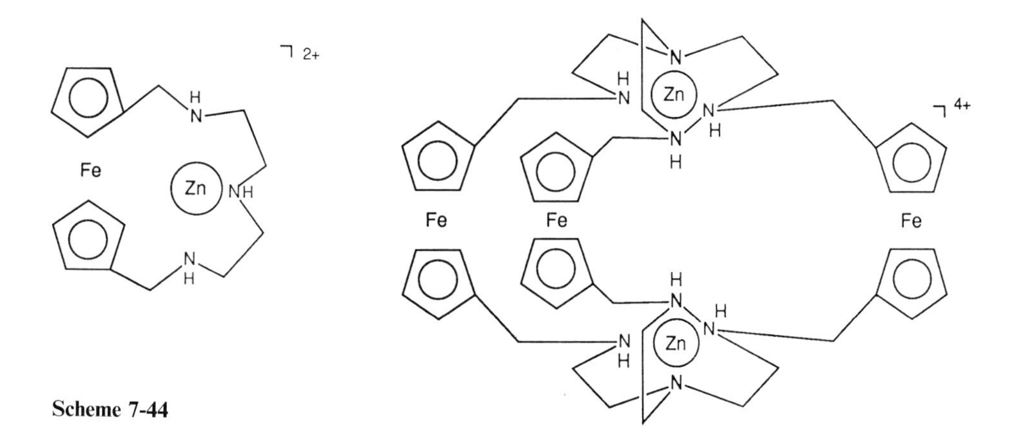

Scheme 7-44

7.2.9.2 Cadmium Complexes

On the basis of recent ^{57}Fe Mössbauer spectroscopic data, the 1,1′-bis(diphenylphosphino)ferrocene derivative $[Fe(\eta^5\text{-}C_5H_4PPh_2)_2]CdCl_2$ likely possesses square planar geometry [204]. As for the corresponding zinc(II) complex, it also undergoes an irreversible one-electron oxidation in 1,2-dichloroethane [151].

Parallel to the behavior previously described for the corresponding cobalt, nickel, copper, and zinc complexes, the thiosemicarbazonate complex $[(\eta5\text{-}C_5H_5)Fe(\eta^5\text{-}C_5H_4CHN(N)=C(S)NH_2)]_2Cd$ also undergoes a reversible, ferrocenyl-centered, two-electron oxidation ($E^{\circ\prime} = +0.46$ V) in acetonitrile [120].

7.2.9.3 Mercury Complexes

Figure 7-76 shows the cyclic voltammetric response exhibited by $[(\eta^5\text{-}C_5H_5)Fe(\eta^5\text{-}C_5H_4)]_2Hg$ [205].

This complex undergoes a chemically reversible two-electron oxidation at $E^{\circ\prime} = +0.29$ V. Quite similar data have been previously obtained by polarographic techniques in acetonitrile solution ($E^{\circ\prime}$ (0/2+) = +0.28 V) [206].

Figure 7-77 illustrates the molecular structure of bis[2-(ferrocenylmethyleneamino)benzenethiolato]mercury(II) [149]. In contrast to the N_2S_2 four-coordination of the nickel(II), palladium(II) and zinc(II) analogs previously described, the mercury atom assumes a linear HgS_2 coordination [149].

In acetonitrile the latter mercury complex undergoes one reversible oxidation centered on the ferrocenyl units ($E^{\circ\prime} = +0.67$ V).

Scheme 7-45 shows a series of 1,1′-bis(diphenylphosphino)ferrocene-mercury complexes, to which a tetrahedral geometry has been assigned [151, 204, 207].

It has been reported that the present complexes undergo polarographically reversible oxidation in acetonitrile at potentials significantly higher than that of the free $Fe(C_5H_4PPh_2)_2$ ligand (Table 7-34) [207]. Indeed, more recent electrochemical investigations on $[Fe(C_5H_4PPh_2)_2]HgCl_2$ in 1,2-dichloroethane solution raises doubts about the reversibility of these electron transfers [151].

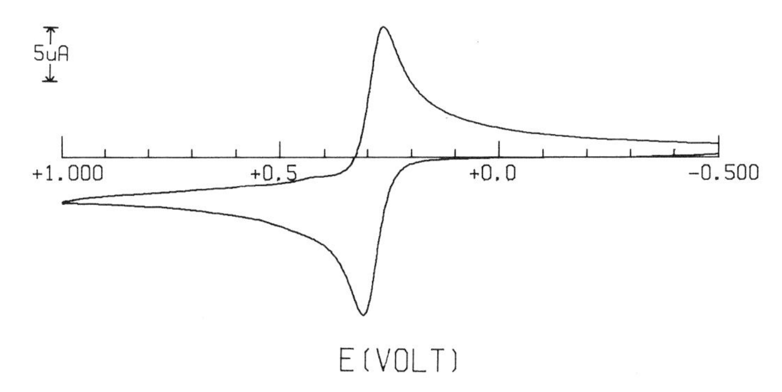

Fig. 7-76. Cyclic voltammogram recorded at a platinum electrode on a dichloromethane solution of $[(C_5H_5)Fe(C_5H_4)]_2Hg$. Scan rate 0.05 V s^{-1}.

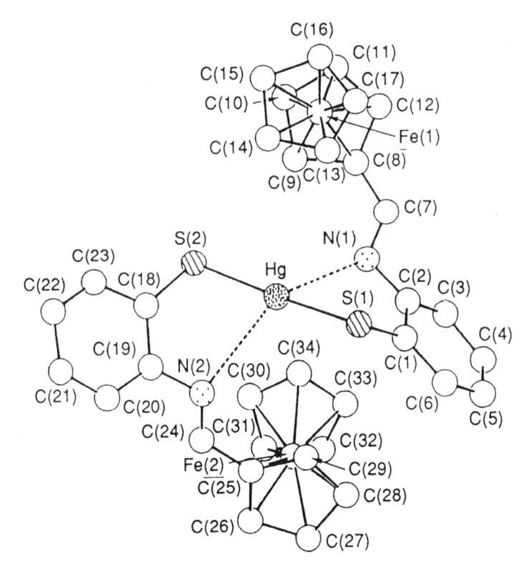

Fig. 7-77. Perspective view of $\{(\eta^5\text{-}C_5H_5)Fe[\eta^5\text{-}C_5H_4\text{-}CH{=}N(C_6H_4)S]\}_2Hg$. Hg-S, average 2.33 Å; Hg···N, average 2.8 Å (reproduced by permission of the Royal Society of Chemistry).

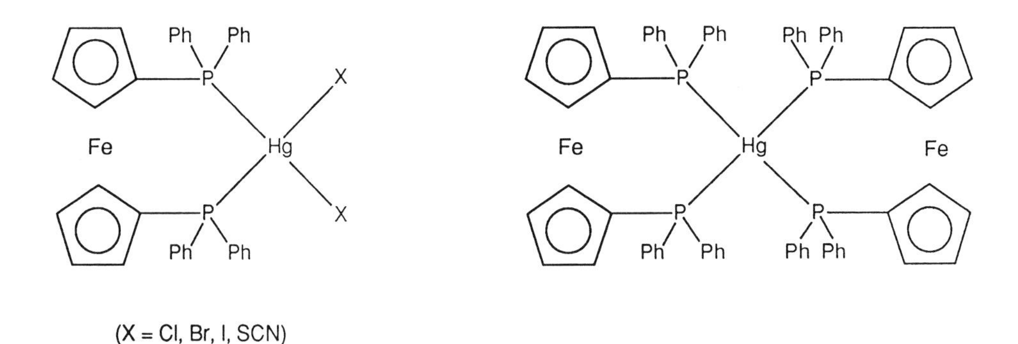

Scheme 7-45

Table 7-34. Polarographic half-wave potentials (*vs.* SCE) for the oxidation processes exhibited by the mercury complexes illustrated in Scheme 7-45. Acetonitrile solution [207]

Complex	$E_{1/2}$, V
$[Fe(C_5H_4PPh_2)_2]HgCl_2$	+0.91
$[Fe(C_5H_4PPh_2)_2]HgBr_2$	+0.89
$[Fe(C_5H_4PPh_2)_2]HgI_2$	+0.74
$[Fe(C_5H_4PPh_2)_2]Hg(SCN)_2$	+1.00
$[\{Fe(C_5H_4PPh_2)_2\}_2Hg]^{2+}$	+1.07[a]
$Fe(C_5H_4PPh_2)_2$	+0.57

a Two-electron step.

7.3 Ferrocene Complexes of Heterometallic Fragments

7.3.1 Trinuclear Complexes

Scheme 7-46 shows the structure assigned to a series of heterometallic assemblies of bis(diphenylphosphino)ferrocene [49].

Scheme 7-46 M = Cr, Mo, W

These complexes exhibit in dichloromethane solution a first ferrocene-based oxidation with features of chemical reversibility, followed by a second irreversible oxidation, which is thought to be centered on the metal-carbonyl fragment. A poorly reproducible oxidation of the Au^I moiety is also observable [49]. The relevant redox potentials are summarized in Table 7-35. With respect to the oxidation of free 1,1′-bis(diphenylphosphino)ferrocene ($E^{\circ\prime} = +0.66$ V) [49], the multimetallic complexation makes oxidation of the ferrocene fragment significantly more difficult (by about 0.4 V).

Table 7-35. Redox potentials (*vs.* SCE) for the electron removals exhibited by the complexes $(ClAuPPh_2C_5H_4)Fe[C_5H_4PPh_2M(CO)_5]$ in dichloromethane solution [49]

M	$E^{\circ\prime}$ (Fc/Fc^+), V	E_p (M^0, M^I), V
Cr	+1.07	+1.44
Mo	+1.06	+1.43
W	+1.06	+1.44

7.3.2 Pentanuclear Complexes

Another spectacular series of heteropolymetallic complexes containing bis(diphenylphosphino)ferrocene as a ligand is shown in Scheme 7-47 [49].

M = Cr, Mo, W

Scheme 7-47 M' = Pd, Pt

Figure 7-78 illustrates the molecular structure of $[\{(CO)_5W(PPh_2C_5H_4)Fe(C_5H_4PPh_2)\}PtCl_2\{(PPh_2C_5H_4)Fe(C_5H_4PPh_2)W(CO)_5\}]$ [49].

It appears as a linear aggregate of five metal atoms linked by $C_5H_4PPh_2$ bridges. The central platinum atom possesses square planar geometry, with the two lateral $(PPh_2C_5H_4)Fe(C_5H_4PPh_2)W(CO)_5$ subunits assuming a *trans* arrangement. In the ferrocene moieties, the cyclopentadienyl rings are nearly eclipsed and parallel.

Like the trinuclear complexes discussed in the preceding section, the main redox processes exhibited by the actual pentanuclear complexes in dichloromethane solution are constituted by a first, nearly reversible, ferrocene-centered oxidation followed by a second Group 6 metal-centered irreversible oxidation [49]. The relevant redox potentials are reported in Table 7-36.

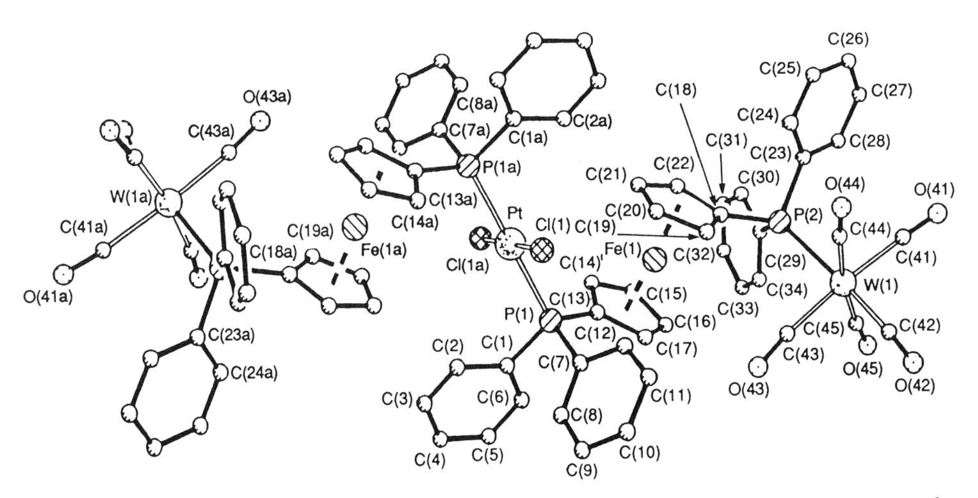

Fig. 7-78. Perspective view of $PtCl_2[(\eta^5\text{-}C_5H_4PPh_2)Fe(\eta^5\text{-}C_5H_4PPh_2)W(CO)_5]_2$. Pt-Cl, 2.30 Å; Pt-P, 2.32 Å; W-P, 2.55 Å; W-C, average 2.03 Å; Fe-C (cyclopentadienyl), 2.04 Å (reproduced by permission of the Royal Society of Chemistry).

Table 7-36. Redox potentials (*vs.* SCE) for the electron removals exhibited in dichloromethane solution by the complexes $M'Cl_2[Fe(C_5H_4PPh_2)_2M(CO)_5]_2$ [49]

M'	M	$E^{\circ\prime}$ (Fc/Fc$^+$), V	E_p (M^0/M^I), V
Pd	Cr	+0.95	+1.44
	Mo	+0.95	+1.41
	W	+0.98	+1.45
Pt	Cr	+0.94	+1.40
	Mo	+0.94	+1.43
	W	+0.95	+1.48

In this case too, complexation of bis(diphenylphosphino)ferrocene with metal fragments makes the electron removal step notably difficult (by about 0.3 V).

7.4 Conclusions

As illustrated above, the presence of peripheral redox-active ferrocene centers in transition metal complexes enhances their chemical reactivity. Thus, exploring the access to different oxidation states of heterometallic complexes may provide an opportunity to monitor the transmission of electronic and stereochemical effects along the molecular framework. In particular, the metallocene centers offer the unique opportunity to evaluate the stereodynamic effects accompanying redox changes, not only on the usual basis of bonding angles and distances, but through two geometrical parameters peculiar to their assembly of cyclopentadienyl rings: (i) the rotation from an eclipsed to a staggered configuration, and vice versa; (ii) their tilting from the common parallel disposition. Both these motions tend, in fact, to relieve molecular steric strain induced by electronic perturbation.

Unfortunately, in spite of the fact that electrochemistry allows access to redox congeners of many heterometallic complexes that could be isolated, we were unable to find examples of solid-state structures of redox congeners. So the main goal of judging quantitatively the effects of electron-transfer processes on these types of complexes could not be fulfilled.

We must note that in many cases the electrochemical recognition has been limited to the simple evaluation of the redox potentials of the electron transfer steps, together with a superficial examination of the chemical stability of the complexes in various oxidation states. This is of little help to synthetic chemists, who would have to prepare redox congeners for complete chemical, physico-chemical, and structural characterization. Controlled potential coulometry and macroelectrolysis tests must be routinarily performed, both to define the number of electrons involved in each redox change and to obtain redox congeners, even if in low quantity, at least for a preliminary determination of their stability and of their spectroscopic properties.

Appendix (Notes added in proof, August 1994)

A significant number of important papers has appeared after submission of the manuscript. A critical selection is given below [208–239]. The numbers in round brackets indicate both the section that the reference refers to and its content according to the following classification:

(A) X-ray structural and electrochemical information;
(B) electrochemical information;
(C) X-ray structural information.

References

[1] W. E. Geiger, *J. Organomet. Chem. Libr.* **1990**, *22*, 142–172.
[2] P. Seiler and J. D. Dunitz, *Acta Cryst.* **1979**, *B35*, 1068–1074.
[3] R. Martinez and A. Tiripicchio, *Acta Cryst.* **1990**, *C46*, 202–205.
[4] D. P. Freyberg, J. L. Robbins, K. N. Raymond, and J. C. Smart, *J. Am. Chem. Soc.* **1979**, *101*, 892–897.
[5] J. Pickardt, H. Schumann, and R. Mohtachemi, *Acta Cryst.* **1990**, *C46*, 39–41.
[6] H. Schumann, C. Janiak, R. D. Köhn, J. Loebel, and A. Dietrich, *J. Organomet. Chem.* **1989**, *365*, 137–150.
[7] M. D. Rausch, W.-M. Tsai, J. W. Chambers, R. D. Rogers, and H. G. Alt, *Organometallics* **1989**, *8*, 816–821.
[8] P. Zanello, A. Cinquantini, S. Mangani, G. Opromolla, L. Pardi, C. Janiak, and M. D. Rausch, *J. Organomet. Chem.* **1994**, *471*, 171–177.
[9] N. Dowling, P. M. Henry, N. A. Lewis, and H. Taube, *Inorg. Chem.* **1981**, *20*, 2345–2348.
[10] B. McCulloch, D. L. Ward, J. D. Woolins, and C. H. Brubaker, Jr., *Organometallics* **1985**, *4*, 1425–1432, and references therein.
[11] T. M. Miller, K. J. Ahmed, and M. S. Wrighton, *Inorg. Chem.* **1989**, *28*, 2347–2355.
[12] J. C. Calabrese, L.-T. Cheng, J. C. Green, S. R. Marder, and W. Tam, *J. Am. Chem. Soc.* **1991**, *113*, 7227–7232.
[13] S. E. Sherman and S. J. Lippard, *Chem. Rev.* **1987**, *87*, 1153–1181.
[14] C. K. Mirabelli, D. T. Hill, L. F. Faucette, F. L. McCabe, G. R. Girard, D. B. Bryan, B. M. Sutton, J. O. Bartus, S. T. Crooke, and R. K. Johnson, *J. Med. Chem.* **1987**, *30*, 2181–2190.
[15] P. Köpf-Maier and H. Köpf, *Chem. Rev.* **1987**, *87*, 1137–1152.
[16] W. R. Cullen and J. D. Woollins, *Coord. Chem. Rev.* **1981**, *39*, 1–30.
[17] L. N. Zakharov, Y. T. Struchkov, V. V. Sharutin, and O. N. Suvorova, *Cryst. Struct. Commun.* **1979**, *8*, 439–444.
[18] J. C. Kotz, W. Vining, W. Coco, R. Rosen, A. R. Dias, and M. H. Garcia, *Organometallics* **1983**, *2*, 68–79.
[19] P. Zanello and F. T. Edelmann, unpublished results.
[20] J. B. Flanagan, S. Margel, A. J. Bard, and F. C. Anson, *J. Am. Chem. Soc.* **1978**, *100*, 4248–4253.
[21] R. Broussier, A. Da Rold, B. Gautheron, Y. Dromzee, and Y. Jeannin, *Inorg. Chem.* **1990**, *29*, 1817–1822.
[22] A. Da Rold, Y. Mugnier, R. Broussier, B. Gautheron, and E. Laviron, *J. Organomet. Chem.* **1989**, *362*, C27–C30.
[23] R. Broussier, Y. Gobet, R. Amardeil, A. Da Rold, M. M. Kubicki, and B. Gautheron, *J. Organomet. Chem.* **1993**, *445*, C4–C5.
[24] F. H. Köhler, W. A. Geike, P. Hofmann, U. Schubert, and P. Stauffert, *Chem. Ber.* **1984**, *117*, 904–914.

[25] M. Herberhold, M. Schrepfermann, and A. L. Rheingold, *J. Organomet. Chem.* **1990**, *394*, 113–120.
[26] S. P. Gubin and V. S. Khandkarova, *J. Organomet. Chem.* **1970**, *22*, 449–460.
[27] C. Degrand and A. Radecki-Sudre *J. Organomet. Chem.* **1984**, *268*, 63–72.
[28] C. Degrand, J. Besancon, and A. Radecki-Sudre, *J. Electroanal. Chem.* **1984**, *160*, 199–207.
[29] C. Degrand, A. Radecki-Sudre, and J. Besancon, *Organometallics* **1982**, *1*, 1311–1315.
[30] K. H. Pannell, H. K. Sharma, F. Cervantes Lee, and P. Zanello, manuscript in preparation.
[31] C. G. Zoski, D. A Sweigart, N. J. Stone, P. H. Rieger, E. Mocellin, T. F. Mann, D. R. Mann, D. K. Gosser, M. M. Doeff, and A. M. Bond, *J. Am. Chem. Soc.* **1988**, *110*, 2109–2116.
[32] M. K. L. Lloyd, J. A. McCleverty, D. G. Orchard, J. A. Connor, M. B. Hall, I. H. Hillier, E. M. Jones, and G. K. McEwen, *J. Chem. Soc., Dalton Trans.* **1973**, 1743–1747.
[33] J. A. McCleverty, D. G. Orchard, J. A. Connor, E. M. Jones, J. P. Lloyd, and P. D. Rose, *J. Organomet. Chem.* **1971**, *30*, C75–C77.
[34] E. O. Fischer, M. Schluge, J. O. Besenhard, P. Friedrich, G. Huttner, and F. R. Kreißl, *Chem. Ber.* **1978**, *111*, 3530–3541.
[35] E. O. Fischer, M. Schluge, J. O. Besenhard, *Angew. Chem. Int. Ed. Engl.* **1976**, *15*, 683–684.
[36] E. O. Fischer, F. J. Gammel, J. O. Besenhard, A. Frank, D. Verhalten, *J. Organomet. Chem.* **1980**, *191*, 261–282.
[37] P. Zanello, in I. Bernal (Ed.) *Sterochemistry of Organometallic and Inorganic Compounds*, Elsevier, Amsterdam, Vol. 4, **1990**, pp. 181–366.
[38] T. El-Shihi, F. Siglmüller, R. Herrmann, M. F. N. N. Carvalho, and A. J. L. Pombeiro, *J. Organomet. Chem.* **1987**, *335*, 239–247.
[39] M. E. N. P. R. A. Silva, A. J. L. Pombeiro, J. J. R. F. da Silva, R. Herrmann, N. Deus, T. J. Castilho, and M. F. C. G. Silva, *J. Organomet. Chem.* **1991**, *421*, 75–90.
[40] J. C. Kotz, C. L. Nivert, J. M. Lieber, and R. C. Reed, *J. Organomet. Chem.* **1975**, *84*, 255–267.
[41] Y.-P. Wang, J.-M. Hwu, and S.-L. Wang, *J. Organomet. Chem.* **1989**, *371*, 71–79.
[42] Y.-P. Wang, J.-M. Hwu, and S.-L. Wang, *J. Organomet. Chem.* **1990**, *390*, 179–192.
[43] T. S. A. Hor, L.-T. Phang, L.-K. Liu, and Y.-S. Wen, *J. Organomet. Chem.* **1990**, *397*, 29–39.
[44] Y.-P. Wang, J.-M. Hwu, S.-L. Wang, and Y.-J. Wu, *J. Organomet. Chem.* **1991**, *414*, 33–48.
[45] J. C. Barnes, W. Bell, C. Glidewell, and R. A. Howie, *J. Organomet. Chem.* **1990**, *385*, 369–378.
[46] I. R. Butler, W. R. Cullen, F. W. B. Einstein, and A. C. Willis, *Organometallics* **1985**, *4*, 603–604.
[47] D. Albagli, G. Bazan, M. S. Wrighton, and R. R. Schrock, *J. Am. Chem. Soc.* **1992**, *114*, 4150–4158.
[48] J. C. Kotz and C. L. Nivert, *J. Organomet. Chem.* **1973**, *52*, 387–406.
[49] L.-T. Phang, S. C. F. Au-Yeung, T. S. A. Hor, S. B. Khoo, Z.-Y. Zhou, and T. C. W. Mak, *J. Chem. Soc., Dalton Trans.* **1993**, 165–172.
[50] J. C. Kotz, C. L. Nivert, J. M. Lieber, and R. C. Reed, *J. Organomet. Chem.* **1975**, *91*, 87–95.
[51] I. R. Butler, W. R. Cullen, T.-J. Kim, S. J. Rettig, and J. Trotter, *Organometallics* **1985**, *4*, 972–980.
[52] D. L. Du Bois, C. W. Eigenbrot, Jr., A. Miedaner, J. C. Smart, and B. C. Haltiwanger, *Organometallics* **1986**, *5*, 1405–1411.
[53] G. Pilloni, B. Longato, and B. Corain, *J. Organomet. Chem.* **1991**, *420*, 57–65.
[54] A. Houlton, J. R. Miller, R. M. G. Roberts, and J. Silver, *J. Chem. Soc., Dalton Trans.* **1990**, 2181–2184.
[55] B. J. Coe, C. J. Jones, J. A. McCleverty, D. Bloor, P. V. Kolinsky, and R. J. Jones, *J. Chem. Soc., Chem. Commun.* **1989**, 1485–1487.
[56] B. J. Coe, C. J. Jones, J. A. McCleverty, K. Wieghardt, and S. Stötzel, *J. Chem. Soc., Dalton Trans.* **1992**, 719–721.
[57] P. D. Beer, S. M. Charsley, C. J. Jones, and J. A. McCleverty, *J. Organomet. Chem.* **1986**, *307*, C19–C22.
[58] P. D. Beer, C. J. Jones, J. A. McCleverty, and R. P. Sidebotham, *J. Organomet. Chem.* **1987**, *325*, C19–C22.
[59] R. P. Sidebotham, P. D. Beer, T. A. Hamor, C. J. Jones, and J. A. McCleverty, *J. Organomet. Chem.* **1989**, *371*, C31–C34
[60] C. J. Jones, J. A. McCleverty, B. D. Neaves, S. J. Reynolds, H. Adams, N. A. Bailey, and G. Denti, *J. Chem. Soc., Dalton Trans.* **1986**, 733–741.

[61] P. Chaudhuri, K. Wieghardt, Y.-H. Tsai, and C. Krüger, *Inorg. Chem.* **1984**, *23*, 427–432.
[62] W. F. Little, C. N. Reilley, J. D. Johnson, K. N. Lynn, and A. P. Sanders, *J. Am. Chem. Soc.* **1964**, *86*, 1376–1381.
[63] J. Breimair, M. Wieser, B. Wagner, K. Polborn, and W. Beck, *J. Organomet. Chem.* **1991**, *421*, 55–64.
[64] O. Orama and J. Hietala, *XIII ICOC*, Torino (Italy), **1988**, p. 327.
[65] C. Cordier, M. Gruselle, J. Vaissermann, L. L. Troitskaya, V. I. Bakhmutov, V. I. Sokolov, and G. Jaouen, *Organometallics* **1992**, *11*, 3825–3832.
[66] P. Zanello and H. Plenio, unpublished results.
[67] W. B. Rybakov, A. I. Tursina, L. A. Aslanov, S. A. Yeremin, H. Schrauber, and L. Kutschabsky, *Z. Anorg. Allg. Chem.* **1982**, *487*, 217–224.
[68] M. Herberhold, H. Kniesel, L. Haumaier, A. Gieren, and C. Ruiz-Perez, *Z. Naturforsch.* **1986**, *41b*, 1431–1436
[69] M. Herberhold, H. Kniesel, L. Haumaier, and U. Thewalt, *J. Organomet. Chem.* **1986**, *301*, 355–367.
[70] S. C. Crawford, C. B. Knobler, and H. D. Kaesz, *Inorg. Chem.* **1977**, *16*, 3201–3207.
[71] S. Onaka, *Bull. Chem. Soc. Jpn.* **1986**, *59*, 2359–2361.
[72] S. Onaka, T. Moriya, S. Takagi, A. Mizuno, and H. Furuta, *Bull. Chem. Soc. Jpn.* **1992**, *65*, 1415–1427.
[73] S. Onaka, H. Furuta, and S. Takagi, *Angew. Chem. Int. Ed. Engl.* **1993**, *32*, 87–88.
[74] S. Onaka, A. Mizuno, and S. Takagi, *Chem. Lett.* **1989**, 2037–2040.
[75] T. S. Hor and L.-T. Phang, *J. Organomet. Chem.* **1990**, *390*, 345–350.
[76] T. S. A. Hor, H. S. O. Chan, K.-L. Tan, L.-T. Phang, Y. K. Yan, L. K. Liu, and Y.-S. Wen, *Polyhedron* **1991**, *10*, 2437–2450.
[77] P. Braunstein, L. Douce, F. Balegroune, D. Grandjean, D. Bayeul, Y. Dusausoy, and P. Zanello, *New. J. Chem.* **1992**, *16*, 925–929.
[78] P. Zanello, manuscript in preparation.
[79] A. C. Macdonald and J. Trotter, *Acta Cryst.* **1964**, *17*, 872–877.
[80] G. M. Brown, T. J. Meyer, D. O. Cowan, C. Le Vanda, F. Kaufman, P. V. Roling, and M. D. Rausch, *Inorg. Chem.* **1975**, *14*, 506–511.
[81] K. H. Pannell, J. B. Cassias, G. M. Crawford, and A. Flores, *Inorg. Chem.* **1976**, *15*, 2671–2675.
[82] K. L. Bieda, A. L. Kranitz, and J. J. Grzybowski, *Inorg. Chem.* **1993**, *32*, 4209–4213.
[83] L.-K. Liu, S.-K. Yeh, and C.-C. Lin, *Bull. Inst. Chem., Academia Sinica* **1988**, *35*, 45–52.
[84] T.-J. Kim, K.-H. Kwon, S.-C. Kwon, J.-O. Baeg, S.-C. Shim, and D.-H. Lee, *J. Organomet. Chem.* **1990**, *389*, 205–217.
[85] T.-J. Kim, S.-C. Kwon, Y.-H. Kim, N. H. Heo, M. M. Teeter, and A. Yamano, *J. Organomet. Chem.* **1991**, *426*, 71–86.
[86] I. R. Butler, W. R. Cullen, and S. J. Rettig, *Organometallics* **1987**, *6*, 872–880.
[87] A. L. Rheingold, U. T. Mueller-Westerhoff, G. F. Swiegers, and T. J. Haas, *Organometallics* **1992**, *11*, 3411–3417.
[88] G. J. Small and J. Trotter, *Can. J. Chem.* **1964**, *42*, 1746–1748.
[89] M. Watanabe and H. Sano, *Bull. Chem. Soc. Jpn.* **1990**, *63*, 777–784.
[90] M. G. Hill, W. M. Lamanna, and K. R. Mann, *Inorg. Chem.* **1991**, *30*, 4687–4690.
[91] S. P. Gubin and A. A. Lubovich, *J. Organomet. Chem.* **1970**, *22*, 183–194.
[92] A. F. Diaz, U. T. Mueller-Westerhoff, A. Nazzal, and M. Tanner, *J. Organomet. Chem.* **1982**, *236*, C45–C48.
[93] N. Dowling and P. M. Henry, *Inorg. Chem.* **1982**, *21*, 4088–4095.
[94] M. Herberhold, W. Feger, and U. Kölle, *J. Organomet. Chem.* **1992**, *436*, 333–350.
[95] H. Lehmkuhl, R. Schwickardi, C. Krüger, and G. Raabe, *Z. Anorg. Allg. Chem.* **1990**, *581*, 41–47.
[96] Y. Kasahara, Y. Hoshino, M. Kajitani, K. Shimizu, and G. P. Sato, *Organometallics* **1992**, *11*, 1968–1971.
[97] P. D. Beer, O. Kocian, and R. J. Mortimer, *J. Chem. Soc., Dalton Trans.* **1990**, 3283–3288.
[98] M. I. Bruce, I. R. Butler, W. R. Cullen, G. A. Koutsantonis, M. R. Snow, and E. R. T. Tiekink, *Aust. J. Chem.* **1988**, *41*, 963–969.
[99] C. R. S. M. Hampton, I. R. Butler, W. R. Cullen, B. R. James, J.-P. Charland, and J. Simpson, *Inorg. Chem.* **1992**, *31*, 5509–5520.

[100] W. R. Cullen, A. Talaba, and S. J. Rettig, *Organometallics* **1992**, *11*, 3152–3156.
[101] A. A. Koridze, A. I. Yanovsky, and Yu. T. Struchkov, *J. Organomet. Chem.* **1992**, *441*, 277–284.
[102] W. R. Cullen, S. T. Chacon, M. I. Bruce, F. W. B. Einstein, and R. H. Jones, *Organometallics* **1988**, *7*, 2273–2278.
[103] S. T. Chacon, W. R. Cullen, M. I. Bruce, O. B. Shawkataly, F. B. Einstein, R. H. Jones, and A. C. Willis, *Can. J. Chem.* **1990**, *68*, 2001–2010.
[104] M. I. Bruce, P. A. Humphrey, O. B. Shawkataly, M. R. Snow, E. R. Tiekink, and W. R. Cullen, *Organometallics* **1990**, *9*, 2910–2919.
[105] W. R. Cullen, S. J. Rettig, and T. C. Zheng, *Organometallics* **1992**, *11*, 853–858.
[106] S. M. Draper, C. E. Housechroft, and A. L. Rheingold, *J. Organomet. Chem.* **1992**, *435*, 9–20.
[107] A. J. Arce, P. A. Bates, S. P. Best, R. J. H. Clark, A. J. Deeming, M. B. Hursthouse, R. C. S. McQueen, and N. I. Powell, *J. Chem. Soc., Chem. Commun.* **1988**, 478–480.
[108] W. R. Cullen, S. J. Rettig, and T. C. Zheng, *Organometallics* **1992**, *11*, 277–283.
[109] W. R. Cullen, S. J. Rettig, and T. C. Zheng, *Organometallics* **1993**, *12*, 688–696.
[110] W. R. Cullen, S. J. Rettig, and T. C. Zheng, *Organometallics* **1992**, *11*, 928–935.
[111] W. R. Cullen, S. J. Rettig, and T. C. Zheng, *J. Organomet. Chem.* **1993**, *452*, 97–103.
[112] W. R. Cullen, S. J. Rettig, and T. C. Zheng, *Can. J. Chem.* **1992**, *70*, 2215–2223.
[113] W. R. Cullen, S. J. Rettig, and T. C. Zheng, *Organometallics* **1992**, *11*, 3434–3439.
[114] W. R. Cullen, S. J. Rettig, and T. C. Zheng, *Can. J. Chem.* **1992**, *70*, 2329–2334.
[115] M. D. Rausch, F. A. Higbie, G. F. Westover, A. Clearfield, R. Gopal, J. M. Troup, and I. Bernal, *J. Organomet. Chem.* **1978**, *149*, 245–264.
[116] J. Kotz, G. Neyhart, W. J. Vining, and M. D. Rausch, *Organometallics* **1983**, *2*, 79–82.
[117] A. Ohkubo, T. Fujita, S. Ohba, K. Aramaki, and H. Nishihara, *J. Chem. Soc., Chem. Commun.* **1992**, 1553–1555.
[118] M. A. Murguia, D. Borchardt, and S. Wherland, *Inorg. Chem.* **1990**, *29*, 1982–1986.
[119] M. Bracci, C. Ercolani, B. Floris, M. Bassetti, A. Chiesi-Villa, and C. Guastini, *J. Chem. Soc., Dalton Trans.* **1990**, 1357–1363.
[120] W.-D. Fleishmann and H. P. Fritz, *Z. Naturforsch.* **1973**, *28b*, 383–388.
[121] P. Braunstein, D. G. Kelly, A. Tiripicchio, and F. Uguzzoli. *Inorg. Chem.* **1993**, *32*, 4845–4852.
[122] P. Braunstein and P. Zanello, work in progress.
[123] M. Adachi, M. Kita, K. Kashiwabara, J. Fujita, N. Iitaka, S. Kurachi, S. Ohba, and D. Jin, *Bull. Chem. Soc. Jpn.* **1992**, *65*, 2037–2044.
[124] S. B. Colbran, L. R. Hanton, B. H. Robinson, W. T. Robinson, and J. Simpson, *J. Organomet. Chem.* **1987**, *330*, 415–428.
[125] G. H. Worth, B. H. Robinson, and J. Simpson, *Organometallics* **1992**, *11*, 501–513.
[126] S. Colbran, B. H. Robinson, and J. Simpson, *J. Chem. Soc., Chem. Commun.* **1982**, 1361–1362.
[127] S. Colbran, B. H. Robinson, and J. Simpson, *Organometallics* **1983**, *2*, 943–951.
[128] G. H. Worth, B. H. Robinson, and J. Simpson, *Organometallics* **1992**, *11*, 3863–3874.
[129] S. Colbran, B. H. Robinson, and J. Simpson, *Organometallics* **1983**, *2*, 952–957.
[130] S. Colbran, B. H. Robinson, and J. Simpson, *Organometallics* **1984**, *3*, 1344–1353.
[131] W. H. Watson, A. Nagl, S. Hwang, and M. G. Richmond, *J. Organomet. Chem.* **1993**, *445*, 163–170.
[132] J. Borgdorff, E. J. Ditzel, N. W. Duffy, B. H. Robinson, and J. Simpson, *J. Organomet. Chem.* **1992**, *437*, 323–346.
[133] M. T. Ahmet, R. A. Brown, R. M. G. Roberts, J. R. Miller, J. Silver, and A. Houlton, *Acta Cryst.* **1993**, *C49*, 1616–1619.
[134] E. S. Shubina, L. M. Epstein, Y. L. Slovokhotov, A. V. Mironov, Y. T. Struchkov, V. S. Kaganovich, A. Z. Kreindlin, and M. I. Rybinskaya, *J. Organomet. Chem.* **1991**, *401*, 155–165.
[135] M. Herberhold, G.-X. Jin, A. L. Rheingold, and G. F. Sheats, *Z. Naturforsch.* **1992**, *47b*, 1091–1098.
[136] P. Zanello, M. Casarin, L. Pardi, J. Silver, M. Herberhold and G.-X. Lin, manuscript in preparation.
[137] F. Estevan, P. Lahuerta, J. Latorre, E. Peris, S. Garcia-Granda, F. Gómez-Beltrán, A. Aguirre, and M. A. Salvadó, *J. Chem. Soc., Dalton Trans.* **1993**, 1681–1688.
[138] F. Estevan, J. Latorre, and E. Peris, *Polyhedron* **1993**, *12*, 2153–2156.

[139] W. R. Cullen, F. W. B. Einstein, C.-H. Huang, A. C. Willis, and E.-S. Yeh, *J. Am. Chem. Soc.* **1980**, *102*, 988–993.
[140] W. R. Cullen, T.-J. Kim, F. W. B. Einstein, and T. Jones, *Organometallics* **1983**, *2*, 714–719.
[141] W. R. Cullen, T.-J. Kim, F. W. B. Einstein, and T. Jones, *Organometallics* **1985**, *4*, 346–351.
[142] G. J. Lamprecht, J. C. Swarts, J. Conradie, and J. G. Leipoldt, *Acta Cryst.* **1993**, *C49*, 82–84.
[143] U. Casellato, B. Corain, R. Graziani, B. Longato, and G. Pilloni, *Inorg. Chem.* **1990**, *29*, 1193–1198.
[144] R. B. Bedford, P. A. Chaloner, and P. B. Hitchcock, *Acta Cryst.* **1993**, *C49*, 1614–1616.
[145] G. De Santis, L. Fabbrizzi, M. Licchelli, P. Pallavicini, and A. Perrotti, *J. Chem. Soc., Dalton Trans.* **1992**, 3283–3284.
[146] A. De Blas, G. De Santis, L. Fabbrizzi, M. Licchelli, C. Mangano, and P. Pallavicini, *Inorg. Chim. Acta* **1992**, *202*, 115–118.
[147] G. De Santis, L. Fabbrizzi, M. Licchelli, C. Mangano, P. Pallavicini, and A. Poggi, *Inorg. Chem.* **1993**, *32*, 854–860.
[148] S. Sahami and M. J. Weaver, *J. Solution Chem.* **1981**, *10*, 199–207.
[149] T. Kawamoto and Y. Kushi, *J. Chem. Soc., Dalton Trans.* **1992**, 3137–3143.
[150] U. Casellato, D. Ajó, G. Valle, B. Corain, B. Longato, and R. Graziani, *J. Cryst. Spectr. Res.* **1988**, *18*, 583–590.
[151] B. Corain, B. Longato, G. Favero, D. Ajó, G. Pilloni, U. Russo, and F. R. Kreissl, *Inorg. Chim. Acta* **1989**, *157*, 259–266.
[152] A. Louati and M. Huhn, *Inorg. Chem.* **1993**, *32*, 3601–3607.
[153] D. Matt, M. Huhn, J. Fischer, A. De Cian, W. Kläui, I. Tkatchenko, and M. C. Bonnet, *J. Chem. Soc., Dalton Trans.* **1993**, 1173–1178.
[154] C. G. Pierpont and R. Eisenberg, *Inorg. Chem.* **1972**, *11*, 828–832.
[155] T. Hayashi, M. Konoshi, Y. Kobori, M. Kumada, T. Higuchi, and K. Hirotsu, *J. Am. Chem. Soc.* **1984**, *106*, 158–163.
[156] C. E. Housecroft, S. M. Owen, P. R. Raithby, and B. A. M. Shaykh, *Organometallics* **1990**, *9*, 1617–1623.
[157] M. Sato, H. Asano, and S. Akabori, *J. Organomet. Chem.* **1993**, *452*, 105–109.
[158] M. Sato, M. Sekino, and S. Akabori, *J. Organomet. Chem.* **1988**, *344*, C31–C34.
[159] M. Sato, M. Sekino, M. Katada, and S. Akabori, *J. Organomet. Chem.* **1989**, *377*, 327–337.
[160] M. Sato, H. Shigeta, M. Sekino, and S. Akabori, *J. Organomet. Chem.* **1993**, *458*, 199–204.
[161] A. Louati, M. Gross, L. Douce, and D. Matt, *J. Organomet. Chem.* **1992**, *438*, 167–182.
[162] J. C. Kotz, E. E. Getty, and L. Lin, *Organometallics* **1985**, *4*, 610–612.
[163] R. A. Holwerda, J. S. Kim, T. W. Robinson, R. A. Bartsch, B. P. Czech, *J. Organomet. Chem.* **1993**, *443*, 123–129.
[164] B. Longato, G. Pilloni, G. Valle, and B. Corain, *Inorg. Chem.* **1988**, *27*, 956–958.
[165] M. Onishi, K. Hiraki, S. Wada, Y. Ohama, *Polyhedron* **1987**, *6*, 1243–1245.
[166] P. D. Harvey and L. Gan, *Inorg. Chem.* **1991**, *31*, 3239–3241.
[167] T. Ukai, H. Kawayura, Y. Ishii, J. J. Bonnet, and J. A. Ibers, *J. Organomet. Chem.* **1974**, *65*, 253–266.
[168] C. G. Pierpont and M. C. Mazza, *Inorg. Chem.* **1974**, *13*, 1891–1895.
[169] D. Seyferth, B. W. Hames, T. G. Rucker, M. Cowie, and R. S. Dickson, *Organometallics* **1983**, *2*, 472–474.
[170] M. Cowie and R. S. Dickson, *J. Organomet. Chem.* **1987**, *326*, 269–280.
[171] T. Hayashi, M. Kumada, T. Higuchi, and K. Hirotsu, *J. Organomet. Chem.* **1987**, *334*, 195–203.
[172] C. López, J. Sales, X. Solans, and R. Zquiak, *J. Chem. Soc., Dalton Trans.* **1992**, 2321–2328.
[173] T. Hayashi, A. Yamamoto, M. Hojo, K. Kishi, Y. Ito, E. Nishioka, H. Miura, and K. Yanagi, *J. Organomet. Chem.* **1989**, *370*, 129–139.
[174] P. Braunstein, T. M. G. Carneiro, D. Matt, F. Balegroune, and D. Grandjean, *J. Organomet. Chem.* **1989**, *367*, 117–132.
[175] D. A. Clemente, G. Pilloni, B. Corain, B. Longato, M. Tiripicchio-Camellini, *Inorg. Chim. Acta* **1986**, *115*, L9–L11.
[176] O. Carugo, G. De Santis, L. Fabbrizzi, M. Licchelli, A. Monichino, and P. Pallavicini, *Inorg. Chem.* **1992**, *31*, 765–769.
[177] C. E. L. Headford, R. Mason, P. R. Ranatunge-Bandarage, B. H. Robinson, and J. Simpson, *J. Chem. Soc., Chem. Commun.* **1990**, 601–603.

[178] S. Akabori, T. Kumagai, T. Shirahige, S. Sato, K. Kawazoe, C. Tamura, and M. Sato, *Organometallics* **1987**, *6*, 526–531.
[179] W. R. Cullen, S. V. Evans, N. F. Han, and J. Trotter, *Inorg. Chem.* **1987**, *26*, 514–519.
[180] P. F. Kelly, A. M. Z. Slawin, D. J. Williams, and J. D. Woollins, *Polyhedron* **1988**, *7*, 1925–1930.
[181] R. E. Hollands, A. G. Osborne, R. H. Whiteley, and C. J. Cardin, *J. Chem. Soc., Dalton Trans.* **1985**, 1527–1530.
[182] A. L. Bandini, G. Banditelli, M. A. Cinellu, G. Sanna, G. Minghetti, F. Demartin, and M. Manassero, *Inorg. Chem.* **1989**, *28*, 404–410.
[183] B. S. Haggerty, C. E. Housechroft, A. L. Rheingold, and B. A. M. Shaykh, *J. Chem. Soc., Dalton Trans.* **1991**, 2175–2184.
[184] A. L. Abuhijleh and C. Woods, *J. Chem. Soc., Dalton Trans.* **1992**, 1249–1252.
[185] P. Zanello, unpublished results.
[186] P. D. Beer, J. E. Nation, S. L. McWhinnie, M. E. Harman, M. B. Hursthouse, M. I. Ogden, and A. H. White, *J. Chem. Soc., Dalton Trans.* **1991**, 2485–2492.
[187] G. Pilloni, B. Corain, M. Degano, B. Longato, and G. Zanotti, *J. Chem. Soc., Dalton Trans.* **1993**, 1777–1778.
[188] A. Benito, J. Cano, R. Martinez-Manez, J. Soto, J. Paya, F. Lloret, M. Julve, J. Faus, and M. D. Marcos, *Inorg. Chem.* **1993**, *32*, 1197–1203.
[189] U. Casellato, R. Graziani, and G. Pilloni, *J. Cryst. Spectr. Res.* **1993**, *23*, 571–575.
[190] G. Pilloni and B. Longato, *Inorg. Chim. Acta* **1993**, *208*, 17–21.
[191] M. R. Churchill, Y.-J. Li, D. Nalewajek, P. M. Schaber, and J. Dorfman, *Inorg. Chem.* **1985**, *24*, 2684–2687.
[192] M. A. El-Sayed, A. Ali, G. Davies, S. Larsen, and J. Zubieta, *Inorg. Chim. Acta* **1992**, *194*, 139–149.
[193] A. Togni, G. Rihs, and R. E. Blumer, *Organometallics* **1992**, *11*, 613–621.
[194] A. N. Nesmeyanov, Y. T. Struchkov, N. N. Sedova, V. G. Andrianov, Y. V. Volgin, and V. A. Sazonova, *J. Organomet. Chem.* **1977**, *137*, 217–221.
[195] J. C. Medina, T. T. Goodnow, M. T. Rojas, J. L. Atwood, B. C. Lynn, A. E. Kaifer, and G. W. Gokel, *J. Am. Chem. Soc.* **1992**, *114*, 10583–10595.
[196] A. N. Nesmeyanov, N. N. Sedova, Y. T. Struchkov, V. G. Andrianov, E. N. Stakheevam, and V. A. Sazonova, *J. Organomet. Chem.* **1978**, *153*, 115–122.
[197] T. S. A. Hor, S. P. Neo, C. S. Tan, T. C. W. Mak, K. W. P. Leung, and R.-J. Wang, *Inorg. Chem.* **1992**, *31*, 4510–4516.
[198] D. T. Hill, G. R. Girard, F. L. McCabe, R. K. Johnson, P. D. Stupik, J. H. Zhang, W. M. Reiff, and D. S. Eggleston, *Inorg. Chem.* **1989**, *28*, 3529–3533.
[199] A. Houlton, R. M. G. Roberts, J. Silver, and R. V. Parish, *J. Organomet. Chem.* **1991**, *418*, 269–275.
[200] A. Togni, S. D. Pastor, and G. Rihs, *J. Organomet. Chem.* **1990**, *381*, C21–C25.
[201] V. G. Andrianov, Y. T. Struchkov, and E. R. Rossinskaya, *J. Chem. Soc., Chem. Commun.* **1973**, 338–339.
[202] A. N. Nesmeyanov, E. G. Perevalova, K. I. Grandberg, D. A. Lemenovskii, T. V. Baukova, and O. B. Afanassova, *J. Organomet. Chem.* **1974**, *65*, 131–144.
[203] P. D. Beer, O. Kocian, R. J. Mortimer, and P. Spencer, *J. Chem. Soc., Chem. Commun.* **1992**, 602–604.
[204] A. Houlton, S. K. Ibrahim, J. R. Dilworth, and J. Silver, *J. Chem. Soc., Dalton Trans.* **1990**, 2421–2424.
[205] P. Zanello and F. T. Edelmann, work in progress.
[206] W. H. Morrison, Jr., S. Krogsrud, and D. N. Hendrickson, *Inorg. Chem.* **1973**, *12*, 1998–2004.
[207] K. R. Mann, W. H. Morrison, Jr., and D. N. Hendrickson, *Inorg. Chem.* **1974**, *13*, 1180–1185.
[208] B. J. Coe, C. J. Jones, J. A. McCleverty, D. Bloor, P. V. Kolinsky, and R. J. Jones, *Polyhedron* **1994**, *13*, 2107–2115 (7.2.3.2, B; 7.2.3.3, B).
[209] R. H. Cayton, M. H. Chisholm, J. C. Huffman, and E. B. Lobkovsky, *J. Am. Chem. Soc.* **1991**, *113*, 8709–8724 (7.2.3.3, B).
[210] M. Sekino, M. Sato, A. Nagasawa, and K. Kikuchi, *Organometallics* **1994**, *13*, 1451–1455 (7.2.3.3, B).
[211] H. Schulz, K. Folting, J. C. Huffman, W. E. Streib, and M. H. Chisholm, *Inorg. Chem.* **1993**, *32*, 6056–6066 (7.2.3.3, C).

[212] M. Sato, Y. Hayashi, H. Shintate, M. Katada, and S. Kawata, *J. Organomet. Chem.* **1994**, *471*, 179–184 (7.2.5.1, B).
[213] E. C. Constable, A. J. Edwards, R. Martínez-Máñez, P. R. Raithby, and A. M. W. C. Thompson, *J. Chem. Soc., Dalton Trans.* **1994**, 645–650 (7.2.5.1, B; 7.2.5.2, B; 7.2.6.1, B).
[214] E. C. Constable, R. Martínez-Máñez, A. M. W. C. Thompson, and J. V. Walker, *J. Chem. Soc., Dalton Trans.* **1994**, 1585–1594 (7.2.5.1, B; 7.2.5.2, B; 7.2.6.1, B; 7.2.7.1, B; 7.2.8.1, B).
[215] M. Sato, H. Shintate, Y. Kawata, M. Sekino, M. Katada, and S. Kawata, *Organometallics* **1994**, *13*, 1956–1962 (7.2.5.2, A).
[216] B. Farlow, T. A. Nile, J. L. Walsh, and A. T. McPhail, *Polyhedron* **1993**, *12*, 2891–2894 (7.2.5.2, B).
[217] M. C. B. Colbert, S. L. Ingham, J. Lewis, N. J. Long, and P. Raithby, *J. Chem. Soc., Dalton Trans.* **1994**, 2215–2216 (7.2.5.2, B; 7.2.5.3, A).
[218] W. R. Cullen, S. J. Rettig, and T. C. Zheng, *Can. J. Chem.* **1994**, *71*, 399–409 (7.2.5.3, C).
[219] K. Onitsuka, X.-Q. Tao, W.-Q. Wang, Y. Otsuka, K. Sonogashira, T. Adachi, and T. Yoshida, *J. Organomet. Chem.* **1994**, *473*, 195–204 (7.2.6.1, C).
[220] M. Sawamura, R. Kuwano, and Y. Ito, *Angew. Chem.* **1994**, *106*, 92–93; *Angew. Chem. Int. Ed. Engl.* **1994**, *33*, 111–113 (7.2.6.2, C).
[221] A. Togni, C. Breutel, M. C. Soares, N. Zanetti, T. Gerfin, V. Gramlich, F. Spindler, and G. Rihs, *Inorg. Chim. Acta* **1994**, *222*, 213–224 (7.2.6.2, C; 7.2.7.2, C; 7.2.7.3, C).
[222] A. M. Allgeier, E. T. Singewald, C. A. Mirkin, and C. L. Stern, *Organometallics* **1994**, *13*, 2928–2930 (7.2.6.2, C).
[223] P. D. Beer, Z. Chen, M. G. B. Drew, J. Kingston, M. Ogden, and P. Spencer, *J. Chem. Soc., Chem. Commun.* **1993**, 1046–1048 (7.2.7.1, B; 7.2.8.1, B; 7.2.9.1, B).
[224] G. De Santis, L. Fabbrizzi, M. Licchelli, C. Mangano, and P. Pallavicini, *Inorg. Chim. Acta* **1993**, *214*, 193–196 (7.2.7.1, B).
[225] M. Sato, K. Suzuki, H. Asano, M. Sekino, Y. Kawata, Y. Habata, and S. Akabori, *J. Organomet. Chem.* **1994**, *470*, 263–269 (7.2.7.2, A; 7.2.7.3, A).
[226] B. Jedlicka, C. Kratky, W. Weissensteiner, and M. Widhalm, *J. Chem. Soc., Chem. Commun.* **1993**, 1329–1330 (7.2.7.2, C).
[227] K. Hamamura, M. Kita, M. Nonoyama, and J. Fujita, *J. Organomet. Chem.* **1993**, *463*, 169–177 (7.2.7.2, C).
[228] R. Bosque, C. López, J. Sales, X. Solans, and M. Font-Bardía, *J. Chem. Soc., Dalton Trans.* **1994**, 735–745 (7.2.7.2, C).
[229] K. V. Katti, Y. W. Ge, P. R. Singh, S. V. Date, and C. L. Barnes, *Organometallics* **1994**, *13*, 541–547 (7.2.7.2, C).
[230] T. B. Baumann, J. W. Sibert, M. M. Olmstead, A. G. M. Barrett, and B. M. Hoffman, *J. Am. Chem. Soc.* **1994**, *116*, 2639–2640 (7.2.7.2, C).
[231] M. V. Russo, A. Furlani, S. Licoccia, R. Paolesse, A. Chiesi Villa, and C. Guastini, *J. Organomet. Chem.* **1994**, *469*, 245–252 (7.2.7.3, C).
[232] P. R. R. Ranatunge-Bandarage, B. H. Robinson, and J. Simpson, *Organometallics* **1994**, *13*, 500–510 (7.2.7.3, C).
[233] P. R. R. Ranatunge-Bandarage, N. W. Duffy, S. M. Johnston, B. H. Robinson, and J. Simpson, *Organometallics* **1994**, *13*, 511–521 (7.2.7.3, C).
[234] B. Delavaux-Nicot, R. Mathieu, D. de Montauzon, G. Lavigne, and J.-P. Majoral, *Inorg. Chem.* **1994**, *33*, 434–443 (7.2.8.1, B).
[235] A. L. Abuhijleh, J. Pollitte, and C. Woods, *Inorg. Chim. Acta* **1994**, *215*, 131–137 (7.2.8.1, C).
[236] S.-P. Neo, T. S. A. Hor, Z.-Y. Zhou, and T. C. W. Mak, *J. Organomet. Chem.* **1994**, *464*, 113–119 (7.2.8.2, C).
[237] M. C. Gimeno, A. Laguna, C. Sarroca, and P. G. Jones, *Inorg. Chem.* **1993**, *32*, 5926–5932 (7.2.8.3, A).
[238] R. D. Rakhimov, K. P. Butin, and K. I. Grandberg, *J. Organomet. Chem.* **1994**, *464*, 253–260 (7.2.8.3, B).
[239] A. Houlton, D. M. P. Mingos, D. M. Murphy, D. J. Williams, L.-T. Phang, and T. S. A. Hor, *J. Chem. Soc., Dalton Trans.* **1993**, 3629–3630 (7.2.8.3, C).
[240] L.-T. Phang, T. S. A. Hor, Z.-Y. Zhou, and T. C. W. Mak, *J. Organomet. Chem.* **1994**, *469*, 253–261 (7.2.8.3, C).
[241] S. Q. Huo, Y. J. Wu, Y. Zhu, and L. Yang, *J. Organomet. Chem.* **1994**, *470*, 17–22 (7.2.9.3, C).

Received: November 12, 1993

Part 3. Materials Science

8 Ferrocene-Containing Charge-Transfer Complexes. Conducting and Magnetic Materials

Antonio Togni

8.1 Introduction

The phenomena of electricity and, even more, of magnetism have captivated the imagination and fascination of mankind since their discovery. Whereas it is a commonplace to say that modern society would be inconceivable without the products derived from exploiting these phenomena, the search for new materials having unprecedented properties is nowadays a very active field of research, both in industry and academia. Thus, the discovery, during the past decade, of molecular materials displaying properties usually connected with the metallic phase, such as superconductivity [1] and ferromagnetism [2], has invigorated the activity in the field and inspired many research groups. A possible practical criterion justifying research in this field is that molecular materials, as compared with metals, are in general lighter, soluble, transparent, and may have particular optical properties. Among others, these aspects could open new, unique, and more ecomic opportunities for the processing of these materials. In other words, molecular materials constitute new perspectives, in particular for the electronic (optotronic) industry, which is aiming at nano- and molecular scale devices [3].

Charge-transfer (CT) complexes containing a donor (D) and an acceptor component (A) capable of forming stable radical cations and anions, respectively, and displaying a stacked structure in the solid state, are, depending on the type of stacking (i.e., segregated A and D stacks *vs.* alternating stacks, see Sect. 8.2), electrical conductors with potential metallic character, or may possess particular magnetic properties. Such CT complexes, because of their structure, constitute low-dimensional solids and therefore their physical properties will show considerable anisotropy. Since the discovery in 1973 of the high electrical conductivity and metallic behavior of the one-dimensional CT complex formed by the donor tetrathiafulvalene (TTF) and the acceptor tetracyano-*p*-quinodimethane (TCNQ) [4], research in the field has boomed [5]. In 1980 there followed the discovery of the first molecular compounds becoming superconducting at a temperature around 1 K and even at ambient pressure: the tetramethyltetraselenofulvalene (TMTSF) salts [6]. After these milestones (the interested reader may gather more information about the development of the field from the excellent reviews available [1, 5]), efforts have been concentrated on two main structural types [7]. Those containing the C_2E_4-fragment (E = S, Se), with the parent TTF donor and the ligand dmit as the most prominent prototypes, and the quinoid structural types dominated by TCNQ and dicyanoquinodiimine (DCNQI) [8] (representative examples are shown in Scheme 8-1). Both structural elements lead to planar molecules allowing for the formation of segregated stacks of donors and acceptors in the solid state, this being the basis for high conductivity. Synthetic chemists have focused their attention on variations and combinations of these basic structures.

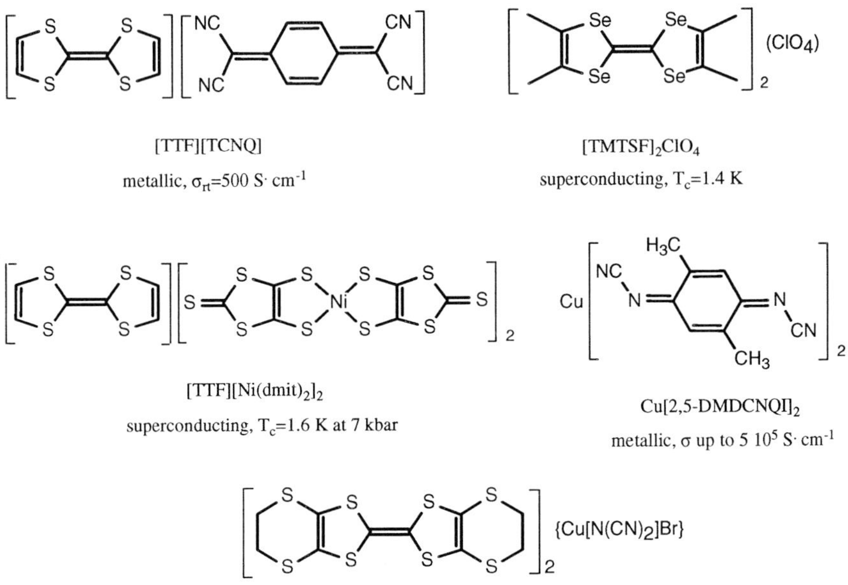

[TTF][TCNQ]
metallic, σ_{rt}=500 S· cm^{-1}

$[TMTSF]_2ClO_4$
superconducting, T_c=1.4 K

$[TTF][Ni(dmit)_2]_2$
superconducting, T_c=1.6 K at 7 kbar

$Cu[2,5\text{-}DMDCNQI]_2$
metallic, σ up to 5 10^5 S· cm^{-1}

κ-$(BEDT\text{-}TTF)_2Cu[N(CN)_2]Br$
superconducting, T_c=11.6 K

Scheme 8-1. Selected examples of molecular conducting and superconducting materials.

What appears to be a new trend, from a synthetic point of view, is the preparation of derivatives that may be considered as oligomers of the parent compounds [9], but completely different structural types have not yet been the object of systematic studies. In particular, there are only few reports in the literature about the use of organometallics for the preparation of conducting CT complexes [10].

As far as organometallics are concerned, and important exception is represented by the (per)alkylated metallocenes and the related bis(arene) complexes of various transition metals. These compounds have been intensively studied in connection with the magnetic properties of their CT complexes, mainly by Miller and co-workers [11]. These studies led to the discovery of the first organometallic compound displaying bulk ferromagnetic properties (vide infra).

The aim of this chapter is to review the known ferrocene-containing CT complexes from a chemical-structural point of view. Therefore, the main criteria for a classification will be the number and nature of the substituents attached to the ferrocene core, as well as the nature of the acceptor partners in the CT complexes.

For tutorials concerning the fundamental physical bases of electrical conductivity [12] and magnetic behavior [13], the reader is referred to other reviews available in the literature.

8.2 Chemical and Structural Requirements

From a structural point of view, a CT complex may display one of two main types of relative arrangement of donor and acceptor molecules in the solid state. These are either segregated in separate stacks ($...D^+D^+D^+D^+D^+...; ...A^-A^-A^-A^-A^-...$), or they form alternate stacks ($...D^+A^-D^+A^-D^+A^-D^+A^-...$). In the former case, the corresponding compounds are likely to be electrically conducting, their specific properties possibly varying from weakly conducting, semiconductor type, or metallic in character. However, for the attainment of high electrical conductivity, the simple geometric constraint of having separate stacks is not yet sufficient. The charge-density distribution in the stacks should be uniform, but the degree of charge-transfer should not be total. In other word the nominal charge of each ion in the stacks should be less than 1. If complete charge-transfer occurs, electron-transport will be hindered because of the intermediate formation of energetically disfavoured, highly charged species, and the material will be at best a semiconductor. Furthermore, the view of a regular distribution of molecules in the donor and acceptors columns is somewhat oversimplified. In reality, besides regular stacking, one also observes much more complex structures. Thus, a variety of irregular stacks of parallel dimers, trimers, and even tetramers, but also deformed stacks, as well as orthogonal dimers, have been observed [14].

On the other hand, alternate stacks will generally lead to insulating materials. As will be seen later, it is this type of structure that afforded ferri- and ferromagnetic compounds.

In order to form CT complexes displaying any type of stacking in the solid state, as described above, donor and acceptor molecules should meet several criteria [15], briefly summarized here. (a) For the sake of optimum geometric matching, the molecules should be planar or composed of planar fragments, thus allowing a parallel arrangement. It has been suggested that a matching of donor and acceptor sizes also favors high conductivity [16]. (b) They should form stable radical species in which the energy gap between HOMO and LUMO is reasonably small. This is attained, for example, by the incorporation of heteroatoms that also enhance molecule polarizability. Since the formation of CT complexes is a redox process, the redox potential of the two partners should match, i.e., the difference between the potential E_{1D} and E_{1A} (for the reactions $D^+ + e^- \leftrightarrow D^0$ and $A^0 + e^- \leftrightarrow A^-$, respectively) should not be too large (≤ 0.25 V). (c) The molecules should possess extended π-systems and should be able to approach one another at distances closer than the sum of their van der Waals radii, in order to increase intermolecular overlap.

How does ferrocene and its derivatives meet these constraints? First, from a geometric point of view and as opposed to most known organic donors (or acceptors), ferrocene is a more compact, cylindrical molecule, composed of *two* planar cyclopentadienyl rings held together by strong covalent interactions with the iron atom, and occupying *two parallel planes*. The distance of ca. 3.3 Å between the two cyclopentadienyl (Cp) rings, encountered in most ferrocene derivatives, will have an important impact on the structure of the corresponding CT complexes, because it will influence the spacing between the CT partners (vide infra). Ferrocenes are prone to one-electron reversible oxidation (depending on the substituents, a two-electron reversible oxidation is also possible), and the corresponding ferrocenium salts obtained are stable. Thus, ferrocenes, as long as they do not contain too bulky substituents that will not be able to adopt a coplanar arrangement with the Cp rings, are suitable donor compounds for the formation of CT complexes. An important advantage of these classical organometallic complexes is that their electron-donor ability may be fine-tuned by the choice of number and nature of the substituents. As revealed by the data collected in Table 8-1 [17], the redox potential of ferrocenes can be varied over a span of more than 1 V. It is also clear that ferrocene itself is relatively difficult to oxidize (by ca. 0.5 V as compared with

Table 8-1. Redox potentials (vs. SCE) for representative ferrocene derivatives and for some typical acceptors

Compound	E^1, V[a]	Ref.
$[Cp_2Fe]$	+0.44	[17c]
$[(CpCOOH)(CpCOOMe)Fe]$	+0.85	[17c, d]
$[(PhCp)CpFe]$	+0.37	[17c]
$[(PhCp)_2Fe]$	+0.37	[17c]
Biferrocene	+0.31 (+0.64)[b]	[58]
$[(MeCp)_2Fe]$	+0.24	[17b]
1,1′-Biferrocenylene	+0.13 (+0.72)[b]	[58]
$[Cp^*CpFe]$	+0.12	[17a]
$[(C_5Et_5)_2Fe]$	−0.11	[17a]
$[Cp^*_2Fe]$	−0.12	[17a]
TTF	+0.30	[17e]
TCNE	+0.15	[17e]
TCNQ	+0.17	[17e]
$[TCNQ]^-$	−0.37	[17e]
$TCNQ(CN)_2$	+0.65	[17e]
$TCNQF_4$	+0.53	[17e]
$[TCNQF_4]^-$	+0.02	[17e]
$TCNQI_2$	+0.35	[17e]
DCNQI	+0.19	[35]
$[Ni\{C_2S_2(CF_3)_2\}_2]$	+0.92	[17h]
$[Pt\{C_2S_2(CF_3)_2\}_2]$	+0.82	[17e]
$[Ni\{C_2S_2(CN)_2\}_2]$	+1.02	[17h]
$[Ni\{C_2S_2(CN)_2\}_2]^-$	+0.23	[17h]
$[Mo\{C_2S_2(CF_3)_2\}_3]$	+0.95	[17h]
$[Mo\{C_2S_2(CF_3)_2\}_3]^-$	+0.36	[17h]

a For the reactions $D^+ + e^- \leftrightarrow D^0$, or $A^0 + e^- \leftrightarrow A^-$, respectively, most commonly in acetonitrile.

b Potential in brackets is for the reaction $D^{2+} + e^- \leftrightarrow D^+$.

decamethyl ferrocene), thus it will form only very weak CT complexes with most acceptors. For comparison, Table 8-1 also includes the redox potential of some of the most common organic and inorganic acceptors encountered in the following sections.

8.3 Charge-Transfer Complexes of Polyalkylated Ferrocenes

Before turning our attention to alkylated ferrocenes, it is worthwile to discuss some of the few CT complexes reported for the parent compound, Cp_2Fe. In a 1962 paper [18], Webster et al. reported that the reaction between ferrocene and tetracyano-ethylene (TCNE) in the solid state at 60 °C afforded a green 1 : 1 CT complex. This complex was found to be soluble in acetonitrile, where ESR spectroscopy indicated the presence of the TCNE anion radical and the ferrocenium cation radical. Addition of cyclohexane, however, was sufficient to reverse the equilibrium, restoring the neutral species and indicating the weak nature of such a CT complex. Further studies by Rosenblum and co-workers provided more detailed insight into the physical properties of the compound, showing it to be paramagnetic in the solid state, but having a modest degree of electron-transfer [19]. The crystal structure was reported in 1967 and showed the typical (but at that time still rare) alternating ...ADADAD... arrangement in the solid state (Fig. 8-1) [20]. The C_2 axis passing through the midpoint of the TCNE molecule is virtually coincident with the C_5

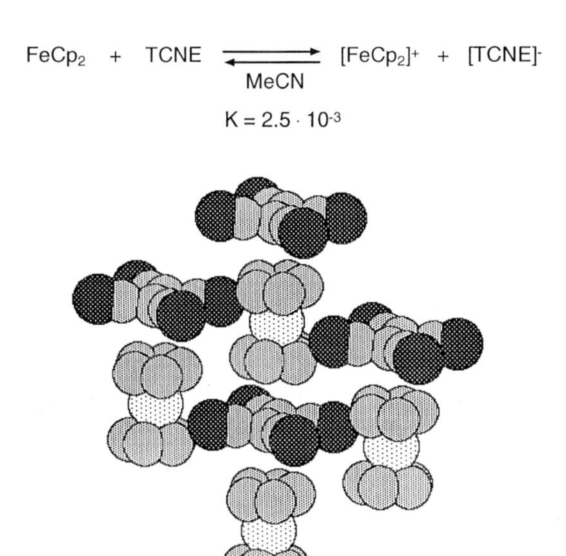

Fig. 8-1. The equilibrium between $[Cp_2Fe]$ and TCNE in solution and the solid state structure of the CT complex $[Cp_2Fe][TCNE]$ (adapted from ref. [20]).

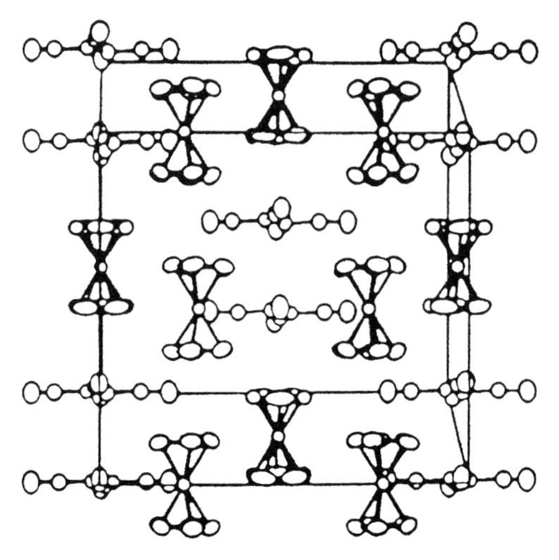

Fig. 8-2. The solid state structure of the CT complex $[Cp_2Fe]^+_{1.5}[(NC)_2C{=}C(CN)O]^-$ (reprinted with permission from ref. [21], copyright 1983 American Chemical Society).

axis of the Cp ring and each stack is surrounded in a hexagonal way by six others.

The CT complex $[Cp_2Fe][TCNE]$ was found to decompose in solution, not only affording the neutral congeners, but also other species. This is due to the complex chemistry of the TCNE radical anion [18]. For example, when equivalent amounts of ferrocene and TCNE are allowed to react in warm ethyl acetate, $[Cp_2Fe]^+_{1.5} \cdot [(NC)_2C{=}C(CN)O]^-$ is formed in good yield [21]. This material was characterized by X-ray crystallography and found to contain the anion tricyanoethenolate, which was shown to be formed from $[TCNE]^-$ in the presence of oxygen (Fig. 8-2). The stoichiometry of this CT complex is also quite unusual and it is best described as a ternary phase of the type $[D^+][D_{0.5}A^-]$.

In view of the low degree of electron-transfer in $[Cp_2Fe][TCNE]$, Miller, Reiff and co-workers have used the stronger electron acceptor 7,7,8,8-tetracyanoperfluoro-*p*-quinodimenthane($TCNQF_4$) and prepared two corresponding CT complexes of 1 : 1 and 2 : 3 stoichiometry, respectively [22]. In $[Cp_2Fe]^+[TCNQF_4]^-$ cations and anions are segregated in separate stacks, where alternating short/long interplane distances (3.225 and 3.675 Å) in the acceptor stacks are observed, i.e., $[TCNQF_4]^-$ forms $[TCNQF_4]_2^{2-}$ dimers, a feature previously detected in other salts and also for TCNQ [23]. On the other hand, the 2 : 3 CT complex contains anion dimers as well as neutral acceptor molecules arranged in an orthogonal manner to each other, this being a rare structural motif for CT complexes containing TCNQ, whereas cations and anions occupy segregated chains. The assignment of the formal charge of this type of acceptor, as reported by Ward and Johnson [24], is usually achieved using IR spectroscopy. Indeed, the CN stretching frequency is very sensitive to the state of charge of the molecule. Both complexes were reported not to show any signal in their ESR spectra. This behavior was attributed to fast paramagnetic relaxation of the $S = \frac{1}{2}\,[Cp_2Fe]^+$ cation, which couples with the $[TCNQF_4]_2^{2-}$ triplets

($S = 1$), thus broadening the signals beyond observation. No further electric or magnetic properties of these two CT complexes were reported.

An interesting and somewhat esthetically appealing CT complex of ferrocene and C_{60} fullerene was recently reported by Kroto and co-workers [25]. A $C_{60}[Cp_2Fe]_2$ stoichiometry was found when black crystals were isolated from concentrated solutions containing a 1 : 2 ratio of the two components. As illustrated in Fig. 8-3, this material consists of layers of C_{60} molecules arranged in a closed-packed manner and stacked one above the other. The space between the layers is occupied by the ferrocene molecules, which show both a side-on and a face-to-face contact with the fullerene. This second interaction takes place between a pentagonal face of C_{60} and one Cp. The two five-membered rings are slipped sideways to each other by 0.8 Å. The compound was formulated as a weak CT complex containing essentially neutral partners.

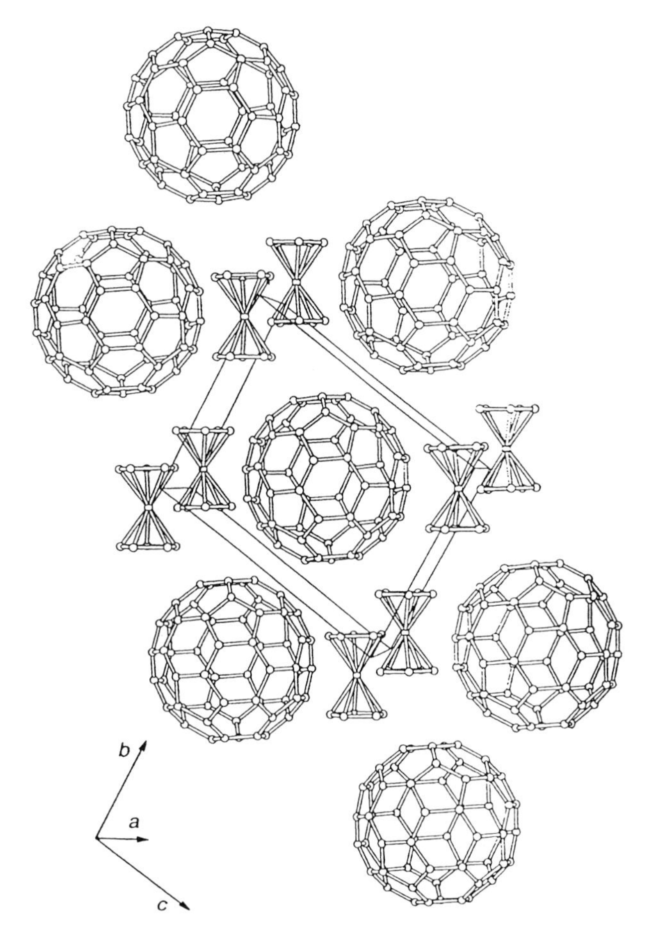

Fig. 8-3. The solid state structure of the CT complex between the fullerene C_{60} and ferrocene, $C_{60}[Cp_2Fe]_2$. View is perpendicular to the *ab* plane (reprinted with permission from ref. [25], copyright 1992 Royal Chemical Society).

8.3.1 Decamethylferrocene

Decamethylferrocene, $[FeCp^*_2]$ ($Cp^* = \eta^5\text{-}C_5Me_5$) [26], is the most common alkylated ferrocene derivative used for the preparation of CT complexes. $[FeCp^*_2]$ and its oxidized form, as is the case for most other decamethylmetallocenes [27], has D_{5d} symmetry in the solid state. For use as an electron-donor it has the important advantage that it is resistant towards substitution reactions, ring exchange, and hydrolysis. It is also much more electron-rich than its parent compound, as reflected by the redox data reported in Table 8-1.

CT complexes containing $[FeCp^*_2]$ have been extensively studied by Miller, Epstein and co-workers in connection with magnetic properties, and the results have been reported in an impressive series of papers, mainly during the course of the last decade [11]. Several types of organic molecules and selected coordination compounds of transition metals have been used as acceptors. The CT complexes were usually fully characterized, in terms of their crystal structure and magnetic behavior. A few other authors also reported the synthesis of CT complexes containing $[FeCp^*_2]$ and the study of their conducting properties, but this topic, in contrast to magnetism, is still to be considered as underdeveloped (vide infra).

8.3.1.1 With Organic Acceptors

Table 8-2 collects examples of CT complexes containing $[FeCp^*_2]$ and a series of organic acceptors whose structure and acronyms are given in Scheme 8-2. It also includes information about their solid state structural motifs and main physical properties.

Table 8-2. Some CT complexes containing $[Cp^*_2Fe]^{\cdot+}$ and organic acceptors

Acceptor	Structure	Stoichiometry	Main behavior[a]	Ref.
$[TCNE]^-$	1D-DADA	1:1	Bulk ferromagnetic	[28]
$[C_3(CN)_5]^-$	1D-DADA	1:1	Antiferromagnetic	[28]
$[C_4(CN)_6]^-$	1D-DADA	1:1	Ferromagnetic	[29]
$[C_3\{C(CN)_2\}_3]^-$	1D-DDAA monoclinic	1:1	Antiferromagnetic	[30]
$[C_3\{C(CN)_2\}_3]^-$	1D-DDAA triclinic	1:1	Paramagnetic	[30]
$[TCNQ]^-$	1D-DADA	1:1	Metamagnetic	[31]
$[TCNQ]^-$	DAAD dimer	1:1	Paramagnetic	[31]
$[(TCNQ)_2]^-$	Segregated stacks	1:2	Conducting	[31]
$[TCNQF_4]^-$	DAAD dimer	1:1	Paramagnetic	[32]
$[TCNQF_4]^{2-}$	DADDAD chains	2:1	[b]	[32]
$[TCNQI_2]^-$	1D-DADA	1:1	Ferromagnetic	[33]
$[DDQ]^-$	1D-DADA	1:1	Ferromagnetic	[35]
$[Me_2DCNQI]^-$	Not determined	1:1	Ferromagnetic	[36]
$[C(CN)_3]^-$	Complex	1:1	Paramagnetic	[37]
$[(PTCI)_2]^-$	Segregated stacks	1:2	Semiconducting	[38]
$[(PDCI)_2]^-$	Segregated stacks	1:2	Semiconducting	[38]
$[(TCIDBT)_2]^-$	Segregated stacks	1:2	Semiconducting	[38]

a Refers to the physical behavior that has been best characterized.
b Probably diamagnetic.

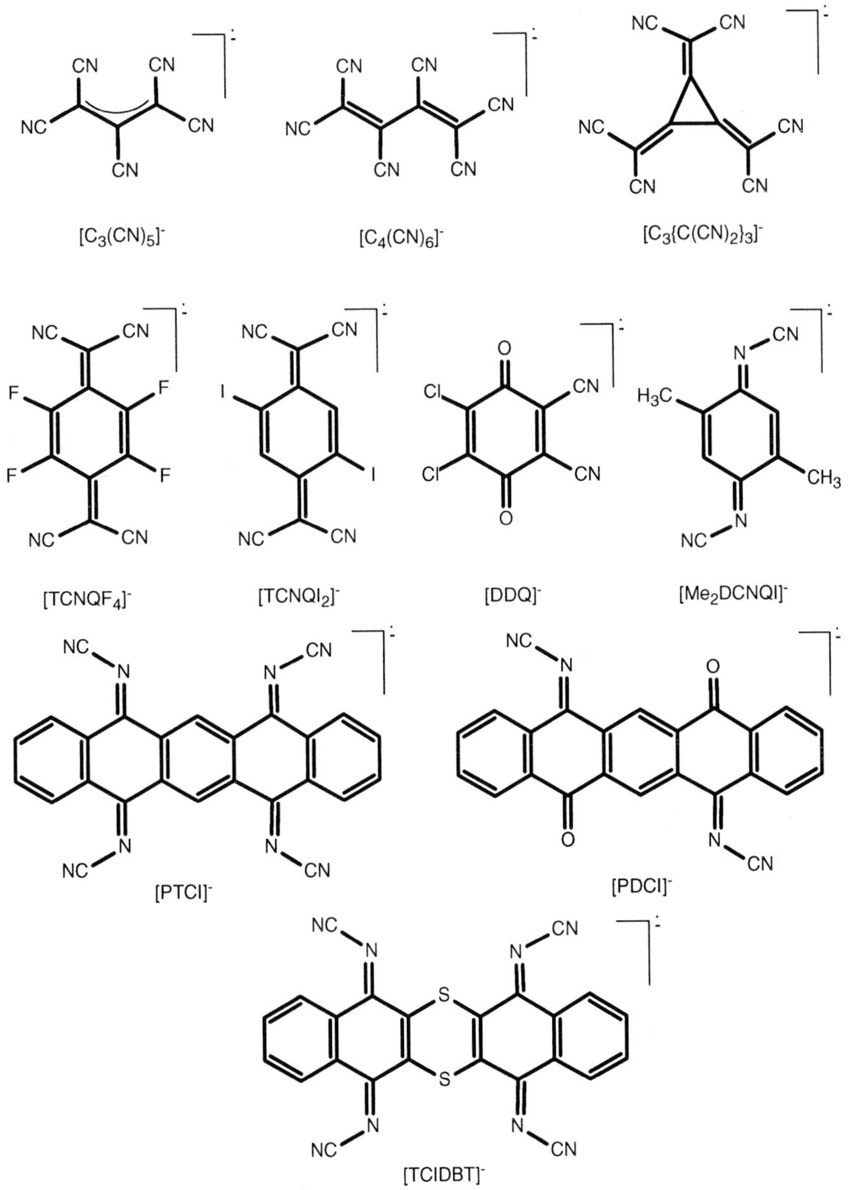

$[C_3(CN)_5]^-$ $[C_4(CN)_6]^-$ $[C_3\{C(CN)_2\}_3]^-$

$[TCNQF_4]^-$ $[TCNQI_2]^-$ $[DDQ]^-$ $[Me_2DCNQI]^-$

$[PTCI]^-$ $[PDCI]^-$

$[TCIDBT]^-$

Scheme 8-2. Some organic electron-acceptors used for the formation of CT complexes.

The CT complexes containing the donor $[FeCp^*_2]$ have been chiefly studied from the point of view of their magnetic properties by the Miller–Epstein group. CT complexes are usually prepared by direct reaction of stoichiometric amounts of $[FeCp^*_2]$ with the desired acceptor under inert conditions, whereupon the product forms as a crystalline material. An alternative procedure is the reaction of

Table 8-3. Some CT complexes containing $[Cp_2^*Fe]^{\cdot +}$ and inorganic acceptors

Acceptor	Structure	Stoichiometry	Main behavior[a]	Ref.
$[Ni\{C_2S_2(CN)_2\}_2]^-$	DAAD dimer	1:1	Paramagnetic	[43]
$[Pt\{C_2S_2(CN)_2\}_2]^-$	DA_2D_2A (α-form) (chains + layers)	1:1	Ferromagnetic	[43]
$[Pt\{C_2S_2(CN)_2\}_2]^-$	$DA_2D[DA]$ (β-form) (chains + dimers)	1:1	Ferromagnetic	[43]
$[Ni\{C_2S_2(CF_3)_2\}_2]^-$	1D-DADA	1:1	Ferromagnetic	[43]
$[Mo\{C_2S_2(CF_3)_2\}_3]^-$	1D-DADA	1:1	Ferromagnetic	[44]
$[Mo\{C_2S_2(CF_3)_2\}_3]^{2-}$	Interpenetrating DADA chains	2:1	Antiferromagnetic	[44]
$[Ni(dmit)_2]^-$	DDAA stacks	1:1	Ferromagnetic	[46]
$[Ni(bds)_2]^-$	DDADDA stacks + A sheets	1:1	Ferromagnetic	[46]
$[Au(dmit)_2]^-$	DDAA stacks	1:1	Insulating	[47]
$0.33\,[Au(dmit)_2]^-$	Not determined	3:1	Conducting	[47]
$[M(C_2B_9H_{11})_2]^-$ M = Cr, Fe, Ni	CsCl-type	1:1	Paramagnetic	[38a, 49]

a Refers to the physical behavior that has been best characterized.

salts containing the oxidized form of the donor and the reduced form of the acceptor.

Research in this field led to the discovery in 1987 of the first organometallic bulk (3D) ferromagnet: $[FeCp_2^*]^{\cdot +}[TCNE]^{\cdot -}$ [28]. This compound is nowadays the prototype CT complex displaying such three-dimensional cooperative effects, which are related to its specific structural features, i.e., the presence of 1D parallel ...DADA... chains constituted by radical cations $D^{\cdot +}$ and radical anions $A^{\cdot -}$. An illustration of the solid state structure of $[FeCp_2^*]^{\cdot +}[TCNE]^{\cdot +}$ is provided by Fig. 8-4.

The physical properties of $[FeCp_2^*]^{\cdot +}[TCNE]^{\cdot -}$ have been thoroughly studied by Miller and Epstein [39], who showed it to display a Curie temperature T_c of 4.8 K. Below this temperature the material exhibits the onset of spontaneous magnetization in zero applied field. This property is consistent with a three-dimensional, bulk ferromagnetic ground state. Further support for bulk ferromagnetism was provided by magnetization measurements versus applied field, carried out on aligned single crystals [40]. These measurements show typical hysteresis loops with a large coercive field H_c of 1 kG at 2 K (compare H_c of 1 G for iron, or H_c of 213 G for Fe_3O_4 at room temperature [11a]). The hysteresis behavior is illustrated in Fig. 8-5.

To account for spin alignment in the bulk of a molecular ferromagnetic material constituted by chains of alternating radical anions and radical cations, the model originally proposed by McConnell has been applied [41a, b]. According to this model, an excited state with $m_s = 1$ formed either by a virtual forward or retro charge-transfer ($D^{\cdot +} + A^{\cdot -} \rightarrow D^{2+} + A^{2-}$, or $D^0 + A^0 \leftarrow D^{\cdot +} + A^{\cdot -}$, respectively) admixes with the ground state, thus stabilizing ferromagnetic coupling. The $m_s = 1$ excited state can arise either from an $m_s = 1$ donor, or an $m_s = 1$ acceptor, but not both. Although this model has been successfully employed by Miller and

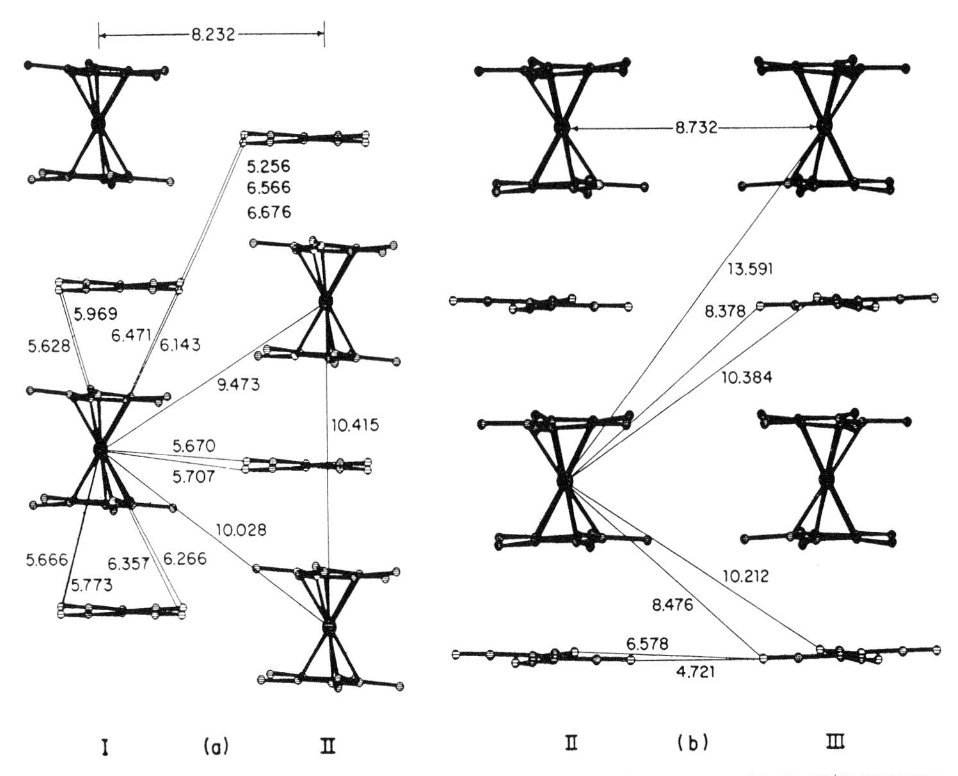

Fig. 8-4. The solid state 1D-DADA structure of the bulk ferromagnet $[FeCp_2^*]^{\cdot+}[TCNE]^{\cdot-}$. Inregistry chains (I–II) and out-of-registry chains (II–III), as well as selected interatomic distances are shown (reprinted with permission from ref. [28], copyright 1987 American Chemical Society).

Epstein [40b] to predict ferromagnetic coupling for several CT complexes, it has recently been challenged by Collmar and Kahn [42]. These authors suggested an alternative mechanism to explain ferromagnetism in $[FeCp_2^*]^{\cdot+}[TCNE]^{\cdot-}$, involving configurational mixing of a singly excited configuration of the ferrocenium cation, which does not need to invoke donor-acceptor charge-transfer states [42b].

TCNE is one among several compounds belonging to the class of poly- or percyanated olefinic acceptors. Indeed, CT complexes of $[FeCp_2^*]$ with, e.g., pentacyanopropenide ($[C_3(CN)_5]^-$) [28], hexacyanobutadienide ($[C_4(CN)_6]^-$) [29], hexacyanotrimethylenecyclopropanide ($[C_3\{C(CN)_2\}_3]^-$) [30], as well as with the simplest representative tricyanomethanide ($[C(CN)_3]^-$) [36] have been reported. The respective 1:1 CT complexes with $[FeCp_2^*]$ have been structurally characterized and their magnetic properties studied.

The compound with pentacyanopropenide was found to possess a 1D-DADA structure in the solid state, but the main magnetic behavior was of the antiferromagnetic type, as the material obeys the Curie–Weiss law ($\chi = C/T - \theta$) governing the temperature dependence of the magnetic susceptibility with $\theta = -1.2$ K (compare with $\theta = +30$ K for $[FeCp_2^*]^{\cdot+}[TCNE]^{\cdot-}$, above 60 K) [28].

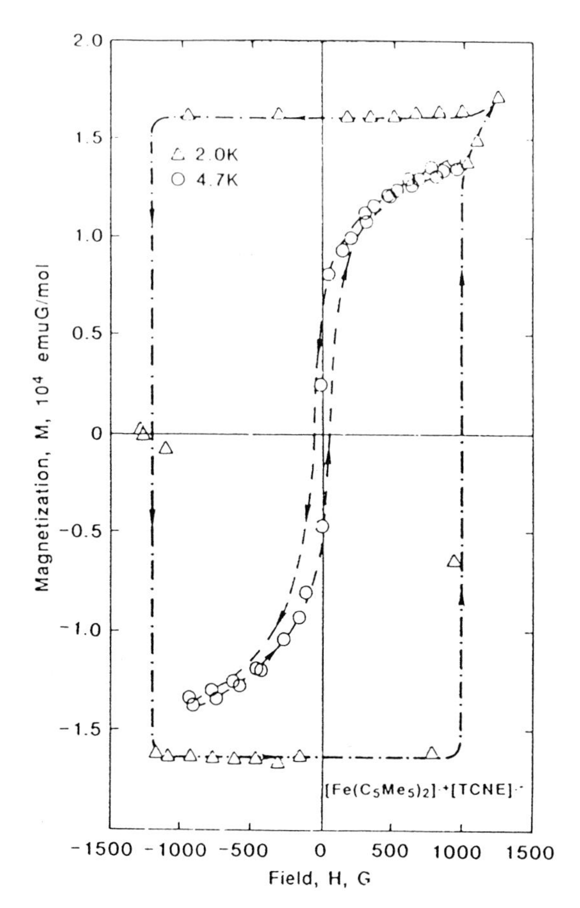

Fig. 8-5. Hysteresis behavior of a single crystal of $[FeCp_2^*]^{\cdot+}[TCNE]^{\cdot-}$ with the stacking axis oriented parallel to the applied field (reprinted with permission from ref. [11b], copyright 1988 American Chemical Society).

Again, the same 1D-DADA structure was found for the hexacyanobutadienide CT complex, as illustrated in Fig. 8-6. This material also displays in-registry and out-of-registry interactions between chains, as is the case for $[FeCp_2^*]^{\cdot+}[TCNE]^{\cdot-}$. Furthermore, it shows dominant ferromagnetic interactions, as indicated by the Curie–Weiss constant θ +35 K [29].

Two different crystalline phases of the 1:1 CT complex with hexacyanotrimethylenecyclopropanide have been found [30]. Both the monoclinic and triclinic modifications possess a 1D structure with segregated chains of anions and cations, which, in principle, could ensure electrical conductivity, but no related experimental evidence was reported. The spacing between the parallel planes of the anions is uniform in both modifications and corresponds to 3.224 Å (monoclinic) and 3.20 Å (triclinic). The magnetic properties were studied in detail for the monoclinic form only and indicate an antiferromagnetic coupling. This behavior is interpreted on the basis of a singlet ground state of the anion chains due to their relatively short separations.

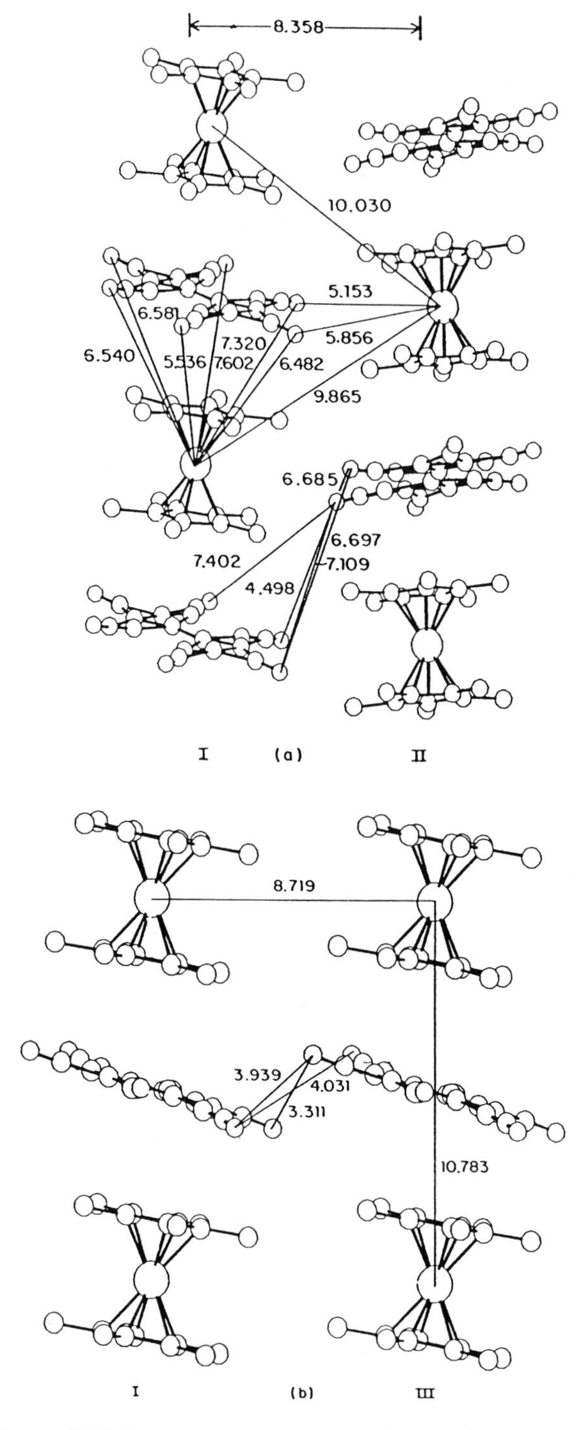

Fig. 8-6. The solid state 1D-DADA structure of $[FeCp^*_2]^{\cdot +}[C_4(CN)_6]^{\cdot -}$. In-registry chains (I – II) and out-of-registry chains (I – III), as well as selected interatomic distances are shown (reprinted with permission from ref. [29], copyright 1987 American Chemical Society).

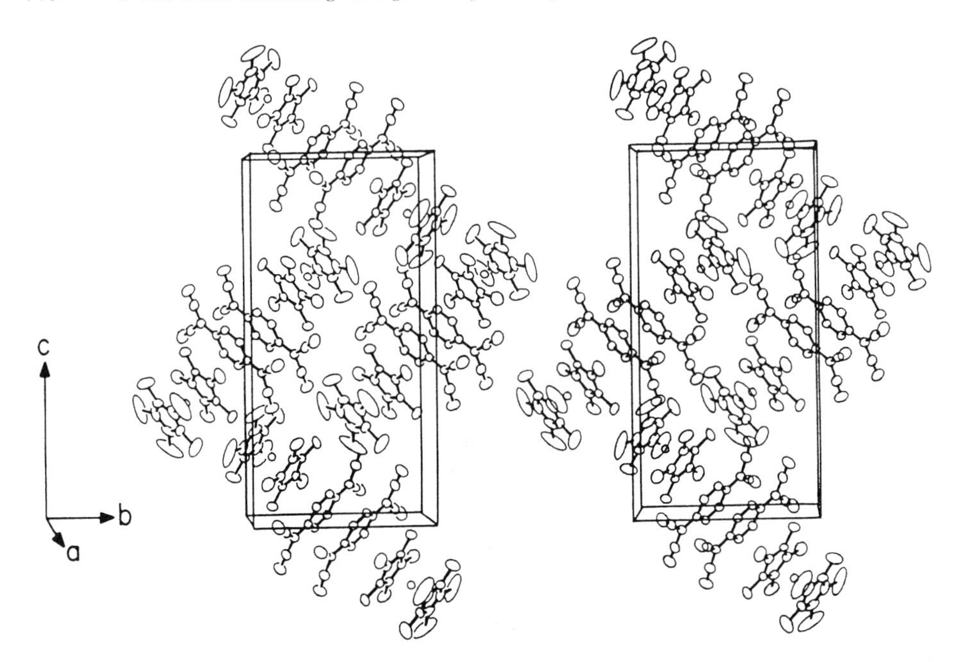

Fig. 8-7. A view of the unit cell of the thermodynamic 1:1 phase of $[FeCp^*_2]_2^{\cdot +}[(TCNQ)_2]^{2-}$ showing the DAAD dimeric structure (reprinted with permission from ref. [31c], copyright 1987 American Chemical Society).

The classical acceptor TCNQ was found to undergo formation of three different polymorphs, the preparation conditions for each of these phases differing only very slightly from one another [31]. The 1:2 phase is electrically conducting and is obtained by carefully controlling the stoichiometry during crystallization. The other two forms are both 1:1 phases and can be best described as the kinetic and thermodynamic forms. The latter is a DAAD dimeric form composed, in the solid state, of discrete stacks of such dimeric units, as depicted in Fig. 8-7. Because of this type of structure, strong antiferromagnetic coupling occurs within the anion dimers and negligible magnetic coupling is observed between the distant ferrocenium cations. This results in the overall paramagnetic behavior of the material. The kinetic phase, obtained by fast crystallization from warm dichloroethane or acetonitrile solution, belongs to the 1D-DADA type structure. As such it compares with $[FeCp^*_2]^{\cdot +}[TCNE]^{\cdot -}$ and $[FeCp^*_2]^{\cdot +}[C_4(CN)_6]^{\cdot -}$, as it contains 1D chains of alternating donors and acceptors, displaying in-registry and out-of-registry interactions. The magnetic behavior is best described as metamagnetic which, similarly to ferromagnetism, is characterized by the so-called Néel temperature ($T_N = 2.5$ K). Metamagnetism is encountered when a ferromagnetic state is generated from an antiferromagnetic state in an applied magnetic field.

The analogous, but much stronger, acceptor $TCNQF_4$ forms a 1:1 CT complex with $[FeCp^*_2]$ that is isostructural to that described above, i.e., displaying a DAAD dimeric arrangement of donors and acceptors, and also showing very similar magnetic properties [32]. $TCNQF_4$ also undergoes double reduction, thus leading to a 2:1 phase. This is composed of parallel DADDAD chains, characterized by three unique pairwise interactions, all of the out-of-registry type, as illustrated in Fig. 8-8 [32]. This material was studied in particular because it offered the opportunity to compare experimental and theoretical data concerning the structure of the acceptor depending on its state of charge. No magnetic properties were reported for this material, but it is expected that this 2:1 CT complex should be diamagnetic, analogous to $\{[CoCp_2^*]^{\cdot +}\}_2[TCNQI_2]^{2-}$, which possesses a very similar structure [33].

The 1D-DADA CT complex $[FeCp_2^*]^{\cdot +}[TCNQI_2]^{\cdot -}$ displays a positive Curie–Weiss constant ($\theta = +9.5$ K), indicating a dominant ferromagnetic coupling, but this behavior becomes less important at a temperature below 60 K. The lack of 3D ferromagnetic interactions, as demonstrated by the nonobservation of hysteresis, was ascribed to the reduced symmetry and increased size of the acceptor, as compared with, for example, TCNE. CT complexes of $[FeCp_2^*]$ with other substituted TCNQs have been reported, but no improvement of the magnetic properties could be achieved [34].

Yet another CT complex with the same type of 1D-DADA structure containing both radical anions and radical cations is that containing the acceptor DDQ (2,3-dichloro-5,6-dicyanobenzoquinone) [35]. The nature of the acceptor in the solid state was first erroneously described as the diamagnetic anion DDQH [35b], but further studies of this material indicated the presence of DDQ radical anions, indicating that the main magnetic behavior should be of the ferromagnetic type.

A very recent report disclosed the properties of the 1:1 CT complex with the acceptor Me_2DCNQI (2,5-dimethyl-*N*,*N'*-dicyanoquinonediimine) [36]. Although the crystal structure could not be determined, a dominant ferromagnetic coupling was found, as indicated by a positive Curie–Weiss constant $\theta = +10.8$ K, but no 3D ferromagnetic interactions could be deduced from the measurements so far carried out.

From the discussion above it is apparent that ferromagnetic order in CT complexes of the type $[FeCp_2^*]^{\cdot +}[A]^{\cdot -}$ is determined by a series of interplaying factors that are extremely difficult to control and to design rationally. One of these factors appears to be symmetry, and indeed materials containing highly symmetric acceptor molecules have a better chance to afford the desired magnetic properties by ensuring the decisive 1D-DADA structure. An exception to this rule of thumb is constituted by the D_{3h} symmetrical tricyanomethanide anion. Besides the fact that this anion is diamagnetic, the structure of the corresponding CT compound is very complex and does not compare with any other material of this class [37].

A completely different class of organic acceptors for the formation of conducting CT complexes was studied by a research group at CIBA [38]. The acceptors used are the polycyclic aromatic compounds shown in Scheme 8-2, containing *N*-cyanoimine functionalities: 5,7,12,14-tetrakis-(cyano-imine)pentacene (PTCI), 5,12-bis(cyanoimine)pentacene-7,14-dione (PDCI), and 5,7,12,14-tetrakis(cyanoimine)-dibenzo[*b*,*i*]thianthrene (TCIDBT). 2:1 CT complexes were obtained in all cases from the reaction of these acceptors with $[FeCp_2^*]$, typically in hot 1,2-dichloroethane. The materials thus prepared were found to display a significant room temperature conductivity as polycrystalline samples (pressed pellets) of 1.0 S cm^{-1} (PTCI),

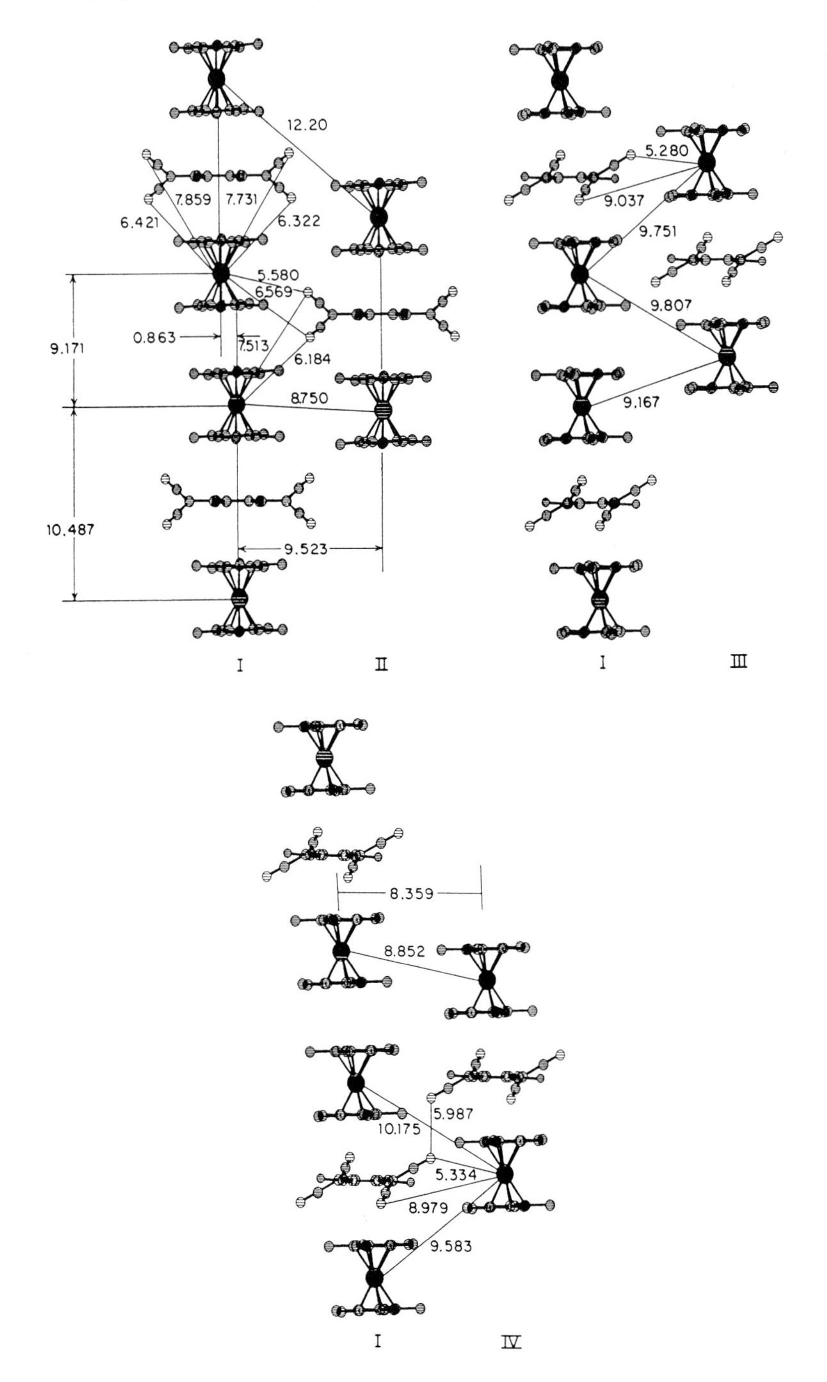

Fig. 8-8. The solid state structure of the 2:1 CT complex $[FeCp^*_2]_2^{\cdot+}[TCNQF_4]^{2-}$ showing the out-of-registry interactions between chains of the DADDAD type (reprinted with permission from ref. [32], copyright 1989 American Chemical Society).

0.52 S cm^{-1} (PDCI), and 0.01 S cm^{-1} (TCIDBT). Although no crystal structures were determined, the relatively high conductivities indicate that these materials should be composed of segregated stacks of donors and acceptors in the solid state. A strong dependence of the electrical conductivity on the number of methyl substituents on the ferrocene was observed (see Sect. 8.3.2).

8.3.1.2 With Inorganic Acceptors

Decamethylferrocene has also been used as a donor for the formation of CT complexes with inorganic acceptors. Such acceptors are mainly composed of late transition metal complexes containing planar ligands. Some of these acceptors are illustrated in Scheme 8-3. As for their organic counterparts, the inorganic partners have at least two reversibly accessible oxidation states. The reduced form present in the CT complex is usually a radical anion.

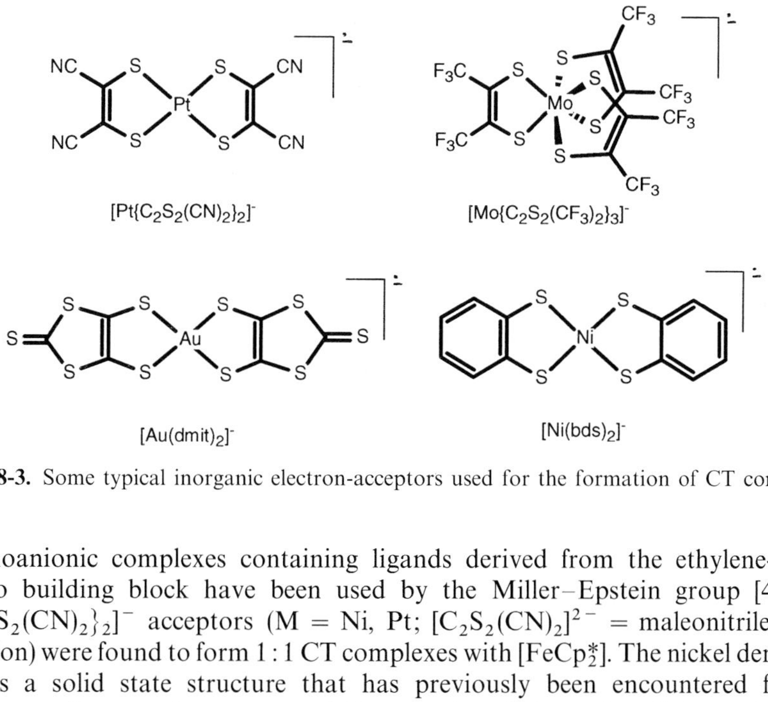

$[Pt\{C_2S_2(CN)_2\}_2]^{\cdot-}$ $[Mo\{C_2S_2(CF_3)_2\}_3]^{\cdot-}$

$[Au(dmit)_2]^{\cdot-}$ $[Ni(bds)_2]^{\cdot-}$

Scheme 8-3. Some typical inorganic electron-acceptors used for the formation of CT complexes.

Monoanionic complexes containing ligands derived from the ethylene-1,2-dithiolato building block have been used by the Miller–Epstein group [43, 44]. $[M\{C_2S_2(CN)_2\}_2]^-$ acceptors (M = Ni, Pt; $[C_2S_2(CN)_2]^{2-}$ = maleonitriledithiolate anion) were found to form 1 : 1 CT complexes with $[FeCp^*_2]$. The nickel derivative displays a solid state structure that has previously been encountered for CT complexes with organic acceptors (vide supra) and the unit cell is made up of isolated DAAD dimers [43]. The compound was studied from the point of view of its magnetic properties and was found to exhibit simple paramagnetic behavior. This was interpreted on the basis of the presence of eclipsed $S = 0\,[A_2]^{2-}$ anions and nearly isolated $S = \frac{1}{2}[FeCp^*_2]^{\cdot+}$ cations. The corresponding Pt derivatives exist in two different forms with more complex structures. Both the α- and β-form are comprised of orthogonal chains and layers and the best overall description is DA_2D_2A (α-form) and $DA_2D[DA]$ (β-form). In the β-form one can identify

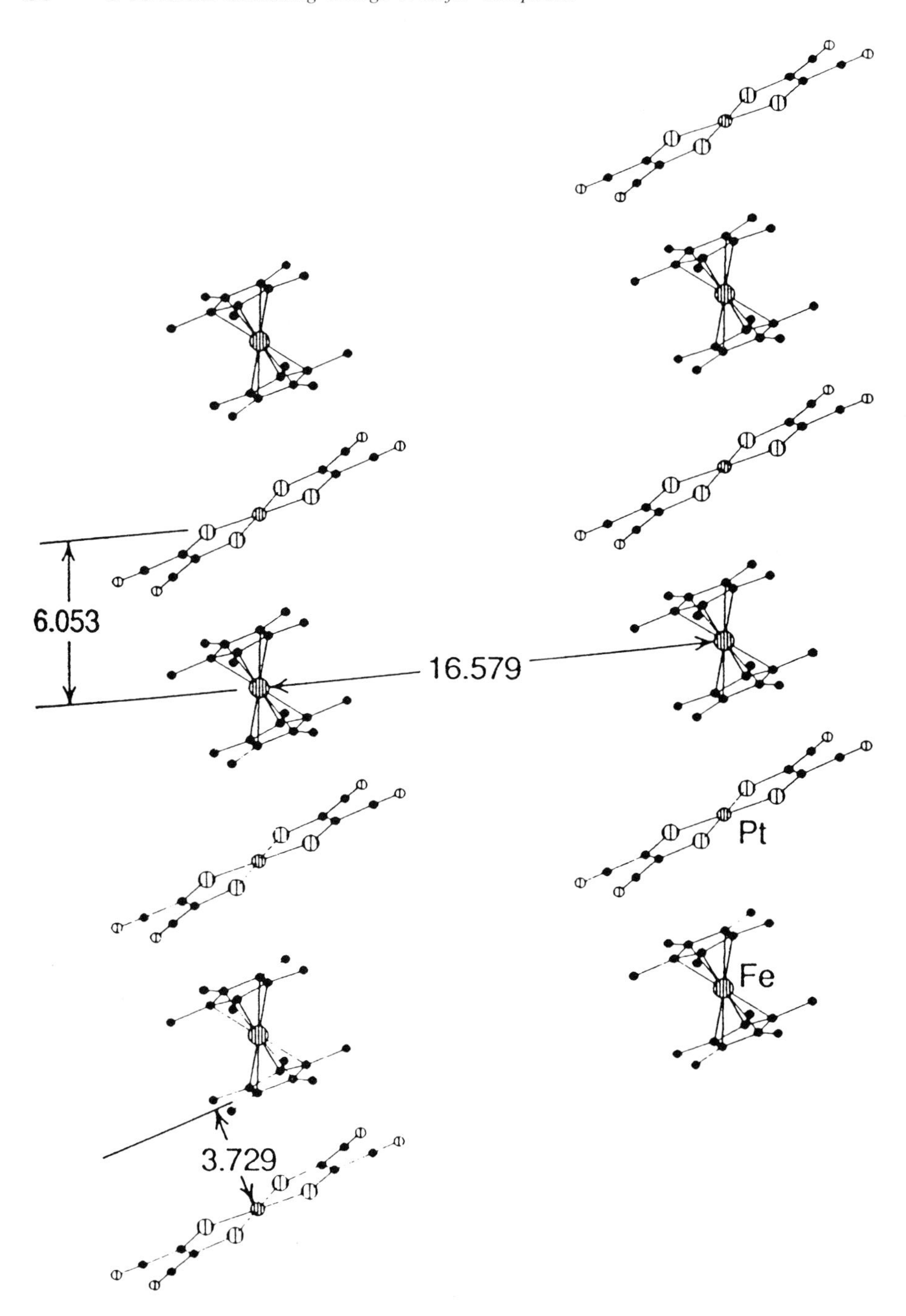

Fig. 8-9. The solid state structure of the 1D-DADA CT complex β-$[FeCp^*_2][Pt\{C_2S_2(CN)_2\}_2]$ (reprinted with permission from ref. [43], copyright 1989 American Chemical Society).

1D-DADA chains of alternating donors and acceptors, as depicted in Fig. 8-9. The two materials exhibit positive values of the Curie–Weiss constant ($\theta = +6.6$ K and $\theta = +9.8$ K, respectively) and their main magnetic coupling is of the ferromagnetic type, although no 3D cooperative effects could be observed [43].

Switching to the slightly more powerful acceptor $[Ni\{C_2S_2(CF_3)_2\}_2]^-$ ($[C_2S_2(CF_3)_2]^{2-}$ = bis(trifluoromethyl)ethylene-1,2-dithiolate) leads to a CT complex with a 1D-DADA structure [43]. This material has a high Curie–Weiss constant $\theta = +15$ K, a large effective moment, as well as a marked field dependence of the magnetic susceptibility, i.e., it shows a ferromagnetic coupling, although bulk ferromagnetism was not observed. This supports the contention that a 1D-DADA structure is an important structural feature for a ferromagnetic behavior, although it is not a prerequisite. Ferromagnetic CT complexes displaying different and more complex solid state structures have been reported (vide infra).

The same ligand $[C_2S_2(CF_3)_2]^{2-}$ coordinated to molybdenum forms pseudo-octahedral (the actual structure is intermediate between octahedral and trigonal prismatic) complexes of the type $[Mo\{C_2S_2(CF_3)_2\}_3]^{x-}$, which have been employed as acceptors [44]. Two CT complexes with different stoichiometries with $[FeCp^*_2]$ have been described. The 1 : 1 material shows in the solid state a 1D-DADA type structure and turns out to be ferromagnetic ($\theta = +8.4$ K, but no bulk ferromag-

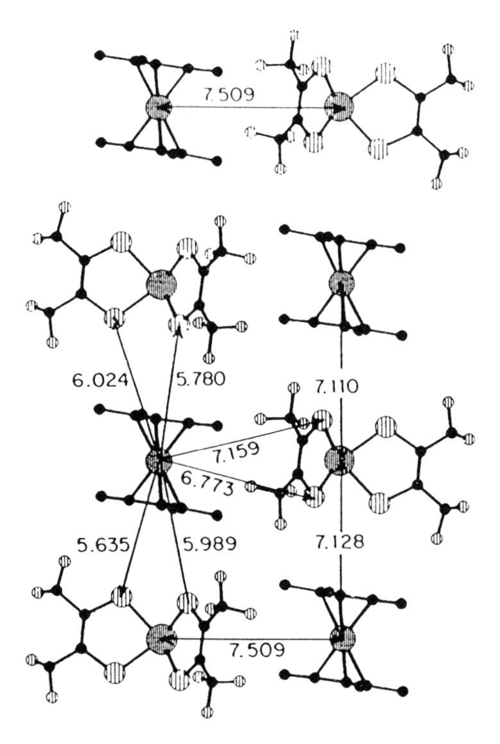

Fig. 8-10. The solid state structure of the 1D-DADA 1 : 1 CT complex $[FeCp^*_2][Mo\{C_2S_2(CF_3)_2\}_3]$. One ligand at molybdenum has been omitted for clarity (reprinted with permission from ref. [44], copyright 1990 American Chemical Society).

netism). This is somewhat surprising in view of the fact that the anion is not a planar entity, and because its shape and size prevent optimal interchain interaction between anions and cations. The 1D-DADA structure of this material is illustrated in Fig. 8-10.

The 2 : 1 phase, prepared by mixing equivalent amount of $[FeCp^*_2]$ with the neutral form of the oxidizing molybdenum complex, is composed in the solid state of two distinct interpenetrating DADA chains of alternating donors and acceptors [44]. The material, which is made up of diamagnetic $[Mo\{C_2S_2(CF_3)_2\}_3]^{2-}$ dianions and $[FeCp^*_2]^{\cdot+}$ cations, shows a weak antiferromagnetic coupling, as indicated by the small negative Curie–Weiss constant $\theta = -3.2$ K.

Complexes of the well-known and ubiquitous ligand dmit (1,3-dithiole-2-thione-4,5-dithiolate) [45] have been used for the formation of $[FeCp^*_2]$-containing CT compounds. Hoffman and co-workers reported a 1 : 1 ferromagnetic (but no bulk ferromagnet) CT complex containing the acceptor $[Ni(dmit)_2]^-$ [46]. Because of its rather unusual solid state structure, this material somewhat challenges the view requiring a 1D-DADA arrangement of donors and acceptors in order to obtain

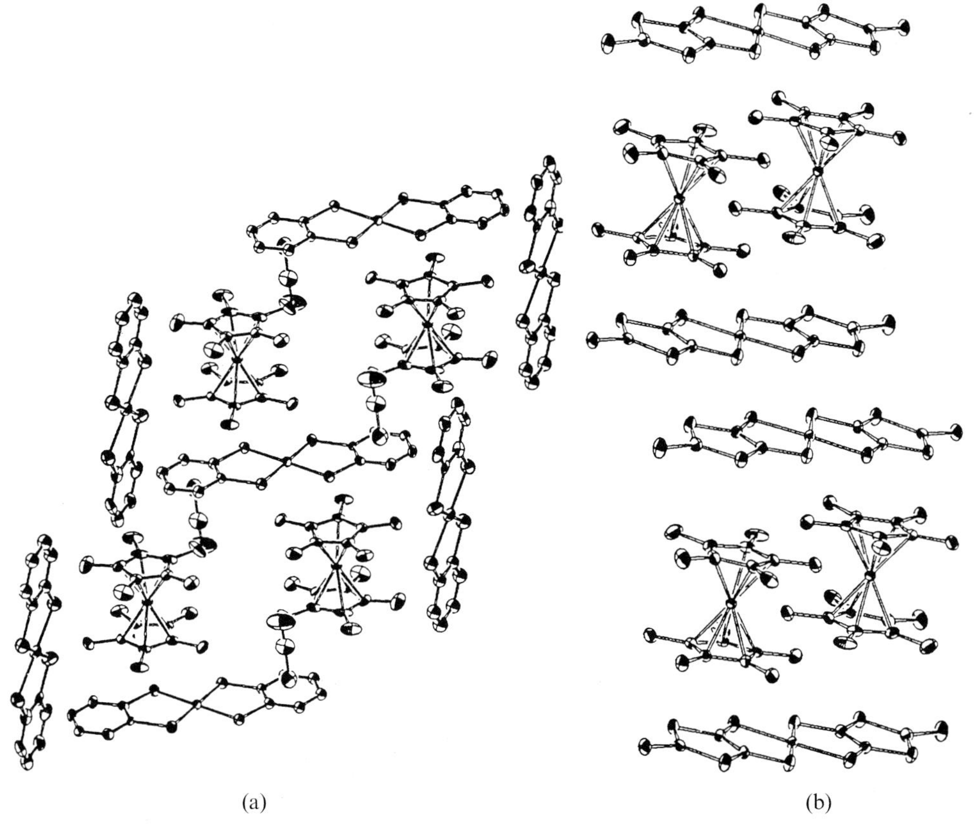

Fig. 8-11. DDADDA stacks and part of the flanking A sheets in the 1 : 1 CT complex $[FeCp^*_2]^{\cdot+}[Ni(bds)_2]^{\cdot-}$ (a) and DDAA stacks of the 1 : 1 CT complex $[FeCp^*_2]^{\cdot+}[Ni(dmit)_2]^{\cdot-}$ (b) (reprinted with permission from ref. [46], copyright 1989 American Chemical Society).

ferromagnetic coupling (vide supra). $[FeCp_2^*]^{\cdot +}[Ni(dmit)_2]^{\cdot -}$ is made up of DDAA stacks in which pairs of cations, arranged in a side-by-side manner, alternate with pairs of anions organized in a face-to-face arrangement, as illustrated in Fig. 8-11. At very low temperature (< 2 K) this material shows a crossover to antiferromagnetic behavior, possibly arising from favorable interactions between anions.

The corresponding 1 : 1 CT complex with $[Au(dmit)_2]^-$ has been studied from the point of view of its electric properties, but no details of its magnetic behavior have been disclosed [47]. Due to a solid state structure very similar to that of the nickel derivative (DDAA stacks), this material turns out to be essentially insulating, its conductivity being even smaller than that of $[NBu_4][Au(dmit)_2]^-$ ($\sigma_{rt} = 2.1 \times 10^{-8}$ *vs.* 4.9×10^{-8} S cm^{-1}). On the other hand, a CT complex obtained by electrocrystallization and displaying a 1 : 3 stoichiometry ($\{[FeCp_2^*]^{\cdot +}\}_{0.33}[Au(dmit)_2]^-$) has an appreciable electrical conductivity ($\sigma_{rt} = 0.093$ S cm^{-1}) and typical semiconducting character (activation energy 0.061 eV) [47]. No crystal structure of this latter compound was reported, but conduction pathways are likely to arise from stacking of the partially oxidized $[Au(dmit)_2]$ units ensuring relatively short S-S intermolecular contacts. In this context it is interesting to note that ferrocene also forms CT complexes with $[Au(dmit)_2]$ of 1 : 3 and 1 : 4 stoichiometries similar to those of $[FeCp_2^*]$ and that such materials possess very similar electrical properties [48].

An acceptor related to $[Ni(dmit)_2]^-$, $[Ni(bds)_2]^-$ (bis(benzene-1,2-dithiolato)nickelate), has been shown to form a 1 : 1 CT complex having an interesting solid state structure [46]. This can be best described as consisting of DDADDA stacks forming two-dimensional layers. These layers occupy the *ab* planes of the triclinic crystals and are separated by sheets of acceptors molecules located in the *ac* planes. Despite the unusual structure, this material is characterized by a dominant ferromagnetic coupling, although again no indications for bulk ferromagnetism were found. The

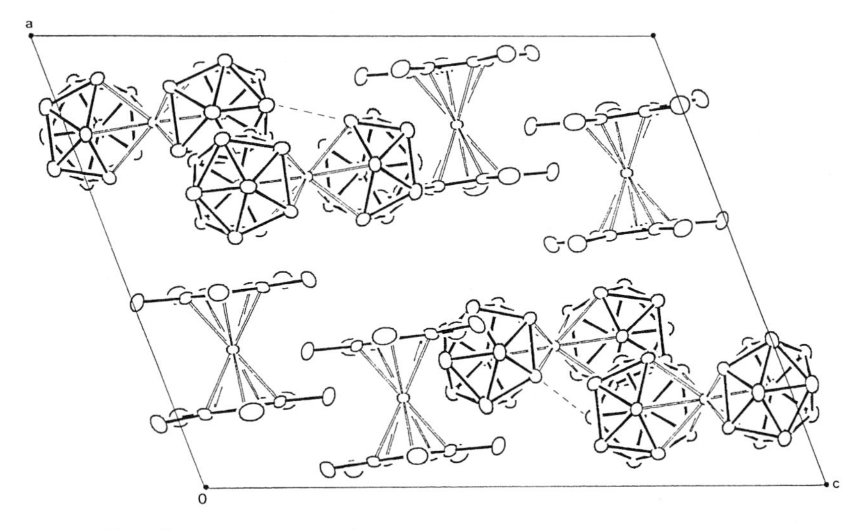

Fig. 8-12. Packing diagram of $[FeCp_2^*]^{\cdot +}[Ni(C_2B_9H_{11})_2]^-$ (ref. [38a]).

ferromagnetic interactions were interpreted on the basis of the previously discussed McConnell model.

A completely different type of acceptors that have been used very recently for the formation of CT complexes are the metallocarborane sandwich compounds. Mingos and co-workers [49] and, independently, Chetcuti [38a] have prepared 1 : 1 CT salts of $[FeCp_2^*]$ containing bis-(3)-1,2-dicarbollide metallate anions ($[M(C_2B_9H_{11})_2]^-$, M = Cr, Fe, Ni). These complexes were found to be isostructural, displaying a distorted CsCl-type structure in the solid state (see Fig. 8-12). The low values of the Curie–Weiss constant shown by these compounds were taken as an indication that no significant magnetic couplings (either ferromagnetic or antiferromagnetic) are operative. The materials are therefore best described as paramagnetic.

8.3.2 Other Alkylferrocenes

The synthesis of pentaethylcyclopentadiene, and hence of decaethylferrocene ($[Fe(\eta^5\text{-}C_5Et_5)_2]$), has recently been reported independently by two groups [50, 51]. The corresponding CT complexes with the acceptors TCNE and TCNQ were described, and that with the latter acceptor was structurally characterized [50]. The complex adopts a 1D-DADA solid state structure similar to that of the analogous $[FeCp_2^*]$ derivative and displays in-registry, as well as out-of-registry interchain interactions. Due to the increased bulk of the $[Fe(\eta^5\text{-}C_5Et_5)_2]$ donor, the intra- and interchain interactions are now characterized by longer intermolecular separations. An illustration of the solid state structure of this compound is given in Fig. 8-13. Thus the changed structural features lead to a significantly decreased ferromagnetic coupling ($\theta = +6.8$ K, *vs.* $+11.6$ K for the decamethyl derivative) and to a complete absence of bulk magnetic ordering.

The analogous CT complex with the donor $TCNQF_4$ also displays a 1:1 stoichiometry, but a completely different solid state structure. It consists namely of discrete stacks of donors and acceptors molecules aligned with the crystallographic *b*-axis [51]. Pairwise parallel organization of the anions in dimers (interplane distance 3.13 Å) and an approximate side-to-side arrangement of the cations is observed. This type of structure is responsible for the observed paramagnetic behavior of this material, which follows the Curie law ($\theta \approx 0$) over the whole temperature range for which magnetic susceptibility measurements have been carried out.

Octamethylferrocene ($[Fe(\eta^5\text{-}C_5Me_4H)_2]$ has been reported to undergo formation of 1:1 CT complexes with the classical acceptors $[TCNE]^-$, $[TCNQ]^-$, $[C_4(CN)_6]^-$, $[C_3\{C(CN)_2\}_3]^-$, $[DDQ]^-$, and $TCNQF_4$ [52]. The former two materials have been structurally characterized and were thereby shown to possess the classical 1D-DADA solid state structure similar to that reported for the corresponding $[FeCp_2^*]$ compounds. However, their magnetic behavior, as for the other materials mentioned above, is at best described as weakly ferromagnetic, as indicated by the slightly positive values of the Curie–Weiss constant. The drastic change in magnetic behavior has been attributed to the reduced symmetry of the donor molecule, as

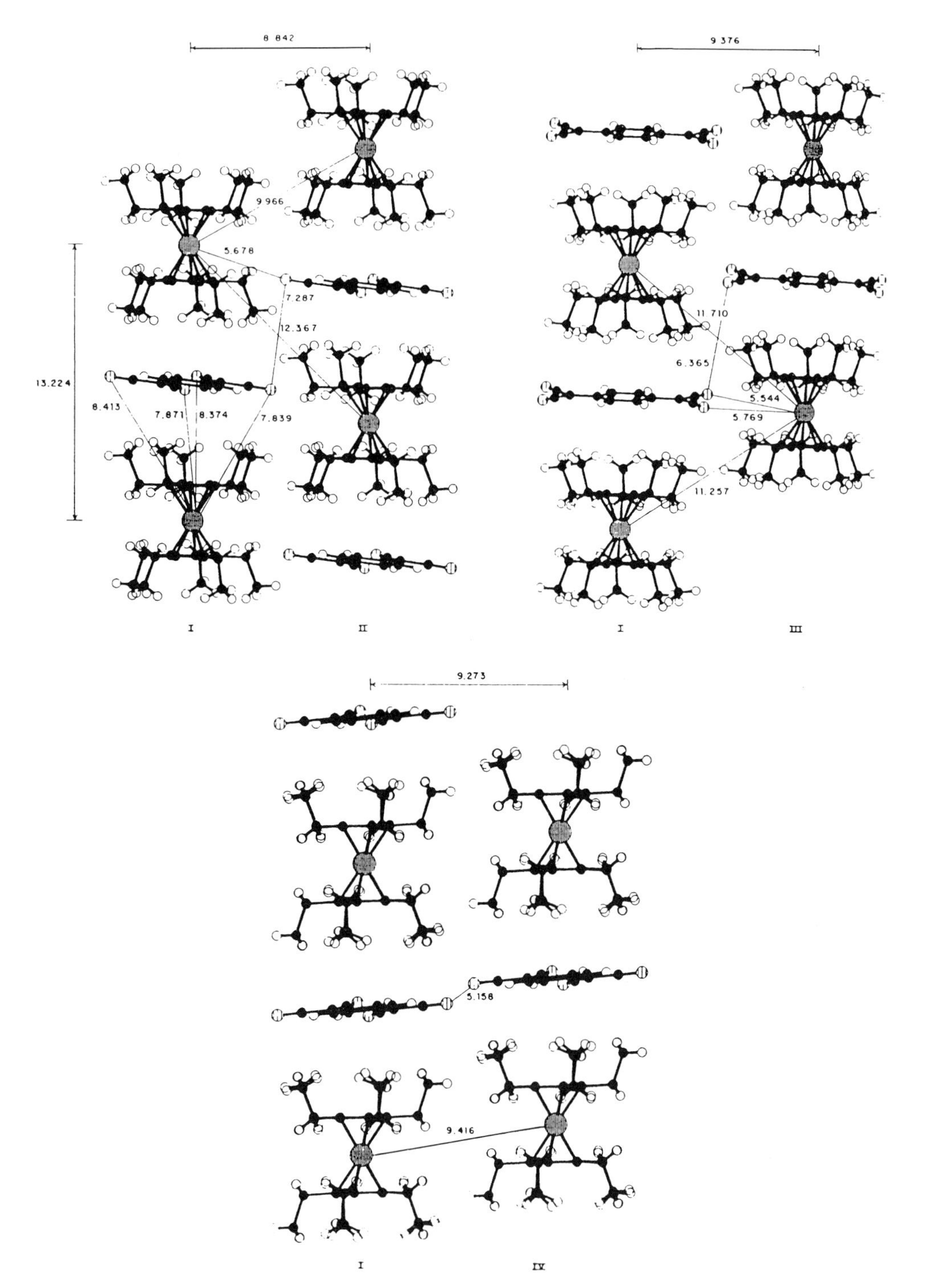

Fig. 8-13. The solid state structure of the 1D-DADA 1 : 1 CT complex $[Fe(\eta^5\text{-}C_5Et_5)_2]^{\cdot+}[TCNQ]^{\cdot+}$, showing out-of-registry (I – II and I – III) and in-registry (I – IV) interactions between chains (reprinted with permission from ref. [50], copyright 1991 American Chemical Society).

compared to [FeCp$^*_2$] [53]. As a consequence the charge-transfer excited state could be a singlet and not a triplet, as required by the McConnell model used to interpret ferromagnetism in such compounds (vide supra). The alternative view postulating weaker interactions between the chains being responsible for weaker magnetic coupling seems less probable.

The TCNQ and $TCNQF_4$ CT complexes of octaethylferrocene ([Fe(η^5-$C_5Et_4H)_2$] have been described by Sitzmann and co-workers [54]. Whereas the former material adopts a 1D-DADA structure in the solid state, the second is found to comprise DDAA stacks. These are constituted by a side-by-side arrangement of the cations and by parallel anion dimers. Interestingly, the cation pairs correspond to pairs of enantiomeric conformations of the octaethylferrocene, related by a crystallographic inversion center. No magnetic properties of these two materials have been reported.

The mixed cyclopentadiene ferrocene [Cp*CpFe] (1,2,3,4,5-pentamethylferrocene) undergoes formation of CT complexes with TCNE in two different stoichiometries [17a]. The 2:3 phase contains both neutral TCNE molecules and $[(TCNE)_2]^{2-}$ dimer anions, as well as THF of crystallization. In contrast, the 1:1 phase is constituted by 1D-DADA chains. The latter compound was shown to display a weak ferromagnetic coupling.

Several ferrocenes differing in their degree of methylation have been used for the preparation of conducting 1:2 CT complexes with the organic polycyclic acceptors PTCI and TCIDBT (see Scheme 8-2) [38a]. Table 8-4 shows pertinent conductivity data for polycrystalline samples at room temperature as a function of the number of methyl groups present in the donor molecule.

It is interesting to note that substantial conductivities vary with the nature of the donor, although electron transport very likely is exclusively taking place along the organic stacks. Thus, the degree of substitution of the ferrocene is possibly exerting a fine-tuning effect on the structure of the anion stacks, thereby influencing the intramolecular interactions. The exact nature of this effect could not be determined, since only one of the compounds was structurally characterized. Figure 8-14 shows the packing of the $[(MeCp)_2Fe]^{\cdot+}[(PTCI)_2]^-$. For this material single crystal conductivity vs. temperature was determined, showing a semiconducting behavior with

Table 8-4. Conductivity data for several 1:2 CT complexes of methylated ferrocenes with the organic acceptors PTCI and TCIDBT[a]

Donor	Acceptor	Conductivity,[b] σ S cm^{-1}
$[(MeCp)_2Fe]$	PTCI	1.8
$[(1,3,5\text{-}Me_3Cp)_2Fe]$	PTCI	5.0
$[(1,2,3,4\text{-}Me_4Cp)_2Fe]$	PTCI	3.0
$[Cp_2Fe]$	TCIDBT	0.52
$[(MeCp)_2Fe]$	TCIDBT	2.13
$[(1,2\text{-}Me_2Cp)_2Fe]$	TCIDBT	0.90
$[(1,3,5\text{-}Me_3Cp)_2Fe]$	TCIDBT	0.35

a See Scheme 8-2.
b Pressed pellets at room temperature.

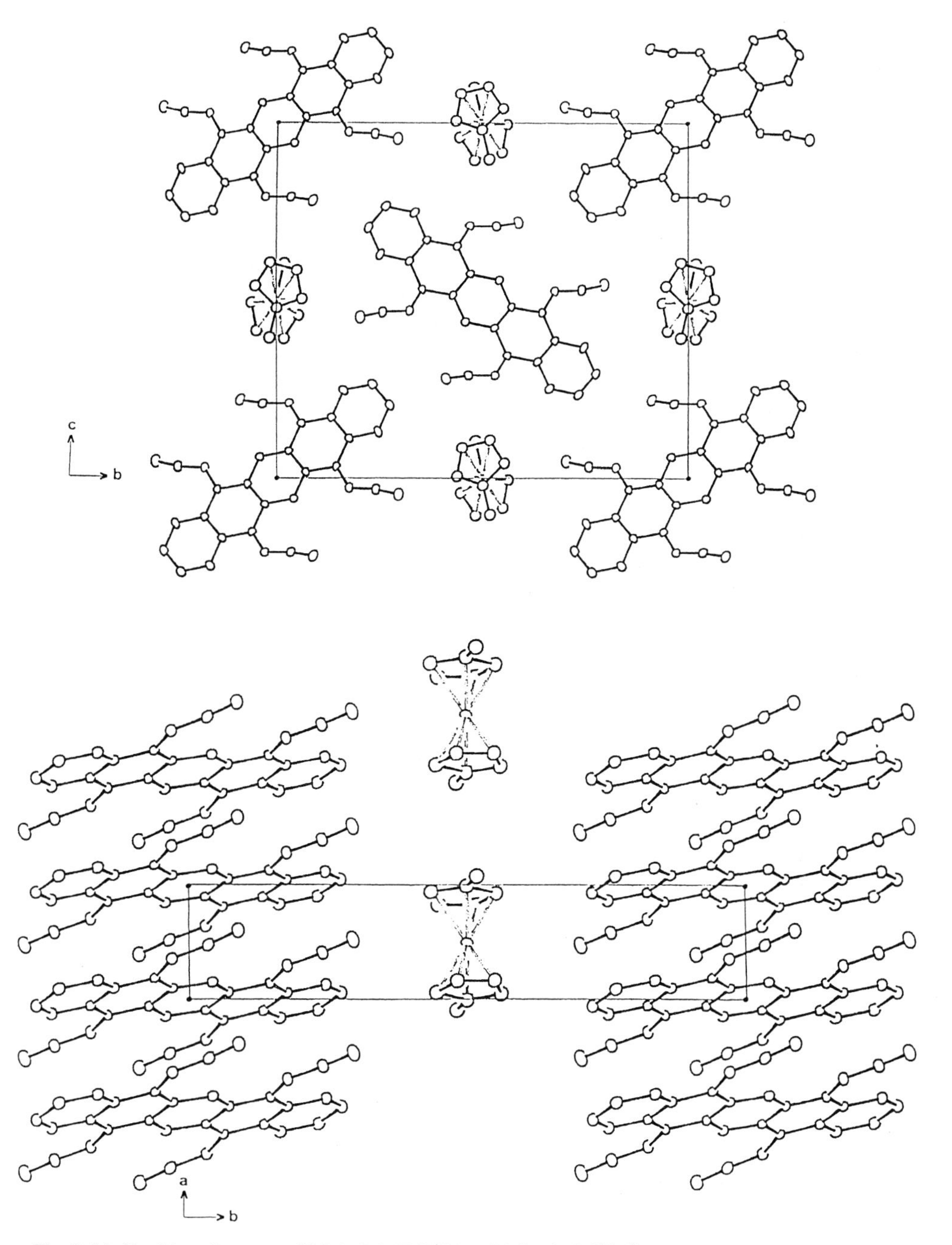

Fig. 8-14. Packing diagram of $[(MeCp)_2Fe]^{\cdot +}[(PTCI)_2]^-$ (ref. [38a]).

some metallic character. Finally, one notes that [$FeCp_2^*$] gives, with both acceptors, the CT complexes with the lowest conductivity, the differences being larger for the TCIDBT acceptor (see Sect. 8.3.1.1).

8.3.3 Miscellaneous Ferrocenes

A 1:2 CT complex of bis(η^5-tricyclo[5.2.1.$0^{2,6}$]deca-2,5,8-trien-6-yl)iron(II) **1** (depicted below) with TCNQ has recently been reported [55]. The solid state structure is characterized by segregated stacks of donor and acceptor ions, in which the TCNQ columns form a zig-zag array of equally spaced molecules. This type of arrangement is in principle well suited for the purpose of electrical conductivity, but in this case no measurements thereof have been carried out.

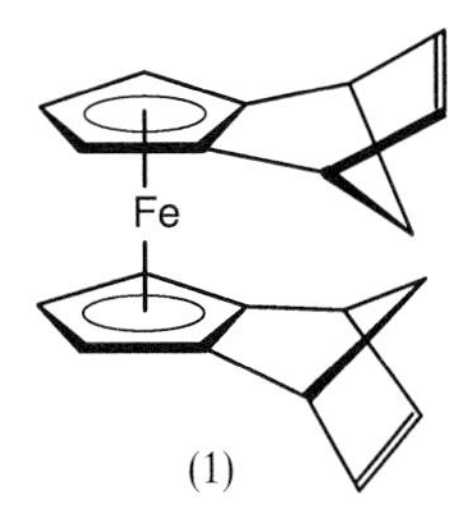

(1)

An interesting approach to the construction of ferromagnetic ferrocene-containing CT complexes, inferred from the McConnell model, has been reported by Iwamura and co-workers [56]. The idea consisted of combining a ferrocene moiety with a stable radical in order to create, after formation of the CT complex, a stable triplet diradical, in analogy to the virtual triplet prescribed by the McConnell model. The actual compound **2** that was prepared contains a nitronyl-nitroxide fragment and is shown below. Such a molecule should generate, upon oxidation, a diradical that could exist in a triplet state (ferromagnetic coupling) or as a singlet (antiferromagnetic coupling). The corresponding 1:1 CT complex with DDQ was shown to display a complex temperature-dependent magnetic behavior, but mainly of the antiferromagnetic type, thus indicating that in the oxidized form of the donor the two spins are not independent. It is conceivable that this approach could lead to the discovery of new ferromagnetic materials in the future.

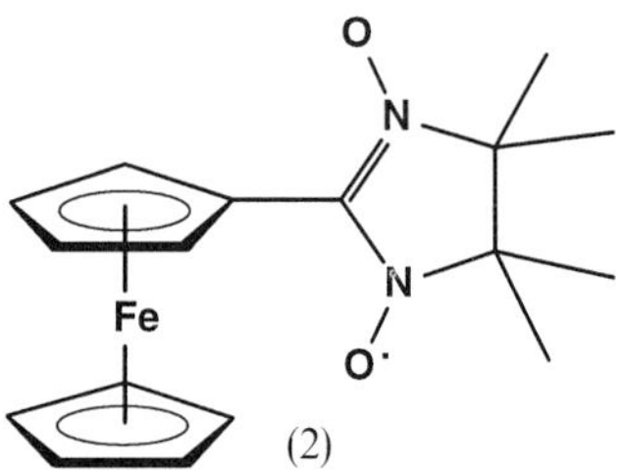

(2)

8.4 Biferrocenes

Biferrocene **3** and 1,1′-biferrocenylene **4** (shown below) have attracted much attention already more than two decades ago, in particular because of their easy formation of mixed-valence Fe(II)-Fe(III) species [57]. In this context, it is interesting to note that the latter is much more easily oxidized to the monocationic form than biferrocene (see Table 8-1) [58].

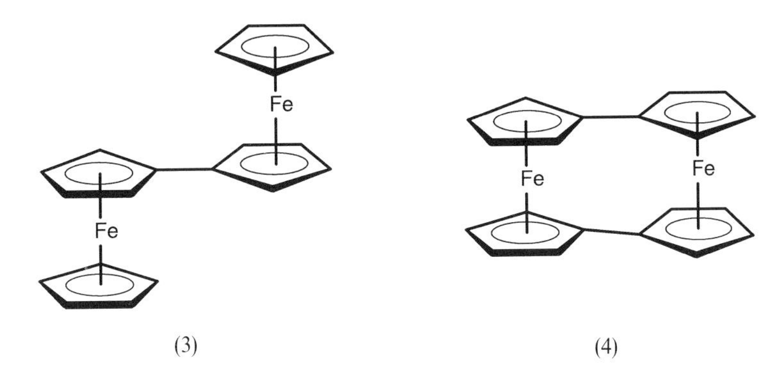

(3) (4)

The biferrocene[Fe(II)-Fe(III)] picrate was one of the first ferrocene-containing CT complexes to be characterized from the point of view of its electrical conductivity [59]. Although the room temperature conductivity of single crystals of such a semiconducting material were found to be rather modest ($\sigma_{rt} = 2.3 \times 10^{-8}$ S cm^{-1}, $E_a = 0.43$ eV), it could be recognized that the mixed valence system was four to six orders of magnitude more conducting than the corresponding ferrocenium compounds.

The related 1,1′-biferrocenylene was reported in 1972 by two groups independently to form a 1:2 CT complex with TCNQ [60a, b]. In contrast to ferrocene itself, the electron transfer from 1,1′-biferrocenylene to TCNQ in solution, even at very low concentration, is essentially an irreversible process. Polycrystalline samples of the CT complex were found to have a relatively high room temperature conductivity ($\sigma_{rt} = 10$ S cm^{-1}).

The differences between biferrocenium and 1,1′-biferrocene[Fe(II)-Fe(III)] cations have been previously discussed [57]. It is sufficient to note here that the crystal structure of 1,1′-biferrocene[Fe(II)-Fe(III)] picrate [61a] reveals a shorter Fe-Fe distance than that in the neutral compound [61b] (3.64 *vs.* 3.98 Å) indicating increased interaction between the two metal centers in the cation. Possibly because a high yield synthesis of 1,1′-biferrocenylene is not available, and because the derivative chemistry of this compound is essentially undeveloped, no other relevant studies directed toward the preparation of molecular materials based on it have appeared in the literature.

In very recent years derivatives of biferrocene have witnessed a revival of interest. Thus, alkylated biferrocenium salts have been studied mainly from the point of view of their intramolecular electron-transfer rates [62] and charge localization [63]. Thus, the decamethylbiferrocene **5** is readily oxidized by iodine and forms the corresponding I_3^- monocationic salt [63]. The crystal structure of this complex reveals two dimensionally different ferrocene moieties (e.g., the distances between the planes of the Cp rings are 3.32 and 3.41 Å). Charge localization has also been confirmed by Mössbauer spectroscopy down to 4.2 K. Furthermore, the dioxidized species was found to display a weak antiferromagnetic coupling ($\theta = -5.6$ K), thus indicative of an $S = 0$ ground state.

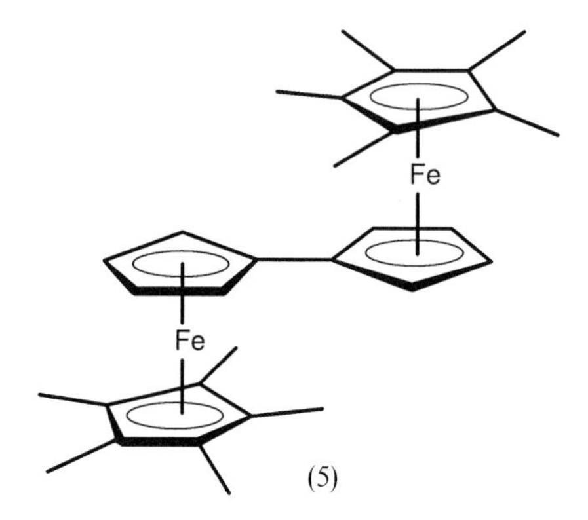

(5)

8.5 Charge-Transfer Complexes of Ferrocenes Containing Conjugated Heteroatom Substituents

Due to their structural and electrochemical properties, it would appear almost obvious to combine ferrocene-containing fragments and derivatives of tetrathiafulvalene to construct new donors for conducting CT complexes. This approach would lead to new multistage redox systems that are likely to display different solid state properties to those of their congeners. However, only very few derivatives of this type have been so far reported. What appears to be the first compound belonging to this class was prepared by Ueno et al. in 1980 [64]. Bis(ferrocenyl)tetrathiafulvalene was obtained as a *trans*/*cis* isomeric mixture (**6** and **7**, respectively) and was shown to form 1 : 1 CT complexes with TCNQ and DDQ. These materials possess

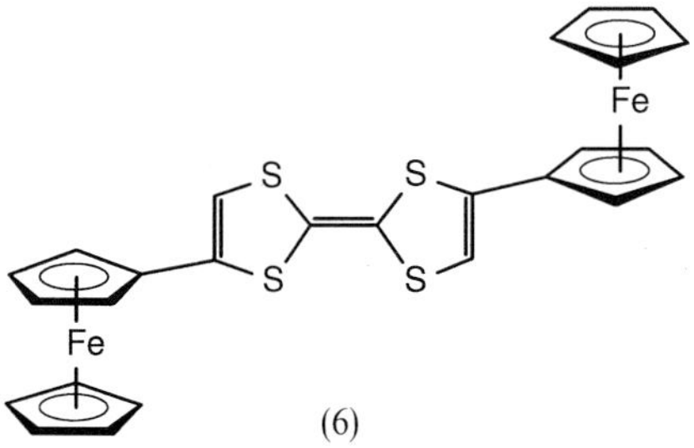

(6) 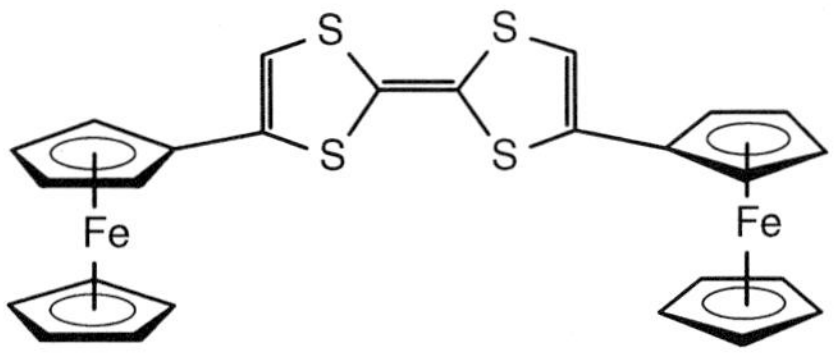 (7)

a modest electrical conductivity (σ_{rt} of pressed pellets for the DDQ derivative: 1.2×10^{-3} S cm^{-1}). On the basis of electronic spectra it was concluded that these CT complexes are to be regarded as tetrathiafulvalenium and not ferrocenium salts. Since no structural study was reported, how donor and acceptor molecules are arranged in the solid state and how to explain the observed conductivity, remains a matter of speculation. Furthermore, the two isomers were not separated, although it would have been of interest to examine the differences in solid state properties due to isomerism.

A recent report from our laboratory dealt with the synthesis of 1,1′-disubstituted ferrocene derivatives as novel donors for the preparation of CT complexes [65]. The approach was to attach two donor moieties structurally and electronically related to the TTFs and which are conjugated with the ferrocene fragment. The underlying concept was (a) to create a multistage redox system by virtue of combining three electron donors in the same molecule, and (b) to examine the possibility that such systems would lead to intramolecular stacking by preferentially displaying the eclipsed conformation of the ferrocene fragment.

It was found that the two compounds depicted in Fig. 8-15, i.e., 1,1′-bis[(5,6-dihydro-1,3-dithiolo[4,5-*b*][1,4]dithiin-2-ylidene)-methyl]ferrocene and 1,1′-bis[(1,3-benzodithiol-2-ylidene)methyl]ferrocene are very similar, as far as their redox behavior and CT complexes are concerned. Thus, they both display three subsequent oxidation processes, only the first two exhibiting features of chemical reversibility. Furthermore, they form 1:2 CT complexes with TCNQ that are electrically

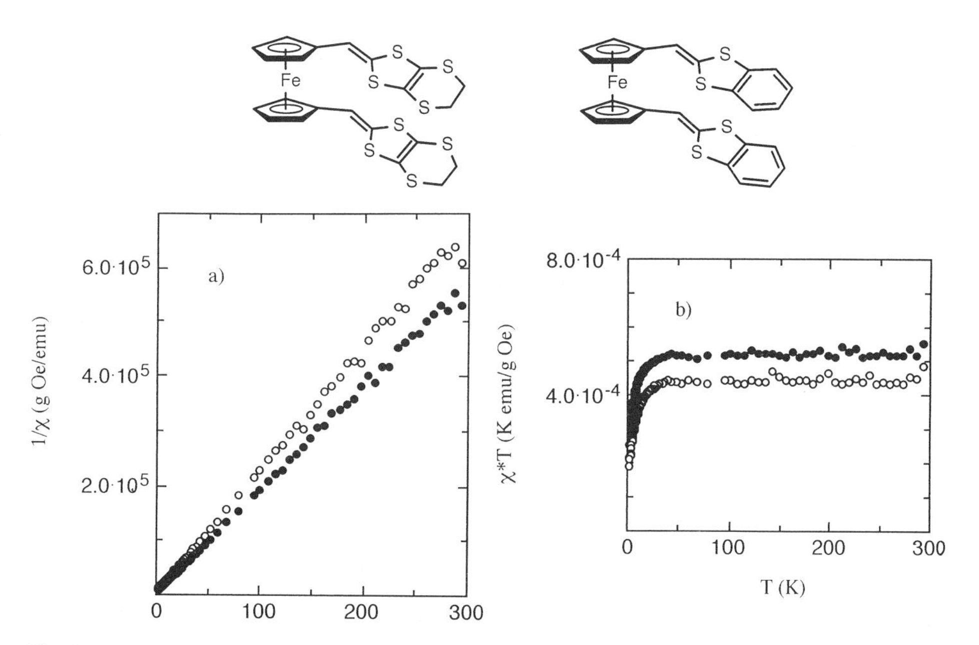

Fig. 8-15. Two 1,1′-disubstituted ferrocenes and the temperature dependence of the magnetic susceptibility at constant applied field of their 1:2 CT complexes with TCNQ (open and closed circles, respectively) (ref. [65]).

conducting (σ_{rt} of pressed pellets: 0.26 and 0.20 S cm^{-1}, respectively). Their magnetic behavior was characterized by susceptibility measurements to be weakly antiferromagnetic ($\theta = -3$ and -4 K, respectively). For samples cooled below 50 K, an abrupt drop in χT *vs.* T was observed (Fig. 8-15). This is indicative of a phase transition to an antiferromagnetically ordered state, which is reached below the Néel temperature of 5 K.

No crystals suitable for X-ray diffraction studies could be obtained for these two CT complexes. On the other hand, the structure of the donor in its neutral form is known and, as it is shown in Fig. 8-16, it displays an approximately C_s symmetric conformation with the two substituents lying on top of each other, thus indicating that the intramolecular stacking can be realized. However, the actual conformation of the molecule is distorted, such that the two heterocyclic substituents are not perfectly parallel, and are found to significantly diverge from coplanarity with the respective Cp ring. If such a distorted conformation is also adopted by the cationic form in the CT complex, then it will turn out to be detrimental to optimum intermolecular interactions in the solid state. Whether or not stacks of such cations significantly contribute to conductivity is at present not known.

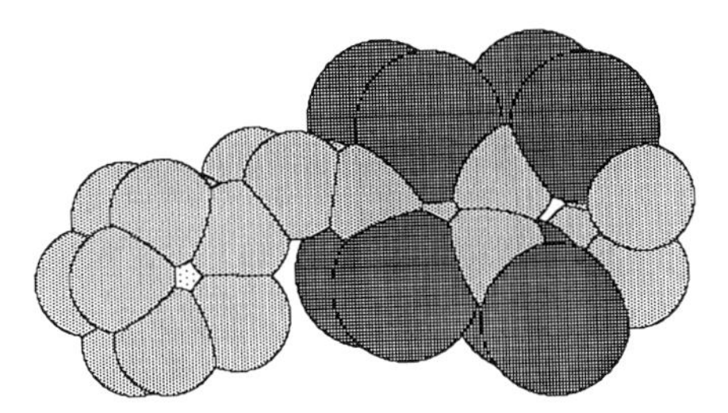

Fig. 8-16. The eclipsed structure of 1,1′-bis[(5,6-dihydro-1,3-dithiolo[4,5-*b*][1,4]-dithiin-2-ylidene)-methyl]ferrocene (ref. [65]).

A very similar type of donor molecule has recently been reported by Bryce and co-workers [66]. 1,1′-Bis[1-(1,3-dithia-cyclopenten-2-ylidene)ethyl]ferrocene **8** and 1,1′-Bis[1-(4,5-dimethyl-1,3-dithia-cyclopenten-2-ylidene)ethyl]ferrocene **9** (see be-

(8) (9)

low) were prepared in an analogous way to the above described derivatives. These two derivatives have very similar properties compared to those of the compounds above, but their 1:2 CT complexes with TCNQ display slightly lower conductivities (σ_{rt} of pressed pellets 0.20 and 0.10 S cm^{-1}, respectively). The former of these two donors was characterized by X-ray analysis and in the solid state was found to possess a structure corresponding schematically to that depicted. Thus, the molecule assumes a chiral conformation with a crystallographic C_2 symmetry, as opposed to C_s symmetry for the derivative described above, and a remarkable twisting of the substituents with respect to the plane of the Cp rings. The main overall difference in conformation could be responsible for the lower conductivity observed, when one assumes that the same conformation is also present in crystals of the CT complexes. Among the possible factors determining these conformational differences there could be the 1-methyl groups or the absence of the second ring in the conjugated substituents in the C_2 symmetric derivatives.

A last example to be discussed in this section is particularly instructive from a structural point of view. In the case where the two substituents in the 1,1′ position are (4-methylthio-phenyl)-2-*E*-ethenyl, the corresponding ferrocene derivative (see Fig. 8-17) still acts as a donor and forms a 1:2 CT complex with TCNQ, but the degree of electron transfer is very weak [65]. This material is non-conducting (σ_{rt} of pressed pellets $< 10^{-8}$ S cm^{-1}), shows very weak signals in both ESR spectra and magnetic susceptibility measurements, and displays a CN stretching frequency typical for neutral TCNQ. As illustrated in Fig. 8-17, the solid state structure of this CT complex is characterized by DA_2 stacks, whereby the donor in the DA_2 unit assumes an antiperiplanar conformation, thus being able to accomodate the

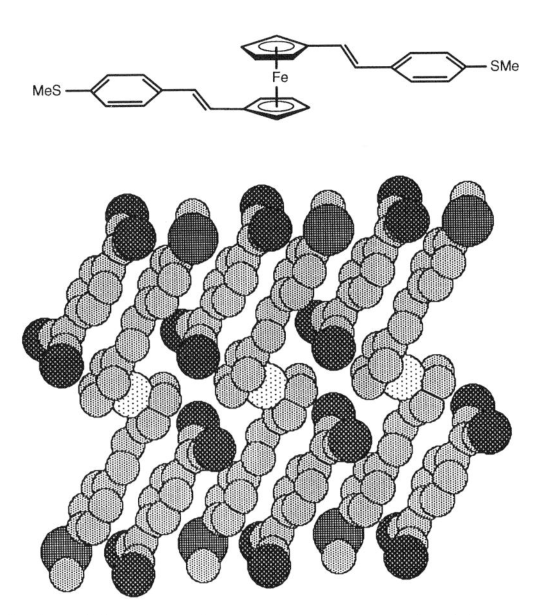

Fig. 8-17. A 1,1′-disubstituted ferrocene donor adopting an antiperiplanar conformation and the structure of its 1:2 CT complex with TCNQ (ref. [65]).

two TCNQ molecules in approximate parallel position to the Cp rings. This leads to an overall DADA structure type. Despite its uninteresting physical properties, the structural motif of this compound should be taken into consideration as a new possibility for the construction of highly symmetrical CT complexes. Assuming a high degree of electron transfer (which could be obtained, for example, by fine-tuning the nature of the substituents), this type of structure should be suited for warranting optimal intermolecular interactions, needed for the observation of magnetic coupling.

8.6 Comparison with Other Organometallic Charge-Transfer Complexes

Organometallic precursors as donors for the preparation of CT complexes have not been limited to ferrocene derivatives and it would be misleading not to mention, at least briefly, what has been achieved with other types or classes of compounds.

As far as magnetic properties are concerned, an important aspect is related to the spin S of the species involved in the CT complex. Given a bulk ferromagnetic system, the critical Curie temperature T_c is proportional to $S(S + 1)$ and to the exchange integral J [67]. Therefore, for an isostructural system based on components with higher values of S, a higher value of T_c should be observed. This reasoning led to the discovery of the bulk ferromagnet $[MnCp_2^*]^{\cdot +}[TCNE]^{\cdot -}$, which is isostructural with the corresponding iron compound, but has a T_c of 8.8 K (*vs.* 4.8 K) [68]. Decamethylmanganocenium is a d^4 high-spin ($S = 1$) system, as opposed to the d^5 low-spin ($S = 1/2$) ferrocenium cation. Bulk ferromagnetism has also been reported for the decamethylchromocene/TCNQ [69] CT complex and for $[MnCp_2^*]^{\cdot +}[TCNQ]^{\cdot -}$ [70]. These two latter materials have shown that the interpretation of their ferromagnetism cannot be based on the simple McConnell model, which invokes a configurational mixing of a virtual CT state (vide supra).

Several CT complexes of metal bis(arene) compounds containing, among others, the familiar organic acceptor TCNQ have been reported [71]. None of these materials has been shown to display physical properties superior to those of the metallocene systems. One very important exception is related to the reaction of bis(benzene) vanadium (which is isoelectronic with $[MnCp_2^*]$) with TCNE, which affords a material with unprecedented properties [72]. The amorphous material obtained displays bulk ferromagnetism at room temperature and its T_c even exceeds the decomposition temperature of the sample of about 350 K. The compound is no longer an arene complex as it has an empirical composition corresponding to $V(TCNE)_2 \cdot 1/2\,(CH_2Cl_2)$.

The Cp*Ru(II) fragment, easily available as the tris(acetonitrile) complex salt $[Cp^*Ru(NCMe)_3]CF_3SO_3$, has been reported as a very well suited building block for the formation of cationic mixed Cp*-arene donor systems [73]. This constitutes an alternative to ruthenocene derivatives that do not afford the corresponding ruthenocenium salts on oxidation, but usually readily decompose [74]. Since the Cp*Ru(II) fragment has a high affinity for arenes, a number of mono-, di-, and polynuclear complexes and their $(TCNQ)_x$ salts have been obtained. Such derivatives are not so easily accessible for the corresponding iron fragment. Among others, the dinuclear complex $[(Cp^*Ru)_2(\eta^6,\eta^6\text{-}[2_2](1,4)\text{-cyclophane})]^{2+}$ (**10**, depicted below) forms a $[TCNQ]_4$ salt that displays a substantial electrical conductivity.

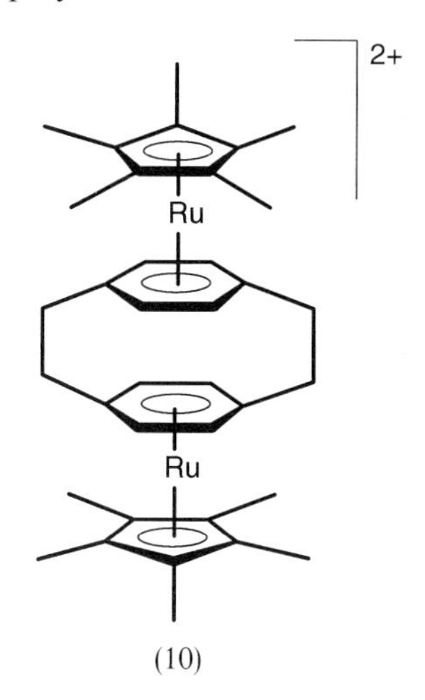

(10)

Last but not least, mention should be made of the use of redox active organometallic clusters for the formation of CT complex. An early example was reported by Eisenberg, Miller and co-workers who exploited the trinuclear niobium cluster $[Nb_3(\mu\text{-}Cl)_6(\eta^6\text{-}C_6Me_6)_3]^+$ and prepared the $[TCNQ]_2^{2-}$ salt, upon oxidation to the dicationic form. This material was found to be paramagnetic and to display a relatively modest conductivity ($\sigma_{rt} = 0.001$ S cm^{-1}, semiconductor) [75]. Finally, the pentanuclear trigonal bipyramidal vanadium cluster $[(MeCp)_5V_5(\mu^3\text{-}S)_6]$ was shown to undergo CT complex formation with TCNQ affording $[(MeCp)_5V_5(\mu^3\text{-}S)_6]^{\cdot+}[(TCNQ)_2]^-$ [76]. This material, which was structurally characterized, is an n-type semiconductor with a band gap of 0.25 eV.

8.7 Conclusions and Perspectives

Ferrocene and some of its derivatives are well suited for the formation of CT complexes and offer an entry into solid state organometallic chemistry. Because such CT complexes contain radical species, the consideration of their physical properties is of fundamental importance and now constitutes an interdisciplinary research field. The systematic study of the magnetic properties of CT complexes of, notably, decamethylferrocene, by the Miller–Epstein group has led to the discovery of bulk ferromagnetic molecular materials and to new insights into the mechanisms of magnetic coupling. Such results have shown that 1D materials, as most CT complexes are, can display 3D properties if the molecular components are matched to each other in an optimal manner. Despite this success, the rational design of organic/organometallic ferromagnetic materials with given structure and given physical properties still remains an essentially unattained goal. Furthermore, research directed toward the synthesis of organometallic conductors (or superconductors) is even less developed.

Ferrocene, its derivatives, and, more in general, organometallic sandwich compounds capable of existing in different oxidation states of the central metal atom, convey to their CT complexes an important structural bias due to their cylindrical "biplanar" nature. What has been achieved with decamethylferrocene indicates that we are learning how to match the structure of the acceptors in order to possibly predict, for example, a 1D-DADA structure of the CT complexes. On the other hand, the well established and extremely rich derivative chemistry of ferrocene has placed synthetic chemists in a position to further manipulate ferrocenes and to design better molecular components incorporating, e.g., several donor (or acceptor) fragments, thus leading to well defined multistage redox systems. Research in this direction could lead to the discovery of new materials with improved magnetic or electrical properties in the future.

References

[1] See, for example, (a) J. M. Williams, J. R. Ferraro, R. J. Thorn, K. D. Carlson, U. Geiser, H. H. Wang, A. M. Kini, M.-H. Whangbo, *Organic Superconductors. Synthesis, Structure, Properties, and Theory*, Prentice Hall, Englewood Cliffs, New Jersey, **1992**; (b) J. M. Williams, H. H. Wang, T. J. Emge, U. Geiser, M. A. Beno, P. C. W. Leung, K. D. Carlson, R. J. Thorn, A. J. Schultz, *Prog. Inorg. Chem.* **1987**, *35*, 51–218.

[2] See, for example, *Magnetic Molecular Materials*, D. Gatteschi, O. Kahn, J. S. Miller, F. Palacio (Eds.), NATO ASI Series, Series E: Applied Sciences, Vol. 198, Kluwer, Dordrecht, **1991**.

[3] See, for example, *Nanotechnology. Research and Perspectives*, B. C. Crandall, J. Lewis (Eds.), The MIT Press, Cambridge, Massachusetts, **1992**.

[4] (a) J. Ferraris, D. O. Cowan, V. V. Walatka, J. H. Perlstein, *J. Am. Chem. Soc.* **1973**, *95*, 948–950; (b) J. P. Ferraris, T. O. Poehler, A. N. Bloch, D. O. Cowan, *Tetrahedron Lett.* **1973**, 2553–2556.

[5] For reviews, see: (a) A. Graja, *Low-dimensional Organic Conductors*, World Scientific, Singapore, **1992**; (b) M. R. Bryce, *Chem. Soc. Rev.* **1991**, *20*, 355–390; (c) P. Cassoux, L. Valade in *Inorganic Materials* (Eds.: D. W. Bruce, D. O'Hare), Wiley, Chichester, **1992**, pp. 2–58; (d) P. Cassoux, L. Valade, H. Kobayashi, A. Kobayashi, R. A. Clark, A. E. Underhill, *Coord. Chem. Rev.* **1991**, *110*, 115–160.

[6] (a) D. Jérome, A. Mazaud, M. Ribault, K. Bechgaard, *J. Phys. Lett. (Paris)* **1980**, *41*, L95–L98; (b) K. Bechgaard, C. S. Jacobsen, K. Mortensen, H. J. Pedersen, N. Thorup, *Solid State Commun.* **1980**, *33*, 1119–1125.

[7] For other classes of compounds, see, for example, (a) B. Tiecke, A. Wegmann, W. Fischer, B. Hilti, C. W. Mayer, J. Pfeiffer, *Thin Solid Films* **1989**, *179*, 233–238; (b) E. Günther, S. Hünig, J.-U. von Schütz, U. Langohr, H. Rieder, S. Söderholm, H.-P. Werner, K. Peters, H. G. von Schnering, H. J. Lindner, *Chem. Ber.* **1992**, *125*, 1919–1926, and references cited therein.

[8] See, for example, (a) A. Aumüller, P. Erk, G. Klebe, S. Hünig, J.-U. von Schütz, H.-P. Werner, *Angew. Chem.* **1986**, *98*, 759–761; *Angew. Chem. Int. Ed. Engl.* **1986**, *25*, 740–741; (b) S. Hünig, P. Erk, *Adv. Mater.* **1991**, *3*, 225–236.

[9] For examples, see, (a) M. R. Bryce, G. Cooke, A. S. Dhindsa, D. J. Ando, M. B. Hursthouse, *Tetrahedron Lett.* **1992**, *33*, 1783–1786; (b) A. Izuoka, R. Kumai, T. Sugawara, *Chem. Lett.* **1992**, 285–288; (c) M. Iyoda, Y. Kuwatani, N. Ueno, M. Oda, *J. Chem. Soc. Chem. Commun.* **1992**, 158–159; (d) T. Tachikawa, A. Izuoka, R. Kumai, T. Sugawara, Y. Sugawara, *Solid State Commun.* **1992**, *82*, 19–22.

[10] For scattered reports, see, for example, (a) U. T. Müller-Westerhoff, P. Eilbracht, *J. Am. Chem. Soc.* **1972**, *94*, 9272–9274; (b) S. Z. Goldberg, B. Spivack, G. Stanley, R. Eisenberg, D. M. Braitsch, J. S. Miller, M. Abkowitz, *ibid.* **1977**, *99*, 110–117; (c) Y. Ueno, H. Sano, M. Okawara, *J. Chem. Soc. Chem. Commun.* **1980**, 28–30; (d) C. M. Bolinger, J. Darkwa, G. Gammie, S. D. Gammon, J. W. Lyding, T. B. Rauchfuss, S. R. Wilson, *Organometallics* **1986**, *5*, 2386–2388.

[11] For reviews, see: (a) J. S. Miller, A. J. Epstein, *Angew. Chem. Int. Ed. Engl.* **1994**, *33*, 385–416; (b) J. S. Miller, A. J. Epstein, W. M. Reiff, *Chem. Rev.* **1988**, 88, 201–220; (c) J. S. Miller, A. J. Epstein, W. M. Reiff, *Acc. Chem. res.* **1988**, 21, 114–120; (d) J. S. Miller, A. J. Epstein, W. M. Reiff, *Science* **1988**, 240, 40–47; (e) J. S. Miller, A. J. Epstein in *Research Frontiers in Magnetochemistry* (Ed.: C. J. O'Connor), World Scientific, Singapore, **1993**, pp. 283–302.

[12] See, for example, (a) ref. [5a], Chapter 2; (b) ref. [1a], Chapter 4.

[13] See, for example, (a) ref. [11a, b]; (b) O. Kahn, Y. Pei, Y. Journaux in *Inorganic Materials* (Eds.: D. W. Bruce, D. O'Hare), Wiley, Chichester, **1992**, pp. 59–114; for a standard treatise on magnetochemistry, see (c) R. C. Carlin, *Magnetochemistry*, Springer, Berlin, **1986**.

[14] For a classification of the different types of stacks in conducting CT complexes, see: P. Delhaès in *Lower-Dimensional Systems and Molecular Electronics* (Eds.: R. M. Metzger, P. Day, G. C. Papavassiliou), NATO ASI Series, Series B: Physics, Vol. 248, Plenum, New York, **1991**, pp. 43–65.

[15] For a discussion on "Design constraints for organic metals and superconductors", see: D. O. Cowan, J. A. Fortkort in *Lower-Dimensional Systems and Molecular Electronics* (Eds.: R. M. Metzger, P. Day, G. C. Papavassiliou), NATO ASI Series, Series B: Physics, Vol. 248, Plenum, New York, **1991**, pp. 1–22.

[16] R. C. Wheland, *J. Am. Chem. Soc.* **1976**, 98, 3926–3930. See also: J. M. Williams, H. H. Wang, A. M. Kini, M. A. Beno, U. Geiser, A. J. Schultz, K. D. Carlson, J. R. Ferraro, M.-H. Whangbo in *Lower-Dimensional Systems and Molecular Electronics* (Eds.: R. M. Metzger, P. Day, G. C. Papavassiliou), NATO ASI Series, Series B: Physics, Vol. 248, Plenum, New York, **1991**, pp. 91–95.

[17] (a) J. S. Miller, A. J. Epstein in ref. [2], pp. 151–158; (b) P. Lemoine, M. Gross, P. Braunstein, F. Mathey, B. Deschamps, J. H. Nelson, *Organometallics* **1984**, 3, 1303–1307; (c) M. M. Sabbatini, E. Cesarotti, *Inorg. Chim. Acta* **1977**, 24, L9–L10; (d) H. Grimes, S. R. Logan, *ibid.* **1980**, 45, L223–L224; (e) R. C. Wheland, J. L. Gillson, *J. Am. Chem. Soc.* **1976**, 98, 3916–3925; for the synthesis of substituted TCNQ's, see: (f) R. C. Wheland, E. L. Martin, *J. Org. Chem.* **1975**, 40, 3101–3109; for an early reference on redox properties of (mainly mono) substituted ferrocenes, see: (g) G. L. K. Hoh, W. E. McEwen, J. Kleinberg, *ibid.* **1961**, 83, 3949–3953; for the synthesis and properties of metal ethylenedithiolato complexes, see: (h) J. A. McCleverty, *Prog. Inorg. Chem.* **1968**, 10, 49–221.

[18] O. W. Webster, W. Mahler, R. E. Benson, *J. Am. Chem. Soc.* **1962**, 84, 3678–3684.

[19] M. Rosenblum, R. W. Fish, C. Bennett, *J. Am. Chem. Soc.* **1964**, 86, 5166–5170.

[20] E. Adman, M. Rosenblum, S. Sullivan, T. N. Margulis, *J. Am. Chem. Soc.* **1967**, 89, 4540–4542.

[21] B. W. Sullivan, B. M. Foxman, *Organometallics* **1983**, 2, 187–189.

[22] J. S. Miller, J. H. Zhang, W. M. Reiff, *Inorg. Chem.* **1987**, 26, 600–608.

[23] R. M. Metzger, N. E. Heimer, D. Gundel, H. Sixl, R. H. Harms, H. J. Keller, D. Nothe, D. Wehe, *J. Chem. Phys.* **1982**, 77, 6203–6214.

[24] M. D. Ward, D. C. Johnson, *Inorg. Chem.* **1987**, 26, 4213–4227, and references cited therein.

[25] J. D. Crane, P. B. Hitchcock, H. W. Kroto, R. Taylor, D. R. M. Walton, *J. Chem. Soc. Chem. Commun.* **1992**, 1764–1765.

[26] Decamethyl ferrocene was first reported by King and co-workers: R. B. King, M. B. Bisnette, *J. Organomet. Chem.* **1967**, 8, 287–297.

[27] For a comparison of several permethylmetallocenes with respect to their electronic structure and physical properties, see: J. L. Robbins, N. Edelstein, B. Spencer, J. C. Smart, *J. Am. Chem. Soc.* **1982**, 104, 1882–1893; for decamethylosmocene, see: D. O'Hare, J. C. Green, T. P. Chadwick, J. S. Miller, *Organometallics* **1988**, 7, 1335–1342.

[28] J. S. Miller, J. C. Calabrese, H. Rommelmann, S. R. Chittipeddi, J. H. Zhang, W. M. Reiff, A. J. Epstein, *J. Am. Chem. Soc.* **1987**, 109, 769–781.

[29] J. S. Miller, J. H. Zhang, W. M. Reiff, *J. Am. Chem. Soc.* **1987**, 109, 4584–4592.

[30] J. S. Miller, M. D. Ward, J. H. Zhang, W. M. Reiff, *Inorg. Chem.* **1990**, 29, 4063–4072.

[31] (a) G. A. Candela, L. J. Schwartzendruber, J. S. Miller, M. J. Rice, *J. Am. Chem. Soc.* **1982**, 101, 2755–2756; (b) A. H. Reis, Jr., L. D. Preston, J. M. Williams, S. W. Peterson, G. A. Candela, L. J. Schwartzendruber, J. S. Miller, *ibid.* **1982**, 101, 2756–2758; (c) J. S. Miller, J. H. Zhang, W. M. Reiff, D. A. Dixon, L. D. Preston, A. H. Reis, Jr., E. Gebert, M. Extine, J. Troup, A. J. Epstein, M. D. Ward, *J. Phys. Chem.* **1987**, 91, 4344–4360.

[32] D. A. Dixon, J. C. Calabrese, J. S. Miller, *J. Phys. Chem.* **1989**, 93, 2284–2291.

[33] J. S. Miller, J. C. Calabrese, R. L. Harlow, D. A. Dixon, J. H. Zhang, W. M. Reiff, S. Chittipeddi, M. A. Selover, A. J. Epstein, *J. Am. Chem. Soc.* **1990**, 112, 5496–5506.

[34] For a more complete list of known compounds, see, for example, ref. [11e].

[35] (a) J. S. Miller, P. J. Krusic, D. A. Dixon, W. M. Reiff, J. H. Zhang, E. C. Anderson, A. J. Epstein, *J. Am. Chem. Soc.* **1986,** 108, 4459–4466; (b) E. Gebert, A. H. Reis, Jr., J. S. Miller, H. Rommelmann, A. J. Epstein, *ibid.* **1982**, 104, 4403–4410.

[36] J. S. Miller, C. Vasquez, R. S. McLean, W. M. Reiff, A. Aumüller, S. Hünig, *Adv. Mater.* **1993,** 5, 448–450.

[37] D. A. Dixon, J. C. Calabrese, J. S. Miller, *J. Am. Chem. Soc.* **1986**, 108, 2582–2588.

[38] (a) P. Chetcuti, G. Rist, G. Rihs (CIBA-Geigy Ltd., Basel, Switzerland), unpublished results. We thank CIBA for permission to disclose these results. For acceptors syntheses, see: (b) L. L. Miller, C. A. Liberko, *Chem. Mater.* **1990**, 2, 330–340; (c) S. F. Rak, C. A. Liberko, L. L. Miller, *Synthetic Metals* **1991**, 41–43, 2365–2375; (d) M. Matsuoka, A. Iwamoto, T. Kitao, *J. Heterocyclic Chem.* **1991**, 28, 1445–1447; (e) H. Bock, D. Jaculi, *Phosphorus, Sulfur, and Silicon* **1991**, 61, 289–304.

[39] (a) A. J. Epstein, J. S. Miller in ref. [2], pp. 159–169; (b) A. J. Epstein, J. S. Miller, *Mol. Cryst. Liq. Cryst.* **1989,** 176, 359–368.

[40] S. R. Chittipeddi, K. R. Cromack, J. S. Miller, A. J. Epstein, *Phys. Rev. Lett.* **1987**, 58, 2695–2698.

[41] (a) H. M. McConnell, *Proc. R. A. Welch Found. Chem. Res.* **1967**, 11, 144; for a specific treatment of CT complexes analogous to $[FeCp^*_2]^{\cdot +}[TCNE]^{\cdot -}$, see, (b) J. S. Miller, A. J. Epstein, *J. Am. Chem. Soc.* **1987**, *109*, 3850–3855; (c) D. A. Dixon, A. Suna, J. S. Miller, A. J. Epstein in ref. [2], pp. 171–190; theoretical models have been put forward also by other groups: (d) A. L. Tchougreeff, *J. Chem. Phys.* **1992**, *96*, 6026–6032; (e) A. L. Tchougreeff, I. A. Misurkin, *Phys. Rev. B* **1992**, *46*, 5357–5365; (f) Z. G. Soos, P. C. M. Williams, *Mol. Cryst. Liq. Cryst.* **1989**, *176*, 369–380; (g) K. Yamaguchi, H. Namimoto, T. Fueno, T. Nogami, Y. Shirota, *Chem. Phys. Lett.* **1990**, *166*, 408–414.

[42] (a) C. Kollmar, O. Kahn, *J. Am. Chem. Soc.* **1991**, *113*, 7987–7994; (b) C. Kollmar, M. Couty, O. Kahn, *ibid.* **1991**, *113*, 7994–8005; see also (c) W. E. Broderick, B. M. Hoffman, *ibid.* **1991**, *113*, 6334–6335.

[43] J. S. Miller, J. C. Calabrese, A. J. Epstein, *Inorg. Chem.* **1989**, *28*, 4230–4238.

[44] W. B. Heuer, P. Mountford, M. L. H. Green, S. G. Bott, D. O'Hare, J. S. Miller, *Chem. Mater.* **1990**, *2*, 764–772.

[45] For a discussion of the chemistry of dmit, see: R.-M. Olk, B. Olk, W. Dietzsch, R. Kirmse, E. Hoyer, *Coord. Chem. Rev.* **1992**, *117*, 99–131.

[46] W. E. Broderick, J. A. Thompson, M. R. Godfrey, M. Sabat, B. M. Hoffman, *J. Am. Chem. Soc.* **1989**, *111*, 7656–7657.

[47] G. Matsubayashi, A. Yokozawa, *Inorg. Chim. Acta* **1992**, *193*, 137–141.

[48] G. Matsubayashi, A. Yokozawa, *J. Chem. Soc. Dalton Trans.* **1990**, 3535–3539.

[49] J. M. Forward, D. M. P. Mingos, A. V. Powell, *J. Organomet. Chem.* **1994**, *465*, 251–258.

[50] K.-M. Chi, J. C. Calabrese, W. M. Reiff, J. S. Miller, *Organometallics* **1991**, *10*, 688–693.

[51] D. Stein, H. Sitzmann, R. Boese, E. Dormann, H. Winter, *J. Organomet. Chem.* **1991**, *412*, 143–155.
[52] J. S. Miller, D. T. Glatzhofer, D. M. O'Hare, W. M. Reiff, A. Chakraborty, A. J. Epstein, *Inorg. Chem.* **1989**, *28*, 2930–2939.
[53] For a discussion of structure – magnetic behavior relationships in this type of CT complexes, see: J. S. Miller, A. J. Epstein, *Mol. Cryst. Liq. Cryst.* **1989**, *176*, 347–358.
[54] D. Stein, H. Sitzmann, R. Boese, *J. Organomet. Chem.* **1991**, *421*, 275–283.
[55] J. C. Gallucci, G. Opromolla, L. A. Paquette, L. Pardi, P. F. T. Schirch, M. R. Sivik, P. Zanello, *Inorg. Chem.* **1993**, *32*, 2292–2297.
[56] Y. Nakamura, N. Koga, H. Iwamura, *Chem. Lett.* **1991**, 69–72.
[57] (a) U. T. Müller-Westerhoff, *Angew. Chem.* **1986**, *98*, 700–716 (*Angew. Chem. Int. Ed. Engl.* **1986**, *25*, 702–716); (b) D. O. Cowan, C. Le Vanda, J. Park, F. Kaufman, *Acc. Chem. Res.* **1973**, *6*, 1–7.
[58] (a) W. H. Morrison, Jr., S. Krogsrud, D. N. Hendrickson, *Inorg. Chem.* **1973**, *12*, 1998–2004; (b) T. Matsumoto, M. Sato, A. Ichimura, *Bull. Chem. Soc. Jpn.* **1971**, *44*, 1720.
[59] (a) D. O. Cowan, F. Kaufman, *J. Am. Chem. Soc.* **1970**, *92*, 219–220; (b) F. Kaufman, D. O. Cowan, *ibid.* **1970**, *92*, 6198–6204.
[60] (a) D. O. Cowan, C. Le Vanda, *ibid.* **1972**, *94*, 9271–9272; (b) U. T. Müller-Westerhoff, P. Eilbracht, *ibid.* **1972**, *94*, 9272–9274.
[61] (a) M. Hillman, Å. Kvick, *Organometallics* **1983**, *2*, 1780–1785; (b) M. R. Churchill, J. Wormald, *Inorg. Chem.* **1969**, *8*, 1970–1974.
[62] For a review, see: (a) D. N. Hendrickson, S. M. Oh, T.-Y. Dong, T. Kambara, M. J. Cohn, M. F. Moore, *Comments Inorg. Chem.* **1985**, *4*, 329–349; (b) T.-Y. Dong, C.-K. Chang, C.-H. Huang, Y.-S. Wen, S.-L. Lee, J.-A. Chen, W.-Y. Yeh, A. Yeh, *J. Chem. Soc. Chem. Commun.* **1992**, 526–528; (c) T.-Y. Dong, C. Y. Chou, *ibid.* **1990**, 1332–1334.
[63] M.-H. Delville, F. Robert, P. Gouzerh, J. Linarès, K. Boukheddaden, F. Varret, D. Astruc, *J. Organomet. Chem.* **1993**, *451*, C10–C12, and references cited therein.
[64] Y. Ueno, H. Sano, M. Okawara, *J. Chem. Soc. Chem. Commun.* **1980**, 28–30.
[65] A. Togni, M. Hobi, G. Rihs, G. Rist, A. Albinati, P. Zanello, D. Zech, H. Keller, *Organometallics* **1994**, *13*, 1224–1234.
[66] A. J. Moore, P. J. Skabara, M. R. Bryce, A. S. Batsanov, J. A. K. Howard, S. T. A. K. Daley, *J. Chem. Soc. Chem. Commun.* **1993**, 417–419.
[67] For a discussion, see, for example, ref. [13c].
[68] G. T. Yee, J. M. Manriquez, D. A. Dixon, R. S. McLean, D. M. Groski, R. B. Flippen, K. S. Narajan, A. J. Epstein, J. S. Miller, *Adv. Mater.* **1991**, *3*, 309–311.
[69] W. E. Broderick, B. M. Hoffman, *J. Am. Chem. Soc.* **1991**, *113*, 6334–6335.
[70] W. E. Broderick, J. A. Thompson, E. P. Day, B. M. Hoffman, *Science* **1990**, *249*, 401–403.
[71] For examples, see: (a) D. O'Hare, M. D. Ward, J. S. Miller, *Chem. Mater.* **1990**, *2*, 758–763; (b) M. D. Ward, D. C. Johnson, *Inorg. Chem.* **1987**, *26*, 4213–4227; (c) D. O'Hare, M. Kurmoo, R. Lewis, H. Powell, *J. Chem. Soc. Dalton Trans.* **1992**, 1351–1355.
[72] J. M. Manriquez, G. T. Yee, R. S. McLean, A. J. Epstein, J. S. Miller, *Science* **1991**, *252*, 1415–1417.
[73] (a) P. J. Fagan, M. D. Ward, J. C. Calabrese, *J. Am. Chem. Soc.* **1989**, *111*, 1698–1719; (b) M. D. Ward, P. J. Fagan, J. C. Calabrese, D. C. Johnson, *ibid.* **1989**, *111*, 1719–1732; (c) S. Li, H. S. White, M. D. Ward, *Chem. Mater.* **1992**, *4*, 1082–1091.
[74] For a discussion of the different chemical and redox properties of decamethylferrocene, -ruthenocene, and -osmocene, and their oxidized forms, see: D. O'Hare, J. C. Green, T. P. Chadwick, J. S. Miller, *Organometallics* **1988**, *7*, 1335–1342, and references quoted therein.
[75] S. Z. Goldberg, B. Spivack, G. Stanley, R. Eisenberg, D. M. Braitsch, J. S. Miller, M. Abkowitz, *J. Am. Chem. Soc.* **1977**, *99*, 110–117.
[76] C. M. Bolinger, J. Darkwa, G. Gammie, S. D. Gammon, J. W. Lyding, T. B. Rauchfuss, S. R. Wilson, *Organometallics* **1986**, *5*, 2386–2388.

9 Ferrocene-Containing Thermotropic Liquid Crystals

Robert Deschenaux and John W. Goodby

9.1 Introduction

The design and synthesis of molecular units (building blocks), capable of forming supramolecular architectures (micelles, vesicles, mono- and multi-layers, lyotropic and thermotropic liquid crystals) exhibiting a unique combination of new properties have generated enthusiastic studies at the frontiers of chemistry, physics and biology. First, the beauty of organized molecular assemblies fascinated scientists and led them to take more and more interest. Second, this interdisciplinary field of research was predestined to provide fundamental results, the importance of which had to play a key role towards better understanding of natural processes and the development of a new technology: the nanometer scale technology.

Since the description of liquid crystallinity for cholesteryl benzoate and cholesteryl acetate at the end of the 19th century by Reinitzer [1], an intense activity has been devoted to thermotropic liquid crystals, especially since the early 1970s, owing to the fabrication and application of liquid crystal displays in electronic technology.

Most of the research was focused on purely organic materials and today several thousand compounds are known [2]. During this period, only minor interest was shown in metal-containing liquid crystals (metallomesogens). Indeed, although the first thermotropic metal-containing liquid crystals were reported by Vorländer [3] in 1910 (he discovered that alkali-metal carboxylates formed classical lamellar phases), the era of metallomesogens began only in 1977 when Giroud and Mueller-Westerhoff [4] described the liquid crystal behavior of some Ni(II)dithiolene complexes.

Metal-containing liquid crystals opened the way to new geometries and new topologies, in comparison with wholly organic materials, and allowed further investigation of the relationship between structure and mesomorphic properties. Mononuclear (Ni, Cu, Rh, Pd, Pt, V, Ag, Au, Zn, rare earths) and dinuclear (Rh, Mo, Ru, Ni, Cu and Pd) coordination complexes containing liquid crystals were prepared and studied during the last decade [5].

Much less attention was devoted to organometallic liquid crystals, even though the first examples, diarylmercury derivatives, were described as early as 1923 by Vorländer [6]. Recently, Ziminski and Malthête [7] reported the first mesomorphic(butadiene)iron-tricarbonyl derivatives.

This contribution presents the results published up to June 1993 on ferrocene-containing thermotropic liquid crystals and should be considered as a complement to four outstanding reviews recently published on metal-containing liquid crystals [5].

To help readers who are not familiar with liquid crystals several terms are defined below. Those who are interested in learning the basic concepts of these materials in more detail are referred to the references [1, 5c, 8].

A thermotropic liquid crystal (mesogen) is a compound that, on heating the crystal or on cooling the isotropic liquid, gives rise to mesomorphism. Liquid crystallinity occurs between the crystal and isotropic liquid states. The intermediate phases, or mesophases, can be either enantiotropic, i.e., thermodynamically stable, or monotropic, i.e., thermodynamically unstable. The solid to mesophase transition is referred to as the melting point, while the mesophase to isotropic liquid transition is referred to as the clearing point.

Liquid-crystalline phases are characterized to some degree by the shape of the molecules and by their packing arrangements and ordering in the mesomorphic state. Typically, molecules can have either disc- or rod-like shapes and can form discotic or calamitic mesophases, respectively. Ferrocene liquid crystal systems that have so far been synthesized tend to have molecular structures that are lath- or rod-like in shape, and consequently the phases observed are calamitic. However, this does not preclude the possibility that a polysubstituted ferrocene could be prepared where the molecular shape is disc-like, thereby holding out the prospect of possibly producing discotic/columnar phases.

Calamitic phases are essentially defined as a set of mesophases that occur between the breakdown of the long-range periodic, translational ordering of the crystal and the loss of long range orientational order at the transition to the isotropic liquid. In this context six mesophases exist: the nematic phase and five smectic phases labelled A, B, C, F and I. Other closely related soft crystal phases, in which the molecules have long-range periodic order but are themselves rotationally disordered, also exist in concert with liquid crystal phases. This second set of mesophases are labelled $B_{(cryst)}$, E, J, G, H, and K, and are essentially the crystal versions of the above smectic phases. In ferrocene systems, not very many materials have been found that exhibit these crystal mesophases, and most liquid-crystalline behaviour has been defined as being nematic, smectic A or smectic C. This may be due to the size of the ferrocene unit, which sterically interferes with the packing of the molecules, thereby preventing the build up of long-range order that would result in the formation of a soft crystal phase (or even the slightly more ordered liquid crystal phases I and F).

In the least ordered mesophase, the nematic modification, the molecules have long range orientational order, but no positional order. The molecular correlations are very short range, extending only over a few molecular centers. Thus, the nematic phase can be considered as a one-dimensionally ordered elastic fluid. The long axes of the molecules are, however, aligned on average parallel to one another, this direction being called the director of the phase, which is usually given the symbol *n*. The degree to which the long axes are aligned is called the order parameter, which is defined by the equation

$$S = 1/2\langle 3\cos^2\theta - 1\rangle$$

where θ is the angle made between the long axis of each individual rod-like molecule and the director. The brackets in the equation indicate that this is an average taken over a very large number of molecules.

An order parameter of zero implies that the phase has no order at all (it is liquid-like) whereas a value of one indicates that the phase is perfectly ordered, i.e., all the long axes of the molecules are parallel to one another and to the director, n. For a typical nematic phase the order parameter has a value in the region 0.4 – 0.7, indicating that the molecules are considerably disordered. The order parameter has the same symmetry properties as the nematic phase, in that it is unchanged by rotating any molecule through an angle of 180°. Thus, in the bulk nematic phase, there are as many molecules pointing in one direction relative to the director as there are pointing in the opposite direction (a rotation of 180°), i.e., the molecules have a disordered head-to-tail arrangement in the mesophase. Thus, the phase has rotational symmetry relative to the director. Furthermore, the rod-like molecules in the nematic phase are free to rotate about their short axes and to some degree about their long axes, however the relaxation times for rotations about the long axes are much longer ($\approx 10^{-6}\ s^{-1}$) than those about the short axes ($\approx 10^{-11}\ s^{-1}$). The structure of the nematic phase is depicted in Fig. 9-1 (top).

As noted above, a number of liquid-crystalline smectic phases exist in which the molecules are arranged in lamellae. The structures of these phases are dependent on the packing arrangements and orientations of the molecules within the lamellar structure. Thus, the molecules can pack in layers in which their long axes are either tilted (smectics C, I and F) or orthogonal (smectics A and B) to the layer planes. Locally the rod-like molecules are on average hexagonally close-packed, but the extent of the in-plane ordering is dependent on the mesophase type. In the smectic A and smectic C phases the positional ordering of the molecules is only short range extending over 15 – 20 Å at most, whereas in smectics B, I, and F it extends over a few hundred Å. In contrast, the positional ordering in the soft crystal phases, J, G, $B_{(cryst)}$, E, H, and K, is long range, and both in and out of the planes of the layers.

On passing from the least ordered smectic A phase down to the more ordered crystal H and K phases the layer planes tend to sharpen up. In the smectic A and C phases the layers are therefore very diffuse, and can be thought of as one-dimensional density waves relative to the director. Thus, locally these two phases are very similar in structure to the nematic phase.

In the smectic A phase the molecules are arranged so that their long axes are on average perpendicular to the diffuse layer planes (Fig. 9-1, middle). The average layer spacing is usually fractionally shorter than the fully extended molecular length, indicating some degree of disordering in the layers. As with the nematic phase the molecules undergo rapid rotation about both their short and long axes. One of the more important smectic phases exhibited by ferrocene systems is the smectic C phase (Fig. 9-1, bottom). In this phase the long axes of the molecules are tilted at a temperature-dependent tilt angle θ with respect to the layer planes. The temperature dependence of the tilt angle takes the form

$$\theta_T = \theta_0(T_c - T)^{\alpha}$$

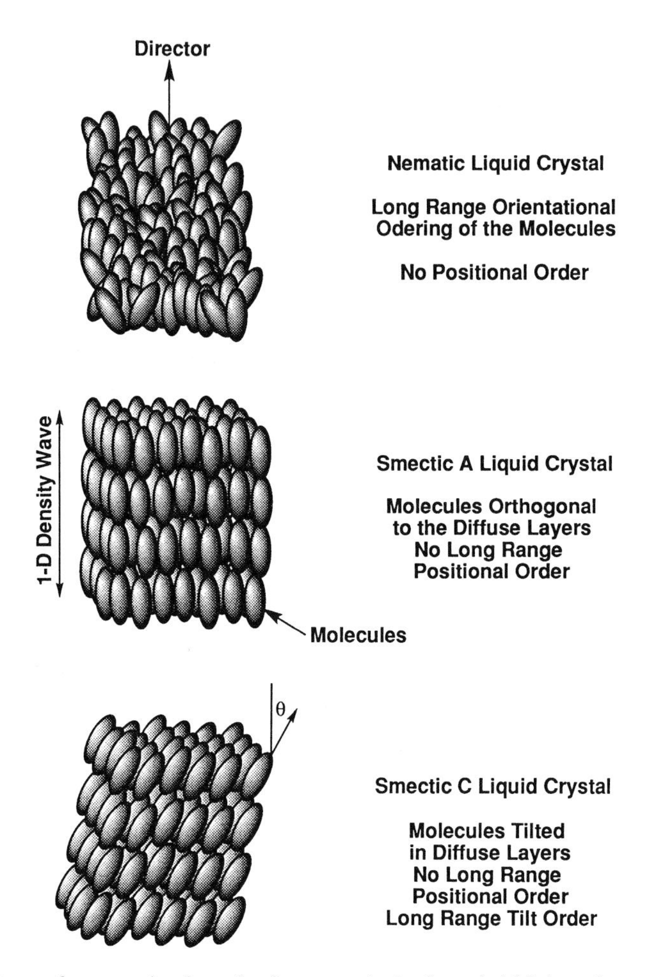

Fig. 9-1. Structure of a nematic phase (top), a smectic A phase (middle), and a smectic C phase (bottom).

where θ_T is the tilt angle at temperature T, θ_0 is a constant, T_c is the smectic A to smectic C transition temperature, T is the temperature and α is an exponent equal to 0.5.

The smectic state, apart from exhibiting a rich variety of modifications, is also thermodynamically ordered. Thus, there is a well-defined sequence of phases formed on cooling from the isotropic liquid. For instance, the following transition sequence would be observed for a ferrocene liquid crystal exhibiting enantiotropic smectic C, smectic A and nematic phases:

$$\text{Crystal} \xrightarrow{T_1} \text{Smectic C} \underset{}{\overset{T_2}{\rightleftharpoons}} \text{Smectic A} \overset{T_3}{\rightleftharpoons} \text{Nematic} \overset{T_4}{\rightleftharpoons} \text{Isotropic liquid}$$

where T_1 (melting point) $< T_2 < T_3 < T_4$ (clearing point).

It is interesting to consider substitution patterns in respect of the ferrocene unit, and their effect on the phase morphology described above. Typically, calamitic or rod-like systems can be produced by monosubstitution, and by 1,2-, 1,3- or 1,1′-disubstitution.

Monosubstitution, as in the case of conventional calamitics, is expected to lead to a predominance of nematic and smectic A phases, with possibly nematic phases being favored because of the repulsive steric effects of the ferrocene unit causing a reduced ability of the molecules to pack in layers. Disubstitution, on the other hand, may be expected to lead to smectic polymorphism because the overall shape of the molecule can be extended and varied from being linear to being slightly bent. For instance, let us first consider disubstitution in one of the ferrocene rings. In 1,2-disubstituted materials, the mesogenic units will have a relatively acute angle between them making the overall molecular structure bent like a hair-pin. This geometry will disfavor both intra- and inter-molecular interactions between the mesogenic moieties, which will lead to very weak mesomorphic tendencies. The structural situation in 1,3-disubstituted systems is, however, much more conducive to phase formation as the molecules will be less bent. Apart from a slight bend in structure, 1,3-disubstituted systems will also have a lateral ferrocene unit protruding from the side, as shown schematically in Fig. 9-2a. It would be unexpected for such molecules to be capable of packing together in a parallel arrangement in order to form orthogonal phases, because of the steric interference caused by the ferrocene moiety and the overall bent structure that is expected to predominate in the liquid-crystalline state. Rather, it is more likely that such molecules would have a staggered packing arrangement, which is favored by smectic C phases (depending on the mesogenic groups).

For substitution in both of the rings of the ferrocene unit, i.e., the 1,1′-system, the molecule is more linear in nature, either in a geometry in which the mesogenic groups of the same molecule overlie one another (Fig. 9-2b), or in the fully extended structure (Fig. 9-2c). For the elongated structure, it can be seen that there is a step in the overall shape. Deformities of this nature encourage tilted packing arrangements of the molecules as theorized by Wulf [9]. Therefore it is not surprising that these systems might favor smectic C phases. Alternatively, it is possible for the mesogenic groups on the opposing cyclopentadienes to lie almost one on top of another (Fig. 9-2b), and in this case parallel packing might be expected to be preferred, thereby giving rise to smectic A phases. However, it is doubtful that the smectic B phase would be favored under such conditions because of the steric bulk of the ferrocene unit, which will prevent close packing of the type required to form a hexagonal matrix.

In all cases however, it should be remembered that ferrocene is a relatively bulky group to incorporate into a mesogenic system. Its lateral width, no matter what the substitution pattern is, will cause steric repulsion of neighboring molecules leading to a reduction in transition temperatures relative to other units such as phenyl, cyclohexyl, pyrimidyl, etc. As a consequence, possibly the best way of promoting mesogenic behavior is to extend the molecular length in systems that are relatively linear. Some of these guidelines for creating mesogenic ferrocenes will become apparent in the following discussion of materials.

(a) 1,3 - disubstituted ferrocenes - bent form

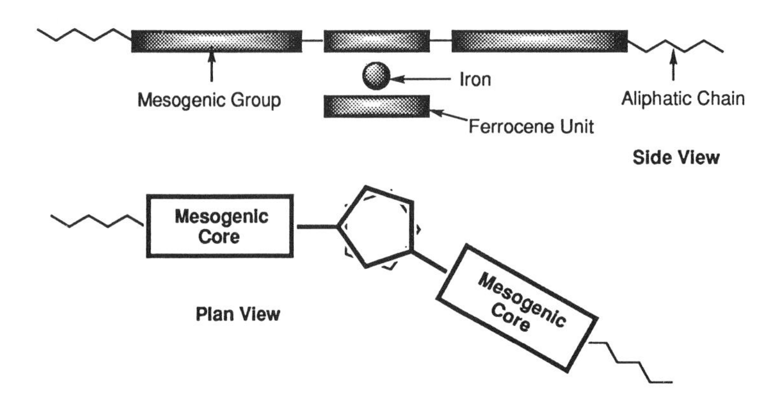

(b) 1,1'- disubstituted ferrocenes - bent form

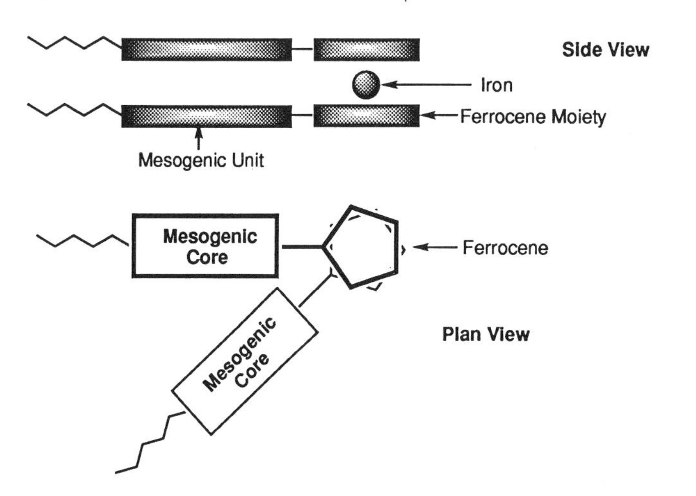

(c) 1,1'- disubstituted ferrocenes - elongated form

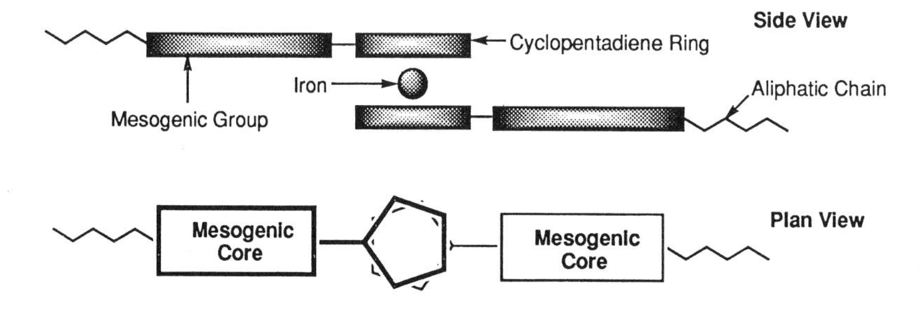

Fig. 9-2. Stereochemistry about the ferrocene unit.

9.2 Monosubstituted Ferrocene Derivatives

The first ferrocene-containing thermotropic liquid crystals were reported by Malthête and Billard [10] in 1976 (Fig. 9-3). These metal-containing liquid crystals were specially designed to undertake Mössbauer studies in a mesomorphic medium formed from a Mössbauer metal active marker. The ferrocene derivatives **1** ($n = 8$) (C · 153 · N · 167 · I)[1] and **1** ($n = 10$) (C · 143 · N · 159 · I) gave enantiotropic nematic phases with liquid crystalline domains of 14 °C and 16 °C, respectively. Compounds **2** ($n = 5, 6, 8$), bearing and alkyl chain instead of an alkyloxy tail as in **1**, and those carrying a -CH_2-CH_2- spacer between the ferrocenyl moiety and the rigid organic core **3** ($n = 8, 10$) showed monotropic nematic phases. The ferrocenyl Schiff-bases **1**–**3** are the first well characterized organotransition metallomesogens. In previous reviews [5], compounds **1**–**3** have been erroneously described as smectogenic liquid crystals.

Another example of a ferrocenyl Schiff-base derivative **4** (Fig. 9-3) was described by Galyametdinov [11], and gave rise to an enantiotropic nematic phase between 140 °C and 184 °C. Interestingly, the additional OH group, located next to the imine function, allowed the complex formation with a Cu(II) ion to produce the CuL_2 species **5** (Fig. 9-3). This trimetallic derivative retained the nematogenic character (C · 221 · N · 230 · I) of ligand **4**. Formation of the Cu(II) complex led to an increase in both the melting and clearing points and to a narrowing of the liquid crystal range (44 °C for **4**, and 9 °C for **5**).

The last example in this section is the fluorinated derivative **6** [12] (Fig. 9-3), which exhibited an enantiotropic nematic phase with relatively low melting (114.9 °C) and clearing (128.9 °C) points. Its molecular structure revealed an eclipsed conformation for the two cyclopentadienyl rings.

9.3 Symmetrically Disubstituted Ferrocene Derivatives

9.3.1 1,1′-Disubstituted Ferrocene Derivatives

The first disubstituted ferrocene derivatives were reported in 1988 [13]. The ferrocenyl core was linked to various 4-alkoxy-4′-biphenols by ester functions. Three compounds exhibited a monotropic mesophase: compounds **7** ($n = 5$) (I · 140 · SmC · 129 · C) and **7** ($n = 6$) (I · 139 · SmC · 132 · C) gave rise to smectic C phases, and **7** ($n = 11$) (I · 141 · SmA · 132 · C) to a smectic A phase (Fig. 9-4). X-Ray diffraction studies indicated that **7** ($n = 5$) adopted a *trans* conformation ("S" shape) in the crystalline and in the mesomorphic states [14].

1 The following abbreviations are used throughout the text: C = Crystal, SmC = Smectic C phase, SmA = Smectic A phase, N = nematic phase, I = Isotropic liquid.

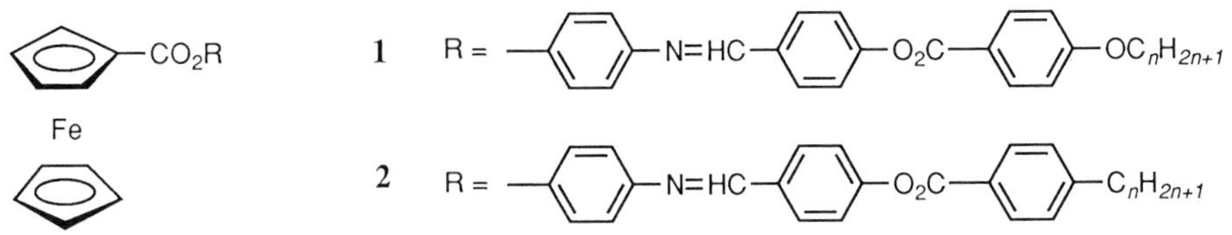

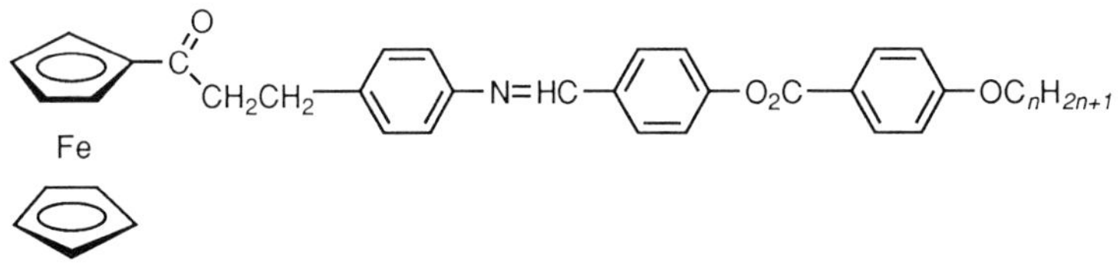

CO_2R

Fe

1 R = $-C_6H_4-N{=}HC-C_6H_4-O_2C-C_6H_4-OC_nH_{2n+1}$

2 R = $-C_6H_4-N{=}HC-C_6H_4-O_2C-C_6H_4-C_nH_{2n+1}$

O, C, CH_2CH_2, N=HC, O_2C, OC_nH_{2n+1}

Fe

3

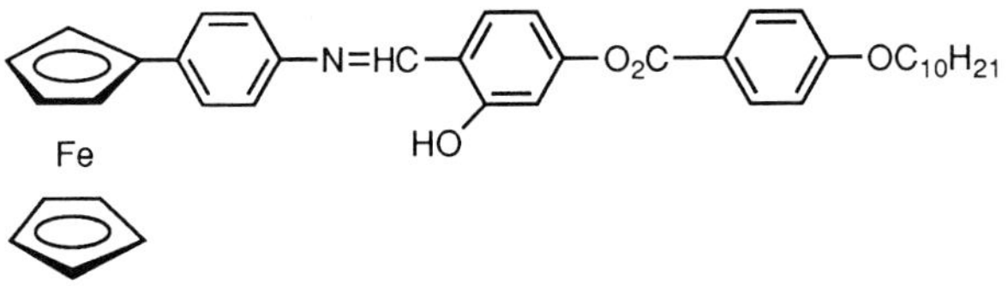

N=HC, O_2C, $OC_{10}H_{21}$, HO

Fe

4

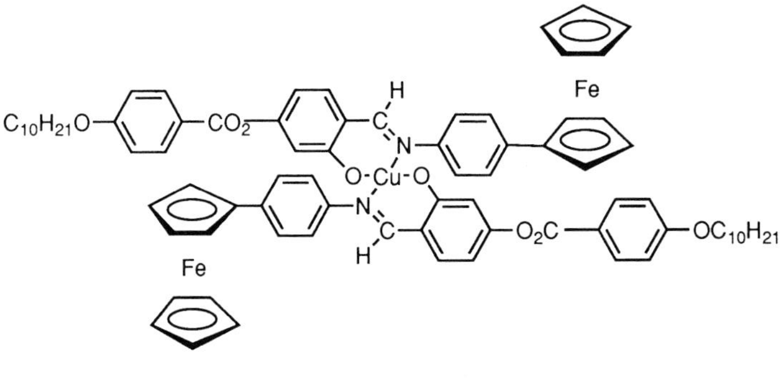

Fe

H

$C_{10}H_{21}O$, CO_2, C, N, O-Cu-O, N, C, H, O_2C, $OC_{10}H_{21}$

Fe

5

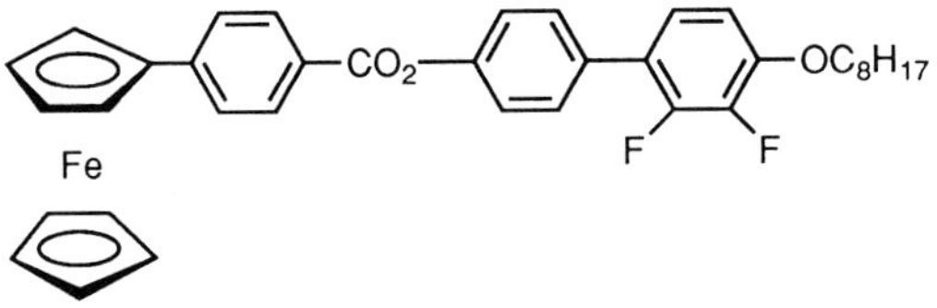

CO_2, OC_8H_{17}, F, F

Fe

6

Fig. 9-3. Monosubstituted ferrocene-containing thermotropic liquid crystals.

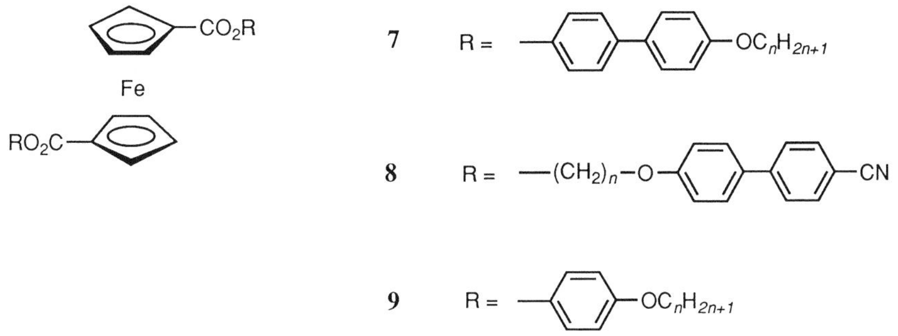

Fig. 9-4. 1,1′-Ferrocene diesters.

Two papers published in 1991 described the thermal properties of the diester ferrocene derivatives **8** [15] and **9** [16] (Fig. 9-4). Complexes **8** (n = 3, 4, 6, 8) were found to be non-mesomorphic and **8** (n = 10) seemed to give rise to a smectic phase (unidentified) between 85 °C and 88 °C during the second heating cycle. When **9** (n = 7) was cooled from the isotropic melt at a rate of 2 °C/min, a liquid crystalline texture was observed (transition temperatures and liquid crystalline range not given) that, according to the authors, was indicative of a smectic phase. The latter was not identified. Derivative **9** (n = 10) formed a nematic phase between 78 °C and 79 °C. This mesophase did not reform on cooling from the isotropic melt.

The three families **10** [17], **11** [18] and **12** [19] were described the following year (Fig. 9-5). Compounds **10** resemble **8** except that the flexible chains were attached to the ferrocenyl moiety without functional groups. This structural change caused an improvement in the liquid crystal properties. An enantiotropic smectic C phase was observed for **10** (n = 6) (C · 109.5 · SmC · 118.2 · I), **10** (n = 8) (C · 102.5 · SmC · 108.5 · I), and **10** (n = 11) (C · 104.5 · SmC · 109.3 · I). For all these compounds, a very narrow anisotropic range was detected (**10** (n = 6): 8.7 °C, **10** (n = 8): 6°C, **10** (n = 11): 4.8 °C). On cooling, the isotropic melts supercooled and reformed the smectic C phase, except for **10** (n = 8), which gave rise to a monotropic smectic A phase (I · 71 · SmA · 33 · C).

Interesting liquid crystal properties resulted from the Schiff-base derivatives **11** [18]. All compounds exhibited mesomorphic behavior (Fig. 9-6). The first members of the series gave rise to enantiotropic nematic phases and the long chain derivatives exhibited enantiotropic smectic A phases. The intermediate chain length derivatives presented monotropic nematic or smectic A phases.

Ferrocene derivatives **12** presented a nematic phase [19]. The transition temperatures were reported for two homologues, i.e., **12** (n = 3) (C · 210 · N · 220 · I) and **12** (n = 8) (C · 166 · N · 197 · I). Rapid decomposition in the nematic phase was observed for the former compound. The molecular structure of **12** (n = 3) showed a *cis* conformation, in contrast to that of **7** (n = 5).

We started our research on disubstituted ferrocene derivatives late in 1989. At this time, only the paper describing compounds **7** [13] was known.

First, from considerations of CPK models, we strongly felt that the position of the two substituents, namely the 1,1′-, the 1,2-, and the 1,3-positions, would strongly

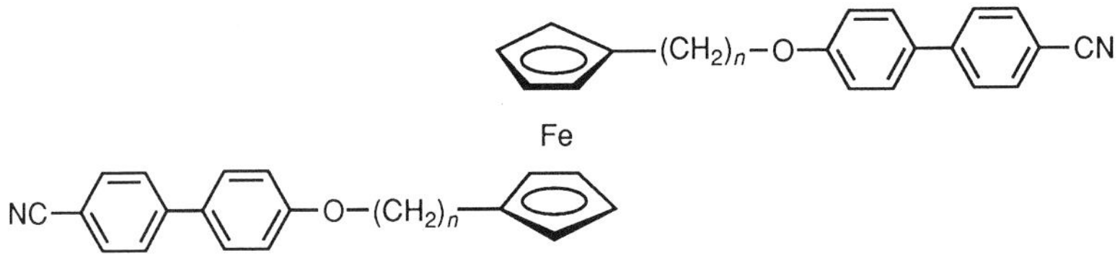

10

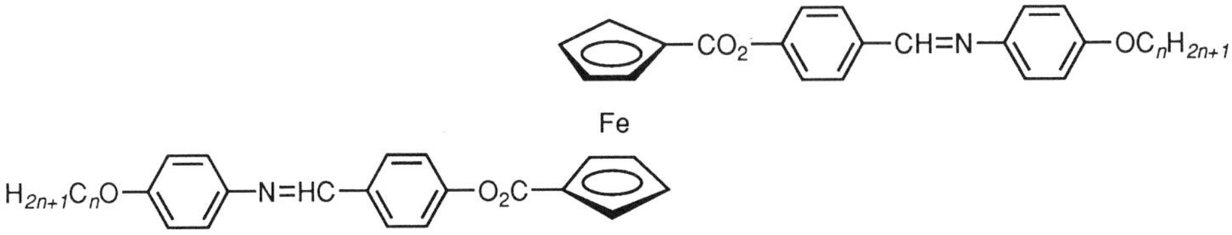

11

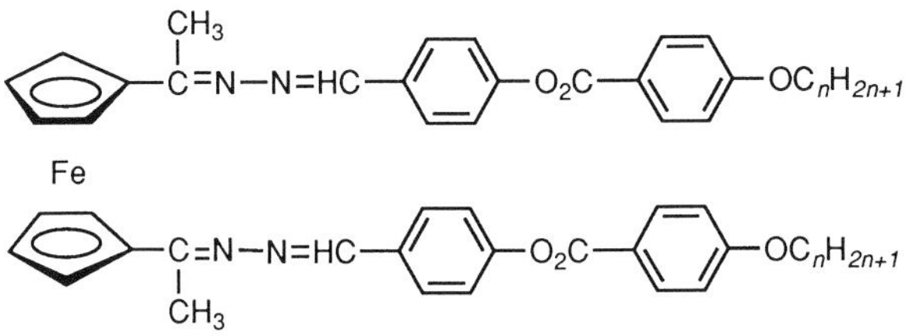

12

Fig. 9-5. 1,1′-Disubstituted ferrocene derivatives reported in 1992.

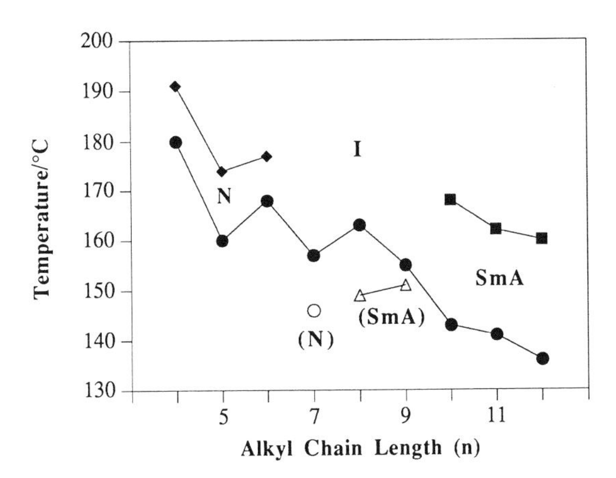

Fig. 9-6. Phase diagram of ferrocenes **11**. ● Melting point; ◆ nematic/isotropic liquid transition; ■ smectic A/isotropic liquid transition; △ isotropic liquid/smectic A transition; ○ isotropic liquid/nematic transition.

influence the mesomorphic properties of disubstituted ferrocene derivatives. Second, stereoelectronic factors were also considered as a potential parameter that could influence liquid crystal formation and mesophase stability. Therefore, we undertook a systematic study to understand how structural and electronic factors could be used to engineer mesomorphic properties of disubstituted ferrocene-containing liquid crystals.

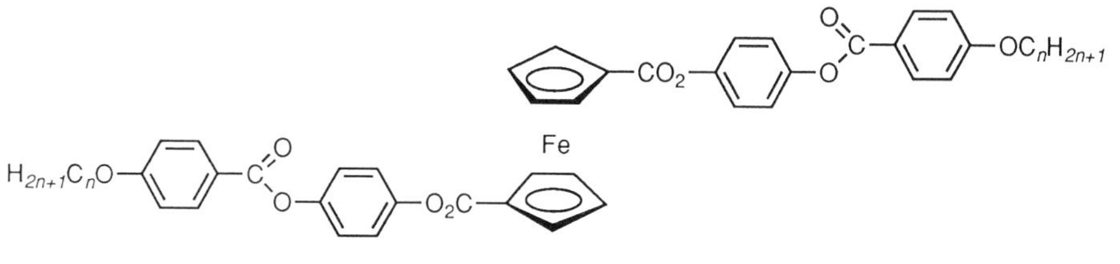

CO_2 O C O OC_nH_{2n+1}
Fe
$H_{2n+1}C_nO$ C O O O_2C

13

CO_2 C O O OC_nH_{2n+1}
Fe
O C O $H_{2n+1}C_nO$ O O_2C

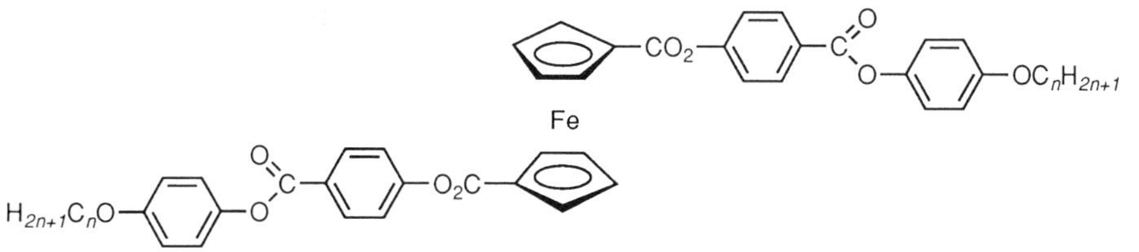

14

Fig. 9-7. Isomeric 1,1′-disubstituted ferrocene derivatives which differ in the orientation of the external ester functions.

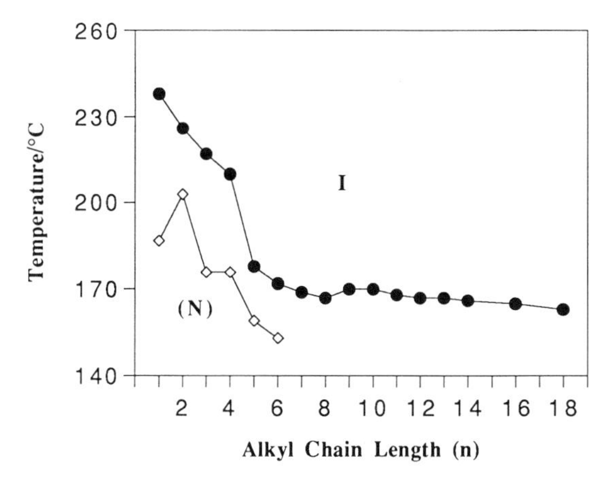

Fig. 9-8. Phase diagram of ferrocenes **13**. ● Melting point; ◇ isotropic liquid/nematic transition.

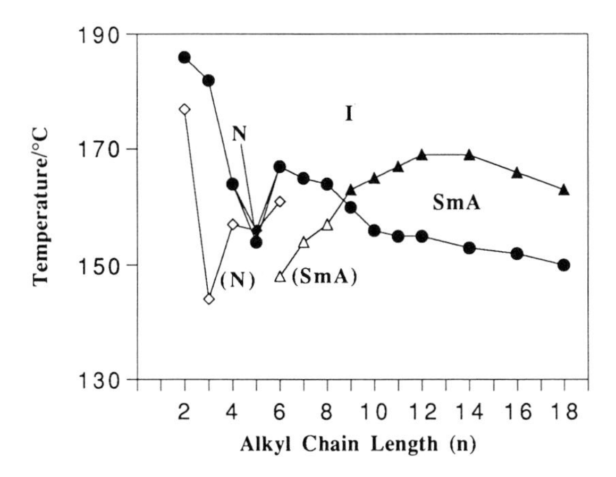

Fig. 9-9. Phase diagram of ferrocenes **14**. ● Melting point; ◆ nematic/isotropic liquid transition; ▲ smectic A/isotropic liquid transition; ◇ isotropic liquid/nematic transition; △ isotropic liquid or nematic/smectic A transition.

We prepared and studied the two isomeric families **13** and **14** (Fig. 9-7), which differ in the orientation of the external ester functions [20]. Their phase diagrams are reported in Fig. 9-8 and 9-9, respectively. (In a recent review [5d], the structure of **13** ($n = 6$) was incorrectly drawn.)

None of the ferrocene derivatives **13** showed liquid crystal properties on heating. They all melted into an isotropic melt. When cooled from the isotropic liquid, the first members of the series ($n = 1-6$) exhibited a monotropic nematic phase. A representative example of a nematic *schlieren* texture is shown in Fig. 9-10.

More interesting mesomorphic properties resulted from series **14**. The derivatives with short alkyl chains ($n = 2-4$) gave rise to monotropic nematic phases. Comparison of the crystal-to-liquid transition temperatures between **13** and **14** revealed that the members of series **14** always melted at a lower temperature than that of the corresponding isomers of family **13.** During the first heating, the compound with $n = 5$ showed an enantiotropic nematic phase over a very short range (2 °C). The derivatives with $n = 6-8$ exhibited monotropic nematic and/or smectic A phases. Then, the long alkyl chain derivatives ($n \geq 9$) showed enantiotropic smectic A phases with reasonable liquid crystal domains. Figure 9-11 illustrates a typical focal-conic texture of a smectic A phase.

The results observed for **13** and **14** clearly showed the strong influence of the orientation of the ester linking function on the mesomorphic properties. An explanation of this influence was attempted from stereoelectronic and structural considerations (Scheme 9-1). In **A**, the organic unit used to prepare **13**, electron delocalization occurs from the O-atom of the alkoxy chain to the ester function. Thus, mesomerism takes place in the external part of the organic fragment. However, in **B**, from which series **14** was built up, electron delocalization appears in the interior of the organic core, and in the opposite direction. Consequently, the O-atom of the ether group is more polar in **A** than in **B.** Furthermore, examination of CPK

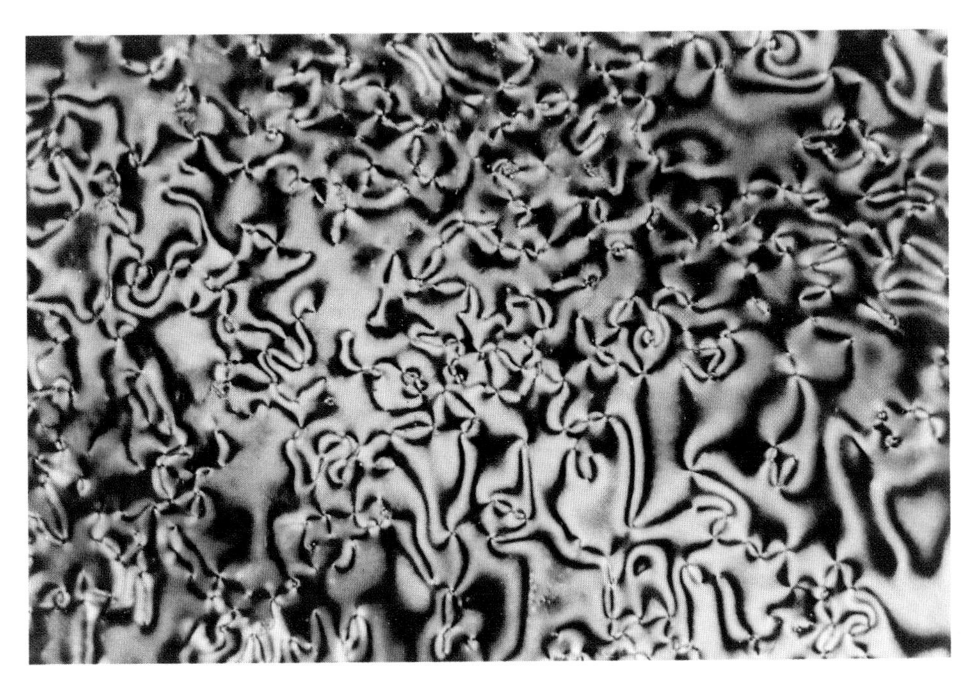

Fig. 9-10. Representative thermal polarized optical micrograph of the nematic *schlieren* texture displayed by **13** ($n = 5$) on cooling from the isotropic melt to 159 °C.

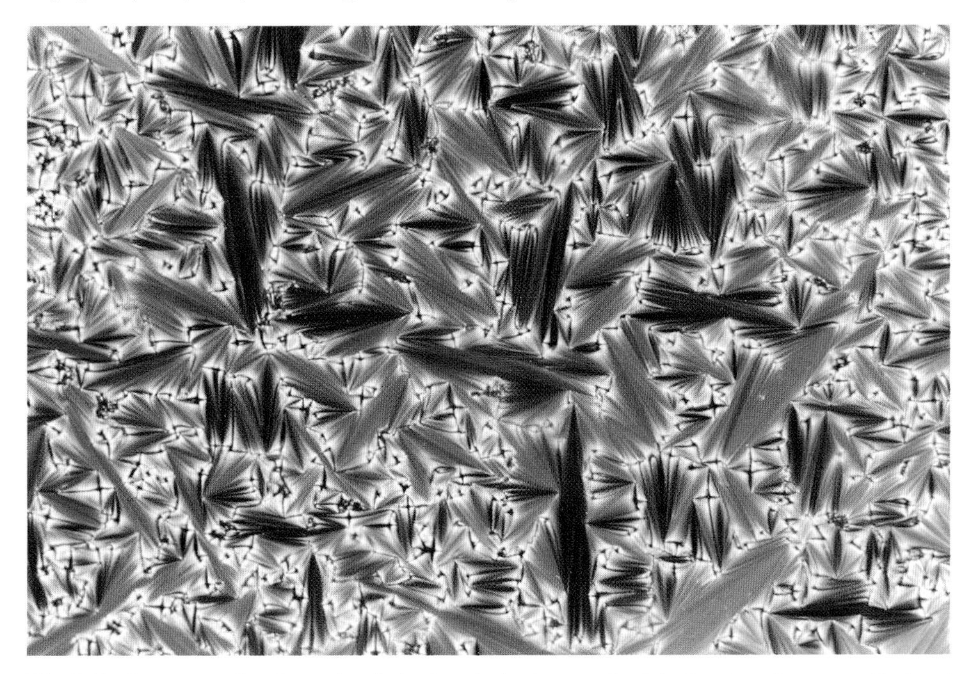

Fig. 9-11. Representative thermal polarized optical micrograph of the focal-conic texture displayed by **14** ($n = 12$) in the smectic A phase on cooling from the isotropic liquid to 164 °C.

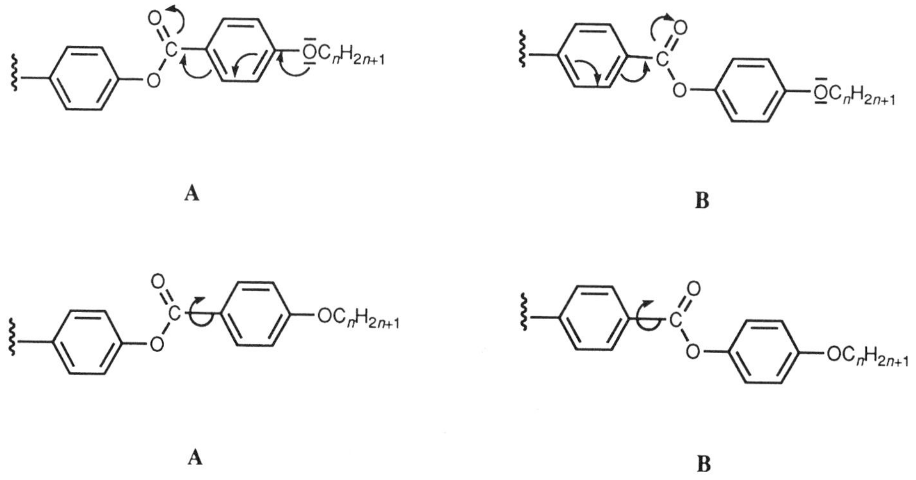

Scheme 9-1

models indicated that rotation could occur around some C-C bonds: in **B**, the rotation is probably more restricted than in **A** as it requires the motion of a larger molecular fragment. A combination of both electron delocalization and rotational motion is probably the origin of the different mesomorphic behavior observed between **13** and **14.**

The thermal characteristics exhibited by the series **11** [18] and **14** [20] proved that interesting liquid crystal properties can be obtained with symmetrically 1,1′-disubstituted ferrocene derivatives. The structure of the organic unit, which is combined with the ferrocenyl moiety, must be carefully chosen.

9.3.2 1,3-Disubstituted Ferrocene Derivatives

Compounds **15** [21, 22] and **16** [21] represent the first 1,3-disubstituted ferrocene-containing liquid crystals (Fig. 9-12). (In a recent review [5d], the structures of **15** ($n = 6-8$) and **16** were incorrectly drawn.)

Ferrocene derivatives **15** exhibited remarkable liquid crystal properties (Fig. 9-13). Indeed, they all gave rise to enantiotropic mesophases. Structures with $n = 1$ to 11 showed nematic phases. From $n = 12$ a smectic C phase formed. The latter was monotropic only for **15** ($n = 12$). The smectic C domain increased from $n = 13$ to $n = 16$, and, inversely, the nematic range narrowed. The last member of this series ($n = 18$) presented one smectic C phase between 159 °C and 179 °C. A nematic to smectic C transition and a focal-conic texture of a smectic C phase are presented in Figs. 9-14 and 9-15, respectively.

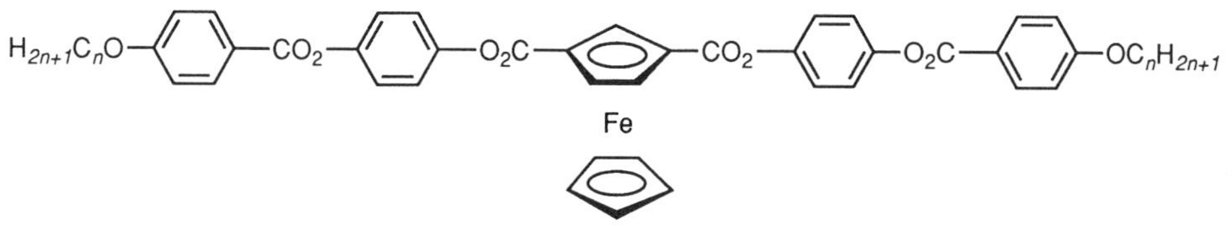

15

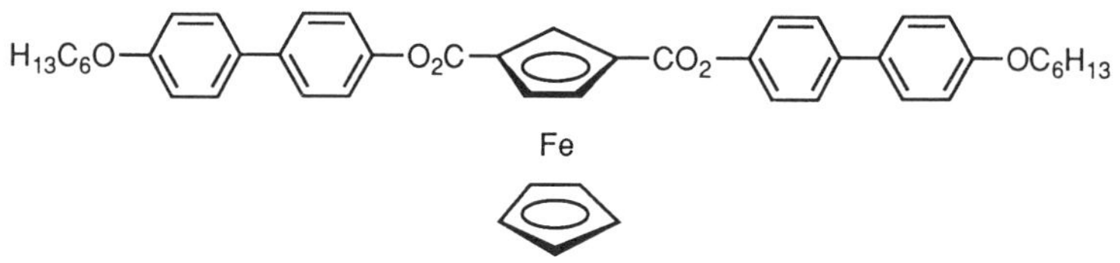

16

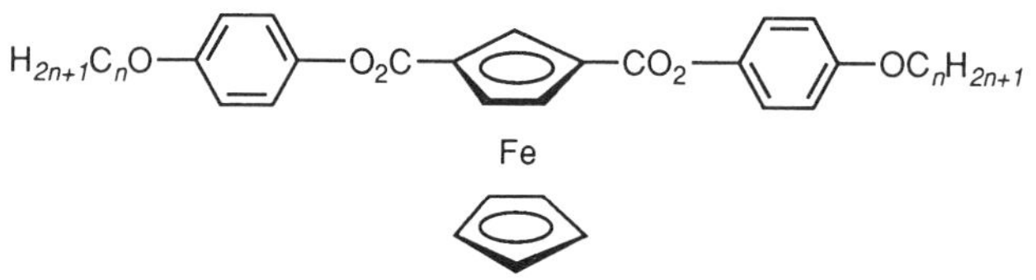

17

Fig. 9-12. First 1,3-disubstituted ferrocene-containing liquid crystals (**15** and **16**) and non mesomorphic 1,3-disubstituted ferrocene derivatives **17**.

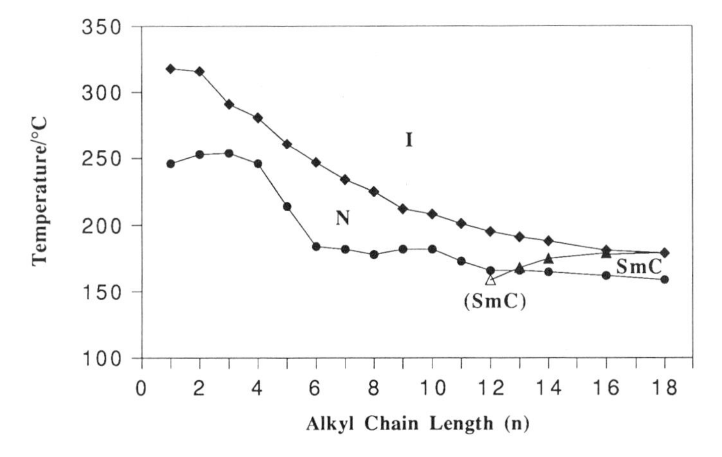

Fig. 9-13. Phase diagram of ferrocenes **15**. ● Melting point; ◆ clearing point; ▲ smectic C/nematic transition; △ nematic/smectic C transition.

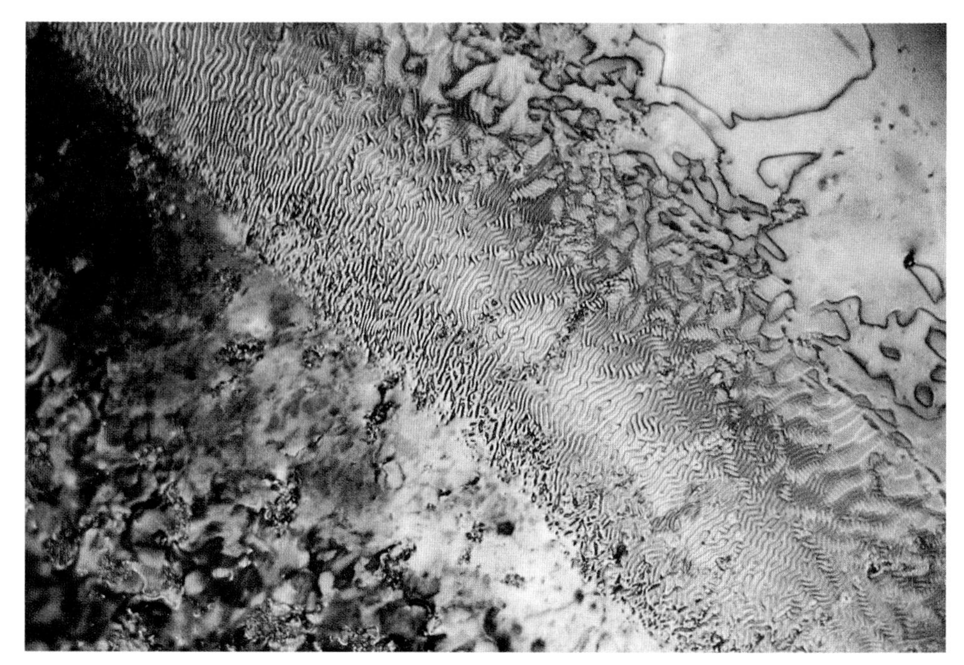

Fig. 9-14. Representative thermal polarized optical micrograph of the nematic to smectic C transition displayed by **15** (n = 13) on cooling at 169 °C.

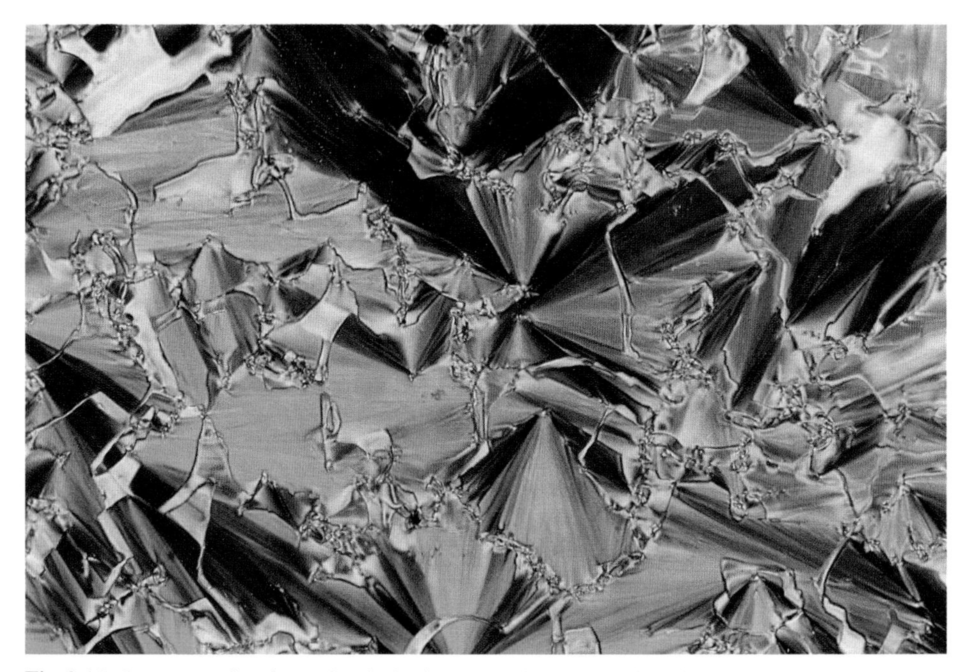

Fig. 9-15. Representative thermal polarized optical micrograph of the focal-conic texture displayed by **15** (n = 18) in the smectic C phase on cooling from the isotropic liquid to 165 °C.

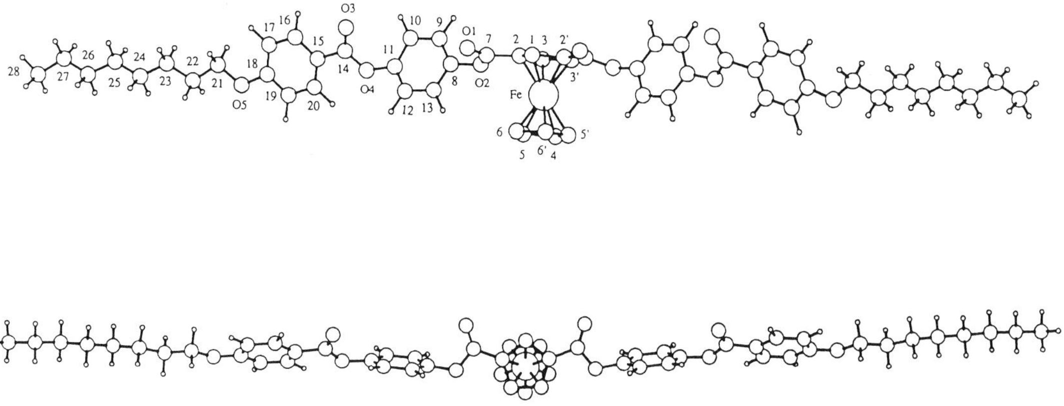

Fig. 9-16. Molecular structure of **15** ($n = 8$).

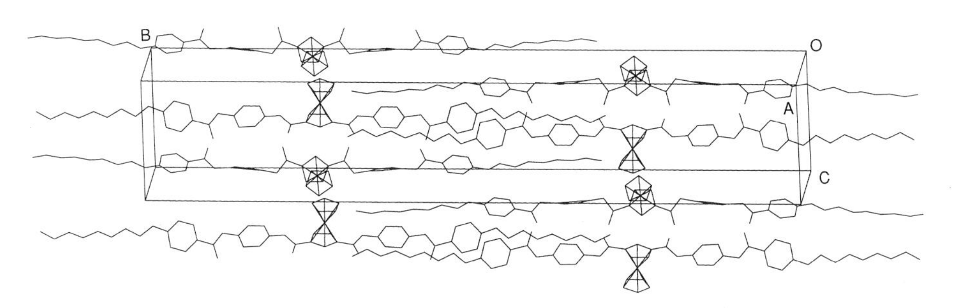

Fig. 9-17. Crystal structure of **15** ($n = 8$).

The molecular (Fig. 9-16) and crystal (Fig. 9-17) structures of compound **15** ($n = 8$) were determined and revealed interesting features [22]. First, despite the angle of 142° between the two substituents, a linear shape with a highly anisometric structure was observed. Second, the crystal packing indicated that a molecular arrangement allowing favorable intermolecular attactions was obtained. These findings demonstrated that compound **15** ($n = 8$) and its homologues had the required structural characteristics to show liquid crystal properties.

Ferrocene derivative **16** showed an enantiotropic nematic phase between 204 °C and 235 °C. This result indicated that a biphenyl system, when associated with the ferrocenyl moiety substituted in the 1,3-positions, was also propitious to liquid crystal formation (compare **16** with its 1,1′-isomeric analogue **7** ($n = 6$)).

No mesomorphic behavior resulted from family **17** [22] (Fig. 9-12). This observation is of interest, as it showed that a limit exists for obtaining liquid crystallinity even in the case of 1,3-disubstituted ferrocene derivatives.

The cyclopentane ring has already been used to prepare thermotropic liquid crystals [23]. In **17**, the bulky ferrocene core, due to its depth, acts as a spacer separating the aromatic rings from each other. Consequently, the intermolecular attractions are too weak to give rise to mesomorphism. The lowering of the melting points on going from **15** and **16** to **17** supports this interpretation.

Therefore, the l/d ratio (l = length of the rigid core, d = distance between the two cyclopentadienyl rings) is a limiting factor for observing liquid crystal properties. The following length values of the rigid segments have been determined from either crystallographic data or CPK models: 27.5 Å, 23.7 Å and 15.1 Å, for **15**, **16** and **17**, respectively. The depth of the ferrocene being ca. 3.3 Å [24], l/d ratios of 8.3, 7.2, and 4.6 are obtained for **15**, **16** and **17**, respectively. These values suggest that liquid crystal behavior can be expected when the l/d ratio is greater than 5–7.

9.3.3 1,2-Disubstituted Ferrocene Derivatives

Some 1,2-disubstituted ferrocene derivatives, **18** and **19**, have been prepared [25] (Fig. 9-18). None of these structures showed liquid crystal properties. The 1,2-isomeric structures always gave the lowest melting point: compare, for example, **18** ($n = 6$) (*m.p.* = 149 °C) with **15** ($n = 6$) (*m.p.* = 184 °C) and **13** ($n = 6$) (*m.p.* = 172 °C), or **18** ($n = 12$) (*m.p.* = 113 °C) with **15** ($n = 12$) (*m.p.* = 166 °C) and **13** ($n = 12$) (*m.p.* = 167 °C), or **19** (*m.p.* = 149 °C) with **16** (*m.p.* = 204 °C) and **7** ($n = 6$) (*m.p.* = 181 °C).

The fact that the 1,2-isomeric structures exhibit the lowest melting points is certainly due to steric hindrance caused by the bulky substituents located adjacent to each other. Consequently the intermolecular attractions are reduced, compared

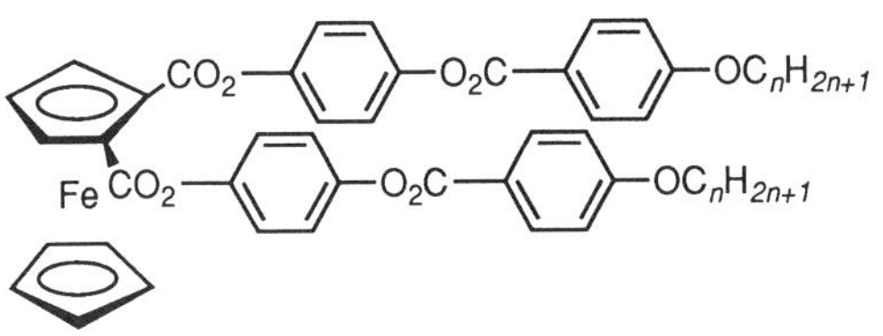

18

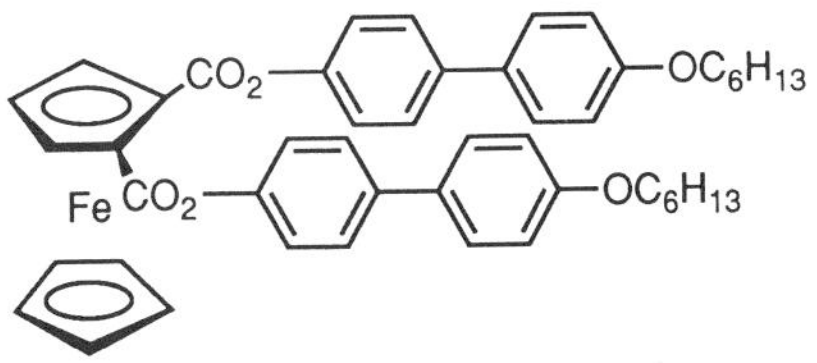

19

Fig. 9-18. Non mesomorphic 1,2-disubstituted ferrocene derivatives.

with their 1,1′- and particularly to their 1,3-isomers, and become too weak to allow mesomorphism to develop.

Our systematic study conducted on isomeric series demonstrated that liquid crystal tendency follows the series: 1,3- $>$ 1,1′- $\gg$ 1,2-isomeric structures.

9.4 Unsymmetrically 1,1′-Disubstituted Ferrocene Derivatives

Unsymmetrically disubstituted complex **20** [16], obtained during the preparation of **9** ($n = 10$) [16], seemed to show a nematic phase. The latter did not reform during the cooling process. A mixture of the two derivatives **20** (4 parts) and **9** ($n = 10$) (1 part) gave rise to a broad nematic phase (C · 110 · N · 170 · I).

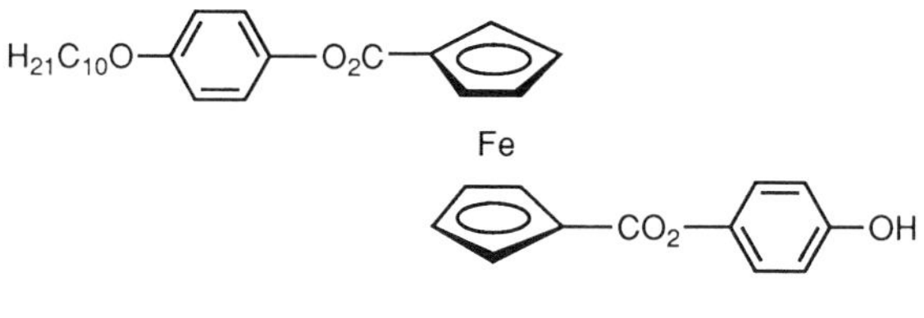

20

We recently reported a collection of unsymmetrically 1,1′-disubstituted ferrocene derivatives (Fig. 9-19) [26]. The basic 1-alkyl-1′-substituted ferrocene ester derivatives **21**–**24** have been constructed from various rigid organic frameworks and alkyl chain lengths. Additionally, several structural parameters have been subtly modified, such as the length of the rigid segment ($m = 1$ or 2) in **22b**, and the nature of the terminal substituent in **22a** and **23.** Complexes **24a** and **24b** are alkyl derivatives of Malthête's Schiff-base ferrocene-containing liquid crystals (see **1**, Fig. 9-3). The thermal properties of compounds **21**–**24** were carefully examined and compared with those of their analogous structures incorporating either a cyclohexyl or a phenyl ring instead of the ferrocene core.

This work showed that the replacement of a 1,4-disubstituted benzene ring with a 1,1′-disubstituted ferrocene unit led to a moderate decrease of the melting points but to a dramatic reduction of the clearing points. The order of promotion of nematic phase stability was found to be: cyclohexyl $>$ benzene $\gg$ ferrocene. This general effect is depicted graphically in Fig. 9-20 for the nematic to isotropic liquid transition of one particular family of related "four ring" materials. It can be seen that the clearing points for the analogs of 4-[(4′-octyloxybiphenyl-4-yl)-oxycarbonyl]phenyl-1′-hexylferrocene-1-carboxylate, in which the ferrocene unit has been replaced by either phenyl or cyclohexyl moieties, are over 150 °C higher than for the parent system. This demonstrates that ferrocene is a fairly inefficient unit for the promotion of mesogenic properties. This type of result can be attributed to a number of factors: (a) the ferrocene moiety is bulky and is set more towards one end of the molecule, (b) 1,1′-substitution creates a step in the overall structure

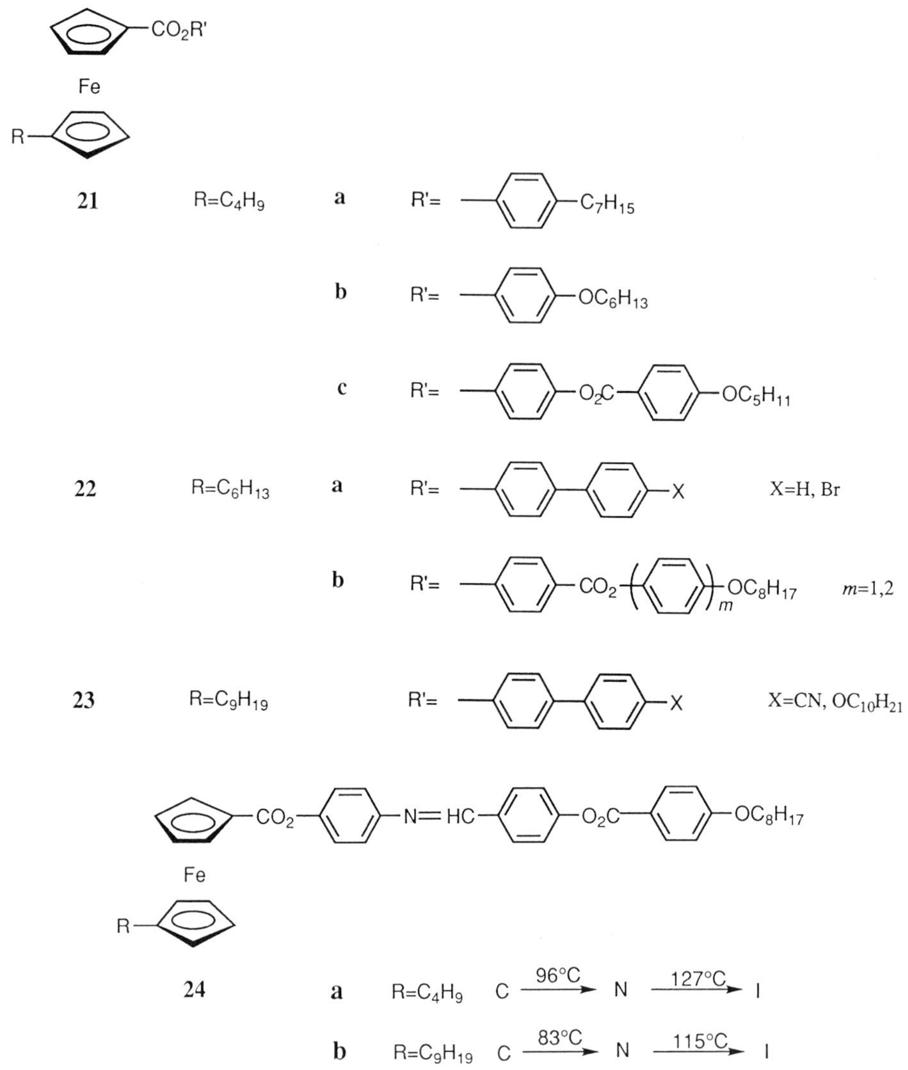

21 $R=C_4H_9$ **a** $R'=$ –C₆H₄–C_7H_{15}

b $R'=$ –C₆H₄–OC_6H_{13}

c $R'=$ –C₆H₄–O_2C–C₆H₄–OC_5H_{11}

22 $R=C_6H_{13}$ **a** $R'=$ –C₆H₄–C₆H₄–X X=H, Br

b $R'=$ –C₆H₄–CO_2–(C₆H₄)$_m$–OC_8H_{17} *m*=1,2

23 $R=C_9H_{19}$ $R'=$ –C₆H₄–C₆H₄–X X=CN, $OC_{10}H_{21}$

24 **a** $R=C_4H_9$ C $\xrightarrow{96°C}$ N $\xrightarrow{127°C}$ I

b $R=C_9H_{19}$ C $\xrightarrow{83°C}$ N $\xrightarrow{115°C}$ I

Fig. 9-19. Unsymmetrically 1,1′-disubstituted ferrocene derivatives reported by Thompson, Goodby and Toyne.

leading to a lack of coplanarity between the substituents in the 1,1′-positions, and (c) 1,1′-substitution leads to a slightly bent structure, whereas the phenyl and cyclohexyl analogues are essentially linear. Overall, the length to breadth ratio of the ferrocene system is much less than for the other two conventional liquid-crystalline materials. As a consequence, much lower transition temperatures should be expected for ferrocene-containing systems in comparison with traditional material formulations and therefore it is necessary to extend the molecular length in order to achieve length to breadth ratios that are more conducive to phase formation.

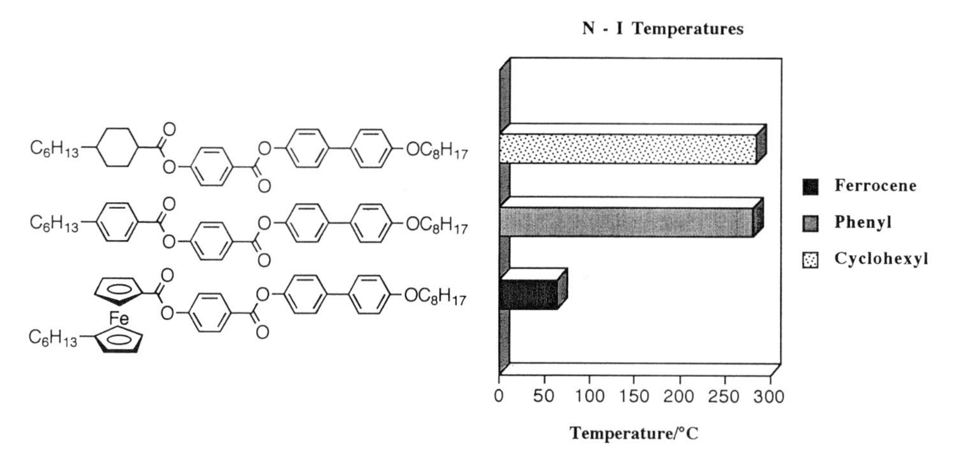

Fig. 9-20. Comparison of the nematic to isotropic liquid transition of related systems.

In conclusion the ferrocene unit is a poor component of a mesogen but can be tolerated if the molecule is long enough to compensate for the reduction of mesogenic properties induced by its inclusion. This statement is in agreement with the findings obtained for the 1,3-disubstituted ferrocene derivatives (compare **15** (l/d = 8.3) and **16** (l/d = 7.2) with **17** (l/d = 4.6)).

We designed and studied [27] a family of unsymmetrically 1,1'-disubstituted ferrocene derivatives obtained by combining the organic units **A** and **B** (above), used to prepare the families **13** and **14**, respectively within the same molecular framework. Structures **25** (Fig. 9-21) led to remarkable mesomorphic properties. All derivatives exhibited liquid crystal properties. Compound **25** (n = 11) gave rise to an enantiotropic smectic A phase. Complex **25** (n = 12) showed an enantiotropic

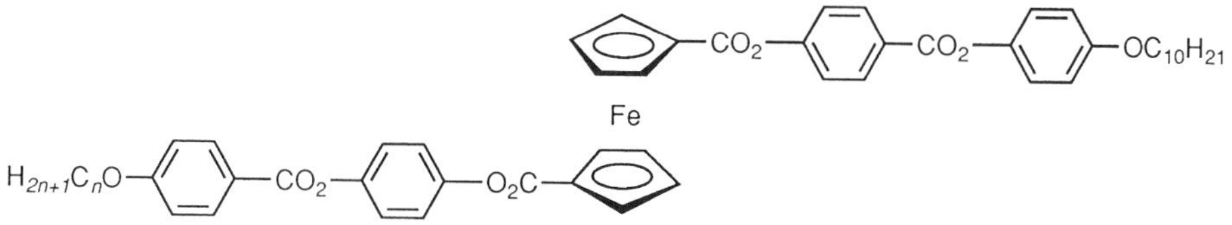

CO_2 — CO_2 — $OC_{10}H_{21}$

Fe

$H_{2n+1}C_nO$ — CO_2 — O_2C

25

n=11 C $\xrightarrow{144°C}$ SmA $\xrightarrow{149°C}$ I

n=14 C $\xrightarrow{135°C}$ SmC $\xrightarrow{142°C}$ SmA $\xrightarrow{152°C}$ I

n=15 C $\xrightarrow{132°C}$ SmC $\xrightarrow{145°C}$ SmA $\xrightarrow{151°C}$ I

Fig. 9-21. Unsymmetrically 1,1'-disubstituted ferrocene-containing thermotropic liquid crystals and examples of their transition temperatures.

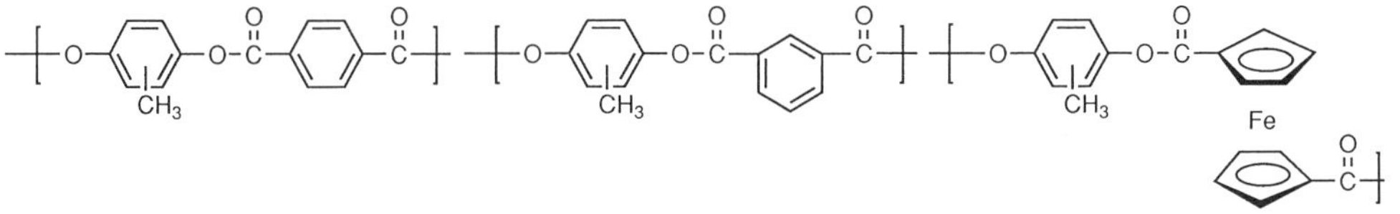

26

smectic A phase and a monotropic smectic C phase. Ferrocene derivatives **25** ($n = 13-16$) exhibited enantiotropic smectic C and smectic A phases. An explanation of the different thermal properties observed between the ferrocene symmetrically (**13** and **14**) and unsymmetrically (**25**) disubstituted is not straightforward since two structural modifications must be taken into consideration: (a) the alkyl chain length and (b) the orientation of the external ester function. Further investigations, which are currently in progress in our laboratory, are necessary to fully understand the effect of symmetry reduction on the mesomorphic properties.

However, one can already mention that the introduction of dissymmetry had two positive effects on the mesomorphic properties: a depression of the melting points was observed and ferrocene derivatives exhibiting smectic C and smectic A phase were obtained for the first time.

9.5 Ferrocene-Containing Liquid Crystal Polymers

Polyesters **26** represent the first ferrocene-containing liquid crystal polymers [28]. These copolymers were found to be insoluble in THF, toluene, dichloromethane, chloroform, *p*-chlorophenol, and tetrachloroethane. The terephthaloyl chloride/isophthaloyl chloride ratio was maintained constant (7:3), but the content of the ferrocene unit was varied from 0 to 100%.

Except for the structure containing 100% of ferrocene unit, which decomposed before melting, all the organometallic copolymers exhibited birefringent melts. Nematic textures were identified by means of polarized optical microscopy and, in one case, by X-ray diffraction studies. For comparison purposes, a polymer without ferrocene unit was prepared, but showed no mesomorphism. The authors deduced that the ferrocene framework was contributing to the liquid crystallinity of the ferrocene-containing polymers.

After our investigations on low molar mass ferrocene derivatives, we examined the thermal properties of isomeric polymeric structures.

Polycondensates **27** and **28** (Fig. 9-22), incorporating either the 1,3- or 1,1′-disubstituted ferrocene unit, were prepared by solution polymerisation. Polymers **27** [25] were obtained by reaction of the novel monomer **29** [25] (Fig. 9-23) with the desired bis(acid chloride) **31** (Fig. 9-23) in refluxing CH_2Cl_2 in the presence of triethylamine. Polymers **28** [25] were prepared following the same procedure from bis-phenol **30** [16] (Fig. 9-23).

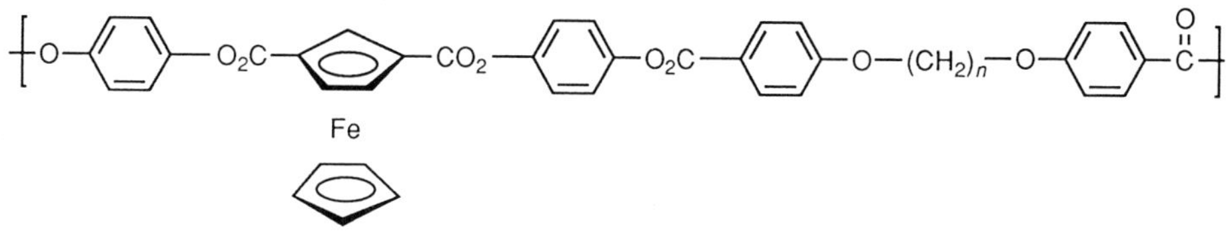

O, O–, O₂C–, –CO₂–, –O₂C–, –O–(CH₂)n–O–, C=O

Fe

27

–CO₂–, –O₂C–, –O–(CH₂)n–O–, C=O

Fe

O–, O₂C–

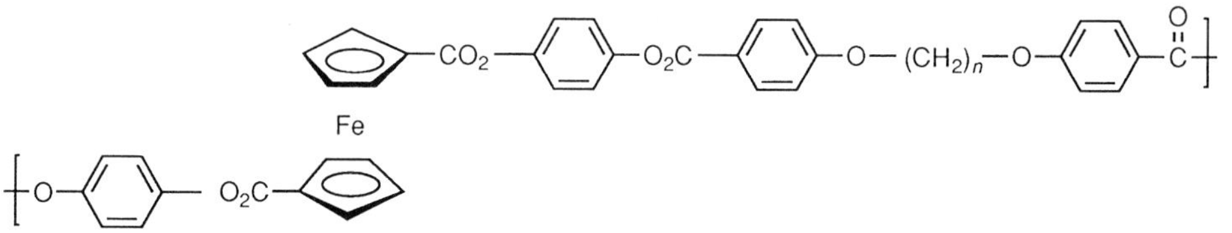

28

Fig. 9-22. Isomeric ferrocene-containing polyesters.

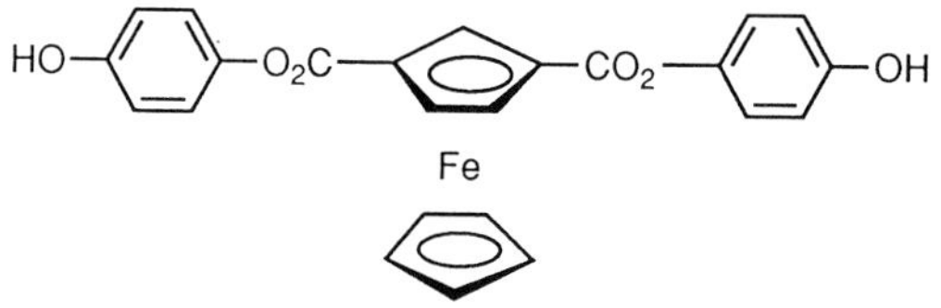

HO–, –O₂C–, –CO₂–, –OH

Fe

29

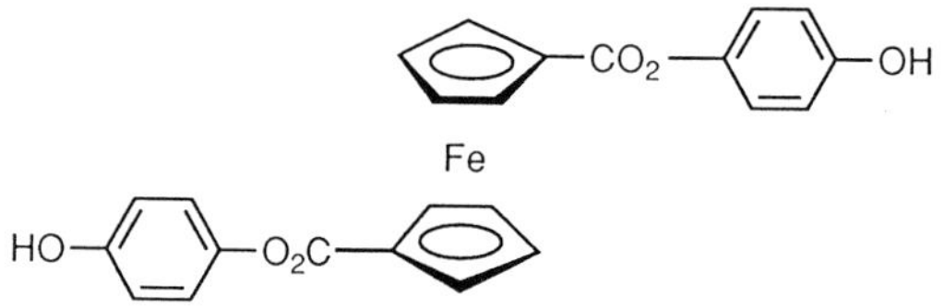

–CO₂–, –OH

Fe

HO–, –O₂C–

30

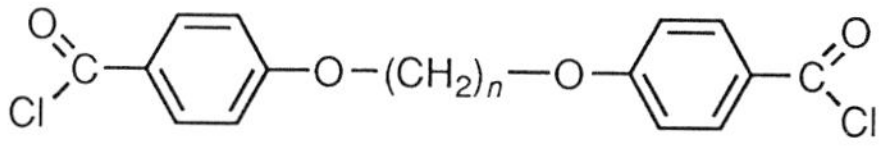

O, C, Cl, –O–(CH₂)n–O–, C, O, Cl

31

Fig. 9-23. Monomers used for preparing polymers **27** and **28.**

Ferrocene-containing polymers **27** and **28** were soluble in dichloromethane and chloroform, and partly soluble in THF. Analysis by gel permeation chromatography (GPC) indicated molecular weights in the range 10000 – 20000, and revealed, therefore, an oligomeric nature for **27** and **28.**

Materials **27** and **28** decomposed above ca. 200 °C. Regarding the high thermal stability of the model compounds, this result was unexpected and was attributed to the presence of unreacted end-groups. None of the 1,1′-isomeric structures **28** exhibited liquid crystal properties before decomposition. Interestingly, oligomers **27** clearly showed nematic phases between ca. 180 – 200 °C.

The trend in liquid crystal properties in oligomeric structures, and certainly in polymeric structures, is the same as that in low molar mass models.

9.6 Concluding Remarks

The research on ferrocene-containing thermotropic liquid crystals has concentrated, so far, on design, synthesis, and investigations of thermal properties. These efforts, and more particularly the results obtained, could be a source of inspiration for the development of new metallomesogens. As an example, we recently reported the first liquid crystalline ruthenocene derivatives [29].

A challenging problem is the design and synthesis of liquid crystalline ferrocene derivatives exhibiting mesomorphism near room temperature. If this goal can be reached, ferrocene-containing liquid crystals would represent a promising class of metallomesogens that might be used for the development of liquid crystal devices in nanoscale technology.

References

[1] F. Vögtle, *Supramolecular Chemistry*, Wiley, Chichester, **1991**, pp. 231 – 281.
[2] (a) D. Demus, H. Demus, H. Zaschke, *Flüssige Kristalle in Tabellen*, VEB Deutscher Verlag für Grundstoffindustrie, Leipzig, **1974**; (b) D. Demus, H. Zaschke, *Flüssige Kristalle in Tabellen II*, VEB Deutscher Verlag für Grundstoffindustrie, Leipzig, **1984.**
[3] D. Vorländer, *Ber. Dtsch. Chem. Ges.* **1910**, *43*, 3120 – 3135.
[4] A.-M. Giroud, U. T. Mueller-Westerhoff, *Mol. Cryst. Liq. Cryst.* **1977**, *41*, 11 – 13.
[5] (a) A.-M. Giroud-Godquin, P. M. Maitlis, *Angew. Chem. Int. Ed. Engl.* **1991**, *30*, 375 – 402; (b) P. Espinet, M. A. Esteruelas, L. A. Oro, J. L. Serrano, E. Sola, *Coord. Chem. Rev.* **1992**, *117*, 215 – 274; (c) D. W. Bruce in *Inorganic Materials*, (Eds. D. W. Bruce and D. O'Hare), Wiley, Chichester, **1992**, pp. 405 – 490; (d) S. A. Hudson, P. M. Maitlis, *Chem. Rev.* **1993**, *93*, 861 – 885.
[6] D. Vorländer, *Z. Phys. Chem. Stoechiom. Verwandschaftsl.* **1923**, *105*, 211 – 254.
[7] L. Ziminski, J. Malthête, *J. Chem. Soc., Chem. Commun.* **1990**, 1495 – 1496.
[8] G. W. Gray, J. W. Goodby, *Smectic Liquid Crystals*, Leonard Hill, London, **1984.**
[9] A. Wulf, *Phys. Rev. A* **1975**, *11*, 365 – 375.

[10] J. Malthête, J. Billard, *Mol. Cryst. Liq. Cryst.* **1976**, *34*, 117–121.
[11] Yu. G. Galyametdinov, O. N. Kadkin, I. V. Ovchinnikov, *Izv. Akad. Nauk., Ser. Khim.* (Russia) **1990**, *10*, 2462–2463; Yu. G. Galyametdinov, O. N. Kadkin, I. V. Ovchinnikov, *Izv. Akad. Nauk., Ser. Khim.* (Russia) **1992**, *2*, 402–407.
[12] C. Loubser, C. Imrie, P. H. van Rooyen, *Adv. Mater.* **1993**, *5*, 45–47.
[13] J. Bhatt, B. M. Fung, K. M. Nicholas, C.-D. Poon, *J. Chem. Soc., Chem. Commun.* **1988**, 1439.
[14] M. A. Khan, J. Bhatt, B. M. Fung, K. M. Nicholas, E. Wachtel, *Liq. Cryst.* **1989**, *5*, 285–290.
[15] J. Bhatt, B. M. Fung, K. M. Nicholas, *J. Organomet. Chem.* **1991**, *413*, 263–268.
[16] P. Singh, M. D. Rausch, R. W. Lenz, *Liq. Cryst.* **1991**, *9*, 19–26.
[17] J. Bhatt, B. M. Fung, K. M. Nicholas, *Liq. Cryst.* **1992**, *12*, 263–272.
[18] K. P. Reddy, T. L. Brown, *Liq. Cryst.* **1992**, *12*, 369–376.
[19] A. P. Polishchuk, T. V. Timofeeva, M. Yu. Antipin, Yu. T. Struchkov, Yu. G. Galyametdinov, I. V. Ovchinnikov, *Kristallografiya* **1992**, *37*, 705–711.
[20] R. Deschenaux, J.-L. Marendaz, J. Santiago, *Helv. Chim. Acta* **1993**, *76*, 865–876.
[21] R. Deschenaux, J.-L. Marendaz, *J. Chem. Soc., Chem. Commun.* **1991**, 909–910.
[22] R. Deschenaux, I. Kosztics, J.-L. Marendaz, H. Stoeckli-Evans, *Chimia* **1993**, *47*, 206–210.
[23] L. A. Karamysheva, T. A. Geyvandova, I. F. Agafonova, K. V. Roitman, S. I. Torgova, R. KH. Geyvandov, V. F. Petrov, A. Z. Rabinovich, M. F. Grebyonkin, *Mol. Cryst. Liq. Cryst.* **1990**, *191*, 237–246.
[24] C. Elschenbroich, A. Solzer, *Organometallics*, VCH, Weinheim, **1989.**
[25] R. Deschenaux, J.-L. Marendaz, unpublished results.
[26] N. J. Thompson, J. W. Goodby, K. J. Toyne, *Liq. Cryst.* **1993**, *13*, 381–402.
[27] R. Deschenaux, M. Rama, J. Santiago, *Tetrahedron Lett.* **1993**, *34*, 3293–3296.
[28] P. Singh, M. D. Rausch, R. W. Lenz, *Polym. Bull.* **1989**, *22*, 247–252.
[29] R. Deschenaux, J. Santiago, *J. Mater. Chem.* **1993**, *3*, 219–220.

Received: September 28, 1993

10 Synthesis and Characterization of Ferrocene-Containing Polymers

Kenneth E. Gonsalves and Xiaohe Chen

10.1 Introduction

Metal-containing polymers have emerged as an important category of polymeric materials [1]. The impetus for developing these materials is based on the premise that polymers containing metals are expected to possess properties significantly different from those of conventional organic polymers. Examples of these properties include: electrical conductivity, magnetic behavior, thermal stability, non-linear optical (NLO) effects, and possibly superconductivity. Perhaps it is not presumptuous to infer, based on the considerable amount of research conducted recently in developing these types of polymers, that an area of research termed organometallic polymer science has been established. This field of activity can be seen as crossing the conventional boundaries of chemical sub-disciplines such as inorganic, organic, and polymer chemistry, and also materials science.

Since its serendipitous discovery [2], ferrocene has intrigued inorganic, organic, and polymer chemists [1]. Again the motivation for this, both in academic and in industrial laboratories, has been directed towards using the potential advantages of incorporating the ferrocene moiety in a polymer, in order to investigate novel properties such as those mentioned above. It is noteworthy to mention that significant efforts have been aimed at the use of ferrocene-containing polymers as materials for electrode coatings [3] and more recently as non-linear optical materials [4]. This chapter is intended to provide a general framework for the more recent accomplishments in ferrocene polymer chemistry and materials science, i.e., within the past decade. For earlier work the reader is encouraged to refer to the excellent reviews included in the bibliography at the end of this chapter. The main thrust is on synthetic strategies of ferrocene-containing polymers. However, it is important to mention that there have been significant recent advances in the synthesis of organometallic monomers [1]. Therefore, in addition to ferrocene, a number of comparable monomers are now available. Some of these monomers (**1**–**12**) are listed in Scheme 10-1.

In general, there are two basic routes for the synthesis of metal-containing polymers [1]. One approach is to form derivatives of organic polymers with organometallic groups. The second involves the synthesis of organometallic compounds that contain polymerizable functional groups. These monomers can then be homo- or co-polymerized with conventional organic monomers. Both of these approaches have been employed in the synthesis of ferrocene-containing polymers. However, in this chapter we will emphasize the latter approach.

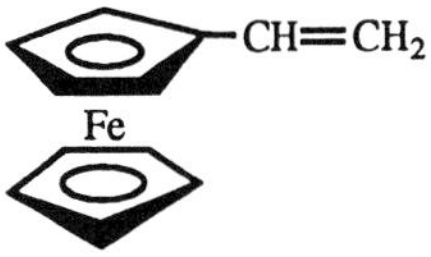

$CH{=}CH_2$ Fe

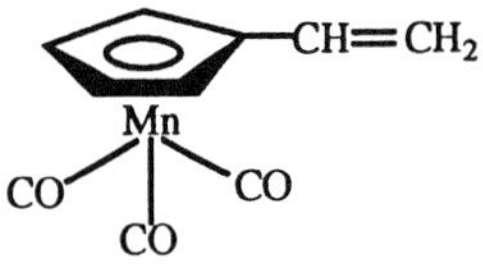

1

$CH{=}CH_2$ Mn CO CO CO

2

$CH{=}CH_2$ Cr ON CO CO

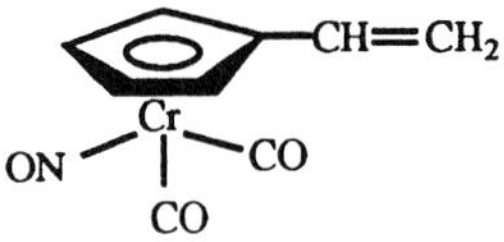

3

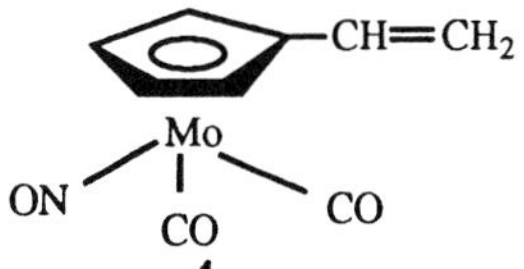

$CH{=}CH_2$ Mo ON CO CO

4

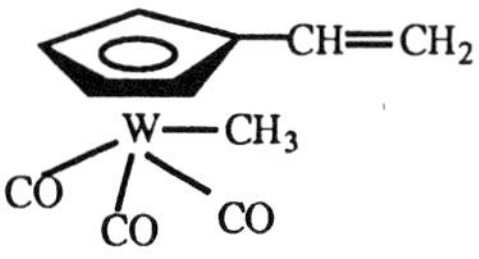

$CH{=}CH_2$ W — CH_3 CO CO CO

5

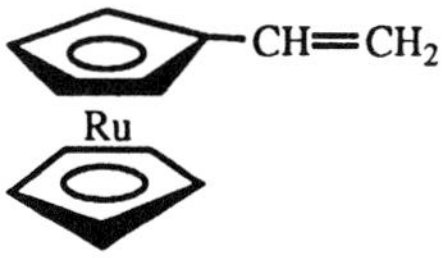

$CH{=}CH_2$ Ru

6

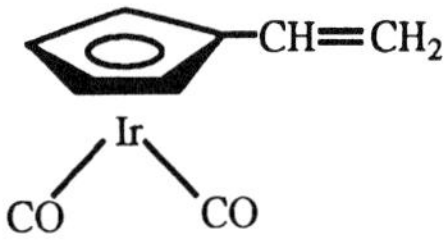

$CH{=}CH_2$ Ir CO CO

7

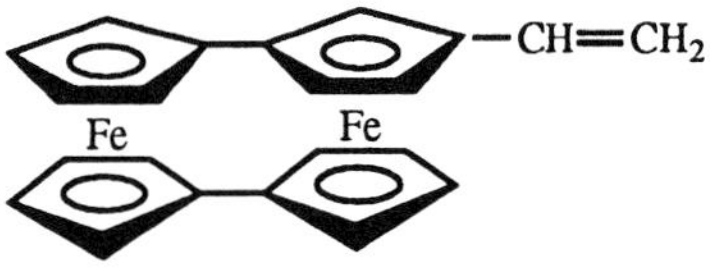

$CH{=}CH_2$ Fe Fe

8

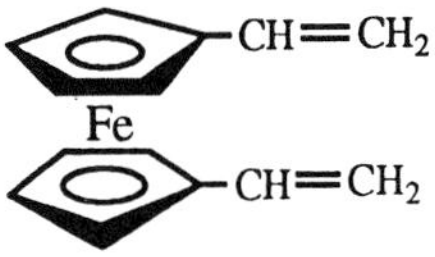

$CH{=}CH_2$ Fe $CH{=}CH_2$

9

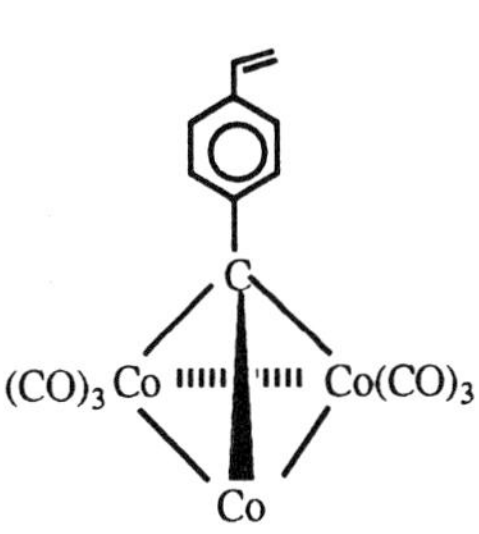

C $(CO)_3Co$ $Co(CO)_3$ Co

10

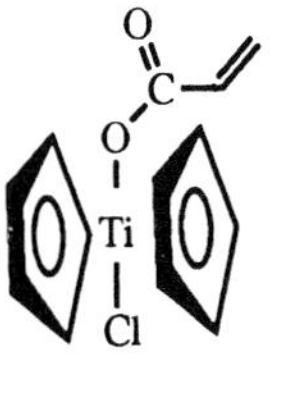

O C O Ti Cl

11

Ph_3P — Pd — PPh_3 Cl

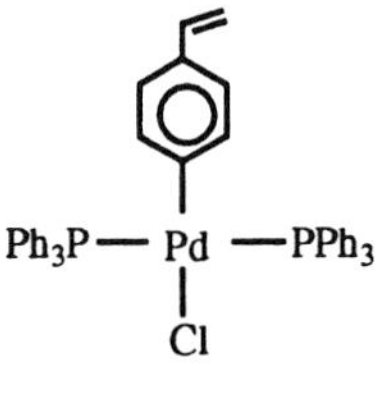

12

Scheme 10-1

10.2 Addition Polymers

Addition polymerization of ferrocene-containing vinyl monomers is probably the earliest preparation method that has been widely studied. Vinylferrocene **1** [5], the first organometallic monomer, was synthesized in 1955 and its polymerization behavior has been extensively studied under radical [5, 6], cationic [7], and Ziegler–Natta conditions [7]; it is inert to anionic initiation [8].

The effect that organometallic functions might exert in vinyl polymerizations is beginning to become clear [9]. As expected, the transition metal fragment with its various accessible oxidation states and large steric bulk is expected to exert unusual electronic and steric effects during the polymerization process. Within this perspective, the vinylferrocene monomer **1** has been studied in great detail and important aspects of its homopolymerization initiated under radical conditions will be considered first. Vinylferrocene undergoes oxidation with peroxide initiators, so that azo initiators such as AIBN have been used exclusively. Unlike most vinyl monomers, the molecular weight of poly(vinylferrocene) does not increase with a decrease in initiator concentration [10]. This is a consequence of the anomalously high chain transfer constant of vinylferrocene ($C_m{}^* = 8 \times 10^{-3}$ versus 6×10^{-5} for styrene at 60 °C) [10]. Finally, the rate law (see Eq. 10-1) for vinylferrocene homopolymerization is first-order in initiator in benzene [11], which indicates intramolecular termination (see also Fig. 10-1).

$$V_p{}^* = 5.64 \times 10^{-4}\,[\text{vinylferrocene}]^{1.12}[\text{AIBN}]^{1.1} \qquad (10\text{-}1)$$
$$(\text{units: } V_p\text{: mol L}^{-1}\text{s}^{-1};\ k\text{: mol}^{-1}\ \text{s}^{-1})$$

Mößbauer studies [11] support the mechanism shown in Scheme 10-2, which involves electron transfer from the iron atom to the growing chain radical to give a zwitterion, which terminates the chain and further results in a high spin Fe(III) complex [11].

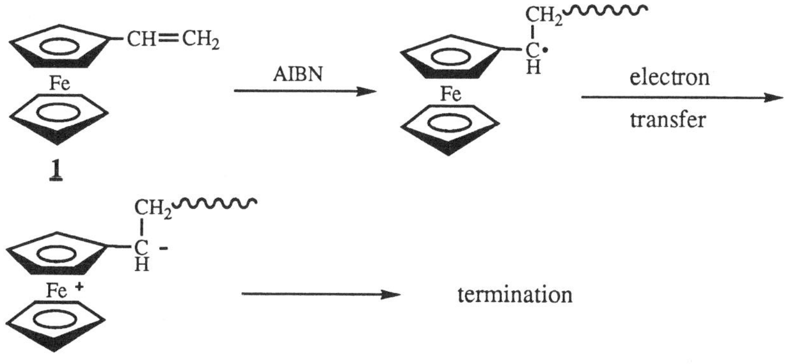

Scheme 10-2

* For the definitions of these basic parameters for chain polymerization, i.e., C_m, V_p, r_1, r_2, Q, e, see G. Odian, *Principles of Polymerization* (3rd ed.), Chap. 3 and 6, Wiley, New York, **1991**.

The organometallic acrylates and methacrylates containing the ferrocene nucleus undergo ready radical-initiated homo- and copolymerization. Unlike the unusual kinetic behavior of vinylferrocene, the homopolymerization of ferrocenylethyl acrylate **13** (Scheme 10-4) and ferrocenylethyl methacrylate **14** (Scheme 10-4) was found to be first-order in monomer and half-order in initiator, similar to that of their organic analogs. In these monomers, the vinyl groups are removed from the influence of the ferrocene nucleus. Monomers developed using this concept will also be discussed later.

The electron richness of vinylferrocene as a monomer has been demonstrated in its copolymerization with maleic anhydride, in which 1 : 1 copolymers were obtained over a wide range of feed ratios and $r_1 \cdot r_2{}^* = 0.003$ [13]. Subsequent copolymerization of vinylferrocene with classic organic monomers, such as styrene [13], *N*-vinyl-2-pyrrolidone [15], methyl methacrylate [13] and acrylonitrile [13] were carried out and the Alfrey–Price Q^* and e^* parameters [16] determined. The value of e is a semiempirical measure of the electron richness of the vinyl group. The best value of e for vinylferrocene is about -2.1, which, when compared with the e values of maleic anhydride ($+2.25$), *p*-nitrostyrene ($+0.39$), styrene (-0.80), *p*-*N*,*N'*-dimethylaminostyrene (-1.37) and 1,1'-dianisylethylene (-1.96), again emphasizes the electron rich nature of the vinyl group in vinylferrocene.

The latest vinylferrocene monomer $\{\eta^5\text{-}C_5H_4CH_2O_2CC(CH_3){=}CH_2\}\{\eta^5\text{-}C_5H_4CH{=}C(CN)CO_2Et\}Fe$ **15** that undergoes radical polymerization has been prepared as shown in Scheme 10-3 [17]: Copolymerization of the monomer with methyl methacrylate produced copolymer **16**, via radical initiation using AIBN in benzene. The ethyl α-cyanoacrylate moiety on the ferrocene remained intact through the polymerization process. The thermal behavior of **16** was similar to that of polymethyl methacrylate: glass transition temperature, $T_g{}^*$ 120 °C, melt transition

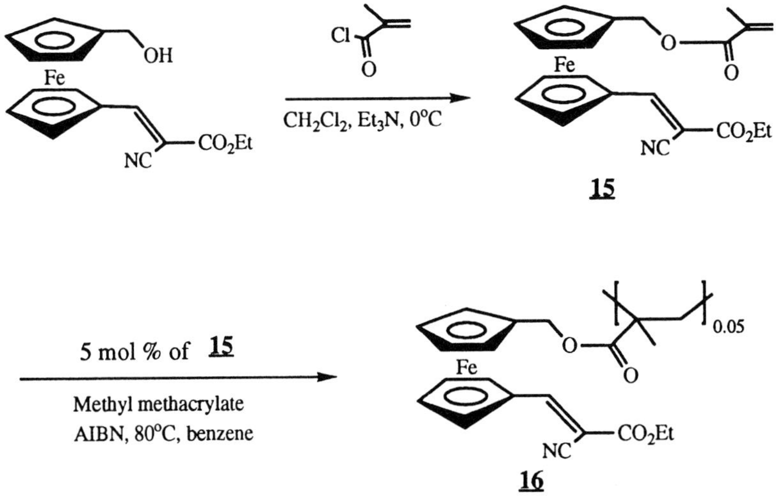

Cl; O; CH_2Cl_2, Et_3N, 0°C; OH; Fe; NC; CO_2Et; O; **15**

5 mol % of **15**; Methyl methacrylate; AIBN, 80°C, benzene; Fe; O; NC; CO_2Et; 0.05; **16**

Scheme 10-3

* For thermal transition behavior of polymers, such as glass transition and melt transition, refer to F. W. Billimeyer, Jr., *Textbook of Polymer Science* (3rd ed.), Wiley, New York, **1984**.

temperature, T_m* 225 °C, and M_n of ca. 30000 by GPC (relative to polystyrene standards). This polymer exhibited unique non-linear optical properties such as a second harmonic efficiency, approximately 4 times that of the quartz standard.

The focus will now shift from radical addition polymerization of ferrocene vinyl type monomers to cationic-initiated polymerization processes. The *Q-e* values of vinylferrocene (as mentioned above) have shown that the metallocene group is strongly electron donating and it is known that the isopropenyl group has a substantial advantage over the vinyl group in cationic polymerization reactions, as seen in the relative polymerizabilities of α-methylstyrene and styrene. For this reason, a wide range of η^5-cyclopentadienyl-metal monomers that contain isopropenyl functional groups have been synthesized (Scheme 10-4), as it was anticipated that these new organometallic monomers will be potentially attractive candidates for cation initiated polymerization. The results for the cation initiated polymerization of isopropenyl ferrocene **17** are outlined here. Compound **17** was synthesized by a three step procedure involving the acetylation of ferrocene, conversion of the resulting product to 2-ferrocenyl-2-propanol, and subsequent dehydration of the carbinol formed. The monomer was homopolymerized under various cationic conditions using the initiators $BF_3 \cdot OEt_2$, anhydrous $AlCl_3$, $SnCl_4$, or $Ph_3C^+SbCl_6^-$, mainly in dichloromethane as solvent at a temperature varying from −78 °C to ambient

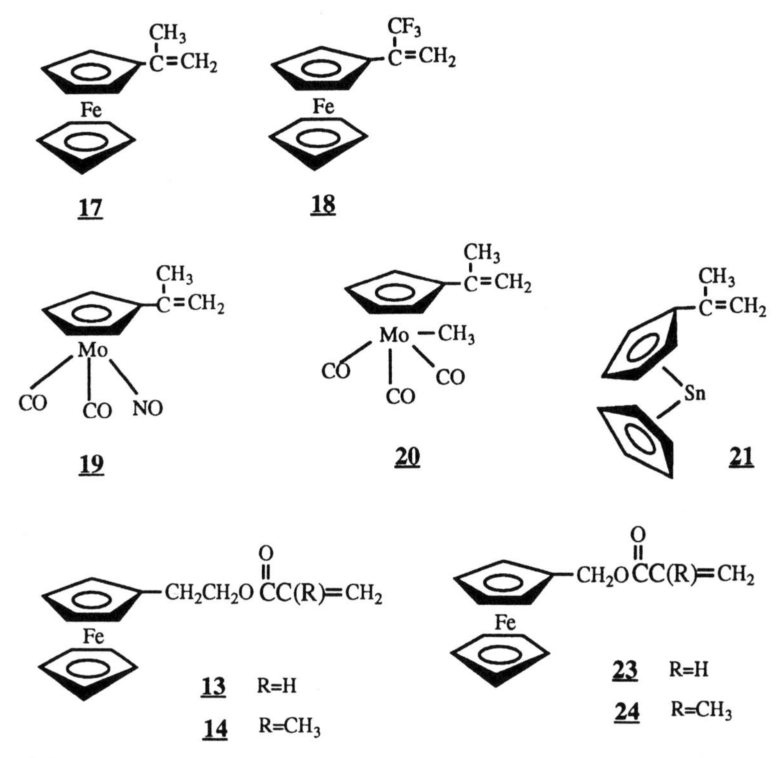

Scheme 10-4

[18]. Only low molecular weight oligomers were obtained. Copolymerization of isopropenyl ferrocene with styrene and *p*-methoxy-α-methylstyrene also gave only low molecular weight products. The formation of the latter can be ascribed to the unusually high stability [19] of the α-ferrocenyl carbocation and the subsequent effect on its propagation reactions, in which it acts as a chain terminator. The propagating carbocation undergoes electrophilic ring substitution on the unsubstituted cyclopentadienyl ring of the penultimate group to form a bridged system (Fig. 10-1). Evidence for this has been provided by analysis of the end groups using ^1H NMR spectroscopy [19]. Similar results for vinylferrocene had been reported earlier by Kunitake et al. [20]. Only low molecular weight polymers were formed in methylene chloride at 0 °C and 24 h using $Et_2AlCl \cdot tBuCl$ or $BF_3 \cdot OEt_2$ as the cationic initiators.

Copolymerization of isopropenylferrocene with styrene, initiated by $BF_3 \cdot OEt_2$ in CH_2Cl_2 at 0 °C, resulted in the incorporation of styrene when the styrene content of the comonomer feed exceeded 90 wt.% [18]. This result also points to the stable ferrocenyl carbocation acting as an inhibitor in the polymerization. The low level of incorporation of styrene could be attributed to the greater reactivity of the isopropenylferrocene, as also determined for vinylferrocene by Aso and Kunitake, who showed that *r* for vinylferrocene was greater than that of styrene [20].

The cation initiated polymerizations of 1,1′-divinylferrocene and 1,1′-diisopropenylferrocene have been extensively studied by Russian [21, 22] and Japanese [23] groups. However, Jablonski and Chisti [24] recently reinvestigated the homopolymerization of 1,1′-diisopropenylferrocene **22**. They used $BF_3 \cdot OEt_2$ and CF_3COOH as the initiators at elevated temperature and reaction periods of 24 h. Polymer yields greater than 90% were achieved, but the products were of low molecular weight ($8100 < M_n < 14700$). Spectroscopic evidence established an initial protonation of the monomer to give a tertiary monocarbocation that did not readily undergo intramolecular cyclization. Analysis of isolated homopolymers by ^1H and ^{13}C NMR spectroscopy revealed significant residual unsaturation, consistent with facile chain transfer. The homopolymers were postulated to have a mixed structure more complicated than that proposed earlier by Sosin et al. [22]. The types of processes probably occurring in the cationic homopolymerization of **22** as proposed by Chisti

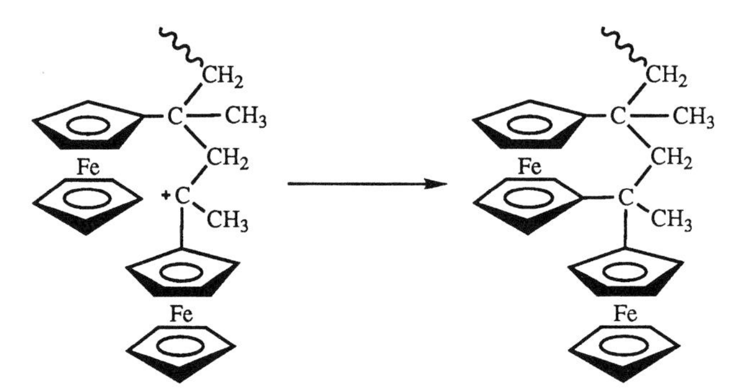

Fig. 10-1. Chain termination in the cationic polymerization of isopropenylferrocene.

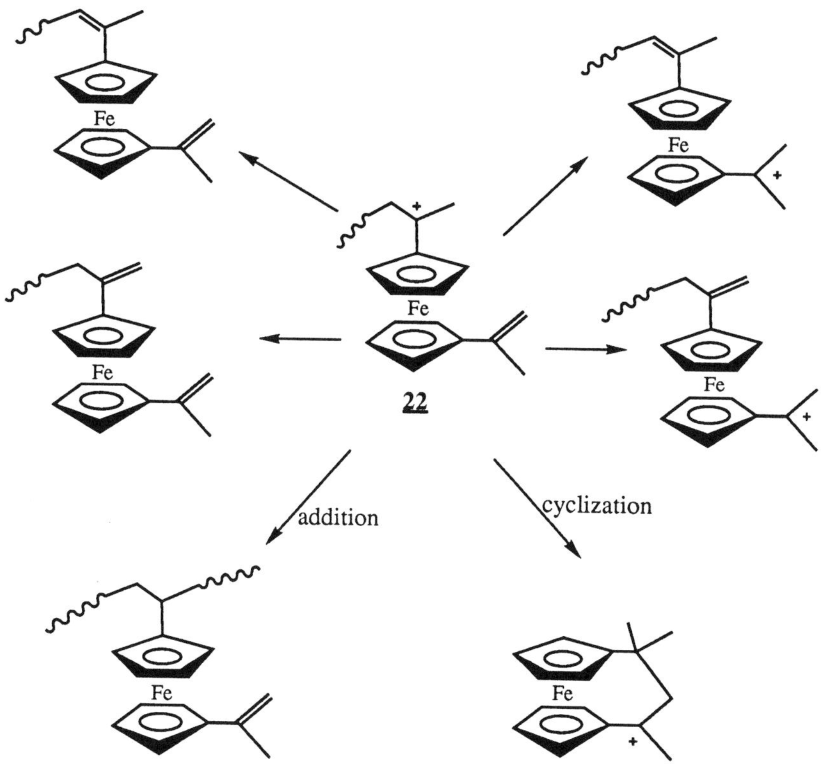

Fe Fe Fe Fe 22 Fe Fe addition cyclization Fe Fe

Scheme 10-5

and Jablonski are outlined in Scheme 10-5. Thus, the cationic polymerization of **22** is extremely complex.

Although vinylferrocene does not undergo anionic polymerization, ferrocenylmethyl methacrylate **23** and -acrylate **24** do [25], using $LiAlH_4$/TMEDA in THF. By varying the mole ratio monomers: $LiAlH_4$ from 17 to 300 the M_n/M_w increased from 3000/5400 to 277000/724000. Block copolymers of ferrocenylmethyl methacrylate and methyl methacrylate and acrylonitrile have been reported, which is indicative of living polymerization,* characteristic of anionic-type polymerization of conventional organic monomers. It should be pointed out here that by placing the reactive vinyl functional group away from the influence of the ferrocene nucleus, the ferrocenylmethyl methacrylate and ferrocenylmethyl acrylate act as conventional organic monomers. This is extremely important as the removal of the functional groups from the ferrocene nucleus reduces the electronic and steric effects of the bulky metallocene. This will be shown more convincingly in the condensation type polymerizations to be considered next.

---

* For the characteristics of living polymerization, see G. Odian, *Principles of Polymerization* (3rd ed.), Chapter 5, Wiley, New York, **1991**.

10.3 Condensation Polymers

Metallocene methylene polymers have also been prepared via the preparation of an α-metallocenylcarbocation with an acidic or Lewis acid initiator, followed by polycondensation involving electrophilic substitution of the metallocene nucleus [26]. However, earlier investigations on the polycondensation of various ferrocenylcarbinols have generally been conducted under drastic conditions, such as high temperature reactions [27] or in heated sealed tubes [28]. The products were reported to contain ferrocenylmethylene units with a random sequence distribution of units possessing both homo- and heteroannularly disubstituted ferrocenylene groups. Recent studies have investigated the self condensation of α-hydroxyisopropylferrocene **25** under mild conditions using cationic initiators [29]. The monomer α-hydroxyisopropylferrocene was synthesized using an abnormal Grignard reaction outlined in Scheme 10-6. Critical to the synthesis of **25** was the ratio of acetylferrocene to the Grignard reagent.

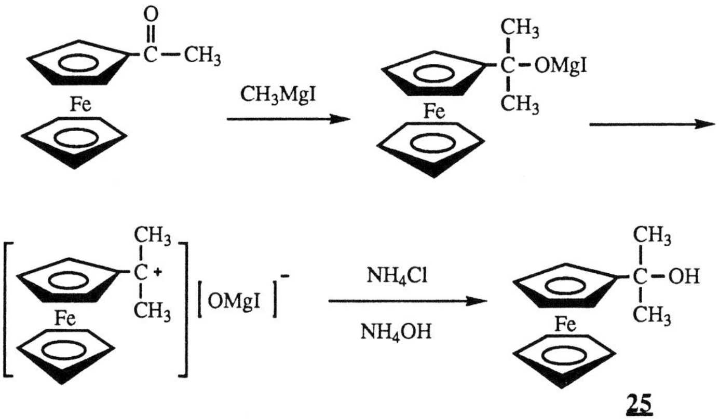

Scheme 10-6

The best yield of **25** was obtained with the ratio 1 : 4.0 in THF at 25 °C for 16 h. The monomer **25** was polymerized at 20 °C with $SnCl_4$ or $BF_3 \cdot OEt_2$ in CH_2Cl_2 for 24–48 h. The polymerization proceeded by the self-alkylation of the stable ferrocenylcarbocation on the cyclopentadienyl ring to form oligomers that contained both homoannular and heteroannular [29] links (Scheme 10-7). In this work the extreme stability of the isopropylferrocenylcarbocation was demonstrated by synthesis and isolation of the α-isopropylferrocenyl carbenium fluoroborate form **25** and using it to initiate the polymerization of styrene. Initiation was successful at 20 °C and at 0 °C but no polymerization occurred at −78 °C. The condensation of ferrocene and acetone in the presence of $AlCl_3$ also gave oligomers with structures very similar to those obtained from the cationic polymerization of isopropenylferrocene [29].

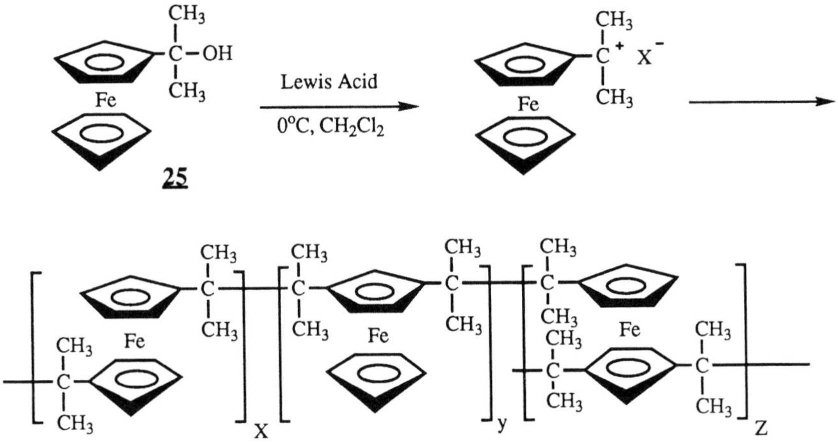

25

Lewis Acid

0°C, CH_2Cl_2

Scheme 10-7

Conventional condensation polymerization [30] of organometallic monomers has generally been conducted at elevated temperature and the resulting products have frequently been poorly characterized [31]. Recently, ferrocene-containing polyamides and polyureas have been synthesized at ambient temperature using interfacial methods [32]. In some instances, film formation has been observed at the interface. Analogous polyesters and polyurethanes have also been prepared [32]. The monomers used were 1,1′-bis(β-aminoethyl)ferrocene **26** and 1,1′-bis(β-hydroxylethyl)-ferrocene **27** [32]. These monomers were synthesized from ferrocene, using modifications of procedures outlined by Sonoda and Moritani [33] and Ratajczak et al. [34]. The intermediate diacid, 1,1′-ferrocenedicarboxylic acid, was synthesized according to the more convenient procedure of Knobloch and Rausher [35]. Monomer **26** was vacuum distilled prior to use (*b.p.* 120 °C, 133 Pa). Details of the synthetic route are given in Scheme 10-8. It should be emphasized that in contrast to previous ferrocene-containing monomers [36], in **26** and **27** the reactive amino and hydroxyl functional groups are located two methylene units from the ferrocene nucleus. This feature minimizes steric effects and also enables **26** and **27** undergo the Schotten–Baumann reaction readily without the classical α-metallocenyl carbocation providing any constraints [19]. Polyamide formation was observed to be vigorous, exothermic, and instantaneous. Interfacial or solution polycondensation, with or without stirring, was the general procedure employed for the preparation of the polyamides and polyureas. Details are provided in Table 10-1. An important point to be noted is that in the unstirred interfacial condensation polymerization of **27** with sebacoyl chloride or terephthaloyl chloride in the organic phase and triethylamine as the acid acceptor, film formation occurred immediately at the interface. Attempts to determine the molecular weight of these iron-containing polyamides in *m*-cresol were unsuccessful, due to the limited solubility of the materials in

* For the definition of intrinsic viscosity, [η], see R. J. Young and P. A. Lovell, *Introduction to Polymers* (2nd ed.), Chapman & Hall, London, **1991**.

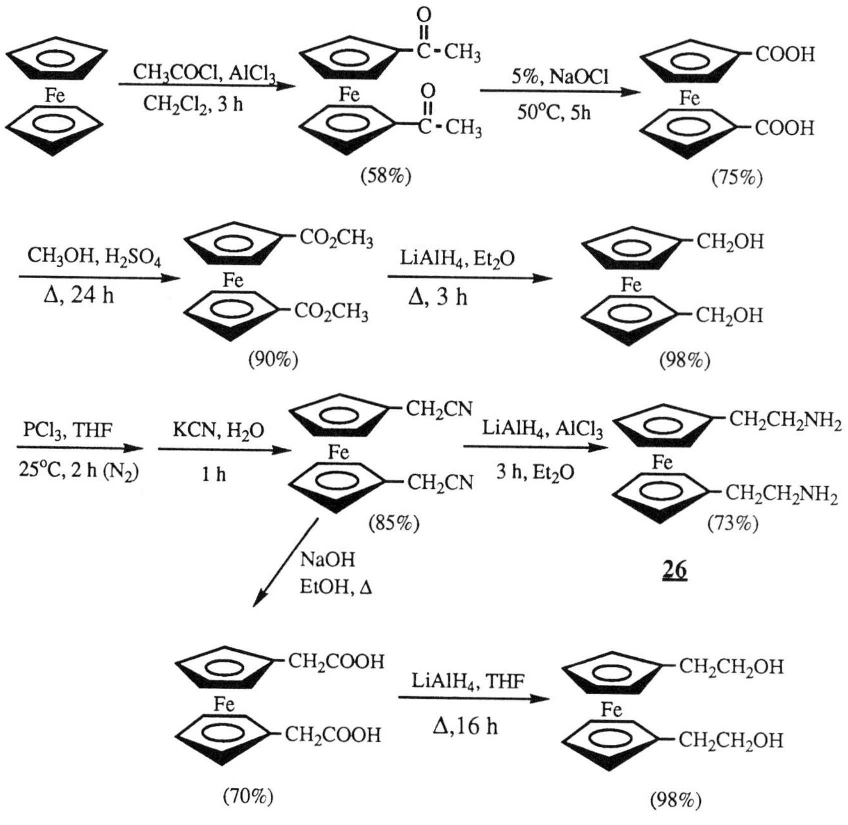

Fe $\xrightarrow[CH_2Cl_2,\ 3\ h]{CH_3COCl,\ AlCl_3}$ Fe(C(O)CH$_3$)$_2$ (58%) $\xrightarrow[50^\circ C,\ 5h]{5\%,\ NaOCl}$ Fe(COOH)$_2$ (75%)

$\xrightarrow[\Delta,\ 24\ h]{CH_3OH,\ H_2SO_4}$ Fe(CO$_2$CH$_3$)$_2$ (90%) $\xrightarrow[\Delta,\ 3\ h]{LiAlH_4,\ Et_2O}$ Fe(CH$_2$OH)$_2$ (98%)

$\xrightarrow[25^\circ C,\ 2\ h\ (N_2)]{PCl_3,\ THF}$ $\xrightarrow[1\ h]{KCN,\ H_2O}$ Fe(CH$_2$CN)$_2$ (85%) $\xrightarrow[3\ h,\ Et_2O]{LiAlH_4,\ AlCl_3}$ Fe(CH$_2$CH$_2$NH$_2$)$_2$ (73%) **26**

NaOH, EtOH, Δ

Fe(CH$_2$COOH)$_2$ (70%) $\xrightarrow[\Delta,16\ h]{LiAlH_4,\ THF}$ Fe(CH$_2$CH$_2$OH)$_2$ (98%) **27**

Scheme 10-8

organic solvents. The intrinsic viscosity $[\eta]$* of the polymers was 1.5–0.10 dL/g. The low values of $[\eta]$ can be attributed to the premature precipitation from solution. This is a significant drawback in the synthesis of high molecular weight organometallic polymers. The polyamides and polyureas exhibited broad, intense N–H stretches at ca. 3300 cm^{-1}. A strong carbonyl stretching vibration was present at 1630 cm^{-1}. The amide II band was evident near 1540 cm^{-1}. In addition, sp^2 C–H stretches occurred at about 3100 cm^{-1} and symmetric and asymmetric stretches at 2950 and 2860 cm^{-1}, respectively. The polyurethane showed the carbonyl absorption near 1700 cm^{-1} and C–O stretches in the vicinity of 1220 cm^{-1}. Similar absorptions were present in the polyester. The polyamides and polyureas are thus thought to have the structures outlined in Scheme 10-9. Further elucidation of these reactions and structures has been provided by the synthesis and characterization of model analogs [32].

The monomers **26** and **27** have also been used to synthesize segmented poly(ether urethane) from poly(propylene glycol) (PPG) and 4,4′-methylene-bis(phenylisocyanate) (MDI) [37]. In the prepolymer method employed, MDI (2 equiv.) and PPG (1 equiv.) were allowed to react at 60 °C in the presence of 1 mol% dibutyltin

Table 10-1. Polycondensation of 1,1′-bis(β-amino-ethyl)ferrocene **26** and (β-hydroxyethyl)ferrocene **27** with diacid chlorides and diisocyanates (adapted from [1 d])

Monomer (M_1)	Monomer (M_2)	Process (base used)	yield, %	η^e dL/g
26[a]	Terephthaloyl chloride (CH_2Cl_2)	UI[b] (Et_3N)	72	1.50
26[a]	Sebacoyl chloride (CCl_4)	UI (Et_3N)	85	0.37
26[a]	Sebacoyl chloride (CCl_4)	UI (NaOH)	39	0.59
26[a]	Sebacoyl chloride (CCl_4)	I[b]	51	1.09
26[a]	Adipoyl chloride (CCl_4)	UI (Et_3N)	47	0.53
26	Terephthaloyl chloride (CH_2Cl_2)	S[b] (Et_3N)	45	0.80
27	Terephthaloyl chloride (*m*-xylene, reflux)	S (pyridine)	51	0.16
27	TDI[c] (Me_2SO, 115 °C)	S	46	0.20
26	TDI ($CHCl_3$)	UI	58	0.16
26	TDI ($CHCl_3$)	S	53	0.10
26	MDI[d]	S	67	f

a Monomer in aqueous phase.
b UI, unstirred interfacial. S, solution, I, stirred interfacial.
c TDI: tolulyene 2,4-diisocyanate (80%) + 2,6-isomer (20%).
d MDI: methylene-bis(4-phenylisocyanate).
e Intrinsic viscosity determined in *m*-cresol at 32 °C.
f Insoluble in *m*-cresol.

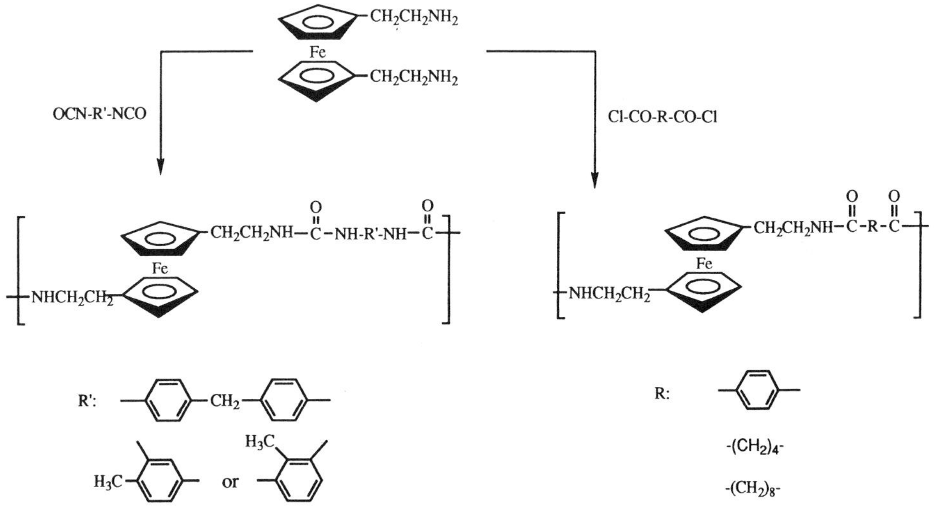

Scheme 10-9

dilaureate as catalyst in the melt. The course of the polymerization was followed by monitoring the intensity of the -NCO peak in the IR. After the MDI and PPG had reacted to form the prepolymer, **26** (1 equiv.) in dry DMF was added as the chain extender and curing continued at ambient temperature, as outlined in Scheme 10-10. Similar curing of the prepolymer with **27** as the chain extender at 60 °C also produced a polymer film. On the basis of their IR, NMR, and mass spectra

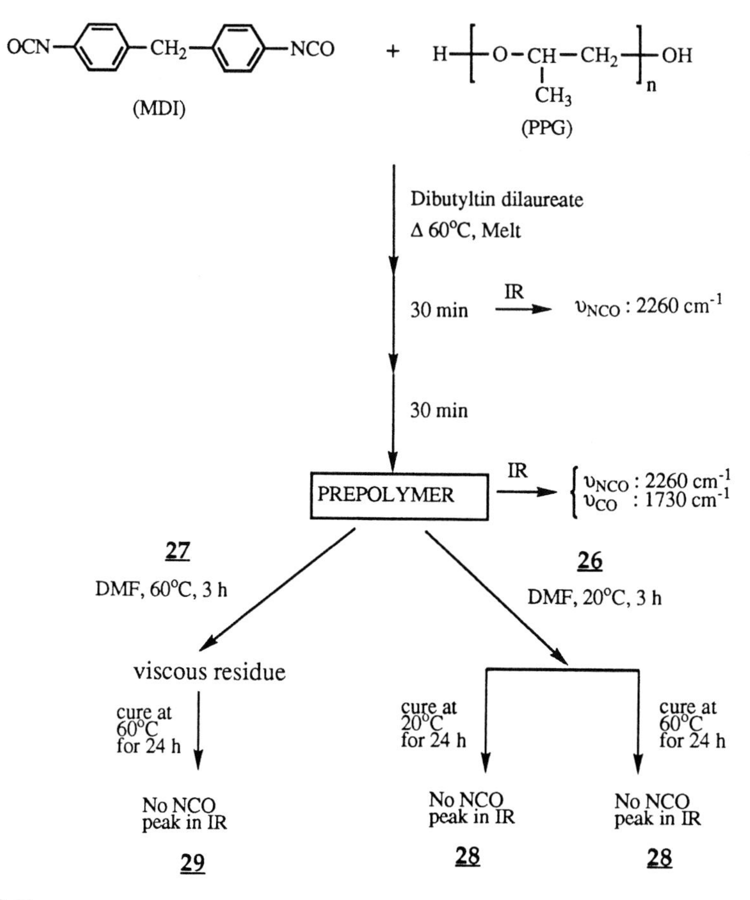

Scheme 10-10

the polymers were assigned the structures **28** and **29** (Fig. 10-2). The molecular weight distribution (MWD) of the polymers was determined by GPC. The solvent used was THF and the instrument calibrated with narrow MWD polystyrenes. Polymer **28** had M_w 56000 (M_n = 12500) and **29** had M_w 97000 (M_n = 9100). The amount of iron in **28** and **29** was determined to be 1.9 and 1.2 wt.% respectively, corresponding to 6.7% and 5% ferrocene units in the copolymers.

Ferrocene-containing liquid crystalline polymers **30** have been reported from the solution polymerization of 1,1′-bis(chlorocarbonyl)ferrocene, terephthaloyl chloride, and methylhydroquinone in refluxing dichloromethane [38], as indicated in Scheme 10-11. With one exception, these ferrocene containing copolyesters were reported to have birefringent melts. The presence of liquid crystallinity was verified by differential scanning calorimetry (DSC), polarized light microscopy, and X-ray diffraction studies.

There have been reports of the synthesis of organometallic arylidene polyesters containing ferrocene derivatives in the main chain. Interfacial polymerization of

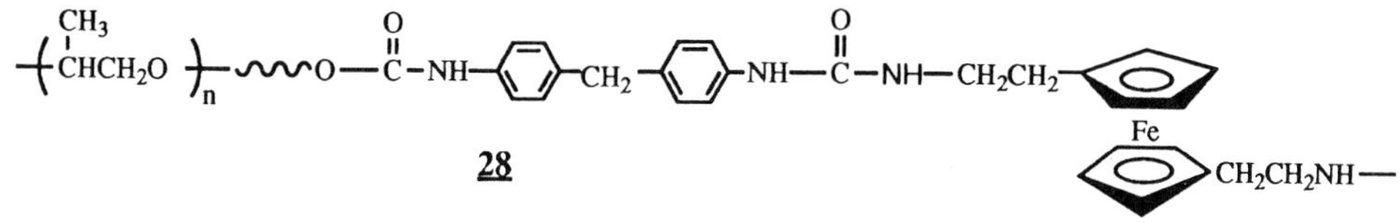

28

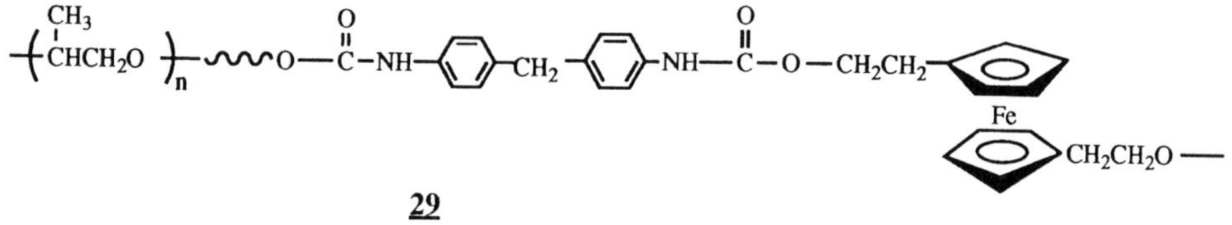

29

Fig. 10-2. Ferrocene-containing polyurethanes.

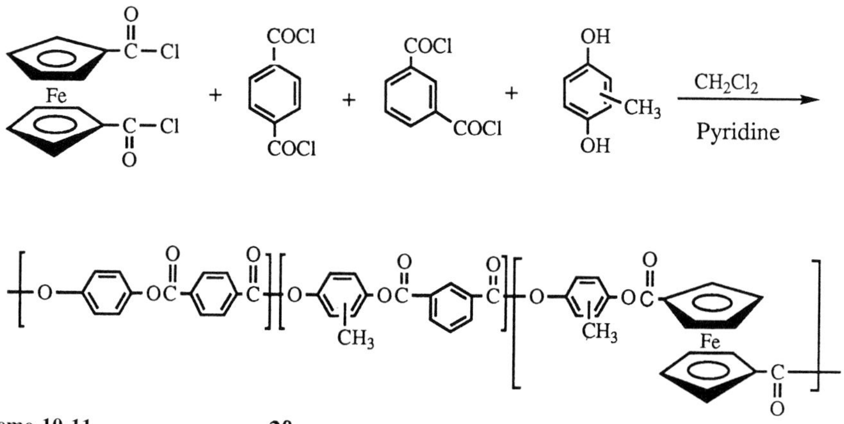

Scheme 10-11

30

1,1′-dichlorocarbonylferrocene or 1,1′-dichlorocarbonyl-4,4′-diiodoferrocene with diarylidenecycloalkanones, as shown in Scheme 10-12, at ambient temperature afforded polyesters **31** and **32**, respectively. The intrinsic viscosity [η] of the polymers was low, 0.39 – 0.64 dL/g. These polyester syntheses are essentially extensions of earlier work by Pittman et al., who prepared a series of polyesters containing ferrocene from the reaction of 1,1′-dichlorocarbonylferrocene and various diols using high temperature solution polymerization techniques [39]. Analogous polyesters had also been reported earlier by Knobloch and Rausher [35].

As mentioned above, it has recently been demonstrated that ferrocene systems, especially polymers, are potentially excellent NLO materials [4]. Two examples of such materials obtained by polycondensation are described here. New monomers, namely **33** and **34**, have been prepared (as shown in Scheme 10-13) by selective functionalization of the cyclopentadienyl rings of ferrocene [40].

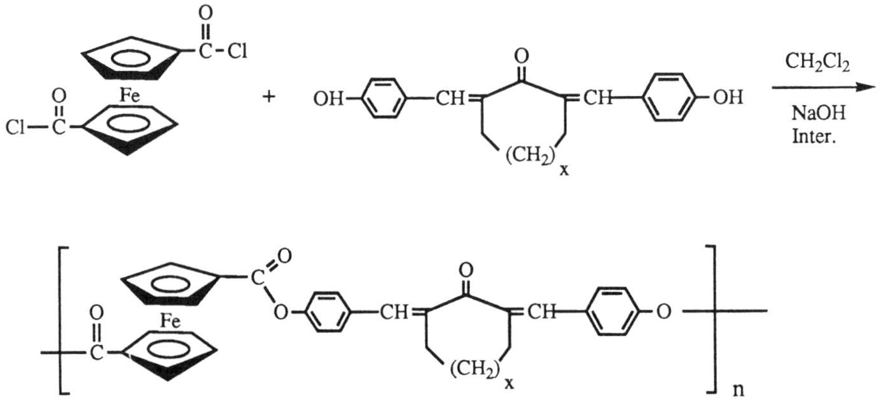

CH_2Cl_2

NaOH

Inter.

31a,b

a=(X=0); b(X=1)

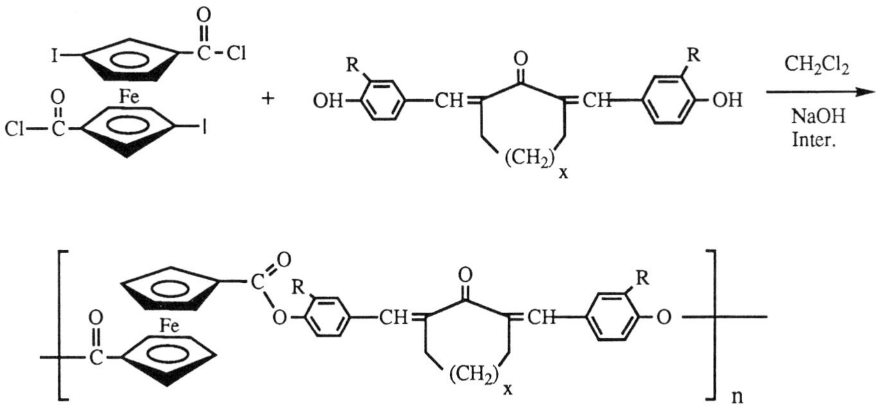

CH_2Cl_2

NaOH

Inter.

32a-e

a (X=0, R=H);
c (X=1, R=H);
e (X=2, R=H);

b (X=0, R= OCH_3)
d (X=1, R= OCH_3)

Scheme 10-12

Monomer **33** was made to undergo transesterification polymerization using $Ti(OC_4H_9)_4$, while monomer **34** was appropriate for a Knoevenagel polycondensation. The transesterification polymerization resulted in the formation of an intractable material of unknown structure. Homopolymerization of **34** by the Knoevenagel technique afforded polymer **35** with a low molecular weight (M_n 6800). A major byproduct in this polymerization was a macrocyclic lactone, formed via an intramolecular Knoevenagel condensation (Scheme 10-14).

OH
Fe
SnBu3
1.NaH, THF
2. I-(CH2)6OSi(Me)2(t-Bu)
3.n-BuLi, DMF
Fe
O
(CH2)6
OSi(Me)2(t-Bu)
CHO
NBuF4
Fe
O-(CH2)6-OH
CHO
CNCH2CO2Et
K2CO3, THF
40°C
CNCH2CO2H,
DCC, CH2Cl2, 0°C
O-(CH2)6-OH
Fe
CHO
33
O
CN
O-(CH2)6-O
Fe
CHO
34

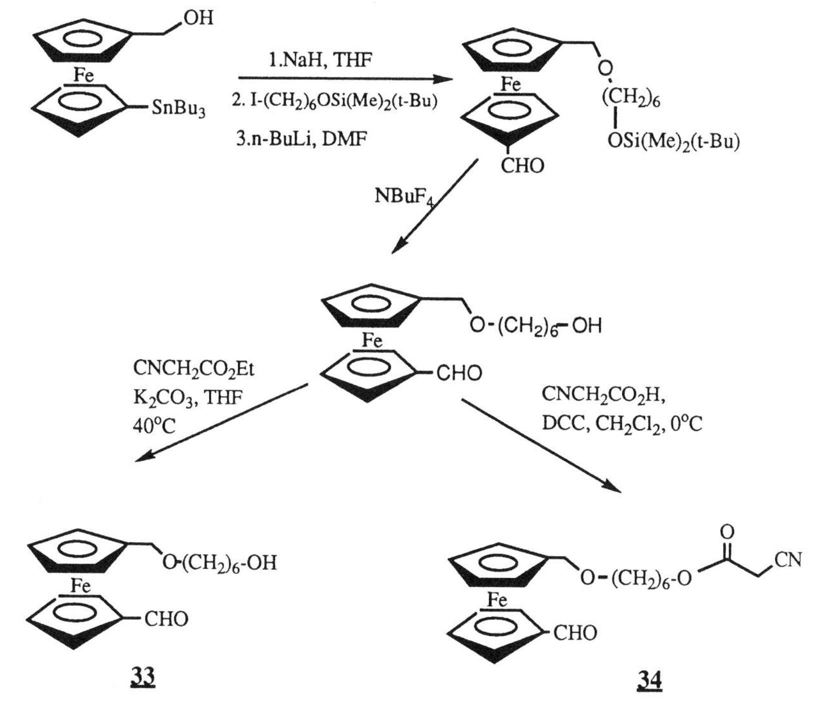

Scheme 10-13

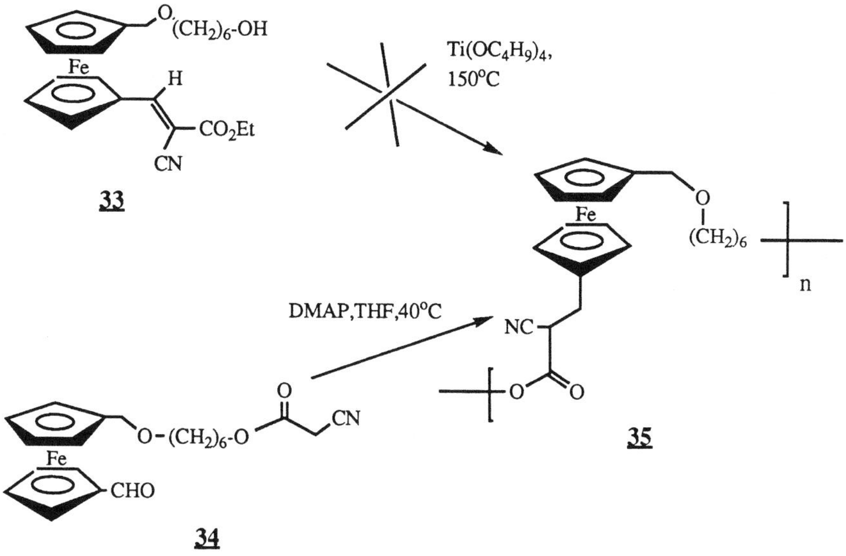

O
(CH2)6-OH
Fe
H
CO2Et
CN
33
Ti(OC4H9)4,
150°C
Fe
O
(CH2)6
n
DMAP,THF,40°C
NC
O
O
35
O
CN
O-(CH2)6-O
Fe
CHO
34

Scheme 10-14

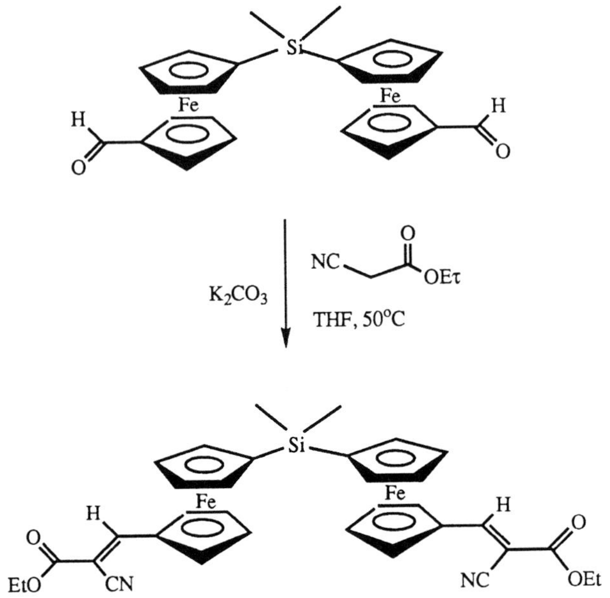

K_2CO_3

NC–CH₂–C(O)OEt

THF, 50°C

Scheme 10-15

36

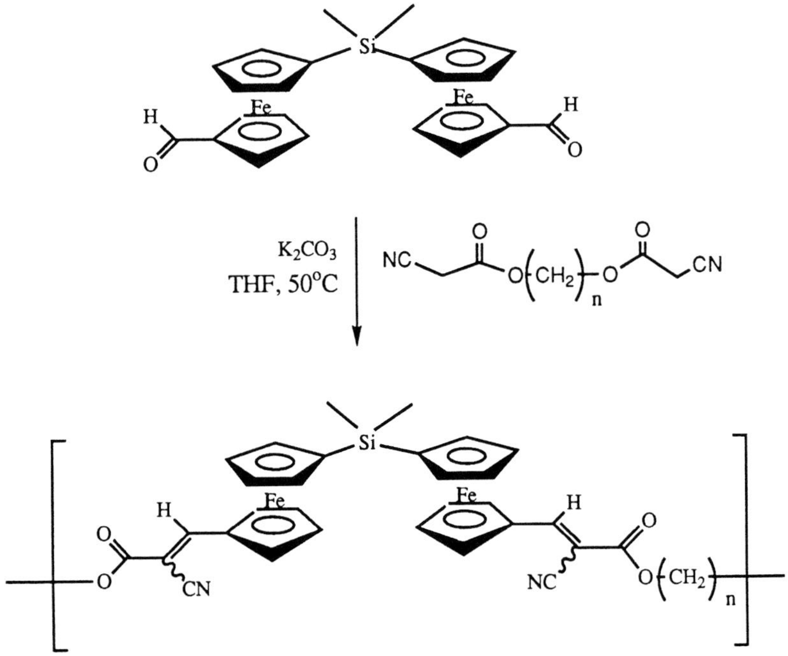

K_2CO_3
THF, 50°C

Scheme 10-16

37 a (n=4)
37 b (n=6)
37 c (n=8)

The bridged ferrocenyl monomer **36** was prepared in two steps as shown in Scheme 10-15 [41]. Knoevenagel polycondensation of **36** and $CNCH_2CO_2(CH_2)_n$-O_2CCH_2CN (n = 4, 6, 8; K_2CO_3, THF, 50 °C) afforded a new class of organometallic "according type" copolymers **37**, as depicted in Scheme 10-16. The value of M_n was 9100 – 26000 with a polydispersity of approx. 2.1. The polymers were soluble in organic solvents. The polycondensation, however, appeared to produce a mixture of E and Z isomers (9/1, respectively) and the isomerically enriched polymer **37** (i.e., 98% E isomer) was obtained by selective precipitation in methanol.

Ferrocene has been reported to be very effective as a soot reducing agent in combustion [42 – 44]. Thus, when ferrocene compounds are incorporated in a fire retardant polymer, such as a phenolphthalein-based polymer and poly(phosphate ester)s, they have shown added advantages in that they promote extinction and reduce smoke formation by accelerated char reduction [45, 46]. The synthesis of such ferrocene-containing poly(phosphate ester)s was achieved by interfacial polycondensation using a phase transfer catalyst [47]. Accordingly, 1,1′-bis(p-hydroxyphenylamido)ferrocene and 1,1′-bis(p-hydroxyphenylcarbonyl)ferrocene underwent condensation with various aryl phosphoroic acid dichlorides to yield two series of ferrocene-containing polymers, i.e., poly(amide-phosphate ester)s **38a** and poly(ester-phosphate ester)s **38b** respectively, as shown in Scheme 10-17.

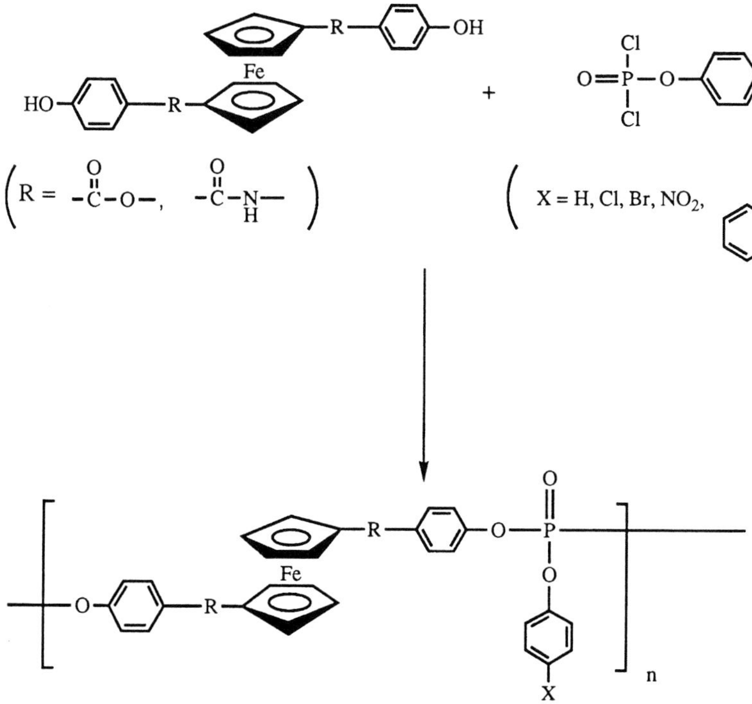

Scheme 10-17

The M_n of these polymers, determined by end group analysis using ^{31}P NMR spectral data, was 3270–4550. These polymers were suggested to have improved thermal stability and flame resistance.

10.4 Ring-Opening Polymerization

Ring-Opening polymerization has also been involved in the synthesis of ferrocene-containing polymers. Several very recent examples deal with the ring-opening polymerization of strained, ring-tilted ferrocenylsilanes [48] (Fig. 10-3), ferrocenylgermanes [49], ferrocenylphosphines [49], and ring-opening desulfurization of trithiaferrocenophanes [50].

The ferrocenylsilanes were obtained from strained ferrocenophanes, themselves prepared by the reaction of dilithioferrocene with the appropriate dichloroorganosilane [48]. On slight heating beyond the melting point of the monomer these monomers yielded high molecular weight polymers **39** (Scheme 10-18a) as shown in Table 10-2. The polymers did not exhibit a T_m. Values of T_g were dependent on the alkyl substituent and were found to be in the range $-26\,°C - +33\,°C$. An interesting feature of these polymers was that their UV/VIS spectra were found to be consistent with an essentially localized electronic structure for the polymer backbone [2]. Ferrocenophanes also undergo ring-opening polymerization on heating at 300 °C in sealed tubes. These polymers were intractable and insoluble. However, polymers obtained from [2] ferrocenophanes with a methyl group on each cyclopentadienyl ring were soluble with a molecular weight of the order of M_n = 80000 [48a] (Scheme 10-18b).

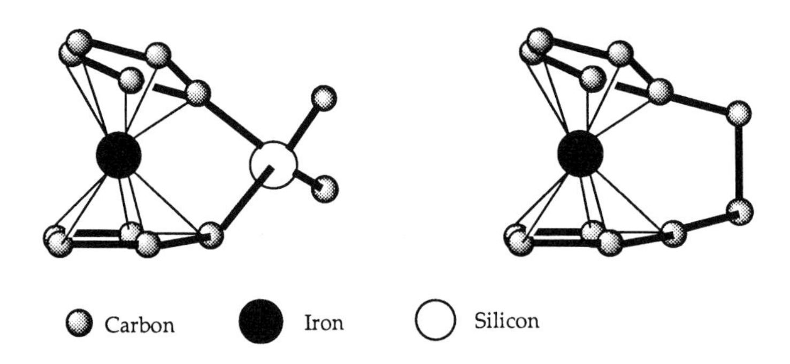

•Hydrogen atoms not shown

•Si- and C_2-bridged ferrocenophanes have same ring-tilt angle (21°)

•Source : X-ray crystallographer Alan Lough, University of Toronto

Fig. 10-3. Structures of Si- and C_2-bridged ferrocenophanes. (Adapted from Chemical and Engineering News **1993**, *71*, 22).

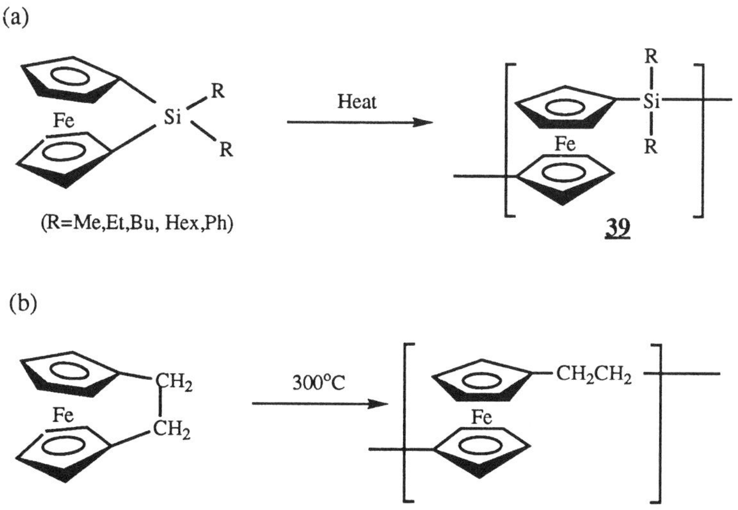

(a)

(R=Me,Et,Bu, Hex,Ph)

Heat

39

(b)

300°C

Scheme 10-18

Table 10-2. Visible spectroscopic, molecular weight, and glass transition data for poly(ferrocenylsilanes) **39** (adapted from [48a])

	R	λ_{max} (ε, M^{-1} cm^{-1})	M_w [a]	M_n [a]	PDI	T_g [d], °C
39[a]	Me	430 (190)	5.2×10^5	3.4×10^5	1.5	33/25[e]
39[b]	Et	444 (200)	7.4×10^5	4.8×10^5	1.6	22
39[c]	Bu	450 (160)	8.9×10^5	3.4×10^5	2.6	3
			1.7×10^5 [b]	9.3×10^4 [b]	1.9	
39[d]	Hex	450 (160)	1.15×10^5	7.6×10^4	1.5	−26/−27[e]
39[e]	Ph		5.1×10^4 [c]	3.2×10^4 [c]	1.6[c]	

a Determined by GPC.
b Sample used for light scattering.
c Soluble polymer fraction extracted with hot THF.
d T^g values obtained by DMA experiments.
e T^g values obtained by DSC.

Analogous ferrocenophanes containing a single germanium atom, prepared by reaction of dilithioferrocene with dichlorodialkyl germane have also been reported [49]. Again, on heating beyond the melting point, rapid exothermic ring-opening polymerization to ferrocenylgermanes occurred (Scheme 10-19). The molecular weight of the polymers **40**, determined by GPC, ranged between $5.2 \times 10^5 - 2 \times 10^6$ (M_w) to $5.2 \times 10^4 - 8.5 \times 10^5$ (M_n) for R = R′ = Me and 3.4×10^4 (M_w) to 2.9×10^4 (M_n) for R = R′ = Et. The polymers are reported to be stable in air both in the solid state and in solution. Cyclic ferrocenylphosphines [49] were obtained by the reaction of dilithioferrocene · TMEDA with dichlorophosphines. When heated in the melt, they yielded a polymer. The molecular weights were not reported.

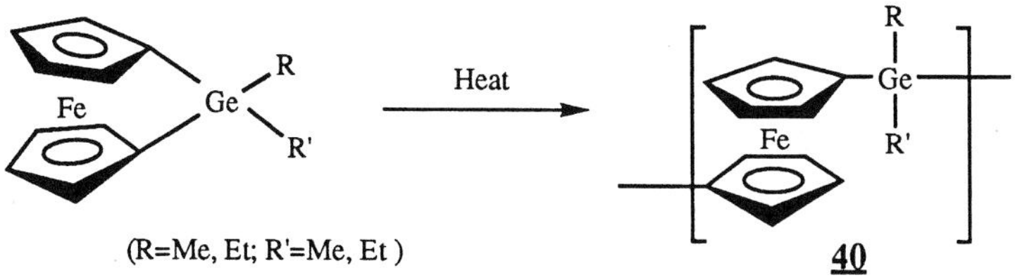

(R=Me, Et; R'=Me, Et)

40

Scheme 10-19

The trithiaferrocenophanes monomers $(tBuC_5H_3)Fe(C_5H_4)S_3$ **41** and $(tBuC_5H_3)_2$-FeS_3 **42** were prepared from the *tert*-Bu-substituted ferrocene by lithiation (*n*BuLi), followed by treatment with elemental sulfur [50]. These monomers polymerized in the presence of Bu_3P at room temperature in 48 h to yield polymers **43** with M_n 3700, M_w 26000 and M_n 2800, M_w 250000, respectively. The broad MWD is indicative of a competing and complex polymerization mechanism (Scheme 10-20).

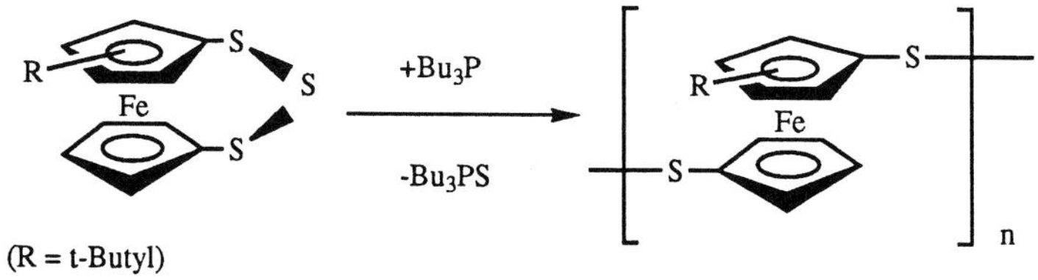

(R = t-Butyl)

41 43

Scheme 10-20

Living ring-opening polymerization (ROMP)* of norbornenes and norbornadienes [51, 52] appear to be promising, because functionalized derivatives are relatively easy to synthesize in a wide variety using the Diels-Alder reaction. In addition, well characterized ROMP catalysts have been developed that tolerate a variety of functional groups [52, 53]. Ferrocene has been incorporated into the norbornene skeleton **44**, as shown in Scheme 10-21. This monomer was polymerized quantitatively, employing **44a** as the initiator [54]. The polymer **45** was cleaved from the catalyst component with pivaldehyde or *p*-trimethylsilylbenzaldehyde and purified by precipitation from a polar solvents such as hexane, or pentane. GPC analysis showed that the resulting polymers (Table 10-3) had a polydispersity of 1.13, expected for polymers prepared by living systems. The ^1H and ^{13}C NMR spectra are indicative of polymers containing *cis/trans* olefinic linkages, *endo/exo* and head-to-tail isomers. Block copolymers containing norbornadienes (NBE) were also reported [54], as shown in Scheme 10-22, the molecular weights of which are also included in Table 10-3.

* See G. Odian, *Principles of Polymerization* (3rd ed.), Chap. 7, Wiley, New York, **1991**.

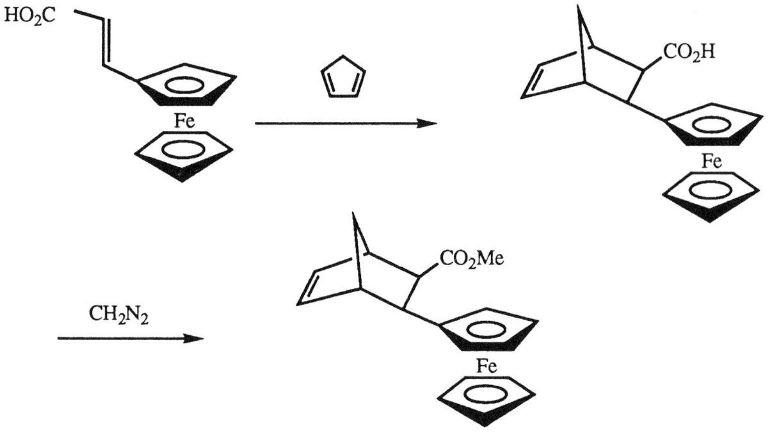

HO_2C Fe CO_2H Fe CH_2N_2 CO_2Me Fe

Scheme 10-21 **44**

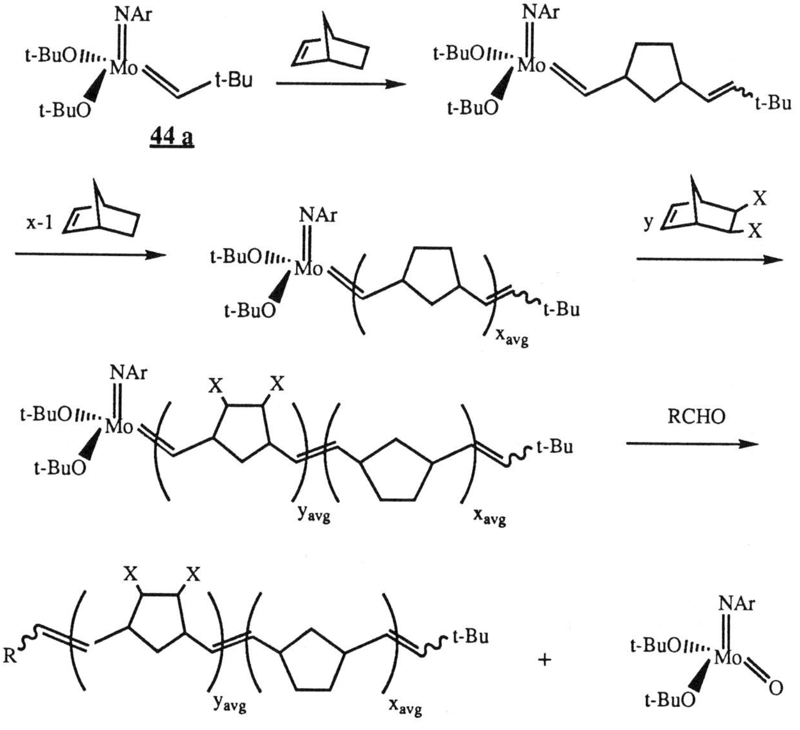

NAr t-BuO Mo t-Bu t-BuO **44 a** NAr t-BuO Mo t-BuO t-Bu

x-1 NAr t-BuO Mo t-BuO t-Bu x_{avg} y X X

NAr t-BuO Mo t-BuO X X t-Bu y_{avg} x_{avg} RCHO

R X X t-Bu y_{avg} x_{avg} + NAr t-BuO Mo O t-BuO

45

Scheme 10-22

Table 10-3. Characterization of redox-active polymers and block copolymers (adapted from [54])

Polymers[a]	PDI[b]	M_n[c]	M_W (theory)[d]	T_g[e], °C
45$_{15}$	1.13	5090	5183	143
45$_{30}$	1.13	9030	10226	150
$(NBE)_{15}$**45**$_{15}$[f]	1.05	10460	6688	60, 123
$(NBE)_{60}$**45**$_{15}$[f]	1.07	16190	10925	43, 107

a Polymers were prepared with **44a** as the initiator and capped by addition of pivaldehyde, unless otherwise noted.
b Polydispersity was determined by gel permeation chromatography in dichloromethane versus polystyrene standards. For **45**$_{15}$ the PDI measured by FD-MS was 1.06.
c M_n = weight average molecular weight as determined by GPC *vs.* polystyrene standards.
d Theoretical molecular weight calculated on the basis of stoichiometry.
e Determined by DSC (scan rate 20 °C/min).
f The end group is $CH(p\text{-}C_6H_4SiMe_3)$.
[Note: Polymer **45**$_x$ is where x represents the number of equivalents of monomer **45** added to the initiation].

The versatility of polyphosphazenes is evident by the scope of applications that have been reported for these materials [55]. If the phosphazene ring is subjected to ring strain by the presence of a transannular ferrocenyl group, as in structure **46**, under these circumstances polymerization takes place to give **47** (Fig. 10-4). In this case reaction occurs, although no halogen atoms are attached to the phosphorus atoms. However, the reaction is accelerated by the presence of catalytic quantities of $(NPCl_2)_3$ [56].

The linkage of ferrocenyl side groups in phosphazenes **48, 49** has also been accomplished by the method shown in Scheme 10-23 [57].

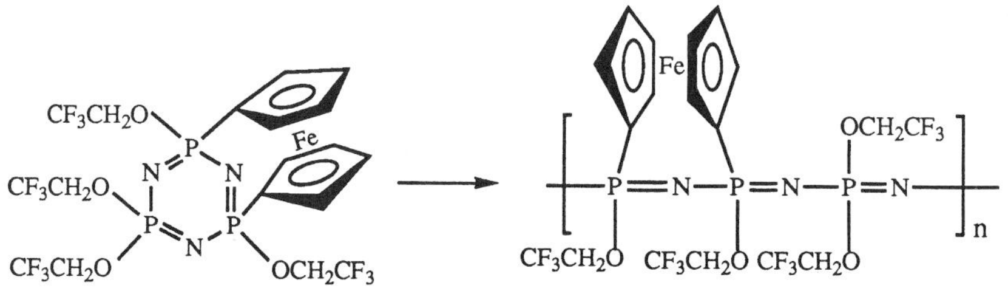

46

47

Fig. 10-4. Ferrocene-containing polyphosphazenes.

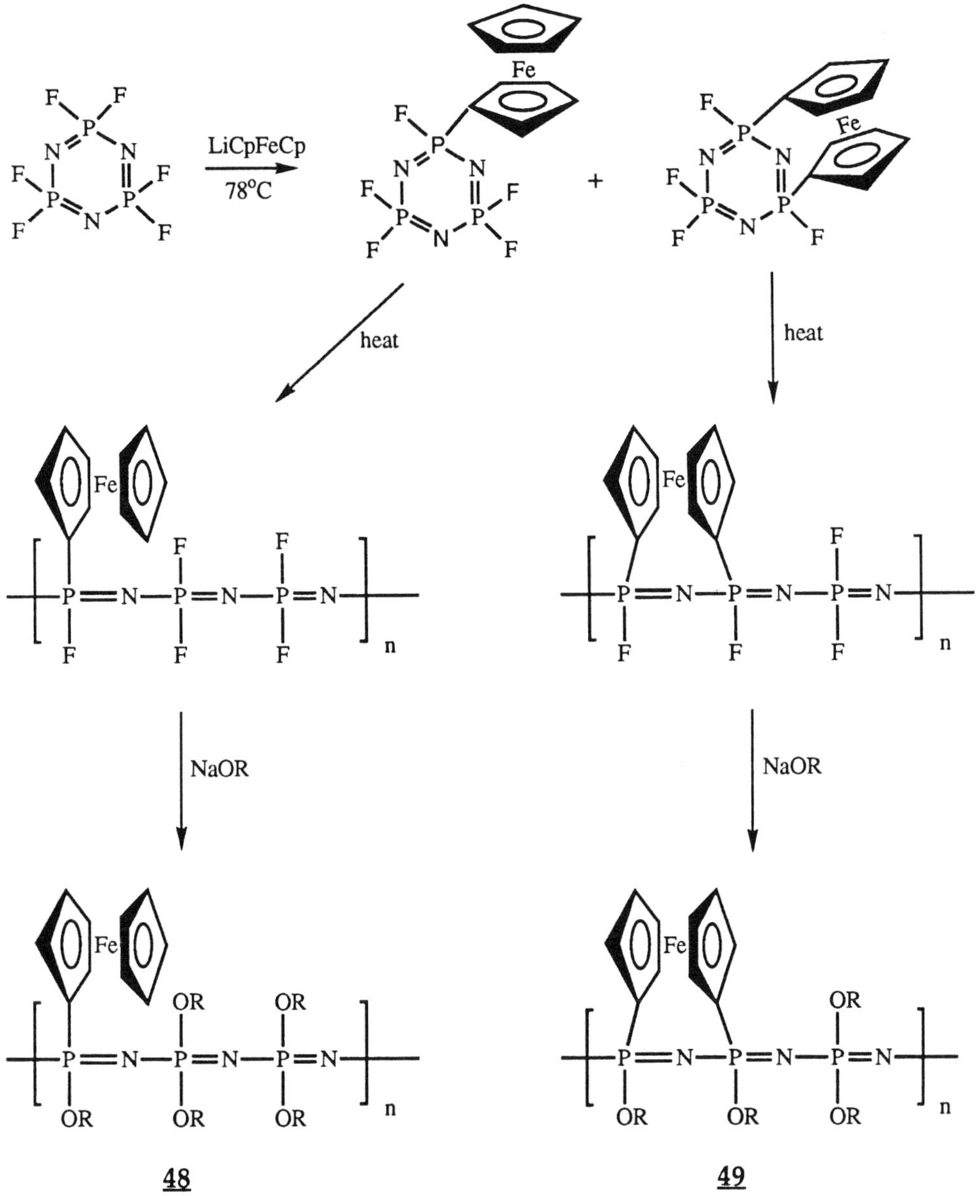

LiCpFeCp
78°C
heat
heat
NaOR
NaOR

48

49

Scheme 10-23

10.5 Derivatization of Preformed Organic Polymers [58]

A wide range of conventional organic type polymers has been used for modification with the ferrocene functional group. Treatment of polystyrene with lithioferrocene or dilithioferrocene led to incorporation of the ferrocene as a pendant group [59]. Another interesting example is the preparation of a ferrocene-containing ionomer

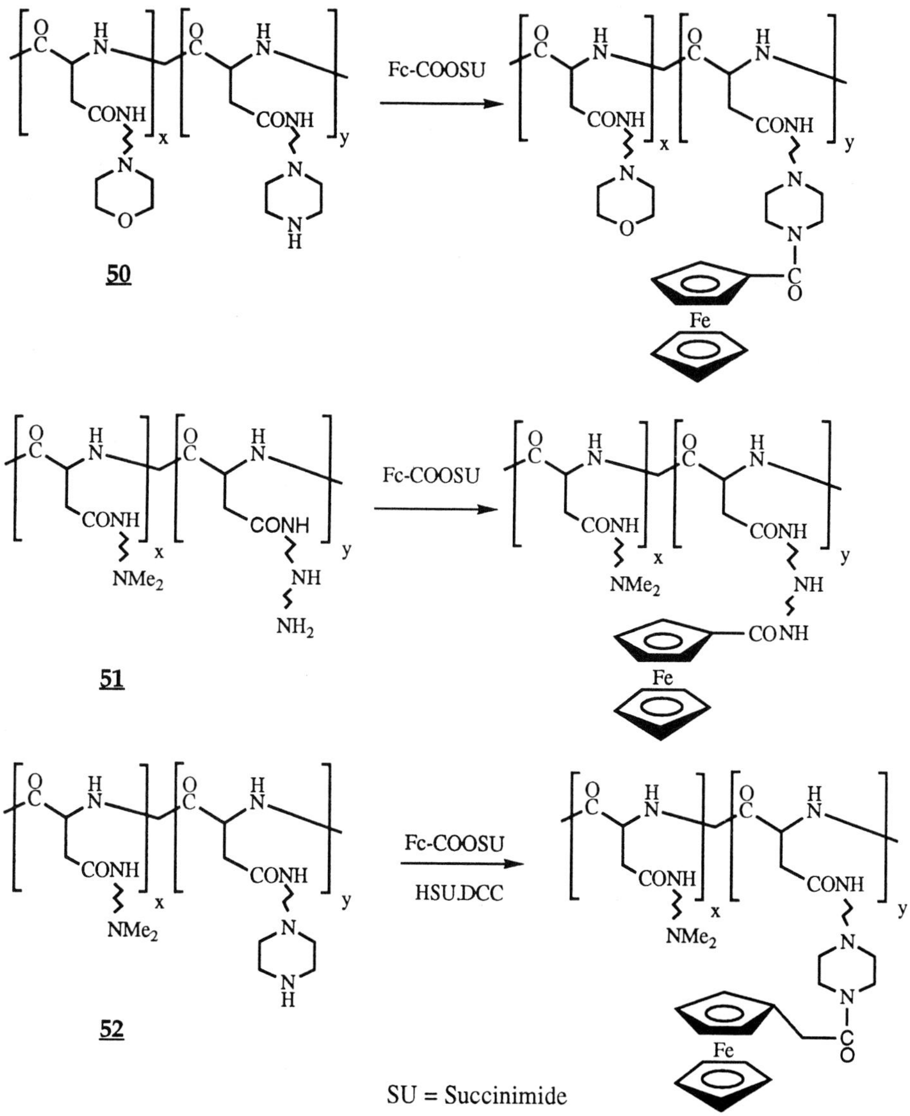

Fc-COOSU

50

Fc-COOSU

51

Fc-COOSU

HSU.DCC

52

SU = Succinimide

Scheme 10-24

by the reaction of water-soluble poly(vinylpyridine) with ferrocenylmethyltrimethyl ammonium iodide followed by reaction with chloroacetic acid. This ionomer is a polycationic material containing ferrocene as a redox center that can be oxidized at a platinum electrode [60].

The ferrocenyl fragment has also been bound covalently to a variety of copolyaspartamides via spacer groups attached to the aspartamide nitrogen atoms [61]. Attachment of ferrocene to carrier polymers, for example **50**, **51**, and **52**, was brought about by reaction of the amine functions attached to the spacer with ferrocenecarboxylic acids and -carboxaldehydes. The reactions shown in Scheme 10-24 are representative of these modifications.

By incorporating *tert*-amine groups the polymers are rendered water soluble, despite the lipophilic character of the anchored metallocene, as the amine groups are hydrosolubilizing functionalities. The inherent viscosity was typically 10 – 15 mL/g and the degree of ferrocene incorporation was 20 – 75% of the theoretical value, depending on the reactant ratios employed. These polymers are of interest as conjugates for biomedical applications requiring solubility in aqueous media.

Methyl-(β-ferrocenylethyl)- and methyl-[β-(1′,3′-dimethylferrocenyl)ethyl]siloxane polymers **53** and **54**, respectively were prepared by the hydrosilylation of vinylferrocene and 1,1′-dimethylferrocene-3-vinylferrocene with methyl hydrosiloxane (molecular weight was originally reported to be 2270) or methylhydrosiloxane-dimethylsiloxane copolymer (molecular weight was originally reported to be 2000 – 2100) in the presence of chloroplatinic acid as a catalyst. The synthetic route is given in Scheme 10-25 [62]. The reaction was monitored by IR spectroscopy until the complete disappearance of the Si-H absorption at 2161 cm^{-1}.

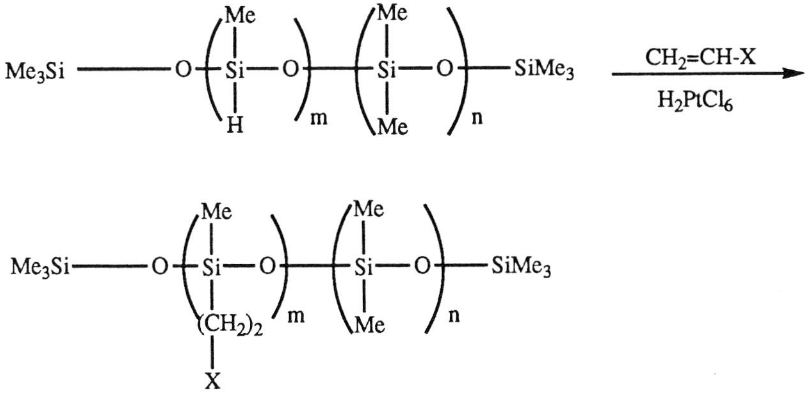

$$Me_3Si{-}O{-}(Si(Me)(H){-}O)_m{-}(Si(Me)(Me){-}O)_n{-}SiMe_3 \xrightarrow[H_2PtCl_6]{CH_2{=}CH{-}X} Me_3Si{-}O{-}(Si(Me)((CH_2)_2X){-}O)_m{-}(Si(Me)(Me){-}O)_n{-}SiMe_3$$

53a. X= ferrocenyl, m:n = 1:0;
53b. X= ferrocenyl, m:n = 1:2;
54. X= 1',3-dimethylferrocenyl, m:n = 1:2

Scheme 10-25

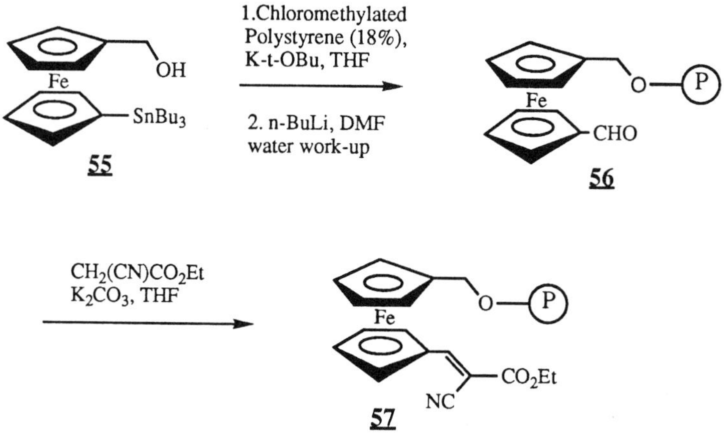

Scheme 10-26

The ferrocene complex **55** has also been attached to chloromethylated polystyrene (Scheme 10-26) under basic conditions [63] to form **56** and then polymer **57**.

In poly(methylphenylphosphazene), $[Ph(Me)P{=}N]_n$ **58**, both the phenyl and methyl substituents are potential sites for formation of derivatives [65]. Deprotonation of ca. half of the methyl substituents on this polymer was carried out in THF at $-78\,^{\circ}C$ using *n*-BuLi. On treatment of the intermediate polymer anion with ferrocenyl ketones and subsequent quenching with a mild proton source, phosphazenes **59** containing the OH functional group were prepared. The amount of substitution, determined by ^1H NMR and elemental analysis, was found to be 45 and 36%, respectively, for polymers **59a** and **59b** (Scheme 10-27). For these substituted polymers M_w was 187000 and 154000, respectively, and no degradation of the parent polymer occurred.

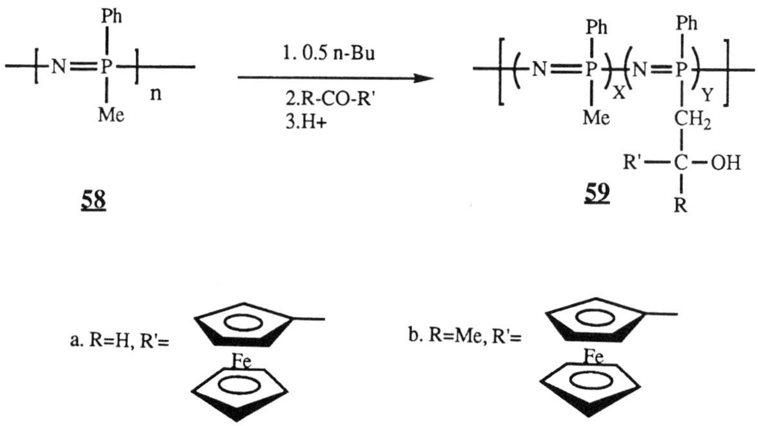

Scheme 10-27

10.6 Electropolymerization

Ferrocene-containing polymers are interesting materials with which to modify electrode surfaces [3]. Electropolymerization of monomers such as pyrrole or thiophene, which contain pendant ferrocene groups or its derivatives, leads to the rapid and direct deposition of the resulting polymers onto the electrode surface. This technique offers a unique approach to control of the characteristics of the resulting polymer, including conductivity, permeability, redox properties, selectivity toward other species, and solubility. By using different ferrocene derivatives or by varying the chain length of the polymer, it is the possible to "tune" E°, the standard oxidation potential of the probe and to obtain information on the effect of distance on the rate of electron transfer for the electrode system. In addition, electropolymerization may be expected to produce ultrathin films of conducting polymers in a controlled fashion. This is highly desirable for producing a reproducible electrode, because the thinner the film, the higher the conductivity [66].

Polypyrroles and polythiophenes are particularly adaptable to such studies. They can be polymerized by an oxidative polymerization mechanism that proceeds through an electrochemically accessible radical cation intermediate [66].

In the synthesis of ferrocene-functionalized polypyrrole systems, numerous derivatives are readily obtained by substitution at either the 3-position of the ring or on the nitrogen of the pyrrole prior to polymerization. It was reported that polymers with both metallic properties and enhanced solubility may be synthesized by judicious derivatization and choice of polymerization conditions, such as a variety of processing solvents [67, 68].

One such synthesis dealt with the preparation of redox-active polymer thin films of polypyrrole with ferrocene covalently attached [68]. The anodic electropolymerization of an *N*-substituted ferrocenylpyrrole monomer, *N*,*N'*-bis-[3-(pyrrol-1-yl)propyl]ferrocene-1,1'-dicarboxamide, was carried out in acetonitrile. This monomer was prepared in 75% yield by the reaction of *N*-(3-amino-propyl)pyrrole with ferrocene-1,1'-dicarbonyl chloride under dry argon in the presence of triethylamine. The monomer concentration in acetonitrile was 2×10^{-3} *M*, together with the presence of Bu_4NClO_4 or Bu_4NBF_4. Golden films of polymer were deposited on the electrode surface (Pt, glass carbon, or indium–tin oxide), either by repeated cycling between -0.1 and $+1.3$ V or by controlled potential electrolysis at $+1.3$ V. Thicker films were obtained on longer electrolysis, with an apparent molar coverage of 3.8×10^{-8} mol cm^{-2} on a glassy carbon electrode. Because this polymer showed electroactivity decay, which might suggest degradation of the polymer under the measurement conditions, copolymerization of the above monomer was attempted with *N*-methylpyrrole. Electropolymerization of a 1 : 1 monomer ratio with 0.1 *M* Bu_4NClO_4 gave a golden film that exhibited a persistent and stable cyclic voltammogram response in monomer-free electrolyte. Growth of the copolymer film was monitored by cyclic voltammetry, which showed an increase in the intensity of the response with the growth of apparent molar coverage. The electroactive polymer behaved almost ideally with rapid charge transfer kinetics instead of a decay type.

The electropolymerization of polythiophene with pendant ferrocene groups has also been investigated in detail [65]. The choice of the thiophene monomer stems from many useful properties of thiophene. In addition, the synthesis of derivatives with substituents at the 3-position is generally easier with thiophene than with pyrrole, the ring nitrogen of which must be protected [69]. Anodic electropolymerization of thiophene is possible in acetonitrile. Furthermore, long-chain substituents at the 3-position of the thiophene ring have been shown to form a polymer that is soluble in common organic solvents.

2-(3-Thienyl)ethanol as a starting material will give monomers with an ether linkage in the substituent at the 3-position. Such monomers, once polymerized, have exhibited the ability to complex cations such as Li^+ in a loose crown ether type structure [70]. This in turn leads to enhanced conductivity of the polymer when such cations are part of the supporting electrolyte. An added benefit of electropolymerization of polythiophene originates from the fact that sulfur has a tendency to physisorb to metals such as gold and platinum, which are electrode materials. Hence they may enhance the adsorption of polymer to the electrode and thus improve the physical stability of the system, as well as the extent of polymer/electrode interaction. The synthesis of these type of monomers (e.g., **60**) is shown in Scheme 10-28.

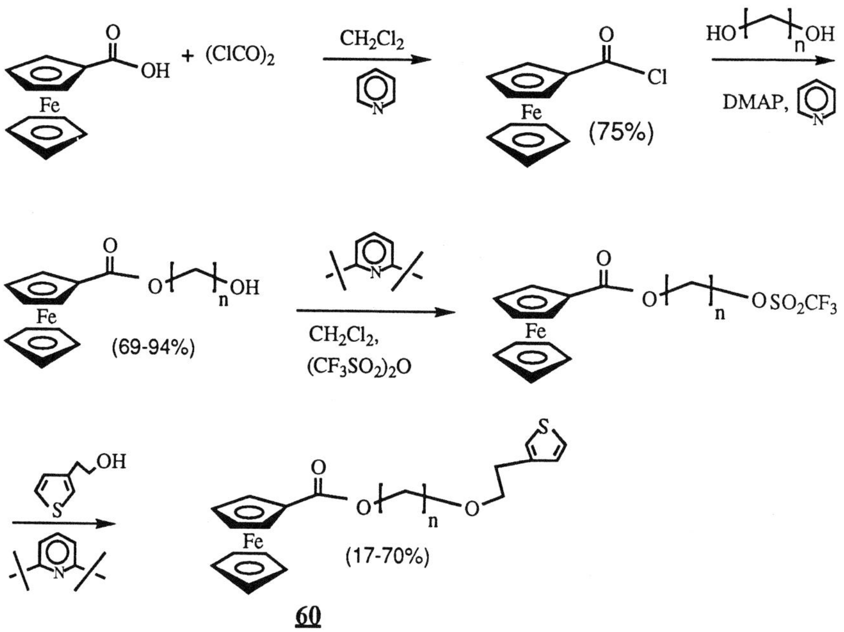

Scheme 10-28

Reaction of ferrocene acid chloride with appropriate diols in pyridine, with a catalytic amount of (dimethylamino)pyridine (DMAP), gave the corresponding ferrocene alcohol compounds. Formation of bis(ferrocene) compounds was minimized by working at high dilution with a 3-fold excess of diol. The ferrocene alcohols were then converted to the corresponding triflates, and the triflate moiety was displaced by 2-(3-thienyl)ethanol to yield a series of ferrocene-derived thiophenes. A series of monomers, consisting of a ferrocene ester unit attached to a 2-(3-thienylethoxy) group by an alkyl chain, with chain length of 6–16 carbon atoms, has been synthesized. The typical yield of these monomers was about 70%.

The electropolymerization of ferrocene/thiophene conjugates [65] was conducted by oxidation on a Pt electrode and led to the deposition of a monolayer of poly(thiophene). The electropolymerization was performed from several solution systems, such as tetrabutylammonium hexafluorophosphate/acetonitrile and lithium perchlorate/acetonitrile, at a concentration of 0.1 *M*. Constant potential experiments (+2.0 V) for a definite time were used to effect polymerization. Polymerization was also attempted using cyclic voltammetry (repeatedly sweeping from 0.0 to +2.5 V) and pulse potential (potential stepped from 0.0 to 2.0 V and back to 0.0 V).

The polymer-modified electrodes were examined by cyclic voltammetry and chronocoulometry to characterize the electrochemistry of the ferrocene and the polymer itself. Chronocoulometry reveals that material oxidizable at +1.0 V adheres to the Pt electrode after each polymerization experiment. This is presumably the ferrocene oxidation, but could include a component of the oxidation of the polythiophene backbone. Cyclic voltammetry indicates the presence of a surface wave centered at about +600 mV, corresponding to the oxidation and reduction of the ferrocene ester. These observations confirm that the electropolymerization leads to the localization of ferrocene ester moieties onto the electrode surface. An estimate of surface coverage using the geometrical electrode area indicates that a surface coverage of about 100 $Å^2$ per ferrocene moiety is attained with a 10 m*M* bulk solution; this is of the order of monolayer coverage. Less concentrated solutions (2–10 m*M*) lead to only a fraction (30–40%) of this coverage. Therefore it was concluded that the ferrocene-derived thiophenes do not favor formation of multilayers or large quantities of homopolymer. The electrochemical deposition of a ferrocene-derived thiophene monomer onto a Pt electrode precoated with a conducting film of 3-methylthiophene is also limited to monolayer equivalent coverage. The result is similar to that of the homopolymer deposited directly on the Pt electrode surface.

In the characterization of electropolymerization products from ferrocene/thiophene conjugates, however, it was shown that they are relatively unstable to repeated potential cycling between 0.0 and +1.0 V and are quickly destroyed at potentials of +2.0 V or more. The system appears to have an intrinsic limit to the amount of polymer that may be deposited onto the electrode surface. This may be rationalized to arise from the cationic state of ferrocenium when the oxidation potential of thiophene monomer is reached. Surface coverage calculations show that this limit corresponds to about a monolayer of ferrocene groups.

On the other hand, in order to affix more ferrocene groups onto the electrode surface, attempts were made to copolymerize a ferrocene-derived thiophene with 3-methylthiophene. This results in a film that has a greater than monolayer equivalent coverage. The number of ferrocene units on the electrode surface, calculated by the total charge beneath the cyclic voltammogram, is approximately 10^3 monolayer equivalents of ferrocene. The optimum monomer ratio was also studied and was found to be ca. 1:10. An increase in the monomer ratio, e.g., to 1:4, produced material with a similar number of ferrocene units per thiophene unit. This was explained by the ferrocenium ions inhibiting, but not preventing, further polymerization, as their concentration in the polymer increases. Copolymerization experiments with a 1:20 ratio produced material in which ferrocene was virtually undetectable.

Although many aspects are presently still under study, all of these studies provide an insight into the deposition mechanism of the ferrocene-containing poly(thiophene) conjugates onto an electrode surface and establish a methodology for deliberately producing a monolayer coverage of conducting polymers.

In another study, the electropolymerization of ferrocene monomers with aniline and phenol substituents has been described [71]. The synthesis involved the ferrocene compounds $CpFe(C_5H_4CH_2NHR^n)$, where $R^n = C_6H_5$, $-C_6H_4NH_2$, $-C_6H_4OH$, and $-C_6H_5NH(CH_2)_2$. The monomers were dissolved in a CH_3CN solution of Bu_4NClO_4 with no other additive and deaerated with nitrogen prior to polymerization. Generally, polymerization was carried out under a cycling electrode potential between 0 and about 0.75 V at 50 mV s^{-1}. The color of the film depended on its thickness, ranging from golden for thin to blue for thick films. All compounds underwent polymerization on electrochemical oxidation of the phenyl portion of the complex. The phenol-substituted complex shows unique polymerization behavior; adding NEt_3 to the electrolysis solution greatly enhances the rate of film deposition. Electrochemical characterization of the film shows well-defined couples for the ferrocene but no indication that the polymer backbones are electroactive. Spectroelectrochemical measurements show changes in the visible region of the spectrum characteristic for the formation of ferrocenium cations, $\lambda_{max} = 620$ nm, on oxidation.

The electrochemical polymerization of these ferrocene monomers was thought to occur through the pendant aniline or phenol groups, because no polymer is formed if these moieties are not oxidized. Attack of the aromatic amine on the cyclopentadienyl rings is unlikely, since E° for each polymer is shifted very slightly compared to that of the respective monomer. It was also revealed that substitution on the aromatic amine shows that both the amine nitrogen and the *para* position are involved in film formation, as the monomers $CpFe(C_5H_4CH_2NH(CH_3)C_6H_5)$ and $CpFe(C_5H_4CH_2NHC_6H_4OCH_3)$, do not undergo electropolymerization.

A number of desirable properties were exhibited in this work, which include ease of monomer synthesis, mild positive electropolymerization potential, polymer stability to continuous potential cycling, and stability to storage under ambient conditions. Unfortunately, the nature of the polymer backbone could not be definitely assigned. Nevertheless, the utility of pendant phenol and aniline groups for anchoring metal complexes to an electrode surface is a method worth further investigation.

10.7 Summary

Developments in the synthesis and characterization of ferrocene-containing polymers within the past ten years have provided a great deal of exciting and significant results. These include the synthesis of new monomers, the development of novel synthetic routes to obtain the desired polymeric products, the use of a better and broader range of analytical methodology for characterization of the polymers, and increasing exploration of possible areas of application. The potential use of ferrocene-containing polymers as specialty materials such as those offering unusual electronic, optical, and magnetic properties, is still attractive to both academic and industrial interests. Therefore the continuous development of synthetic strategies to prepare processable high molecular weight polymers, such as novel ring-opening polymerization from strained cyclic monomers, appears to be very promising [72].

Acknowledgement. We acknowledge discussions with Prof. M. D. Rausch at the University of Massachusetts at Amherst.

References

[1] For a general review of organometallic polymers see: (a) *Organometallic Polymers* (Eds.: C. E. Carraher, Jr., J. E. Sheats, C. U. Pittman, Jr.), Academic Press, NY, **1978**; (b) *Metal-Containing Polymeric Systems*, (Eds.: J. E. Sheats, C. E. Carraher, Jr., C. U. Pittman, Jr.), Plenum, NY, **1985**; (c) *Inorganic and Metal-Containing Polymeric Materials* (Eds.: J. E. Sheats, C. E. Carraher, Jr., C. P. Pittman, Jr., M. Zeldin, B. Currell), Plenum, NY, **1985**; (d) *Inorganic and Organometallic Polymers*, (Eds.: M. Zeldin, K. J. Wynne, H. R. Allcock), *ACS Symp. Ser. 360*, **1988**; (e) B. M. Cullbertson, & C. U. Pittman, Jr., *New Monomers & Polymers*, Plenum, NY, **1984**; (f) E. W. Neuse, H. Rosenberg, *Metallocene Polymers*, Marcel-Dekker, NY, **1970**; (g) J. E. Sheats in *Kirk-Othmer Encyclopedia of Chemical Technology*, 3rd ed., Vol. 15, Wiley, NY, **1981**.

[2] (a) T. J. Kealy, P. L. Pauson, *Nature* **1951**, *168*, 1039–1040; (b) S. A. Miller, J. A. Tebboth, J. F. Tremaine, *J. Chem. Soc.* **1952**, 632–635; (c) For a general review of ferrocene chemistry see M. Rosenblum, *Chemistry of Iron Group Metallocenes*, Wiley, NY, **1965**; (d) W. E. Watts, *Organomet. Chem. Rev.* **1967**, *2*, 231–254.

[3] For a general view on ferrocene polymers used for electrode surface modification, see: (a) P. D. Hale, T. Inagaki, H. I. Karan, Y. Okamoto, T. A. Skotheim, *J. Am. Chem. Soc.* **1989**, *111*, 3482–3484; (b) T. Inagaki, H. S. Lee, T. A. Skotheim, Y. Okamoto, *J. Chem. Soc., Chem. Commun.* **1989**, 1181–1183; (c) H. Nishihara, M. Noguchi, K. Aramaki, *Inorg. Chem.* **1987**, *26*, 2862–2867; (d) A. R. Hillman, D. A. Taylor, A. Hamnett, S. J. Higgins, *J. Electroanal. Chem.* **1989**, *266*, 423–435; (e) S. Nakahama, R. W. Murray, *J. Electroanal. Chem.* **1983**, *158*, 303–322; (f) A. H. Schroeder, F. B. Kaufman, V. Patel, E. M. Engler, *J. Electroanal. Chem.* **1980**, *113*, 193–208; (g) P. A. Peerce, A. J. Bard, *J. Electroanal. Chem.* **1980**, *108*, 121–125; **1980**, *112*, 97–115; **1980**, *114*, 89–115.

[4] For non-linear optical polymeric materials and those specific to ferrocene, see (a) M. E. Wright, E. G. Toplikar, *Macromolecules* **1992**, *25*, 1838–1839; (b) M. E. Wright, E. G. Toplikar, *Macromolecules* **1992**, *25*, 6050–6054; (c) M. E. Wright, M. S. Sigman, *Macro-*

molecules **1992**, *25*, 6055–6058; (d) P. N. Prasad, D. R. Ulrich, *Nonlinear Optical and Electroactive Polymers*, Plenum Press, New York, **1988**; (e) *Materials for Nonlinear Optics: Chemical Perspectives*, ACS Symp. Series no. 455, ACS, Washington, DC, **1991**; (f) *Organic Materials for Non-linear Optics*, RSC, London, Spec. Publ. no. 69, **1989**;(g) *Organic Materials for Nonlinear Optics II*, RSC, London, Spec. Publ. no. 91, **1991**; (h) For the first synthesis of an NLO ferrocene compound, see M. L. H. Green, S. R. Marder, M. E. Thompson, J. A. Bandy, D. Bloor, P. V. Kolinsky, R. J. Jones, *Nature* **1987**, *330*, 26, 360–362; (i) For a theoretical treatment of organometallic NLO materials, see D. R. Kanis, M. A. Ratner, T. J. Marks, *J. Am. Chem. Soc.* **1990**, *112*, 8203–8204; (j) For $\chi^{(3)}$ active organometallic polymers: M. H. Chisholm, *Angew. Chem., Int. Ed. Engl.* **1991**, *30*, 673; (k) For ferrocene complexes with large experimental β values: S. R. Marder, J. W. Perry, B. G. Tiemann, *Organometallics* **1991**, *10*, 1896–1901.

[5] (a) F. S. Arimoto, A. C. Haven, Jr., *J. Am. Chem. Soc.* **1955**, *77*, 6295–6297; (b) K. Schlogl, A. Mohar, *Monatsh. Chem.* **1961**, *92*, 219.

[6] J. C. Lai, T. Rounsefell, C. U. Pittman, Jr., *J. Polym. Sci. A-1* **1971**, *9*, 651–662.

[7] C. Aso, T. Kunitake, T. Nakashima, *Makromol. Chem.* **1969**, *124*, 232–240.

[8] C. U. Pittman, Jr., C. C. Lin, *J. Polym. Sci. Polym. Chem. Ed.* **1979**, *17*, 271–275.

[9] See *"Metal-Containing Polymers: An Introduction"* by C. E. Carraher, Jr., C. U. Pittman, Jr., in *Metal-Containing Polymeric Systems* (Eds.: J. E. Sheats, C. E. Carraher, Jr., C. U. Pittman, Jr.), Plenum, New York, **1985**.

[10] Y. Sasaki, L. L. Walker, E. L. Hurst, C. U. Pittman, Jr., *J. Polym. Sci. Polym. Chem. Ed.* **1973**, *11*, 1213–1224.

[11] (a) G. F. Hayes, M. H. George, "Participation of the Ferrocene Nucleus in the Polymerization of Vinylferrocene and its Effects on Polymer Properties" in *Organometallic Polymers*, (Eds.: C. E. Carraher, Jr., J. E. Sheats, C. U. Pittman, Jr.), Academic Press, New York, **1978**; (b) M. H. George, G. F. Hayes, *J. Polym. Sci. Polym. Chem. Ed.* **1975**, *13*, 1049–1070.

[12] C. U. Pittman, Jr., R. L. Voges, W. R. Jones, *Macromolecules* **1971**, *4*, 291–297.

[13] C. U. Pittman, Jr., R. L. Voges, J. Elder, *Polym. Lett.* **1971**, *9*, 191–194.

[14] J. C. Lai, T. Rounsefell, C. U. Pittman, Jr., *J. Polym. Sci. A-1* **1971**, *9*, 651–662.

[15] C. U. Pittman, Jr., P. C. Grube, *J. Polym. Sci. A-1* **1971**, *9*, 3175–3186.

[16] (a) T. Alfrey, Jr., C. C. Price, *J. Polym. Sci.* **1947**, *2*, 101–106; (b) T. Alfrey, Jr., L. J. Young, "The *Q-e* Scheme", Chap. II in *Copolymerization* (Ed.: G. E. Ham), Wiley-Interscience, New York, **1964**.

[17] M. E. Wright, E. G. Toplikar, R. F. Kubin, M. D. Seltzer, *Macromolecules* **1992**, *25*, 1838–1839.

[18] K. E. Gonsalves, L. Zhan-Ru, R. W. Lenz, M. D. Rausch, *J. Polym. Sci. Polym. Chem. Ed.* **1985**, *23*, 1707–1722

[19] (a) M. Cais, A. Eisenstadt, *J. Org. Chem.* **1965**, *30*, 1148–1154; (b) S. Lupan, M. Kapon, M. Cais, F. H. Herbstein, *Angew. Chem. Int. Ed.* **1972**, *11*, 1025; (c) W. E. Watts, *Organomet. Chem. Rev.* **1979**, *7*, 399.

[20] C. Aso, T. Kunitake, T. Nakashima, *Makromol. Chem.* **1969**, *124*, 232–240.

[21] S. L. Sosin, V. V. Korshak, T. M. Frunze, *Dokl. Akad. Nauk SSSR* **1968**, *179*, 345–347.

[22] (a) S. L. Sosin, L. V. Dzhashi, B. A. Antipova, V. V. Korshak, *Vysokomol Soedin., Ser. B* **1970**, *12* (9), 699; (b) S. L. Sosin, L. V. Dzhashi, B. A. Antipova, V. V. Korshak, *Vysokomol Soedin., Ser. B* **1974**, *16* (5), 347.

[23] (a) T. Kunitake, T. Nakashima, C. Aso, *J. Polym. Sci., A-1* **1970**, *8*, 2853–2859; (b) T. Kunitake, T. Nakashima, C. Aso, *Makromol. Chem.* **1971**, *146*, 79–90; (c) T. Kunitake, T. Nakashima, *Bull. Chem. Soc. Jpn.* **1972**, *45*, 2892–2895.

[24] A. S. Chisti, C. R. Jablonski, *Makromol. Chem.* **1983**, *184*, 1837–1848.

[25] (a) C. U. Pittman, Jr., C. C. Liu, *J. Polym. Sci. Polym. Chem. Ed.* **1979**, *17*, 27–37; (b) C. U. Pittman, Jr., A. Hirao, *J. Polym. Sci. Polym. Chem. Ed.* **1977**, *15*, 1677; *ibid.* **1978**, *16*, 1197–1209.

[26] E. W. Neuse, H. Rosenberg, *Metallocene Polymers*, Marcel Dekker, New York, **1970**, p. 45–53.

[27] (a) E. W. Neuse, O. S. Trifari, *J. Am. Chem. Soc.* **1963**, *85*, 1952–1958; (b) E. W. Neuse, K. Koda, *Bull. Chem. Soc. Jpn.* **1966**, *39*, 1502; (c) H. J. Lorkowski, *Fortschr. Chem. Forsch.* **1967**, *9*, 207; (d) A. Wende, H. J. Lorkowski, *Plaste Kautsch* **1963**, *10*, 32.

[28] (a) E. W. Neuse, E. Quo, *Bull. Chem. Soc. Jpn.* **1966**, *39*, 1508–1514; (b) E. W. Neuse, US Patent, 3 341 495, **1967**.

[29] Z. R. Lin, K. E. Gonsalves, R. W. Lenz, M. D. Rausch, *J. Polym. Sci. Polym. Chem. Ed.* **1986**, *24*, 347–358.

[30] P. W. Morgan, *Condensation Polymers by Interfacial and Solution Methods*, Wiley, New York, **1965**.

[31] (a) C. E. Carraher, Jr. in *Interfacial Synthesis. Volume II. Polymer Applications and Technology* (Eds.: F. Millich, C. E. Carraher, Jr.), Marcel Dekker, New York, **1988**; (b) E. W. Neuse, H. Rosenberg, *Metallocene Polymers*, Marcel Dekker, New York, **1970**.

[32] (a) K. E. Gonsalves, R. W. Lenz, M. D. Rausch, *Appl. Organomet. Chem.* **1987**, 81; (b) K. E. Gonsalves, Z. R. Lin, M. D. Rausch, *J. Am. Chem. Soc.* **1984**, *106*, 3826–3863.

[33] A. Sonoda, I. Moritani, *J. Organomet. Chem.* **1971**, *26*, 133–140.

[34] A. Ratajczak, B. Czech, L. Drobek, *Synth. React. Inorg. Metal.-Org. Chem.* **1982**, *12*, 557–563.

[35] F. W. Knobloch, W. K. Rausher, *J. Polym. Sci.* **1961**, *54*, 651–656.

[36] C. U. Pittman, Jr., *J. Polym. Sci. Polym. Chem. Ed.* **1968**, *6*, 1687–1695.

[37] K. E. Gonsalves, M. D. Rausch, *J. Polym. Sci.: Part A: Polym. Chem.* **1986**, *24*, 1599–1607.

[38] P. Singh, M. D. Rausch, R. W. Lenz, *Polym. Bull.* **1989**, *22*, 247–252.

[39] M. M. Abd-Alla, M. F. El-Zohry, K. I. Aly, M. M. M. Abd-el. Wahab, *J. Appl. Poly. Sci.* **1993**, *47*, 323–329.

[40] M. E. Wright, E. G. Toplikar, *Macromolecules* **1992**, *25*, 6050–6054.

[41] M. E. Wright, M. S. Sigman, *Macromolecules* **1992**, *25*, 6055–6058.

[42] J. B. A. Mitchell, *Combustion and Flame* **1991**, *86*, 179–184.

[43] J. B. A. Mitchell, D. J. M. Miller, M. Sharpe, *Combust. Sci. Technol.* **1991**, *74*, 63–66.

[44] K. E. Ritrievihe, J. P. Longwell, A. F. Sarifim, *Combustion and Flame* **1987**, *70*, 17–31.

[45] S. K. Brauman, *J. Fire Ret. Chem.* **1980**, *7*, 161–171.

[46] T. Tajima, Japan Kokai 7 851 249; *Chem. Abstr.* **1978**, *89*, 111482r; T. Tajima, Japan Kokai 7 856 299; *Chem. Abstr.* **1978**, *89*, 111523r.

[47] K. Kishore, P. Kannan, K. Iyangar, *J. Polym. Sci.: Part A: Polym. Chem.* **1991**, *29*, 1039–1044.

[48] (a) A. Foucher, R. Ziembinski, B. Tang, P. M. Macdonald, J. Massey, C. R. Jaeger, G. J. Vancso, I. Manners, *Macromolecules* **1993**, *26*, 2878–2884; (b) I. Manners, J. M. Nelson, H. Rensel, *J. Am. Chem. Soc.* **1993**, *115*, 7035–7036.

[49] C. Honeyman, D. A. Foucher, O. Mourad, R. Rulkens, I. Manners, *Polym. Preprints* **1993**, *34* (1), 330–331.

[50] D. L. Compton, T. B. Rauchfuss, *Polym. Preprints* **1993**, *34* (1), 351–352.

[51] R. H. Grubbs, W. Tumas, *Science* **1989**, *243*, 907–915.

[52] R. R. Schrock, *Acc. Chem. Res.* **1990**, *23*, 158–165.

[53] (a) G. C. Bazan, R. R. Schrock, H. N. Cho, V. C. Gibson, *Macromolecules* **1991**, *24*, 4495–4502; (b) G. C. Bazan, J. H. Oskam, H. N. Cho, L. Y. Park, R. R. Schrock, *J. Am. Chem. Soc.* **1991**, *113*, 6899–6907.

[54] D. Albagli, G. Bazan, M. S. Wrighton, R. R. Schrock, *J. Am. Chem. Soc.* **1992**, *114*, 4150–4158.

[55] J. E. Mark, H. R. Allcock, R. West, *Inorganic Polymers*, Prentice Hall, New Jersey, **1992**.

[56] I. Manners, G. H. Riding, J. A. Dodge, H. R. Allcock, *J. Am. Chem. Soc.* **1989**, *111*, 3067–3069.

[57] I. Manners, G. H. Riding, K. D. Laving, *Macromolecules* **1987**, *18*, 1340–1345; *ibid.* **1987**, *20*, 6–10.

[58] General reviews on the derivatization of preformed organic polymers (a) C. U. Pittman, Jr. in *Comprehensive Organometallic Chemistry*, (Eds.: G. Wilkinson, F. G. A. Stone, E. W. Abel), Pergamon Press, Chapter 55, **1992** p. 553; (b) C. U. Pittman, Jr. in *Polymer Supported Reactions in Organic Chemistry* (Eds.: P. Hodge, D. C. Sherrington), Wiley, New York, Chapter 5, **1980**, p. 249; (c) Y. Chauvin, D. Coumerenc, F. Dawans, *Prog. Polym. Sci.* **1977**, *5*, 95.

[59] I. R. Butler, W. R. Cullen, N. F. Han, F. G. Herring, N. R. Jagannathan, J. Li, *Appl. Organomet. Chem.* **1988**, 2.

[60] R. Tabakovic, I. Tabakovic, A. Davidovic, *J. Electroanal. Chem.* **1992**, *332*, 297–301.

[61] E. W. Neuse, C. W. N. Mbonyyana, "Synthesis of Polyaspartamide-Bound Ferrocene Compounds" in *Inorganic and Metal-Containing Polymeric Materials*, (Eds.: J. E. Sheats, C. E. Carraher, Jr., C. U. Pittman, Jr., M. Zeldin, B. Currell, Plenum, New York, **1990**.

[62] T. Inagaki, H. S. Lee, T. A. Skotheim, Y. Okamoto, *J. Chem. Soc., Chem. Commun.* **1989**, 1181–1183.

[63] M. E. Wright, S. A. Svejda in *Materials for Nonlinear Optics: Chemical Perspectives*, Chap. 39, ACS Symposium Series no. 455, ACS, Washington, DC, **1991**.

[64] P. Wisian-Neilson, R. R. Ford, *Macromolecules* **1989**, *22*, 72–75.

[65] R. Back, R. B. Lennox, *Langmuir* **1992**, *8*, 959–964.

[66] E. M. Genies, G. J. Bidan, *J. Electroanal. Chem.* **1983**, *149*, 101–113.

[67] See, for example, (a) A. C. Chang, R. L. Blankespoor, L. L. Miller, *J. Electroanal. Chem.* **1987**, *236*, 239–252; (b) R. L. Elsenbaumer, K. Y. Jen, R. Oboodi, *Synth. Met.* **1986**, *15*, 169; (c) R. L. Blankespoor, L. Miller, *J. Chem. Soc., Chem. Commun.* **1985**, 90–92; (d) T. Inagaki, M. Hunter, X. Q. Yang, T. A. Skotheim, Y. Okamoto, *J. Chem. Soc., Chem. Commun.* **1988**, 126–127.

[68] J. G. Eaves, R. M. Mirrazaei, D. Parker, H. S. Munro, *J. Chem. Soc., Perkin Trans. II* **1989**, 373–376.

[69] D. Delabouglise, F. Garnier, *Adv. Mater.* **1990**, *2* (2), 91.

[70] J. Roncali, R. Garreau, D. Delabouglise, F. Garnier, M. Lemaire, *J. Chem. Soc., Chem. Commun.* **1989**, 679–681.

[71] (a) C. P. Horiwitz, G. C. Dailey, *Chem. Mater.* **1990**, *2*, 343–346; (b) C. P. Horiwitz, N. Y. Suhu, G. C. Dailey, *J. Electroanal. Chem.* **1992**, *324*, 79–91.

[72] *C & E News* **1993**, *71* (31), 22–23.

Received: December 2, 1993

Index